PEDIATRIC DRUG DEVELOPMENT

PEDIATRIC DRUG DEVELOPMENT
CONCEPTS AND APPLICATIONS

Edited by

Andrew E. Mulberg MD
Steven A. Silber MD
Johnson and Johnson Pharmaceutical Research and Development, LLC,
Established Products

John N. van den Anker MD, PhD
Children's National Medical Center

WILEY-BLACKWELL
A JOHN WILEY & SONS, INC., PUBLICATION

Published by John Wiley & Sons, Inc., Hoboken, New Jersey
Published simultaneously in Canada

Wiley-Blackwell is an imprint of John Wiley & Sons, Inc., formed by the merger of Wiley's global Scientific, Technical, and Medical business with Blackwell Publishing.

Library of Congress Cataloging-in-Publication Data:

Pediatric drug development: concepts and applications / [edited by] Andrew Mulberg, Steven Silber, John van den Anker.
p. ; cm.
Includes bibliographical references and index.
ISBN 978-0-470-16929-2 (cloth)
1. Pediatric pharmacology. 2. Drug development. I. Mulberg, Andrew. II. Silber, Steven. III. Van den Anker, John N.
[DNLM: 1. Child. 2. Drug Evaluation- -methods. 3. Age Factors. 4. Clinical Trials as Topic- -standards. 5. Drug Design. 6. Drug Evaluation- -ethics. 7. Drug Evaluation- -standards. QV 771 P371 2008]
RJ560.P3875 2008
615′.190083- -dc22

2008019010

Printed in the United States of America

10 9 8 7 6 5 4 3 2 1

Andrew E. Mulberg
To my children, Nathaniel and Rebecca and my wife, Elyse for giving me Love, strength and courage; to the memories of Richard, Matthew and Kristen who remind me daily of a doctor's need for compassion and empathy; and to the world's children who have needs to ensure drug therapy will be made with them in mind.

Steven A. Silber
Dedicated to my mother, Carol, who knew the meaning of unconditional love. . .at least when applied to her children; and to Solomon and Abraham, the best sons a father could have.

John van den Ander
To the children of the world: For their patience in waiting for the safe and effective medicines they deserve!

CONTENTS

PART X CASE STUDIES: SUCCESSES FOR CHILDREN

PREFACE

As an outgrowth of my own passion for pediatrics this book has been born to develop a textbook devoted to the concepts and applications of pediatric drug development. Understanding the similarities and differences of drug development of pharmaceutical based therapeutics from adults targeted to children has been needed since Shirkey described children as "therapeutic orphans." This is the first textbook devoted to an understanding of the issues in pediatric drug development, is an outgrowth of a need to fill this void and attributed to the shared vision of many esteemed colleagues whose similar passions have contributed to this textbook.

I have been fortunate to have mentors throughout my career who are integral to my success in achieving this important milestone: especially Kurt Hirschhorn, MD (Mount Sinai School of Medicine). Fayez Ghishan, MD (Vanderbilt, University of Arizona), both who started me on the path to pediatrics and pediatric gastroenterology; to Steven Altschuler, MD, and William Schwartz, MD (Children's Hospital of Philadelphia) for never failing to recognize my potential and set me off to authorship; to Richard Grand, MD (New England Medical Center, Harvard) who brought me the knowledge that only he can teach and always a friend; to Stephen Spielberg, MD, PhD (JJPRD, Dartmouth), the founder (to me and many) of pediatric drug development; and to Steven A. Silber, MD, a gentleman, a scholar and a teacher of issues beyond that of pharmaceutical development.

As the senior editor and on behalf of my co-editors, John and Steven, I wish to thank our many colleagues of academia, from the multiple pharmaceutical industry partners and functional area experts who contributed to individual chapters and for their outstanding efforts for this book. These chapters have provided a keen insight into all of the functional areas needed for a proper and deep understanding of the issues and the population related to pediatric drug development.

For Steven, this book is dedicated to his mother Carol, who lived her life with devotion to and unconditional love for her children. For John, the dedication is devoted to neonates, infants, children and adolescents who are in need of safe and effective medicines wherever they are in the world.

I would like to thank my parents, Nathan (of blessed memory) and Betty, who always encouraged me to pursue the most noble of professions; to my wife, Elyse, for her tireless tolerance of my shortcomings and my insane desire to complete this book, to my loving children, Nathaniel and Rebecca, for whom I adore and wonder daily how fortunate I am to be part of their lives and their love and who have patiently tolerated my occasional absence in order to complete this project.

To the memories of many special children, including Kirsten who taught me of the value of being a physician as a young intern, and Matthew who so young but whose spirit continues to foster how important it is not to lose sensitivity as a physician; and to the special memory of my friend, Richard S, son of Beth and Richard, a young 15 year old, who died so young from

Ewing's sarcoma despite his courage and strength of fight, an event that too often occurs because we aren't at a place where therapeutics for children could have brought him benefit. To their individual and collective spirits, which continue to teach humanity to all, I am indebted.

This book is dedicated in honor of all of the world's children, and to Richard's memory and the **Go4TheGoal** Foundation (http://www.go4thegoal.com). May this book in some small way help to stimulate the youngest and oldest of minds to pursue the paths of their dreams, never to allow their energies to be diminished by naysayers and may their dreams become reality for the betterment of mankind.

With much respect and admiration, for my mentors and colleagues, who helped to motivate and to contribute to its successful launch.

Cherry Hill, NJ ANDREW E. MULBERG, MD, FAAP, CPI

CONTRIBUTORS

Peter C. Adamson, MD Chief, Division of Clinical Pharmacology & Therapeutics, Director, Office of Clinical and Translational Research, The Children's Hospital of Philadelphia, University of Pennsylvania School of Medicine, Philadelphia, PA 19104
[email: adamsonp@mail.med.upenn.edu]

Pieter P. Annaert, PhD Laboratory for Pharmacotechnology and Biopharmacy, Department of Pharmaceutical Sciences, Katolieke Universiteit Leuven, 3000 Leuven, Belgium
[email: pieter.annaert@pharm.kuleuven.be]

Gerri R. Baer, MD Community Neonatal Associates, 1500 Forest Glen Road, Silver Spring, MD 20910
[email: gbaer73@gmail.com]

Graham P. Bailey, BSc Reproduction Toxicology, Global Preclinical Development, Johnson & Johnson Pharmaceutical Research and Development, Turnhoutseweg 30, B-2340 Beerse, Belgium
[email: gbailey1@prdbe.jnj.com]

Martin Otto Behm, MD Merck and Co., Inc., Clinical Pharmacology – MRL, 351 North Sunneytown Pike, North Wales, PA 19454
[email: martin_behm@merck.com]

Katia Boven, MD Global Clinical Research & Development, Tibotec Inc., 1020 Stony Hill Road, Yardley, PA 19067
[email: kboven@tibus.jnj.com]

Carolyn A. Campen, PhD, PMP Johnson & Johnson, Pharmaceutical Global Strategic Marketing, Raritan, NJ 08869
[email: ccampen@psmus.jnj.com]

Anne F. Clothier 108 West Springfield Ave, Philadelphia, PA 19118
[email:anne_clothier@verizon.net]

Emmett Clemente, PhD Manchester Consulting, Inc., 23 Loading Place Road, Manchester, MA 01944
[email: elclem88@verizon.net]

Nadine Cohen, PhD Head of Pharmacogenomics and Senior Research Fellow, Johnson & Johnson Pharmaceutical Research and Development, 1000 Route 202, Raritan, NJ 08869
[email: ncohen2@prdus.jnj.com]

Timothy P. Coogan, PhD, DABT Senior Director, Toxicology and Investigational Pharmocology, Centocor Research & Development, Inc., 145 King of Prussia Road, Radnor, PA 19087
[email: tcoogan@its.jnj.com]

Sandra Cottrell, PhD Therapeutic Area Head, Regulatory Affairs, Novo Nordisk, Inc., Princeton, NJ 08540
[email: sctr@novonordisk.com]

Loeckie L. de Zwart, PhD Global Preclinical Development, Johnson & Johnson Pharmaceutical Research and Development LLC, Turnhoutseweg 30, B-2340 Beerse, Belgium
[email: ldzwart@prdbe.jnj.com]

Luc Dekie, PhD Director of Pharmaceutical Operations, Biomedical Systems, Chaussée de Wavre 1945, 1160 Brussels, Belgium
[email: ldekie@biomedsys.com]

Leonard R. Friedland, MD GlaxoSmithKline Biologicals, Executive Director, Vaccines, Clinical Research and Development and Medical Affairs, NA, 2301 Renaissance Blvd, King of Prussia, PA 19406
[email: leonard.r.friedland@gsk.com]

Brahm Goldstein, MD, MCR Professor of Pediatrics, University of Medicine and Dentistry of New Jersey—Robert Wood Johnson Medical School, New Brunswick, NJ 08901
[email: brahm.goldstein@yahoo.com]

Magali Haas, MD, PhD Johnson & Johnson Pharmaceutical Research and Development, Turnhoutseweg 30, B-2340 Beerse, Belgium
[email: mhaas8@prdbe.jnj.com]

Seth V. Hetherington, MD Icagen, Inc., Durhan, NC 27703 and Professor of Pediatrics, University of North Carolina School of Medicine, Chapel Hill, NC 27579
[email: shetherington @icagen.com]

Matthew Hill, JD Attorney, Novo Nordisk, Inc., Princeton, NJ 08540
[email: mhi@novonordisk.com]

Steven Hirschfeld, MD, PhD CAPT USPHS, Associate Director for Clinical Research, National Institute of Child Health and Human Development, 31 Center Drive, Bethesda, MD 20814
[email: hirschfs@mail.nih.gov]

Michael R. Hoy, PhD Johnson & Johnson Pharmaceutical Research and Development, LLC, 920 Route 202 South, Raritan, NJ 08869
[email: mhoyphd@prdus.jnj.com]

Karen M. Kaplan, MD Director, Safety Surveillance, Benefit Risk Management, Division of Johnson & Johnson Pharmaceutical Research & Development LLC,.100 Tournament Drive, Titusville, NJ 08560
[email: kkaplan@brmus.jnj.com]

Leah Kleinman, DPH Senior Research Scientist, Center for Health Outcomes Research, United BioSource Corporation, 2601 4th Ave, Seattle, WA 98121
[email: leah.kleinman@unitedbiosource.com]

Hui Kimko, PhD Advanced Pharmacokinetic/Pharmacodynamic Modeling & Simulation, Johnson & Johnson Pharmaceutical Research and Development LLC, Raritan, NJ 08869
[email: hkimko@prdus.jnj.com]

Greg Koski, MD, PhD Senior Scientist, Institute for Health Policy. Massachusetts General Hospital, Associate Professor of Anesthesia, Harvard Medical School, President, Academy of Pharmaceutical Physicians and Investigators, Association of Clinical Research Professionals, Boston, MA 02114
[email: gkoski@partners.org]

Donald P. Lombardi President and CEO, Institute for Pediatric Innovation, One Broadway, 14th Floor, Cambridge, MA 02142
[email: donald.lombardi@pediatricinnovation.org]

Samuel D. Maldonado, MD, MPH, FAAP Vice President and Head, Pediatric Drug Development Center of Excellence, Johnson & Johnson Pharmaceutical Research & Development LLC, 920 Route 202 South, Raritan, NJ 08869
[email: smaldon3@prdus.jnj.com]

Gerard P. McNally, PhD Vice President, Research & Development, McNeil Consumer Healthcare, 7050 Camp Hill Road, Fort Washington, PA 19034
[email: gmcnall@mccus.jnj.com]

Ursula Merriman, RN Clinical Management, Ortho-McNeil Janssen Scientific Affairs, LLC, Titusville, NJ 08560
[email: umerrim@omjus.jnj.com]

Jacqueline M. Miller, MD GlaxoSmithKline Biologicals, Director, Vaccines Clinical Research and Development and Medical Affairs, NA, 2301 Renaissance Blvd., King of Prussia, PA 19406
[email: jacqueline.m.miller@gsk.com]

Christopher-Paul Milne, DVM, MPH, JD Associate Director, Tufts Center for the Study of Drug Development, Tufts University, 75 Kneeland St., Boston, MA 02111
[email: christopher.milne@tufts.edu]

Geert Molenberghs, PhD Professor and Head, International Institute for Biostatistics and Statistical Bioinformatics, Universiteit Hasselt, Agoralaan 1, B-3590 Diepenbeek, Belgium
[email: geert.molenberghs@uhasselt.be]

Johan G. Monbaliu, PhD Global Preclinical Development, Johnson & Johnson Pharmaceutical Research and Development LLC, Turnhoutseweg 30, B-2340 Beerse, Belgium
[email: jmonbali@prdbe.jnj.com]

Andrew E. Mulberg, MD, FAAP, CPI Senior Director, Internal Medicine Established Products, Johnson and Johnson Pharmaceutical Research and Development LLC, 1125 Trenton Harbourton Road,Titusville, NJ 08560, Associate Professor of Pediatrics, University of Pennsylvania School of Medicine, Attending Physician, Gastroenterology and Nutrition, Children's Hospital of Philadelphia, Philadelphia, PA
[email: amulberg@its.jnj.com]

Alexander Cvetkovich Muntanola, MD INC Research, Barcelona, Spain 0811
[email: acvetkovich@incresearch.com]

Hidefumi Nakamura, MD, PhD National Center for Child Health and Development, 2-10-1 Ookura, Setagaya-ku, Tokyo 157-8535, Japan
[email: nakamura-hd@ncchd.go.jp]

Robert M. Nelson, MD, PhD Pediatric Ethicist, Office of Pediatric Therapeutics, Food and Drug Administration, Rockville, MD 20857
[email: robert.nelson@fda.hhs.gov]

Seth L. Ness, MD, PhD Medical Director, CNS, Pain, and Translational Medicine, Johnson & Johnson Pharmaceutical Research & Development LLC, 1125 Trenton-Harbourton Road, Titusville, NJ 08560
[email: sness@prdus.jnj.com]

Jennifer Niesz Senior Specialist, IMPACT Group, Johnson & Johnson Pharmaceutical Research & Development LLC, 1125 Trenton-Harbourton Road, Titusville, NJ 08560
[email: amulberg@its.jnj.com]

Robin Norris Clinical Instructor, Division of Oncology, Clinical Pharmacology & Therapeutics, Children's Hospital of Philadelphia, Philadelphia, PA 19146
[email: norrisr@chop.edu]

Jeffrey S. Nye, MD, PhD Chief Medical Officer, REDEC, Johnson & Johnson Pharmaceutical Research & Development LLC, Welsh and McKean Roads, Spring House, PA 19477
[email: jaye@its.jnj.com]

Shunsuke Ono, PhD Graduate School of Pharmaceutical Sciences, The University of Tokyo, Tokyo 113-0033 Japan

Camille Orman, PhD Johnson & Johnson Pharmaceutical Research and Development, 1125 Trenton-Harbourtown Road, Titusville, NJ 08560
[email: corman@prdus.jnj.com]

Kate Owen, BS Project Management, Novodisk, Inc., Princeton, NJ 08540
[email: kaow@novonordisk.com]

Cindy Levy-Petelinkar, MBA, RAC Director, Portfolio Performance Analysis & Benchmarking, Global Project and Portfolio Management, GlaxoSmithKline Research & Development, 709 Swedeland Road, King of Prussia, PA 19406
[email: cindy.levy@gsk.com]

Mary Pipan, MD Children's Hospital of Philadelphia, Division of Child Rehabilitation and Developmental Medicine, Philadephia, PA 19104
[email: pipan@email.chop.edu]

A. Procaccino Senior, Director, Technical Training, Johnson & Johnson Pharmaceutical Research & Development LLC, 1125 Trenton-Harbourton Road, Titusville, NJ 08560
[email: aprocacc@its.jnj.com]

Wayne Rackoff, MD Vice President, Clinical Oncology, Ortho Biotech Oncology Research & Development, Johnson & Johnson Pharmaceutical Research and Development, 920 Route 202 South, Raritan, NJ 08869
[email: wrackoff@ordus.jnj.com]

Aniruddha M. Railkar, PhD Principal Scientist, Johnson & Johnson Pharmaceutical Research and Development, LLC, Welsh and McKean Roads, Spring House, PA 19477
[email: gmcnall@mccus.jnj.com]

Natella Y. Rakhmanina, MD Divisions of Pediatric Clinical Pharmacology and Infectious Diseases, Children's National Medical Center, Washington, DC 20010
[email: nrakhman@cnmc.org]

Edward J. Roche, PhD Director of Science and Technology, Business Development, McNeil, Johnson & Johnson, 420 Delaware Drive, Ft. Washington, PA 19034
[email: eroche2@mccus.jnj.com]

Klaus Rose, MD, MS Head Pediatrics, F. Hoffmann-La Roche Ltd, CH-4070 Basel, Switzerland
[email klaus.rose@roche.com]

Margaret Rothman, PhD Senior Director, World Wide Patient Reported Outcomes Center of Excellence, Johnson & Johnson Pharmaceutical Services LLC, 301 South Alexander Avenue, Washington, GA 30673
[email: mrothman@psmus.jnj.com]

Alisha J. Rovner, PhD Division of Epidemiology, Statistics and Prevention, Eunice Kennedy Shriver National Institute of Child Health and Human Development, 6100 Executive Blvd, Rockville, MD 20852
[email: rovneral@mail.nih.gov]

Mahesh N. Samtani, PhD Advanced Pharmacokinetic/Pharmacodynamic Modeling & Simulation, Johnson & Johnson Pharmaceutical Research and Development LLC, Raritan, NJ 08869
[email: msamtani@prdus.jnj.com]

Luc M. De Schaepdrijver, DVM, PhD Toxicology/Pathology, Global Preclinical Development, Johnson & Johnson Pharmaceutical Research and Development, Turnhoutseweg 30, B-2340 Beerse, Belgium
[email: ldschaep@prdbe.jnj.com]

Euguene Schneider, MD Medical Director, Clinical Research Amicus Therapeutics, 6 Cedar Brook Drive, Cranbury, NJ 08512
[email: eschneider@amicustherapeutics.com]

Steven A Silber, MD Vice President, Established Products, Johnson and Johnson Pharmaceutical Research and Development LLC, 1125 Trenton Harbourton Road, Titusville, NJ 08560
[email: ssilber@mccus.jnj.com]

M. Renee Simar, PhD INC Research, Inc., Austin, TX 78746
[email: rsimar@hughes.net]

Tibor Sipos, PhD Digestive Care, Inc., Bethlehem, PA 18017
[email: tibsipos@fast.net]

Robbyn E. Sockolow, MD Division of Pediatric Gastroenterology and Nutrition, New York Presbyterian Hospital, Weill Cornell Medical College, New York, NY 10021
[email: deyongpc@aol.com]

Aliza B. Solomon, DO Division of Pediatric Gastroenterology and Nutrition, New York Presbyterian Hospital, Weill Cornell Medical College, New York, NY 10021
[email: als9047@med.cornell.edu]

Stephen P. Spielberg, MD, PhD Professor of Pediatrics, and of Pharmacology and Toxicology, Dartmouth Medical School and College 1 Rope Ferry Rd., Hanover, NH 03755
[email: stephen.p.spielberg@dartmouth.edu]

Bert Suys, MD, PhD Congenital and Pediatric Cardiology, University Hospital Antwerp, Wilrijkstraat 10, 2650 Edegem (Antwerp), Belgium
[email: bert.suys@uza.be]

John N. van den Anker, MD, PhD Division of Pediatric Clinical Pharmacology, Children's National Medical Center, Professor of Pediatrics, Pharmacology and Physiology, George Washington University School of Medicine and Health Sciences, 111 Michigan Avenue, NW, Washington, DC 20010
[email: jvandena@cnmc.org]

Gigi Veereman-Wauters, MD, PhD Division of Pediatric Gastroenterology, Hepatology, and Nutrition, Queen Paola Children's Hospital ZNA and University Hospital Antwerp, 2020 Antwerp, Belgium
[email: gveereman@skynet.be]

Babette Zemel, PhD Division of Gastroenterology, The Children's Hospital of Philadelphia, University of Pennsylvania School of Medicine, 3535 Market Street, Philadelphia, PA 19104
[email: zemel@email.shop.edu]

PART I

PAST, PRESENT, AND FUTURE OF PEDIATRIC DRUG DEVELOPMENT

CHAPTER 1

A New Model for Children

ANDREW E. MULBERG, MD

Internal Medicine Portfolio, Johnson and Johnson Pharmaceutical Research and Development, LLC, 1125 Trenton Harbourton Road, Titusville, New Jersey 08560

JOHN N. VAN DEN ANKER, MD, PhD

Children's National Medical Center, 111 Michigan Avenue, N.W., Washington, DC 20010

STEVEN A. SILBER, MD

Johnson and Johnson Pharmaceutical Research and Development, LLC, 1125 Trenton Harbourton Road, Titusville, New Jersey 08560

Januscz Korczak (1878–1942), a noted author, founder of modern day orphanages in Poland, and posthumously honored as "pediatrician" by the American Academy of Pediatrics, has long been known as a children's advocate. Korczak spoke of a Declaration of Children's Rights long before any such document was drawn up by the Geneva Convention in 1924 or the United Nations General Assembly in 1959. Excerpts of his writing as a call to action include many famous aphorisms including: "Love the child, not just your own" and "Who asks the child for his opinion and consent? As the years pass, the gap between adult demands and children's desires becomes progressively wider."

The concept for this book grew out of my 20-year career as a pediatrician and pediatric specialist and became an even more urgent endeavor since I am now involved in drug development for children at Johnson & Johnson. The development of a book devoted to the critical issues involving children, of all age groups from preterm to adolescents, has been a personal mission since I have entered the pharmaceutical industry. From partnership and colleagueship, this book is a multidimensional text of all aspects and stages of the pediatric drug development process, coedited with academic and industry colleagues, John N. van den Anker and Steven A. Silber. Given the rapid evolution of regulatory policy governing drug development for children, it is a timely book that should be a valuable resource to industry, regulatory agencies, academia, and investigators worldwide.

Even in the so-called developed world, there are significant unmet medical needs for both adults and children. Moreover, the history of drug development is clearly one of neglecting the specific needs of children, by assuming, incorrectly of course, that data generated in

Pediatric Drug Development: Concepts and Applications
Edited by Andrew E. Mulberg, Steven A. Silber, and John N. van den Anker

adults can be applied directly to children. Thus, this new textbook addresses the significant and unmet needs of infants and children regarding proper drug development. For decades, the needs of infants, children, and adolescents have been ignored in the drug development process. For decades, Shirkey[1] has described infants and children as "therapeutic orphans," attesting to the fact that drugs are not often developed for their specific and unmet medical needs. The awareness of the differences between the pediatric patient and the adult patient has in large part not been adequately addressed by the pharmaceutical industry. Drugs are used in children every day with little guidance on appropriate dosing based on a lack of understanding of the specific pathobiology, metabolic and physiological differences, and developmental changes that characterize the differences from the adult subject. Without understanding the differences between the child and the adult, there have been many examples of therapeutic misadventures, including thalidomide, elixir of sulfanilamide, and chloramphenicol. These three of many examples have led to the death of infants and children due to the lack of appreciation of the developmental differences of infants and children from the adult subject. The potential adverse impact on the pediatric patient without understanding the individuality of the pediatric subject is not acceptable. Far beyond the deaths, however, are the many potentially preventable adverse drug reactions or decreased efficacy in children, which occur because of over- or underdosing and unrecognized drug–drug interactions because the clinical information was never developed in pediatric populations.

The scope of this book addresses the unmet medical needs of all stakeholders to develop a new model of collaboration for the benefit of children who require therapeutic options supported by data generated in appropriate populations. This book addresses the scientific background of the differences between the pediatric and adult patient, the ethics of exploring these differences in clinical development programs, the business case for proper development of drugs for children, the technical feasibility, and the process that is necessary for a comprehensive pediatric drug development program. The applications of these approaches will benefit all stakeholders because it will result in better and safer drugs for the pediatric population.

The chapter flow represents the importance of this new model, integrating the needs of all stakeholders in the drug development process: government, academia, parents/patient organizations, and pharmaceutical industry. We start with a historical perspective of pediatric therapeutics, outline the population demographics, develop the business case for proper pediatric drug development, and review the ethics demanding a new model, and then we present the unique functional areas for which unique expertise is required. These functional areas represent the cornerstone of the pediatric drug development process: regulatory directives from Food and Drug Administration (FDA), the European Medicines Evaluation Agency (EMEA), and Japan (U.S., EU, and Japanese perspectives), preclinical/developmental toxicology, clinical pharmacology, clinical and operational development including safety, CMC/formulation development and case examples of successes.

Why have these issues been ignored for so long and what accounts for the changing environment? The economics of drug development together with a narrow interpretation of the "ethics" of drug development for children dictated that, despite rare occasions of specific agents developed for niche diseases, first indications for new chemical entities (NCEs) would be in adults. Approved drugs would then be "downstreamed" for children, without, in many cases, additional clinical trials or even pharmacokinetic (PK) studies. Change has come largely through the actions of governmental (FDA) incentives, including the Best Pharmaceuticals for Children Act and the Pediatric Research Equity Act. Due to a "perfect storm" of regulatory guidances, the interest of pediatric experts from academia, and an evolving work force in industry who understood the business case for proper drug development for drugs for children, this book has been born.

It is with great honor and pride that this mission has now been completed. Targeting the needs of children is an important task for all societies, in both developed and "developing" countries, where the impact of utilizing proper therapies proven to be safe and effective has very significant medical and economic consequences.

Nelson Mandela stated that "there can be no keener revelation of a society's soul than the way in which it treats its children." It is with pride that John, Steve, and I sincerely hope that this textbook brings luster to this issue and harnesses the critical and relevant topics in pediatric drug development.

Good reading!

REFERENCE

1. Shirkey H. Therapeutic orphans. *Pediatrics* 1968;72:119–120.

CHAPTER 2

History of Pediatric Drug Development and Therapeutics

STEPHEN P. SPIELBERG, MD, PhD

Dartmouth Medical School and Dartmouth College, Hanover, New Hampshire 03755

"It was the best of times, it was the worst of times; it was the age of wisdom, it was the age of foolishness; it was the epoch of belief, it was the epoch of incredulity; it was the season of Light, it was the season of Darkness; it was the spring of hope, it was the winter of despair; we had everything before us, we had nothing before us; we were all going directly to Heaven, we were all going the other way." Dickens wrote these lines in 1859, yet how they resonate today in the 21st century, particularly for the lives of children. We have witnessed, in but a short lifetime, remarkable advances in the health and well-being of children, as well as the devastating impacts of war, political unrest, poverty, and "new" diseases such as HIV/AIDS. The lives of children have never been better in some venues, and never worse in others.

Since I was a child, infant mortality (deaths per 1000 live births to age 1 year) has decreased from approximately 30 to less than 7 in the United States (it was ~100 in 1900). And yet, it is more than double that among African Americans, and more than 100 times that in most of sub-Saharan Africa. Much of the improvement here is attributable to public health measures—water purity, milk pasteurization, improved hygiene, and education. In fact, the best correlate of an individual child's health internationally and in the United States is maternal education level and literacy. Health is not the province of medicine and medicines alone. Neonatal mortality rates (in the first hours to 7 days of life) have fallen dramatically, due in part to better maternal health as well as remarkable advances in neonatal care, including premature intensive care units and attendant therapeutics including surfactant. Immunization has had a profound impact. Pertussis, measles, polio, mumps, and diphtheria are close to diseases of the past; all can regain their negative impact on children (and adults) with failure of vigilance, and assurance of herd and individual immunity. When I was a pediatric house officer in the early 1970s, meningitis and epiglottitis were two of our most feared admitting diagnoses, now rare with *Haemophilus* influenzae, pneumococcal, and meningococcal vaccines. And yet, HIV/AIDS continues to ravage so much of the world, the "miracle" of prevention of vertical transmission from mother to fetus by medicines unimaginable only 25 years ago barely

Pediatric Drug Development: Concepts and Applications
Edited by Andrew E. Mulberg, Steven A. Silber, and John N. van den Anker

impacting sub-Saharan Africa (only perhaps 100 cases of neonatal HIV infection in the United States per *year* versus 1600 per *day* in Africa).

Diseases that were uniformly fatal when I was a child, or even when I began my career in pediatrics, are now treatable. We have seen remarkable success with many types of congenital heart disease, metabolic disorders including cystic fibrosis, many types of childhood cancer, and extreme prematurity. Children with what once were fatal conditions are now living into adolescence and adulthood. We are beginning to see chronic conditions never before seen in adults. This challenges our health care system. Who will treat these patients who are too old for pediatric care but have ongoing health needs only poorly understood and not taught to internists? What is the long-term prognosis of "survivors of pediatric success"? What new interventions will these patients need to live successful, meaningful lives? And, as well, we are faced with new child health issues of childhood obesity and of behavioral and learning problems that increasingly are challenging our knowledge and our ability to intervene. Despite our unbelievable increase in knowledge of human biology, we need to be humble, balanced by our honest recognition that we are just beginning to scratch the surface. It is apparent in all this that those who argue that we know enough and that all we need to do is to assure access are dismissing the needs of all those patients with conditions we poorly understand and for whom we have inadequate interventions, while those who argue that all we need is more science are turning their backs on all who do not have optimal access to the care we have available today. All children (and all people) deserve our dedication to both—assurance that science and medicine advance through research (fundamental, basic, translational, clinical, and outcome) and that the advances of human wisdom are available to all in need. If one measure of a society is how it treats its most vulnerable members, including children, we have a long way to go to assure that as knowledge and health care advance, children are at the forefront.

The context of the improved overall health status of children in the United States, the impact of public health, or "low tech" interventions, such as bicycle helmets and child restraints in cars, can make the issues in this book seem like not such "big" problems. That is why the perspective above is so important. Yes, the lives of most children in the United States in the 21st century are vastly better than at any time in the past ("the best of times"), yet the research needs of sick children, truly sick children for whom we lack adequate therapy, often are not addressed with the same urgency as in adult medicine. Some simple realities apply. Children do not vote; they require advocates in the political decision-making processes to assure their needs are not forgotten. The good news is that most children in the United States are indeed "healthy" and do not need complex therapeutic interventions, and those with serious and chronic diseases are not distributed among a few highly prevalent illnesses. However, this happy reality has major consequences for drug development for children.

In adults, heart disease, hypertension, Alzheimer's disease, many types of cancer, and many other diseases have high prevalence in the population. If a drug candidate is developed to treat one of these diseases, there are many patients to enroll in clinical trials, and once the drug is approved, the marketplace is large. For children, there are relatively few patients with any specific diagnosis (beyond conditions such as asthma, upper respiratory infections, etc.), and thus fewer patients to study, making clinical trials difficult. Furthermore, "children" are not one group. Studies might be needed in newborns, infants, toddlers, older children, and adolescents. Each group may have different dosing needs, different responses to medications, and different adverse reaction profiles and may need different dosage forms for safe and accurate administration. The cost and complexity of studies go up and, once the studies are completed, the marketplace may be so small that it is economically not viable to produce and distribute the product.

Traditionally, this has left children as "therapeutic orphans," and pediatricians have been forced to use medicines with little hard data about proper choice of medicine and doses and about what to expect in terms of efficacy and side effects. Tragic outcomes from gray baby syndrome from chloramphenicol and deaths from drug excipients, such as diethylene glycol in elixir of sulfanilamide (1936) and benzyl alcohol in intravenous drugs used in newborns (1980s), are well-documented examples of the need for scientifically and ethically driven pediatric drug development and therapeutics. "Experiments" such as the Best Pharmaceuticals for Children Act have attempted to raise the priority of pediatric studies through financial incentives and have been remarkably successful over the last years. There has been a major response to the Act by the pharmaceutical industry, working with academic pediatricians and the FDA. We have developed much needed new data about medicines in children and have learned that old assumptions about doses, efficacy, and side effects of many drugs were incorrect. Efforts by the National Institute of Child Health and Development (NICHD) to assure pediatric investigation as part of the National Institutes of Health (NIH) Roadmap initiative in Clinical and Translational Science Awards and in creating the Pediatric Pharmacology Research Unit (PPRU) Network have been vital in advancing the field of developmental and pediatric pharmacology. These issues are adequately covered in this textbook by authors who have led the way in proving their utility. Similarly, efforts within the FDA to improve pediatric drug development have been pivotal to moving the agenda forward. Improving the "therapeutic lives" of sick children requires ongoing advocacy and public–private partnerships, bringing together all those who support and implement pediatric clinical investigation.

As we move forward in the United States, we need to remember that birth rates in the "developed" world are falling, and most children live in regions of the globe plagued by poverty and political and social instability and are afflicted by diseases not prevalent in the developed world. There are signs of hope in the latter sphere with initiatives such as the Gates Grand Challenges in Global Health, seeking novel public–private partnerships to address the needs of children and adults internationally. More "out of the box" creative initiatives will be needed to focus on the needs of those so long neglected.

One of the critical elements in all human investigation is assurance of the very highest ethical standards. Indeed, some of the elements of human investigation ethical assurance, particularly those derived from "respect for persons" and human autonomy leading to concepts of voluntary informed consent, had been viewed as impediments to having children participate in clinical trials. Issues were specifically raised in trials where the prospect of direct benefit to the individual trial was absent or unlikely. We have struggled with these issues over the years, and while there always should be ongoing discussion and consideration of ethical issues as context, science, and society change, there is now a reasonable grounding to allow for pediatric investigation both in the United States and internationally. In fact, as the science of clinical investigation has advanced to allow us to "technically" address key issues in therapeutics in children, we have come to realize that use of medicines in clinical practice not guided by data from good studies *in children* places children at greater risk than participation in well-designed, ethically conducted clinical trials. The advances in outcomes of children with cancer are a tribute to the value of capturing clinical results of therapies in structured clinical trials. Guidance can be found in statements from the American Academy of Pediatrics Committee on Drugs (*Pediatrics* 1995; 95:294), the U.S. government (NIH and FDA: 45 CFR 46, 21 CFR 50), the Institute of Medicine (mandated by the Best Pharmaceuticals for Children Act to review current guidelines: Ethical Conduct of Clinical Research Involving Children, Institute of Medicine of the National Academies, 2004), and international regulatory documents (International Conference on Harmonisation, ICH E11).

The future of pediatric therapeutics will depend on our expanding knowledge of disease pathogenesis, the discovery of new "drug targets" and new medicines, the skillful and ethical evaluation of those medicines in children, and more innovative approaches to "driving" the development of medicines specifically for pediatric diseases and for assuring that all medicines that are used in children are properly evaluated. We will need more and better trained pediatric clinical investigators and clinical pharmacologists. We will need more networks of centers skilled at working together, in sharing large volumes of data. We need to be sure that the needs of children are kept at the forefront of thinking in academic, industry, and regulatory settings.

We are approaching an era of increased emphasis on "personalized medicine," providing the right drug to the right patient based on improved diagnostic specificity and understanding of the net benefit of a specific therapy for a specific patient. We have yet to develop creative approaches from a health care delivery and financing perspective that will ultimately lead to a significant change in the use and development of medicines. We may be seeing a gradual move away from traditional development of "blockbuster" drugs to more targeted therapeutics, but the costs of drug development keep rising, and regulatory change is necessary. If we are correct that this approach ultimately will lead to improved outcomes for our patients and improved prediction of effectiveness and safety, then as pediatricians we should be partnering with others struggling to make this happen. Pediatric subpopulations, by age and by disease state, should—*must*—be included in thinking about how to optimize therapeutics. I believe that this may represent a real opportunity for children. Ultimately what matters in therapeutics is how specific medicines work in individual patients. Given human heterogeneity, we all are "therapeutic orphans," seeking optimum, individualized treatment. Recognizing that, children do not really seem so different conceptually but still will need our advocacy, wisdom, focus, and dedication. It is with these sentiments that the textbook herein developed can partially fulfill the needs of the pediatric population involved in clinical research and development.

CHAPTER 3

Perspectives on Pediatric Clinical Trials: The Good, the Bad, and the Ugly

GREG KOSKI, MD, PhD
Harvard Medical School, Boston, Massachusetts 02114

OVERVIEW

Using children as the subjects of scientific study is a topic that never fails to stimulate debate, whether broached in scientific circles, in the ethics domain, or in the public media. No one questions the need or desire to know more about conditions affecting the health and well-being of children, whether from the physiological, psychological, or pathological vantage point. And yet, even with everyone in agreement on this fundamental point, the avenue forward soon splits into many different paths, some less well traveled than others, some highly risky if not treacherous, and most seemingly fraught with obstacles at every turn.

Navigating the complexities of this domain is unquestionably challenging. The several excellent contributions in this volume offer invaluable guidance for those wishing to try. The intent of this chapter is to offer an overview of research involving children considered from several perspectives, including those of investigators, sponsors, contract research organizations, physicians, regulators, ethicists, and, most importantly but least well appreciated, from the perspective of children themselves to the extent that we as adults can truly know and understand their points of view or appreciate their experiences.

The use of the term *children* to describe the remarkably diverse population of individuals in question is a gross oversimplification. Usual demographic descriptors of children, such as sex, age, weight, height, newborn, infant, toddler, adolescent, teenager, and so on, cannot begin to adequately encompass the vast range of individual variability in physical maturation, cognitive development, intellectual capacity, emotional maturity, and life experience within any given group. Accordingly, generalizations, even if well intended as a means to simplify the discussion, are more likely than not to be inaccurate and misleading.

At the same time, among the many goals espoused for the clinical research endeavor in children and those engaged in it, there are several competing interests that come to bear,

Pediatric Drug Development: Concepts and Applications
Edited by Andrew E. Mulberg, Steven A. Silber, and John N. van den Anker

including those of the children as individuals, their families, science, industry, and society at large. Competing interests often become conflicting interests, particularly when the prospect of financial gain or loss is added to the mix, drawing into question the motives of those who are empowered, rightly or wrongly, to make decisions on behalf of children, most of whom are denied the right to decide for themselves what will be done to them, with them, or for them.

Knowing all of this up front, those who choose to venture into this difficult area must be either intrepid or ignorant—this ground is not tread lightly. Some of the assertions and opinions offered in this chapter will probably be controversial at best, or simply wrong at worst, and yet excusable in light of the many complexities noted. At least, they are based on a broad foundation of training and experience in the domains relevant to the topic and certainly well intended.

Those who engage in clinical research do so with a bias that knowledge is a good thing, and to the extent that research is the pathway to knowledge, research is good and good research is ethical. What makes research ethical has been addressed by others and need not be reiterated here, save for the three fundamental principles on which ethical research is founded—respect for persons, beneficence, and justice. Many include a fourth principle, autonomy, derived from the principle of respect for persons. Autonomy is perhaps the most vexing aspect of research involving children. Any individual or group of individuals whose autonomy is limited is vulnerable to the will of others and easily victimized. Such is the case of children. Whether the autonomy of children is limited by their age, stage of development, or through laws passed by a society sometimes unwilling to grant children the respect and autonomy they may rightfully deserve, the fact that children generally either cannot or may not make decisions about their own destiny is often the root cause of the dilemmas faced when scientists endeavor to study them.

Perspective is always changing in a dynamic world; what one sees depends not only on the direction in which one is looking, but on where one is standing at the moment. In research involving children, perspectives vary widely, from the very positive to the extremely negative, depending in large measure on how one views motivation for the work, management of risk, approach to recruitment and oversight, and the specifics of the populations being studied.

THE GOOD

While empirical evidence regarding public attitudes toward research in children is limited, a common view about using children as subjects holds that *good* research is not *too* risky and has a high likelihood of yielding information that will truly benefit children as a whole or, at least, specific subpopulations of children with a condition that compromises their health or well-being. This perspective is widely accepted among scientists, ethicists, and parents. While some might argue that it is improper and unethical ever "to use children as guinea pigs," tenants of this view are often considered extremists by the mainstream of modern society within which a more pragmatic view prevails. Just as some would argue that animals should never be used for research, others will take the same view regarding children as research subjects, but this view is widely considered to be unjustified and irrational. A fairly concise and practical guideline for what constitutes good health research in children is set forth in the Australian National Statement on Health Research (http://www.nhmrc.gov.au/publications/humans/part4.htm):

Research is essential to advance knowledge about children's and young peoples' well-being, but research involving children and young people should only be conducted where:

(a) The research question posed is important to the health and well being of children or young people;
(b) the participation of children or young people is indispensable because information available from research on other individuals cannot answer the question posed in relation to children or young people;
(c) the study method is appropriate for children or young people; and
(d) the circumstances in which the research is conducted provide for the physical, emotional and psychological safety of the child or young person.

The Australian position is perhaps unique in that it does not even mention the word *risk* in relation to research in children. Even if Australians as a whole may have a somewhat different perspective on acceptance of risk than other nationalities, as Bill Bryson observed in his narrative about traveling *In a Sunburned Country* (Broadway Book, New York, 2000), the ability to characterize *good* research as research that *should* be done in the absence of a nebulous risk/benefit calculus is laudable, as it serves to underscore the benefits of acquiring knowledge rather than the danger of doing so.

Most reasonable people would probably agree with this view. Whether the research is being designed, proposed for review, conducted, or published, studies that fall within the comfortable bounds set by these guidelines will rarely be objectionable on either scientific or ethical grounds. Importantly, the guideline works equally well in the biomedical industry.*

From the perspective of one in industry charged with designing a *good clinical trial*, the specific points in the Australian statement can be translated into a useful operational guideline. Any drug, device, or biologic that is not important to the health and well-being of children is unlikely to be profitable even if it were to be marketable, so its development would be questionable from both an ethical and business perspective. Similarly, if the study methods are not appropriate for the population to be studied, the likelihood of successfully recruiting sufficient numbers of subjects to participate in the study in a timely fashion is small, and unless the physical, emotional, and psychological safety of the participants is reasonably assured, few parents would be inclined to consent.

When it comes to the issue of consent, the pathway to best practice is more obscure despite agreement in principle that consent for a child's participation in research must be obtained from the child, the child's parent(s) or guardian, or another individual or organization as prescribed by law. The perspectives of children and their families often differ very significantly from those of sponsors and investigators when it comes to consent, and the differences are very important. Consent, as employed in most clinical trials, is a legal construct that requires comprehension, understanding, and volition, along

* The American Academy of Pediatrics has issued *Guidelines for the Ethical Conduct of Studies to Evaluate Drugs in Pediatric Populations* (http://aappolicy.aappublications.org/cgi/content/abstract/pediatrics;95/2/286?maxtoshow=&HITS=&hits=&RESULTFORMAT=&fulltext=Guidelines + for + the + Ethical + Conduct + of + Studies + to + Evaluate + Drugs + in + Pediatric + Populations&searchid=1169123798554_184), something that everyone involved in the development of pediatric clinical trials ought to study, along with the U.S. federal regulations related to permissible research involving children as subjects, 45 CFR 46, Subpart D (http://www.hhs.gov/ohrp/humansubjects/guidance/45cfr46.htm).

with documentation. For most clinical research teams today, the process of consent is merely a prelude to enrollment, and some see it as part of the recruitment process. When practiced in this way—a practice most appropriately described as "consenting the patient" rather than seeking a potential subject's consent—consequences can be dire when something goes wrong. A more appropriate model for which we should strive is that of *informed decision making*: not a process that seeks to acquire consent as the preferred outcome, but one that values a well-informed voluntary affirmation to participate as the best outcome, whether or not the potential subject enrolls.

Parents and children often find the consent process to be intimidating and legalistic, probably because it is too often just that—intimidating and legalistic. From their perspective, a more desirable approach is a compassionate, honest dialogue with someone who really knows and understands what is involved and who is viewed as having the best interest of the child in mind. This set of conditions is not easily achieved in many clinical trial settings, but an investment in training appropriate personnel and monitoring their performance can make a very significant difference in how the process is actually conducted.

A major challenge for the investigator's studying children is that our legal system has stripped children of their right to consent. Legally, all children, even teenagers, lack the capacity to consent for themselves not because they are incapable of making a decision, but because the law does not recognize the legality of their decision making. In our society, individuals are considered to have the capacity to consent unless the legal system, whether through legislation or judicial action, declares them incompetent. In working with children less than the age of majority in the jurisdiction of record, affirmative assent is the requirement, but willing, uncoerced participation is really the desired goal. Much has been written on this topic, and many approaches have been described to try to achieve this goal (see Chapter 30 by Simar in this textbook for further description).

For newborns, infants, and toddlers, parents must make decisions for their children and many would consider it inconceivable that parents would not act in the best interest of their children; a *good* clinical trial would never put them in a position to do so! When the formula described earlier is used to develop a trial, it is a given that the trial *is* in the best interest of the children who might participate. Thus, the burden of ensuring the safety and well-being of the children falls not to the parents at the time of consent but to the investigator and the sponsor at the conception and development of the trial.

This construct of what constitutes a *good clinical trial* springs from a concordance between a statement of ethical principles and what is recognizable as good business strategy. The simple implication is unmistakable—in the clinical trials arena, good ethics is good business! In a world that every day becomes more cynical about the pharmaceutical industry, and where *business ethics* is often cited as an oxymoron, no message can be more critically important. The best approach to developing clinical trials in children is to identify, first and foremost, opportunities to improve the health and well-being of children rather than to improve corporate profits and shareholder equity.

As unrealistic and antithetical to the views of pharmaceutical development and marketing as this approach may seem to some, its legitimacy has been substantiated in practice, both inside and outside the bioscience industry. The often cited and discussed handling by Johnson & Johnson of the Tylenol tampering case in 1982 is considered by some to be a primary example of how ethics-driven decision making can and should apply in the business world. The more recent example of actions taken by the nation's largest mortgage lender, Countrywide Financial, to refinance loans for the benefit of individuals caught in the subprime mortgage crisis rather than foreclose on their homes, if it plays out as many hope that it will, offer some evidence that a company can do well by doing good. Of course, this is sometimes more easily said than done, but it should not be in the case of clinical trials, in either adults or children.

Arguably, the biosciences industry is granted special consideration by society, allowing the use human beings as a means toward its end, drug and device development, and in the end, industry may reap huge profits as a result. Some argue that profits in the pharmaceutical sector have been excessive and that price controls are warranted. Indeed, such controls exist in one form or another in virtually every developed nation in the world *except* the United States, and the prospect of such controls being adopted here are a serious concern for the industry. Public cries for controls become louder and more widespread whenever the industry is characterized as deceiving, bilking, or otherwise abusing the public, and when clinical trials go awry, especially when children are involved, the impact can be substantial. Regrettably, there are too many examples that can be cited to make this case.

THE BAD

Recent surveys indicate that while Americans generally hold research in high esteem, few know much about it, and when the term "research" is associated with the pharmaceutical industry rather than a university, positivity of public opinion declines (http://www.ljshealthcare.com/fileadmin/ljs pharma/ClinicalTrials_Web2.pdf). Again, few are surprised by this observation, but it underscores the disproportionate impact that bad press can have to undermine public opinion when the public is not well informed and the issues are so complex when it comes to a problem with a clinical trial.

So, what is a *bad clinical trial* when children are involved? This question is readily and simply answered within the Australian National Statement cited earlier: "An HREC must not approve, and consent cannot be given for, research which is contrary to the child's or young person's best interests." The reference here is to a Health Research Ethics Committee. According to this view, a *bad clinical trial* involving children is one that is contrary to a child's best interests. This is not the same as saying that a trial is too risky and instead places the trial in a framework of what is really good for the child—it is a question of priority and balance.

For many years, consistent with policies and practices that allow, and even encourage, physicians to use any approved and marketed drug as they see fit, drugs used in children have rarely been well studied in appropriate pediatric populations. More commonly, even though the cardinal rule in pediatrics is that children are not just little adults, drugs are used in pediatrics as if children *are* just little adults—and usually with successful outcomes and acceptable toxicity despite the lack of hard scientific evidence for safety and efficacy. Sponsors thus have little incentive to conduct difficult, expensive, time-consuming trials in children when there is not requirement that they do so before they can be used in children.

As a result, most drugs are prescribed for children through extrapolation of data from trials conducted with adults, complemented by pharmacokinetic and pharmacodynamic studies in pediatric patients receiving the drugs regardless of whether or not they are known to be either safe or effective. To remedy this situation, policymakers in the United States sought to provide incentives for drug makers to perform clinical trials to formally evaluate safety and efficacy of medicines in children. In the language of the pharmaceutical industry, this translates to *exclusivity*, or at least an extension of exclusivity, for pediatric indications of prescription drugs established through rigorous clinical trials.

The program has generally been effective despite widespread criticism that the industry is "trolling" for new pediatric indications for drugs they have already been developed, not because they are necessarily good for children but because they can increase the return on their initial investment in research and development of the drug by increasing patent life and profits. Whether or not this is a *bad* approach depends on the justification for the trials. And again, it comes down to the determination of whether or not the trial is really in the best interest of children.

While the simplicity of this concept of a *bad clinical trial* as stated by the Australians is appealing, the wording chosen is unfortunate. No doubt, an HREC or institutional review board (IRB) ought never to approve a trial in children that is not in the best interests of children; but, more importantly, no manufacturer, sponsor, or investigator should ever design or propose such a study in the first place.

An undesirable and yet undeniable consequence of the regulatory and oversight process for clinical trials in the United States is that IRB approval is considered to mean that a clinical trial is *good*—well, at least okay. Operationally, if a sponsor or investigator can get a protocol approved by an IRB, that is, *any* IRB, it is assumed to be ethically and scientifically sound. This simply is not the case. The process through which IRBs and ethics committees make judgments has never been well studied, but two points are substantiated. First, IRBs approve almost every study proposal they receive, despite the observation that many IRBs may lack sufficient expertise to make appropriate decisions; and second, they behave in highly idiosyncratic ways, often taking multiple and divergent views when reviewing the same protocol. Just as there are good and bad clinical trials, there are good and bad IRBs.

As long as ethics committees and IRBs are permitted to operate largely behind closed doors and without better understanding of how and why they make decisions, and whether or not those decisions are appropriate to ensure the safety and well-being of research participants, the public is potentially exposed to clinical trials that may be unsound both ethically and scientifically.

Another perspective of the public regarding the biosciences industry is the pervasiveness of conflicts of interest. Because industry sponsors have significant interests in the proprietary financial consequences of their research and development efforts, conflicts of interest are unavoidable. History has taught an important lesson: problems are more likely to arise when marketing and science are commingled. When studies are objectively designed to determine whether a new entity is safe and effective, or to determine whether one treatment is more effective than another, they are generally good. On the other hand, studies designed to show that one company's product is better than the competitor's, or solely to create or enlarge a market for an entity, are not science. Sometimes the line between the two is fine or poorly defined.

Once again, intent is important. A close colleague and I once had an intense discussion about this issue. He, a vice president for global research and development for one of the world's pharmaceutical giants, claimed that without vigorous marketing, the public would never have access to promising new drugs. I argued that safe and effective drugs ultimately sell themselves because they make people's lives better. We ultimately agreed that we were both right, but that in the absence of effective marketing, companies would probably sell less of their products and make less money. Where we were unable to agree, however, was around the issue of clinical trials proposed or designed by a company's marketing department. I continue to believe that such studies undermine the integrity of science and the public's confidence in clinical research. When studies designed under the provisions of the pediatric exclusivity extension fall into the marketing realm, we do an injustice to children and their families. I believe we can and should do better. Unfortunately, we can also do worse and, on occasion, we have.

THE UGLY

It is tragic when something goes wrong in a clinical trial, and if the trial happens to involve children, emotions generally run high and reaction is strong. The disclosures of the studies of hepatitis transmission done at the Willowbrook State School and the radio-iodine lacing of milk given to mentally retarded inmates at the Fernald School during Cold War radiation experiments in the 1960s are historically among the frequently cited examples of unethical use of children in research. More recent examples include studies on lead paint abatement performed in Baltimore under the auspices of the Kennedy–Krieger Institute and HIV/AIDS studies conducted at New York City's Incarnation Children's Center (ICC).

ICC is a group home exclusively for HIV-positive foster children near Harlem, established in 1989. In 2004, allegations arose that, between 1995 and 2002, more than 100 ICC children, mostly black and Latino babies and children from poor families, were illegally used as test subjects in Phase I and II clinical trials funded by the NIH and administered by physician investigators from Columbia–Presbyterian Medical Center.

According to published reports (http://www.altheal.org/texts/house.htm), many of the children were taken from the custody of their HIV-positive mothers by the city's Administration for Children's Services and were used in trials of vaccines and drugs, including AZT, protease inhibitors, and combinations supplied by several major pharmaceutical manufacturers. Freelancer Liam Scheff, based on interviews with ICC's medical director and several parents of children at the home, reported that "when the children refuse the drugs, they're force-fed. If the kids continue to refuse, they're given a surgery to implant a plastic tube through their abdomen into their stomach. The drugs are then injected directly into their stomachs—no refusing." Scheff alleged that several children had suffered severe treatment side effects and that some had died as a result.

Many have questioned whether the ACS was acting in the best interest of the children entrusted to its custody when it gave consent for these experiments. The prospect that this research can be characterized as "commercially driven" is part of the basis for concern. That view, together with the concern of using populations of convenience, children lacking parents who would have to consent to their children's participation, cast the research in an unflattering light and it dimmed even further by the public's sagging perception of the trustworthiness of the industry (http://www.pharmaceutical-business-review.com/article_-feature.asp?guid=DE5ED0BA-28DB-4011-9FFA-781D96BBD6EB).

Among the ugliest of recent episodes was the alleged collusion between the pharmaceutical industry and the U.S. Food and Drug Administration to "cover up" the increased risk of suicide in children taking selective serotonin reuptake inhibitors (SSRIs) for major depressive disorder (http://www.newmediaexplorer.org/sepp/2005/01/01/eli_lilly_knew_prozac_causes_suicides_violence_fda_closed_both_eyes.htm). Subsequent to public hearings on the topic, the FDA issued warnings regarding the use of these drugs in children (http://www.fda.gov/cder/drug/advisory/mdd.htm), but not until after internal documents were released amidst allegations that the agency's own internal safety analysis had been suppressed by high-level FDA officials (http://www.antidepressantsfacts.com/mosholder-ssri-suicidal-events-report-barred.htm). SSRIs are now widely considered to be contra-indicated in children, and the credibility of both the industry and the FDA has been damaged as a result of the episode and its coverage by the media. Some child advocacy groups have complained that these events, when juxtaposed with increasingly widespread screening programs for depressive disorder in children and adolescents (http://www.ahrq.gov/clinic/3rduspstf/depression/depressrr.htm), with industry support, amount to "disease mongering" and market manipulation aimed at children (http://www.mindfreedom.org/kb/psych-drug-corp/drugs-in-schools/view). However one views the events that have transpired in this case, the overall impact is negative and pervasive. Much has been written about these events, how they came about, who was to blame, and what should be done. The simple truth is that these events should never have occurred, and steps should be taken to ensure that they do not recur. Greater openness and transparency here and in similar cases would go a long way in that regard, but additional changes in attitude and actions are needed.

THE WAY FORWARD

Whatever the specifics may be in the above referenced cases, these several examples illustrate the kinds of situations and activities that contribute to a growing perception that the pharmaceutical industry cannot be trusted, and this concern has the potential to spread to clinical trials involving children (http://www.usatoday.com/tech/news/techpolicy/2003-05-28-kids-usat_x.ht). Justified or not, the association between corporate profits and efforts to expand clinical trials in children, however well intended, is perceived negatively by many, and unless this perception can be effectively countered, skepticism is likely to continue. Toward this end, several specific courses of action may be helpful.

First, education is critical, but it must be provided by independent sources whose intent to promote and protect the interests and well-being of children is unquestioned. Several groups, including the FDA (http://www.fda.gov/consumer/updates/pediatrictrial101507.html), the American Academy of Pediatrics, and the American Academy of Family Practice, offer information and guidance for parents considering enrolling their children in clinical trials. The need to better educate the public about clinical research in general is great across the board, but the challenge with children and parents is complicated by the special considerations already discussed. The concurrent need to better educate those engaged in research activities involving children is further addressed below.

The second step is for the industry to openly acknowledge and address the skepticism that surrounds essentially any activity in which it engages where money is an issue. Today, even if a company wishes to donate resources to a worthy cause, if there is any potential that the company might benefit financially, a public that has become cynical and distrusting will view its actions and motivation with suspicion. With respect to clinical trials in children, sponsors need to be more cognizant of and sensitive to concerns over competing and conflicting interests, and should take steps to ensure that only studies determined by independent third parties to be in the best interests of children are conducted.

As an example, rather than donate monies to underwrite screening programs intended to identify more children with conditions for which their drugs can be prescribed (i.e., "market building"), or to undertake clinical trials intended primarily to identify new conditions in children for which existing drugs can be marketed (i.e., "disease mongering"), companies should instead focus their research on known serious problems affecting children and establish rigorous policies and procedures for internal and independent external review of proposed studies involving children, and this review ought to incorporate input from parents and children with the condition in question. These policies could include firewalls between their marketing and research divisions, so that no clinical study is ever proposed with the specific intent of marketing to children rather than finding better ways to treat and prevent conditions affecting children. Companies can continue to provide a useful public service by donating funds to support independent, objective educational programs about clinical research involving parents and children, and if this were done collectively by a group of industry leaders, all would benefit. To be effective, however, such an initiative must be free of commercial intent. Finally, all sponsors of clinical research involving children should ensure (1) that every individual involved in the endeavor is properly qualified and rigorously trained to conduct the research; (2) that such studies be conducted only with the review approval and oversight of an institutional review board or ethics committee that is accredited by an independent accrediting organization, such as the Association for Accreditation of Human Research Protection Programs; and (3) that such studies be performed only at sites that have been intensely scrutinized to ensure proper facilities and resources are available.

At the present time, no such requirements exist, and individual investigators with little or no training in pediatrics are permitted to conduct clinical trials involving children without demonstration of their competency to do so. Sponsors, working in concert with professional organizations and oversight agencies such as the FDA, could dramatically improve the current situation overnight by simply demanding that all investigators and members of their research teams, whether selected through internal mechanisms or by outside contract research organizations, are properly trained and qualified and meet the highest professional standards, including third party certification.

No one can deny that the only way to truly expand our knowledge about better ways to prevent and treat conditions involving children is through research, or that this research must be done right. When it comes to clinical trials involving children, doing it *right* means putting the interests and well-being of children ahead of *all* other considerations. Improving health, quality of life, and disease prevention are valuable goals, even when the research needed to reach those goals involves a measure of risk. If we ask every research question with this in mind, design every trial as if our own children would be enrolled, and engage both parents and their children in an open, compassionate discussion aimed at making an informed decision about participation instead of simply acquiring consent to enroll, the goodness of what we do will be enhanced, and its future success assured.

CHAPTER 4

Population Dynamics, Demographics, and Disease Burden of Infants and Children Across the World

CHRISTOPHER-PAUL MILNE, DVM, MPH, JD

Tufts Center for the Study of Drug Development, Tufts University, Boston, Massachusetts, 02111

INTRODUCTION

Despite the fact that children comprise about 40% of the population worldwide and childhood extends for one-quarter of the human life span, pediatric uses are not routinely assessed in the development of biopharmaceuticals intended for the general population without the impetus of incentive programs or regulatory requirements. Among the reasons why children are not routinely included in clinical trials are that the practical difficulties of conducting clinical trials in children are usually not recompensed by the return on investment. The pediatric patient pool, already considered a subpopulation, is further fragmented, both in terms of the market and clinical trial participation. In clinical trials, older children are sometimes included with adult participants, whether that is advisable or not from a scientific standpoint. Infants and toddlers are generally excluded from mainstream clinical trials because of the developmental differences from the adult participants, and their need for special formulations. In the marketplace, there are a number of medicines specifically formulated or developed for children, but the majority of pharmacological needs in pediatric practice are met by using adult medicines off-label. The purpose of this chapter is to explore features of the pediatric population in the United States and worldwide that affect the pediatric market for medicines. In this context, the number of children that are potential patients in various regions of the world and how and why those proportions will change over the next few decades are important considerations. Similarly, the specific and general types of diseases that children experience and their contribution over time and by region to the overall disease burden will affect the viability of the pediatric clinical trials enterprise.

Pediatric Drug Development: Concepts and Applications
Edited by Andrew E. Mulberg, Steven A. Silber, and John N. van den Anker

DEMOGRAPHICS

United States

By 2003, there were 73 million children aged 0–17 in the United States, or 25% of the population, down from a peak of 36% at the end of the baby boom (1964). This proportion is expected to decline only slightly to 24% by 2020. The racial and ethnic mix was 60% of children white-alone/non-Hispanic, 16% black-alone, 4% Asian-alone, and 19% Hispanic (this group having the most dramatic change over time, more than doubling from 9% in 1980).[1]

While the proportion of the youngest segment declines, the proportion of the oldest segment of the population increases. By 2030, the proportion of Americans 65-and-older is expected to almost double from what it was in 2000. This demographic trend has implications for the healthcare system. In 1987, an average of $1033 was spent on healthcare for children under age 6, and $3858 for persons over 65. Ten years later, the inflation-adjusted spending on children was $905 and on the elderly it was $6265. There is tension among competing priorities. For example, some states have chosen not to provide a recent vaccine for the prevention of bacterial meningitis and pneumonia to children due to cost. Medicaid too is considering tightening the eligibility requirements for children or eliminating some benefits. The drug approval process has also been highlighted as an area where children are disenfranchised.[2]

World

The proportion of the world population under age 15 was 29% of a world population approximated at 6,555,000,000 in mid-2006, growing to 7,940,000,000 in 2025.[3] According to another source, the proportion of children approaches 40%, when the age range is expanded to 18 or 19.[4] The proportion of the population that children comprise is extremely variable depending on the region of the world, ranging from just under 50% in Africa to half that in Europe in the 1990s for children 0–14 years of age.[5] Even within regions or continents, the proportions are variable. Egypt, for example, is in North Africa but its population under 15 years of age in 1998 was only 36%, with a growth rate of that population slowing to 1.9% annual average increase from 2.6% in the period 1980–1995.[6] In contrast to growing youth populations in Asia generally, the percent of children (under age 15) in Japan is anticipated to decrease from 14.7% in 2000 to 13.7% in 2020.[7]

The World Health Organization (WHO) projected that the proportion of the world population of 0–9 year olds was expected to decrease to 31% in 2025 from 39% in 1995. In more developed countries, it is expected to decrease from 27% to 24% over the same time period; in emerging economies from 40% to 30%; and in lesser developed countries from 50% to 43%.[8] Another source predicts that the percentage of the world population formed by youth aged 10–24 will go down from 27% in mid-2006 to 23% in 2025.[4] Corroborating evidence for this age distribution transformation comes from the United Nations Population Division as depicted in Figure 4.1.[9] It shows that as of 2000 the population of 0–3 year olds was still substantially on the rise only in sub-Saharan Africa and South Asia. Asia currently accounts for 77% of the world's pediatric population.[10] For example, the population of children in China alone is nearing 300 million,[11] while that of the United States and Europe combined, representing most of the countries of the developed world, is just over half that number at 175 million. While the population of infants is flattening in the Middle East, North

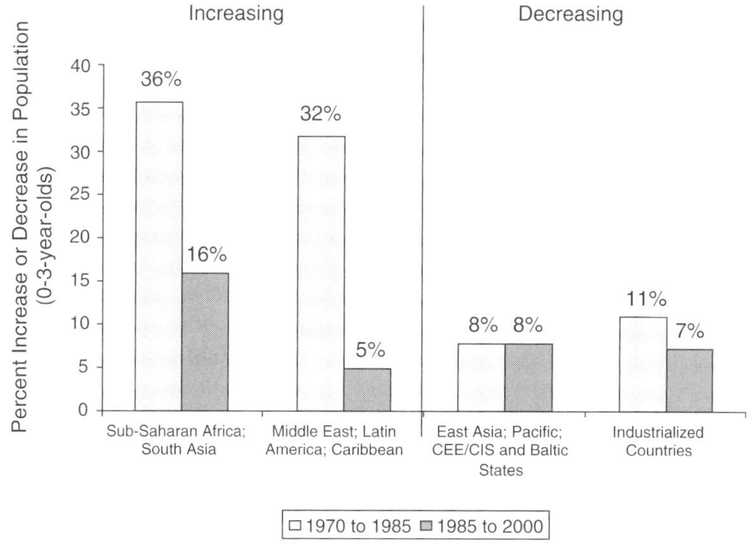

FIGURE 4.1 Trends in child population by region: 0–3 year olds, 1970–1985 and 1985–2000. (*Source*: Adapted from Ref. 9.)

Africa, and the Latin America/Caribbean region, over the next few years, nine in every ten people in regions of Africa, Asia, and Latin America will be in the pediatric age range.[12] Lastly, the population of infants is actually declining in East Asia/Pacific, CEE/CIS, Baltic States, and industrialized countries.[9]

GENERAL DISEASE BURDEN

Chronic Illness in the United States

In 2004, 13–23% of the 0–17 population experienced special healthcare needs, such as asthma (12%), learning disabilities (8%), and attention deficit hyperactivity disorder (ADHD) (7%).[13] Approximately 20 million children suffer from one of 200 chronic conditions and disabilities,[14] accounting for 42.1% of total medical care costs.[15] In addition, five of the top ten leading causes of death in children were chronic diseases: malignant neoplasms; congenital malformations, deformations, and chromosomal abnormalities; diseases of the heart; chronic lower respiratory tract diseases; and cerebrovascular diseases. Chronic conditions resulted in 66 million restricted-activity days and 27 million days lost from school.[15] Consequently, drug utilization trends reflect this disease burden. In 2005, there were over 9.7 million children (13%) in the United States who had a problem for which prescription medicines had been taken regularly for at least 3 months,[16] up from 9% or 6.6 million children in 1998.[17]

Chronic Illness Worldwide

Although current global public health concerns focus on acute illnesses, especially in children under age 5, that toll is expected to decline by half by 2025. In contrast, the proportion of chronic diseases, such as depression and cardiovascular illness, throughout the world is expected to rise from 55% in 1990 to 73% in 2020, with the most rapid increase expected in the developing world.[18]

Acute Diseases in the United States

Appendicitis, intestinal infection, noninfectious gastroenteritis, abdominal pain, esophageal disorders, and congenital digestive anomalies accounted for 75% of the gastrointestinal (GI) discharge diagnoses.[19]Respiratory syncytial virus (RSV) is the most common cause of hospitalization from acute respiratory disease among children age 1 and under. Annually, an estimated 51,000–82,000 young children are hospitalized, and approximately 2% die.[20] Acute illnesses were the third leading cause of hospitalizations (after normal newborn infants and conditions related to pregnancy) in 1997 with 330,000 pediatric discharges, costing $2.6 billion in hospital charges.[19]

Acute Diseases Worldwide

Deaths in children under age 5 result from neonatal causes (37%) including diarrhea, acute respiratory infections (19%), diarrhea deaths during the postneonatal period (17%), HIV/AIDS (3%), measles (4%), malaria (8%), and other causes (13%).[21] More than 10 million children under age 5 die each year, mostly in developing countries, and account for approximately one-sixth of annual deaths in the population worldwide. Yet, 63% of these deaths in children under age 5 could be prevented by universal coverage with available preventative and treatment interventions, and about half of these interventions are pharmaceuticals.[22]

SPECIFIC DISEASE BURDEN

Allergies

United States Twelve percent of U.S. children under age 18 suffered from respiratory allergies, 11% from hay fever, and 13% from other allergies in 2005 (essentially unchanged from 1998),[16,17] although the prevalence of allergic rhinitis is believed to have increased substantially during the last 20 years.[23] The total economic burden of allergic rhinitis and its complications in the United States is estimated at $6 billion per year.[24]

World According to the WHO, 10–20% of children will develop allergies by adolescence, but rates differ by country depending on disease definition, diagnosis criteria, and type of population studied.[25] There is a growing consensus that asthma and allergic rhinitis are linked diseases. For example, European studies indicate that treatment of allergic rhinitis may prevent onset of asthma.[23]

Asthma

United States By 2005, over 9 million U.S. children under 18 years of age (13%) have ever been diagnosed with asthma, and 6.5 million children (9%) still have asthma.[16] Asthma is the leading cause of school absences for a chronic illness,[26] accounting for 14 million lost schooldays and an annual total cost to the U.S. economy of $16.1 billion, $11.5 direct and

$4.6 indirect.[20] According to a study of Medicare/Medicaid patients, the cost breakdown consists of 43% inpatient hospitalization, 30% prescription medication, 14% physician-related services, 8% ER visits, and 5% outpatient hospital services.[27] Two decades ago, asthma affected only half as many children under 18 years of age and cost approximately eightfold less according to the American Lung Association.[28] However, there has been an increase of only 1% in the prevalence of children ever diagnosed with asthma from 1998 to 2005.[16]

World The prevalence of asthma (including wheezing) in 0–14 year olds in selected European countries is somewhat variable: United Kingdom, 13.7%; Germany, 7.1%; France, 6.1%; Italy, 6%; and Spain, 4.8%.[25] Yet, it ranges even more widely in the rest of the world: India,[29] 29.5%; Latin America,[30] 27–32%; Western Pacific, 50%; and Australia,[31] 40%. The prevalence has doubled in the developed world in the past 20 years and is expected to do so in the developing world. Among asthmatics in Europe, 18% have severe persistent asthma, 19% have severe moderate asthma, 19% have mild persistent asthma, and 44% have intermittent asthma.[25] The annual costs for all asthma patients in the EU is US $16.3 billion,[25] while in Australia, it's between US $500 million and US $1 billion each year.[31]

Cancer

United States In 2005, approximately 9510 children under age 15 were diagnosed with cancer. Some 1585 died. Among the 12 major types, hematologic and central nervous system (especially brain) cancer account for over one-half of new cases. The most common type of leukemia is acute lymphoblastic leukemia and the most common solid tumors are gliomas and medulloblastomas. Over the past 20 years, the incidence of all forms of invasive cancer in children has increased from 11.5/100,000 in 1975 to 14.6/100,000 in 2002, but 5 year survival rates have increased dramatically from 55.9 in 1974–1976 to 78.6 in 1995–2001, due to significant advances in treatments.[32]

World The burden of pediatric cancer in the developing world will dwarf that of developed countries. For example, a conservative projection of 45,000 new cases of pediatric cancer are expected in China. Low survival rates will exacerbate cancer's impact on child mortality. Some 60% of children treated for acute lymphocytic leukemia in Honduras are unlikely to survive compared to developed countries, in which 80% survive. The most common reasons for failure are abandonment of treatment (23%) and death caused by treatment (20%), whereas in Europe, abandonment of treatment is virtually unknown and only 2% die from treatment.[33]

Diabetes/Obesity

United States Type 1 diabetes (also called juvenile diabetes, caused by inadequate production of insulin) is one of most common chronic pediatric diseases (1 in 600) with an incidence of more than 400,000 new cases reported every year, up to age 24 (although typically first diagnosed in teens), and a prevalence of 1 million.[34]

Type 2 diabetes (insulin resistant) now accounts for up to 45% of all newly diagnosed diabetes in pediatric patients.[35] A study of 370,000 insured children, aged 10–19, from 2001 to 2006 showed an increase of 167% in girls taking drugs for type 2 diabetes, while boys increased 33%.[36] The major predisposing factor is being overweight, and especially being obese (BMI > 30). Some 2 million adolescents ages 12–19, or 1 in 6 overweight adolescents, have pre-diabetes.[20] A child born in the year 2000 has a 30–40% lifetime risk of developing diabetes unless current trends are reversed.[37] The total economic cost of diabetes in 2002 was an estimated $132 billion, or $1 out of every $10 spent in the United States on healthcare.[20]

At the root of the epidemic is the change over time in the weight status of the pediatric population. Approximately, one-quarter to one-third of children and teens are now overweight, an increase of two- to threefold during the past two to three decades.[38] This may be an underestimate. A study at one children's hospital in Pittsburgh found that two-thirds of pediatricians failed to note that a child was overweight. It found that only 7% of pediatricians ordered lab tests to screen for weight-related problems and only 15% noted children's activity levels and hours spent watching television.[39] Childhood obesity is associated with a variety of adverse consequences, such as risk factors for cardiovascular disease as well as symptoms such as headaches, snoring, daytime somnolence, abdominal pain, hip pain or limp, urinary frequency, nocturia, polydipsia, polyuria, and irregular menses or amenorrhea. Conditions associated with being overweight, such as sleep apnea and gallbladder disease, tripled in children and adolescents from 1979–1981 to 1997–1999.[35] The most common obesity-related comorbidities account for over one-third of pediatric hospitalizations.[40]

World Worldwide, 150 million are affected with diabetes—type 2, 90% and type 1, 10%. The expected patient population will increase to 250 million during the next 10 years.[25] The annual incidence rate in Europe of type 1 diabetes is quite variable among countries, ranging from 3.2 cases per 100,000 in Macedonia to 40.2 cases per 100,000 in Finland.[41] At the same time, the incidence and prevalence of type 2 diabetes in the pediatric population is increasing dramatically in Europe and worldwide according to EMEA.[42] Asian children too are affected. For example, a 1999 urine test screening of 2.8 million Taiwanese children aged 6–18 found that more than 15,000 tested positive for diabetes.[43] Type 2 diabetes incidence almost doubled in Japan from 7.3/100,000 to 13.9/100,000 in junior high children from the late 1970s to early 1990s.[44]

In Europe, about half a million children have metabolic syndrome (high blood pressure, elevated cholesterol, and insulin resistance) with 2000–10,000 children diagnosed with type 2 diabetes.[45] As seen in Figure 4.2, the percentage of children between the ages of 6 and 17 who are overweight in various European countries ranges from less than half to slightly more than the percent of children overweight in the United States.[46] Nearly half the children in North and South America and 38% in Europe could be overweight by 2010 if the current trend continues.[47]

Epilepsy

United States Some 77% of first seizures occur in patients aged 0–19.[48] Many patients with continuing seizure disorders are subsequently diagnosed with epilepsy. In fact, by age 20, 1% of the population is expected to have been diagnosed with epilepsy.[49] Of the 2.3 million persons with epilepsy in the United States, 14% are age 14 or under, and there are around 45,500 new cases of seizure and epilepsy in this age group every year.[50] Epilepsy costs the United States $412.5 billion in medical costs and lost wages.[51]

World The prevalence of adults and children with epilepsy is 50 million worldwide.[52] About 20% present before age 5 and 50% before age 25.[53] Some 80% of cases are found in developing countries,[8] while approximately the same percentage of cases go untreated.[53]

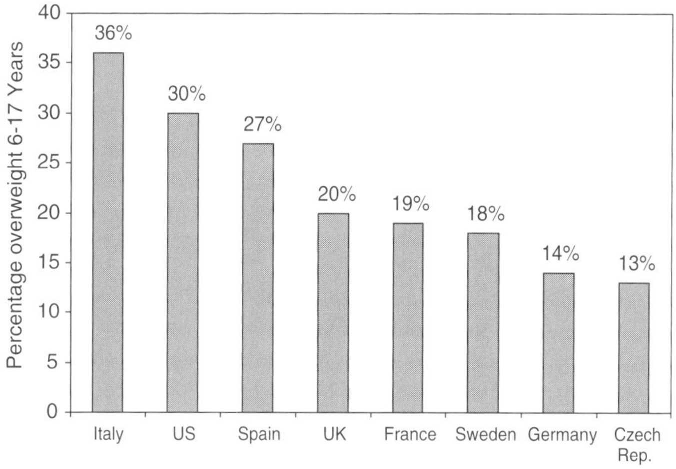

FIGURE 4.2 Percent of children aged 6–17 who are overweight in select European countries compared to the United States. (*Source*: Adapted from Ref. 46.)

However, a recent study showing that the inexpensive and widely available drug phenobarbital is just as effective for treating epilepsy as carbamazine, which is more commonly prescribed, could present a promising approach to address the economic barrier to increasing availability of treatment.[54]

Genetic Disorders

United States There are many genetic disorders that afflict the world's population, the vast majority understandably first manifesting in children. Two such disorders that typify the public health and healthcare economic burden of such diseases are cystic fibrosis and sickle cell anemia.

Cystic fibrosis (CF) is the second most common life-shortening, childhood onset inherited disorder.[20] CF is diagnosed in about 1000 Americas each year, most frequently in Caucasians. It is a life-shortening, genetic disorder characterized by chronic respiratory infections and gradual loss of lung function, typically accompanied by digestive disorder and malnutrition.[55] More than 80% are diagnosed by age 3. Today, nearly 40% of the CF population is 18 and older.[20]

Sickle cell disease (SCD) is a genetic disorder that affects 2000 infants born in the United States each year. This population has a high susceptibility to severe bacterial infection that can cause meningitis, pneumonia, septicemia, and death. Thirty years ago, about half of children with SCD did not reach age 18. Today, about 85% of children with SCD will reach adulthood, thanks to newborn screening, prophylactic penicillin, and the increasing use of disease-modifying therapies such as blood transfusions.[56]

World Cystic fibrosis (CF) is the most common lethal genetic disease in Caucasians, which affects 70,000 people worldwide.[57] Sickle cell disease is most common in West and Central Africa, where as many as 25% of people have sickle cell trait and up to 2% of all babies are born with a form of the disease.[20] Lack of treatment options in the developing world leave these patient populations where U.S. populations were a few decades ago with high disability, low productivity, and shortened life spans.

Infectious Disease

United States Generally, infectious diseases are not among the major causes of child mortality in the United States but do cause considerable morbidity. The annual cost of medical care from treating infectious diseases in the United States is estimated to exceed $120 billion.[20] Some specific diseases have become serious public health threats at different points in time, but the risk and severity have been reduced by advances in treatment. For example, there were an estimated 11,000 HIV-infected children[58] in 2005; however, perinatal transmission declined dramatically with the introduction of the zidovudine regimen for pregnant women in 1994, dropping 70% from 1992 to 1998.[59]Respiratory syncytial virus (RSV) caused 73,400–126,300 children to be hospitalized with bronchiolitis or pneumomia according to a 1999 study by Dr. David K. Shay at the Centers for Disease Control (CDC), with 500–4500 estimated deaths.[60] However, the National Center for Health Statistics reports that the number of pediatric ER visits has declined, especially for respiratory emergencies.

In 2002, children under age 19 made more than 29 million visits to emergency departments (EDs) in the United States according to the National Hospital Ambulatory Medical Care Survey, accounting for 30% of all ED visits. The majority of pediatric ED visits involve children over age 5, and the most frequent diagnoses for young children (under age 10) in the ED are upper respiratory infection and otitis media.[61] Rotavirus is the most common cause of diarrhea in children under age 5, resulting in approximately 55,000 hospitalizations in the United States every year.[20] However, in the United States, these infections result in relatively few deaths and a vaccine has recently reached the marketplace.

World In 1995, of more than 11 million deaths of children under age 5, 9 million were due to infectious diseases, and 25% were preventable through vaccination.[8] Almost 10 years later, there is little improvement. In rural Africa, for instance, the risk of any bacteremic disease for children was 2.5% by the time they were 5 years old.[62]*Plasmodium falciparum* malaria is still a leading cause of morbidity and mortality, resulting in 1.5–2.7 million deaths, mostly of children under 5 years old in sub-Saharan Africa.[63] Rotavirus causes the deaths of 600,000 children annually worldwide.[20] While more than 400,000 children under age 1 die every year from such diseases in Latin America, the rate varies country-by-country (around 10/1000 in Cuba, Chile, and Costa Rica, but 50/1000 in Bolivia and Haiti). Some of this variation is due to public health measures. Chile, for example, achieved a 72% reduction in infant mortality due to pneumonia and bronchopneumonia, and a 48% reduction in neonatal mortality due to respiratory diseases by expanding immunization, physiotherapy, inhaled steroids and bronchodilators, brief one-day hospitalizations, and rational use of antibiotics.[64] Developing and emerging economies are not the only regions at risk. For example, research indicates that 22% of young infants and a few older children admitted to London hospitals with respiratory difficulties actually had whooping cough despite the fact that of the children who actually had the illness, whooping cough was suspected in only 28% of infant cases and 75% of older cases.[65]

Inflammatory Diseases

United States Atopic dermatitis (eczema) is a chronic inflammatory skin disorder that affects 9% of the population. It usually begins during the first year of life and almost always presents within the first 5 years of life.[20] Some 300,000 children suffer some form of arthritis or rheumatic disease. Juvenile rheumatoid arthritis is the most common form in children, affecting 70,000–100,000 under age 16,[20] doubling from a decade earlier.[48] It is estimated that 1 in every 800 children has Crohn's disease, a chronic inflammation of the intestinal wall, which usually appears between the ages of 12 and 16.[20]Ulcerative colitis (UC), a chronic inflammation of the large intestine with ulceration, usually starts between ages 15 and 30 and the prevalence in the United States is 229/100,000.[25]

World Some 17.5% of schoolchildren suffer from atopic dermatitis (eczema).[66] The majority of cases of Crohn's disease start between the ages of 14 and 24, with roughly equal occurrence between the sexes. The overall prevalence is 144/100,000; Japan has six cases per 100,000; Europe ranges from 20 to 76 per 100,000 (United Kingdom, Italy, France, Germany, Spain from highest to lowest).[25] Ulcerative colitis prevalence in Europe runs the gamut from 30 to 122 per 100,000 (United Kingdom but also Italy, Spain, France, and Germany from highest to lowest are in that range), and Japan's prevalence is 18/100,000.[25]

Mental Illness

United States An estimated one in ten children and adolescents suffer from mental illness severe enough to cause impairment, but in any given year fewer than one in five children receive needed treatment.[67] In 1998, the direct cost of treatment was $11.75 billion or $173 per child.[20] In the mid-2000s, the overall costs of mental illness including both direct (treatment-related) and indirect (lost productivity) expenses may exceed $150 billion per year.[68]

The prevalence of the most common behavioral disorder, ADHD, is 5–8% of children and adolescents with impairment that lasts into adulthood, with the disorder three to four times more common in boys than girls.[20] About half of children diagnosed with ADHD receive stimulant agents (methylphenidate, dextroamphetamine) and related agents including stimulants (modafinil) and nonstimulants (atomoxetine). In the United States, a switch-over to long-acting agents (effects lasting more than 8 hours) began in 2000.[69] In the United States alone, 2.5 million children, including as many as 10% of boys aged 10–12, take stimulants.[70] The 9-year median medical care costs for children/adolescents with ADHD were double those of children without the disorder.[71] More than half of primary care pediatricians have prescribed drugs for sleep problems and three-quarters have recommended over-the-counter (OTC) medicines. Sleep problems affect 20–25% of children.[72] Use of prescription sleeping pills doubled to 0.3% of boys and 0.44% of girls.[36]

More serious psychiatric problems are prevalent in the U.S. pediatric population. Approximately 3% of 9–17 year olds had an anxiety disorder that caused serious impairment.[73] Obsessive–compulsive disorder prevalence was 268,000 (under age 18) in 2000.[48] The disease burden of autism was 300 expected cases per 100,000 by age 21.[74] Ten years ago, autism was diagnosed in 1 out of 10,000 children, with a recent increase to 1 out of 166. Of the 1.5 million people living with autism, 450,000 are children.[20] Depression affects 3–8% of children and adolescents. Up to one-third of the 3.4 million children and adolescents with depression may actually be experiencing early onset bipolar disorder.[20] Schizophrenia disease burden was 300 expected cases per 100,000 (risk by age 21).[74]

World Neuropsychiatric disorders are responsible for 31% of disability in the world, with 450 million people suffering a mental or neurological disorder.[52] Nearly two-thirds of people with a known mental disorder never seek help.[8] The impact on children's disease burden was demonstrated by a mental health survey of children in England, Scotland, and Wales indicating that 10% of 5–15 year olds have some type of mental disorder.[75]

According to the WHO's Atlas Project, which conducted a survey of 181 countries, 43% had no mental health policy, 23% had no mental health legislation, 38% had no community care facilities, and only 41% had treatment available for severe mental disorders at the primary care level.[76] About two-thirds of countries spend 1% or less of their healthcare budget on mental health, and 25% of countries do not have access to the three most commonly prescribed drugs to treat schizophrenia, depression, and epilepsy at the primary care level.[8]

Regarding specific behavioral disorders, ADHD is the most studied. Although some countries, like the United Kingdom, with different diagnostic traditions, report lower prevalence rates, overall ADHD appears in multiple nations and cultures at prevalence rates similar to the United States. In 1993, 31 countries had adopted the use of ADHD medications, and by 2003 the number had grown to 55. The usage of ADHD medications increased 274% from 1993 to 2003, with the U.S. share actually decreasing slightly. Among Organization for Economic Cooperation and Development (OECD) countries, use of ADHD medications is positively correlated with increasing gross domestic product (GDP). However, Canada and Australia show higher than expected use, while Italy, Ireland, Austria, Japan, Sweden, and Finland use less.[69]

Among psychiatric diseases, depression is the most significant with a prevalence of 121 million cases worldwide.[52] In the early 2000s, the antidepressants were considered a panacea for this problem, and there was a significant increase in use of these medications in many regions of the world during 2000–2002.[77] Despite the urgency of the problem, however, treatment options have recently become more limited. For example, prescriptions for antidepressants jumped in Britian from 400,000 to 700,000 (2000–2002), but then fell after an increased risk of suicidal ideation was posited. Unfortunately, doctors have been left without alternative treatment options because of the 6–9-month waiting list for psychotherapy.[78]

PAIN/PALLIATIVE CARE

United States Every day more than 39,000 children present to an emergency department, doctor's office, or clinic seeking treatment for an injury. Between 20% and 25% of all youngsters sustain an injury serious enough to warrant medical attention.[79] Adolescents are one of the highest risk groups for traumatic brain injury. One million children in the United States have life-threatening conditions (e.g., AIDS, cancer, trauma) requiring palliative care, and each year 54,000 die, most often in hospitals.[80] Pain is a fact of life for patients with many chronic diseases as well. Children with vaso-occlusive episodes of sickle cell anemia may have to be treated with opioids to achieve adequate pain relief.[81] Eleven percent of children aged 5–15 years and 28% of 15–19 year olds suffer chronic headaches. Forty percent of headache-prone children experience daily symptoms.[82] Approximately 10–15% of children complain of recurrent abdominal pain.[83]

World More therapeutic options for children are clearly needed. In the United Kingdom, 33% of prescriptions for pain management are off-label (i.e., lacking a specific indication for use in children).[84] According to a study published in the *American Journal of Gastroenterology* (September 2005), childhood abdominal pain is a common complaint and may progress to adult irritable bowel syndrome (IBS). In a study from the University of

Dunedin in New Zealand, a history of abdominal pain was documented in 18% of children, peaking at age 7–9 in boys but remaining stable among females. By age 26, IBS was two to three times more common in subjects with a history of childhood abdominal pain.[85]

DEMOGRAPHIC FACTORS AND DISEASE BURDEN

The disease burden is not uniform across the pediatric population and varies by age, region, socioeconomic status (SES), and ethnicity/culture. These factors are complex and interact with other factors such as political upheaval, population displacement, and natural disasters. The following discussion is an overview. Figures and examples are intended to be illustrative, not an exhaustive list.

Ethnicity/Culture

In the United States, 20% of children account for 80% of healthcare spending on children. Skin color predicts a child's life expectancy.[86] The prevalence of overweight adolescents in the United States was 15.5% for children aged 12–19. Overall, this represents an increase of about 4% on average in fewer than 10 years. For non-Hispanic, African-American, and Mexican-American adolescents, the percentage of overweight children was 23% in 1999–2000, having increased ten percentage points in fewer than 10 years.[87] Even though certain ethnic groups appear to suffer more from certain chronic diseases, they can be less likely to use medicines consistently, in a manner more closely associated with ethnicity than socioeconomic status. For example, despite having worse asthma than Caucasian children, African-American and Hispanic children with similar insurance and sociodemographic characteristics are 31% and 42% less likely, respectively, to use inhaled anti-inflammatory medication.[88] Although the prevalence of major depression is believed to be the same as in the United States, sales of antidepressants in Asia are miniscule due to reliance on alternative healers and the stigma of mental illness.[89] Such a cultural bias can persist even when subpopulations relocate and despite greater susceptibility that should engender greater vigilance. Despite the fact that suicide rates among Asian-American adolescents and young adults are higher than for Caucasians, and the fact that they are vulnerable to posttraumatic stress disorders and depression associated with adverse experiences in their countries of origin, these children are less likely to use mental health services than Caucasians.[90]

Immigration

An estimated 1.5 million legal and illegal immigrants arrive in the United States each year. Some 43% of immigrant children live in low-income families, and almost one-third do not have health insurance. Each year 750,000 children are born to immigrant women.[91] U.S. citizens adopted 21,616 foreign children in 2003, mostly from China and Russia, according to the State Department. The Centers for Disease Control (CDC) reports that infectious diseases have been found in as many as 60% of children adopted from abroad, depending on country of origin, and many infections show no symptoms.[92] More than half of the immigrants coming into the United States each year come from countries with endemic intestinal parasitosis. Some 39% of all tuberculosis (TB) cases reported in the United States were found in foreign-born patients (in California it was 67%); one-half of all measles cases in the United States are "imported" from other countries. This problem is not just a U.S. public health issue. More than 2 million people cross national borders every day, 500,000 yearly by air travel alone. Yet we know that one-third of the world population is latently infected with tuberculosis. In fact, the WHO has developed guidelines to help minimize the risk of contracting TB during airline flights over 8 hours. One recent study found that 40% of asymptomatic children returning to the United Kingdom from sub-Saharan Africa were subsequently diagnosed with a tropical diseases.[93]

Poverty

In 2003 in the United States, 18% of all children aged 0–17 lived in poverty and 89% had health insurance coverage at some point during the year.[1] In 2000, 4 million U.S. children younger than 6 years of age (approximately one in five), lived in poverty.[94] Worldwide, approximately 70% of the world's children live in disadvantaged countries,[33] and roughly 40% of the global poor are younger than 15 years of age.[95]

Poverty impacts the susceptibility to diseases in both children and adults, as well as decreasing the likelihood of treatment initiation and maintenance. For example, minority populations living in poverty are at greater risk for epilepsy.[51] In 2005, the percentage of children with government health insurance (16%) who took a medicine for more than 3 months was more than double that of children with no insurance (7%).[16] In New Zealand, a study from Dunedin shows that, even in a developed modern society, the socioeconomic conditions in which children are raised influenced their health in adult life.[96] However, this effect is not linear or absolute. Among OECD countries, there is no obvious relationship between per capita GDP and child well-being. For example, the Czech Republic scored higher than France, Austria, the United Kingdom, and the United States in overall rank despite having lower socioeconomic status.[97]

The impact of poverty also affects the outcomes of disease and treatment for specific disease subpopulations. Low-income children with sickle cell disease were dispensed an average of only 148 days (41%) of an expected 365-day supply of prophylactic antibiotics. Ten percent received none. Evidence at the time suggested that an 84% reduction in infection could be expected when young children took daily penicillin.[98] This association has been documented outside the United States as well. In Scotland, adverse socioeconomic circumstances in childhood have a specific influence on mortality from stroke and stomach cancer in adulthood, which was not due to the continuity of social disadvantage throughout life.[99]

Age Group

Different pediatric groups have their own age-specific susceptibility to certain types of disease. Although the health outcomes range from premature mortality to poor school performance, the costs to national healthcare systems and productivity are similar in magnitude. The following are some examples.

Neonates In the United States, about 71% of babies are born and discharged without problems. However, conditions identified in the neonatal period are among the most expensive diagnoses for children; prematurity, cardiac and circulatory birth defects, other birth defects, respiratory distress syndrome, and other neonatal respiratory problems account for $4.6 billion in costs or 10% of total dollars spent on hospital stays for children and adolescents.[100] Worldwide, of 10.8 million deaths occurring in children under age 5, 3.9 million occur in low-income communities during the neonatal period due to the following causes: 24% by severe infection, 29% by birth asphyxia, 24% by complications of prematurity, and 7% by tetanus.[101]

Infants In the United States, infectious diseases account for five or six of the top ten diagnoses for infants and 1–5 year olds.[100] Worldwide, about 90% of deaths of children younger than 5 years old occur in just 42 countries. Nearly two-thirds of deaths in these 42 countries and 57% worldwide are due predominantly to infectious diarrhea and pneumonia, in addition to other neonatal disorders, with malaria and AIDS playing an important part in some areas of the world, such as sub-Saharan Africa.[101]

Children Allergies are a common ailment that develop in childhood and the incidence seems to be increasing worldwide. Food allergy to peanuts in children tested at one center in Isle of Wight tripled from 1989 to 2001. Half of these children had a history of asthma, and nearly all had eczema.[102] Allergic rhinitis is the most common chronic condition in children and is estimated to affect up to 40% of all children. It is usually diagnosed by the age of 6. Preliminary data indicate that a high proportion of children aged 6–14 years with allergic rhinitis will develop asthma within 2–5 years. A major impact of allergic rhinitis is comorbidity—sinusitis, otitis media with effusion, and/or bronchial asthma.[103]

Adolescents For children aged 13–17, affective disorders are the most common cause of hospitalization, beside nonneonatal or nonpregnancy related conditions.[100] The chance that a teenager has a treatable psychiatric illness such as anxiety, mood, or addictive disorder is nearly 21%. Median age of onset of any mental disorder is 14 years.[104] Half of all individuals who have a mental illness during their lifetime report that the onset of disease occurred by age 14 and three-quarters by age 24.[105]

CONCLUSION

Worldwide, there are significant unmet medical needs; many of the 1500 unmet medical needs identified by the WHO are discussed in this chapter. There is an increased recognition of the need to treat chronic diseases in children; but generally, except in a few countries, there are insufficient resources or prioritization to do so. However, there is an increasing appreciation in the biopharmaceutical sector for the potential value of developing pharmaceutical solutions for niche markets and emerging economies. For example, the value of vaccines is being factored into national healthcare budgets, and there are increasing incentives for vaccine research and development (R&D) due to substantial donations from the charitable sector, the steady income that vaccines can provide to biopharma as a hedge against the vagaries of the blockbuster market, and advocacy efforts directed at increasing R&D for certain diseases. Incentive programs from the public sector or charitable foundations are needed to jump-start R&D for certain neglected diseases such as AIDS, malaria, tuberculosis, tropical infectious diseases, and parasitism. Another positive trend is the increasing awareness of the globalization of public health. Health concerns in one region of the world no longer necessarily remain confined to a specific geographic locale for indefinite periods of time. Public health makes good neighbors of us all.

ACKNOWLEDGMENT

The Tufts Center is supported in part by unrestricted grants from pharmaceutical firms, biotechnology companies, and related service providers. No companies were involved with the production of the author's work.

REFERENCES

1. Federal Interagency Forum on Child and Family Statistics. *America's Children: Key National Indicators of Well-Being, 2005*. Federal Interagency Forum on Child and Family Statistics, Washington DC: US Government Printing Office. Available at http://childstats.gov/americaschildren/highlights.asp. (Accessed October 10, 2007.)
2. Freed GL, Fant K. The impact of the "aging of America" on children. Commentary. *Health Affairs*. March/April 2004;23(2):168–173.
3. World Population Data Sheet. Population Reference Bureau (PRB). 2006. Available at http://www.prb.org. (Accessed September 15, 2007.)
4. Ladley E. Safety, economics converge. *MED AD NEWS*. May 1, 2006;25(5):1, 50.
5. Global Population Profile: 2002. US Census Bureau, International Programs Center. International Database. Available at http://www.census.gov/ipc/prod/wp02/wp-02.pdf. (Accessed October 10, 2007.)
6. Ghazal A. Egypt emerges as clinical trials location. *Appl Clin Trials*. September 1999;8(9):86.
7. Epidemiology—Japan. In: Smith RC, ed. *The Medical & Healthcare Marketplace Guide*, 18th ed. Philadelphia, PA: Dorland Healthcare Information; 2003: I-12.
8. WHO 2000: Population by age and sex, 1995 and 2025. Available at http://www.who.int/whr/1998/fig13e.jpg. (Accessed September 15, 2007.)
9. UN Population Division, World Population Prospects: The 1998 Revision. Available at http://www.unicef.org (Accessed September 15, 2007.)
10. Study sees fast growth for pediatric drug markets. *Pharma Marketletter*. July 5, 2004;31(27): 26.
11. Ribeiro RC, Pui C-H. Saving the children—improving childhood cancer treatment in developing countries. *N Engl J Med*. May 26, 2005;352(21):2158–2160.
12. Global pediatric drug sales to top $46B. *Pharma Marketletter*. January 23, 2006;33(4):16.
13. AHRQ Pub. No. 05-PO11. Selected Findings on Child and Adolescent Health Care from the 2004 National Healthcare Quality/Disparities Reports. Fact Sheet, March 2005. Available at www.ahrq.gov. (Accessed September 15, 2007.)
14. Children's health highlights: Chronic Illness. In: *Child Health Research Findings, Jan 2001–Dec 2004*. Agency for Healthcare Research and Quality (AHRQ). Available at http://www.ahrq.gov/child/highlts/chhigh1.htm # Chronic. (Accessed October 10, 2007.)
15. DeAngelis CD, Zylke, JW. Theme Issue on chronic diseases in infants, children, and young adults: call for papers. *JAMA*. October 11, 2006;296(14):1780.
16. Summary Health Statistics for US Children: National Health Interview Survey, 2005. Series 10, Number 231. Hyattsville, MD: DHHS, CDC, National Center for Health Statistics; December 2006. DHHS Publ. No. (PHS) 2007-1559.
17. Summary Health Statistics for US Children: National Health Interview Survey, 1998. Series 10, Number 208. Hyattsville, MD: DHHS, CDC, National Center for Health Statistics; October 2002. DHHS Publ. No. (PHS) 2002-1536.
18. Epidemiology—The World Health Report. In: *Medical and Healthcare Marketplace Guide*, 2000: Section I:15–16.

19. GI disorders are a leading cause of hospitalization in children. Child Health Research Findings; Jan 2001–Dec 2004. AHRQ grant HS 11826. (Also published by Guthery, Hutchings, Dean, and Hoff, *J Pediatr*. 2004;144:589–594. Agency for Healthcare Research and Quality (AHRQ). Available at www.ahrq.gov. (Accessed September 15, 2007.)
20. More than 200 medicines are in testing to meet the needs of children. April 2007 Report, Pharmaceutical Research and Manufacturers of America (PhRMA). Available at www.pharma.org. (Accessed October 10, 2007.)
21. WHO, World Health Report, 2005. Available at http://www.unicef.org. (Accessed September 15, 2007.)
22. Jones G, Steketee RW, Black RE, et al. How many child deaths can we prevent this year? *Lancet*. July 5, 2003;362:65–71.
23. Plaut M. Immune-based, targeted therapy for allergic diseases. *JAMA*. December 19, 2001;286 (23):3005–3006.
24. Casale TB, Condemi J, LaForce C, et al. Effect of Omalizumab on symptoms of seasonal allergic rhinitis: a randomized controlled trial. *JAMA*. December 19, 2001;286(23):2956–2967.
25. 2006 Innovative Medicines Initiative, European Commission and European Federation of Pharmaceutical industries and Associations. The Innovative Medicines Initiative (IMI): Strategic Research Agenda. Creating Biomedical R&D Leadership for Europe to Benefit Patients and Society. September 15, 2006;62–63: 71, Appendix 8.3. Available at http://www.efpia.org/4_pos/SRA.pdf. (Accessed September 15, 2007.)
26. Malik R, Hampton G. Counseling hospitalized pediatric patients with asthma. *Am. J. Health-System Pharm.* 2002;59:1829, 1833. Available at http://www.medscape.com. (Accessed December 15, 2002.)
27. Hegner RE. Bad news/good news in the asthma epidemic: prevalence jumps sharply, but symptoms can be controlled. In: *National Health Policy Forum, The George Washington University*, September 2000. Available at http://www.nhpf.org/pdfs_bp/BP_AsthmaEpidemic_9-00.pdf. (Accessed October 10, 2007.)
28. AHCPR seeks improved respiratory disease care. December 3, 1996. Available at http://www.ahrq.gov/news/press/respdis.htm. (Accessed September 15, 2007.)
29. Paramesh H. Epidemiology of asthma in India. *Indian J. Pediatr.* 2002;69:309–312.
30. Mallol J, Sole D, Asher I, et al. Prevalence of asthma symptoms in Latin America: the international study of asthma and allergies in childhood (ISSAC). *Pediatr Pulmonol.* December 1, 2000;30(6):439–444.
31. King D. Asthma vaccine "in years" with kid tests. *The Australian*. June 29, 2004.
32. National Cancer Institute Research on Childhood Cancer. NCI Fact Sheet (reviewed April 22, 2005). US National Institutes of Health, National Cancer Institute. Available at www.cancer.gov. (Accessed October 10, 2007.)
33. Lilleyman J. Simple deliverable therapy needed for childhood leukemia. Commentary. *Lancet*. 30 August 2003;362:676.
34. Overcoming juvenile diabetes. *FDA Consumer*. July/August 2000:28–32.
35. Dietz WH, Robinson TN. Overweight children and adolescents. *N Engl J Med.* May 19, 2005;352 (20):2100–2109.
36. Medco Health Drug Trend Report 2007. Available for purchase from Medco Health Solutions, Inc.
37. Dooren JC. Health advisers call for action to battle childhood obesity. *Wall Street Journal.* October 1, 2004: B2.
38. Cara JF. The epidemic of type 2 diabetes mellitus in children. 2002 Annual Meeting AAP Conference Coverage. Available at http://www.medscape.com. (Accessed January 15, 2003.)

39. Study: pediatricians can miss obesity. AP, August 2, 2004. Available at http://www.mslive.com. (Accessed October 17, 2004.)
40. Feigin RD. Propects for the future of child health through research. *JAMA*. September 21, 2005;294(11):1373–1379.
41. Variation and trends in incidence of childhood diabetes in Europe. (EURODIAB ACE Study Group. Correspondence to Dr. CC Patterson). *Lancet* 2000;355(9207): 873–876.
42. CPMP guidance on diabetes & paediatrics. *SCRIP*. 4 October 2000;2580:6.
43. Taiwan study raises child diabetes fears. CNN.com. September 10, 2003. Available at http://cnn.health.printthis. (Accessed October 17, 2003.)
44. Fagot-Campagna A, Venkat Narayan KM, Imperatore G. Type 2 diabetes in children. *BMJ*. February 17, 2001;322:377–378.
45. Many European children suffer health problems tied to obesity. *Wall Street Journal*. June 2, 2005: D3. Summary available at https://www.aibonline.org/resources/bibliography/Obesity.htm. (Accessed October 10, 2006.)
46. Ball D. Swedish kids show difficulty fighting fat. *Wall Street Journal*. 2 December 2B1.
47. Study: child obesity expected to soar worldwide. March 6, 2006. Available at http://www.msnbc.com. (Accessed September 15, 2006.)
48. Pediatric drug research: substantial increase in studies of drugs for children, but some challenges remain. Statement of Janet Heinrich. United States General Accounting Office. May 8, 2001. GAO-01-705T.
49. Myshko D. The little patient. *R&D Directions*. November/December 1997;3(6):20–26.
50. Trileptal gets US paediatric monotherapy indication. *SCRIP*. August 13, 2003;2875: 21.
51. Epilepsy: targeting a chronic medical condition through policy and research. *Public Health News. State Health Notes*. September 9, 2002:3.
52. Brundtland GH. Mental health: new understanding, new hope. *JAMA*. November 21, 2001;286 (19):2391.
53. Therapeutic categories outlook. SG Cowen, October 2001:153–155.
54. Small change could stop epilepsy seizures for millions. *Epilepsy News*. June 7, 2007. Available at http://www.epilepsy.com/newsfeed/pr_1181309455.html. (Accessed October 10, 2007.)
55. Letherman S, McCarthy D. Monitoring and evaluation for cystic fibrosis. Quality of health care for children and adolescents. UNC Program on Health Outcomes, The University of North Carolina at Chapel Hill. Commonwealth Fund. April 2004:50. Available at www.cmwf.org. (Accessed September 15, 2007.)
56. Quinn CT, et al. Survival of children with sickle cell disease. *Blood*. 2004;103:4023.
57. In Brief. *Pharma Marketletter*. April 16, 2007:28.
58. Harwell JI, Obaro SK. Antiretroviral therapy for children: substantial benefit but limited access. *JAMA*. July 19, 2006;296(3):330–331.
59. NDAs for perinatal HIV prevention may refer to foreign data—Cmte. *Pink Sheet*. October 11, 1999;61(41):31.
60. Petersen M. Ad campaign has parents asking for a costly drug. *New York Times*. January 31, 2001. Available at http://www.nytimes.com. (Accessed September 17, 2001.)
61. Committee on the Future of Emergency Care in the United States Health System. *Emergency Care for Children: Growing Pains*. Washington, Institute of Medicine, National Academies Press; 2007:18. Available at http://books.nap.edu/openbook.php?record_id=11655&page=18. (Accessed October 10, 2007.)
62. Mulholland EK, Adegbola RA. Bacterial infections—a major cause of death among children in Africa. *N Engl J Med*. January 6, 2005;351(1):75–77.

63. John CC, Idro RI. Cerebral malaria in children. *Infect Med.* 2003;20(1):53–58.

64. Jiménez J, Romero MI. Reducing infant mortality in Chile: success in two phases. *Health Affairs.* March/April 2007;26(2):458–465.

65. Boseley S. Whooping cough still a threat to babies. *The Guardian.* 25 August 2003. Available at http://www.guardian.co.uk. (Accessed September 18, 2005.)

66. Martinez B. Novartis fights eczema drug's cancer warning. *Wall Street Journal.* April 8, 2005: B1, B2.

67. Surgeon general releases a national action agenda on children's mental health. Press release. January 31, 2001. HHS News. Office of the Surgeon General. DHHS. Available at http://www.surgeongeneral.gov/news/pressreleasechildren.htm. Report available at http://www.surgeongeneral. gov/cmh/. (Accessed September 15, 2007.)

68. *2004 Survey: Medicines in Development for Mental Illness.* Pharmaceutical Research and Manufacturers of America (PhRMA). Available at www.pharma.org. (Accessed October 10, 2007.)

69. Scheffler RM, Hinshaw SP, Modrek S, Levine P. The global market for ADHD medications. *Health Affairs.* March/April 2007;26(2):450–457.

70. Harris G. Panel advises disclosure of drugs' psychotic effects. *New York Times.* March 23, 2006. Available at http://www.nytimes.com. (Accessed September 15, 2007.)

71. Leibson CL, Katusic SK, Barbaresi WJ, et al. Use and costs of medical care for children and adolescents with and without attention-deficit/hyperactivity disorder. *JAMA.* January 3, 2001;285 (1):60.

72. Morris BR. Lullabies in a bottle: prescribing for children. *New York Times.* May 13, 2003. Available at http://nytimes.com. (Accessed October 18, 2003.)

73. Coyle JT. Drug treatment of anxiety disorders in children *N Engl J Med.* April 26, 2001;344 (17):1326–1327.

74. Kaiser J. Everything you wanted to know about children, for $2.7 billion. *Science.* July 11, 2003;301:162–163.

75. Yamey G. Survey finds that 1 in 10 children has a mental disorder. *BMJ.* December 4, 1999;319 (7223):1456. Available at www.ons.gov.uk. (Accessed January 15, 2000.)

76. Mental disorders set to rise. *SCRIP.* April 13, 2001;2634:16.

77. Increased prescribing trends of paediatric psychotropic medications. *Arch Dis Child.* 2004;89: 1131–1132.

78. Alvazez L. Therapy? Or pills? A quandary in Britain. *New York Times.* December 21, 2004. Available at http://www.nytimes.com/2004/12/21/health/psychology/21depr.html. (Accessed October 10, 2007.)

79. Philadelphia program lets kids shoot cameras, not guns. For your information: *State Health Notes.* April 7, 2003;24(393):back page.

80. Hospitals seek to help children with life-threatening conditions. For your information: *State Health Notes.* October 22, 2001;22(358):8.

81. Berde CB, Sethna NF. Analgesics for the treatment of pain in children. *N Engl J Med.* October 3, 2002;347(14):1094–1103.

82. Chronic headaches cause emotional damage to kids. *Detroit Free Press.* July 7, 2003. Available at www.freep.com. (Accessed August 7, 2003.)

83. Elliott VS. Researchers seek tool for children's stomach trouble. February 21, 2005. Available at http://www.amednews.com. (Accessed September 15, 2006.)

84. Better formulations and information needed for children, says UK NSF. *SCRIP.* September 22, 2004;2989.

85. Kids' abdominal pain may become adult IBS. Reuters. Available at http://www.healthypages.co.uk/newsitem.php?news=5238. (Accessed October 10, 2007.)

86. Simpson L. Lost in translation? Reflections on the role of research in improving health care for children. *Health Affairs*. 2004;23(5):125–130.

87. Ogden CL, Flegal KM, Carroll MD, Johnson CL. Prevalence and trends in overweight among US children and adolescents, 1999–2000. *JAMA*. October 9, 2002;288(14):1728–1732.

88. Child health highlights: asthma. AHRQ grant HS09935 Child Health Research Findings; Jan 2001–Dec 2004. Agency for Healthcare Research and Quality (AHRQ). Available at http://www.ahrq.gov/child/highlts/chhigh1.htm - Asthma. (Accessed October 10, 2007.) (Also published by Lieu, Lozano, Finkelstein, et al. *Pediatrics* 2002;109(5):857–865.)

89. Mental illness: drugmakers struggle in Asian markets. Online newsletter. January 12, 2001.

90. Abright AR, Chung R. Depression in Asian Amercian children. *West J Med*. 2002;176 (4):244–248.

91. Mullan F. Immigration pediatrics. *Health Affairs*. November/December 2005;24 (6):1619–1623.

92. Fiondella F. Catch-up care is essential for foreign adoptees. *Wall Street Journal*. May 11, 2004: D 60

93. Milne C-P. The health of the world's children: what goes around, comes around. *Drug Information J*. 2000;34(1):213–221.

94. Zigler E, Styfco SJ. Extended childhood intervention prepares children for school and beyond. *JAMA*. May 9, 2001;285(18):2378–2380.

95. Gwatkin DR, Guillot M, Heuveline P. The burden of disease among the global poor. *Lancet*. August 14, 1999;354:586–589.

96. Power C. Childhood adversity still matters for adult health outcomes. *Lancet*. November 23, 2002;360:1619–1620.

97. Child poverty in perspective: an overview of child well-being in rich countries. Innocenti Research Centre. *UNICEF*. 2007: 7.

98. Letherman S, McCarthy D. Quality of health care for children and adolescents. Prescription of antibiotics to prevent infection among Medicaid-insured young children with sickle cell disease. UNC Program on Health Outcomes, The University of North Carolina at Chapel Hill. Commonwealth Fund. April 2004: 48. Available at www.cmwf.org.

99. Smith GD, Hart C, Blane D, Hole D. Adverse socioeconomic conditions in childhood cause specific mortality: prospective observational study. *BMJ*. May 30, 1998;316:1631.

100. Owens PL, Thompson J, Elixhauser A, et al. AHRQ. Care of children and adolescents in US hospitals. Healthcare Cost and Utilization project (HCUP) Fact Book No. 4. 2004: 6. Available at http://www.ahrq.gov/data/hcup/factbk4/factbk4.pdf. (Accessed October 10, 2007.)

101. Black RE, Morris SS, Bryce J. Where and why are 10 million children dying every year? *Lancet*. June 28, 2003;361(9376):2226–2234.

102. Study: more kids allergic to peanuts. November 18, 2002. Available at http://msnbc.com/news. (Accessed October 18, 2003.)

103. Galant SP, Wilinson R. Clinical prescribing of allergic rhinitis medication in the preschool and young school-age child: what are the options? *BioDrugs*. 2001;15(7):453–463.

104. Friedman RA. Uncovering an epidemic—screening for mental illness in teens. *N Engl J Med*. 28 December 2006;355(26):2717–2719.

105. Kuehn BM. Mental illness takes heavy toll on youth. *JAMA*. July 20, 2005;294(3):293–295.

CHAPTER 5

Pharmaceutical Economics and Applications to Pediatrics: Business Case Development

CHRISTOPHER-PAUL MILNE, DVM, MPH, JD

Tufts Center for the Study of Drug Development, Tufts University, Boston, Massachusetts, 02111

INTRODUCTION

The research and development (R&D) enterprise for medicines is a daunting one. Typically, it costs on average $1.2–1.3 billion, and 10 years to bring a drug or biological product to market.[1] This figure includes both out-of-pocket and opportunity costs, as well as the impact of failures, of which there are many, both technically and commercially. Only approximately 10–20% of drugs and biologicals survive from lab bench to pharmacy shelf. Of those, 34% of approved products actually have present value above the average costs of R&D; that is, only one-third of new chemical entities that reach the market can pay the freight for the time and costs that it took to get them there.[2]

The environment is increasingly challenging. The basic raw materials for making medicines are promising molecules, investigators, study participants, facilities, and funding. Each decade has seen clinical trial costs increase while product approvals decrease proportionately. The number of investigational compounds brought forward into the clinic has increased in the last few years, but this only serves to make the funding crunch and competition for resources more critical, and late-stage failure rates are going up instead of down. At the same time, experienced investigators are leaving the R&D sector and recruiting patients has become more and more difficult. For example, in the United States, fewer than 6% of eligible patients participate in clinical trials and there are 80,000 clinical trials per year.[3]

The prospects for bringing a drug to market for a pediatric indication are considered even more formidable, owing to a number of disincentives, such as liability and ethical concerns, limited population for certain diseases, difficulties in conducting trials in pediatrics (from practical to technical), scientific disagreement that dosing could be determined by

Pediatric Drug Development: Concepts and Applications
Edited by Andrew E. Mulberg, Steven A. Silber, and John N. van den Anker

weight-based calculations ("little adults" fallacy), the lack of accepted endpoints and validated pediatric assessment tools, and limited marketing potential compared to adults. Yet, in 1991, there was a seminal meeting hosted by the U.S. National Academy of Science's Institute of Medicine, which focused on the lack of prescription medicines for children. In the United States, two-thirds of medicines being used in pediatrics were off-label, and evidence later confirmed that adverse events were in fact more frequent with off-label use. Similarly, in Europe 50% of medicines used in children were being used off-label or off-license. In the European Union (EU), 20% of medical products were being sold to treat children, but only 7% of clinical trials involved children. In Japan, just 16% of package inserts have sufficient dosage and usage information for children.[4] Data provided by the Paediatric Clinical Trial Office of the MedChild Institute indicates that on a global level "three quarters of all medications marketed today do not carry drug regulatory agency approved labeling for use in neonates, infants, children and adolescents."[5] Moreover, the evidence base supporting much of pediatric use of adult medications was sparse and child-friendly formulations were generally not available.

MAJOR ECONOMIC DISINCENTIVES TO PEDIATRIC R&D

Three major obstacles to conducting clinical trials for pediatric indications of adult drugs confronted those who wanted to see change. All of them impacted on the risk and resources required to undertake the additional burden of investigating the use of drugs approved for adults in what amounted to four subpopulations: neonates, infants, children, and adolescents. Each of these age groups had their own particular difficulties, yet together they did not comprise a sufficiently robust market to justify even a marginal return on investment. Drug development has became increasingly complex in the United States with the expansion of evidentiary requirements for regulatory approval. In addition, the clear business imperative to address unmet medical needs that offer the greatest potential return on investment and the race to the major disease marketplace required prioritization of company resources that created significant economic disincentives to address the unmet medical needs of children.

Infrastructure

Resource constraints and competition were crucial factors that had to be addressed to build an environment in which successful pediatric drug development programs could flourish. Although there are approximately 50,000 pediatricians in the United States, few were willing to participate as investigators and even fewer were trained in Good Clinical Practice (GCP), particularly GCP as it related to pediatrics, perhaps as few as 2500.[4] Trained pediatric investigators had multiple, ongoing trials at any given time, many of which were competing for the same patient population. There was also competition with other ongoing trials, whether or not the drugs are direct competitors, since a standard exclusion was participation in any other ongoing trial. There were few contract research organizations (CROs) with specific pediatric experience from which to choose, and fewer hospitals equipped and staffed with personnel trained to handle pediatric patients. Also, there were few qualified pharmacokinetic sites capable of conducting pediatric trials, and few centralized laboratories (regulatory compliant) with the ability to handle microsized blood and serum samples from neonates and infants. With a small patient pool that was often concentrated in specific clinics and academic centers, meeting recruitment needs was challenging and costly. Even stratification by age and inclusion of children within an adult trial may increase the sample size required, thus increasing the cost.

Formulations

The development of formulations suitable for children was, and still is, a major challenge that separates pediatric drug development from mainstream R&D. The FDA understands that it may be necessary for a manufacturer to begin developing a pediatric formulation of a new molecular entity (NME) before initiating clinical trials—in some cases, before Phase I studies are completed for the parent compound. By the FDA's own estimates, the financial impact of new formulation development is considerable and the agency recognizes that the difficulty and cost of producing a formulation can vary greatly based on such factors as the compound's solubility and taste. The International Conference on Harmonisation (ICH) guidelines regard pediatric formulation needs as potentially extensive, suggesting that several formulations, such as liquids, suspensions, chewable tablets, patches, or suppositories with varied concentrations, may be needed for pediatric patients in different age groups. For injectables, the concentration must be appropriate for the doses administered, including doses for small premature infants or when fluid restrictions are relevant for extremely small patients. The ICH recommends that, since the development of pediatric formulations can be difficult and time consuming, it is important that this process be considered early in medicinal product development.

Not only do adult dosage forms not provide for the needs of children due to developmental/behavioral differences from adults, noncompliance, and caregiver convenience, but also different pediatric age groups have variable needs. For example, according to the Physician Drug & Diagnosis Audit conducted by Scott Levin in 1997, four times as many 16 year olds were able to take a solid dose formulation than could 6 year olds. Some dosage forms will be particularly challenging in regard to stability, impurities, degradation products, and solubility profiles. The development of age-appropriate formulations is technically challenging and time consuming and requires a number of steps, including the identification of problem drugs, the application of specialized formulation technology (e.g., tastemasking and transdermal enhancement), and the completion of dosing and pharmacokinetic studies.[4]

There are also considerations related to specific therapeutic area requirements (e.g., inhalers for respiratory disease), taste (and aftertaste), texture, tolerability, excipients (safety and tolerability), total volume (oral liquids and IV solutions), and so on. The selection of inert ingredients requires greater care, since children can have adverse reactions to preservatives, colorings, and flavoring agents routinely used in adult formulations. Taste is particularly crucial for compliance in children—a survey of 500 parents conducted by Ascent Pediatrics indicated that approximately 50% of children refuse to take their medication at some time and that, for 75% of the noncompliers, the reason was due to taste. Unique disease-specific requirements may also impact a formulation's tolerability. For instance, autistic children often have an aversion to sweet foods or specific flavors, which will make many current pediatric formulations unpalatable.[4]

Liability

A review by the Tufts Center for the Study of Drug Development (Tufts CSDD) of tort litigation cases from the late 1960s to the mid-1990s, which were compiled in a 1998 legal compendium, indicates that of 147 active moieties listed by the FDA in early 2001 as possible candidates for pediatric studies, 19 had been the subject of litigation. The average amount awarded in adult cases was $671,813, while in cases involving children the average award was $898,100. The defendants were not manufacturers by-and-large (less than 10%), but doctors and hospitals. So, while manufacturers are justifiably leery of litigation because of the size of the awards against them (an average of $33.6 million compared to $798,000 for other plaintiffs), liability fears affect all levels of healthcare, from medical students to liability insurers. Because of the lag time involved in cases reaching the courts, it is likely that few drugs approved in the 1990s are represented here, and some of the cases are 20–30 years old, when awards were generally for lower amounts. For example, some of the award amounts sought by plaintiffs in more recent cases that are still pending before the courts were in the neighborhood of $100–155 million. Moreover, 90% of actions never reach court, some state cases are unreported, and settlement amounts are often undisclosed as part of the settlement agreement. Thus, the actual litigation costs from cases involving these active moieties could be many-fold higher.[6] In addition to actual litigation expenses and plaintiff awards, liability can have secondary effects, such as causing manufacturers to overestimate risks and jump into crisis mode whenever a new risk is identified, causing insurers and reinsurers to withhold protection, making investors nervous, and inducing companies to reprioritize resources to less risky endeavors.[7]

GOVERNMENT PROGRAMS FOR OVERCOMING ECONOMIC DISINCENTIVES TO PEDIATRIC DRUG TRIALS

By the late 1990s, the EU was still in the process of organizing its overarching regulatory authority, now called the European Medicines Agency (EMEA), and pediatric study measures were largely relegated to whatever programs existed within individual member nations. Meanwhile, in Japan, two programs to facilitate pediatric indications of adult drugs were in place by the late 1990s: (1) the notification on "Medicinal Drug Use with Off-label Indications" states that if there is substantial evidence and experience with a drug approved outside Japan and the drug is recommended by the relevant physicians' association, the drug may be approved without domestic trials (1999); and (2) the Ministry of Health, Labor, and Welfare "Ordinance on Postmarketing Surveillance" provides that, when a company conducts clinical trials for a pediatric indication, the drug will have its reexamination period extended for up to 10 years (a form of protection from generic drug competition).[8] By this time in the United States, the FDA and pediatric health advocates were frustrated at the lack of success of regulatory policies to encourage drug firms to include children in the drug development process. While the FDA prepared a pediatric assessment regulation with enforcement provisions for newly developed drugs, congressional sponsors considered options for incentivizing pediatric clinical trials for already marketed drugs. Two programs started in the 1980s are examples of successful models. The Orphan Drug Act of 1983 was a model for the type of incentive that might command drug firms' attention as it provided for a 7-year period of market protection against both generic and brand-name competitors, in addition to whatever relevant patents were applicable. The Hatch–Waxman Act of 1984 was a model for a compromise piece of legislation that provided both positive and negative incentives to drug firms to innovate by awarding additional periods of market protection to developers of new drugs while lowering the barriers for their generic competitors to enter the market.

In late 1997, the FDA Modernization Act (FDAMA) was passed and among its many provisions was an incentive program for the pharmaceutical industry to conduct pediatric studies, requested by the FDA, in exchange for an award of an additional 6 months of market protection against generic competition for all the products containing the active ingredient studied (referred to as "pediatric exclusivity"). This FDAMA provision was reauthorized in January 2002 as the Best Pharmaceuticals for Children Act (BPCA) and was reauthorized again in September 2007 as Title V of the Food and Drug Administration Amendment Act (FDAAA). In late 1998, the FDA issued a regulation mandating pediatric assessment of new drugs (or already-marketed drugs under certain circumstances), which was later codified as the Pediatric Research Equity Act of 2003 (PREA) and also reauthorized in September 2002 as Title V of the FDAAA. The "carrot-and-stick" approach was firmly in place.

Success of the Program

Evidence of the success of the pediatric exclusivity program comes from multiple sources: the FDA's pediatric drug development website, congressional testimony at the 2001 FDAMA reauthorization hearings, the Pharmaceutical Research and Manufacturers of America (PhRMA) surveys of member companies conducting pediatric R&D, and surveys conducted by CenterWatch and Tufts CSDD. Industry invested considerable resources into the pediatric exclusivity program because, as one industry spokesperson pointed out during the reauthorization hearings, the pediatric exclusivity incentive was sufficiently attractive to raise "the priority of pediatric studies among competing programs within a company."[9] Coincident with the pediatric studies initiative, overall interest in pediatric research grew. While the number of drugs being studied for pediatric indications (typically along with adult indications) grew by 28% from 1990 to 1997, it increased by 49% from 1997 to 2000, or twice the increase in half the time.[10] Not only did the prioritization flag get raised in the private sector but within the public sector as well. Initially, the FDA allocated 65 staff members from other jobs within the agency to work on the pediatric studies initiative. By the end of the first 5 years of the FDAMA legislation, the FDA remained resource-challenged, yet the pediatric studies program was one of only four agency programs to actually receive increased funding (along with food safety, generic drugs, and bioterror programs).

In a period of less than 10 years, the pediatric exclusivity program led to more than 500 pediatric studies by 70 companies on 120 diseases and conditions affecting children. This, in turn, has resulted in the labeling of over 130 new or already approved drugs for use in children (the majority with significant new information that will assist prescribers and caregivers). The yearly economic benefits to U.S. society from lowering healthcare expenditures from major disease categories have been estimated to be $7.13 billion, while costs to industry have been approximately $375 million, and to society $695 million.[6] The program has been emulated by the European Commission (EC), which adopted a similar proposal in 2004 that was implemented in 2007. Similarly, recent testimony at government hearings before Japan's Ministry of Health, Labor and Welfare from the Japan Pediatric Society and other concerned organizations proposed that a similar program should be established in Japan.[11]

The therapeutic benefits for pediatric patients in the United States and worldwide are considerable. As seen in Figure 5.1,[12] newly labeled products were spread across a number of

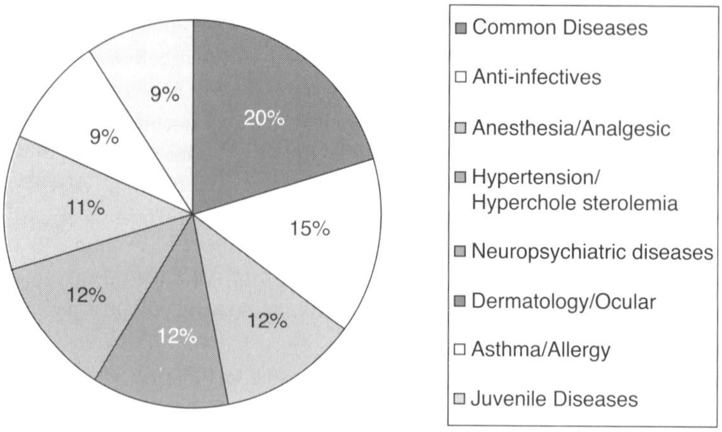

FIGURE 5.1 Percentage of drugs newly labeled for use in children under the U.S. Pediatric Exclusivity Program by therapeutic area. (*Source*: Reference 12.)

therapeutic categories including diseases common to both adults and children (20%), such as GERD and cancer but with different clinical impacts. Fully 15% of the new labels contained information for anti-infective medications against such global public health threats as malaria, HIV/AIDS, influenza, antibiotic-resistant bacterial strains as well as hepatitis B and C. The lack of labeled medicines for hospital use in pediatric AIDS patients was an early impetus for the pediatric studies initiative.[13] Another 12% of the newly labeled products were for neuropsychiatric diseases. About 7% of pediatric hospital stays are for mental disorders, with affective disorders (primarily depression) accounting for 74,000 hospital stays in 2000.[14] Furthermore, 11% of the new labels reflect information on pediatric use for dermatological and ocular conditions, which are prevalent in children but previously not studied widely in pharmaceutical research programs. For example, 15 million people in the United States are affected by atopic dermatitis, including as many as 20% of infants and young children. While topical corticosteroids are the standard of care, they are used less frequently in children due to safety concerns, which are now being addressed by pediatric studies.[15] Another 12% of the newly labeled products provided information on a number of drugs largely used in the hospital care setting such as anesthesia, analgesia, postoperative care, and inpatient treatment (e.g., hemodialysis), which are critically needed but a difficult area for clinical investigations in children. Of the 36 million U.S. hospital stays in 2000, about 18% of these were for children, 17 years and younger.[16] Nearly another 10% of products labeled under the program were for allergic rhinitis and asthma, which greatly impact healthcare expenditures. In fact, three respiratory problems—pneumonia, acute bronchitis, and asthma—are responsible for nearly $3 billion in charges or nearly 7% of the total U.S. healthcare bill for children and adolescents.[16] More than 10% of school-age children have seasonal allergic rhinitis, which contributes to missed school days, poor school performance, and impaired function on standardized tests.[17] Several well-known medical journals highlighted the problem of the lack of therapeutic options for early cardiovascular disease. One stated that cardiovascular risk factors in early teenage years are associated with permanent damage to the arterial wall. It continued to say that the difficulty of modifying lifestyles of teenagers in the current environment should not be underestimated; however, the safety and efficacy of pharmacological therapy for youths was not well established.[18] Another article laments that starting therapy for heterozygous familial hypercholesterolemia at 18 may be too late but there is a lack of evidence base for prescribing statins for children.[19] These shortcomings have been addressed by the pediatric exclusivity program by virtue of 15 drugs (12%) approved for hypertension or hypercholesterolemia.

In terms of addressing the major economic disincentives, there was considerable success, but obstacles remain. For example, regarding formulation development, a recent *JAMA* article that examined the results of the pediatric exclusivity program found that 31% of the drugs studied under the pediatric exclusivity program that have completed labeling now have child-friendly formulations (newly approved drugs, newly reformulated drugs, or new recipes for extemporaneous formulations).[20]

However, not only the cost of developing a formulation or a group of age-appropriate formulations has to be factored into resource considerations, but also the cost of keeping a formulation in channels of trade (manufacturing and distribution to wholesalers, distributors, and pharmacies). Often a company will develop a pediatric formulation, but then it is not commercialized due to cost constraints and small commercial value of the pediatric market. Formulation needs remain high not only in the United States, but also in the developed and developing world. For example, the French Regulatory Agency notes that pediatric formulations are needed in many therapeutic areas such as HIV/AIDS, attention deficit hyperactivity disorder (ADHD), rheumatology, osteoporosis, hypertension, cardiac, cancer, and hematology.[21] Dr. Phillipa Musoke, who works at a children's hospital in Uganda, says that antibiotics, antimalarials, and pain medications are very difficult to get in the appropriate formulations, because syrups are expensive and harder to store, and tablets need to be crushed in order to be properly administered.[22]

The impact of the incentive programs (i.e., pediatric exclusivity) on building a pediatric clinical trial infrastructure is harder to measure. Nonetheless, among 436 investigative sites, the number with pediatric expertise increased from 13% to 22%. Similarly, a group of nonprofit academic site management organizations (SMOs), called the Pediatric Pharmacology Research Units (PPRUs), which is a network of academic medical centers specializing in pediatric research and organized under the aegis of the National Institutes of Health (NIH), doubled its number of participating institutions in 1999 from 7 to 13 in reaction to the increased workload from the pediatric research initiative, and were working on as many as 70 industry-sponsored studies. A Tufts CSDD survey of 35 CROs revealed that of the 21 respondents, 13 were working on pediatric studies, 113 studies involving 7000 patients. Half of these CROs, which noted that pediatric research was responsible for most or some of the additional workload they had experienced in the last 2 years, attributed it to the FDAMA's pediatric provision. Thus, the effect on pediatric capacity ramp-up was immediate in the wake of the FDAMA incentive. Just this sample of CROs and SMOs (representing roughly about 5% of their respective outsourcing service sectors) were working on nearly 200 pediatric studies themselves in 2001, compared to fewer than 150 pediatric studies involving drugs executed by drug companies and the outsourcing industry as a whole in 1997.[6]

In the longer term, the direction and configuration of pediatric clinical trial capacity is still evolving. About 40–50% of sponsors have reported that they relied on outsourcing to help conduct the pediatric studies under the exclusivity program.[23] However, in 2000 outsourcing was the only mechanism by which sponsors addressed the additional workload, while in 2006 one-third of sponsors reported that they also added full time equivalents (FTEs), started a discrete pediatric research unit, or created an in-house pediatric advisory council. The number of medical centers and research institutions identifying themselves as having specific capacity in pediatric and neonatology studies grew five-fold from 1997 to 2007 (from 58 to 267).[24]

Liability was not specifically addressed by the pediatric exclusivity program; however, the potential return on investment (ROI) of the pediatric exclusivity award along with the need for regulatory compliance were apparently sufficient to overcome concerns about potential liability. Nonetheless, what the impact of several recent events involving selective serotonin reuptake inhibitors (SSRIs) and vaccines may be on the risk of liability remains to be seen. Considerable consensus was reached on several challenging ethical issues such as definition of minimal risk and guidelines for assent and consent that could help to mitigate potential liability issues. Nonetheless, liability remains an unresolved issue. There are examples where conducting pediatric trials has led to specific label changes and even black box warnings. These are often handled first on a drug-by-drug basis, before there are class changes. For approved products where the sales are primarily in adults, such a warning (even though it is restricted to pediatrics) has an enormous impact on the sale of that product and may impact its competitive advantage. It is also worth noting that prospective litigation costs may be passed along to consumers and taxpayers in the form of higher drug prices. One study found that liability risk is responsible for one-third to one-half of the price differential between the United States and Canada, where drugs are substantially lower priced.[25] Thus, any liability avoided by better labeling for drugs used in children, especially since they typically involved more costly awards in this sample of cases, will be beneficial across all segments of society.

NONGOVERNMENT APPROACHES FOR OVERCOMING ECONOMIC DISINCENTIVES TO PEDIATRIC DRUG TRIALS: CANCER AND CYSTIC FIBROSIS AS SPECIAL CASE EXAMPLES

Cancer

There is a need for less toxic and more efficacious cancer drugs. Around 20–25% of children do not respond to currently available chemotherapeutic agents,[26] and there are few available. Small patient groups (i.e., only about 10,000 children diagnosed with cancer out of a total of 1.37 million new cancer cases annually) [27] create economic disincentives and clinical trial recruitment challenges.[28] Despite these obstacles, however, there is progress in that some 50% of children with cancer enroll in clinical trials compared to only 3% of adults.[29] Pediatric oncologists and advocates were able to accomplish this by leveraging available public sector funding through consortium-building between Children's Oncology Group (COG) and National Cancer Institute (NCI), which formed a 238-insitution consortium with 212 of the institutions in the United States,[30] which is looking to expand its global reach. In 2003, just 15 of 100 drugs approved by the FDA's Division of Oncology Drug Products had pediatric use information.[31] Yet, the COG–NCI consortium together with an industry partner in May 2003 brought a new pediatric indication of a cancer drug (Gleevec) to the marketplace for the first time in 13 years, using extrapolated information from adults for efficacy and proof-of-concept in children.[32] On the international level, twinning is a variation on the consortium approach in which institutions in developed countries assist those in underdeveloped countries. Twinning programs for childhood cancer treatment have been in place in Central and South America, northwest Africa, and Southeast Asia for as along as 10 years.[33]

Cystic Fibrosis

The advocates of R&D for cystic fibrosis (CF) have taken a different tack. The CF Foundation (CFF) was established in 1955 and realized in the 1990s that the small market (about 30,000 living with CF in the United States) was a problem for industry. In response, in 1997, the foundation started the Therapeutics Development Program to channel private funding to biopharmaceutical companies and partner with the existing clinical trials infrastructure, overseen by the foundation's nonprofit affiliate, the CF Foundation Therapeutics Inc. Pulmozyme went through the CFF-mediated (clinical trial) process in 1993. Later, because of the need to increase identification of promising new agents, the program expanded to include a Therapeutics Discovery Component that supports combinatorial chemistry and high-throughput screening. Most CF funding comes directly from donations from the public with additional support from corporations and others. Typically, the foundation spreads funding seeds around in the amount of $1.5–2 million, but it has committed itself to a 5-year, $46.9 million investment with Aurora Biosciences for high-throughput screening.[34]

Success of the Approach

As a consequence of the efforts targeting pediatric cancer and CF advocates, these two diseases were among the top five areas of pediatric drug testing in 2001: cancer, 32 trials; infectious diseases, 20 trials; cardiovascular disease, 19 trials; psychiatric disorders, 14 trials; and cystic fibrosis tied with asthma, 13 trials each.[35] They both are still very active areas of pediatric R&D according to PhRMA's 2007 survey of medicines in development. Cancer is number one with 39 projects, cystic fibrosis comprises half of the projects in the third leading category (genetic disorders, $N=26$).

In contrast, spinal muscular atrophy (SMA), a disease that strikes about 1 in 6000 Americans, was attracting only about $1 million in research funding compared to $117 million spent on cystic fibrosis in 2002. One reason was highlighted when SMA advocates, seeking congressional sponsors on Capitol Hill, were continuously asked: "Do you have a celebrity?"[37] Similarly, research on obesity, diabetes, and cardiology are greatly underfunded,[38] despite the urgent public health concerns, because they had no organized advocacy programs.

EMERGING AND REMAINING ECONOMIC BARRIERS AND BURDENS

Although considerable progress has been made to address the disincentives to pediatric studies, some problems remain more than a decade since the initiative was launched, and new problems have emerged, such as the unforeseeable nature of some clinical responses, the possibility of catastrophic unanticipated reactions, the increased recognition of the possibility of long-term effects on growth and development, and the difficulty in predicting dose–response or concentration–response relationships by extrapolation from adults.[39] Moreover, pediatric studies have become more expensive, more complex, and more resource-intensive as evidenced by comparing the results of two Tufts CSDD surveys in 2000 and 2006 on the experience of drug sponsors conducting pediatric clinical trials for exclusivity in response to a Written Request (WR) from the FDA.[23]

Scope

The most notable difference between 2000 and 2006 was the increase in the category of efficacy/safety trials, basically Phase III studies, from 25% to 40%. These are the most time-consuming and expensive type of study. There was commensurate decrease from 2000 to 2006 in pharmacokinetic (PK) studies that typically involve small numbers of participants, as well as less time and money. The expansion in the scope of the pediatric studies to meet the requirements of the Written Request (WR) can readily be appreciated by the difference between 2000 and 2006 reports on participants per study and studies per WR. The number of patients per study increased over 2.5-fold while the number of studies per WR nearly doubled. Adding to the expansion in the scope of pediatric studies in the 2006 cohort compared to 2000 was a doubling (from 17% to 34%) of the number of programs required to perform long-term follow-up studies (>6 months) on growth and development. In terms of developing specific research techniques and tools, one-third of pediatric exclusivity projects reported the development and validation of new clinical assessment tools and sampling and analytical techniques by the end of 2000, and an additional 27% of sponsors reported similar activity in a follow-up Tufts CSDD survey in 2006.[23]

Time

Earlier work by the Tufts CSDD indicated that pediatric study programs are typically in progress 15 months before the WR was actually listed by the FDA, suggesting that most of the WRs, whose sponsors were surveyed, were in progress from 4 to 6 years by the time of the survey. Increases in the complexity and scope of the studies (in addition to other possible factors such as availability of patients, investigators, and facilities, access to FDA staff, or sponsor allocation of resources) are likely to have contributed to the 2.5-fold increase in the time required to complete the WR program (i.e., protocol development to submission of pediatric study reports) from 2000 to 2006. The intense negotiation and interaction between FDA and sponsor are reflected by the fact that the majority of sponsors had two to four formal meetings with the FDA, nearly 40% negotiated written agreements with the FDA, and nearly 90% amended their WRs an average of two or three times, according to the 2006 survey.[23]

Success Rates

A small number of WRs from the FDA to conduct pediatric studies are refused by sponsors. The reasons cited were business decisions, unrealistic study design, or inability to make appropriate pediatric formulations. As of 4–5 years from the date that these WRs were listed, only half the sponsors had submitted pediatric study reports, and of those only about half had been accepted, while the rest had either not been accepted or were still awaiting an FDA determination.[23]

Cost

According to the Tufts CSDD surveys, there was an eightfold increase in the overall average of self-reported costs for completing a WR from 2000 to 2006. This is not an unexpected result given that, as already noted, there has been a 2.5-fold increase in the time to conduct the studies, as well as significant changes in the scope of the pediatric studies in terms of numbers of patients, numbers of studies, types of studies, and the proportion of studies requiring formulation development and long-term studies. Moreover, based on previous work by Tufts CSDD on drug development in general, there would also be an annual clinical cost inflation rate of about 10% per year.[23]

A study published by Li et al.[40] at the same time as the Tufts CSDD study reported a median cost figure of $12.34 million, compared to a median cost figure of $20 million for the Tufts CSDD study.[23] The study by Li and colleagues was not based on reported costs but estimates based on data available in FDA files. These did not include data on the cost of formulations. According to the Tufts CSDD surveys of pediatric study programs, formulation costs can be quite significant, although widely variable, ranging from $500,000 to $15,000,000. This variability is not unexpected since the formulation challenges ranged from having to develop five new formulations of one product for various age groups, to simply doing stability testing for a recipe based on extemporaneous compounding. Among the products requiring formulations, the range was from one to five pediatric-specific formulations, including suspensions, sprinkles, oral solutions, coated granules for reconstitution in water, and adult pumps converted to deliver smaller doses. The authors of the Li study further state that another limitation was the use of software designed for adult trials, although the authors recognize that pediatric trials are "usually more expensive."[40] In fact, the Tufts CSDD 2006 survey found average pediatric per-patient costs that were more than double those of average reported adult per-patient costs of $26,000 for Phase III trials.[41] Furthermore, the Li cohort consisted of pediatric studies done on average 2 years earlier than the Tufts CSDD cohort, so the clinical cost inflation rate would increase their spending estimate to about $15 million for the same time frame as the Tufts study.

Other cost estimates have been published, often without needed detail as to what they encompass or the source of the information and methodology used to derive them. The National Institute of Child Health and Human Development and the European Commission estimated costs for pediatric studies in 2004 at $10 million (6.3 million euros) and $5.4 million (3.4 million euros), respectively.[42,43]

SPECIAL ECONOMIC ISSUES

Cost Effectiveness

Most of the leading diseases of adulthood have their roots in health as a child. Effectively treating those diseases would significantly decrease expenditures on healthcare in the future and create a healthier future as well. Yet, there is considerable debate as to the role of pharmacotherapy in addressing these unmet medical needs as well as the trade-offs required in healthcare expenditures. The issues are complex and vary disease-by-disease and region-by-region. For example, according to the Centers for Disease Control (CDC), 79% of children diagnosed with cancer under the age of 15 live 5 years or more, compared to 56% in the mid-1970s.[44] However, 30 years after diagnosis, 40% of childhood cancer survivors have a serious health problem, and one-third have multiple problems. Problems may be due to chemotherapy, radiation treatment, drugs used to treat infections in immune-suppressed patients, and/or earlier onset of traditional age-related problems.[45] An estimated 1 in 250 adults under age 40 will be survivors of childhood cancer by 2010 according to the American Academy of Pediatrics (AAP).[46] Although patient populations are small, the costs to the healthcare system are large, especially as a greater proportion live to adulthood. As cancer treatment options for children get better, however, long-term prognoses are likely to improve along with productivity and quality of life.

Another example involves a Dutch study of children suffering from familial hypercholesterolemia, which found that they could be successfully treated with statins (e.g., pravastatin) without side effects related to growth, maturation, hormone levels, or muscle and liver enzymes.[47] However, a previous study from Europe suggests that although dietary manipulation is poor as a cholesterol-lowering treatment, and that the only way to lower cholesterol is with drugs, there is no evidence that a possible benefit from cholesterol lowering from a young age may balance possible side effects from long-term drug use.[48] New recommendations published by the American Heart Association in March 2007 appear to reconcile this debate by taking the approach that, if needed, statins should be considered as first-line treatment for children at risk, if the source of the risk is from high-risk lipid abnormalities or a family history of premature coronary heart disease, rather than just being overweight or obese.[49]

Chronic pain is far more common than previously thought in children; experiences of pain in early life may lead to long-term consequences, both physiological and psychological.[50] Acute pain is one of the most common symptoms experienced by children, occurring as a result of injury, illness, and necessary medical procedures, and is associated with increased anxiety, avoidance, somatic symptoms, and parent distress.[51] There has been a history of inadequate treatment of pain in children and infants.[52] Pain is undertreated due to lack of training in proper evaluation and fear of side effects of potent analgesics. Yet, even children under 2 can be treated with "properly dosed" narcotics, and combining it with acetaminophen or ibuprofen can minimize the amount needed.[53] However, continued concerns about diversion to the "black market" and abuse indicate a need to develop new approaches to the treatment of pain. In the United States, nearly 3 million adolescents ages 12–17 use prescription drugs for recreational purposes or nonmedical reasons at least once, and among the most commonly abused drugs are opioid analgesics and tranquilizers.[54]

The UK's Joint Committee on Vaccination and Immunization suggests that the number of people falling ill in the general population would go down by a fifth if 60% of children aged 6 months to 2 years were immunized against common flu strains A and B; yet the government health service declined to do so.[55] In the United States, about 85% of child vaccines are administered in private offices. Although state and federal programs pay for 55% of those vaccines, the costs of purchasing, administering, and storing vaccines are becoming prohibitive. If a child receives all recommended doses by age 18, he/she would get 37 shots and 3 oral doses at a cost exceeding $1600. Some states that once provided free vaccines, like North Dakota, have abandoned that practice. Spending by the federal Vaccines for Children Program, which pays for immunization for Medicaid children and some others, has grown from $500 million in 2000 to $2.5 billion.[56]

Over 2 million U.S. adolescents became obese and an additional 1.5 million remained obese ($BMI \geq 30$) as they matured into adulthood from 1996 to 2001.[57] Childhood overweight is associated with a variety of adverse consequences, such as risk factors for cardiovascular disease and type 2 diabetes. Childhood onset overweight accounts for 25% of adult obesity. The economic costs of obesity are second only to tobacco use at an annual cost estimated to be $117 billion in direct and indirect costs.[58] There are two weight-loss drugs approved for teens—sibutramine and orlistat (also adolescents)—but as with other chronic diseases, it is likely to require ongoing drug therapy.[59] Orlistat in conjunction with diet, exercise, and behavioral modification statistically significantly improves weight management in obese adolescent subjects. However, the article cautions that without data on the long-term risks and benefits of orlistat and other pharmacotherapies and treatment settings, its use should not be stand-alone, but should take place in a setting that offers comprehensive assessment and management.[60] A Dutch study published in the

Journal of Clinical Endocrinology and Metabolism (April 2007) stated that adolescents put on a low-calorie diet and exercise program do not seem to derive any additional benefit from taking the drug sibutramine.[61] An FDA draft guidance again appears to take the middle road, stating that lifestyle modification is the cornerstone of weight management. Because of inherent risks of drugs, use of weight-management products should be undertaken only after a sufficient trial of lifestyle management has failed, and the risks of excess adiposity and the anticipated benefits of weight loss are expected to outweigh known and unknown risks of treatment with a particular product.[62]

Pediatric Market

The pediatric medicines market is difficult to characterize as either boom or bust for incentivizing pediatric R&D. On the one hand, the U.S. market does not look sufficiently strong to carry the entire subsector. The value of the pediatric prescription market in the United States was placed at somewhere between $13 billion and $18 billion in 2002.[63] The breakdown by therapeutic area appears as follows: 25% vaccines, 23% allergy/respiratory, 21% anti-infectives, and 32% for all other disease categories.[63] Upon consideration of this therapeutic market breakdown, the need for follow-on pediatric indications from adult drugs is easy to appreciate, because the range of indications supported by pediatric use alone is limited.

On the other hand, predictions for the worldwide market look more promising. Pediatric prescription drug sales reached $36.4 billion in 2005, up from about $18 billion in 1996,[65] and are expected to grow annually by 6.2% to $46 billion by 2009.[64] By major therapeutic category, sales break down as follows: anti-infectives, 42%; allergy and respiratory drugs, 18%; CNS drugs, 16%; hormone drugs, 8%; and GI, cardiovascular, cancer, and other drugs, 16%.[65] Meanwhile, the world vaccine market reached an estimated $10 billion in 2007 with pediatric vaccines having a 56% share according to Kalorama.[66]

Despite the variability regarding the actual size of the market, there is general agreement that although the market is small (generally thought to be less than 10% of the total prescription drug market), it is growing. A recent report asserts that the under-19 patient population in the United States is the fastest growing segment of the prescription drug market with dollar volume increasing 85% from 1997 to 2001, and that children are using medicines 34% longer.[67] However, this trend is counterbalanced by the fact that pediatric drugs are typically cheaper and used for short durations, and thus generate less revenue. In 2004, an analysis by Frost and Sullivan found the worldwide market for pediatric pharmaceuticals generated revenues of $8.73 billion in 2002, which is estimated to increase to $14.48 billion by 2010.[68]

Parental Trust

Parents' and caregivers' trust in the health industry is decreasing and their confidence in mainstream medicine has been increasingly challenged. A poll of several thousand adults in the United States indicated that 61% believe ADHD drugs are prescribed too often for children under age 13 and only 2% not enough.[69] Yet use of ADHD drugs has actually leveled off in girls and dropped in boys to 8%.[70] Another example is a study that found that 21% of parents said they had treated their children with alternative therapies in the past year, such as herbs and specific vitamins, but only 36% admitted to telling their pediatricians about it.[71] Yet another found that nearly half (46%) of the predominantly white, well-educated parents of children with cancer in this study used complementary therapy (CT), ranging from acupuncture and magnets to dietary supplements and herbal remedies, and 33% began using a new CT following their child's cancer diagnosis.[72] Ultimately, this environment affects the way parents feel about clinical trials, and thus patient recruitment becomes ever more difficult. For example, a recent survey tested parental willingness to consider allowing children to participate in trials: 25% responded Yes; 30% said No; and 45% Not Sure. Answers changed depending on circumstances: from 75% (you thought the drug would cure your child) down to 21–26% (your child was healthy and your doctor/specialist was not conducting the trial); and, depending on the illness, 79% were likely to allow children to participate for cancer trials, but only 53% if the trial was for ADHD.[73]

Pharmacogenomics/Preventive Medicine

We are entering the era of predictive medicine, yet ethical questions about "red-flagging" individuals at risk and economic pressures may slow its implementation. Newborn screening programs already are commonly in place for nearly 10 diseases, including cystic fibrosis. However, at present, testing for adult-onset diseases is not being implemented broadly based on decisions by two committees of the American Academy of Pediatrics (AAP).[74] The American College of Medical Genetics' newborn screening expert group recommended that states screen all 4 million infants born annually for 29 rare disorders that affect about 5000 individuals per year and that are all treatable if discovered early, yet few do.[75] In the near future, personalized medicine based on genotype will affect treatment selection and preventive healthcare advice but will not be widely adopted by clinicians until insurance coverage is assured for individuals with preexisting conditions.[76] Unfortunately, children rely disproportionately on state financing, but states are cutting costs by erecting additional barriers to enrollment or eligibility or both. Unfortunately, doctors are also dependent on government financing. Pediatricians and family practitioners have the lowest incomes and already are more likely to practice in undercapitalized settings than other providers do. Child health providers and organizations often do not reap direct benefit from preventive service investments because the positive outcomes may not be captured until decades later and thus, initially, they can lose money; for example, hospitals investing in programs for asthmatic children will have fewer pediatric hospitalizations and shorter lengths of stay.[77]

FUTURE

In the final analysis, statutory and regulatory incentives in the United States and Europe may help to build a global pediatric research infrastructure with a sufficient economy of scale and performance to permit pediatric medicine development to become a sustainable sector of the drug and biological products industry. The growing movement toward regulatory harmonization and the internationalization of clinical research, together with the growth of multinational CROs, should facilitate the conduct of pediatric trials in multicountry settings, and should lay the groundwork for a global pediatric research infrastructure. Long-term sustainability is likely only to be achieved with the expansion to global markets in which close to 30–40% of the population are children. In this respect, it may be a harbinger of the direction in which research and development is heading as a whole. Pharmacogenomics, consumer empowerment, public policy pressures for price controls, formulary requirements, regulatory incentive programs, and competitive forces may compel the industry to subdivide research and development, marketing, and manufacturing units along subpopulation lines as well as by therapeutic areas or geographic sectors. Some members of "big pharma" are already moving in this direction by taking actions, such as embracing the concept of personalized medicines (and thus the concept of smaller patient markets); investing heavily in pharmacogenomics, early-stage biotechnology projects, and unmet medical needs; moving away from the primary care market to the specialty care market; and targeting future growth from sales in the developing world.

In the short-term, however, incentives remain critical to maintain and enhance a growing but fragile pediatric clinical trials infrastructure. Unfortunately, there is currently some political momentum in the United States to roll back incentives, despite previous success. Instead, new initiatives should be focused on (1) maintaining current incentives to ensure continued success and (2) addressing unmet needs in pediatrics, where the current incentives do not necessarily reach (e.g., formulation development, resistant organisms, and cancer). Some approaches suggested are as follows:

1. Continuous prioritization process for public sector programs based on, for example emerging unmet medical needs, changing market forces, and feasibility of increasing access to imports through current laws or proposed legislation.
2. Continuous refinement and international harmonization of standards and guidelines for extemporaneous compounding to supplement formulation development needs.
3. Promotion of international partnerships and consortiums for specific disease areas such as cancer and global public health threats such as malaria and AIDS including data sharing up to the point of "specific utility" when relevant intellectual property rights accrue.
4. Establishing a funding mechanism for public sector programs to develop formulation technology and pediatric trials of unpatented drugs such as a small royalty percentage of revenues from sales.
5. Establishing incentive programs for economically unattractive areas of pediatric R&D (neonates, etc.) such as transferable pediatric exclusivity or priority review status.
6. Harmonization of pediatric assessment regulations internationally to avoid redundancy and encourage consistency.
7. Limit liability through legislative solutions modeled on the PREP Act for bioterror countermeasures or the National Vaccine Injury Compensation Program.[4]

ACKNOWLEDGMENT

The Tufts Center is supported in part by unrestricted grants from pharmaceutical firms, biotechnology companies, and related service providers. No companies were involved with the production of the author's work.

REFERENCES

1. DiMasi JA, Hansen RW, Grabowski HG. The price of innovation: new estimates of drug development costs. *J Health Economics*. March, 2003;22(2):151–185.
2. Grabowski HG, Vernon J, DiMasi JA. Returns on research and development for 1990s new drug introductions. *PharmacoEconomics*. 2002;20 (Suppl 3):11–29.
3. Reach subjects through targeted recruiting. *Clin Trials Administrator*. May 2005;3(5):49–60. Available at www.ahcpub.com. (Accessed September 15, 2007.)
4. Milne C-P, Bruss JB. The economics of pediatric formulation development for off-patient drugs. *Therapeutics*. 2008;30(11):1–13.
5. Children and drugs—about the qualification of the use of drugs in children. Pediatric Clinical Trial Office, MedChild Institute. Available at http://www.medchild.org/eng/modules/progetti/index.asp? CNT_Area=P&CNT_ID=27. (Accessed October 10, 2007.)
6. Milne C-P, The pediatric studies incentive: equal medicines for all. Tufts Center for the Study of Drug Development, Tufts University, Boston MA. White Paper, April, 2001.
7. Mello MM, Brennan TA. Legal concerns and the influenza vaccine shortage. *JAMA*. 2005;294:1817–1820.
8. Uchiyama A. Pediatric clinical studies in Japan: regulations and current status. *Appl Clin Trials*. July, 2002;11(7):57–59.
9. Better Pharmaceuticals for Children: Assessment and Opportunities, Hearing Before the Senate Comm. On Health, Education, Labor, and Pension, 107th Cong. 36 (2001) (quoting testimony of Stephen P. Spielberg, MD, PhD, then vice president of pediatric drug development for Janssen Pharmaceuticals).
10. US pediatric studies incentive led to new labeling for nearly 100 drugs. In: Kaitin KI, ed. *Tufts CSDD Impact Report*. July/August, 2005;7 (4).
11. Japan to discuss pediatric drug use. *Pharma Marketletter*. December 18, 2006;33(15):17.
12. Unpublished data. Tufts Center for the Study of Drug Development, 2007.
13. Drug Development and the Pediatric Population: Report of a Workshop 1. Institute of Medicine, National Academy of Sciences, 1991.
14. Care of Children and Adolescents in US Hospitals. Agency for Health Care Research and Quality, DHHS. Executive Summary, October, 2003. Aavailable at http://www.ahrq.gov/data/hcup/factbk4/factbk4.pdf. (Accessed October 10, 2007.)
15. National Institute of Arthritis and Musculoskeletal and Skin Diseases. Handout on Health: Atopic Dermatitis. 2003. Available at http://www.niams.nih.gov/Health_Info/Atopic_Dermatitis/default.asp. (Accessed October 10, 2007.)
16. Owens PL, Thompson J, Elixhauser A, et al. Care of children and adolescents in US hospitals. Healthcare Cost and Utilization Project (HCUP), Agency for Hearthcare Quality and Research (AHRQ). 2004; Fact Book No. 4:3–6.
17. Antihistamines for the treatment of allergic rhinitis in children. *Pediatr Pharmacotherapy*. September 1995;1:(9). Available at http://www.people.virginia.edu/~smb4v/pedpharm/v1n9.html. (Accessed August 15, 2005.)
18. McGill HC, McMahan CA. Starting earlier to prevent heart disease. *JAMA*. November 5, 2003;290(17):2320–2322.
19. Bradbury J. Should children with familial hypercholesterolaemia be prescribed statins? *Lancet*. 2002;360:1077.
20. Roberts R, Rodriquez W, Murphy D, et al. Pediatric drug labeling: improving the safety and efficacy of pediatric therapies. *JAMA*. 2003;290(7):905–911.
21. French publish list of paediatric medicine needs. *SCRIP*. April 2, 2004;2940:3.

22. Tobias LD, Harkness J. Briefing paper on paediatric medicines and clinical research. International Alliance of Patients'. Organizations (IAPO). 2006:17.
23. Pediatric study costs increased 8-fold since 2000 as complexity level grew. In: Kaitin KI, ed. *Tufts CSDD Impact Report*. March/April, 2007;9(2):1–4.
24. Profiles of centers conducting clinical research in pediatrics/neonatology. In: *CenterWatch Clinical Trials Listing Service*. Boston, MA: Thompson Centerwatch. Available at http://www.centerwatch.com. (Accessed September 15, 2007.)
25. Manning RL. Products liability and prescription drug prices in Canada and the United States. *J Law Economics*. April, 1997;XL(5):203–243.
26. Marcus AD. Testing "smart drugs" on children with cancer. *Wall Street Journal*. May 31, 2005: D1–D7.
27. Seltzer J.Drugmakers fail to address cancer in children. Reuters. May 9, 2004. Available at http://www.cancerpage.com/news/article.asp?id=7053. (Accessed October 10, 2007.)
28. An interview with Division of Oncologic Drug Products Director, Richard Pazdur, MD, *US Regul Reporter*. March, 2000;16(9): 7.
29. Couzin J. Tight budget takes a toll on US-funded clinical trials. *Science*. March 2, 2007; 315:1202–1203.
30. Global consortium would facilitate early pediatric oncology trials—cmte. *Pink Sheet*. July 21, 2003;65(29):32.
31. Pediatric oncology labeling should reflect exploratory approach—Pazdur. *Pink Sheet*. March 10, 2003;65(11):14.
32. What's new? In: CDER, FDA website. Available at http://www.fda.gov.cder/cancer/whatsnew.htm. (Accessed September 15, 2007.)
33. Ribeiro RC, Pui C-H. Saving the children—improving childhood cancer treatment in developing countries. *N Engl J Med*. May 26, 2005;352(21):2158–2160.
34. Adamson PC, Weiner SL, Simone JV, Gelband H, eds. *Making Better Drugs for Children with Cancer*. Committee on Shortening the Time Line for New Cancer Treatments, Institute of Medicine and the National Research Council. Box 8-A: independent entities for drug development, 2005.
35. Zimmerman R. Desperately seeking kids for clinical trials. *Wall Street Journal*. May 29, 2002: D1.
36. More than 200 medicines are in testing to meet the needs of children. Pharmaceutical Research and Manufacturers of America (PhRMA). April 2007 Report. Available at www.pharma.org. (Accessed October 10, 2007.)
37. O'Connor A. A deadly disease of infants attracts new research money. *New York Times*. October 28, 2003. Available at www.nytimes.com. (Accessed September 15, 2007.)
38. Nelson R. US paediatric researchers say funding is uneven. *Lancet*. 2003;362(9377):50.
39. Peds subcommittee hears oncology update (Report on Pediatric Subcommittee Meeting). DIA Dispatch (online newsletter). Washington, DC: Drug Information Association; October 21, 2005.
40. Li JS, Eisenstein EL, Grabowski HG, et al. Economic return of clinical trials performed under the pediatric exclusivity program. *JAMA*. February 7, 2007;297(5):480–488.
41. Phase 3 clinical trial costs exceed $26,000 per patient (report of survey by Cutting Edge Information). Yahoo.news, October 26, 2006.
42. Progress in Implementing the Best Pharmaceuticals for Children Act (BPCA). US Department of Health and Human Services, National Institutes of Health (NIH), National Institute of Child Health and Development (NICHD). Available at http://www.nichd.nih.gov/bpca/documents/progress_implementing_BPCA.pdf. (Accessed September 15, 2007.)

43. Arlett P. Regulation on medicines for children: frequently asked questions. In: European Commission website. October 28, 2004. Available at http://ec.europa.eu/enterprise/pharmaceuticals/paediatrics/docs/paeds_qa_october_28.pdf. (Accessed October 10, 2007.)
44. Fox M. New partnership needed for child cancer—report. Reuters. April 18, 2005; Available at http://www.reuters.co.uk. (Accessed August 15, 2005.)
45. Emery G. Child cancer survivors face worse health as adults. Reuters. October 12, 2006. Available at http://www.reuters.co.uk. (Accessed November 15, 2006.)
46. Marcus AD. Parents face painful choice in treating childhood cancers. *Wall Street Journal*. March 1, 2005:D1–D6.
47. Dutch study backs cholesterol drug for children. Reuters. July 20, 2004. Available at http://www.reuters.co.uk. (Accessed August 15, 2005.)
48. Ravnskov U. Prevention of atherosclerosis in children (Letter). *Lancet*. January 1, 2000;355:69.
49. Heart association backs statin use for at-risk kids. In: healthfinder.gov website, 2007. Available at http://www.hhs.gov. (Accessed October 10, 2007.)
50. Howard, RF. Current status of pain management in children. *JAMA*. November 12, 2003;290:(18):2464–2469.
51. AAP, Committee on Psychosocial Aspects of Child and Family Health and American Pain Society, Task Force on Pain in Infants, Children and Adolescents. The assessment and management of acute pain in infants, children and adolescents. *Pediatrics*. September, 2001;108(3):793–797.
52. Berde CB, Sethna NF. Analgesics for the treatment of pain in children. *N Engl J Med*. October 3, 2002;347(14):1094–1103.
53. Cox News Service. Pain assessment a priority for kids. November 17, 2003. Available at www.intelihealth.com. (Accessed November 15, 2005.)
54. Kumar A. Prescription drug abuse soars; youth at forefront. *LA Times*. January 17, 2003. http://www.latimes.com. (Accessed February 15, 2003.)
55. Roberts M. Child flu jab "veto" questioned. In: *BBC News Health Reporter*. February 9, 2007. Available at http://newsvote.bbc.co.uk. (Accessed October 10, 2007.)
56. Pollack A. Pediatricians voice anger over costs of vaccines. *New York Times*. March 24, 2007. Available at http://www.nytimes.com. (Accessed October 10, 2007.)
57. Stein R. Severe obesity rises sharply. *Washington Post*. October 14, 2003:A08.
58. *2004 Survey: Medicines in Development for Mental Illness*. Pharmaceutical Research and Manufacturers of America (PhRMA). Available at www.pharma.org. (Accessed October 10, 2007.)
59. Dietz WH, Robinson TN. Overweight children and adolescents. *N Engl J Med*. May 19, 2005; 352(20):2100–2109.
60. Joffe A. Pharmacotherapy for adolescent obesity: a weighty issue. *JAMA*. June 15, 2005; 293(23):2932–2394.
61. Diet drug may be of little benefit in obese teens. Yahoo.news. May 3, 2007. Available at http://news.yahoo.com. (Accessed October 10, 2007.)
62. FDA says no to metabolic syndrome; new obesity guidance takes hard line. *Pink Sheet*. February 19, 2007;69(8):21–22.
63. Durrant D. (Market analysis based on Kalorama and IMS data.) Presentation, Pediatric Market United States March 2003. Information on file at Tufts Center for the Study of Drug Development.
64. Global pediatric drug sales to top $46B. *Pharma Marketletter*. January 23, 2006;33(4):16.
65. Myshko D. The trials of youth: bringing children into clinical research. *Pharmavoice*. September 2003;3(9):42–48.

66. World vaccine market to exceed $15 billion in 2012. *Pharma Marketletter*. February 19, 2007; 34(8):27.
67. 2002 Medco Health Drug Trend Report. *Formulary*. January 2003;38:28.
68. Konopka A. Shifting the pediatric paradigm. *PharmaVoice*. March 2006;6(3):24–30.
69. Many believe drugs to treat ADHD are prescribed too often, poll finds. ADHD opinion poll: WSJ online/Harris Interactive Healthcare. In: WSJ online. April 18, 2006. Available at http://online.wsj.com. (Accessed September 15, 2007.)
70. *Medco Health Drug Trend Report 2007*. Available for purchase from Medco Health Solutions, Inc.
71. Elias M. Doctors caution parents on using herbal remedies for kids. *USA Today*. October 24, 2001:9D. Available at http://www.usatoday.com. (Accessed November 15, 2001.)
72. Parents' decision-making preferences in pediatric oncology: the relationship to health care involvement and complementary therapy use. Child Health Research Findings; Jan 2001–Dec 2004. Agency for Healthcare Research and Quality (AHRQ grant T32 HS00063). (Also published by Gagnon EM, Recklitis CJ. *Psycho-Oncology* 2003;12:442–452.) Available at www.ahrq.gov. (Accessed October 10, 2007.)
73. Parental willingness to have children participate in clinical trials depends on many different factors. *WSJ Health Care Poll*. October 13, 2004;3(20). Harris Interactive. Available at http://online.wsj.com.
74. Khoury MJ, McCabe LL, McCabe ERB. Population screening in the age of genomic medicine. *N Engl J Med*. January 2, 2003;348(1):50–58.
75. Brink S. Rare but deadly. *US News and World Report*. May 30, 2005. Available at http://www.usnews.com/usnews/health/articles/DJ050530/30child.htm. (Accessed October 10, 2007.)
76. Feigin RD. Propects for the future of child health through research. *JAMA*. September 21, 2005;294(11):1373–1379.
77. Simpson L. Lost in translation? Reflections on the role of research in improving health care for children. *Health Affairs*. Sept./Oct. 2004;23(5):125–130.

CHAPTER 6

Pediatric Market Dynamics

ANNE F. CLOTHIER
Philadelphia, Pennsylvania 19118

MARKET SIZE AND TRENDS

Currently, there are approximately 74 million children under 18 years of age in the United States, which accounts for 25% of the total U.S. population. Over the past 5 years, there has been little growth in this segment of the population.[1] Most children in the United States report very good or excellent health, with only 2% suffering from fair or poor health. Unfortunately, children living in poverty account for a disproportionate share of children with poor health.[2]

The most frequent pediatric illnesses are otitis media, upper respiratory infection, attention deficit hyperactivity disorder (ADHD), asthma, dermatitis, allergic rhinitis, viral infections, fungal infections, conjunctivitis, infant diarrhea, and esophagitis secondary to gastroesophageal reflux disease. Pediatric diagnoses that have grown significantly over the past several years include ADHD, dermatitis/eczema, bipolar affect/emotional disorders, and esophagitis. Among children under 2 years of age, the fastest growing diagnosis is acute gastritis. For children between the ages of 3 and 11, diagnoses of bipolar disease affect and emotional disturbances have become more frequent with an annual growth rate of 30–40% per year. Among young teenagers between the ages of 12 and 16, diagnoses of abnormal weight gain/obesity, affective psychoses, and bipolar and other emotional disorders have skyrocketed in recent years.[3]

The pediatric prescription drug market worldwide, which is growing at a rate of 6.2% per year, is projected to exceed $46 billion by 2009.[4] Medco, which is a major pharmacy benefit manager in the United States, reported that spending on pediatric prescriptions grew 85% over the period from 1997 to 2002 due to increased drug utilization in this segment, particularly with use of high-priced medicines in certain therapeutic areas. Asthma, allergies, and anti-infectives represent the largest areas of prescription drug spending for children. Drugs for treatment of ADHD and proton pump inhibitors for gastrointestinal use were among the fastest growing areas of drug spending for children during this period.[5]

Pediatric Drug Development: Concepts and Applications
Edited by Andrew E. Mulberg, Steven A. Silber, and John N. van den Anker

PEDIATRIC DRUG UTILIZATION

In 2003, there were over 9 million children in the United States (13%) who had a problem for which prescription medication had been taken regularly for at least 3 months. Youths aged 12–17 years were more likely to have been on regular medication for at least 3 months (17%) than children ages 5–11 years (13%) or children under 5 years of age (8%). Children with private (13%) or public (16%) health insurance coverage were almost twice as likely as children with no health insurance coverage (6%) to have been on regular medication.[6]

Almost two-thirds of the dollars spent in the pediatric prescription drug market are for asthma, allergies, infections, and central nervous system disorders. While overall growth in spending on pediatric prescriptions has been trending around 7% per year, there is wide variation across pediatric market segments. Sales of specialty therapeutics for treating growth hormone deficiency, respiratory syncytial virus (RSV), and cystic fibrosis grew at an average rate of 14%. The fastest growing pediatric market segments in prescription spending include treatments for ADHD, asthma, dermatologic conditions, and psychoses. In 2004, spending on anti-infectives and antidepressants declined 12% and 8%, respectively, which was primarily due to lower drug utilization in these categories.[7]

PEDIATRIC MARKET ECONOMICS

The economics of the pediatric market are a primary driver impacting both development and prescribing of medicines for children. Although the majority of U.S. children have either private or public health insurance, approximately 7 million children (10%) have no health insurance coverage. Children with health coverage are two times more likely to be on regular medications.[8] These statistics suggest that reimbursement for medicines is one variable influencing utilization of prescription medicines in children.

The financial incentives of drug development companies are perhaps the most significant variable resulting in the relatively limited availability of medicines for children. The average cost of developing a new drug is estimated at $800 million and continues to skyrocket.[9] Likewise, pediatric development costs for drugs that have previously been approved for use in adults has increased from $4 million in 2000 to $31 million in 2006.[10] When making choices about which drugs to advance through the final and most costly stages of development, biopharmaceutical companies are likely to evaluate financial metrics such as the net present value (NPV) and internal rate of return (IRR) of different options. Key factors that determine financial value include the projected revenues, costs, time, and risk to develop and commercialize a new drug in development. Since overall prescription drug utilization rates are significantly higher among adults than children, particularly for chronic medications, biopharmaceutical companies invest more resources in development and commercialization of drugs for adults.

Furthermore, unless a new medicine is being developed for treating (1) a condition primarily or exclusively in the pediatric population or (2) a serious or life-threatening disease in the pediatric population, initial drug development programs are designed for adult populations. *FDA Guidance for Industry for Clinical Investigations of Medicinal Products in the Pediatric Population,* which was released in 2000, states that "it is important to carefully weigh benefit/risk and therapeutic need in deciding when to start pediatric studies." In addition, biopharmaceutical companies usually need to develop specific pediatric formulations, which adds to drug development costs and time.[11] These issues are more fully discussed in other chapters in this textbook, including the regulatory framework in the United States and European Union as well as formulation requirements for pediatrics by Hoy, Roche, and others.

Due to these financial hurdles, in the absence of incentives extending market exclusivity and regulations requiring pediatric drug development programs, very few medicines have been developed and approved for use in the pediatric populations. As a result, healthcare providers have been left to prescribe medicines to children with little or no information about dosing, efficacy, and safety in children. Some estimates suggest that 50–75% of all medicines prescribed to children do not have pediatric labeling.[12] In order to secure approved pediatric labeling, biopharmaceutical companies need to conduct appropriate studies in this population.

PEDIATRIC DRUG DEVELOPMENT

While many opportunities remain for drug development that would address unmet needs in the pediatric population, the first major stimulus for pediatric drug development came in 1997 with passage of the FDA Modernization Act (FDAMA), which included provisions for pediatric exclusivity. Pediatric exclusivity granted 6 months of additional market exclusivity for a prescription medicine if the biopharmaceutical company completed and submitted the necessary studies detailed in a Written Request approved by the FDA. Pediatric exclusivity provided a substantial financial incentive to biopharmaceutical companies, particularly for "blockbuster" medicines with annual sales of $1 + billion. As a result, pediatric drug development began to finally receive a boost for innovation in the United States.

In addition to providing financial incentives through pediatric exclusivity, between 1998 and 2003 the U.S. government passed several new regulations with requirements for pediatric drug development. The Pediatric Rule was approved in 1998, which enabled the FDA to mandate pediatric studies and, where necessary, to develop a new formulations for use in the pediatric population. In 2002, the Best Pharmaceuticals for Children Act (BPCA) authorized the extension of pediatric exclusivity and established processes for development of off-patent drugs that needed pediatric formulations and/or pediatric labeling. In addition, BPCA required publication of results from pediatric studies conducted for pediatric exclusivity as well as public review of safety reporting for products granted pediatric exclusivity. The Pediatric Research Equity Act (PREA) was approved in late 2003 and reauthorized in 2007. This legislation codified the Pediatric Rule by requiring pediatric studies of certain drugs in order to obtain pediatric labeling for new indications, new dosage forms, new route of administration, new dosing regimens, and/or new active ingredients. The PREA added criteria with requirements for pediatric studies in situations where there is (1) need for "meaningful therapeutic benefit" over existing alternatives or (2) need for additional options.[13] These regulatory and financial issues are more fully discussed by Milne (Chapters 4 and 5) and Maldonado (Chapter 12) and others in chapters related to finance and regulatory issues in the textbook.

The results of these financial incentives and regulatory requirements have been noteworthy. In 2007, the Tufts Center for the Study of Drug Development (TCSDD) reported that less than 10 years after passing the first legislation in the United States to support pediatric drug development, pediatric studies increased tenfold with 58 completed studies by 2000 and 568 completed studies by 2006. Findings from these studies have resulted in new pediatric labeling for 115 medicines. During this time period, pediatric studies have become larger, take longer, and are more costly due to a significant increase in the number of efficacy/safety studies (40% in 2006 compared to 25% in 2000) as well as an increased number of studies and patients required.[14]

PEDIATRIC MARKET OPPORTUNITIES

Despite the dramatic increase in pediatric drug development over the past decade, a number of unmet needs have not been addressed in the pediatric market for drug treatments. Opportunities remain to improve treatments for the largest pediatric market segments, which include anti-infectives, asthma, allergies, and CNS disorders. Although multiple treatment options are available in each of these categories that have been approved for use in children, pediatric patients could benefit from improved formulations, drug delivery, dosing frequency, efficacy, tolerability, and safety. For products that are studied in children to obtain pediatric exclusivity, drug manufacturers should evaluate the pediatric market potential and the commercial value of bringing the new indication to market.

While some pediatric markets have very small patient populations, the high burden of illness for diseases that occur in children, such as cystic fibrosis, sickle cell disease, respiratory distress syndrome, and growth hormone deficiencies, could be eased by breakthrough new treatments. In turn, biopharmaceutical companies that are successful in developing new treatments for pediatric diseases with the greatest demands will reap financial rewards that more than offset their investments in drug development.

Medimmune is one example of a company that delivered and captured value by addressing a significant unmet need in the pediatric market. Synagis® (palivizumab) is the only monoclonal antibody approved by the FDA to prevent serious respiratory infections caused by respiratory syncytial virus (RSV) in high-risk babies. Before Synagis was introduced to the market in 1998, infections caused by RSV resulted in approximately 125,000 hospitalizations per year.[15] By the end of 2006, product sales totaled $1.1 billion.[16]

Many untapped opportunities remain in the pediatric market for medicines. As biopharmaceutical companies face continued challenges in developing and bringing new products to market, perhaps there will be further consideration of the pediatric market. Companies that have invested in pediatric development programs in order to meet regulatory requirements should take time to evaluate pediatric market needs and identify how they can maximize their returns on investment. In doing so, children would benefit from more medicines being studied and developed appropriately for their needs.

REFERENCES

1. US Census Bureau, Population Division: Annual Estimates of the Population by Selected Age Groups and Sex for the United States: April 1, 2000 to July 1, 2006.
2. National Children's Health Survey, 2003. Available at www.CDC.gov.
3. *IMS National Disease and Therapeutic Index*, 2004.
4. *The Worldwide Market for Prescription Drugs*, 2nd ed. Kalorama Information; 2006.
5. *Medco Drug Trend Report*. May 2004;6:15–16.

6. National Children's Health Survey, 2003. Available at www.CDC.gov.
7. *Medco Drug Trend Report*. May 2005;7:20–21.
8. National Children's Health Survey, 2003. Available at www.CDC.gov.
9. DiMasi JA, Hansen RW, Grabowski HG. The price of innovation: new estimates of drug development costs. *Health Economics*. 2003;22:151–185.
10. Pediatric study costs rose 8-fold since 2000 as complexity level grew. *Tufts Center for the Study of Drug Development Report,* March/April 2007;9(2):1–4.
11. *FDA Guidance for Industry for Clinical Investigations of Medicinal Products in the Pediatric Population*, 2000.
12. Roberts R, Rodriguez W, Murphy D, Crescenzi T. Pediatric drug labeling: improving the safety and efficacy of pediatric therapies. *JAMA*. 2003;290:905–911.
13. Murphy D. Impact of pediatric initiatives. Presentation to Drug Information Association Annual Meeting, June 2004.
14. Pediatric study costs rose 8-fold since 2000 as complexity level grew. *Tufts Center for the Study of Drug Development Report,* March/April 2007;9(2):1–4.
15. Herper M. Is there hidden value in MedImmune? 2004. Available at www.forbes.com.
16. MedImmune Annual Report, 2006.

CHAPTER 7

Industry Benchmarks in Pediatric Clinical Trials

CINDY LEVY-PETELINKAR, MBA, RAC
GlaxoSmithKline, King of Prussia, Pennsylvania 19406

CAROLYN A. CAMPEN, PMP, PhD
Johnson & Johnson, Raritan, New Jersey 08869

INTRODUCTION

In 1995, the American Academy of Pediatrics' Committee on Drugs stated that there is a "moral imperative to formally study drugs in children so that they can enjoy equal access to existing as well as new therapeutic agents".[1] This statement was made in response to the paucity of Food and Drug Administration (FDA) approved drugs possessing labeling for children, and the ethical dilemma this placed on the treating physician. The importance of clinical trials in children is increasingly recognized by governmental and professional bodies worldwide such as the Food and Drug Administration, the National Institutes of Health (NIH), American Academy of Pediatrics (AAP), the Medical Research Council (MRC), the Royal College of Paediatrics and Child Health (RCPCH), the European Agency for the Evaluation of Medicinal Products (EMEA), and the European Commission. The FDA, through the enactment of the Food and Drug Administration Modernization Act (FDAMA, 1997), its 2002 reauthorization as the Best Pharmaceuticals for Children Act (BPCA), and the Pediatric Research Equity Act (PREA) in 2003, has the ability to either incentivize (by providing 6 months of additional marketing exclusivity) or require the pharmaceutical industry to conduct clinical studies of approved drugs in the pediatric population. This legislation was renewed in October 2007 and is discussed separately in Chapter 12 (by Maldonado) and in Chapter 15 (by Owen et al.). Since the passage of the FDAMA, the FDA has issued over 340 Written Requests, resulting in label changes for nearly 130 drugs.[2–4] Clearly, this set of legislative actions has advanced useful clinical information in the pediatric population compared to the pre-FDAMA environment.

Pediatric Drug Development: Concepts and Applications
Edited by Andrew E. Mulberg, Steven A. Silber, and John N. van den Anker

TABLE 7.1 Comparison of Characteristics[a] for Pediatric-Only, Adult-Only, and Elderly-Only Studies, 2003–2005

	Total Number of Companies			Total Number of Studies		
Group	2003	2004	2005	2003	2004	2005
Pediatric only	15	13	13	35	30	30
Adult (including adolescents)	24	29	30	332	422	263
Elderly	8	6	8	19	11	10

[a] Data are shown for Phase II (including Phase Ip) and Phase III studies that were completed between 2003 and 2005. Pediatric-only studies include patients between the ages of 0 to ≤14 years only. Adult studies include patients between the ages of >14 years and <65 years only. Elderly studies include patients ≥65 years. Data provided by Centre for Medicines Research International, Thomson Scientific (CMR), from 2006 Global Clinical Performance Metrics Programme.

IMPACTS

As these pediatric laws are reevaluated and reenacted over time, it is of interest to gauge their impact. The Tufts Center for the Study of Drug Development reported that the cumulative number of pediatric studies for 32 drug products rose from 58 at the end of 2000 to an estimated 568 in 2006.[5] However, despite these initial successes resulting from the implementation of BPCA and PREA, there is evidence that their full impact has yet to be achieved. A recent report from the United States Government Accountability Office (GAO) stated that approximately two-thirds of drugs that are prescribed for children have not been studied or labeled for pediatric use.[6] In a retrospective cohort study, analyzing data from 31 U.S. tertiary-care pediatric hospitals, Shah et al.[7] reported that at least one drug was used off-label in 78.7% of hospitalized patients 18 years or younger. In addition, for those pharmaceutical companies that participate in Centre for Medicines Research (CMR) surveys, the last few years have shown little change in the number of companies conducting pediatric trials as well as the actual number of pediatric trials (Table 7.1).

While it does appears that pediatric clinical trial activity has not changed over the past few years, Figure 7.1 shows that the pharmaceutical industry is exploring a larger diversity of

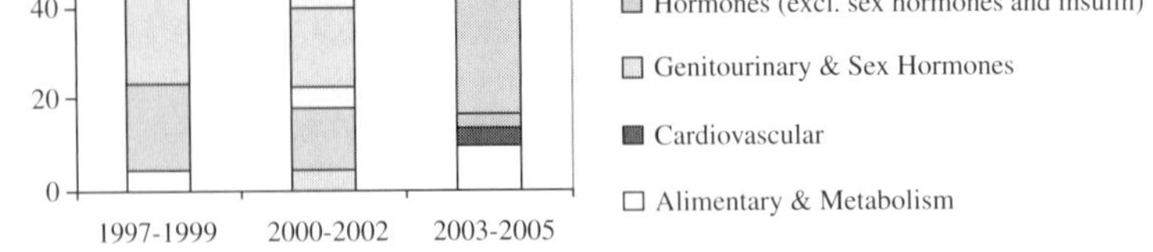

FIGURE 7.1 Number of pediatric trials, by therapeutic area, from 1997 through 2005 by 3-year increments. Data are shown for Phase II (including Phase Ip) and Phase III studies that were active between "first patient enrolled" and "last patient enrolled." Pediatric-only studies include patients between the ages of 0 to ≤14 years only. Data provided by Centre for Medicines Research International, Thomson Scientific (CMR), from 2006 Global Clinical Performance Metrics Programme.

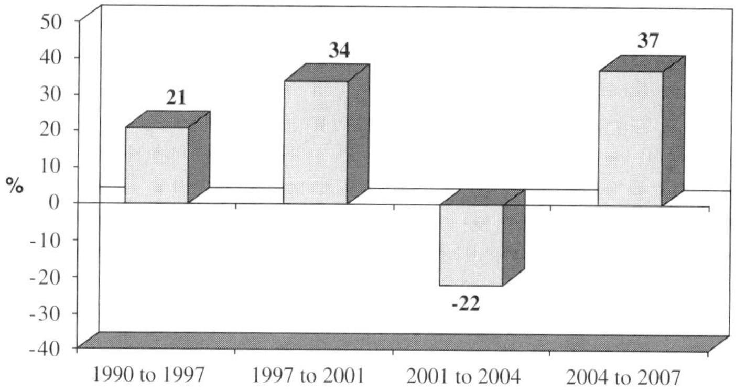

FIGURE 7.2 The changes in the number of drugs being studied for pediatric indications. (Source: PhRMA surveys; Tufts CSDD presented at the DIA 43rd Annual Meeting.)

therapeutic areas within the pediatric population, notably in the areas of musculoskeletal, anticancer, and alimentary and metabolism. The impact of government policy can also be demonstrated in Figure 7.2, which shows that the only time the number of drugs being tested in the pediatric population actually declined occurred during the 2001–2004 time period, when several key pediatric drug development policies were due for review and reenactment, potentially creating an environment of pediatric policy uncertainty.[8]

THE FUTURE FOR PEDIATRIC TRIALS

Although there are policies to promote the inclusion of children in clinical trials,[9] this continues to be difficult due to a myriad of issues. The regulations enacted appear to have initially achieved their goal to inspire drug development for the pediatric population, but perhaps this momentum has slowed in recent years. Continued positive reinforcement of the healthcare industry, more aggressive governmental policies, balanced media coverage, and an increase in true public awareness are possible keys to the further development of medicines for this vulnerable population.

REFERENCES

1. American Academy of Pediatrics' Committee on Drugs. Guidelines for the ethical conduct of studies to evaluate drugs in pediatric populations. *Pediatrics.* February 1995;95(2): 286–294.
2. http://www.fda.gov/cder/pediatric/wrstats.htm.
3. Pediatric studies lead to more information on drug labels. *AAP News* 2007; 28:20–25.
4. www.fda.gov/cder/pediatric/labelchange.htm.
5. Milne CP, Faden L. Pediatric study costs increased 8-fold since 2000 as complexity level grew. *Tufts Center for the Study of Drug Development Report* 2007;9(2):1–4.
6. US GAO Pediatric Drug Research: Studies Conducted Under Best Pharmaceuticals for Children Act, GAO-07-557.
7. Shah SS, Hall M, Goodman DM, et al. *Arch Pediatr Adolescent Med.* 2007;161:282–290.

8. Milne CP. Pediatric Exclusivity Program: What Hath Congress Wrought? Tufts Center for the Study of Drug Development. DIA 43rd Annual Meeting.
9. NIH Policy and Guidelines on the Inclusion of Children as Participants in Research Involving Human Subjects, March 6, 1998. Available at http://grants.nih.gov/gra Cents/guide/notice-files/not98-024.html.

CHAPTER 8

Novel Organizational Strategies for Advancing Pediatric Products: Business Case Development

DONALD P. LOMBARDI

Institute for Pediatric Innovation, Cambridge, Massachusetts 02142

INTRODUCTION

> I recently treated a young child who had a urinary tract infection. The child was extremely uncomfortable. The pain medication I prescribed was only available, as a large tablet, which contained too much drug for a small child. Yet it could not be easily divided. I nervously advised the parent on how to crush the tablet, conceal it in juice, place it in the child's sippy-cup and hope for the best. This situation is very common when prescribing medications to children. For the physician, there are two less-than-ideal choices: withhold the medication because it can't be administered in the proper dose (in which case the child is left in pain) or give the medication, hoping that the child will not receive an overdose. In either case, the child could suffer. Cases such as these regularly show me the need for pediatric drug development has never been more acute.
>
> —Michael W. Shannon, MD, MPH, Professor of Pediatrics, Harvard Medical School and Chief Emeritus and Chair, Division of Emergency Medicine, Children's Hospital Boston

As cited elsewhere in this book pediatricians often need to prescribe drugs that have not been fully tested and labeled for use with and administration to children. In this chapter, we focus on a subset of the products needed for treating children: namely, existing drugs that have previously received an FDA approval but require reformulation into different dosages or delivery modalities for optimal administration to children. We outline a novel organizational strategy for stimulating development of such products and enabling their adoption by commercial companies that will make them available in the marketplace. The strategy relies on forming a consortium of leading centers of pediatric care and research. The consortium will provide data, clinical expertise, and innovation resources needed to define the most critical clinical needs. Clinicians from consortium hospitals will also participate in developing products for commercial adoption that are optimized to meet these needs. We include a

Pediatric Drug Development: Concepts and Applications
Edited by Andrew E. Mulberg, Steven A. Silber, and John N. van den Anker

case study to illustrate one novel pathway for advancing a reformulated drug to commercial realization, and conclude with an action plan.

THE NEED FOR BETTER-FORMULATED DRUGS IN TREATING CHILDREN

It is commonly understood in the pediatric community, including the pediatric divisions of the Food and Drug Administration (FDA) and National Institutes of Health (NIH), that most drugs are administered to children off-label. Less attention has been paid to the fact that many drugs that pediatricians and pediatric specialists wish to administer to children are not available in a dosage or delivery form that is optimized for treating their patients' medical needs. All professional caregivers, as well as parents or others who care for children at home, are aware of the difficulty of administering many pharmaceutical products to children. Recent inquiries to leading pediatric institutions revealed that their pharmacies compound, reformulate, or repackage 60–80% of the prescriptions they fill, at a cost of 20–30% of total pharmacy operations cost. The inherent variability of ad hoc reformulation, coupled with the lack of rigorous clinical test data dealing with the unique physiological environment of growing children, places patients and their caregivers at risk. Some of the risks are inconsistent dosing, poor patient compliance, low efficacy, adverse drug reactions and events, developmental disorders, and death.

MARKET OPPORTUNITIES FOR BETTER PRODUCTS

This gap in availability of drug formulations specifically adapted for children provides the opportunity to develop and to market a robust portfolio of new, differentiated, value-added formulations of known compounds. Although many of these products may address markets only in the \$20–100 million/year range, they can be commercially feasible under the right circumstances. By focusing on off-patent compounds that already have a clinical record of use in pediatrics and using well-established formulation modalities, a company may be able to launch new products in a 2–4-year time frame with a total development cost in the range of \$2–7 million per product. Interviews with pediatric clinicians at one hospital yielded more than 40 products they desired for use in different therapeutic areas that may fit this profile.

There are at least three potential commercial avenues for such compound development:

- A large pharmaceutical company that has a proprietary compound for a particular pediatric specific therapeutic area could acquire several reformulated products directed to the same specialty therapeutic area. These products would serve as product line extenders, providing the sales force with additional products to serve the same customers and provide added value.
- A smaller niche-oriented specialty pharmaceutical company could acquire a group of reformulated products for a range of clinical areas to be sold through a dedicated pediatric sales and marketing force.
- A company that has a proprietary drug delivery platform with particular applicability for pediatric administration could develop a portfolio of products using compounds for which their technology was well adapted. This could provide significant therapeutic value to pediatric patients.

SPECIAL CONDITIONS REQUIRED FOR MAKING REFORMULATED DRUGS FOR PEDIATRIC CARE COMMERCIALLY VIABLE

Given the relatively small market size for pediatric pharmaceutical formulations, the risks and obstacles to develop these products need to be reduced to stimulate and to facilitate making the products available. Some of the ways that newly formulated products for pediatrics can become commercially feasible are as follows:

- Clinical needs for new formulations have to be carefully qualified and quantified in relation to compelling potential patient outcomes.
- Initial product candidates need to be selected that have a short development and approval cycle.
- The people who provide patient care, including parents as well as clinicians, need to be involved in the details of product design in order to appropriately meet the interests of all stakeholders.
- Clinical information on the prior pediatric use of the compound needs to be assessed in order to discuss potential reduction in the clinical testing requirements and burden for the reformulated version with the FDA.
- In some cases, a product's projected sales may be sufficient to sustain the product in the market, but not great enough to attract investment. In these cases, funding sources that do not require a return on investment may need to be accessed to support some or all of product development. Examples of such sources are Federal Small Business Innovation (SBIR) grants and grants from foundations focused on specific diseases or patient populations. This would apply especially to orphan drugs and needs of children with rare diseases.
- Patient populations will need to be aggregated efficiently across multiple sites in order to conduct clinical trials for the product approval process.

Accomplishing all of these development tasks effectively will require engaging clinical organizations nationally, securing participation of clinical and industry experts of the highest caliber, participating in the development of a full portfolio of product opportunities, and facilitating placement of these products into appropriate companies for full commercial sale.

A NOVEL MODEL FOR PUBLIC–PRIVATE FOR-PROFIT–NONPROFIT COLLABORATION FOR DEVELOPING PEDIATRIC DRUG PRODUCTS

Individual reformulated drugs will need to be able to sustain themselves in the market in order to impact pediatric care broadly. However, the process and infrastructure to bring together clinical experts, product innovation specialists, biomedical companies, foundations, and government experts to collaborate on a product development program must also be financed. The experience of the pediatric market suggests that this infrastructure cannot be established and sustained solely on the basis of anticipated return on investment from product revenues. No major corporate players are serving the market broadly. Ascent Pediatrics developed several useful products in the early 1990s but was then sold several times and no longer exists. In 2006, Johnson & Johnson reorganized McNeil Consumer & Specialty Pharmaceuticals into McNeil Pediatrics as a subsidiary that is now focused primarily on the issues relevant to the pediatric population from a marketing perspective. PediaMed Pharmaceuticals, founded in 1999, developed several products but is now also out of business. On the contrary, there are a few, truly very few, small companies that are pediatric centric, but they do exist. Most business strategies are not pediatric centric, but there is a small, albeit nonsubstantial, activity in this area. There are some small companies still around as well as other companies, large and small, that have organized a pediatric department. For example, Pediatric Pharmaceutical Inc., Alliant Pharmaceuticals, or foundations such as the Pediatric Cancer Foundation do exist in this regard, as well as Topaz Pharmaceuticals, which is developing a topical head lice treatment for children.

Institute for Pediatric Innovation

The Institute for Pediatric Innovation (IPI) was founded as a new nonprofit structure that can translate the needs for new pediatric products into viable commercial opportunities that companies will adopt. The organizational structure includes a consortium of pediatric care and research centers, a resource network of people and organizations with expertise in commercial development of medical products, and an action plan focused on near-term product opportunities.

Pediatric Hospital Consortium

IPI is convening a Pediatric Hospital Consortium consisting of a geographically and organizationally diverse group of the most innovative U.S. pediatric hospitals, representing a microcosm of the pediatric market. As of this writing, University Hospitals Rainbow Babies and Children's Hospital in Cleveland, Lucile Packard Children's Hospital at Stanford, and Children's Mercy Hospitals and Clinics in Kansas City have joined together to initiate the consortium. These hospitals have agreed to provide access to their clinical staff and clinical operations to determine which products are most needed to make substantive improvements in patient care, with focus on products that would impact the quality of care parameters, such as improved outcomes, reduced risk of morbidity and mortality, better patient compliance, and higher satisfaction for patients, clinicians, and parents. In addition, staff from the participating hospitals will take part in defining the clinical requirements of the products selected for advancement. Intellectual property agreements are being established to assure fair and effective management of both existing and newly created inventions. At a later stage, when products are ready for clinical validation, the hospitals will have the first option to participate in funded clinical studies, subject to appropriate conflict of interest considerations and the need for independent verification. In addition to their specialty clinics, most of the participating hospitals also have internal primary care programs, and several have affiliated community hospitals and extensive networks of community-based pediatric and family practices. Along with their affiliates, the collaborating pediatric institutions will be able to help define products needed for most segments of the pediatric market. This is an opportunity to develop a broad and diverse think tank to address unmet medical needs for the pediatric population.

Resource Network

IPI is building a network of people and organizations interested in contributing their expertise to the commercial development of pediatric products. The group includes angel investors, product development and regulatory experts, leaders in pediatric care and medical innovation, design engineers, formulation companies, entrepreneurs, strategy consultants, and corporate executives. Members of this group have demonstrated the willingness to provide general advice and contacts on a pro bono basis and are also available to participate as consultants or collaborators to execute substantive projects.

Consortium Collaboration Strategy

IPI and the founding members of the Pediatric Hospital Consortium have committed to an implementation plan for the first year that will focus on assessing and organizing the consortium's innovation resources and identifying, characterizing, and qualifying selected product opportunities. The plan progresses through the following steps:

- Complete an innovation audit of each founding member of the consortium, resulting in an *innovation profile*.
- Assist each founding consortium member in formalizing its processes and increasing its capacity for developing product innovations.
- Identify and analyze the areas of clinical care that can most benefit from new devices or reformulated drugs specific to pediatric care.
- Conduct a formalized *needs assessment* within targeted clinical specialties.
- Based on the advice of the consortium clinical members, select products for development that (1) can significantly improve children's care in hospitals, (2) hold promise of near-term commercialization, and (3) can leverage the resources of the member base.
- Access the resource network of product development experts, corporate collaborators, investors, and entrepreneurs to assist in formulating a plan for the development of each product.
- Create *pediatric product opportunity analyses* for the selected products that detail product clinical requirements, technology characteristics, forecasted adoption rates, marketing channels, product development and regulatory path, intellectual property requirements, timeline, budget, and target sponsor.

We expect that each product opportunity will enter one or more of the following next stages: (1) direct financing for the next phase of development from a foundation or other development funding source interested in the clinical area, (2) collaboration with a formulation company to develop the formulation and/or jointly apply for a federal or other grant, or (3) direct licensing to a commercial entity that has the resources to both develop the formulation and make it available to the pediatric market.

A Product Case Study

A project started in 2004 by the Intellectual Property Office at Children's Hospital Boston illustrates an example of a product development strategy that can be applied to other clinical needs for diverse therapeutic areas. Because this project is still in progress, much information is confidential, so only the general approach will be described.

Dr. Michael Shannon and his colleagues at Children's Hospital Boston had treated more than 5000 children for lead poisoning over 30 years using penicillamine, an off-patent drug. The drug had previously been approved for different adult indications. Currently, there is no FDA-approved treatment for moderate levels of lead poisoning. Published retrospective studies indicated safety and efficacy of penicillamine to treat this clinical need. The compound proved substantially more effective in reducing blood lead levels than the only approved drug treatment for children. In fact, the dosing had been reduced over the years, effectively yielding dose-ranging data that would typically come from a Phase II clinical trial. However, the drug is available only as a large pill that cannot be administered to small children. The pill must be crushed by a parent or other caregiver and resuspended in applesauce or other food. Unfortunately, in this form it gives off a foul smell, as it is a sulfur-containing compound, and is incompatible with milk. This makes the taste and odor harder to mask. The result is significant difficulty in administration and palatability for children, uncertain dosage, and poor compliance. This difficulty is exaggerated by a course of treatment that lasts several weeks.

In order to develop an appropriate reformulation of the drug, a risk-sharing arrangement was made with a formulation company. The company agreed to develop a new formulation for the compound in exchange for a share of any proceeds the hospital would receive from licensing the new formulation. In collaboration with the hospital, the formulation company developed a powder that could be constituted into a dose-adjusted, stable, and palatable liquid formulation. The hospital and the company filed a joint patent application on the new formulation. An angel investor with experience in pharmaceutical product development agreed to determine the risks and costs of conducting the clinical validation that would be required for product approval. He concluded that, given the hospital's long history with the drug and its demonstrated efficacy profile, the clinical study could be completed and the product approved in 2 years with an investment of less than $1.5 million. He also concluded that the addressable market was upwards of $40 million annually in the United States. He created a syndicate for the needed funds, established a virtual company, obtained a license for the formulation and patent data, and will manage the pivotal trial. After FDA approval, the investors would plan to sell or license the resulting product to a commercial entity at a multiple of the investment. This makes the opportunity attractive to the angel investors in the syndicate.

This case worked because development began with a need rather than an invention; the project employed a nontraditional codevelopment mode; the problem was solved through creative reengineering, not research; and the resulting product, while addressing a small market, can be developed efficiently at sufficiently small cost to offer a return to the investors.

The Need for Alternative Development and Commercial Pathways

The nonprofit platform for developing product opportunities can bring together diverse constituencies that are interested in advancing products for pediatric care. Clinicians participate first in helping to define the needs, then to determine required clinical characteristics for the products, and later to conduct clinical validation trials to support a registration pathway for the FDA. People with product and regulatory development skills need to be involved, initially as advisors in helping to qualify the product opportunities, and later in managing the detailed development of specific product opportunities. To reduce cash investment requirements, some of this assistance may be pro bono or, alternatively, provided on a risk basis. In the long term, as the practices for conducting these analyses and preliminary formulation studies become more established, collaborations with a graduate professional school, such as a school of pharmacy, under mentorship by experienced pharmaceutical industry drug development experts, may help to limit development costs.

In some cases, funding for the subsequent stages of product development may come from investors, as in the case study presented previously. Alternatively, depending on the primary indication for the product, the project may be suitable for a Small Business Innovation Research grant from a division of the NIH or other government entity, or for a grant from a disease-focused foundation. In some cases, a company may see strategic advantage in leveraging the resources of the Pediatric Hospital Consortium by providing a development grant to scout proactively for a portfolio of product opportunities in a given clinical specialty area or employing a drug delivery platform well suited to pediatrics.

First-Year Program Plan

In the first year of its Pediatric Pharmaceutical Reformulation Program, IPI and its collaborators will:

1. Convene an advisory committee with expertise and experience in the disciplines required to successfully develop pediatric reformulated pharmaceutical products. This committee will provide assistance in identifying target compounds, contacts, potential collaborators, selection processes and criteria, and other program needs.
2. Identify a broad list of compounds that pediatric clinicians believe need to be reformulated for optimal treatment of children, gathering data from participating hospitals, their affiliated primary care networks, national organizations concerned with the effective care of pediatric patients, the FDA, and the National Institutes of Health.
3. Classify and set priorities on the identified compounds according to potential for clinical impact, formulation development requirements, clinical testing and regulatory pathway and risks, availability of resources for product design and clinical testing, potential sources of development financing, and commercialization options.
4. Prepare a detailed product reformulation development business plan to finance and initiate at least two product development projects per year for the next 4 years.
5. Prepare a complete product opportunity analysis for the first selected product and secure the financing and collaboration commitments necessary to initiate the development stage for this product in the beginning of the second year.
6. Conduct a conference that will present and analyze UNSUCCESSFUL examples of pediatric pharmaceutical development projects and UNSUCCESSFUL prior efforts to create commercial entities to develop, produce, and sell pediatric drugs. This conference will serve to guide further development of the IPI pharmaceutical development program.
7. Produce an educational video product about the issues, challenges, and opportunities related to making medical products available for treating children. This resource will be aimed at decision-makers in government, industry, and philanthropy and other relevant sectors.

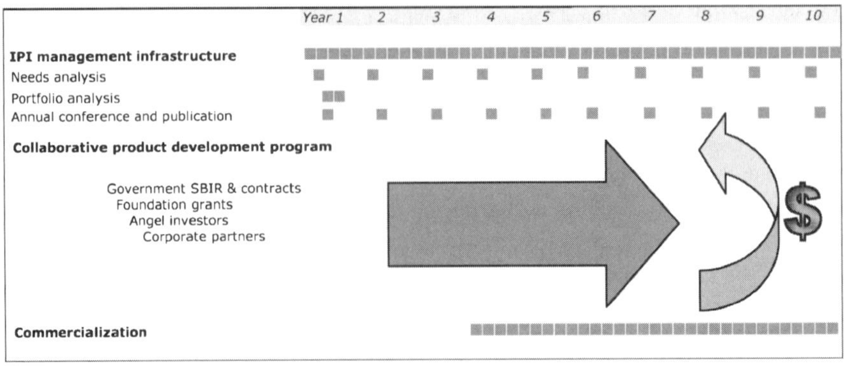

FIGURE 8.1 Multiyear plan.

Multiyear Sustainability

IPI seeks to establish a core infrastructure to manage its pharmaceutical reformulation program. This infrastructure will enable IPI to work with its collaborators to leverage additional future funding from multiple sources for development of products and also to participate actively in facilitating and monitoring the product development projects. Based on securing financing for product development, IPI expects to help generate two new product opportunities per year, and to license these opportunities to commercial entities that will begin introducing the products into the market beginning in the fourth year. Commercialization of the reformulated products will generate transaction fees, royalties, and other consideration that will flow back to IPI and its collaborators. Such funds will help to support the core program infrastructure after the initial 5 year period and could potentially help to finance selected product reformulation development programs. The multiyear plan is illustrated in Figure 8.1.

National Resources Applicable to the Task

IPI has begun to identify national organizations that can provide information and guidance in the development of the program. In some instances, these parties may be able to help identify avenues for financing later stage product development. Key parties identified as of this writing are:

- American Academy of Pediatrics
- American Society of Health System Pharmacists
- Child Health Corporation of America
- Elizabeth Glazer Pediatric AIDS Foundation
- Food and Drug Administration—Office of New Drugs
- Institute for Safe Medication Practices
- National Institutes of Health—Obstetric and Pediatric Pharmacology
- NIH Foundation
- Organization for Rare Disorders
- National Dissemination Center for Children with Disabilities

- Pediatric Pharmacy Advocacy Group
- PhRMA Pediatric Working Group

CONCLUSION

In this chapter, we have outlined a needed area of product development, namely, reformulation of existing drugs for better administration to children. An approach to meeting this need has been described that engages clinical staff from a consortium of pediatric institutions in qualifying needs and in designing and testing reformulated drug products to meet these needs. The clinical expertise provided by consortium members is complemented by a resource network of people and organizations with the other areas of expertise required to advance products to the stage of commercial adoption. The resulting program will produce a portfolio of reformulated drug products that will be available to interested companies.

ACKNOWLEDGMENTS

The author wishes to thank the following people for reviewing drafts of this chapter and offering critiques and suggestions: Emmett Clemente, PhD; Roger Kitterman; Douglas McNair, MD, PhD; Andrew E. Mulberg, MD; Michael Shannon, MD, MPH; Stephen Spielberg, MD, PhD; and Jing Watnick, PhD, MBA.

This work was funded in part by a planning grant from Ewing Marion Kauffman Foundation.

PART II

ETHICAL UNDERPINNINGS

CHAPTER 9

Additional Protections for Children Enrolled in Clinical Investigations

ROBERT M. NELSON, MD, PhD

Food and Drug Administration, Rockville, Maryland 20857

INTRODUCTION

The additional protections for children to be enrolled in a clinical investigation can be divided into four "nested" domains with each protection building on an adequate response to the prior protection. First, the enrollment of children in a clinical investigation should be scientifically necessary before one evaluates whether the research interventions or procedures present the appropriate balance of risk and potential benefit. Second, a clinical investigation must meet the criteria for the appropriate balance of risk and potential benefit found in 21 CFR 50, Subpart D, before considering the role of parental permission and child assent. This chapter addresses the first two of these nested protections: the ethical principle of scientific necessity and the appropriate balance of risk and potential benefit.

THE ETHICAL PRINCIPLE OF SCIENTIFIC NECESSITY

The first protection for children who may be enrolled in a clinical investigation is the ethical principle of "scientific necessity." This principle holds that children should not be enrolled in a clinical investigation unless absolutely necessary to answer an important scientific question about the health and welfare of children. The principle of scientific necessity is grounded in the general regulations governing human subject protections. (See Figure 9.1.) First, the risks to subjects must be minimized by using procedures that do not unnecessarily expose subjects to risk (21 CFR 56.111(a)(1)). In other words, any research procedure that

The opinions, views, and comments expressed in this chapter are solely those of the author and do not necessarily represent the views of the Food and Drug Administration or the Department of Health and Human Services.

Pediatric Drug Development: Concepts and Applications
Edited by Andrew E. Mulberg, Steven A. Silber, and John N. van den Anker

Subpart C--IRB Functions and Operations

21 CFR 56.111 Criteria for IRB approval of research.

(a) In order to approve research covered by these regulations the IRB shall determine that all of the following requirements are satisfied:

(1) Principle of "Scientific Necessity"

(1) Risks to subjects are minimized:

(i) By using procedures which are consistent with sound research design *and*...

which do not unnecessarily expose subjects to risk, *and*

(ii) whenever appropriate, by using procedures already being performed on the subjects for diagnostic or treatment purposes.

(3) Selection of subjects is equitable. In making this assessment the IRB should take into account the purposes of the research and the setting in which the research will be conducted and should be particularly cognizant of the special problems of research involving vulnerable populations, such as children.

(2) Appropriate Balance of Risk and Potential Benefit

(2) Risks to subjects are reasonable in relation to anticipated benefits, *if any*, to subjects, and the importance of the knowledge that may be expected to result.

Apply Additional Safeguards of 21 CFR 50, Subpart D

(b) When some or all of the subjects, such as children, ... are likely to be vulnerable to coercion or undue influence additional safeguards have been included in the study to protect the rights and welfare of these subjects.

(c) In order to approve research in which some or all of the subjects are children, an IRB must determine that all research is in compliance with part 50, subpart D of this chapter.

FIGURE 9.1 General regulations governing human subject protections.

does not contribute to answering a scientific objective of the clinical investigation is unnecessary.* Although not tied directly to the principle of scientific necessity, the risks to subjects must also be minimized, whenever appropriate, by using procedures already being performed for diagnostic or treatment purposes. Second, the selection of subjects must be equitable. Equitable selection should be evaluated in the context of the purposes of the research and the research setting (21 CFR 56.111(b)). More specifically, children should not be enrolled in a clinical investigation unless their involvement is essential. By essential, there should be no other option to answer the scientific objective, such as using nonclinical animal models or adult human subjects.[1] In addition, the scientific objective(s) must be relevant to the health and welfare of children. A corollary of this principle of equitable (or fair) selection is that, where appropriate, subjects who are capable of informed consent (such as adults) should be enrolled prior to older children who are capable of assent. Similarly, older children should be enrolled prior to enrolling children who are not capable of assent.[2] This approach, of course, assumes that there are no significant scientific reasons to enroll younger children preferentially to older children. This principle of the equitable or fair selection of children for participation in clinical investigations is widely supported in international guidelines for pediatric research (e.g., CIOMS, Guideline 14,[3] Medical Research Council guidelines,[4] Section 4.1, and Section 4.8.14 of the ICH Good Clinical Practice guidelines[5]).

Extrapolation

The ethical principle of scientific necessity has been operationalized by the Food and Drug Administration (FDA) in the scientific principle of extrapolation. As described in the Pediatric Research Equity Act of 2007, "if the course of the disease and the effects of the drug are sufficiently similar in adult and pediatric patients, the Secretary may conclude that pediatric effectiveness can be extrapolated from adequate and well-controlled studies in adults, usually supplemented with other information obtained in pediatric patients, such as pharmacokinetic studies."[6] This same principle of extrapolation can be found in the 1998 Pediatric Rule,[7] the 2000 International Conference on Harmonisation guidance on pediatric research,[8] and the Pediatric Research Equity Act of 2003.[9] The need for pediatric studies is assessed by asking a series of questions about the similarity of the adult and pediatric disease, response to treatment, drug exposure–response, and pharmacokinetic and pharmacodynamic measurements that could be used to predict efficacy. (See Figure 9.2.[10])

The FDA approach to extrapolation for determining the necessity of pediatric studies is in contrast to the current policy of the National Institutes of Health (NIH) on the inclusion of children as participants in research involving human subjects. The NIH approach is to include children *unless* there are scientific or ethical reasons *not* to include them, rather than to justify the inclusion of children based on the scientific necessity of obtaining information pertinent to the health and welfare of children.[11] NIH policy suggests that the inclusion of children would allow for the generalization of the overall research results to the pediatric population absent sufficient power to test for group differences between

* A protocol may contain interventions or procedures that are being performed for diagnostic or treatment purposes independent of whether these same interventions or procedures also serve a research purpose. The principle of scientific necessity does not imply that these clinical procedures are unnecessary and should be removed from the protocol. However, the protocol should clearly distinguish the interventions and/or procedures being performed for (1) research purposes only, (2) clinical purposes only, and (3) combined clinical and research purposes. These distinctions are essential to the proper ethical analysis of the protocol.

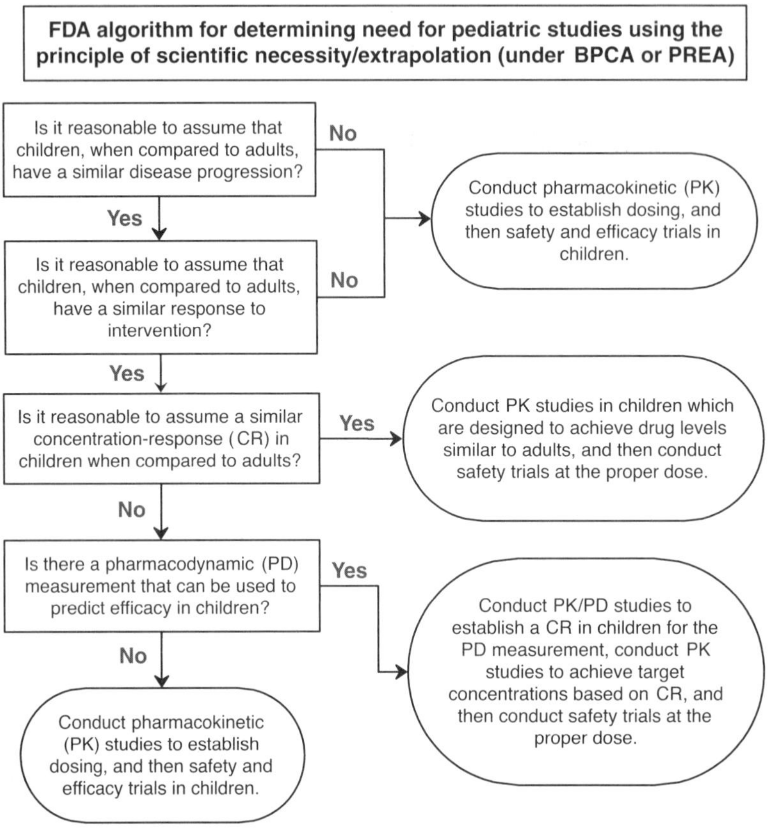

FIGURE 9.2 FDA algorithm for determining need for pediatric studies using the principle of scientific necessity/extrapolation (under BPCA or PREA).

children and adults.[12] In other words, children would be enrolled in research without sufficient numbers to provide for an adequate pediatric sample size to establish any group differences. The overall research results would be generalized to the enrolled children in a manner similar to extrapolation. However, the similarity of a disease between adults and children is used to justify *inclusion* rather than *exclusion* from the research, as the pediatric data would contribute to the overall study results.[12] Oddly, the result is that the scientific possibility of extrapolation would work in reverse, with the required number of adults to achieve an adequate sample size to answer the scientific question reduced by the enrollment of children.[12] Thus, according to NIH policy, the ability to extrapolate from adults to children becomes a reason to include, rather than exclude, children* from a clinical investigation.[12] This approach violates the first protection for the enrollment of

* It should be noted that the NIH definition of a child is an individual under the age of 21. The FDA would not consider individuals between 18 and 21 years of age to be children in need of additional safeguards when participating in a clinical investigation. FDA regulations define children as "persons who have not attained the legal age for consent to treatments or procedures involved in clinical investigations, under the applicable law of the jurisdiction in which the clinical investigation will be conducted" (21 CFR 50.3(o)). In effect, whether or not an individual subject is considered a child for the purposes of the application of Subpart D depends on state law.

children in research, that is, the ethical principle of scientific necessity based on the elimination of unnecessary risk exposure.

THE APPROPRIATE BALANCE OF RISK AND POTENTIAL BENEFIT

Sequential Pediatric Investigational Program

There must be an appropriate balance of risk and potential benefit for *each* of the interventions and procedures included in the trial even if the inclusion of children in a clinical investigation is scientifically necessary. A sequential approach to a pediatric research program may be necessary to generate sufficient data to support either (1) an acceptably low risk of the experimental intervention or procedure absent any prospect of direct benefit, or (2) a sufficient prospect of direct benefit to justify the risks. Such an approach will require the use of nonclinical animal models and/or adult human studies to evaluate the risks of the investigational product and to establish a sufficient "proof of concept" to warrant pediatric studies. Thus, a major challenge for the development of a new product for the treatment of a pediatric disorder or condition is bridging this "gap" between (1) research involving procedures and/or interventions that present only a minor increase over minimal risk (21 CFR 50.53) given the absence of sufficient data to establish a prospect of direct benefit, and (2) the conduct of either "proof of concept" or pivotal trials for dosing, safety, and/or efficacy that offer a sufficient prospect of direct benefit to the enrolled children to justify exposure to greater than a minor increase over minimal risk (21 CFR 50.52).

Incremental Research Risk

Federal research regulations instruct an institutional review board (IRB) to consider "only those risks and benefits that may result from the research (as distinguished from risks and benefits of therapies that subjects would receive even if not participating in the research)" (21 CFR 56.111(a)(2)). A clinical investigation, for example, that compares two approved medications administered according to their labeled dose, indication, and population may be viewed as presenting only a "minimal" incremental risk above the risk of receiving one of those same medications as part of standard clinical care.[13] This situation, however, may be infrequent in pediatrics given the low number of products with pediatric labeling. An IRB should assume that any "off-label" use of an FDA-regulated biologic, drug, or device is not "minimal risk."[†] In addition, the default position for any clinical investigation that involves the "off-label" use of an FDA-regulated product should be that an IND or IDE application should be submitted to the FDA, regardless of the prevalence of "off-label" use in the clinical practice of pediatrics. Absent an IND application, for

[†] 21 CFR 56.110 permits the use of an expedited IRB review process for clinical investigations that present no more than minimal risk and are included on the list of the categories of research. The first category on this list includes studies of FDA-regulated medical or device products that do not require either an investigational new drug (IND) or investigational device exemption (IDE). It should be noted, however, that apart from the use of an expedited review procedure, FDA regulations do not include a waiver of one or more of the elements of informed consent for minimal risk research (other than written documentation) found in 45 CFR 46.

example, a clinician-investigator may not be aware of the lack of appropriate nonclinical juvenile animal testing of the investigational product or may be using an extemporaneous formulation absent sufficient chemistry and/or manufacturing information to assure the safety of that formulation. The IND application offers the FDA the opportunity to inform the clinician-investigator about the need for such studies prior to initiating the proposed clinical pediatric studies, often through the imposition of a clinical hold. The fact that an FDA-regulated product is being used "off-label" in children does not establish that the product has sufficient nonclinical or formulation information to support that use.

Component Analysis and Additional Safeguards for Children Under Subpart D

For adult subjects, the risks of research participation can be justified *either* by the anticipated "direct" benefits to the subjects *or* by the importance of the anticipated knowledge (21 CFR 56.111(a)(2)). This is not the case for clinical investigations involving children, which require the additional safeguards found in 21 CFR 50, Subpart D, to protect the rights and welfare of children (21 CFR 56.111(b) and (c)). For children, the allowable risk exposure for an intervention or procedure not offering the prospect of direct benefit must be restricted to either minimal risk (21 CFR 50.51) or a minor increase over minimal risk (21 CFR 50.53). The allowable risk exposure for an intervention or procedure that does offer the prospect of direct benefit must be restricted to (1) the risk that is "justified" by the anticipated benefit and (2) where the relation of the anticipated benefit to the risk is at least as favorable as that presented by available alternative approaches (21 CFR 50.52). Thus, the individual research interventions and procedures that are contained in an investigational protocol must be categorized and assessed according to whether they do or do not offer the prospect of direct benefit—an approach referred to as "component analysis."

Subpart D links the level of appropriate risk exposure (beyond minimal risk) to whether or not the intervention or procedure offers the "prospect of direct benefit." Thus, the ability to distinguish between components of a protocol that offer a prospect of direct benefit and those that do not offer a prospect of direct benefit is essential to the proper interpretation and application of Subpart D.[2,14,15] Nevertheless, "component analysis" has come under recent criticism for using the norm of "clinical equipoise" as the standard for determining the ethical acceptability of "therapeutic" interventions or procedures.[16,17] The concept of "clinical equipoise" and its applicability to the interpretation of Subpart D will be discussed more fully later. A related (albeit unconvincing) criticism of the "component approach" is directed toward the manner in which the distinction between therapeutic and nontherapeutic procedures is made.[18] All parties to the debate agree on the need to avoid the term "therapeutic research," which may justify (or offset) the risks of nonbeneficial procedures through the inclusion of unrelated beneficial procedures in the same protocol (i.e., the fallacy of the "package deal").[2,4,14]

The common ground in the debate is the ethical requirement to distinguish between interventions or procedures that either offer the prospect of direct benefit, or do not offer the prospect of direct benefit. Wendler and Miller[18] argue that the consequences of the intervention are what matters to the determination of the prospect of direct benefit, rather than the intent of the investigator or the design of the individual intervention. Weijer and

Miller[19,20] argue that the notion of "therapeutic warrant" appeals to the evidence that justifies the "belief" that the intervention offers a prospect of direct benefit. The appeal to sufficient evidence for establishing "therapeutic warrant"[19] and the view that the consequences of the intervention (rather than the investigator's intent) establish the prospect of direct benefit[18] appear, in essence, to be the same position. Prospectively, the intended consequences of an action are evaluated by the structure (or design) of the action itself and not by the investigator's state-of-mind or belief in the therapeutic value of the action. Thus, the determination of the "prospect of direct benefit" is based on the design of the intervention (i.e., choice of dose, duration, method of administration, and so forth) given the available nonclinical and/or adult clinical evidence. In addition, a "benefit" can be considered "direct" if it (1) accrues to the individual subject enrolled in the trial and (2) results from research interventions or procedures that are required to answer the scientific questions posed by the trial (i.e., not from other interventions that are included in the protocol, but are unrelated to the research question).[21] The word "benefit" is often modified by "clinical" to indicate that the "direct benefit" under consideration relates to the health status of the enrolled subject. The purpose of the intervention as "research" is less relevant, as presumably all of the interventions contained in the protocol are for the purposes of answering one or more of the research objectives.

The analysis of a proposed clinical investigation under 21 CFR 50, Subpart D, can be approached either (1) by assessing whether or not each intervention or procedure does or does not offer the prospect of direct benefit, followed by an assessment of the risks of each component, or (2) by assessing the risks of each intervention and procedure, followed by an assessment of the prospect of direct benefit for those components that present greater than minimal risk (as shown in Figure 9.3). An intervention or procedure that presents no more than minimal risk may or may not offer a prospect of direct benefit. As the category under which minimal risk interventions or procedures can be considered does not refer to the prospect of direct benefit, the "minimal risk" category will be discussed under the heading of interventions or procedures that do not offer the prospect of direct benefit.

Interventions or Procedures that Do Not Offer the Prospect of Direct Benefit

Minimal Risk (21 CFR 50.51) Minimal risk is defined as "the probability and magnitude of harm or discomfort anticipated in the research are not greater in and of themselves than those ordinarily encountered in daily life or during the performance of routine physical or psychological examinations or tests" (21 CFR 50.3(k)). This definition is subject to a "relativistic interpretation," which uses the research participants' experiences as the point of reference for determining minimal risk.[14] As such, whether or not any given intervention or procedure is considered minimal risk will vary depending on whether that intervention or procedure would be considered ordinary or routine for the actual subjects to be enrolled in the clinical investigation. The relativistic approach altered the definition proposed by the National Commission which defined minimal risk based on the experience of "healthy" children.[2,22] In addition, the definition of minimal risk includes two different standards: (1) ordinary daily life, and (2) routine physical or psychological examinations or tests. There is well-documented variability in the interpretation and application of the "minimal risk" standard.[23] The standard of the risks of ordinary daily life is more difficult to

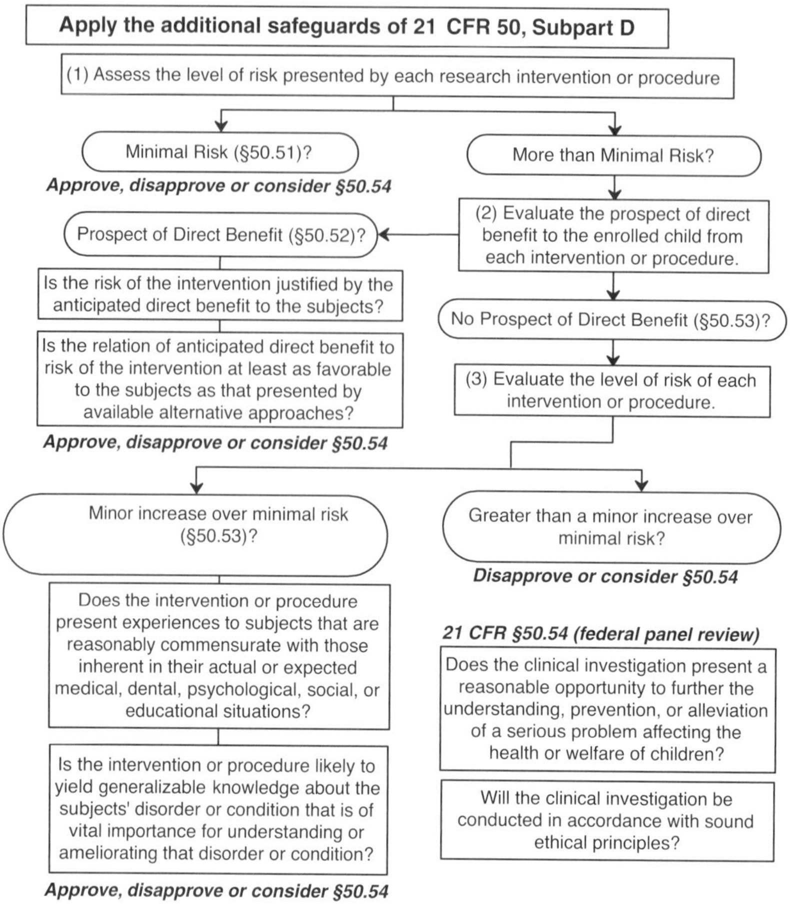

FIGURE 9.3 Analysis of a proposed clinical investigation under 21 CFR 50, Subpart D.

interpret, given the variability in such risks.[14, 24] The standard of "routine physical or psychological examinations or tests" may be subject to less variation, although data are lacking to support this claim.

There is also variability in the definition of minimal risk among international guidelines. CIOMS Guideline 9 omits "ordinary daily life" from the minimal risk standard in favor of "routine medical or psychological examination of such persons" (i.e., individuals incapable of giving informed consent). CIOMS also refers to this as the "low-risk standard."[3] The International Conference on Harmonisation (ICH) Good Clinical Practice Guideline uses the term "low" (without definition), rather than "minimal" risk (Section 4.8.14).[5] The Medical Research Council has a fairly conservative definition of minimal risk, with procedures such as a blood test (which would be classified as minimal risk in the United States) referred to as "low" risk.[4] The Canadian Tri-Council Policy restricts nonbeneficial research to no more than "minimal risk" but adopts a "relative" (rather than "uniform") interpretation of "minimal risk."[25] Building on the 2004 IOM Report,[14] the Secretary's

Advisory Committee on Human Research Protections (SACHRP) recommends the international use of the uniform definition of minimal risk (discussed later).[26]

The uniform standard indexes minimal risk to the normal experiences of average, healthy, normal children such that the level of risk does not vary according to the specific research population.[14] The uniform standard of minimal risk restores the approach recommended by the National Commission[2] and is supported by IOM[14] and SACHRP.[26] In addition, the uniform interpretation of minimal risk was supported by NBAC for all human subjects research, albeit using the phrase "general population" rather than "healthy" persons.[15] The uniform definition of minimal risk serves to protect children with a disorder or condition from research *unrelated* to their disorder or condition that would be considered greater than minimal risk for a healthy child. This approach is grounded on the principle of justice as fairness.[14,24] The interpretation and application of the minimal risk standard should focus on the equivalence of potential harms or discomfort presented by the research interventions or procedures with the harms or discomfort that average, healthy, normal children may encounter in their daily lives or experience in routine physical or psychological examinations or tests.[14] In other words, the actual interventions or procedures included in the clinical investigation do not need to be part of routine examinations or tests in order to be considered minimal risk (e.g., MRI scan without procedural sedation). The risk of harms or discomfort should be interpreted in relation to the ages (and developmental status) of the children to be studied.[2,14,26] The duration, cumulative risks, and reversibility of harm need to be assessed in determining the overall level of risk. For example, a single blood draw may be considered minimal risk while multiple blood draws over a short period of time may be considered greater than minimal risk depending on the age of the child.[14,26]

The role of using empirical data about the current risks of daily life to determine whether any given research intervention or procedure should be considered minimal risk has been the subject of recent debate.[27–32] The IOM criticized the use of the phrase "socially acceptable" as it may imply that the mere fact that a risk occurs in daily life would serve to justify the exposure of children to that risk in research.[14] Although empirical data about the risks of "daily life" or "routine examinations or tests" are a necessary component of an informed evaluation of whether the risks of a research intervention or procedure are "minimal," they are not sufficient. Rather, the concept of "minimal risk" reflects a normative view of parental responsibility (i.e., the "prudent" parent), suggesting that better data about the risks of "daily life" or "routine" interventions (such as vaccinations) may alter our assessment of whether comparable risks should be allowed in the research setting.[30,32]

As examples of minimal risk procedures, the 2001 Preamble of the FDA Subpart D Interim Final Rule lists "clean-catch urinalysis, obtaining stool samples, administering electroencephalograms, requiring minimal changes in diet or daily routine, ... using standard psychological tests, ... [and] a taste test of an excipient or tests of devices involving temperature readings orally or in the ear."[33] The National Commission listed "routine immunization, modest changes in diet or schedule, physical examination, obtaining blood and urine specimens ... developmental assessments ... most questionnaires, observational techniques, noninvasive physiological monitoring, [and] psychological tests and puzzles" as minimal risk.[2] In addition, SACHRP lists a number of physical (e.g., measurement of height, weight, and head circumference; assessment of obesity with skin fold calipers; hearing and vision tests; testing of fine and gross motor development; noninvasive physiological monitoring) and psychological (e.g., child and adolescent intelligence tests; infant mental and motor scales; educational tests; reading and math ability tests; social development assessment; family and peer relationship assessments; emotional regulation scales; scales to detect feelings of sadness or hopelessness) examinations or tests as being no more than minimal risk.[34] Finally, limited exposure to radiation from diagnostic procedures can be viewed as minimal risk.[35] Some of the above procedures may be considered greater than minimal risk depending on the context of the research and the specific population to be enrolled.[34]

Minor Increase over Minimal Risk (21 CFR 50.53) There are three criteria (in addition to adequate provisions for child assent and parental permission) that an intervention or procedure must satisfy in order to be approved under this category: (1) "The risk represents a minor increase over minimal risk;" (2) "the intervention or procedure presents experiences to subjects that are reasonably commensurate with those inherent in their actual or expected medical, dental, psychological, social, or educational situations;" and (3) "the intervention or procedure is likely to yield generalizable knowledge about the subjects' disorder or condition that is of vital importance for the understanding or amelioration of the subjects' disorder or condition." In spite of the central importance of this category of pediatric research, there are no definitions found in the regulations of the following key concepts: (1) "minor increase over minimal risk,"[3] (2) "disorder or condition,"[14] (3) "vital importance,"[33] and (4) "reasonably commensurate."

This category of research provoked the most controversy, with two members of the National Commission dissenting from the majority opinion.[2] In response to the minority criticism, the following three points were made in support of the category. First, the justification for the increased risk exposure above minimal risk is based on scientific necessity, not on an increased baseline risk exposure due to a disorder or condition (e.g., from social or other circumstances).[3,14] Second, parental authority can be viewed as extending to risk exposures that have questionable benefit but which present a minor increase above minimal risk. Third, the National Commission adopted a conservative definition of minimal risk indexed to the experience of a healthy child. In other words, the absolute difference in risk exposure between minimal risk and a minor increase over minimal risk is meant to be "slight."[2]

Concept: "Minor Increase over Minimal Risk" The recommended interpretation by the National Commission,[2] which was reaffirmed by the Institute of Medicine[14] and SACHRP,[26, 34] is that the interpretation of a minor increase over minimal risk should be closely linked to minimal risk. As such, a minor increase over minimal risk represents only a "slight increase in the potential for harms or discomfort beyond minimal risk (as defined in relation to the normal experiences of average, healthy, normal children)."[14] In addition, the interventions or procedures that present only a minor increase over minimal risk should present "no significant threat to the child's health or well-being."[2] The National Commission proposed a "commonsense method of determination" that would draw from a number of data sources.[1] Implicit in the determination that the risks of an intervention or procedure are only "slightly more" than minimal risk is the ability to specify a point estimate and confidence interval (i.e., standard error of the mean) for any given risk. In other words, if the risks of the intervention or procedure are unknown, or have not been adequately characterized (e.g., the confidence interval remains wide), then the risks of the intervention or procedure cannot be considered only a minor increase over minimal risk.

Concept: "Disorder or Condition" The IOM speculated that the variability in defining a "condition" may reflect a narrow interpretation of minimal risk and the subsequent need to define "condition" more broadly in order to approve the research.[14] The two extremes of interpretation should be rejected. On the one hand, a "disorder" should not be limited to an illness, disease, injury, or defect. On the other hand, a more broadly defined "condition" should be linked empirically to a "negative effect on children's health and well-being" such that an answer to the research question would provide information of "vital importance" for understanding or ameliorating that condition. The justification of an increased risk exposure above minimal risk absent the prospect of direct benefit based on the presence of a condition must be evidence based. In other words, the presence of a condition that may negatively affect a child's health or well-being must be based on empirical evidence and not mere speculation.[14]

The IOM recommended that the concept "disorder or condition" be defined as "a specific (or set of specific)... characteristic(s) that an established body of scientific evidence or clinical knowledge has shown to negatively affect children's health and well-being or to increase their risk of developing a health problem in the future."[14] This definition clearly excludes the use of "healthy" children and children with an unrelated disorder or condition in more than minimal risk research that does not offer a prospect of direct benefit to the enrolled children.[3,36] In addition, having a "disorder or condition" is not limited to children with a disease, but may include children who are "at risk" for "manifestations of the disease in the future."[33] The appropriateness of defining a "condition" based on social or environmental characteristics has been a source of controversy. The IOM report pointed out that designating a group of children as having a "condition" based on social or economic circumstances for the purposes of higher risk exposure in nonbeneficial research raises concern about the justice (or fairness) of such a designation.[14]

A recent example illustrating this concern is the environmental study of different lead abatement techniques in Baltimore. Although the media and courts referred to the children involved in the study as "healthy," the majority of the children were "at risk" for lead poisoning and thus could be considered to have a "condition." In addition, the environmental interventions being tested could be considered to offer the prospect of direct benefit to the enrolled children.[37,38] The Baltimore Lead Abatement Study did not involve an FDA-regulated product. However, there have been recent studies of the role of chelation for treating children with an elevated blood lead level (BLL), albeit at levels below existing guidelines for lead chelation therapy.[39] The relationship between BLL and decreased neurodevelopmental performance is linear, making it difficult to define a clear breakpoint at which a child with a measurable BLL is not at risk. Thus, the decision about whether any given child with an elevated BLL is either "at risk" or has a "condition" or has a "disease" that may justify the risks of an investigational intervention (such as chelation) involves a complex balancing of risk and benefit set in the context of available interventions. There are important issues of justice that are raised by the social and economic circumstances that give rise to a child being at risk for lead toxicity; however, these issues should not preclude the search for safe and effective clinical interventions to mitigate the negative impact of lead on children's health and well-being.

Concept: "Vital Importance" The requirement for information of "vital importance" reaffirms the principle of scientific necessity. In essence, a determination that the clinical investigation meets the standard of scientific necessity is both necessary and sufficient to meet the criteria for scientific importance under both 21 CFR 50.53 and 50.54. The question arises whether the standards of "vital importance" (21 CFR 50.53) and "reasonable opportunity" (21 CFR 50.54) are the same, or different. The phrase "reasonable opportunity" should not be interpreted as a lower standard for the importance of the knowledge that may be gained. Both standards reflect the essential requirement for scientific necessity and, as such, do not suggest either a higher or lower standard for the importance of that knowledge. The difference is that the knowledge to be gained in 21 CFR 50.53 is tied closely to the "disorder or condition" of the children to be enrolled in the clinical investigation. Under 21 CFR 50.54, there is no such linkage between the knowledge to be gained and a "disorder or condition" of the enrolled children. In fact, the absence of a "condition or disorder" in the children that would be exposed to a minor increase over minimal risk in a clinical investigation has been the primary reason for the first three referrals to the Pediatric Advisory Committee under 21 CFR 50.54 (see http://www.fda.gov/oc/opt/ethicsmeetings.html). In evaluating whether or not the information may be of "vital importance," the overall developmental program should be taken into consideration. The information gained from the specific protocol under consideration may be an important yet intermediate step leading to further investigations.

Concept: "Reasonably Commensurate" The IOM report implies that the "reasonably commensurate" criterion should be part of risk assessment. However, the discussion of the criterion of "reasonably commensurate" focuses on experiential "familiarity" as a basis for child assent and parental permission, even if the intervention or procedure would occur in the future.[14] Thus, the purpose of this criterion is to improve child assent and parental permission, rather than provide a basis for risk assessment.[2,3] The assessment of the risks of the intervention is independent of whether the child is (or will be) familiar with the intervention. In addition, the requirement for "commensurability" should not be used as a "major criteria" but rather in combination with the other criteria.[26,34] The evaluation of "commensurability" should be from the perspective of the child to be enrolled in the clinical investigation. For example, the child may have experienced procedural sedation with placement of a peripheral intravenous (IV) line and the administration of medication. This experience, however, may or may not have been with the specific medications that are considered to present only a minor increase over minimal risk. This criterion of "commensurability" may be satisfied by the use of educational interventions that familiarize the child with the interventions or procedures to be used in the clinical investigation. For example, some institutions use "mock" MRI scanners to familiarize children with the procedure in the absence of prior or anticipated future experience.

International Guidelines The "minor increase over minimal risk" category of pediatric research does not exist in international guidelines apart from CIOMS.[3] There are arguments in the scientific literature that this category of research should be available in other jurisdictions, such as South Africa.[40–45] The ICH GCP guidelines use the risk category of "low" and appear to allow an exception (under unclear circumstances) to restricting such research to children with a pertinent condition or disorder.[5] The MRC refers to "low" risk procedures (and includes procedures that many would consider minimal risk such as a blood test). The MRC states that such research (greater than minimal risk, no prospect of direct benefit) requires "serious ethical consideration" but provides no specific ethical guidelines (such as found in Subpart D) for the review and approval of such research (such as linking it to the child's condition or disorder).[4] For nontherapeutic research, the 2001 EU Clinical Trial Directive requires that "some direct benefit for the group* of patients is obtained from the clinical trial" and is "essential to validate data obtained in clinical trials on persons able to give informed consent or by other research methods."[36] Although not explicit, this added criterion suggests that a pediatric study should only be performed when such data are necessary to validate or supplement the results of other studies (i.e., the principle of scientific necessity). An additional criterion is that "such research should either relate directly to a clinical condition from which the minor concerned suffers or be of such a nature that it can only be carried out on minors." Thus, the EU Clinical Trial Directive, although based on the ICH GCP guidelines, appears to restrict nonbeneficial research presenting a "low" risk to research that is either related to the child's condition or that can only be done in children.[36]

Examples of procedures that may present a "minor increase over minimal risk (e.g., depending on the research context, the specific population of children to be enrolled, and the skill of the investigator) have included lumbar puncture,[3] bone marrow aspirate, with appropriate procedural sedation,[3,14] placement of a blood-drawing intravenous line, and a single-dose pharmacokinetic (PK) study of a well-characterized drug with limited or sparse sampling. The risk of a single-dose PK study depends on both the approach to blood sampling and on the risks of the drug that is being measured. The risks of the administered drug need to be well-characterized based on prior knowledge, for otherwise the risks could not be assessed as being only a minor increase over minimal risk. Finally, certain approaches to procedural sedation[14] and limited radiation exposure may also be considered a minor increase over minimal risk.[35]

The category of a "minor increase over minimal risk" also serves an important role in the debate over the acceptable risks to which a subject may be exposed by the withholding of known, effective treatment. As discussed later, if the children to be enrolled in a clinical investigation (regardless of group assignment) will not receive a "proven prophylactic, diagnostic or therapeutic method,"[46] the risk exposure related to the withholding of such a method (or intervention) must be limited to no more than a minor increase over minimal risk.

Interventions or Procedures that Offer the Prospect of Direct Benefit (21 CFR 50.52)

This category of research may apply if an intervention or procedure that presents more than minimal risk either "holds out the prospect of direct benefit for the individual subject" or "is likely to contribute to the subject's well-being" (e.g., safety monitoring). In addition to the requirement for adequate provisions for child assent and parental permission, there are two criteria that such interventions or procedures must meet: (1) "the risk is justified by the anticipated benefit to the subjects" and (2) "the relation of the anticipated benefit to the risk is at least as favorable to the subjects as that presented by available alternative approaches." This category of research includes several key concepts that require interpretation: prospect of direct benefit/contribution to well-being, justification of risk, and available alternative approaches.

---

* The language of "direct benefit for the group" (which would be called "indirect" benefit in the United States) was incorporated into the EU Clinical Trial Directive given the passage of an earlier EU resolution that restricted nontherapeutic research on nonconsenting subjects to that which provided direct benefit.

Concept: Prospect of Direct Benefit/Contribution to Well-Being The concept of the prospect of direct benefit from an intervention or procedure has been discussed previously. A monitoring procedure that does not, in and of itself, offer the prospect of direct benefit may be important in evaluating the safety of another intervention that does offer the prospect of direct benefit. There are two ways that the risks of such a monitoring procedure could be evaluated. To the extent that the monitoring procedure is made necessary by the administration of the investigational product, the risks of the monitoring procedure may be justified by the prospect of direct benefit of the experimental intervention. Using this approach, the administration of the investigational product *and* the monitoring made necessary by that administration could both be considered under 21 CFR 50.52.

Alternatively, the monitoring procedure may be viewed as not offering the prospect of direct benefit, and thus considered under either 21 CFR 50.51 (minimal risk) or 21 CFR 50.53 (minor increase over minimal risk). The specific study design may affect which of the two approaches is appropriate. For example, in a concurrently controlled trial, the control group would not receive the investigational product. The comparative balance of risks and potential benefits between the investigational product and the control intervention may be uncertain or unknown (thereby justifying the study as "scientifically necessary"); however, the risks of a necessary monitoring procedure must be justified independently by the prospect of direct benefit for each of the two (or more) comparative interventions (i.e., intervention and control). This distinction is clearest for the use of a placebo or "no treatment" control group. The risks of a necessary monitoring procedure would need to be limited to no more than a minor increase over minimal risk (under 21 CFR 50.53) for the placebo or "no treatment" control group as there would be an insufficient prospect of direct benefit to justify any greater risk exposure. For the group that receives the investigational product (or an active control treatment), the risks of the monitoring procedure potentially could exceed a minor increase over minimal risk if the prospect of direct benefit was sufficient to justify this level of risk exposure under 21 CFR 50.52.

If group assignment is masked, and the risks of the monitoring procedure(s) are not justified by the prospect of direct benefit for one or more of the control interventions, then the risks of the monitoring procedure(s) will need to be limited to no more than a minor increase over minimal risk for the entire study population. If group assignment is not masked, one may be able to justify one or more subject groups having a greater risk exposure for a monitoring procedure made necessary by an experimental intervention; however, the scientific validity of the study may be undermined.

Concept: Justification of Risk The justification of the risks of an experimental intervention is a complex process, involving a mix of quantitative and qualitative judgments.[2,47] The evaluation of risks and potential benefits is similar to that made in clinical practice.[2,3] As discussed earlier, there should be empirical evidence of sufficient probability of direct benefit (i.e., "scientifically sound" expectation of success[2]) to justify exposure to the risks. The justification of risk can include (1) the possibility of avoiding greater harm from the disease and (2) the provision of important anticipated benefit to the individual exposed to risk (i.e., "direct" benefit). Consistent with "component analysis," the risks of an intervention or procedure can only be justified by the benefits to be expected from that same intervention or procedure.[2] The justification of the risks in light of the potential benefits of an experimental intervention needs to be set within the context of the severity of the disease (e.g., degree of disability, life-threatening) and the availability of alternative treatments (e.g., see 21 CFR 312, Subpart E).

Concept: Available Alternative Approaches The requirement that the "relation of the anticipated benefit to the risk is at least as favorable to the subjects as that presented by available alternative approaches" is in addition to the stipulation that the risks of the experimental intervention must be balanced by the potential direct benefits.[2] The underlying ethical reason for considering the "available alternative approaches" is the view that a child's health or welfare should not be placed at a disadvantage by being enrolled in a clinical investigation.[14] The evaluation of risk and potential benefit is similar to that made in clinical practice and must consider alternatives that are available outside the clinical investigation.[2,3] The application of this general principle hinges to a large extent on the interpretation of "available." The National Commission argued that the other approaches that need to be taken into consideration include "any other course of action (or non-action)."[2] However, the modification of "alternative" by "available" raises the question of the extent to which *all* alternatives need to be considered, or only those that are "available" to the subjects to be enrolled in the clinical investigation.[3] In other words, should the range of available alternatives against which the risks and potential benefits of the experimental intervention are compared be all those that are "universally" available, or should the alternatives be limited to those that are "locally" available? This topic is discussed later under the heading "Choice of Control Group."

Clinical (or Research) Equipoise The National Commission's argument in favor of this category of research that offers a prospect of direct benefit (21 CFR 50.52) bears some resemblance to the principle of clinical equipoise.[14] Clinical equipoise is commonly defined as "a genuine uncertainty on the part of the expert medical community about the comparative therapeutic merits of each arm of a clinical trial." As defined, advocates of "clinical equipoise" argue that it "provides a clear moral foundation to the requirement that the health care of subjects not be disadvantaged by research participation."[25] In addition, Weijer argues that "clinical equipoise" is a "core concept of component analysis." In other words, the acceptance of the demarcation of interventions or procedures into those which either do or do not offer the prospect for direct benefit seems to entail the principle of "clinical equipoise."[20] However, there does not appear to be a necessary relationship between the demarcation of beneficial and nonbeneficial interventions or procedures and the principle of "clinical equipoise" as the ethical standard for judging the acceptability of interventions offering the prospect of direct benefit. The ethical requirement that the risks and potential benefits of an investigational intervention should be "justified as they are in medical practice" such that an individual enrolled in a clinical trial is not placed at a disadvantage (i.e., participation is "at least as advantageous")[3] may be consistent with the patient-subject's right to "competent medical care."[48] However, this requirement bears no obvious relationship to "clinical equipoise" interpreted as uncertainty either on the part of an individual clinician-investigator or the community of clinician-investigators.

The concept of equipoise combines two separate principles.[16] The first principle is the scientific principle of "uncertainty" (i.e., the null hypothesis between the investigational product and the comparator or control group). However, this principle is a requirement of *all* ethical research (i.e., the design of the investigation should be capable of answering a question about which there is legitimate scientific uncertainty). The specification and application of this principle of "uncertainty" is complex.[49] The uncertainty of the individual clinician is not decisive, but rather the uncertainty of the relevant community. But whose "uncertainty" determines the ethical appropriateness of a randomized controlled trial (RCT)? The scientific community? The clinical community? The patient community? At what point does that "uncertainty" diminish to where it can no longer serve as a justification for an RCT? The point at which the uncertainty that justifies an RCT can no longer serve in that role varies both within and between different communities. Thus, scientific and/or clinical equipoise appear to be neither necessary nor sufficient justifications for randomization.[16,17,49] In addition, these two forms of equipoise (whether individual or community) are logically separate. The morally problematic area is where sufficient data have been developed such that clinical equipoise is disturbed, but insufficient data exist to justify a scientific (or policy) conclusion.[49,50]

The second principle contained within the concept of equipoise is the ethical norm that no one enrolled in a trial should receive a known inferior treatment (i.e., known effective treatment should be provided). Here "clinical equipoise" is seen as a specification of the "duty of care."[48] From this perspective, the dispute about the role of equipoise is primarily about whether the "duty of care" (carried over from the clinical setting based on the fiduciary duty of a physician to act in a patient's best interest) should be the ethical framework for clinical research.[14] Proponents of equipoise may argue in favor of active-control comparator trials, as such trials may provide more useful clinical information. However, this approach does not solve the tension between having enough information to make an "individual patient decision" and enough to "warrant making a policy decision."[50] The question arises whether a clinical investigator can have a "duty of care" that has a different specification than that based on a physician's fiduciary obligation. Can this "duty of care" be structured within the protocol itself (i.e., in the intervention that offers the prospect of direct benefit, the inclusion and exclusion criteria, and the criteria for withdrawal from the protocol)?

All parties to the debate over equipoise as a guiding principle of clinical research accept that there needs to be a demarcation between interventions or procedures that either offer or do not offer a prospect of direct benefit. Weijer argues that the endorsement of this distinction (which is the basis for component analysis) reflects a patient-subject's entitlement to "competent care," which is secured by the principle of clinical equipoise.[48] In response, Miller argues that the principle of clinical equipoise will not permit even the smallest deviation in "comparable risks and benefits" (such as forgoing the use of known effective treatment as a control group if the risks are no more than a minor increase over minimal risk). Miller offers a different standard (i.e., nonexploitation) to distance his position from the clinical "duty of care" and "clinical equipoise"[16,18] and proposes an alternative to "component analysis" called the "net risks" test.[18] Weijer responds that the "net risks" test would expose patient-subjects to "substandard medical treatments" and reaffirms "component analysis" and the correlative rights to "competent care" (i.e., as specified by "clinical equipoise") and "protection from exposure to undue harm solely in the interests of others" (i.e., limiting nonbeneficial risk exposure to a minor increase over minimal risk).[19] The "net risks" test incorporates the distinction between interventions or procedures that do or do not offer the prospect of direct benefit. As such, it is a form of "component analysis" (a term that Miller avoids in order to distance himself from the principle of "clinical equipoise"). However, both approaches result in the *same* standard for the assessment of nonbeneficial interventions or procedures. The alleged difference is the ability of the "net risks" test to incorporate the category of a "minor increase over minimal risk" as a specification of the limit on risk exposure for patient-subjects who would not receive a known effective treatment as part of a control group.

Subpart D is grounded on the distinction between interventions or procedures that offer the prospect of direct benefit and those that do not. The view that no child should be disadvantaged by participation in a clinical trial bears some resemblance to clinical equipoise. However, an affirmation of the child-patient's (1) right to competent medical care and (2) protection from undue risk of harm when participating in a clinical investigation does not require or entail the principle of clinical equipoise. The right to competent medical care as part of a clinical investigation presenting more than a minor increase over minimal risk should be reflected in the protocol study design. Has the appropriate dose (or range of doses) been selected to optimize the balance of risk and potential benefit? Have the inclusion criteria been crafted so as to enroll children with the best chance of experiencing direct benefit? Have the exclusion criteria been crafted to exclude those children who would be at greatest risk for a severe adverse event absent a sufficient prospect of direct benefit? Are there clear monitoring and safety standards so that children for whom the risks of the investigational intervention are no longer justified by the prospect of direct benefit can be withdrawn from the study? In effect, the patient-subject's right to competent medical care should be operationalized in the structure of the investigational protocol so that ideally there is no conflict between the clinician-investigator's obligation to protect the well-being of the individual subject and the obligation to follow the research protocol.[51] Finally, the concept of equipoise should not be used as the basis for 21 CFR 50.52 in favor of the clear distinction between the concepts of uncertainty and comparable treatment. Rather, the nature of the scientific uncertainty (i.e., question) to be resolved by the study design should be specified, and the comparability of alternatives (i.e., the ethical and scientific argument in favor of the chosen control group) should be justified.

Choice of Control Group The choice of an appropriate control group for a clinical investigation should be approached from two perspectives—scientific and ethical. From a scientific perspective, what is the appropriate comparator (i.e., control group) to use in order to demonstrate the safety and/or efficacy of the investigational intervention? The choice of comparator may depend on the specific scientific objective(s) and the type of study design.[52] The primary focus is on designing the clinical investigation so that any uncertainty (i.e., lack of knowledge) about the research objective(s) is resolved. From an ethical perspective, does enrollment in a clinical investigation place subjects at an "unreasonable" risk (i.e., one that is not compensated by a sufficient prospect of direct benefit)? Are individuals enrolled in the clinical investigation not receiving a treatment that they should otherwise receive as part of "competent medical care"? Although the debate has focused on the withholding of known effective treatment in favor of a placebo, this question should be asked about all subjects to be enrolled in a clinical investigation, including those subjects who would receive the investigational intervention.

The ethics of the choice of control group has been the subject of much debate, focused either on the use of placebo controls or on the choice of a "local standard" (which may or may not be a placebo) as the control group. The starting point of the debate about the use of placebo controls has been Paragraph 29 of the 2000 version of the Declaration of Helsinki, which stated that "a new method should be tested against... the best current prophylactic, diagnostic, and therapeutic methods." According to this standard, a placebo (or no treatment) control could only be used in the absence of a "proven prophylactic, diagnostic or therapeutic method."[46] The ICH published guidance on the choice of control group in 1999, which argued that a placebo-controlled trial may be ethically justified when there would be "no serious harm" from withholding known effective treatment. The scientific concern is the ability of a clinical investigation "to distinguish an effective treatment from a less effective or ineffective treatment" (a property of a clinical investigation referred to as "assay sensitivity").[52] The inclusion of a placebo control arm (distinguished from a "no treatment" control arm by the ability to blind the treatment allocation) would assure one that the finding of "equivalence" would not lead to the false conclusion of effectiveness if an active but ineffective comparator had been used.[3,52] There are other solutions to this problem such as the use of a superiority design with an active comparator. Even if there would be a good scientific reason to withhold a known effective treatment in order to demonstrate the efficacy of a new treatment, the ICH document makes it clear that this would be unethical if the effective treatment "is known to prevent death or irreversible morbidity." Furthermore, "where a placebo-controlled trial is unethical and an active control trial would not be credible, it may be very difficult to study new drugs at all."[52]

In response to debate stimulated by the 1999 ICH document, the World Medical Association clarified Paragraph 29 of the Declaration of Helsinki in 2002. The clarification noted that "a placebo-controlled trial may be ethically acceptable, even if proven therapy is available, under the following circumstances: (1) Where for compelling and scientifically sound methodological reasons its use is necessary to determine the efficacy or safety of a prophylactic, diagnostic or therapeutic method; *or* (2) Where a prophylactic, diagnostic or therapeutic method is being investigated for a minor condition and the patients who receive placebo will not be subject to any additional risk of serious or irreversible harm."[46] The CIOMS guidelines, also published in 2002, corrected a serious oversight with the Declaration of Helsinki.[3] With the use of the word "or," the clarification of the Declaration of Helsinki appears to allow for the withholding of proven therapy "for compelling and scientifically sound methodological reasons" alone.[46] CIOMS clearly states that withholding proven therapy would *only* be ethically acceptable if the "use of placebo would not add any risk of serious or irreversible harm to the subjects" even if the use of an active comparator would undermine the ability of the clinical investigation to produce scientifically sound results.[3] The WMA again revised the Declaration of Helsinki in 2008, modifying and incorporating the clarification of Paragraph 29 into what is now Paragraph 32. The default position remains that "a new intervention must be tested against . . . the best current proven intervention." However, a placebo or no treatment control group would be acceptable even if a current proven intervention exists if such a control group was scientifically necessary and those receiving placebo or no treatment "will not be subject to any risk of serious or irreversible harm."[46] Thus, the current 2008 version of the Declaration of Helsinki and the 1999 ICH guidance on choice of control group in clinical trials are in agreement on the ethical use of placebo or no treatment controls.

Arguably, the clarification of when a placebo control may be used in place of proven effective treatment remains consistent with clinical equipoise. From this perspective, a placebo is acceptable when "patients have provided an informed refusal of standard therapy for a minor condition for which patients commonly refuse treatment and when withholding such therapy will not lead to undue suffering or the possibility of irreversible harm of any magnitude."[25] However, whether or not patient-subjects would commonly refuse standard therapy for a minor condition is not relevant to the ethical justification of the use of a placebo in the absence of "any risk of serious or irreversible harm to the subjects." The withholding of a known effective treatment from children enrolled in a clinical investigation may appear to violate the principle of clinical equipoise.[16,21] Nevertheless, such a violation may be ethically justified if the risk exposure is limited to either minimal risk (21 CFR 50.51) or a minor increase over minimal risk (21 CFR 50.53).[14] In effect, the withholding of known effective treatment can be interpreted as consistent with the Subpart D category of no more than a minor increase over minimal risk provided that (1) such withholding is scientifically necessary *and* (2) does not involve any risk of serious or irreversible harm to the enrolled children.

As a variation in the discussion about placebo (or no treatment) controls, there has been controversy about the use of "local" versus "universal" standards to determine the appropriate control group. In other words, should "proven intervention" only refer to treatments that are *actually available* in the location where the clinical investigation is being conducted? Some argue that the purpose of a clinical investigation is to alter clinical practice. Thus, "it is crucial... to take the study context into account when designing and conducting such studies." Simply, the appropriate control group (or comparator) should be drawn from actual clinical practice in that setting.[20] However, others argue that the withholding of known effective treatment based on the underlying inequities in the distribution of medical care is unjust and exploits those less fortunate.[3]

CONCLUSION

Children should not be enrolled in a clinical investigation unless absolutely necessary to answer an important scientific question about the health and welfare of children. An example of the application of the principle of scientific necessity is the FDA use of extrapolation to determine the need for pediatric studies. Each component of a clinical investigation must be analyzed individually and cumulatively. Absent the prospect of direct benefit, the allowable risk of a research intervention or procedure must be restricted to either minimal risk (21 CFR 50.51) or a minor increase over minimal risk (21 CFR 50.53). The allowable risk exposure for a research intervention or procedure that does offer the prospect of direct benefit must be justified by (1) the anticipated benefit and (2) the relation of anticipated benefit to risk at least as favorable as that presented by available alternatives (21 CFR 50.52). As presented earlier, 21 CFR 50, Subpart D, provides a uniform framework for the ethical evaluation of proposed clinical investigations involving children that is consistent with existing international guidelines.

REFERENCES

1. Office of the Secretary, Department of Health, Education and Welfare (DHEW). Proposed regulations on research involving children. *Fed Regist*. 1978;43 (141):31785–31794.
2. Office of the Secretary, DHEW. Research involving children: report and recommendations of The National Commission for the protection of human subjects of biomedical and behavioral research. *Fed Regist*. 1978;43 (9):2083–2114.
3. Council for International Organizations of Medical Sciences. International ethical guidelines for biomedical research involving human subjects. *Bull Med Ethics*. 2002;182:17–23.
4. Medical Research Council. *Medical Research Involving Children*. 2004.
5. Food and Drug Administration (FDA), Department of Health and Human Services (DHHS). International Conference on Harmonisation; Good Clinical Practice: Consolidated Guideline. *Fed Regist*. 1997;62 (90):25691–25709.
6. Food and Drug Administration Amendments Act of 2007. In: *21 USC 301*. 2007.
7. FDA, DHHS. Regulations requiring manufacturers to assess the safety and effectiveness of new drugs and biologic products in pediatric patients. *Fed Regist*. 1998;63(231):66632–66672.
8. FDA, DHHS. International Conference on Harmonisation; E11:Clinical Investigation of Medicinal Products in the Pediatric Population. *Fed Regist*. 2000;65(71):19777–19781.

9. Pediatric Research Equity Act of 2003. In: *117 Stat*. 2003. 1936–43.
10. FDA, DHHS. *Guidance for Industry: Exposure–Response Relationships—Study Design, Data Analysis, and Regulatory Applications*. April 2003. Available at http://www.fda.gov/cder/guidance/5341fnl.pdf.
11. National Institutes of Health (NIH). *NIH Policy and Guidelines on the Inclusion of Chidren as Participants in Research Involving Human Subjects*. 1998. Available at http://grants1.nih.gov/grants/guide/notice-files/not98-024.html. (Accessed December 31, 2007.)
12. NIH. *Questions and Answers about The NIH Policy and Guidelines on The Inclusion of Children as Participants in Research Involving Human Subjects*. March 1999. Available at http://grants1.nih.gov/grants/funding/children/pol_children_qa.htm. (Accessed December 31, 2007.)
13. Morris MC, Nelson RM. Randomized, controlled trials as minimal risk: an ethical analysis. *Crit Care Med*. 2007;35(3):940–944.
14. Field MJ, Behrman RE, eds. *Ethical Conduct of Clinical Research Involving Children*. Washington DC: The National Academies Press; 2004.
15. National Bioethics Advisory Commission. Ethical and policy issues in research involving human participants. In: *Volume I: Report and Recommendations of the National Bioethics Advisory Commission*. Bethesda, Maryland. August 2001.
16. Miller FG, Brody H. Clinical equipoise and the incoherence of research ethics. *J Med Philos*. 2007;32(2):151–165.
17. Miller FG, Brody H. A critique of clinical equipoise. Therapeutic misconception in the ethics of clinical trials. *Hastings Cent Rep*. 2003;33(3):19–28.
18. Wendler D, Miller FG. Assessing research risks systematically: the net risks test. *J Med Ethics*. 2007;33(8):481–486.
19. Weijer C, Miller, PB. Refuting the net risks test: a response to Wendler and Miller's "Assessing research risks systematically." *J Med Ethics*. 2007;33(8):487–490.
20. Weijer C, Miller PB. When are research risks reasonable in relation to anticipated benefits? *Nat Med*. 2004;10(6):570–573.
21. Miller FG, Wendler D, Wilfond B. When do the federal regulations allow placebo-controlled trials in children? *J Pediatr*. 2003;142(2):102–107.
22. Office of the Secretary, DHEW. Final regulations amending basic HHS policy for the protection of human research subjects. *Fed Regist*. 1981;46(16):8365–8391.
23. Shah S, Whittle A, Wilfond B, et al. How do institutional review boards apply the federal risk and benefit standards for pediatric research? *JAMA*. 2004;291(4):476–482.
24. Kopelman LM. Children as research subjects: a dilemma. *J Med Philos*. 2000;25(6):745–764.
25. Canadian Institutes of Health Research, Natural Sciences and Engineering Research Council of Canada, and Social Sciences and Humanities Research Council of Canada. *Tri-Council Policy Statement: Ethical Conduct for Research Involving Humans*. 1998 (with 2000, 2002, and 2005 amendments) Available at http://pre.ethics.gc.ca/english/policystatement/policystatement.cfm.
26. Prentice ED.SACHRP Chair Letter to HHS Secretary Regarding Recommendations. To The Honorable Michael O. Leavitt, 2005.
27. Wendler D, Belsky AB, Thompson KM, et al. Quantifying the federal minimal risk standard: implications for pediatric research without a prospect of direct benefit. *JAMA*. 2005;294(7):826–832.
28. Wendler D, Varma, S. Minimal risk in pediatric research. *J Pediatr*. 2006;149(6):855–861.
29. Wendler D, Emanuel EJ. What is a "minor" increase over minimal risk? *J Pediatr*. 2005;147(5):575–578.
30. Nelson RM. Minimal risk, yet again. *J Pediatr*. 2007;150(6):570–572.
31. Ross LF, Nelson RM. Pediatric research and the federal minimal risk standard. *JAMA*. 2006;295 (7):759.

32. Nelson RM, Ross LF. In defense of a single standard of research risk for all children. *J Pediatr.* 2005;147(5):565–566.

33. FDA, DHHS. Additional safeguards for children in clinical investigations of FDA-regulated products. *Fed Regist.* 2001;66(79):20589–20600.

34. Secretary's Advisory Committee on Human Research Protections. Appendix B: *SACHRP Chair Letter to HHS Secretary Regarding Recommendations.* 2005. Available at http://www.hhs.gov/ohrp/sachrp/sachrpltrtohhssecApdB.html. (Accessed December 31, 2007.)

35. Nelson RM. Issues in the institutional review board review of PET scan protocols. In: Charron M, ed. *Practical Pediatric PET Imaging.* New York:Springer-Verlag; 2006: 59–71.

36. Directive 2001/20/EC of the European Parliament and of the Council of 4 April 2001 on the approximation of the laws, regulations and administrative provisions of the Member States relating to the implementation of good clinical practice in the conduct of clinical trials on medicinal products for human use. *Official J Eur Communities.* 2001;L121:34–44.

37. Nelson RM. Appropriate risk exposure in environmental health research. The Kennedy–Krieger lead abatement study. *Neurotox Teratol.* 2002;24(4):445–449.

38. Nelson RM. Justice, lead and environmental research involving children. In:Kodish E, ed. *Ethics and Research with Children: A Case-Based Approach,* New York:Oxford University Press; 2005:161–178.

39. Dietrich KN, et al. Effect of chelation therapy on the neuropsychological and behavioral development of lead-exposed children after school entry. *Pediatrics.* 2004;114(1):19–26.

40. Slack C, Strode A, Fleischer T, et al. Enrolling adolescents in HIV vaccine trials: reflections on legal complexities from South Africa. *BMC Med Ethics.* 2007;8:5.

41. Milford C, Wassenaar D, Slack C. Resource and needs of research ethics committees in Africa: preparations for HIV vaccine trials. *IRB: A Review of Human Subjects Research.* 2006;28 (2):1–9.

42. Slack C, Strode A, Grant C, et al. Implications of the ethical–legal framework for adolescent HIV vaccine trials—report of a consultative forum. *S Afr Med J.* 2005;95(9):682–684.

43. Strode A, Slack C, Mushariwa M. HIV vaccine research—South Africa's ethical–legal framework and its ability to promote the welfare of trial participants. *S Afr Med J.* 2005;95 (8):598–601.

44. Slack C, Kruger M. The South African Medical Research Council's Guidelines on Ethics for Medical Research—implications for HIV-preventive vaccine trials with children. *S Afr Med J.* 2005;95(4):269–271.

45. Strode A, Grant C, Slack C, et al. How well does South Africa's National Health Act regulate research involving children? *S Afr Med J.* 2005;95(4):265–268.

46. World Medical Association. *Declaration of Helsinki: Ethical Principles for Medical Research Involving Human Subjects.* 2008. Available at http://www.wma.net/e/policy/pdf/17c.pdf (Accessed November 25, 2008). The 2000 version including the 2002 Note of Clarification on Paragraph 29 can be found at http://www.baskent.edu.tr/tip/helsinkiing.pdf (Accessed November 25, 2008.)

47. Office of the Secretary, DHEW. Belmont Report: Ethical Principles and guidelines for the Protection of Human Subjects of Research. *Fed Regist.* 1979;44(76):23191–23197.

48. Miller PB, Weijer C. Equipoise and the duty of care in clinical research: a philosophical response to our critics. *J Med Philos.* 2007;32(2):117–133.

49. Veatch RM. The irrelevance of equipoise. *J Med Philos.* 2007;32(2):167–183.

50. Gifford F. So-called "clinical equipoise" and the argument from design. *J Med Philos.* 2007; 32(2):135–150.

51. London AJ. Two dogmas of research ethics and the integrative approach to human-subjects research. *J Med Philos* 2007;32(2):99–116.

52. FDA, DHHS. International Conference on Harmonisation: choice of control groups in clinical trials. *Fed Regist.* 1999;64(185):51767–51780.

CHAPTER 10

Ethical Issues in Neonatal Drug Development

GERRI R. BAER, MD

Community Neonatal Associates, 1500 Forest Glen Road, Silver Spring, Maryland 20910

INTRODUCTION

Randomized trials of neonatal therapies are an absolute necessity in order to maximize therapeutic offerings and minimize harms to neonates. However, rigorous scientific evaluation of the most commonly used neonatal pharmacotherapies has not been performed. Historically, clinicians have used drugs that were approved for use in adults or older children, in the absence of efficacy and safety data in neonates. Dosing and formulations for neonatal administration have been improvised from older children and adults as well.

Protection of this vulnerable population has been a primary ethical concern. The ethical principles that are used to protect research participants are drawn from multiple historical documents. Most commonly cited in the United States is the Belmont Report of the National Commission for the Protection of Human Subjects of Biomedical and Behavioral Research, issued in 1978.[1,2] The Belmont Report specified three basic ethical principles to guide human research—respect for persons, beneficence, and justice. Protection of vulnerable subjects (including children) was incorporated as an element of respect for persons. The principles of beneficence and nonmaleficence have been applied to direct investigators to maximize the potential for a child to benefit and minimize potential harm.[1] Finally, the principle of justice classically has been applied to guide the fair distribution of research risks and potential benefits. Vulnerable groups of people (including children) should not be overrepresented as research subjects due to convenience, easy availability, coercion, or undue influence. Rather than overused, neonates are now thought of as an underrepresented population, which leads to another violation of the justice principle. Because neonates (and children) have been understudied, their ability to benefit from the results of research is limited.[1]

The Best Pharmaceuticals for Children Act (BPCA), initially passed in 2002 and reauthorized in 2007, has helped bring this issue to light and has focused attention on the

Pediatric Drug Development: Concepts and Applications
Edited by Andrew E. Mulberg, Steven A. Silber, and John N. van den Anker

need to establish efficacy and safety of drugs in children. Processes were established for the study of both off-patent drugs and on-patent drugs that are used in children.[3] Implementation of the BPCA included the creation of the Newborn Drug Development Initiative (NDDI). The NDDI, a collaborative effort of the National Institute of Child Health and Development (NICHD) and the Food and Drug Administration (FDA), gathered experts in clinical and academic neonatology, industry, and ethics to explore novel approaches to clinical trial design in neonatal medicine.[3,4] Increased attention, and the medical community's appreciation that prescribing untested or poorly tested therapies is less than ideal, should serve to improve neonatal therapeutics in the long term.

NEONATAL RESEARCH: A SCIENTIFIC AND ETHICAL MANDATE

Rigorous testing of treatments on neonates is a scientific and clinical necessity.[1,3,5] Neonates have been harmed by seemingly promising but untested treatments that caused unexpected complications—most prominently the use of 100% oxygen for respiratory distress syndrome and long courses of dexamethasone for bronchopulmonary dysplasia. Yet most of the therapies that are the current standards of care have not been studied in adequately designed, well-controlled randomized trials and/or shown to improve outcomes in neonates with thorough long-term outcome assessment. Neonatologists commonly prescribe medications such as metoclopramide for gastroesophageal reflux and phenobarbital for neonatal seizures, without strong evidence for safety or efficacy.[6,7]

It has often been stated that children are not small adults,[1,8,9]and likewise, neonates (particularly low birth weight infants) are not smaller children. Neonates exhibit unique physiology, different from older children, with additional variation, or "developmental diversity," within the neonatal population.[3] In addition, numerous diseases are unique to the neonatal population, such as perinatal hypoxic-ischemic encephalopathy, respiratory distress syndrome, necrotizing enterocolitis, and apnea of prematurity. Therefore, assumptions of efficacy and safety of drugs untested in the neonatal population may be incorrect.

The current solution to the lack of legitimate evidence-based therapy in neonatology—the common use of off-label therapies[10–12]—is less than ideal. Without well-designed trials demonstrating evidence of safety and efficacy, there is concern that neonatologists are conducting an "uncontrolled experiment" each time an off-label formulation, dosage, or medication is prescribed.[13,14] The incidence of adverse drug reactions (ADRs) in neonates prescribed off-label therapies is not known, due to difficulties with reporting and data gathering, but concerns have been raised regarding increased risks of ADRs.[12] Long-term outcomes related to off-label medications may take years to become apparent, as in the story of dexamethasone and cerebral palsy, or may never come to light without quality long-term outcome data.

From an ethical standpoint, without safety data, we cannot know for certain that a therapy is not harming a neonatal patient. Without demonstrated efficacy, there is no way to determine whether we are giving truly helpful, beneficent treatment that is ultimately of value to the neonatal patient. Finally, off-label treatment of neonates creates a sort of paradox between neonatal research and clinical care. Neonates are treated with therapies off-label regularly as part of clinical care, without any type of parental permission or information exchange, but if an investigator wishes to randomize or standardize the treatment as part of a trial, she must secure institutional review board (IRB) approval as well as informed parental permission.[14] These entrenched off-label practices have exposed countless infants to unproven therapies and hindered the progress of much-needed evidence in neonatology. To address this issue, the NDDI expert working groups have proposed innovative trial designs for commonly used medications that may be acceptable both scientifically and ethically.[6,15–19]

Despite the barrier of ingrained treatment practices, it is clearly possible to execute scientifically sound, well-powered multicenter randomized controlled trials of standard medications in neonates. Methylxanthines, including caffeine, have long been used to treat apnea in preterm infants. Whether caffeine was associated with abnormal neurodevelopmental outcomes and whether treating apnea had any beneficial medium- or long-term effects were until recently unclear.[20]

The recently completed Caffeine for Apnea of Prematurity, or CAP, trial is a prototype for what can be done to advance the medical evidence. The CAP investigators randomized more than 2000 premature infants at risk for apnea of prematurity to caffeine or placebo, with a primary endpoint of death or neurodevelopmental disability at 18–21 months. Infants receiving caffeine had a 37% reduction in bronchopulmonary dysplasia (chronic lung disease of prematurity), and a significant reduction in the primary endpoint of death or disability.[21,22] The evidence of caffeine's safety and efficacy was reassuring to the neonatology community, and the trial was a model for what can and should be done to advance knowledge, even when a largely untested treatment is already "standard."

PROBLEMS WITH THE CURRENT BODY OF EVIDENCE

Suboptimal choices in trial design have led to many inconclusive stories in neonatal therapeutics.[23] The importance of designing neonatal clinical trials with appropriate endpoints, effect size, and sample size cannot be overstated.

When designing a trial, both endpoints and effect sizes help to determine the study's power and sample size. Endpoints and effect sizes must first and foremost be clinically significant to families, patients, and clinicians. Surrogate or short-term endpoints, though important and necessary to report, cannot substitute for long-term outcomes such as disability.[24] The presentation of the potential use and value of surrogate endpoints in pediatric clinical trials is a topic covered in Chapter 38 by Molenberghs and Orman.

A systematic review of 5 years of data from neonatal randomized clinical trials from the 1990s revealed that the majority of trials utilized continuous outcome measures or short-term endpoints. In addition, the trials' median hypothesized relative risk reduction was large[25]—40–50%—a risk reduction that is seldom shown with any therapy and is in excess of what clinicians consider a clinically meaningful effect.[26] Choosing larger effect sizes drives the sample size down, creating a study that is underpowered to detect smaller, but potentially meaningful effects.

Determination of a treatment's desired effect size is complex and involves value judgments about what constitutes a significant improvement. These judgments may incorporate patient or family preferences, costs, social ideologies, moral beliefs, and any number of value-laden factors.[23] Some authors have suggested that effect size may be determined either by clinician consensus and empirical definitions (if they exist) or by existing data from observational studies or pilot trials,[27] but it should not be primarily determined by driving the sample size down.

Planning studies that are underpowered is unacceptable from an ethical standpoint, because (1) investigators involve infants in research (with any degree of potential risk) that is not properly designed to show an effect and (2) investigators do not tend to disclose in the permission process the limited value of the information to be gained by the research.[27] Yet, arguably, much of the evidence on which we base neonatal therapies has been underpowered. (The same issue and concern is related to many pediatric therapeutics.) In an analysis of randomized trials included in Cochrane neonatal systematic reviews prior to 2001, it was noted that the trials' median sample size was 53 participants. A mere 25% of the trials included in Cochrane reviews had sample sizes larger than[28] 102, and these trials constitute the best evidence available in neonatology.

Proposed solutions to the dearth of well-powered neonatal clinical trials include prospective meta-analyses and multicenter trials, and parts of the latter solution already are in place. Multicenter collaborations such as the NICHD Neonatal Research Network, Vermont Oxford Network, the Canadian Neonatal Network, and Pediatrix can help make larger trials feasible.[29,30] In addition, the NICHD has focused further attention on the lack of adequate efficacy and safety data for many commonly used drugs in neonatology. The NICHD working groups proposed trial design strategies and created criteria for prioritizing which drugs to study.[6,15–19,31] A further discussion on the role of the academic community partnership with industry is covered by Goldstein and colleagues in Chapter 36. Appendix II lists the local and national neonatal and pediatric clinical research networks on the Inventory and Evaluation of Clinical Research Networks (IECRN) that participate in Phase I–IV clinical trials (https://www.clinicalresearchnetworks.org/srchnet.asp).

SPECIAL ETHICAL CONSIDERATIONS IN NEONATES

Risk–Benefit Assessment

Subpart D deals specifically with protecting child research subjects. In contrast to Subpart A (which pertains to adults), allowable risk is restricted to either no more than minimal risk or a minor increase over minimal risk for research that does not offer the prospect of direct benefit to its participants (21 CFR 50.51 and 50.53). For research that presents the prospect of direct benefit to participants, risk exposure must be justified by the anticipated benefit; the risk–benefit assessment must be at least as favorable as that presented by available alternatives (21 CFR 50.52), with parent permission and child assent (if appropriate) required as described under 21 CFR 50.55.

In addition to analyzing the entire protocol, research also should be analyzed by examining the risks and benefits of the individual components of the protocol.[1,32,33] Research protocols involving neonatal treatments encompass components that are potentially therapeutic (e.g., administration of study drug) and components that are nontherapeutic (e.g., drawing extra blood samples for pharmacokinetic analysis, or continuation of indwelling catheters for blood sampling). The risk of the therapeutic component should be justified by the prospect of direct benefit to the participant (21 CFR 50.52). Risks of the nontherapeutic component (as it does not offer the participant the prospect of direct benefit) must be justified by the potential to create generalizable knowledge. Nontherapeutic research risk in children must be limited to minimal risk or a minor increase over minimal risk (21 CFR 50.51 and 50.53).[32] Finally, provisions exist so that research not otherwise approvable under 21 CFR 50.51–50.53 can be referred to a federal panel for possible approval (21 CFR 50.54).

When considering the therapeutic risks of a protocol, one cannot help but consider participants in the control or placebo group, who are not receiving the active treatment in question. Placebo effects and general benefits of being involved in research do not equate to the prospect for direct benefit.

Equipoise and Choosing a Control Group

The concept of clinical equipoise was popularly established in 1987 by the late Benjamin Freedman, who defined it as "genuine uncertainty on the part of the expert medical community—not necessarily on the part of the individual investigator—about the preferred treatment."[34] Thus, if there is clinical equipoise, the expert community does not know whether the recipient of the placebo or the control medication will benefit any more or less than the recipient of the active study medication. Assessments of equipoise may also take into account potential risks or toxicities of the therapy in question. Currently, there is a vigorous debate in the bioethics community around the use of the principle of equipoise as a metric in the ethical design of clinical trials.

Franklin Miller and Howard Brody[35] have suggested that equipoise combines two separate principles: (1) the notion of the honest null hypothesis, that is, truly not knowing from a scientific basis which treatment is superior; and (2) the concept of "no inferior treatment," that neither group should be disadvantaged by participation, as an ethical norm driving the design of clinical trials. The two components are not equivalent, and conflating them, according to Miller and Brody, confuses the ethical evaluation of research.[35] Equipoise also has been criticized as a "fragile" concept, hinging on the question of how much scientific evidence is required to prove a treatment effective and safe.[36] A number of authors have argued that clinical equipoise is morally irrelevant, and rather nonexploitation of subjects and voluntary informed consent should be the primary moral guides.[37–39] The philosophical discussion of how and whether to continue to use the principle of equipoise remains active.[40]

Equipoise and control groups have often been stumbling blocks for neonatal randomized trial design. Because of the widespread off-label use of therapies in neonatology, which have become entrenched practices, there is frequently hesitation about the use of placebos in neonatal drug trials.[41–44] The Declaration of Helsinki[45] states that "the benefits, risks, burdens, and effectiveness of a new method should be tested against those of the best current prophylactic, diagnostic, and therapeutic methods." Use of placebo controls can only be permitted when there is "no proven prophylactic, diagnostic, or therapeutic method."[45] Considering what constitutes "proven" therapy may allow for placebo controls when a drug has not been definitively shown to be safe and/or efficacious in large randomized trials. In addition, the apparent lack of assay sensitivity and proof of efficacy to a placebo is often a significant issue in pediatric clinical trials.

The CAP trial used a placebo for its control group, despite extensive use of caffeine and other methylxanthines for apnea of prematurity.[22] Caffeine is not the only treatment for apnea, and control group infants could be treated with rescue treatments such as continuous positive airway pressure or mechanical ventilation. Prior to CAP, there were no randomized trials with long-term neurodevelopmental data to follow up preterm infants who were given caffeine, so although caffeine may have improved the short-term endpoint of apnea, it was unclear whether it altered any important long-term outcomes. Methylxanthines act on receptors in the central nervous system, raising the question of whether they can harm infants. Control group patients could receive open-label caffeine treatment if clinically necessary. (Open-label treatment with caffeine occurred in 9% of the treatment group and 10% of the placebo group ($p = 0.45$).)[21,22]

The International Conference on Harmonisation (ICH) published guidance in 1999 on choosing control groups. It was argued that placebo controls may be used despite the presence of a proven treatment if no serious harm would result from withholding treatment or if the effective treatment is associated with major toxicity.[46] Subsequently, the World Medical Association modified the Declaration of Helsinki to reflect the conditions presented in the ICH guidance report.[45] Both Subpart D and the 1998 Canadian Tri-Council Policy Statement hold that assessment of clinical equipoise requires that a participant's health not be disadvantaged by participating in the research.[47] Therefore, withholding proven treatment in a trial for more than a minor condition would not be allowed under the equipoise requirement of 21 CFR 50.52. But if the risk of withholding the proven treatment creates no more than a minor increase over minimal risk, the research may be approvable under 21 CFR 50.53 for that arm. In the case of the CAP trial, it could be argued either that apnea of prematurity was a minor condition or that the availability of rescue treatment limited risk to no more than a minor increase over minimal. In the absence of strong evidence that it causes irreversible harm to preterm infants, withholding treatment in the placebo arm was ethical.

Innovative trial designs may also help solve the question of appropriate control groups. "Add-on" clinical trials may allow for evaluations of multiple therapies that may be used together. Several new trial frameworks were proposed by the NDDI working groups. However, buy-in from the neonatology community to test drugs that already are used may be difficult to secure. The many years of operating with low-quality evidence have left the field of neonatology with many therapies that are ingrained and comfortable practices, without solid evidence of their efficacy or safety.

The Difficulty of Parental Permission Under Duress

Questions have often been raised about the validity of parental permission for research in the emotionally strained circumstances of neonatal intensive care. Particularly during emergent circumstances, many neonatologists and neonatal researchers have worried that the extreme emotions may impair parents' ability to comprehend, appreciate, and make reasoned decisions about research. Parents may enter trials out of desperation to do anything to save their infant, thus raising the problem of therapeutic misconception when it comes to understanding the purpose of a randomized trial. Simply defined, the therapeutic misconception is the belief that the primary purpose of the research is treatment.[1] There are a number of studies addressing parents' recall and comprehension of informed consent procedures, but these data are often limited by long time periods elapsing between trial participation and recall.

In the United States, parents were found to be inadequately informed after a single-center randomized controlled drug trial—nearly 8% did not recall the trial at all, and though two-thirds of those who recalled the trial recalled its purpose and potential benefits, only 5% could verbalize any risks of the trial. Interviews were carried out between 3 and 28 months after consent forms were signed, with nearly three-quarters of parents interviewed after more than 6 months had passed.[48] Despite being relatively well informed about study details and procedures, 27% of parents of children in a drug study for autism did not understand the random nature of treatment allocation.[49] Interviews of parents who had allowed their infants to be enrolled in the United Kingdom Collaborative ECMO trial also confirmed that many parents do not appreciate the concept of randomization. These data were also limited by years that elapsed between trial participation and parental interview.[50] There remain concerns that incorrect appreciation of randomization and the therapeutic misconception may go hand in hand.

Mason and Allmark[51] interviewed parents of 200 European neonates who had been asked for permission for neonatal trials within the previous year. In 70 percent of cases, there was impairment in one or more consent criteria (competence, information, understanding, and/or voluntariness). Problems with the permission process were more prevalent in parents of children approached for emergency research and research involving more than standard risk.[51] Again, these data are limited by time that elapsed since consent discussions, but they suggest the need for ongoing communication with parents, after the initial discussion, to ensure their voluntary and informed permission.

Several studies have examined parental perceptions about clinical research decisions involving their infants. In a qualitative study from the United Kingdom, 51 parent interviews were conducted after a perinatal research decision had been made. Fear was the dominant emotion associated with research decisions and, in dire situations, often led to parents consenting to a trial for any chance of hope. A number of mothers who were ill or in labor quickly signed permission forms so that the person seeking consent would leave the room. Finally, parents who perceived there were no risks involved in the research were more likely to make rapid decisions about entry into the trial.[52] These are troubling results due to the pervasiveness of fear and physical stress in parents of ill neonates, as well as the fact that virtually no research is without some degree of risk.

Parents of neonates having cardiothoracic surgery in a single U.S. center shared reasons for either participating or not participating in clinical research. The most commonly cited reasons for participation were the potential for societal benefit and personal benefit. One-third of parents were concerned about risks in the research, while one-third perceived no risks to the research.[53] Parents in a Canadian study shared high rates of altruistic and personal motives. Parents who did not give permission to enroll were more concerned about the likelihood of harm, yet there were no obvious differences in perception of illness severity between consenters and nonconsenters. Reassuringly, the vast majority of parents felt "free to decide," did not think the process was "too complex," and were not worried about the implications of not participating.[54] In another Canadian survey study, parents of both infants in the neonatal intensive care unit (NICU) and healthy newborns were noted to believe in the necessity of research, but wanted full disclosure and the power to decide about enrollment.[55]

Some have advocated a modified approach to parental permission for research that does not increase patient risk and have argued that in the presence of equipoise about a treatment, inclusion in a trial of that treatment does not necessarily increase risk.[56] Improvements in the permission process also may be achieved by using an ongoing consent process and official training for consenting clinicians.[57] The Institute of Medicine report also advocates a "careful *process* of communication" with ongoing processes of permission and information exchange incorporating the family's evolving understanding of the child's condition, and a reduction of the emphasis on the consent paperwork.[1]

Institutional Review Boards

Nationally, there appears to be great variability in quality and efficiency of IRB review, a problem that is not unique to neonatal research. The working groups of the NDDI expressed concerns that variability in local and regional standards of care could in part modify opinions about a given trial design's ethical and scientific soundness.[5] It has been shown empirically, in a survey of IRB chairpersons, that there were wide variations in risk/benefit assessments for children, as well as in what constitutes "direct benefit" to a research participant. The data also led to questions about whether IRBs may apply different risk standards to healthy and ill children.[58]

It is unclear what can be done about local and regional differences in standards of care, but it has been suggested that ethical consultation in the protocol-writing phase of a study and regulatory and ethical analysis written into the protocol could help reduce the variability in IRB review.[5]

DOES SUBPART B APPLY TO PREMATURE INFANTS?

A final regulatory question to address surrounds the applicability of Subpart B to research involving premature neonates. Subpart B regulates research on "pregnant women, fetuses, and neonates," including "neonates of uncertain viability," creating the possibility of ambiguity about whether research on very premature neonates is regulated by Subpart D alone, Subpart B, or both.[5,59] If Subpart B were applied to neonatal research, procedures offering no prospect of direct benefit would be required to meet the standard of minimal risk, rather than a minor increase over minimal risk. In addition, no waiver of informed consent for minimal risk research (45 CFR 46.116) could be applied, and the emergency exception from informed consent would be restricted (45 CFR 46).

Viability is defined in Subpart B as "being able, after delivery, to survive (given the benefit of available medical therapy) to the point of independently maintaining heartbeat and respiration"(45 CFR 46.202). Given the definition and the discussion of medical therapy, "independent" must be interpreted as independent from the mother, not from medical technology. Thus, any infant who is able to be resuscitated at birth should be considered viable for the purpose of these regulations. Research on infants who are viable is explicitly covered under Subparts A and D, which are sufficient to protect ill neonates.[5,60] Thus, Subpart B does not apply to neonatal research.

REFERENCES

1. Field MJ, Behrman RE. *Ethical Conduct of Clinical Research Involving Children.* Washington DC: The National Academies Press; 2004.
2. Commission N. *The Belmont Report: Ethical Principles and Guidelines for the Protection of Human Subjects of Research.* Washington DC: US Government Printing Office; 1978.
3. Giacoia GP, Birenbaum DL, Sachs HC, Mattison DR. The Newborn Drug Development Initiative. *Pediatrics.* 2006;117(3):S1–S8.
4. Giacoia GP, Mattison DR. Newborns and drug studies: the NICHD/FDA newborn drug development initiative. *Clin Ther.* 2005;27(6):796–803.
5. Baer GR, Nelson RM. Ethical challenges in neonatal research: summary report of the ethics group of the Newborn Drug Development Initiative. *Clin Ther.* 2006;28(9):1399–1407.
6. Clancy RR. Summary proceedings from the Neurology Group on neonatal seizures. *Pediatrics.* 2006;117(3):S23–S27.
7. Hibbs AM, Lorch SA. Metoclopramide for the treatment of gastroesophageal reflux disease in infants: a systematic review. *Pediatrics.* 2006;118(2):746–752.
8. Bachrach LK. Bare-bones fact—children are not small adults. *N Engl J Med.* 2004;351(9):924–926.

9. Fan LL, Langston C. Pediatric interstitial lung disease: children are not small adults. *Am J Respir Crit Care Med.* 2002;165(11):1466–1467.
10. Barr J, Brenner-Zada G, Heiman E, et al. Unlicensed and off-label medication use in a neonatal intensive care unit: a prospective study. *Am J Perinatol.* 2002;19(2):67–72.
11. Conroy S, McIntyre J, Choonara I. Unlicensed and off label drug use in neonates [see Comment]. *Arch Dis Child Fetal Neonatal Edition.* 1999;80:F142–F145.
12. Cuzzolin L, Atzei A, Fanos V. Off-label and unlicensed prescribing for newborns and children in different settings: a review of the literature and a consideration about drug safety. *Expert Opin Drug Safety.* 2006;5(5):703–718.
13. Chalmers I, Silverman WA. Professional and public double standards on clinical experimentation. *Controlled Clin Trials.* 1987;8:388–391.
14. Tyson JE. Use of unproven therapies in clinical practice and research: How can we better serve our patients and their families? *Semin Perinatol.* 1995;19(2):98–111.
15. Finer NN, Higgins R, Kattwinkel J, Martin RJ. Summary proceedings from the Apnea-of-Prematurity Group. *Pediatrics.* 2006;117(3):S47–S51.
16. Perlman JM. Summary proceedings from the Neurology Group on hypoxic-ischemic encephalopathy. *Pediatrics.* 2006;117(3):S28–S33.
17. Roth SJ, Adatia I, Pearson GD, Members of the Cardiology Group. Summary proceedings from the Cardiology Group on postoperative cardiac dysfunction. *Pediatrics.* 2006;117(3):S40–S46.
18. Short BL, Van Meurs K, Evans JR, Cardiology Group. Summary proceedings from the Cardiology Group on cardiovascular instability in preterm infants. *Pediatrics.* 2006;117(3): S34–S39.
19. Walsh MC, Szefler S, Davis J, et al. Summary proceedings from the Bronchopulmonary Dysplasia Group. *Pediatrics.* 2006;117(3):S52–S56.
20. Stevenson DK. On the caffeination of prematurity. *N Engl J Med.* 2007;357(19):1967–1968.
21. Schmidt B, Roberts RS, Davis P, et al. Caffeine therapy for apnea of prematurity. *N Engl J Med.* 2006;354(20):2112–2121.
22. Schmidt B, Roberts RS, Davis P, et al. Long-term effects of caffeine therapy for apnea of prematurity. *N Engl J Med.* 2007;357(19):1893–1902.
23. Lantos JD. Sample size: profound implications of mundane calculations. *Pediatrics.* 1993;91 (1):155–157.
24. Aschner JL, Walsh MC. Long-term outcomes: What should the focus be? *Clin Perinatol.* 2007;34(1):205–217.
25. Zhang B, Schmidt B. Do we measure the right end points? A systematic review of primary outcomes in recent neonatal randomized clinical trials. *Pediatr.* 2001;138(1):76.
26. Raju TNK, Langenberg P, Sen A, Aldana O. How much "better" is good enough? The magnitude of treatment effect in clinical trials. *Am J Dis Child.* 1992;146(4):407–411.
27. Halpern SD, Karlawish JHT, Berlin JA. The continuing unethical conduct of underpowered clinical trials. *JAMA.* 2002;288(3):358–362.
28. Sinclair JC, Haughton DE, Bracken MB, Horbar JD, Soll RF. Cochrane neonatal systematic reviews: a survey of the evidence for neonatal therapies. *Clin Perinatol.* 2003;30 (2):285–304.
29. Horbar JD. The Vermont-Oxford Neonatal Network: integrating research and clinical practice to improve the quality of medical care. *Semin Perinatol.* 1995;19(2):124–131.
30. Wright L, McNellis D. National Institute of Child Health and Human Development (NICHD)-sponsored perinatal research networks. *Semin Perinatol.* 1995;19(2):112–123.
31. Anand KJS, Aranda JV, Berde CB, et al. Summary proceedings from the Neonatal Pain-Control Group. *Pediatrics.* 2006;117(3):S9–S22.

32. Weijer C. The ethical analysis of risk. *J Law Med & Ethics* 2000;28(4):344–361.
33. Weijer C, Miller PB. When are research risks reasonable in relation to anticipated benefits? *Nat Med.* 2004;10(6):570–573.
34. Freedman B. Equipoise and the ethics of clinical research. *N Engl J Med.* 1987;317(3): 141–145.
35. Miller FG, Brody H. Clinical equipoise and the incoherence of research ethics. *J Med Philos.* 2007;32(2):151–165.
36. Gifford F. So-called "clinical equipoise" and the argument from design. *J Med Philos.* 2007;32 (2):135–150.
37. Miller FG, Brody H. What makes placebo-controlled trials unethical? *Am J Bioethics.* 2002;2(2):3–9.
38. Miller FG, Brody H. A critique of clinical equipoise: therapeutic misconception in the ethics of clinical trials. *Hastings Cent Rep.* 2003;33(3):19–28.
39. Veatch RM. The irrelevance of equipoise. *J Med Philos.* 2007;32(2):167–183.
40. Miller FG, Veatch RM. Symposium on equipoise and the ethics of clinical trials. *J Med Philos.* 2007;32(2):77–78.
41. Gluckman PD, Wyatt JS, Azzopardi D, et al. Selective head cooling with mild systemic hypothermia after neonatal encephalopathy: multicentre randomised trial. *Lancet* 2005;365 (9460):663–670.
42. Shankaran S, Laptook AR, Ehrenkranz RA, et al. Whole-body hypothermia for neonates with hypoxic-ischemic encephalopathy. *N Engl J Med.* 2005;353(15):1574–1584.
43. Blackmon LR, Stark AR, Committee on Fetus and Newborn AAoP. Hypothermia: a neuroprotective therapy for neonatal hypoxic-ischemic encephalopathy. *Pediatrics.* 2006;117(3):942–948.
44. Kirpalani H, Barks J, Thorlund K, Guyatt G. Cooling for neonatal hypoxic ischemic encephalopathy: Do we have the answer? *Pediatrics.* 2007;120(5):1126–1130.
45. Declaration of Helsinki. Tokyo, Japan: World Medical Association; 2000.
46. Food and Drug Administration. International Conference on Harmonisation; choice of control group in clinical trials. *Fed Regist.* 1999;64: 51767.
47. Tri-Council policy statement: ethical conduct for research involving humans. Medical Research Council of Canada, August 1998.
48. Ballard HO, Shook LA, Desai NS, Anand KJ. Neonatal research and the validity of informed consent obtained in the perinatal period. *J Perinatol.* 2004;24(7):409–415.
49. Vitiello B, Aman MG, Scahill L, et al. Research knowledge among parents of children participating in a randomized clinical trial. *J Am Acad Child Adolesc Psychiatry.* 2005;44 (2):145–149.
50. Snowdon C, Garcia J, Elbourne D. Making sense of randomization: responses of parents of critically ill babies to random allocation of treatment in a clinical trial. *Soc Sci Med.* 1997;45(9):1337–1355.
51. Mason SA, Allmark PJ. Obtaining informed consent to neonatal randomised controlled trials: interviews with parents and clinicians in the Euricon study [see Comment]. *Lancet* 2000;356 (9247):2045–2051.
52. Snowdon C, Elbourne D, Garcia J. "It was a snap decision:" Parental and professional perspectives on the speed of decisions about participation in perinatal randomised controlled trials. *Soc Sci Med.* 2006;62(9):2279–2290.
53. Hoehn KS, Wernovsky G, Rychik J, et al. What factors are important to parents making decisions about neonatal research? *Arch Dis Child Fetal Neonatal Edition.* 2005;90:F267–F269.
54. Zupancic JA, Gillie P, Streiner DL, Watts JL, Schmidt B. Determinants of parental authorization for involvement of newborn infants in clinical trials. *Pediatrics.* 1997;99(1):e6.

55. Singhal N, Oberle K, Burgess E, Huber-Okrainec J. Parents' perceptions of research with newborns. *J Perinatol.* 2002;22:57–63.

56. Rogers CG, Tyson JE, Kennedy KA, Broyles S, Hickman JF. Conventional consent with opting in versus simplified consent with opting out: an exploratory trial for studies that do not increase patient risk. *J Pediatr.* 1998;132(4):606–611.

57. Allmark PJ, Mason SA. Improving the quality of consent to randomised controlled trials by using continuous consent and clinician training in the consent process. *J Med Ethics.* 2006;32 (8):439–443.

58. Shah S, Whittle A, Wilfond B, Gensler G, Wendler D. How do institutional review boards apply the federal risk and benefit standards for pediatric research? *JAMA.* 2004;291(4):476–482.

59. Department of Health and Human Services. Subpart B—additional protections for pregnant women, human fetuses and neonates involved in research. *Fed Regist.* November 13, 2001;66: 56778–56780.

60. Nelson RM. Personal communication. 2007.

CHAPTER 11

Ethical Principles of Pediatric Research and Drug Development: A Guide Through National and International Frameworks and Applications to a Worldwide Perspective

KLAUS ROSE, MD, MS

Head Pediatrics, F. Hoffmann-La Roche Ltd., CH-4070 Basel, Switzerland

INTRODUCTION AND METHODOLOGY

Research in children is a rather vague term, addressing a vast array of scientific questions, including studies on psychological and biological maturation of body and mind, child nutrition, use of medical devices in children, safety of children in cars and aircrafts, and many more.[1] While the Belmont Report of 1979 covers biomedical and behavioral research in general,[2] other documents (e.g., the EU Ethical Considerations for Clinical Trials Performed in Children[3]) refer to drug development only. Until the end of the 20th century, research in children was predominantly performed by academic institutions and financed by national governments or other grant systems.[4,5] In consequence, many textbooks address questions of ethics of research with children in the context of national frameworks.[6,7] While academic research addressed more the understanding of child development, we have observed over several decades an explosion of biomedical research that increasingly also addresses and begins to include children.

Since 1997, pediatric legislation to address the deficiencies of drug development lacking for children was first introduced in the United States, followed by comparable legislation in 2006 in Canada and Europe, with further legislation planned in Japan and other countries and regions. For full description of these issues, please see Chapter 14 by Nakamura and Ono, Chapter 12 by Maldonado, and Chapter 13 by Rose in this book. These legislations aim at letting children participate and benefit directly in the undeniable progress of pharmaceutical treatment that we are observing since the 19th century.

While the first U.S. legislation, the Food and Drug Administration Modernization Act (FDAMA) and its unsuccessful predecessors, started with collecting additional data on

Pediatric Drug Development: Concepts and Applications
Edited by Andrew E. Mulberg, Steven A. Silber, and John N. van den Anker

pediatric use of already existing, marketed medications, later legislations target increasingly at including children into the general drug development process and plan at earlier stages.[8,9] Over the past decades, the process of drug development has evolved into a global process performed by pharmaceutical companies that register their medicinal products locally and regionally.[1,10] They always have to comply with local regulations wherever clinical trials or other preclinical, technical, or further tests are physically performed, but they coordinate the development process in a truly global fashion. Each company must navigate through this double framework—national/regional guidelines—as well as international declarations, guidelines, principles, and more. Increasingly, academic centers are in a comparable position as far as they organize global clinical trials across different countries and continents. A knowledge of both the key national and international documents and publications should help all parties involved in international pediatric research to comply with local regulations, in presentations before local ethics committees/institutional review boards (IRBs), in discussions with the respective local clinicians, and in dialogues with interested children once they reach the age of basic understanding of complex issues, with parents, with reimbursement institutions, and with many other involved parties.

This chapter starts with a short historical introduction and then guides the reader through key international and national/regional documents, laws, publications, and organizations that have exerted a considerable influence on discussions on research in children. These documents reflect the fact that there is no world government, that the evolution of science is not a linear process, and that ethical and scientific consensus is a process that evolves over decades of discussion, sometimes slowly, sometimes crystallizing rather fast into action by governments, international bodies, and other institutions. As always, many more documents exist, and the interested reader is provided with a multitude of references for further reading.

OVERVIEW

In the 19th century, the health and welfare of children were the objectives of many conferences and declarations, partially due to the misery caused to children in wars and civil wars, partially due to increasing inclusion of children into industrial production processes. The development of education, health, and social services and the organization of professional societies in many countries led to an increasing number of national and international conferences that addressed the health and welfare of children; for example, ten international congresses of professional workers serving children in European cities between 1883 and 1933 with delegates from Europe and America as well as ten Pan American Child Congresses all over North and South America between 1916 and 1955. Also, the League of Nations requested member states in its Covenant No. 23 to endeavor and maintain fair and human conditions for adult and child workers, to take measures for the prevention and control of disease, and to entrust the League with general supervision over the execution of agreements on the traffic in women and children.[11] However, with the two exceptions of Prussia in 1900 and Germany in 1931,[12] ethics in scientific research in children was not addressed by governments or academic societies before 1950.

Until the end of the first half of the 20th century, the conduct of human experimentation was done by individual investigators, usually within the framework of academic institutions. Research in children was often performed on researchers' own children, servants, and slaves or on orphans.[4,13]

TABLE 11.1 The Ten Points of the Nuremberg Code

1. Voluntary consent.
2. Experiment should aim at fruitful results.
3. Anticipated results should be based on science.
4. Avoidance of unncessary physical and mental suffering.
5. Avoidance of death or disabling injury in the experiment.
6. Risk must not exceed the humanitarian importance of the targeted problem.
7. Protection of participants aganst injury, disability, or death.
8. Trial to be conducted only by scientifically qualified persons.
9. Participants' right to withdraw.
10. Study management to be prepared to terminate trial if risks are perceived for participants.

During and after World War II, medical research expanded at an extraordinary rate. After 1945, the world was outraged by the murders conducted on humans in general and specifically on children by Nazi physicians such as Josef Mengele[4,14] and by Japanese physicians in occupied China.[12] The judges in the trial against Nazi physicians in Nuremberg, Germany, published in 1947 a list of principles that became the "Nuremberg Code" in their critical evaluation of the observed wrongdoings and atrocities.[15,16] The Nuremberg Code is discussed further later on and the critical ten points are listed in Table 11.1.

In 1948, the World Medical Association (WMA) adopted the International Code of Medical Ethics in Geneva,[17] which asked for utmost respect for human life and is quoted completely in Table 11.2. In 1948, in London, the WMA expanded the International Code of Medical Ethics toward duties of the physician in general, who should always be "dedicated to providing competent medical service in full technical and moral independence, with compassion and respect for human dignity."[18] (See Table 11.3).

Also in 1948, the General Assembly of the United Nations adopted and proclaimed the Universal Declaration of Human Rights.[19] Children are mentioned in Article 25(2):

TABLE 11.2 The Ten Principles of the 1948 WMA International Code of Medical Ethics

1. To protect to the maximum extent possible the survival and development of the child, and to recognise that parents (or legally entitled representatives) have primary responsibility for the development of the child and that both parents have common responsibilities in this respect;
2. To ensure that the best interests of the child shall be the primary consideration in health care;
3. To resist any discrimination in the provision of medical assistance and health care from considerations of age, gender, disease or disability, creed, ethnic origin, nationality, political affiliation, race, sexual orientation, or the social standing of the child or her/his parents or legally entitled representatives;
4. To attain suitable pre-natal and post-natal health care for the mother and child;
5. To secure for every child the provision of adequate medical assistance and health care, with emphasis on primary health care, pertinent psychiatric care for those children with such needs, pain management, and care relevant to the special needs of disabled children;
6. To protect every child from unnecessary diagnostic procedures, treatment, and research;
7. To combat disease and malnutrition;
8. To develop preventive health care;
9. To eradicate child abuse in its various forms; and
10. To eradicate traditional practices prejudicial to the health of the child.

TABLE 11.3 The 1948 WMA Declaration of Geneva: The Medical Code of Ethics

AT THE TIME OF BEING ADMITTED AS A MEMBER OF THE MEDICAL PROFESSION:

I SOLEMNLY PLEDGE to consecrate my life to the service of humanity;
I WILL GIVE to my teachers the respect and gratitude that is their due;
I WILL PRACTISE my profession with conscience and dignity;
THE HEALTH OF MY PATIENT will be my first consideration;
I WILL RESPECT the secrets that are confided in me, even after the patient has died;
I WILL MAINTAIN by all the means in my power, the honour and the noble traditions of the medical profession;
MY COLLEAGUES will be my sisters and brothers;
I WILL NOT PERMIT considerations of age, disease or disability, creed, ethnic origin, gender, nationality, political affiliation, race, sexual orientation, social standing or any other factor to intervene between my duty and my patient;
I WILL MAINTAIN the utmost respect for human life;
I WILL NOT USE my medical knowledge contrary to the laws of humanity, even under threat;
I MAKE THESE PROMISES solemnly, freely and upon my honour.

"Motherhood and childhood are entitled to special care and assistance. All children, whether born in or out of wedlock, shall enjoy the same social protection." In 1959, the UN General Assembly proclaimed the Declaration of the Rights of the Child.[20]

In 1964, the Declaration of Helsinki was developed by the World Medical Association (WMA) in Helsinki, Finland, as a set of ethical principles for the medical community regarding clinical research in humans.[21,22] The term "clinical research" had replaced the term "human experimentation" used in the Nuremberg Code. It is widely regarded as the cornerstone document of human research ethics, although, as the Nuremberg Code, it is not a legally binding instrument in international law.

The outrage after World War II did not lead to the application of the newly pronounced Nuremberg Code to experimentation in humans in general and to children in particular in the United States. It took 20 more years until Harvard University trained anesthesiologist Henry K. Beecher criticized in a landslide publication in 1966 twenty-two selected academic research projects that had been published in academic research journals and that were unethical in the view of the author.[23,24]

In 1972, the *Washington Star* broke the story that the U.S. Public Health Service (PHS) had been conducting a study of untreated syphilis on black men in Macon County, Alabama, in and around the county seat of Tuskegee. Peter Buxton, a low-grade employee of the PHS with a major in political science, had heard of this study from a co-worker, had inquired about the study more in depth, and had since 1966 tried to have it terminated. Only when PHS management for several years rebuked his attempts had he approached the press in 1972. The Tuskegee study was terminated almost immediately after this publication.[25] It took 25 more years until President Clinton apologized: "The United States Government did something that was wrong, deeply, profoundly, morally wrong. It was an outrage to our commitment to integrity and equality for all our citizens. We can end the silence. We can stop turning our heads away. We can look at you in the eye and finally say on behalf of the American people what the United States Government did was shameful, and I am sorry."[26]

In 1973, the U.S. Department of Health, Education, and Welfare (HEW), which later changed its name to today the Department of Health and Human Services (DHHS), published a first set of proposed regulations concerning the protection of human subjects in biomedical and behavioral research, which were published in 1974 as specific legal guidelines for clinical investigators (45 CFR 46).[4,27] (CFR stands for the U.S. Code of Federal Regulations.)

In 1974, the U.S. Congress established a National Commission for the Protection of Human Subjects in Biomedical and Behavioral Research, of which two key publications should be mentioned here:

- The 1977 Report and Recommendations: Research Involving Children. This report laid out the case for involving children in research, described the extent of such research, surveyed institutional practices regarding consent for research, and reviewed legal and ethical issues. In contrast to other key publications, neither the 1977 report nor a summary are available online.[28]
- The 1978 Belmont Report: Ethical Principles and Guidelines for the Protection of Human Subjects of Research.[2]

In 1983, 10 years after its first proposals and 6 years after the National Commission report on children, the DHHS issued Subpart D (special protections for children)[29] to 45 CFR 46, after it had supplemented Subpart B (special protections for pregnant women and fetuses)[30] in 1978 by Subpart C (special protections for prisoners).[31]

In 1977, the American Academy of Pediatrics published its first set of professional guidelines on ethics of drug research.[32] The subchapter, "Needs for Ethic Guidelines," is quoted in Table 11.4. These guidelines were updated in 1995.[33] Of interest is also the AAP Statement Before the Institute of Medicine Committee on Clinical Research Involving Children (July 9, 2003).[34] The recommendations of this statement are listed in Table 11.5.

In 1989, the United Nations declared a Convention on the Rights of the Child as an international convention setting out the civil, political, economic, social, and cultural rights of children.[35] The child's right to the enjoyment of the highest attainable standard of health is emphasized. The specific principles are quality of care, freedom of choice, consent and self-determination, access to information, confidentiality, admission to hospital, elimination of child abuse, health education, dignity of the patient, and religious assistance.

In 1990, the International Conference on Harmonisation of Technical Requirements for Registration of Pharmaceuticals for Human Use (ICH) was established as a working

TABLE 11.4 1977 AAP Ethics of Drug Research: "The Need for Ethic Guidelines"

Standards for performance of clinical pharmacologic research in infants and children must be established with the same humane purpose and scientific objectives as standards for clinical practice. Ethical practice requires that treatment modalities available to others be made available to pediatric patients, and that, as for other subjects, appropriate protection be given to pediatric patients when they receive treatment. Poor scientific design or uncontrolled experimentation is unethical.

Ethical, as well as scientific, guidelines are needed for the evaluation of drugs to be used in infants and children and for the development of acceptable clinical research patterns for drug investigations in children. These patterns, when developed by medical-scientific, legal, and social experts (in consultation with informed laymen), will become widely accepted in time and may eventually be acknowledged legally, much as standards of practice activity are accepted by society and legally recognized.

These guidelines have been prepared because there is a need to assure that a balance is maintained between the protection of individual children; the accepted needs of a specific child, group, class of children, or society at large; and societal values in general.

TABLE 11.5 Recommendations from the AAP Statement Before the Institute of Medicine Committee on Clinical Research Involving Children: July 9, 2003

1. Children must be permitted to serve as participants in clinical research so that they may gain from both the personal benefit of participation as well as the benefits that accrue to all children as a group. The alternative of individual practitioners "trying" off-label therapeutic and diagnostic agents outside of the controlled setting of research projects increases the risk for the individual child and slows the acquisition of knowledge that might benefit children as a group.
2. While the AAP supports the necessity at times to conduct research that may involve slightly more than minimal risk, even in the absence of likely personal gain, we find the distinction between the "healthy" child and the child with an illness or condition somewhat artificial and the increment in risk from minimal to a "minor increase above minimal" to be vague. We therefore encourage careful consideration by IRBs of patient personal risk (based in part on prior personal experience with comparable procedures) and conscientious steps to ensure full informed consent and assent.
3. The AAP recommends that the FDA adopt 46.4089(c) of Subpart D (pertaining to waivers of informed consent). Without preempting any local or state law, IRBs should be allowed to give emancipated or mature minors autonomous consent for participation in research protocols.
4. Remuneration of child subjects or their parents should never be so great that it will encourage the parent (or the child) to consent to procedures or place the child at greater risk than they would otherwise do in the absence of such payments. Concerns about remuneration levels are far less important in situations where there is no appreciable risk for participation other than inconvenience. Even in such situations where no significant risk is present, the AAP does not feel that parents should benefit financially from their child's participation in clinical research.
5. The AAP believes that the competence and ethical conduct of the research team is ultimately the most important safeguard for the protection of children in clinical research. Excessive regulatory requirements may discourage participation of potential subjects in appropriate and needed clinical research and increase the risks for children who do participate if excessive regulatory requirements burden IRBs and decrease their ability to evaluate proposals critically. The AAP encourages the academic community to monitor the impact of regulatory requirements on the successful conduct of clinical research, such as recently adopted HIPAA documentation requirements, and to consider new tools such as parent exit surveys that may provide IRBs with useful feedback concerning patient protection.
6. OHRP and FDA should ensure that when protocols are open for public comment, that a vocal minority does not unduly influence decisions regarding the appropriateness of a research project based on the group's desire to promote a particular political or ideological viewpoint.
7. Local IRBs must have proper pediatric expertise and adequate institutional support in order to fulfill the critical role of protecting the rights of child research subjects. Since all research funded by NIH should include pediatric subjects, unless there are specific reasons for their exclusion, then IRBs should require at least one standing member with adequate pediatric expertise.

platform between regulatory agencies and industry associations of Europe, Japan, and the United States, with several further regions represented as observers.[36] Of special interest in the context of this chapter are ICH documents E6 on Good Clinical Practice[37] and E11 on Drug Development in Children.[38]

In 1991, a final "Common Federal Policy for the Protection of Human Subjects" was promulgated as "Common Rule" to integrate and consolidate existing non-DHHS governmental regulations on human subjects. In was accepted by the U.S. Office of Science and Technology as policy and adopted by DHHS and 16 other federal departments and agencies.[39]

In 1997, FDAMA Section 111 introduced the pediatric exclusivity to facilitate pediatric drug research.[40,41] In 1997, the EU Convention on Human Rights and Biomedicine was held in Oviedo, Spain.[42] In Article 1 it emphasized that the convention should "protect the dignity and identity of all human beings and guarantee everyone, without discrimination, respect for their integrity and other rights and fundamental freedoms with regard to the application of biology and medicine."

In 1998, the WMA adopted the World Medical Association Declaration of Ottawa on the Rights of the Child to Health Care.[43] The declaration refers to Article 24 of the 1989 United Nations Convention on the Rights of the Child.[35]

In 2002, the FDAMA was reauthorized as BPCA.[44] In 2003, PREA was introduced as a reintroduction of the pediatric rule that originated from 1998.[45] In 2002, the CIOMS published 21 International Ethical Guidelines for Biomedical Research Involving Human Subjects, among them Guideline 14, "Research Involving Children."[46]

In 2006, the EU **Ethical Considerations for Clinical Trials Performed in Children** was published on the EMEA website[47] for consultation. The consultation period ended in January 2007. It is intended to work in the framework of the EU clinical trials directive,[48] which has been trying since 2001 to harmonize the practice of GCP in all EU countries. The document starts with general undisputed statements, such as that children are not small adults and that as a vulnerable population they need protection against the risks of research, but that this should not lead to denying them the benefits of research. This document continues with a compilation of EU and international documents that regulate clinical research in general and specifically clinical research in children, followed by definitions of key institutions and definitions such as ethics committees, age groups, informed consent, assent, and many more. The main chapters of the EU ethical considerations are given in Table 11.6.

In December 2006, the EU Pediatric Regulation was published in the *EU Official Journal*, with enforcement in January 2007.[49]

In September 2007, both U.S. pediatric legislations BPCA and PREA were reauthorized (see Chapter 12 by Maldonado). In December 2007, the WHO[50] organized an international conference that took place in London, to promote medicines adapted to use in children all over the planet.[51]

KEY DOCUMENTS, PUBLICATIONS, AND INTERNATIONAL ORGANIZATIONS

The Nuremberg Code

The Nuremberg Code is a set of principles for human experimentation that were published in 1947 by the judges who tried German medical doctors for inhumane experiments on adults and children during the Nazi regime. Specifically, the studies performed by Josef Mengele in twin children shocked the international community through their cruelness and complete lack of concern for human and child dignity. The Nuremberg Code consisted of ten points that are summarized in Table 11.1. The Nuremberg Code was never transformed into a legally binding document in American or German law but is broadly recognized as a first internationally binding set of principles that govern studies in humans.

Taken literally, the first principle of the code suggests a prohibition on pediatric research, as children cannot legally give consent to research. This literal interpretation prevailed during the decades following the Nuremberg Code publication and contributed substantially to keeping children out of the scope of biomedical research.

TABLE 11.6 Main Chapters of the EU Ethical Considerations for Clinical Trials Performed in Children

0. Executive Summary
1. Introduction
2. Scope
3. Ethical Principles
4. Legal Context
5. Definitions/Glossary
6. Informed Consent
7. Assent from Children
8. Ethics Committee's Composition in Respect of Paediatric Trials
9. Paediatric Clinical Trial Designs
10. Pain, Distress, and Fear Minimisation
11. Assessment of the Level of Risk
12. Measures of Benefit
13. Assays in Relation to Age/Bodyweight—Blood Sampling
14. Studies in Neonates (Term and Pre-Term)
15. Healthy Children/"Volunteers" Studies
16. Vaccines
17. Paediatric Formulations to Be Used in Paediatric Trials
18. Individual Data Protection
19. Unnecessary Replication of Trials
20. Adverse Reactions and Reporting
21. Inducements Versus Compensation for Children
22. Insurance Issues
23. Trials in Children in Non-EU Countries
24. Ethical Violations, and Noncompliance with GCP
25. Annex 1
26. Annex 2
27. References

The Declaration of Helsinki

The original version of 1964 is short and divided into three chapters on basic principles, clinical research combined with professional care, and nontherapeutic clinical research. It allowed the consent by the legal guardian if the participating person is legally incompetent.

The key points of the Declaration of Helsinki were as follows:

- Clinical research must conform to medical and moral principles and should be based on science.
- Clinical research should be conducted by scientifically qualified personnel.
- Clinical research should be based on careful risk–benefit assessment.
- Participants must be fully informed and freely consent to the research.

- Combination of medical care and clinical research is acceptable to the extent of therapeutic value to the patient.
- Responsibility remains always with the medically qualified person and never with the subject.
- The participant has always the right to withdraw.
- If the investigator sees potential harm for the participant, he/she should discontinue the research.

Since 1964, the Declaration of Helsinki has been revised several times, has become much more lengthy, and has increasingly addressed more and more complex issues such as the need to review any biomedical research project by an institutional review board (in the United States)/ethics committee (other countries), the use of placebo, research in developing versus developed countries, and many more attributes. As of November 2008, six revisions and two notes clarifications have been published by the WMA and express the growing maturity of biomedical research and its increasing complexity within the divergence of a multitude of different interests and situations in a world that has increasingly become global and interlinked.[55]

Henry K. Beecher: Ethics and Clinical Research, 1966

In 1996, the Harvard anesthesiologist Henry K. Beecher listed 22 experiments performed on humans, including several studies performed on children, by U.S. academic institutions that were objectionable ethically. It included, for example, one case of transplantation of malignant melanoma from a daughter to the mother (the daughter died one day, the mother 451 days after the transplantation). In almost all 22 studies, the test subjects were institutionalized or were in other situations that compromised their ability to give free consent, such as being newborns, very elderly, terminally ill, mentally disabled children, soldiers, or charity patients in a hospital. Beecher had first submitted this paper to the *Journal of the American Medical Association* (JAMA), which rejected it. Finally, the *New England Journal of Medicine* (NEJM) accepted the manuscript. Example 16 in his paper describes the artificial induction of hepatitis in an institution for mentally defective children. Beecher's paper, to this day, is frequently quoted as a keystone publication that helped to turn the tide in the need to protect vulnerable patients.

The Belmont Report: Ethical Principles and Guidelines for the Protection of Human Subjects of Research

Published in 1978 by the U.S. National Commission for the Protection of Human Subjects of Biomedical and Behavioral Research, this report identified three basic ethical principles as particularly relevant to the ethics of research on human subjects: the respect for persons, beneficence, and distributive justice.

- *Respect for Persons*: Individuals should be treated as autonomous agents, and persons with diminished autonomy are entitled to protection.
- *Beneficence*: Persons should be treated in an ethical manner not only by respecting their decisions and protecting them from harm, but also by making efforts to secure their well-being. Specifically, research involving children is mentioned under this

principle: "A difficult ethical problem remains, for example, about research that presents more than minimal risk without immediate prospect of direct benefit to the children involved. Some have argued that such research is inadmissible, while others have pointed out that this limit would rule out much research promising great benefit to children in the future. Here again, as with all hard cases, the different claims covered by the principle of beneficence may come into conflict and force difficult choices."

- *Justice*: The benefits of research and its burdens should be addressed in a sense of "fairness in distribution" or "what is deserved." An injustice occurs when some benefit to which a person is entitled is denied without good reason or when some burden is imposed unduly.

In the chapter "Applications," the Belmont Report further discusses the principles of (1) informed consent, (2) assessment of risks and benefits, and (3) selection of subjects. Furthermore, in the final notes the commission declined to make any policy determination on social experimentation and expresses its belief that this problem ought to be addressed by one of its successor bodies.

It should be mentioned here that the three main principles of the Belmont Report have undergone a change in their meaning over the 30 years following their publication. A generation ago, beneficence was still perceived as characterizing the physician's personal characteristics, while today it is more seen as the right to have the best choice within a vast array of therapeutic options. The concept of respect for persons has changed toward patients' insistence to be treated like autonomous persons who, for example, have the right to hear the truth about their medical conditions—something that a generation ago was not expected from the physician.[51]

ICH

The International Conference on Harmonisation (ICH) of Technical Requirements for Registration of Pharmaceuticals for Human Use brings together the regulatory authorities of Europe, Japan, and the United States and experts from the pharmaceutical industry in the three regions to discuss scientific and technical aspects of product registration.[52]

The purpose is to make recommendations on ways to achieve harmonization of technical guidelines and requirements for product registration in order to reduce the need to duplicate work. The objective of such harmonization is a more economical use of human, animal, and material resources, and the elimination of unnecessary delay in the global development and availability of new medicines while maintaining safeguards on quality, safety, efficacy, and regulatory obligations to protect public health. It was at the WHO Conference of Drug Regulatory Authorities (ICDRA), in Paris, in 1989, that specific plans for action began to materialize. Soon afterwards, the authorities approached IFPMA to discuss a joint regulatory–industry initiative on international harmonization, and ICH was conceived. The birth of ICH took place at a meeting in April 1990, hosted by the EFPIA in Brussels. Representatives of the regulatory agencies and industry associations of Europe, Japan, and the United States met, primarily to plan an international conference, but the meeting also discussed the wider implications and terms of reference of ICH. The ICH Steering Committee, which was established at that meeting, has since met at least twice a year, with the location rotating between the three regions.

- In 1996, ICH E6 Guideline on Good Clinical Practice (GCP) was agreed upon.
- In 2000, ICH E11, Clinical Investigation of Medicinal Products in the Pediatric Population, was finalized, adopted by the three agencies EMEA, FDA, and MHLW, and consequently published on their respective websites.

The WMA Declaration of Ottawa

Interestingly, the Declaration of Ottawa emphasized children's "inherent right to life, as well as the right of access to the appropriate facilities for health promotion, the prevention and treatment of illness and the rehabilitation of health." However, at no point does it address the question of where this knowledge comes from. The more the need for clinical and pharmaceutical research in children is discussed in the international literature, the more medical science has to accept that former beliefs in well-founded and eternal knowledge of childhood diseases and their required therapies are in constant evolution—a process that includes continuous revision of dogma that appeared to be written in stone a few decades ago. In other words, the Declaration of Ottawa does not reflect on the origin of therapeutic and medical knowledge. In this sense, the Declaration of Ottawa reflects pediatric care at least until the middle of the 20th century, that is, the deeply rooted conviction that basically medical knowledge of children's therapy was sufficient in the developed country and constituted a well-defined body of wisdom and understanding, where the only problem was a deficiency in the distribution of knowledge.

The only sentence in the entire Declaration of Ottawa that might be interpreted in the sense discussed above would be No. 10 of its principles "to eradicate traditional practices prejudicial to the health of the child," a formulation that probably aimed at malpractices such as routine female circumcision in African and predominantly Muslim cultures. This principle, however, could equally be used to describe the need of continuous revision of medical pediatric practice in the developed world. Examples of successful reviews would be today's right of parents to visit their child in the hospital at any time, in contrast to the exaggerated discipline that dominated pediatric wards a few decades ago, when parents were allowed to see their children at predefined time slots only, irrespective of whether their work schedule allowed visits during these time windows or not. Exceptions were always possible, especially for parents with higher social status. A second example would be the right of children to tasteful nourishment in the hospital, an issue that in many countries is often hampered by institutional barriers that prevent healthy and tasteful food in pediatric as well as adult hospital wards.

Council for International Organizations of Medical Sciences (CIOMS)

CIOMS[53] is an international, nongovernmental, nonprofit organization established jointly by the WHO and UNESCO in 1949. In 2003, the membership of CIOMS included 48 international member organizations, representing many of the biomedical disciplines, and 18 national members, mainly representing national academies of sciences and medical research councils. The main objectives of CIOMS are the following:

- To facilitate and promote international activities in the field of biomedical sciences.
- To maintain collaborative relations with the UN, WHO, and UNESCO.
- To serve the scientific interests of the international biomedical community in general.

The list of CIOMS members is an impressive collection of 16 international members (e.g., the World Medical Association or the International Pediatric Association); of 16 national members (i.e., the national scientific associations of 16 countries from Belgium to Switzerland); and of 25 associated members including the American College of Chest Physicians and the European Forum of Good Clinical Practice (EFGCP).

Among the 21 International Ethical Guidelines for Biomedical Research Involving Human Subjects, their 2002 version of Guideline 14, "Research Involving Children," should be mentioned here. The guideline is short and to the point:

> Before undertaking research involving children, the investigator must ensure that:
>
> - the research might not equally well be carried out with adults;
> - the purpose of the research is to obtain knowledge relevant to the health needs of children;
> - a parent or legal representative of each child has given permission;
> - the agreement (assent) of each child has been obtained to the extent of the child's capabilities; and
> - a child's refusal to participate or continue in the research will be respected.

CIOMS Guideline 14 is then followed by several pages of commentary that are not quoted here.

IFPMA

The International Federation of Pharmaceutical Manufacturers and Associations is a nonprofit, nongovernmental organization (NGO) representing industry associations and companies from both developed and developing countries. Member companies of the IFPMA are research-based pharmaceutical, biotech, and vaccine companies.

In the research and development pipeline, the pharmaceutical industry is working on more than 700 new medicines and vaccines for infectious diseases including HIV/AIDS, cancer, heart disease, and stroke, and diseases that disproportionately affect women, such as osteoporosis.

The main objectives of IFPMA are the following:

- To encourage a global policy environment that is conducive to innovation in medicine, both therapeutic and preventative, for the benefit of patients around the world.
- To contribute industry expertise and foster collaborative relationships and partnerships with international organizations, national institutions, governments, and nongovernmental organizations that are dedicated to the improvement of public health, especially in developing and emerging countries.
- To assure regular contact and experience sharing and coordinate the efforts of its members toward the realization of the above objectives.

CONCLUSION

It is a recent phenomenon that children are perceived by society as entities with principally the same rights as adults. A huge challenge is to balance the needs of children to be protected against the patronizing attitudes of institutions and religious and secular authorities. In the past, children have often survived and overcome treatments that in hindsight were not very pleasant. Children's vitality has in the long term often exceeded the power of institutionalized and encrusted tradition. However, the price to be paid in the form of suffering and lost lives was always too high, and it is our responsibility and that of future generations to help to reduce children's suffering as far as possible.

A patronizing attitude of research toward human subjects in general and specifically children has led to research practices that in today's view are ethically questionable. At the time of their occurrence, however, it required courage to address these issues, and changes evolved over decades, not months. Today, an international framework is evolving that step by step further shapes the place of children in society and research.[54] The pendulum is swinging away from protecting children *against* research toward protecting their health *through* research. Part of the background of this movement is of course the explosion of scientific understanding of adult and pediatric diseases with a first prospective of potential healing. The increase of survival rate of pediatric cancer patients from almost zero to around 80% today is a good example. As we have no world government, this process is not linear. As is human nature, the process becomes more and more complex the more details are addressed on a global level. The original Declaration of Helsinki was a rather short document that was without doubt intended to allow some flexibility in its interpretation. With the following years, it has become longer and more elaborate in order to address the complexity of today's societies. Ethics constitute the body of convictions and assumptions that influence research. Research with children intends to improve the therapeutic armamentarium of pediatricians, physicians, parents, and all child healthcare givers to stabilize and improve children's health and to reduce their suffering when they are ill.

REFERENCES

1. Golombek S, van den Anker J, Rose K, et al. Clinical trials in children—ethical and practical issues. *Int J Pharm Med.* 2007;21(2):121–129.
2. The Belmont Report. Ethical principles and guidelines for the protection of human subjects of research. Available at http://ohsr.od.nih.gov/guidelines/belmont.html.
3. Ethical considerations for clinical trials performed in children. Available at http://ec.europa.eu/enterprise/pharmaceuticals/paediatrics/docs/paeds_ethics_consultation20060929.pdf.
4. Lederer SE, Grodin MA. Historical overview: pediatric experimentation. In: Grodin MA, Glanth LH, eds. *Children as Research Subjects.* New York: Oxford University Press; 1994: 3–25.
5. Grodin MA, Glanth LH, eds. *Children as Research Subjects.* New York: Oxford University Press; 1994.
6. Field MJ, Behrman RW, eds. *Ethical Conduct of Clinical Research Involving Children.* Washington DC: Institute of Medicine of the National Academies, The National Academies Press; 2004.
7. Kodish E, ed. *Ethics and Research with Children.* New York: Oxford University Press; 2005.
8. Rose K, van den Anker JN, eds. *Guide to Paediatric Clinical Research.* Basel, Switzerland: Karger; 2007.
9. Stoetter H. Paediatric drug development—historical background of regulatory initiatives. In: Rose K, van den Anker N, eds. *Guide to Paediatric Clinical Research.* Basel, Switzerland: Karger; 2007: 25–32.
10. Rose, K. Pediatric drug deveopment. Implementation of pediatric aspects into the general drug development process. Applied Clinical Trials 2005 (1). Available at http://actmagazine.

findpharma.com/appliedclinicaltrials/Clinical + Trials/Pediatric-Drug-Development/ArticleStandard/Article/detail/140819.

11. *British Encyclopedia*. 1962. Children's health & welfare.
12. Weyers W. *The Abuse of Man. An Illustrated History of Dubious Medical Experimentation*. New York: Ardor Scribendi; 2003:161–235.
13. Ross LF. *Children in Medical Research: Access Versus Protection*. New York: Oxford University Press; 2006.
14. Mozes-Kor E. The Mengele twins and human experimentation. In: Annas G, Grodin M, eds. *The Nazi Doctors and the Nuremberg Code*. New York: Oxford University Press; 1995.
15. http://ohsr.od.nih.gov/guidelines/nuremberg.html.
16. http://en.wikipedia.org/wiki/Nuremberg_Code.
17. WMA 1948 Declaration of Geneva: International Code of Medical Ethics. Available at http://www.donoharm.org.uk/gendecl.htm.
18. WMA 1949 International Code of Medical Ethics. Available at http://www.cirp.org/library/ethics/intlcode/.
19. UN Universal Declaration of Human Rights, proclaimed 1948. Available at http://www.un.org/Overview/rights.html.
20. http://www.cirp.org/library/ethics/UN-declaration/.
21. http://www.pubmedcentral.nih.gov/picrender.fcgi?artid=1816102&blobtype=pdf.
22. Wikipedia. Declaration of Helsinki. Available at http://en.wikipedia.org/wiki/Declaration_of_Helsinki.
23. Beecher HK. Ethics and clinical research. *N Engl. J Med.* 1966;274(24):1354–1360.
24. Krugman S. The Willowbrook hepatitis studies revisited: ethical aspects. *Rev Infect Dis.* 1986;8:157–162.
25. Jones JH. *Bad Blood. The Tuskegee Syphilis Experiment*. New York: The Free Press, 1981.
26. http://www.pbs.org/newshour/bb/health/may97/tuskegee_5-16.html.
27. http://www.hhs.gov/ohrp/45CFRpt46faq.html.
28. Field MJ, Behrman RW, eds. *Ethical Conduct of Clinical Research Involving Children*. Washington DC: Institute of Medicine of the National Academies, The National Academies Press; 2004: 52–53. The report is available on microfiche, ERIC # ED146763, but not online. See http://www.eric.ed.gov/ERICWebPortal/custom/portlets/recordDetails/detailmini.jsp?_nfpb=true&_&ERICExtSearch_SearchValue_0=ED146763&ERICExtSearch_SearchType_0=no&accno=ED146763.
29. http://www.hhs.gov/ohrp/humansubjects/guidance/45cfr46.htm#subpartd.
30. http://www.hhs.gov/ohrp/humansubjects/guidance/45cfr46.htm#subpartb.
31. http://www.hhs.gov/ohrp/humansubjects/guidance/45cfr46.htm#subpartc.
32. American Academy of Pediatrics (AAP). Guidelines for the ethical conduct of studies to evaluate drugs in pediatric populations. *Pediatrics*. 1977; 60(1):91–101. Available at http://pediatrics.aappublications.org/cgi/reprint/60/1/91.
33. AAP. Guidelines for the ethical conduct of studies to evaluate drugs in pediatric populations. *Pediatrics*. 1995, 95(2):286–294. Available at http://pediatrics.aappublications.org/cgi/reprint/95/2/286.
34. AAP Statement before the Institute of Medicine Committee on Clinical Research Involving Children; July 9, 2003. Available at http://www.iom.edu/Object.File/Master/13/760/AAP%20Statement.pdf.
35. 1989 UN Convention on the Rights of the Child. Available at http://www.cirp.org/library/ethics/UN-convention/.

36. http://www.ich.org/cache/compo/276-254-1.html.
37. http://www.ich.org/LOB/media/MEDIA482.pdf.
38. http://www.ich.org/LOB/media/MEDIA487.pdf.
39. Common Rule. Available at http://www.hhs.gov/ohrp/policy/#common.
40. http://www.fda.gov/cder/fdama/sections.htm.
41. http://www.fda.gov/cder/guidance/2891fnl.htm.
42. EU Convention on Human Rights and Biomedicine, Oviedo, Spain, 1997 Available at http://www.worldcarecouncil.org/pdf/CoE_HRBioMed.pdf.
43. 1998 WMA Declaration of Ottawa. Available at http://www.wma.net/e/policy/c4.htm.
44. Best Pharmaceuticals for Children Act [online]. Available at http://www.fda.gov/cder/pediatric/PL107–109.pdf [Accessed 28 November 2008].
45. Best Pharmaceuticals for Children Act [online]. Available at http://www.fda.gov/cder/pediatric/PL107–109.pdf [Accessed 28 November 2008].
46. CIOMS. International Ethical Guidelines for Biomedical Research Involving Human Subjects. Most updated version 2002. Available at http://www.cioms.ch/frame_guidelines_nov_2002.htm.
47. http://ec.europa.eu/enterprise/pharmaceuticals/paediatrics/docs/paeds_ethics_consultation 20060929.pdf.
48. EU Clinical Trials Directive. Available at http://www.eortc.be/Services/Doc/clinical-EU-directive-04-April-01.pdf.
49. http://ec.europa.eu/enterprise/pharmaceuticals/eudralex/vol-1/reg_2006_1901/reg_2006_1901_en.pdf.
50. WHO. Available at http://www.who.int/childmedicines/en/index.html.
51. Cassell EJ. The principles of the Belmont Report: How have respect for persons, beneficence, and justice been applied in clinical medicine? In: Childress JF, et al. eds. *Belmont Revisited.* Washington DC: Georgetown University Press; 2005.
52. www.ich.org.
53. http://www.cioms.ch/.
54. Neubauer D, Laitinen-Parkkon P, Matthys D. et al. Ethical challenges of clinical research in children. In: Rose K, van den Anker JN, eds. *Guide to Paediatric Clinical Research.* Basel, Switzerland: Karger; 2007.
55. http://wma.net/e/policy/b3.htm.

PART III

REGULATORY GUIDELINES FOR PEDIATRIC DRUG DEVELOPMENT: STIMULATION OF PEDIATRIC DRUG RESEARCH BY REGULATORY AUTHORITIES

CHAPTER 12

United States Perspective

SAMUEL MALDONADO, MD, MPH, FAAP

Vice President, Pediatric Drug Development Center of Excellence, Johnson & Johnson PRD, 920 Route 202 South, Raritan, New Jersey 08869

For nearly four decades the importance of including children in clinical research and meeting the challenges posed by including children in trials testing pharmaceuticals has been part of a public debate led chiefly by those advocating for children's improved access to healthcare. These advocates have clearly articulated how exclusion of children from this process has caused harm. The failure to involve children in this research has resulted in catastrophic misdosing of children based on estimating doses in addition to withholding important therapies in children because the paucity of information does not allow accurate assessment of risk/benefit. In recent years, government and regulatory authorities have created the appropriate infrastructure through effective legislation that mandates as well as provides incentives for industry to include children in clinical development of new products. The most recent examples of these laws in the United States are the Best Pharmaceutical Act for Children (BPCA, better known as "pediatric exclusivity") and the Pediatric Research Equity Act (PREA, also known as "the pediatric rule"). Similar laws have been approved in Europe and Canada and are being explored in Japan, subjects explored extensively in Chapter 13 by Rose and in Chapter 14 by Nakamura and Ono in this book. Based on these legislative actions and the sound scientific rationale and ethics underpinning them, it is clear that involvement of children in clinical development of pharmaceuticals is, and will continue to be, the standard practice in the industry.

Although considered the standard of practice, it can be expected that there will be great challenges over the next decade in meeting the expectations associated with these practices. Including children in clinical trials requires an appropriate investment of resources. We will need to involve experts, both internal and external to companies, who understand the special medical needs of children, experienced clinical operations personnel and investigators who care for children, and appropriate financial resources to carry out pediatric programs.

In the United States, before the initial passage of BPCA, most drugs were not studied in children. However, since the initial approval of the law, there has been an unprecedented surge in the number of pediatric studies. According to a report from the Pharmaceutical

Pediatric Drug Development: Concepts and Applications
Edited by Andrew E. Mulberg, Steven A. Silber, and John N. van den Anker

Research and Manufacturers of America (PhRMA) published in 2007, there are more than 2000 medicines in testing to meet the needs of children.[1] Most pediatric studies have been done in response to U.S. Food and Drug Administration (FDA) issued Written Requests, which is the regulatory mechanism use for the implementation of BPCA. The FDA also has another regulatory and legal mechanism under PREA, but the FDA has not invoked it as often as it has BPCA. BPCA is a voluntary mechanism by which a sponsor submits to the FDA a Proposed Pediatric Study Request (PPSR). The FDA typically modifies the PPSR and turns it around to the sponsor in the form of a Written Request (WR). If the sponsor would like to receive pediatric exclusivity, that sponsor has to reasonably respond to all the requests from the FDA. If the sponsor does not respond to the request or the response is not in accordance with the WR, exclusivity is denied. If the sponsor complies with the request, the FDA makes a determination of granting exclusivity. The value of the pediatric exclusivity is such that it applies to the active moiety in whatever formulation the moiety is present; this includes the formulations for adult patients. Therefore, the exclusivity can be very valuable financially for some drugs. However, there is a wide spread in the financial value of exclusivity. For a minority of drugs, the incentive appears to be disproportionately high but for the majority of drugs it is either commensurate to the investment or may even fall short of the investment made.[2,3] These differences notwithstanding, the incentives created by the government have been able to do exactly what they were intended to do: stimulate pediatric drug development and address the needs of vulnerable population, including preterm neonates and children.

As industry experts and regulators learn more about pediatric drug development and understand some of the limitations and shortcomings of previously conducted clinical trials in children, the need to create more comprehensive pediatric programs is evident. Over time, the FDA has been increasing its demands on the number of studies, the number of patients, and the complexity of the studies they would like to see in response to a Written Request. Assuming that the merits of the FDA demands are scientifically appropriate, the sponsor should expect a more significant investment of resources into the respective pediatric programs. This trend is already obvious to researchers in industry who have done pediatric drug development in the last 10 years. The March/April 2007 Impact Report from Tufts University quantifies the increasing cost of doing pediatric drug development due to the increase in the complexity of such programs. According to this report, the study cost increased 8-fold since 2000.[4] For further information on pharmaceutical economics of pediatric drug development, please review Chapters 4 and 5 by Milne in this book.

When these two pieces of legislation were first enacted as laws in 1997, most sponsors targeted for pediatric studies drug products whose patents were going to expire in the near future. However, as many of those programs completed and appropriate changes were made to their respective labels, the challenge is now to do studies earlier in the life cycle of new drug products. The goal would be to either eliminate or shorten the time that the drug products are in the market without appropriate labeling and information for use in children. The current laws have shortened the time interval between the original New Drug Application (NDA) approval and the supplemental pediatric NDA. Before these laws, the median time between original NDA and pediatric supplement was 7.2 years. After these laws were enacted, the median time decreased to 4.2 years.[5] It is expected that if the trend continues, this time lag will decrease even further. However, as this gap closes, patients, sponsors, and regulators are faced with increased uncertainty and should be willing to accept greater risk in exchange to earlier access to drugs when appropriate. Once the drug is first approved, sponsors and regulators begin to learn the potentials and limitations of the drug products in the "real world." Therefore, initiating pediatric programs early after the first approval of a drug or even before the approval will demand from all researchers a more careful and deliberate action plan. This action plan may also be slower in order to provide all those involved time to analyze emerging data. Therefore, pediatric programs may take longer to be completed than the time they took 10 or 5 years ago. This time factor will also have an impact on the overall resources dedicated to pediatric drug development. It is also possible that in our desire to close this gap, some drugs that will eventually be withdrawn from the market, either by the FDA or sponsors for safety reasons, will already have an ongoing pediatric program or perhaps approved labeling for children. This is a trade-off that should be carefully considered for each drug product.

As Europe, the United States, and other regions of the world continue to create incentives for pediatric drug development, it is expected that the future of pediatric drug development will be even brighter than in the decade since the first U.S. pediatric legislation was introduced. However, competing demands from regulatory bodies around the world will necessitate careful execution of pediatric drug development programs to avoid redundancy and optimize results. Fortunately, as it pertains to Europe, the United States, and Japan, the regulatory framework under the International Conference on Harmonisation provides the necessary platform for such optimization. It is recognized that cultural differences and differences in standard of care between the regions will continue to exist. Muntañola, in Chapter 35 of this book, explores the geographic differences and the impact on operational success and failure of pediatric clinical trials. Therefore, expectations from patients, clinical investigators, practicing clinicians, and regulators may be different across the different regions. However, for drug development in the pediatric population, it would be desirable to create mechanisms to avoid duplication of efforts given the vulnerability, relative scarcity of subjects, and ethical concerns inherent to this population. This harmonization should facilitate the timely execution of pediatric drug development plans by the pharmaceutical industry. This pediatric drug development plan should serve the needs of all regulatory bodies simultaneously.[6] This is of particular concern, since the ethics of performing unnecessary clinical trials and the mandate of acceptability of differences in clinical trial endpoints has already been faced by teams. Please see Chapter 48 (Topiramate Case Study) by Ness, Merriman, and Nye in this book.

The incentives that legislators created in Europe were modeled after the significant success of BPCA in the United States. European legislators have made this law a permanent law. This does not preclude them from optimizing the law in the future if they see the need to do so. Unfortunately, in the United States, the law continues to have a sunset clause that requires a debate and revision of the law every 5 years. The original intent of the sunset clause was laudable. Because many attempts to stimulate pediatric drug development had failed in the previous two decades, the U.S. Congress was not sure whether or not BPCA was going to be the answer they were looking for. Fortunately, for children, this experiment paid off. In 2002, the law was renewed with another sunset clause that expired on September 30, 2007. After 10 years and two reevaluations, it is abundantly clear that BPCA and PREA have not only achieved their intended goals but exceeded them, and millions of sick children and their families have already benefited as a result. In 2012, at the time of the next reevaluation, Congress should consider removing the sunset clauses on BPCA and PREA. By removing the sunset clauses, Congress will remove the uncertainties created every 5 years and encourage the creation of a more sustainable infrastructure for pediatric drug development. Even despite all of the successes of BPCA and PREA in stimulating participation in pediatric drug development across companies of all sizes, the sunset clauses in them remain major hindrances, discouraging companies from formally organizing pediatric infrastructures.

In the absence of a consistent and predictable exclusivity provision, there will remain a considerable and understandable reluctance among companies with countless competing research priorities to devote dedicated resources to formal pediatric divisions. This is especially true as the cost, size, number, and complexity of pediatric studies have increased and the absolute value of the pediatric exclusivity has decreased. By removing the sunset clauses, Congress will convey a powerful message: pediatric drug development is here to stay, and drug safety and effectiveness for children is firmly among the nation's highest priorities. The sunset clauses' removal will also help the advocates of pediatric drug development in industry to encourage their respective institutions to create and sustain the necessary infrastructure to continue improving pediatric therapeutics.[7]

We have made significant and unprecedented progress in the last 10 years in pediatric drug development. However, we still face many challenges for closing the knowledge gap between therapeutics in adults and children. The combined efforts of clinical investigators, drug sponsors, regulators, and government legislators should build a stronger and more sustainable infrastructure that would yield even greater benefits for children in the future.

REFERENCES

1. Tauzin B. More than 200 medicines are in testing to meet the needs of children. In: *PhRMA Report on Medicines in Development for Children*. 2007.
2. Li JS, Eisenstein EL, Grabowski, HG, et al. Economic return of clinical trials performed under the pediatric exclusivity program. *JAMA* 2007;297:480–488.
3. GAO Report to Congressional Committees, March 2007: Studies Conducted Under Best Pharmaceutical for Children Act. GAO-07-557.
4. Pediatric study costs increased 8-fold since 2000 as complexity level grew. Tufts CSDD Impact Report 2007: 9(2):1–4.
5. Schachter AD, Ramoni MF. Paediatric drug development. *Nat Rev*. 2007;6:429–430.
6. Maldonado S, van den Anker JN, Rose K. Conclusions: pediatric drug development in a global context. In: *Guide to Pediatric Clinical Research*. Basel, Switzerland: Karger; 2007: 133–134.
7. Maldonado SD. Testimony before the Senate Health, Education, Labor, and Pensions Committee on March 27, 2007. Available at http://help.senate.gov/Hearings/2007_03_27_b/Maldonado.pdf.

CHAPTER 13

European Perspective

KLAUS ROSE, MD, MS
Head Pediatrics, F. Hoffmann-La Roche Ltd., CH-4070 Basel, Switzerland

INTRODUCTION

Pharmaceutical treatment of children was and is in Europe comparable to the situation in the United States and Canada before the introduction of U.S. pediatric legislation in 1997.[1–3] Numerous publications show that the use of inadequately tested drugs in children ranges from 10% by general practitioners up to 95% in neonatal intensive care.[4–8]

Several issues are underlying as causes of this problem, including skepticism against research and clinical research in children, the small size of the pediatric market as compared with the adult market, and the fact that clinical research itself was a rather young discipline that even for adults had to evolve its rules during the second half of the 20th century.[9,10] In all of them, Europe is comparable to the United States.

The discussion about the need to better adapt pharmaceutical treatment to the specific characteristics of children took place in Europe on the level of individual national countries in the 1980s and 1990s. However, this did not lead to any substantial initiatives. There was, of course, less visible lobbying predominantly by pediatricians and pediatric pharmacologists. A key role was played by the European Society of Developmental Perinatal and Pediatric Pharmacology (ESDP) and a number of other academic and scientific societies dedicated to the promotion of research in children. The debate started to have a European dimension 1997 with a round-table discussion organized by the EU Commission at the EMEA.[11] Due to the different organization and complexity of Europe, the debate took a much longer time than in the United States for resolution of a final plan. Early in the 21st century, the EMEA Committee for Medicinal Products for Human Use (CHMP) established a pediatric expert group (PEG) with the remit to coordinate necessary actions and to advise the EMEA and its organs on development and use of drugs in children. Finally, in December 2006, the European Regulation was published in the *EU Official Journal*, entering into force on January 26, 2007.[12,13] In contrast to other legal entities (e.g., a directive), in the EU a regulation is immediately in force 30 days after its

Pediatric Drug Development: Concepts and Applications
Edited by Andrew E. Mulberg, Steven A. Silber, and John N. van den Anker

publication in all EU member states and does not require a transformation into the multitude of national laws.

After a decade of preparation, the EU regulation helped to put children on the radar screen of many institutions involved in healthcare, including clinicians, regulatory authorities, pharmaceutical companies, insurance companies, parent and patient advocacy groups, and other stakeholders. Furthermore, considerable detailed information is published on the EMEA website[14] and other official EU websites.[15–17]

In the following, the key elements of the EU pediatric regulation are outlined, followed by a short overview over the document itself and a preliminary evaluation of its strengths and weaknesses. In order to avoid converting this chapter into an EU pediatric regulatory affairs handbook, the last section is a guide through the EMEA pediatric website. For in-depth details, the reader should consult the website itself, especially as its contents are reviewed and updated on a regular basis

KEY ELEMENTS OF THE EU PEDIATRIC REGULATION

The two separate U.S. pediatric legislations, FDAMA/BPCA and PREA, are often described as the carrot[18] and the stick.[19] (These issues are discussed more fully in Chapter 12 by Maldonado in this book.) Using this nomenclature, the EU pediatric regulation can be described similarly as a combination of carrot and stick, as it combines mandatory consideration of the use of drugs in children with a reward in the form of 6 months of added market exclusivity in the form of supplementary protection certificate (SPC) prolongation. In EU countries, an SPC is a unique, patent-like, intellectual property right, granted for drugs in compensation for the long development time.[20] A key institution formed by the EU pediatric regulation is the EMEA Pediatric Committee (officially abbreviated as PDCO), which was constituted in July 2007. It is composed of one representative for each EU member state (there are 27 member states in 2008) plus one alternate for each state representative. Furthermore, three representatives from healthcare professions and three representatives from patient advocacy groups became PDCO members during 2008.[21, 22] With the enactment of the EU pediatric regulation, the former EMEA/CHMP Pediatric Expert Group is continued by the PDCO.[23] The PDCO decides about Pediatric Investigation Plans (PIPs), waivers, and deferrals. The composition of the PDCO members aims at representing EU scientific, regulatory, and healthcare professional knowledge of all aspects of pediatric drug development. The pharmaceutical industry is not represented in the PDCO, but regular meetings are arranged with a group of industry regulatory affairs associates through the EU pharmaceutical industry's trade association, EFPIA.[24, 25]

As of July 26, 2008, condition for registering a new drug is pediatric data based on an agreed upon PIP ("the results of all studies performed and details of all information collected in compliance with an agreed pediatric investigation plan"), or a PDCO-approved waiver, or a PDCO-approved deferral. In other words, the EU regulation regards as standard at submission availability of pediatric data reflecting an entire preclinical and clinical development, unless the EMEA has granted a waiver or a deferral before (Article 7). The PIP has to be submitted no later than the end of pharmacokinetic studies in the development program for an individual new chemical entity. As of January 26, 2009, this requirement applies also to submission of a new indication, new pharmaceutical forms, or new routes of administration of already registered drugs, if they are patent protected or protected by an EU supplementary protection certificate (SPC) (Article 8).[26–28]

Compliance with the PIP will be rewarded by 6 months of additional market exclusivity in the form of a prolongation of the SPC.[29] To allow generic industry to prepare for this SPC prolongation, the data generated in compliance with the PIP must be submitted to the EMEA 2 years before SPC expiry. However, until 2012, as a transition measure, the PIP data can be submitted 6 months before SPC expiry.

With the entry of the EU pediatric regulation into force, the EMEA gives free scientific advice for pediatric development questions.[30, 31] Key elements of submitted PIPs are made public on the EMEA website after deletion of commercially confidential data. Details and results of clinical trials performed in children are made public through the EU clinical trials database (EudraCT) after removal of commercially confidential data. The pediatric regulation also enforces existing pharmacovigilance activity in the EU and requires risk-management plans where appropriate.[32]

For orphan drugs, the 10 years of market protection is extended to 12 years if a pediatric development is performed (Article 37). For off-patent drugs, a special pediatric use marketing authorization (PUMA) is introduced. Drugs with a specific pediatric authorization are rewarded with 8 years of data protection and 10 years of market protection (Article 38).[33, 34] All already existing pediatric clinical data on marketed drugs had to be submitted to the EU national competent authorities or to the EMEA before January 26, 2008 (Article 45).

EU PEDIATRIC REGULATION: THE DOCUMENT

Published on December 27, 2006 in the *EU Official Journal,* the English version is 19-page document consisting of a 38-paragraph preamble and 57 articles. The key elements have been given earlier, while additional detailed information will be given later in the guide through the EMEA pediatric website. A critical evaluation of the EU pediatric regulation, a guide for pharmaceutical companies on how to initiate PIP development, and information on how to translate PIPs into practical drug development practice will be given further later, after discussion of the EMEA pediatric website.

A definitive structure and table of contents for the EU pediatric regulation is delineated in Table 13.1.

TABLE 13.1 EU Pediatric Regulation: Structure/Table of Contents

Preamble		(1) – (38)
Title I	Introductory Provisions	
	Chapter 1: Subject Matter and Definitions	Articles 1 and 2
	Chapter 2: Pediatric Committee	Articles 3–6
Title II	Marketing Authorization Requirements	
	Chapter 1: General Authorization Requirements	Articles 7–10
	Chapter 2: Waivers	Articles 11–14
	Chapter 3: Pediatric Investigation Plan	
	• Section 1: Requests for Agreement	Articles 15–19

(*continued*)

TABLE 13.1 (*Continued*)

	• Section 2: Deferrals	Articles 20 and 21
	• Section 3: Modification of a PIP	Article 22
	• Section 4: Compliance with the PIP	Articles 23 and 24
	Chapter 4: Procedure	Articles 25
	Chapter 5: Miscellaneous Provisions	
Title III	Marketing Authorization Procedures	
	Chapter 1: Within the Scope of Articles 7 and 8	Articles 28 and 29
	Chapter 2: Pediatric Use of Marketing Authorization	Articles 30 and 31
	Chapter 3: Identification	Articles 32
Title IV	Post-Authorization Requirements	Articles 33–35
Title V	Rewards and Incentives	Articles 36 – 40
Title VI	Communication and Coordination	Articles 41 – 46
Title VII	General and Final Provisions	
	Chapter 1: General Provisions	
	• Section 1: Fees, Community Funding, Penalties and Reports	Articles 47 – 50
	• Section 2: Standing Committee	Article 51
	Chapter 2: Amendments	Articles 52–57

SCOPE AND LIMITATIONS OF U.S. AND EU PEDIATRIC LEGISLATIONS

Neither the U.S.-based nor the EU-based pediatric regulations change fundamentally the scope of drug development, which over the last five decades has become almost exclusively the domain of the pharmaceutical industry. Pharmaceutical companies continue and will continue to develop new medications for unmet medical needs including specific medications developed for children. In many cases, there was frequent use in children, for example, antibiotics in infectious diseases, Cerezyme for Gaucher disease, Fabrazyme for Fabry disease, pulmozyme (dornase alpha) for cystic fibrosis,[35] and lung surfactant in preterm newborns.[36] The EU regulation, like the U.S. pediatric legislation, only enforces the consideration of pediatric use of medications that are primarily developed for adult patients. In all probability neither the U.S. nor the EU pediatric legislation will lead to the development of new medications for diseases that occur exclusively in children. The U.S. legislation has led to the investigation of adult medications in rare child diseases,[37] and hopefully the EU regulation will have comparable outcomes in areas not covered by the U.S. legislation, specifically biological agents. Furthermore, in both the United States and Europe, orphan disease legislation have successfully stimulated research in rare diseases.[38,39]

A FIRST APPRAISAL OF THE EU PEDIATRIC REGULATION

While in the United States the *voluntary* pediatric legislation has been a success in the view of the FDA, the *mandatory* development was not in force for many years so the FDA only started collecting experience in mandatory development in children since 2003. While discussion and implementation of a pediatric plan is strongly encouraged by the FDA in specific new chemical entity (NCE) development, the pharmaceutical company developing the NCE is the one who decides when to approach the FDA in pediatric issues. The EU regulation goes a step further to promulgate a more definitive and detailed pediatric development program.

For marketed drugs, PIPs must be submitted for new indications, new routes of administrations, and new formulations. If a company wants to register a new formulation, it will have to submit a PIP that covers all licensed indications. For each indication, the company has to explain for each age group if the respective disease exists in children and to what degree a clinical investigation program is feasible. If the drug is currently marketed in tablet form only and the targeted disease also exists in children younger than 6 years old, the EMEA requests as part of the PIP the development of a pediatric formulation. As explained in the EU Commission's PIP guideline,[40] the degree of accuracy expected in the PIP is rather high: the entire preclinical and clinical development is expected to be outlined, unless the PDCO grants a waiver or a deferral. In the case of a deferral, studies and developments should be outlined in the PIP as far as it is possible to describe them at an early stage. Later, an amended PIP is expected to give more details.

For new drugs, companies are expected to submit a first PIP no later than the end of pharmacokinetic studies, as outlined in Article 16 of the EU pediatric regulation, which refers to Section 5.2.3 of Part I of Annex I to Directive 2001/83/EC.[41] Directive 2001/83EC was amended in 2003 by directive 2003/63/EC,[42] and in this document Section 5.2.3 of Part I of Annex I can be found. The relevant passages of the involved documents are given in Table 13.2. At the end of human pharmacokinetics, teams can do their pediatric homework: assessing the frequency of the targeted disease in the different pediatric age groups, and getting an overview of existing alternative treatments.

Many pharmaceutical industry development teams that have reached the end of Phase I studies are in the process of preparing their first PIP. During 2009, we can expect a steep learning curve for both sides (i.e., the EMEA PDCO and pharmaceutical companies) in achieving a balance between an early assessment of pediatric diseases and therapeutic alternatives and a realistic planning of pediatric development that addresses all questions of preclinical testing, the need for juvenile animal testing, pharmacokinetic/pharmacodynamic (PK/PD) studies, pediatric clinical trials, pediatric pharmacovigilance, and many more aspects.

Development projects that were further along than the end of Phase I before June 2008 have some degree of freedom to decide when to submit their PIP to the EMEA. Most probably they were not prepared at the early development stage to perform pediatric development in their specific project.

In conclusion, the EU pediatric regulation forces the industry development teams, at an early stage, to include children in their development programs. This is a huge challenge. Development teams need a basic understanding of the differences between adults and children on the levels of physiological differences, organ maturation, maturation of metabolic pathways, maturation of excretion pathways, and the psychological challenges for parents, children, study personnel, and other involved partners. Every company is free to choose its approach to pediatric development. Some will build their own pediatric departments, while others will outsource as much as possible. But even if most work can be outsourced, key competence is required within the company to overview and supervise the outsourced work. Furthermore, each company needs internal policies, standard operating procedures (SOPs), and guidelines on how to deal with the requirements of pediatric drug development. Since these are now legal requirements, no company has the choice of not

TABLE 13.2 Articles/Chapters Relevant for the Timing of PIP Submissions for New Drugs

EU Paediatric Regulation, Article 16(1)

In the case of the applications for marketing authorization referred to in Articles 7 and 8 or the applications for waiver referred to in Articles 11 and 12, the paediatric investigation plan or the application for waiver shall be submitted with a request for agreement, except in duly justified cases, not later than upon completion of the human pharmacokinetic studies in adults specified in Section 5.2.3 of Part I of Annex I to Directive 2001/83/EC, so as to ensure that an opinion on use in the paediatric population of the medicinal product concerned can be given at the time of the assessment of the marketing authorization or other application concerned.

Directive 2001/83/EC was amended by Directive 2003/63/EC in 2003.

Section 5.2.3 of Part I of Annex I of EU Directive 2003/63/EC of 25 June 2003:

5.2.3. Reports of Human Pharmacokinetic Studies

(a) The following pharmacokinetic characteristics shall be described:
absorption (rate and extent), distribution, metabolism, and excretion.

Clinically significant features including the implication of the kinetic data for the dosage regimen especially for patients at risk, and differences between humans and animal species used in the preclinical studies, shall be described.

In addition to standard multiple-sample pharmacokinetic studies, population pharmacokinetic analyses based on sparse sampling during clinical studies can also address questions about the contributions of intrinsic and extrinsic factors to the variability in the dose–pharmacokinetic response relationship. Reports of pharmacokinetic and initial tolerability studies in healthy subjects and in patients, reports of pharmacokinetic studies to assess effects of intrinsic and extrinsic factors, and reports of population pharmacokinetic studies shall be provided.

(b) If the medicinal product is normally to be administered concomitantly with other medicinal products, particulars shall be given of joint administration tests performed to demonstrate possible modification of the pharmacological action.

Pharmacokinetic interactions between the active substance and other medicinal products or substances shall be investigated.

---

complying with this regulation. (In the editor's opinion, this book will facilitate the accomplishment of these necessary adjustments.)

FDA-TRIGGERED PEDIATRIC RESEARCH AND THE EU PEDIATRIC REGULATION

A lot of energy has been invested in the years preceding the EU pediatric regulation to determine if FDA-triggered pediatric clinical data are eligible for the EU reward. All clinical studies that were finalized before January 26, 2007 (the day the regulation went into force) have to be submitted to the EU competent authorities and are not eligible for the EU reward. Studies that were started before January 26, 2007 and will be finished thereafter will be examined by the EMEA if their outcomes are significant.

Many studies triggered by the U.S. pediatric legislations involved drugs whose patent protections are now near the end. Taking into account the timelines and procedures of the

EMEA, the chance of getting a second EU reward based on new significant study results appears to be limited.

The one area where the EU pediatric regulation will probably yield results rather soon are drugs that (in the United States) are classified as biologics and are therefore not eligible for pediatric exclusivity. Biologics are regarded as drugs in the EU and are eligible for the EU reward just like any other drug.

Another key aspect of the EU regulation is the difference between adult and child indications, where the wording of the regulation is less stringent than the wording of the U.S. pediatric legislation. The U.S. PREA mandates companies to perform pediatric investigations only and exclusively for indications that are registered and planned to be registered in adults. If the FDA wants other uses of the respective medication to be investigated, it can only rely on the willingness of the respective company to perform voluntary pediatric research, which the FDA can decide to reward through pediatric exclusivity. Through this mechanism, a number of research projects into rare diseases have been triggered, as the respective companies had an interest to obtain exclusivity in the main adult market. Examples for this are the use of Alendronate in osteogenesis imperfecta, or the use of tamoxifen in McCune–Albright syndrome.[37]

INTEGRATION OF PEDIATRIC ASPECTS INTO THE GENERAL DRUG DEVELOPMENT PROCESS

Clinical research is in itself a rather young discipline, which began only in the 1960s when companies had to prove safety and efficacy of their drugs, based on data from clinical trials.[1] The U.S. pediatric legislation since 1997 has initiated a process that started with drugs already on the market. For the first 8 years, the FDA had to rely almost exclusively on the voluntary legislation and has been collecting experience with the mandatory PREA only since 2003. The EU process plans for a much higher commitment to pharmaceutical companies to consider the needs of children, not *after* market introduction but *during* the development process. (See Figure 13.1.)

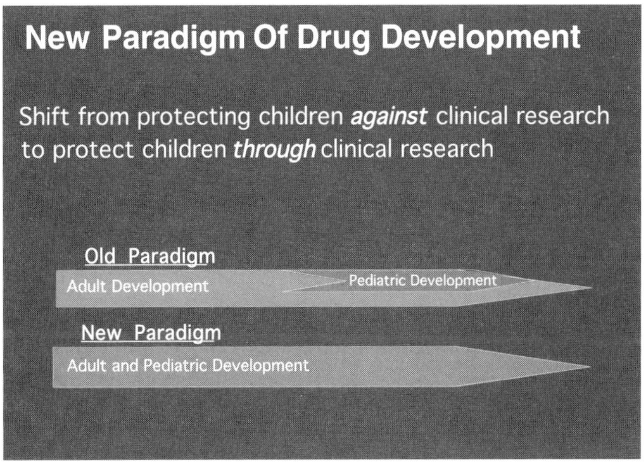

FIGURE 13.1 A change of paradigm.

This is new in the history of drug development. As it was unethical in the past to use drugs that were not properly tested in children, it would be equally unethical to expose children at early development stages to untested substances that potentially could put them in danger. There is a high attrition rate in drug development, and approximately only one in five substances that enter Phase I trials reach the market. To establish a balance that is acceptable to all partners in healthcare will be a lengthy process and will require the goodwill and patience of all involved. It should lead to increasing participation of children in the general pharmaceutical progress without jeopardizing the pharmaceutical progress itself.

EU PEDIATRIC REGULATION

The EU pediatric regulation was preceded by preparation symposia, work delivered by the then EMEA PEG, and many other activities by the EMEA, EU national regulatory authorities and health authorities, and other institutions. Many documents produced by these activities and a lot of additional information is stored on the EMEA pediatric website.[43] Since the EU pediatric regulation is continuously evolving and new information packages are entered on a regulator base, we now explain the structure and main topics of the website. For additional information, the reader should consult the website for additional information posted since the printing of this book. Table 13.3 gives the left drop down list of the EMEA

TABLE 13.3 Table of Contents (Left Drop Down List) of the EMEA Pediatric Website

Introduction
The EU Paediatric Regulation
Paediatric Committee
Guidance for Applicants
- Scientific advice
- Paediatric investigation plans (PIPs), waivers, and modifications
- Paediatric-use marketing authorisations (PUMAs)
- Compliance
- Submission of paediatric studies

Opinions and Decisions on PIP Applications
- Background
- Class waivers
- Product-specific decisions

Paediatric-Related Information
- Inventory of paediatric needs
- Paediatric clinical trials
- Priority list of off-patent medicines
- Scientific guidance
- Presentations

EU Paediatric Network
Global Cooperation
- Member states
- International

Workshops
Related Links

pediatric website, which serves as a table of contents. (The only modification is introduction of bullet points to ease reading. The original UK English spellings have been left unchanged in Table 13.3.)

13.8.1 EMEA Paediatric Website

Introduction[43] A short introduction is given into the scope of the EU pediatric regulation.

The EU Paediatric Regulation[12] This page comprises the direct link to the EU pediatric legislation as well as an amendment, a document on frequently asked questions on the regulatory aspects of the regulation, a joint EU Commission/EMEA implementation plan for the regulation, a website section entitled "Medicines for Children" of the EU Commission's Directorate General Enterprise and Industry, and a document on the history of the EU pediatric initiative.

Paediatric Committee[21] This section of the website contains a short explanation of the PDCO, including a link to an EMEA press release on the occasion of the first meeting of the PDCO as well as a link to the PDCO website and a list of the PDCO meeting dates for the next two years. The PDCO website has the following four sections:

- PDCO role and responsibility: The main responsibility is assessment of PIPs, including applications for waivers and deferrals. Other tasks of the Paediatric Committee include assessment of PIP-triggered data and adoption of opinions on the use of any drug in children; advising EU states on a survey of all existing uses of drugs in children; advising and supporting the EMEA in the creation of a European pediatric research network; advising on pediatric medicines; establishing and regularly updating an inventory of pediatric drug needs; advising the EMEA and EU Commission on communication on pediatric clinical research. The PDCO is not responsible for marketing authorization of drugs for children, which remains fully within the remit of the EMEA CHMP (Committee for Medicinal Products for Human Use). However, the CHMP or any other competent authority may request PDCO's opinion if the clinical data were generated in accordance with an agreed PIP.
- Composition: Gives a link to all PDCO members.[44]
- Rules of procedure: As of November 2008 contains document EMEA/348440/2008 "Rules of Procedure."
- Other related documents: As of November 2008 just one: "Policy on Representation of EMEA Scientific Committees by Their Members."[45]

Guidance for Applicants

Scientific Advice[46] Pediatric scientific advice is offered free of charge.

PIPs, Waivers, and Modifications[47] This section of the website lists documents that are intended to help the companies put together a PIP and submit it to the EMEA:

- EU Commission guideline on format and content of applications for pediatric investigation plans (19 pages).[40] This guideline dates back to January 2007. Despite repeated announcements, an update of this guideline was not published until December 10, 2007.
- Practical aspects on how to submit an application for pediatric investigation plan and requests for waiver and deferral (6 pages).[48]
- Contact details for PDCO members and alternates for sending PIP application: coordinates all PDCO members and their alternates.[44]
- Electronic template for PIP applications (in Adobe Acrobat Reader version 8.0).[49]
- PDCO meeting dates and timelines for submission of applications in 2009.[50,51]
- Template letter of intent: 2 months before submitting a PIP, companies are requested to direct a letter of intent to the EMEA so it can plan its resources.[52]

Paediatric-Use Marketing Authorisations[53] *(PUMAS)* This section gives just a short, very high level description of the PUMA instrument with which the EMA wishes to attract investment into off-patent pediatric medicines: 10 years of data protection is offered. No further documents in this subchapter.

Compliance[54] This section emphasizes that once a PIP has been approved, the submitting company has to follow that plan exactly. It further explains that once the plan is complete, the EMEA or the authorities in the member states will check that all studies and measures required have been performed. This compliance check is necessary before an application for marketing authorization can be considered valid.

Submission of Paediatric Studies[55] All pediatric studies completed by January 26, 2007 had to be submitted by January 28, 2008 to the EMEA PDCO.

Opinions and Decisions on PIP (Applications)

Background[56] At the end, the PDCO gives an "opinion" on any submitted PIP. The company can request a reexamination. The PDCO will reconsider and give a "definitive opinion." The EMEA will then adopt a "decision," based on the PDCO's definitive opinion. All EMEA decisions on PIPs, deferrals, or waivers will be made public after deletion of commercially confidential information.

Class Waivers[57] This section gives the link to a list of class waivers. It will be updated on a regular basis.

Product-Specific Decisions[58] In this section of the website product-specific decisions are published.

Paediatric-Related Information

Inventory of Paediatric Needs[59] This section gives an inventory of pediatric needs in the indication areas including anesthesiology, anti-infective therapy, cardiology, chemotherapy (cytotoxic substances), chemotherapy (supportive therapy), type 1 and 2 diabetes, epilepsy, gastroenterology, immunology, migraine, nephrology, obstructive lung disease, pain, psychiatry, and rheumatology. These therapeutic need inventory lists include off-patent medications as well as patent-protected medications and include which pediatric needs exist in the view of the authors. These lists were produced by the EMEA Pediatric Working Party (PEG), whose responsibilities have now been shifted to the PDCO. Pediatric needs include issues such as pediatric formulations, dose, efficacy and safety data in specific indications and in all appropriate age groups, PK data in children, and more. The inventory also gives a list of medications for which there are no pediatric needs as perceived by the PDCO.

Paediatric Clinical Trials[60] All clinical trials performed in the EU are registered in the EudraCT database, which will now include all pediatric trials performed anywhere in the world if the trial is part of a PIP. This pediatric part of EudraCT will be public. Also, details of the results of pediatric clinical trials, including those terminated prematurely, will be made public by the EMEA from now on.

Priority List of Off-Patent Medicines[61] This section gives the link to a priority list for recommended studies into off-patent drugs used in children,[62] as well as to the EU 7th Framework Programme for off-patent medicines developed for children.[63]

Scientific Guidance[64] This section refers to the EMEA CHMP's scientific guidelines, located in a central repository.[65] They are categorized into quality guidelines, nonclinical guidelines, clinical efficacy and safety guidelines, and multidisciplinary guidelines.

Presentations[66] As of November 2008, this section contains one 27-slide power point presentation on the EU pediatric regulation.

EU Paediatric Network[67] As of November 2008, this section contains a draft implementation strategy for the EMEA network of pediatric networks.

Global Cooperation

Member States[68] This section gives access to the EU Heads of Medicines Agencies (HMA) website and to a document, "Guidance on the Content and Format of Data to Be Collected on All Existing Uses of Medicinal Products in the Paediatric Population."

International[69] This section gives a short history of the U.S. pediatric legislations and links to the FDA pediatric website, as well as to a press release on intensified collaboration between the EMEA and FDA in the area of medicines for children.

Workshops[70] As of November 2008, this section contains the presentations of four EMEA workshops.

Related Links[71] As of November 2008, this section contains three links:

- To the FDA Office of Pediatric Therapeutics.[72]

- To the EU Commission website "Medicines for Children,"[73] which gives access to a multitude of additional documents, including a press release on intensified cooperation between the FDA and EMEA in pediatric drug development and the extended impact assessment of the EU pediatric regulation.
- To the EU European Commission Directorate-General for Research 7th Framework Programme website.[16]

COMMENTS ON THE EU PEDIATRIC REGULATION

SPC Prolongation

It should be noted here that not all drugs have an SPC (e.g., drugs that were developed very fast). Drugs without SPCs are nevertheless expected to comply with all mandatory parts of the pediatric regulation.

PUMA

PUMA highlights a problem for which there is no completely sufficient solution: drugs that do not have any patent protection. Obviously, it is the intention of the EU to attract investment for research into the pediatric use of off-label products. The degree to which the framework offered by PUMA will be successful remains to be seen. PUMA aims at small- and medium-sized companies to develop new pediatric formulations or other programs that will allow a better use of off-label drugs in children. Not all factors that will influence the future success or failure of PUMA are under the direct control of the EU Commission. It will be the individual EU countries and the respective reimbursement institutions that will decide to what degree higher prices for pediatric medications will be accepted.

PEDIATRIC HOMEWORK

The following key questions will have to be answered in the future by each individual development team in each pharmaceutical company at an early development stage toward the end of Phase I trials:

- Is the targeted indication life-threatening or seriously debilitating?
- Does the same or another potential indication exist in children?
- What is the frequency per age groups?
- Is the mechanism of disease different in children versus adults?
- Which therapeutic alternatives exist in children?
- What is the risk/benefit assessment of early start clinical trials in children?

These questions can be called "pediatric homework" and can indeed be answered toward the end of human pharmacokinetics. All further decisions, such as requests for deferrals or waivers, or beginning of physical pediatric development including preclinical work, work on pediatric formulations, and modeling and simulation, will then have to be made on a project-specific level. The pharmaceutical industry as well as the EMEA are at an early stage of experience and will probably both experience a steep learning curve. The dialogue between these two parties will mold the future of the pediatric homework to be performed at early project development stages.

THE PLACE OF EUROPE IN THE GLOBAL PROCESS OF PEDIATRIC DRUG DEVELOPMENT

A first look at the EU pediatric regulation and the EMEA pediatric website shows the deep commitment of EU institutions to promote better medicines for children. It is also obvious that the preparation of a PIP alone will be a major workload for each individual development team. It is to be hoped that this workload will turn into increased benefit for children from pharmaceutical progress. A lot of questions remain open and, as usual, the ongong process seems to pose more open questions with each individual problem resolved. The heavier workload will increase the necessary investment per each individual new medicine before it reaches the market. A continuing dialogue between the partners in healthcare will ensure success for the EU pediatric regulation and will give Europe a new leading role in the development of medicines for children.[74]

REFERENCES

1. Stoetter H. Paediatric drug development—historical background of regulatory initiatives. In: Rose K, van den Anker JN, eds. *Guide to Paediatric Clinical Research*. Basel, Switzerland: Karger; 2007; 25–32.
2. Rose K, van den Anker JN, eds. *Guide to Paediatric Clinical Research*. Basel, Switzerland: Karger; 2007.
3. Editorial: Clinical trials in children, for children. *Lancet* 2006;367(9527): 1953.
4. McIntyre J, Conroy S, Avery A, et al. Unlicensed and off label prescribing of drugs in general practice. *Arch Dis Child*. 2000;83:498–501.
5. Turner S, Longworth A, Nunn AJ, et al. Unlicensed and off label drug use in paediatric wards: prospective study. *BMJ* 1998;316:343–345.
6. Ekins-Daukes S, Helms PJ, Taylor J, et al. Off-label prescribing to children: attitudes and experience of general practitioners. *Br J Clin Pharmacol*. 2005;60(2):145.
7. Conroy S, Choorara I, Impicciatore P, et al. Survey of unlicensed and off-label drug use in paediatric wards in five European countries. *BMJ*. 2000;320:79–82.
8. 't Jong GW, Vulto AG, de Hoog M, et al. A survey of the use of off-label and unlicensed drugs in a Dutch children's hospital. *Pediatrics*. 2001;108:1089–1093.
9. Hilts PJ. *Protecting America's Health*. New York: Alfred A Knopf; 2003.
10. Hawthorne F. *Inside the FDA*. Hoboken, NJ: John Wiley & Sons; 2005.
11. The European paediatric initiative: history of the paediatric regulation. Available at http://www.emea.europa.eu/pdfs/human/paediatrics/1796704en.pdf.
12. Regulation (EC) No 1901/2006 of the European Parliament and of the Council of 12 December 2006 on medicinal products for paediatric use. Available at http://ec.europa.eu/enterprise/pharmaceuticals/eudralex/vol-1/reg_2006_1901/reg_2006_1901_en.pdf.
13. Regulation (EC) No 1902/2006—an amending regulation in which changes to the original text were introduced relating to decision procedures for the European Commission. Available at http://ec.europa.eu/enterprise/pharmaceuticals/eudralex/vol-1/reg_2006_1902/reg_2006_1902_en.pdf.

14. http://www.emea.europa.eu/.
15. http://ec.europa.eu/enterprise/pharmaceuticals/paediatrics/index.htm.
16. EU Commission 7th Framework Cooperation Work Programme: Health. Available at ftp://ftp.cordis.europa.eu/pub/fp7/docs/a_wp_200701_en.pdf.
17. EU Heads of Medicines Agencies (HMA) website. Available at http://www.emea.europa.eu/htms/human/paediatrics/cooperation_ms.htm.
18. http://www.fda.gov/cder/pediatric/presentation/dlb2-DIA-Mar2003/sld004.htm.
19. http://www.fda.gov/cder/pediatric/presentation/DIA_2004June16_Murphy_Shirley/sld002.htm.
20. http://en.wikipedia.org/wiki/Supplementary_protection_certificate.
21. http://www.emea.europa.eu/htms/human/paediatrics/pdco.htm.
22. http://www.emea.europa.eu/pdfs/human/pdco/29568907en.pdf.
23. http://www.emea.europa.eu/pdfs/human/press/pr/25919207en.pdf
24. EFPIA: European Federation of Pharmaceutical Industries and Associations. Available at http://www.efpia.org/.
25. http://www.emea.europa.eu/pdfs/conferenceflyers/EMEA-EFPIA_2Feb2007/1330-N_Seigneuret.pdf.
26. Practical aspects on how to submit an application for paediatric investigation plan and requests for waiver and deferral. Available at http://www.emea.europa.eu/pdfs/human/paediatrics/practical_aspects.pdf.
27. Paediatric investigation plans (PIPs), waivers and modifications. Available at http://www.emea.europa.eu/htms/human/paediatrics/pips.htm.
28. Submission of pediatric studies. Available at http://www.emea.europa.eu/htms/human/paediatrics/studies.htm.
29. Compliance. Available at http://www.emea.europa.eu/htms/human/paediatrics/compliance.htm.
30. Scientific advice. Available at http://www.emea.europa.eu/htms/human/paediatrics/sci_advice.htm.
31. Scientific advice and protocol assistance Available at http://www.emea.europa.eu/htms/human/sciadvice/Scientific.htm.
32. http://www.emea.europa.eu/pdfs/human/phvwp/23591005en.pdf.
33. Paediatric-use marketing authorisations (PUMAs). Available at http://www.emea.europa.eu/htms/human/paediatrics/pumas.htm.
34. Regulation (EC) No 726/2004 of the European Parliament and of the Council of 31 March 2004 laying down community procedures for the authorisation and supervision of medicinal products for human and veterinary use and establishing a European Medicines Agency. Available at http://ec.europa.eu/enterprise/pharmaceuticals/eudralex/vol-1/reg_2004_726/reg_2004_726_en.pdf: Article 14 (11).
35. Aggarwal S. What's fueling the biotech engine? *Nat Biotechnol.* 2007;25:1097–1104. Available at http://www.nature.com/nbt/journal/v25/n10/full/nbt1007-1097.html.
36. Surfactant replacement therapy for respiratory distress syndrome. Available at http://aappolicy.aappublications.org/cgi/content/full/pediatrics;103/3/684.
37. http://www.fda.gov/cder/pediatric/labelchange.htm.
38. http://oig.hhs.gov/oei/reports/oei-09-00-00380.pdf.
39. Fighting rare diseases: 22 new orphan drugs in five years. Available at http://europa.eu/rapid/pressReleasesAction.do?reference=IP/06/844&type=HTML&aged=0&language=EN&guiLanguage=en.

40. European Commission guideline on format and content of applications for paediatric investigation plans. Available at http://ec.europa.eu/enterprise/pharmaceuticals/paediatrics/docs/draft_-guideline_pip_2007-02.pdf.
41. EU Directive 2001/83/EC. Available at http://ec.europa.eu/enterprise/pharmaceuticals/eudralex/vol-1/dir_2001_83/dir_2001_83_en.pdf.
42. EU Directive 2003/63/EC. Available at http://ec.europa.eu/enterprise/pharmaceuticals/eudralex/vol-1/dir_2003_63/dir_2003_63_en.pdf.
43. http://www.emea.europa.eu/htms/human/paediatrics/introduction.htm.
44. http://www.emea.europa.eu/htms/general/contacts/PDCO/PDCO_members.html.
45. http://www.emea.europa.eu/pdfs/general/direct/principle/23147705.pdf.
46. http://www.emea.europa.eu/htms/human/paediatrics/sci_advice.htm.
47. http://www.emea.europa.eu/htms/human/paediatrics/pips.htm.
48. http://www.emea.europa.eu/pdfs/human/paediatrics/practical_aspects.pdf.
49. http://www.emea.europa.eu/pdfs/human/paediatrics/PIP-application-form.pdf.
50. http://www.emea.europa.eu/pdfs/human/paediatrics/26671107en.pdf.
51. http://www.emea.europa.eu/pdfs/human/paediatrics/41346807en.pdf.
52. http://www.emea.europa.eu/htms/human/paediatrics/pips.htm. Click directly on *Template letter of intent*—the word document open in word, no direct hyperlink possible.
53. http://www.emea.europa.eu/htms/human/paediatrics/pumas.htm.
54. http://www.emea.europa.eu/htms/human/paediatrics/compliance.htm.
55. http://www.emea.europa.eu/htms/human/paediatrics/studies.htm.
56. http://www.emea.europa.eu/htms/human/paediatrics/background.htm.
57. http://www.emea.europa.eu/htms/human/paediatrics/classwaivers.htm.
58. http://www.emea.europa.eu/htms/human/paediatrics/decisions.htm.
59. http://www.emea.europa.eu/htms/human/paediatrics/inventory.htm.
60. http://www.emea.europa.eu/htms/human/paediatrics/clinicaltrials.htm.
61. http://www.emea.europa.eu/htms/human/paediatrics/prioritylist.htm.
62. http://www.emea.europa.eu/pdfs/human/paediatrics/19797207en.pdf.
63. ftp://ftp.cordis.europa.cu/pub/fp7/docs/a_wp_200701_en.pdf.
64. http://www.emea.europa.eu/htms/human/paediatrics/sci_gui.htm.
65. http://www.emea.europa.eu/htms/human/humanguidelines/background.htm.
66. http://www.emea.europa.eu/htms/human/paediatrics/presentations.htm.
67. http://www.emea.europa.eu/htms/human/paediatrics/network.htm.
68. http://www.emea.europa.eu/htms/human/paediatrics/cooperation_ms.htm.
69. http://www.emea.europa.eu/htms/human/paediatrics/cooperation_int.htm.
70. http://www.emea.europa.eu/htms/human/paediatrics/workshops.htm.
71. http://www.emea.europa.eu/htms/human/paediatrics/links.htm.
72. http://www.fda.gov/oc/opt/default.htm.
73. http://ec.europa.eu/enterprise/pharmaceuticals/paediatrics/index.htm.
74. Ramet J, Rose K. Europe's path towards better medicines for children. In: Rose K, van den Anker JN, eds. *Guide to Paediatric Clinical Research.* Basel, Switzerland: Karger; 2007: 1–5.

CHAPTER 14

Japanese Perspective

HIDEFUMI NAKAMURA, MD, PhD

National Center for Child Health and Development, Tokyo 157-8535, Japan

SHUNSUKE ONO, PhD

Graduate School of Pharmaceutical Sciences, The University of Tokyo, Tokyo 113-0033, Japan

INTRODUCTION

Off-label, extemporaneous, and/or unlicensed uses of drugs in Japanese children are common as in other parts of the world. After the implementation of ICH E11, the number of clinical trials in children appears to have increased only moderately. Japan does not have "pediatric regulations" equivalent to those in the United States or European Union (EU). Possible incentives for the industries involved in pediatric drug development have been extension of the reexamination period and preferential drug pricing, which appears to have limited effects.

The approval lag of new drugs in Japan has become a serious problem, and the Japanese government has been trying to establish new regulatory schemes to promote drug development in children. For important pediatric drugs that already have been developed in certain countries, the use of foreign data is accepted. The government is also making a tremendous effort to build a competitive environment for clinical trials and to speed up drug approval in the country. There is currently a pediatric clinical trial network being established with the National Center for Child Health and Development, which plays a central role. Discussion on additional regulatory measures to promote pediatric clinical trials is ongoing. With all the efforts combined, further improvements of pediatric drug development are expected in the near future.

PAST AND PRESENT STATUS OF OFF-LABEL, EXTEMPORANEOUS, AND UNLICENSED USE OF DRUGS IN JAPANESE CHILDREN

Off-Label Use of Drugs in Japanese Children

The off-label use of drugs in Japan has been very common, similar to other parts of the world. Morita[1] examined prescriptions of drugs covered by the National Health Insurance during

Pediatric Drug Development: Concepts and Applications
Edited by Andrew E. Mulberg, Steven A. Silber, and John N. van den Anker

1997 and 1998 for pediatric patients in four university hospitals and one affiliated general hospital. For outpatients and inpatients less than 18 years of age, 2032 drugs were prescribed on 531,137 occasions in one year. He and his colleagues examined the package inserts of these medications and found that only 495 drugs (24.4%) had sufficient information for pediatric dosage on the package inserts. Approximately 40% of these had the description "Safety is not established in children." Among these 2032 drugs, approximately 2% were either "contraindicated" or "not recommended in children" for certain age groups, although many of them are commonly used in daily practice. There were also ambiguous statements common in the labels, including "the dose can be adjusted accordingly for age" or no clear guidance on usage in children in approximately 20%. In the opinion of these authors, this situation has not changed over the last 10 years.

Extemporaneous Use of Drugs in Japanese Children

In 2005, Kato and colleagues[2] conducted a survey on the extemporaneous use of drugs in children at 32 Japanese institutions including 18 children's hospitals. All extemporaneous uses of pediatric drugs during the 4-week period from October 17, 2005 to November 13, 2005 were reported. There were a total of 1666 incidences of dosage form changes. In 1227 incidences, age appropriate powders were made either by crushing tablets and/or adding sucrose to make different strengths. On many occasions, Japanese children have powders orally administered as they are or after adding a small amount of water to make them paste-like. This obviously is the reason why the dosage form changes to powder are so common. Most Japanese pharmacies have machines to automatically divide powder evenly and to pack it in sachets.

In 176 cases, tablets were divided into two or more sections; in 50 cases, syrups/suspensions were made from other forms including intravenous solutions (e.g., midazolam for sedation). In 40 cases, the suppositories were cut for appropriate dosages, and in 23 incidences, inhalation solutions were made from other forms including intravenous solutions. The top 10 drugs for dosage form changes on prescription number basis were warfarin potassium, methyl digoxin, enalapril maleate, dantrolene sodium, lisinopril, beraprost sodium, hydrocortisone, baclofen, chloral hydrate, and propranolol hydrochloride.

Unlicensed Pediatric Drugs in Japan

According to the survey by the Japanese Society for Inherited Metabolic Diseases in 2006, "raw" chemical compounds including sodium benzoate, betaine, cysteamine, and glycine are dispensed for certain orphan diseases. Among these, betaine and cysteamine are already approved in some countries, and the Ministry of Health, Labour and Welfare (MHLW) has been seeking for industries that are interested in developing these drugs in Japan following the procedures recommended by the Study Group on Unapproved Drugs.

Patients or physicians may have to import medications, which do not exist in Japan as approved drugs. Especially for patients with oncology and genetic diseases, who have to pay, the cost for these medications can often be more than US $10,000 annually/person. These problems with unlicensed medicines and approval lag (drug lag) have become a focus of social attention, especially in 2004 when the former prime minister, Junichiro Koizumi, ordered the associated cabinet members to take action to improve access to unlicensed medicines and innovative equipment. Approval lag or drug lag is the lag between the time when a drug has been approved in the country and the time when the drug was first approved in another part of the world.

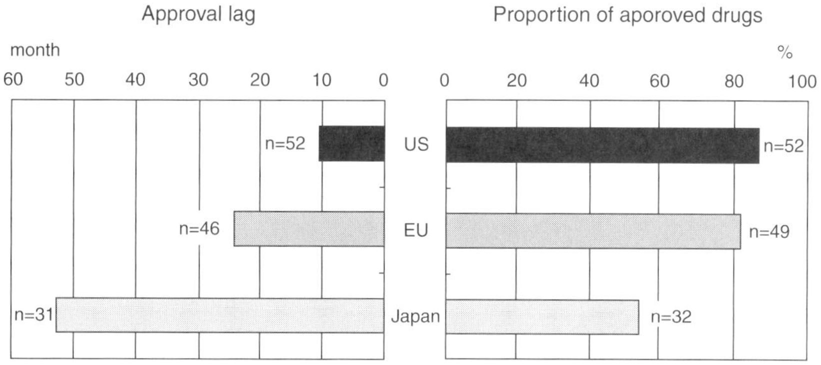

FIGURE 14.1 Proportion of approved drugs and mean approval lag of 60 orphan drugs in the United States (U.S.), Eruopean Union (EU), and Japan. Approval lag was calculated for the approved drugs in each region for which data were available.

Tsuji and Tsutani[3] examined the approval status of the new chemical entities (NCEs) in the United States, EU, and Japan between 1999 and 2005. Out of 334 NCEs approved at least in one of the three regions, 274 (82.6%) were approved in the United States, 262 (78.4%) in the EU, and 181 (54.2%) in Japan. The mean approval lag was 13.5 months in the United States, 13.2 months in the EU, and 46.3 months in Japan. Figure 14.1 shows the mean approval lag and proportion of approved drugs of 60 NCEs designated as orphan drugs (except for anti-HIV drugs). Japan had the lowest number of approvals (32 NCEs or 53.3%) and longer mean approval lag (52.6 months) when compared to the United States or EU.

In this survey, they also examined the development status of unapproved drugs (Figure 14.2a) and unapproved orphan drugs (Figure 14.2b) as of October 31, 2006.

For the NCEs in which they could find the development status, approximately 50% of the total unapproved NCEs in Japan were not under development, whereas 78.6% (22 NCEs)

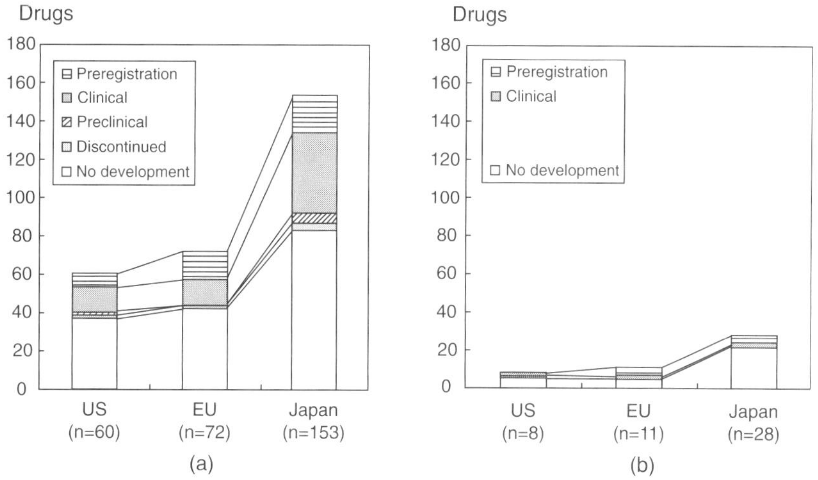

FIGURE 14.2 Development status of unapproved NMEs in the United States (U.S.) European Union (EU), and Japan: (a) for all the unapproved NMEs for which data were available and (b) for all the unapproved NMEs designated as orphan drugs for which data were available.

were not under development as orphan drugs (personal communication). It is believed that one of the reasons why a higher percentage of orphan drugs were left out of the development process is that fewer venture companies are involved in orphan drug development in Japan compared to the United States and EU and there is insufficient support to nourish venture companies in Japan. The need for support for venture companies for drug and device development is considered to be one of the major aims of the 5-year strategy for developing innovative drugs and devices.

Impact of Off-Label, Extemporaneous, or Unlicensed Use of Drugs in Japanese Healthcare

Japan has a nationwide healthcare insurance system, and all the citizens are insured and receive medical services at any time at any institution. For pharmaceutical products, the reimbursement is strictly based on the conditions of approval (i.e., indication(s), dose, regimen, and patient types) under the Pharmaceutical Affairs Law (PAL), the primary law governing drug development in Japan. The Japanese insurance bodies are not allowed to make their own decisions on reimbursement, which is in marked contrast to the diversified decisions by the U.S. insurance bodies. The Japanese healthcare insurance system also does not allow the use of uninsured drugs in combination with the insured practice. If physicians want to use uninsured drugs, the Japanese healthcare insurance system does not cover any of the costs of the medical care during that particular hospital visit.

Off-label, extemporaneous, or unlicensed use of drugs also may mean that the patients who experienced adverse effects cannot receive relief from the Adverse Health Effect Relief Service[4] by the Pharmaceutical and Medical Devices Agency (PMDA). It is also believed that there is significant underreporting of adverse events of off-label, extemporaneously used, or unlicensed drugs.

INVOLVEMENT OF THE JAPAN PEDIATRIC SOCIETY (JPS) FOR FACILITATION OF PEDIATRIC DRUG DEVELOPMENT

The Members of the Committee on Drugs of the JPS organized a working group to study the approval status of pediatric drugs in 1998 as part of the MHLW supported research. Representatives of 17 pediatric subspecialty societies joined the working group and gathered information on off-label drugs, which need to be labeled. The list for these drugs is called the "priority list" and has been used for the selection of drugs for the evaluation by the Study Group on Pediatric Drug Therapy. Some of the working group members have also been involved in setting up the infrastructure for clinical trials in certain areas including neonatology and pediatric nephrology.

The working group also started to accumulate information on unlicensed drugs, which need to be approved urgently in the country. After the drugs are listed, associated pediatric subspecialty societies are responsible to submit request letters to the MHLW for evaluation by the Study Group on Unapproved Drugs.

The Committee also has been involved in several activities to facilitate clinical trials, drug development, and the approval process. For example, at a public hearing in December 2006, the Committee gave requests to the Study Group on Faster Access to Innovative Drugs to consider several issues including the following:

1. Establishment of certain legal obligations for the industry to conduct clinical trials in a timely manner on important pediatric drugs.
2. Appropriate measures to hasten and strengthen the NDA review process for pediatric drugs by the PMDA.
3. Stronger incentives for industry partners to collaborate and more effectively develop opportunities for pediatric drug development.
4. Reinforcement of post market surveillance in cooperation with the JPA.

PAST AND PRESENT GUIDANCES AND NOTIFICATIONS FOR PEDIATRIC DRUG DEVELOPMENTS IN JAPAN

The primary law governing drug development in Japan is the PAL. There are also regulations that set forth the general principles governing drug development and their specific implementing procedures.[5] There are still no regulations equivalent to the U.S. Best Pharmaceuticals for Children Act, the U.S. Pediatric Research Equity Act, or the EU Paediatric Regulation. This section describes the past and present guidances and ordinances for pediatric drug development in Japan.

Impact of ICH E11

Until the implementation of the ICH E11 guideline, "Clinical Investigation of Medicinal Products in the Pediatric Population," in December 2000, there had been no official guideline for pediatric drug development in Japan. After the implementation, the importance of pediatric clinical trials seems to have been recognized more among industry partners and academia, but the number of pediatric clinical trials seems to have increased only moderately. Levy-Petalinkar and Campen discuss these same issues in the United States as presented in Chapter 7 in this book.

Nakagawa reviewed the evaluation reports issued by the PMDA and its former division, the Pharmaceutical and Medical Devices Evaluation Center (PMDEC), from April 2001 to January 2007, for drugs approved under the PAL (pers. comm.). Among 188 new indications between April 2001 and June 2004, 28 (14.8%) included pediatric indications. For these 28 indications, domestic clinical trials were performed for only 10 (35.7%). Among 139 approvals between July 2004 and January 2007, 25 approvals (18.0%) included pediatric indications. For these 25 indications, domestic clinical trials were performed for 18 indications (72.0%). This suggests that the total number of clinical trials in children may have increased recently.

The Official Notification Describing Procedures for Approval of Off-Label Drugs

This ordinance is classified as Kacho-tsuchi, subministerial division notification that is sometimes translated as "guideline."[6] This notification was issued jointly by the Evaluation and Licensing Division (ELD), the Pharmaceutical and Medical Safety Bureau, and the Research and Development Division (RDD), Health Policy Bureau, MHLW.

According to this notification, the RDD will encourage the company to consider pursuing a clinical development strategy for expansion of the already approved indication(s), in cases where there is a formal petition from the academic society to consider an approval AND a recognized medical necessity for the indication. When certain conditions are met, this notification allows off-label indications to be approved with only a literature-based application, waiving some or all domestic registration-directed clinical trials (Chiken), which usually are necessary. If this appears to be a possibility, consultation with the ELD prior to the application submission is recommended. The following are the three situations when domestic Chiken may be waived:

1. The proposed indication has already been approved in a foreign country, which has an equivalent drug approval system to Japan (e.g., the United States and EU), and there is a sizable body of data accumulated from clinical practice. If this is the situation, and the material submitted to the overseas approving authority can be submitted according to the Japanese requirement, domestic Chiken may be waived.
2. In cases when the indication has been approved according to the conditions stipulated above, a sizable body of data from use in clinical practice has been accumulated, AND reliable results are published in internationally recognized journals. To meet this standard, the data should be of high scientific quality.
3. In cases when there exists scientifically reliable data from studies conducted under the auspices of a publicly sponsored organization such as the MHLW. To meet this standard, such studies must have been conducted in accordance with currently accepted international ethical and scientific standards.

According to Nakagawa, among 22 new pediatric indications approved between April 2001 and June 2004, 10 drugs were approved based on this notification without Chiken. Between July 2004 and January 2007, 3 out of 25 approvals of new pediatric indications were based on this notification.

Extension of the Reexamination Period

Extension of the reexamination period was introduced when the revision of the MHLW ordinance on postmarketing surveillance was made in December 2000. In Japan, the reexamination period is applied to all newly approved indications and this was usually 6 years when this "incentive" was introduced. The usual reexamination period was extended to 8 years in April 2007. Data protection is applied during the reexamination period. When a company conducts clinical trials for a pediatric indication during the reexamination period, the reexamination period is extended up to 10 years. This extension is intended to cover the clinical trial period with the reexamination period. It turned out that industries only benefit when their drugs can get a longer reexamination period compared to the patent period. It does not benefit the industry partner whose drug patent period is much longer than the reexamination period.

Ironically, this extension of the reexamination period, which is considered to be an incentive for industry, actually may delay the pediatric drug development process. Based on examination of recent evaluation reports from the PMDA, it appears that many industry partners start postmarketing clinical trials for pediatric indication just before the expiration of the reexamination period so that they can get an extension to cover the clinical trial period. If the industries start the clinical trials too early, they may not get the extension of the reexamination period.

RECENT GOVERNMENT EFFORTS TO INTRODUCE NEW DRUGS TO JAPAN

Many drugs already marketed in the United States and EU are not introduced in Japan for several reasons. Even when they are finally introduced in Japan, it is quite common that there is a significant approval lag in the market introduction. Especially in cancer therapy, the fact that many drugs and therapeutic regimens already approved in Western countries were unavailable or not approved in Japan has caused Japanese policymakers to establish regulatory schemes to bring these drugs to Japan. Lack of domestic approvals in a significant number of drugs and regimens for pediatric treatments has also become a national concern, and there are ongoing MHLW activities to resolve the off-label use and unlicensed use of drugs in children. The information described here is current as of October 2008.

Study Group on Unapproved Drugs (MHLW, January 2005)

This study group started in January 2005 to oversee the situation of the approval lag, to evaluate the need to introduce unapproved drugs in Japan, and to lead them to the clinical development stage. Eighteen formal meetings have been held since its establishment, and development and approval strategies for 43 drugs have been discussed. Of the 43 drugs, 22 are oncology drugs and 15 are pediatric drugs.

The study group also recommends that clinical trials for safety confirmation should be performed for some drugs for which a New Drug Application (NDA) has already been submitted, because such trials could provide opportunities for patients to access these drugs during the NDA review. Compassionate use of investigational or unapproved drugs has not been legitimized in Japan but has been discussed for possible introduction into Japan by the Study Group on Faster Access to Innovative Drugs.

Study Group on Cancer Combination Therapy (MHLW, January 2004 to February 2005)

This study group specifically focused on off-label indications or unapproved regimens of cancer drugs for which Japanese approval was already obtained for some other indication(s) or regimen. To fill the gap between Japan and the United States/EU, this MHLW study group investigated the current Japanese situation and collected clinical evidence available in foreign countries. When the investigation showed that a given drug use was considered "standard" in the therapeutic field, then the MHLW issued approval based primarily on those prior investigations and foreign evidence. In 2005, expanded indications were approved for 30 cancer drugs including 6 pediatric drugs for rhabdomyosarcoma, Ewing's sarcoma, and neuroblastoma.

Committee for Vaccine Industry Vision (MHLW, March 2007)

The objective of the committee is to establish nationwide vaccination programs that are effective and efficient and draw investment in vaccine development from drug companies that are generally hesitant to develop new vaccines because of the very small market size for vaccine products. Due to increasing concerns for possible future pandemics of new type flu viruses, however, the MHLW considers it critical to have an appropriate level of manufacturing capacity for vaccine products in Japan. Activities of this committee will enhance introducing vaccines available in foreign markets.

Study Group on Pediatric Drug Therapy (MHLW, March 2006)

Using the similar scheme of the Study Group on Cancer Combination Therapy, the Study Group on Pediatric Drug Therapy focuses on a solution for the off-label use of pediatric drugs. A list of 99 drugs chosen from the JPS "priority list" was submitted to the study group by pediatric subspecialty societies, and so far 8 drugs have been chosen to be evaluated. The study group investigates the approval status and related information in the United States, United Kingdom, France, and/or Germany, the current Japanese situation, and other clinical evidence available in the literature. When the investigation shows that a given drug use is considered to have "sufficient evidence" for the indication and/or dosage, then the MHLW issues approval based on those investigations and foreign evidence supporting the claim. As of October 2008, expanded indications/dosages have been approved for all the pediatric formulations of acetaminophen, which had rather limited indications and dosage before (e.g., no indication for pain, once daily for suppository). Evaluation of methotrexate and bolulinum toxin type A have been completed, and there is ongoing investigation of the remaining 5 drugs (i.e., flecainide, methylphenidate, ciprofloxacin, cyclophosphamide, and acyclovir). It is most likely that investigation of other drugs will be initiated in the near future.

Compared to 30 cancer drugs approved in 2 years, the progress of this study group has been rather slow. One of the reasons is that there are very scarce published data on the off-label use of drugs in Japan, and it takes some time to gather reliable information on the current status of off-label use. Even after completing the survey on current status, it sometimes becomes clear that there is some discrepancy between the commonly prescribed dose and/or indication in Japan and the approved dose and/or indication in the four foreign countries. Lack of appropriate pediatric formulations for many drugs also can be an obstacle for the forthcoming evaluations. For now, there is minimal or no incentive for the industries to produce new formulations for children, and it will be extremely difficult to have the industry develop new formulations based on the current scheme of the study group.

Drug Pricing for Pediatric Drugs

The best incentive for industries for pediatric drug development is believed to be increasing drug prices. In 2006, preferential drug pricing for pediatric drugs was newly introduced, although the increase of the price was merely 3–10% and only for drugs that have no other pharmacologically similar drugs on the market. In 2008, increase of the drug pricing has further changed to 5–20% and is for drugs whose similar drugs have no preferential pricing. As many off-label drugs have been on the market a long time, and drug prices decrease every year, the drug prices for adult indications of these off-label drugs are cheap. For these kinds of drugs, a 5–20% increase may be too small for industry as an incentive.

Japan does not have "flat price" in the marketplace, and drug prices generally get cheaper, especially drugs used in children smaller than 20 kg. The prices for pediatric dry syrups are generally set to make the price for a 20-kg child equivalent to the price for an adult dose. This rule may become an obstacle when a certain drug is expected mostly to be used in small children.

GOVERNMENT EFFORTS TO PROMOTE CLINICAL TRIALS IN JAPAN

The MHLW and related government agencies have been implementing strategies to boost the number and quality of Japanese clinical trials in the last few years. The outcome of these strategies is expected to be seen in the near future.

Sponsor-Investigator Trials

The revision of PAL in 2002 enabled physicians to conduct sponsor-investigator trials in Japan for the first time. The Committee on Drugs of JPA recommends that JPA members become actively involved in sponsor-investigator trials and utilize them to set up a stronger infrastructure for pediatric clinical trials. There have been seven clinical trials conducted or scheduled.

In August 2007, expanded indication for the use of fentanyl citrate in children including neonates was approved as the first drug ever studied by the sponsor-investigator trials in the country. The drug was contraindicated for 2 year olds and younger children prior to this period. One other clinical trial was completed, another trial is still ongoing, and three more trials started as of October 2008.

Three-Year Clinical Trial Promotion Plan (MHLW and MEXT, 2003–2005)

The MHLW and the Ministry of Education, Culture, Sports, Science and Technology (MEXT) jointly executed a set of programs to promote Japanese clinical trials. Setting up several large networks for clinical trials was one of the important objectives of this plan. Research grants to establish networks and implement some sponsor-investigator trials as model programs were managed by the Center for Clinical Trials of the Japan Medical Association. As of October 2008, there were more than 1400 clinical institutions registered for the networks. Training programs for clinical research coordinators (e.g., research nurses, pharmacists) were provided regularly under this plan. Several symposia were also held to raise public awareness of clinical trials. It is believed that Japanese patients are not necessarily interested in clinical trials, because they can afford medical services without substantial financial burdens.

The national hospitals and research centers directly managed by the MHLW have been promoting harmonization of trial contract documents. These streamlining efforts also are intended to reduce the administrative costs of trials.

New Five-Year Clinical Trial Promotion Plan (MHLW and MEXT, 2007–2011)

In March 2007, the MHLW and MEXT published the 5-year promotion plan that follows the previous 3-year plan. The new plan aims to create several large networks of clinical institutions. Each network has a core hospital that orchestrates the operations with its affiliated hospitals. The MHLW chose 10 core and 30 affiliated major hospitals, based on regional and therapeutic needs as well as their performance in clinical trials. The aim and role of the network of core and affiliated major hospitals described by the MHLW is shown in Figure 14.3.

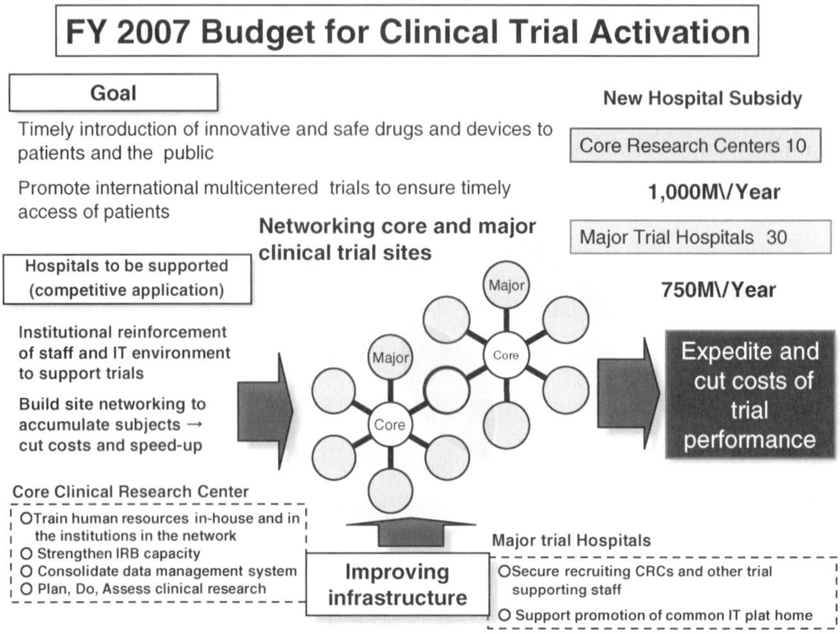

FIGURE 14.3 Role of the network of core and affiliated major hospitals explained by the R&D Division, MHLW.

For a clinical trial network for pediatric drugs, the National Center for Child Health and Development (NCCHD) has been chosen as the core hospital, and the Kanagawa Children's Medical Center, the Osaka Medical Center and Research Institute for Maternal and Child Health, and the Tokyo Metropolitan Kiyose Children's Hospital were chosen as affiliated major hospitals. Networking with other children's hospitals is also expected, and the above four hospitals are to play a major role in conducting networked clinical trials in children.

According to this new 5-year plan, training programs for investigators, research nurses, and pharmacists are provided in more intensive forms than in the previous 3-year plan. In addition, the network hospitals can afford to employ staff, including research nurses, pharmacists, biostatisticians, and data managers through budgetary supports. To attract more industry sponsors, the MHLW promises to remove administrative red tape and streamline operations in the networks, including adoption of information technology. Decrease in the costs of clinical trials is also an alleged goal of the plan. To achieve that goal, however, various factors in addition to costs of clinical trials must be taken into consideration because prices are determined not only by cost components but also by market structure and monopoly, if present.

Study Group on Faster Access to Innovative Drugs (MHLW, 2006–2007)

The MHLW convened an expert panel to discuss the bottlenecks to the introduction of new drugs in Japan in October 2006. This study group focused on the principles of new drug approval, postmarketing safety measures, and enhancement of the PMDA review. It is expected that the results of this will be reflected in upcoming amendments of existing regulations.

In the final report of this study group, issued in July 2007, it is specified that the government should consider stronger incentives for industry and other regulatory measures to facilitate pediatric drug development. It is believed that this statement is in response to the request from the Committee on Drugs of JPA at the public hearing.

"Innovation 25" Strategic Council (Cabinet Office, 2006)

Former Prime Minister Shinzo Abe launched a long-term innovation strategy guideline in his inaugural policy speech in September 2007. The guideline, called "Innovation 25," specifically emphasizes the need to concentrate on several technology areas essential for the solid economic growth of Japan in the coming decades, and pharmaceutical R&D by the drug industry and clinical researchers are regarded as key areas. The Innovation 25 Strategic Council plays the central role in further planning and implementing detailed strategies.

Five-Year Strategy for Developing Innovative Drugs and Devices (MHLW, MEXT, and METI, 2007–2011)

Three ministries made this strategy: the MHLW, Ministry of Education, Culture, Sports, Science and Technology (MEXT), and the Ministry of Economy, Trade and Industry (METI). The major goals of this strategy are to reinforce clinical research infrastructure to ensure safe and secure patient access to new drugs and devices; that is, (1) institutional infrastructure building for core clinical research centers' networks, (2) human resource development for clinical research, and (3) regulatory research promotion and approval system reform. Furthermore, MHLW is planning to establish "medical research clusters" at National Centers in 2008, to accelerate and promote research and development and new partnership with industry. According to this plan, the NCCHD will become involved in establishing the medical research clusters for pediatric drug developments.

PROSPECT FOR THE NEAR FUTURE

Japan has no strong "stick-and-carrot" approach for industry to conduct pediatric clinical trials as the United States and the EU have recently adopted. The Committee of Drugs of JPA will be taking further actions to have governments consider setting up new regulations to promote pediatric drug trials. The MHLW or PMDA does not have the authority to control the patent period. Therefore, the Japanese government may have to consider other incentives, including preferential drug prices, taxation, and fast track review.

As described in this chapter, the Japanese government is taking several critical actions to promote pediatric drug development. The Japan Pharmaceutical Manufacturers Association also has a task force to work on this issue. With all these efforts combined, we are expecting that further promotion of pediatric drug development will occur in the near future.

REFERENCES

1. Morita S. Prescription and package insert description analysis in pediatric drug use (in Japanese). In: *1999 MHLW Project Report for the Research on Current Status of the Off-Label Use of Drugs in Children and Possible Solutions* (PI: Onishi S); 2000: 52–99.
2. Kato Y, Ishikawa I, Kushida K, Nakamura H. Survey of pediatric dosage form changes in pediatric drug therapy Japan. *Paediatric and Perinatal Drug Therapy* 2008.
3. Tsuji K, Tsutani K. Approval of new chemical entities in the US, EU and Japan (in Japanese). *Iryo to Shakai [J Health Care Soc].* 2007;17(2):243–258.
4. Website for the Adverse Health Effect Relief Services. Available at http://www.pmda.go.jp/english/healtheffect.html.
5. Ono S. Ministry of Health, Labour and Welfare (MHLW, Japan). In: D'Agostino RB, Sullivan L, Massaro JM, eds. *Wiley Encyclopedia of Clinical Trials*. Hoboken, NJ: John Wiley & Sons; 2007.
6. Fujiwara Y, Kobayashi K. Oncology drug clinical development and approval in Japan: the role of pharmaceuticals and medical devices evaluation center (PMDEC). *Crit Rev Oncol/Hematol.* 2002;42:145–155.

CHAPTER 15

Regulatory Considerations for Study of Generic Drugs Under Best Pharmaceuticals for Children Act: NICHD and FDA Collaboration

SANDRA COTTRELL, PhD

Novo Nordisk, Inc., Princeton, New Jersey 08540

BRAHM GOLDSTEIN, MD, MCR

University of Medicine and Dentistry of New Jersey–Robert Wood Johnson Medical School, New Brunswick, New Jersey 08901

INTRODUCTION

There is a rich history of U.S. legislative initiatives to achieve adequate pediatric labeling for drugs. To consider the specific implications for generic drugs, it is important to examine briefly the history of the generic drug process, and also the legislative history of pediatric drug development requirements. For further details of these issues as they relate to the EU and Japan, please see Chapters 13 and 14 in this book.

BRIEF SUMMARY OF APPROVAL OF GENERIC DRUGS IN THE UNITED STATES

The 1938 Food, Drug and Cosmetic Act introduced a requirement on each drug's manufacturer to file with the FDA 60 days prior to the marketing of the drug a New Drug Application (NDA) containing safety data. The NDA review was a passive process and in most cases led to marketing at the end of the 60 days. While this requirement technically applied to all versions of the product, between 1941 and 1968 the FDA had a policy of

The opinions, views, and comments expressed in this chapter are solely those of the authors and do not necessarily reflect the opinions or policies of Novo Nordisk, Inc., its directors, officers, employees, agents, or representatives.

Pediatric Drug Development: Concepts and Applications
Edited by Andrew E. Mulberg, Steven A. Silber, and John N. van den Anker

defining certain drugs as no longer "new drugs" but generally recognized as safe (GRAS). With this policy, "generic copies" of drugs essentially escaped FDA review—that is, they had no NDA.

With the 1962 amendment to the 1938 Food, Drug and Cosmetic Act, the definition of new drug was changed to include "generally not recognized as safe and effective" by experts. This change effectively expanded the requirement for safety and effectiveness data in the NDA. An issue quickly arose for the nearly 4000 drugs already approved between 1938 and 1962 under the 1938 Act based on safety data alone (no requirement for efficacy data) in addition to the drugs that were grandfathered under the category of GRAS and were without an NDA on file. To reconcile the drugs approved prior to 1962 with the new standard, a review of each drug's effectiveness was undertaken. This Drug Efficacy Study Implementation (DESI) review, conducted independently by the National Academy of Science and the National Research Council, assessed all versions of these drugs (ineffective, possibly, probably, effective, effective-but). While this review took several years to complete (1967 to 1968; reported in 1969), once a product with the primary NDA was determined as effective, all other copy versions were required to also make a regulatory filing to continue to be marketed.

Beginning in 1970, the FDA allowed generic manufacturers to file Abbreviated New Drug Applications (ANDAs) for these products originally approved between 1938 and 1962. The ANDA made cross-reference to the "pioneer NDA" data for safety and efficacy (animal and human) data but had to provide product-specific manufacturing data and data to prove equivalence to the pioneer product's delivery profile. Blood level and dissolution data comparison between the pioneer and "generic" version was typically used to fulfill this latter requirement. By June 1975, more than 6000 ANDAs had been filed for generic versions of drugs approved prior to 1962. There was, however, no abbreviated application process for a generic version of a product approved after 1962, regardless of whether the patent had expired.[1,2]

Growing pressure for a generic drug process for those drugs approved after 1962 and now off-patent set the stage for the 1984 Drug Price Competition–Patent Term Restoration Act (DPC-PTR Act) (Waxman–Hatch). Under this amendment to the Act, approval was permitted under Section 505(j) of duplicates of approved drugs, thus extending the generic drug application process (abbreviated new drug applications—ANDAs) to drugs approved after 1962. As previously, these ANDA submissions provide information to show that the proposed product is identical in active ingredient, dosage form, strength, route of administration, labeling, quality, performance characteristics, and intended use, among other things, to a previously approved product (cross-referencing to that Reference Listed Drug (RLD) label).

The 1984 DPC-PTR Act, however, also provided for a Suitability Petition process for making modifications, resulting in generic drugs that were "not identical" to the RLD. Such petitioning for modification was limited to one of four specific characteristics, where the resulting labeling would reflect the suitability-approved change. Specifically, the change in the generic drug from the RLD was limited to a change in dosage form, strength, route of administration, or, for a combination drug product, a change in one active ingredient. As will be discussed, this suitability process actually has had some significant implications for subsequent discussions regarding pediatric regulations.

This brief history explains the ANDA process for marketing of generic drug products. It must be emphasized that generic drug manufacturers do not conduct safety/efficacy studies, but rely on cross-referencing to the RDL.

LEGISLATIVE HISTORY OF PEDIATRIC DRUG DEVELOPMENT REQUIREMENTS

The FDA has a history of attempts to address the need for safety/efficacy data specific to pediatric use. As early as 1979, they issued their rule on the Pediatric Use subsection of package inserts. The rule was revised with the 1994 FDA Final Rule on Pediatric Labeling and Extrapolation (effective January 12, 1995), which allowed manufacturers to add pediatric labeling information, but required (and allowed) drugs that had not been tested for pediatric safety and efficacy to bear a disclaimer to that effect.[3,4] However, in light of the disclaimer option, this rule failed to motivate many drug companies to conduct additional pediatric drug trials. Recognizing this shortfall in achieving company motivation to conduct pediatric studies, the FDA and Congress used the 1997 Food and Drug Administration Modernization Act (FDAMA) to pass incentives that gave pharmaceutical manufacturers a 6-month extension of protection on drugs submitted with pediatric trial data.[5]

This legislative approach relied on the incentive of an additional 6-month period of exclusivity to motivate the voluntary fulfillment of conducting studies to generate the requested data. Logistically, the sponsor had to receive and fulfill a Pediatric Written Request that outlined the scope of studies that the FDA required in order to receive the 6 months of exclusivity. It is also important to note that a drug did not need to be proven effective in pediatric subjects to gain the 6 months of exclusivity. Data describing a lack of efficacy or safety limitations for use, which would then be added to the label, were considered equally important. The resulting pediatric exclusivity did not accrue only to the product that was studied in the pediatric population, but attached to all the applicant's formulations, dosage forms, and indications for products with existing marketing exclusivity or patent life that contained the same active moiety. Following the legislative requirements defined under FDAMA, the FDA published the initial prioritized list (Priority List) on May 20, 1998 and subsequently, as required, updated the Priority List annually with those products the FDA felt were especially in need of pediatric data. The list contained products approved for use in adults, but for indications that also occurred in the pediatric population, and were typically being used "off-label" in pediatric subjects. In point, if an interested party wished to have a drug added to the Priority List, they could petition the FDA accordingly. Importantly, however, being on the list did not equate to having a Written Request, and inclusion of a sponsor's drug on the list did not mean that the sponsor was required to take any action, nor in many cases did they—consistent with the voluntary nature of the pediatric aspects of FDAMA. While the 6-month exclusivity was a motivator in some cases, the need to add the 6 months of additional exclusivity onto "something" (existing patent or market exclusivity) was an obvious motivation limitation to the program's success to achieve pediatric labeling.

Accordingly, at the same time, a separate initiative had been underway, and in 1997, the FDA proposed a rule, which became finalized as the 1998 Pediatric Rule (codified as 21 CFR 314.55 regulations), to require pediatric drug trials from the sponsors of New Drug Applications (NDAs).[6–8] Specifically, under the 1998 Pediatric Rule, the FDA required applicants to conduct a pediatric assessment for drug products and indications contained in applications submitted for review. The application could be for a new active ingredient (i.e., a new NDA) or for a new indication, new dosage form, new dosing regimen, or new route of administration for a previously approved product. This approach was the second side of getting sponsors to conduct pediatric studies by requiring a pediatric development plan. Between FDAMA and the 1998 Pediatric Rule, the FDA had created a "carrot-and- stick" approach to achieving pediatric labeling. Of note, as a policy, presuming the pediatric plan under the 1998 Pediatric Rule also met requirements defined in a Written Request, the sponsor could meet the requirement and also get rewarded with 6 months of exclusivity. Importantly, the 1998 Pediatric Rule did not impose any pediatric study requirements on abbreviated applications for exact generic copies of approved drugs (Section 505(j) of the Act). Therefore, relative to the 1998 Pediatric Rule, a clearly defined path for generics to be exempt from the rule was defined. It is ironic then to note that the 1998 Pediatric Rule quickly met with legal challenges by companies well beyond generic considerations, and the 1998 Pediatric Rule was enjoined on October 17, 2002 in federal court as exceeding the FDA's statutory authority. Simply stated, the courts said the FDA could not codify into regulation a law Congress had not written. While this debate regarding the 1998 rule was unfolding, in parallel the 5-year sunset date for FDAMA and its pediatric exclusivity was due for renewal.

As noted earlier, one significant shortfall of the provisions for 6 months of pediatric exclusivity under FDAMA was that the 6 months of exclusivity needed to be "added onto" something (i.e., remaining patent life, or a form of remaining market exclusivity). In cases where there was nothing to "add on to," which was typical for older products including those being used to treat pediatric subjects, the products did not qualify for the incentive to conduct the necessary studies. Moreover, under FDAMA, the created incentive was lost despite the FDA's desperate desire for the label data. Thus, when Congress reauthorized terms for the pediatric exclusivity incentives of FDAMA on January 4, 2002 with the 2002 Best Pharmaceuticals for Children Act (BPCA),[3, 6–8] the new amendment pointedly addressed two important situations, namely, the case of still protected drugs (on-patent or with market exclusivity) for which the sponsor declined doing the Written Request studies, and the case of off-patent products. Thus, while the 2002 BPCA renewed the elements of 6-month exclusivity for on-patent products, it also created a process for obtaining critical information in the pediatric population for still protected products, referring the product to the National Institutes of Health (NIH) for funding of the necessary research as described in the Written Request.

Similarly, for off-patent drugs, the Written Request was sent to all NDA and generic ANDA holders. Should none of these sponsors agree to conduct the studies specified, a process for NIH to publish a request for contract and a request for proposal in the *Federal Register* could be pursued. Furthermore, the BPCA required the FDA to make available to the public a summary of the medical and clinical pharmacology reviews of pediatric studies conducted. The summary was to be made available not later than 180 days after the report on the pediatric studies was submitted to the FDA. These data were expected to lead to label changes by sponsors, or they could face misbranding charges, even if they had previously declined conducting the studies.

The focus for FDA issuance of a Written Request to all holders of approved applications (NDAs and generic ANDAs) for products without patent or market exclusivity protection would be from among the products that were on the Priority List of products lacking adequate pediatric labeling. The list process, created in partnership with NIH/National Institute of Child Health and Human Development (NICHD), was from the onset a long, iterative effort relying on input obtained from Advisory Committees, FDA Divisions, NIH Divisions, American Academy of Pediatrics, United States Pharmacopaeia (USP), and consulted experts. Determining the need for such research was driven by consideration of substantial use (>50,000 U.S. pediatric subjects with the disease/condition) and/or meaningful

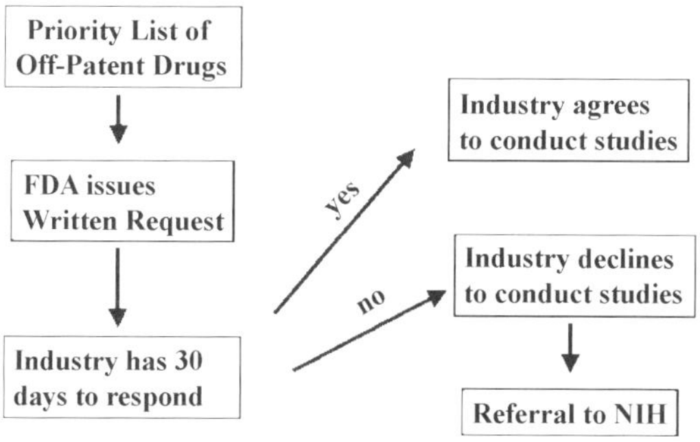

FIGURE 15.1 Process for the study of off-patent drugs.

therapeutic benefit. The first published list of 12 drugs appeared in the *Federal Register* on January 21, 2003.

The process for off-patent products as first described under the 2002 BPCA is displayed in Figure 15.1.

As noted, under the 2002 BPCA, the FDA was able to request NIH sponsoring of studies for pediatric drugs if the marketing authorization holders declined to do the testing. While a step forward, this was still not a perfect solution because these requests were subject to NIH funding constraints.

Moreover, even for the case of products with remaining exclusivity and thus potential incentives, the 2002 BPCA remained voluntary in its primary scope and thus was not an absolute assurance that the necessary studies would be done. Therefore, the FDA still needed a way to mandate pediatric development plans, especially going forward for new products. Accordingly, the FDA needed to trigger a requirement for pediatric assessment for all new products and changes in indication, dosage form, dosing regimen, or route of administration to existing products that had been lost with the enjoinment of the 1998 Pediatric Rule in October 2002. To this end, the concepts defined in the 1998 Pediatric Rule were finally defined within a new amendment to the Act, the Pediatric Research Equity Act (PREA) of 2003, which was passed by Congress and codified by the FDA, and which mandated manufacturer-sponsored pediatric drug trials to be conducted within the context of the drug, disease, and potential applicability to children.[3, 6–8] Essentially, all the points described for the 1998 Pediatric Rule were addressed within the legally allowed framework of the 2003 PREA. PREA legislation did not, however, codify the 2002 BPCA but was to run in parallel, effectively having in place the mandated and voluntary approaches to getting pediatric studies conducted. Unlike BPCA, which was only applicable to products approved under Section 505 of the Act (21 U.S.C. 355), PREA required the conduct of these pediatric studies for drug and biological products, where "drugs" and "drug and biological products" include drugs approved under Section 505 of the Act and also biological products licensed under 351 of the Public Health Service Act (PHSA) (42 U.S.C. 262) that are drugs. Specifically, PREA required new drug applications (NDAs) and biologics licensing applications (BLAs) to have these pediatric data for a new active ingredient, but also for changes in approved products (supplements to applications) leading to a new indication, new dosage form, new dosing regimen, or a new route of administration. The labeling submitted would contain a pediatric assessment (reflecting the study results) unless the applicant had obtained a waiver or deferral. It also authorized the FDA to require holders of applications for previously approved marketed drugs and biological products who are not seeking approval for one of the changes enumerated above to submit a pediatric assessment under certain circumstances, albeit this aggressive approach seems to have less precedence given how sponsors have responded to the legislation. It should be noted that orphan drugs were exempt from the PREA requirements. Similarly, a generic application (ANDA) would not have to include pediatric assessment data unless triggered by the suitability process.

The enactment of PREA,[3] overcoming the enjoinment of the 1998 Pediatric Rule, however, raised again the separate complexity for generics pursuing an ANDA Suitability Petition as a process. Since ANDA Suitability Petitions (except for the case of change in strength) submitted under Section 505(j)(2)(C) for a change in active ingredient, dosage form, or route of administration will trigger pediatric studies based on the definitions under PREA, it means the petition must be denied because "a petition may be denied if investigations must be conducted to show the safety and effectiveness of the change." Effectively, the pediatric studies required under PREA negate the qualification of the Suitability Petition since by definition ANDAs do not contain such clinical investigations. Thus, generic drug applications submitted under a Section 505(j)(2)(C) Suitability Petition for changes in dosage form, active ingredient, or route of administration would trigger pediatric study requirements into their application process, just as a nongeneric product being expanded as part of the product's life cycle management for a new indication, new dosage form, new dosing regimen, or new route of administration would require a pediatric assessment (or a deferral or waiver).

For generic drug products to achieve such changes from the RLD (excluding change in strength), an application under Section 505(b)(2) filings seems the likely recourse. A sponsor seeking to market a generic product but with a change from the RLD would file a Section 505(b)(2) application if the generic applicant is seeking approval of a change to an approved drug that would not be permitted under Section 505(j), because approval will require the review of clinical (pediatric) data. Moreover, a generic manufacturer has always had the discretionary option to submit a Section 505(b)(2) application for a change in a drug product even that was eligible for consideration pursuant to a Suitability Petition under Section 505(j)(2)(C) of the Act, but the impact of PREA seems to now require such a strategy. Importantly, substituting a Section 505(j) generic drug ANDA filing with a Section 505(b)(2) application, while the technical approach to achieving the modification of the RLD, does not negate the need to fulfill the PREA mandated pediatric studies that lead to the Section 505(b) application approach. Therefore, under PREA, generic drug manufacturers are faced with the decision to not make the changes to the RLD label, or to conduct some form of pediatric investigation—clearly outside the activity domain of most generic companies. Given the cost and complexity of doing pediatric studies, PREA may be a barrier to the expansion of generic drug product development previously seeking to modify the dosage form, route of administration, or an active ingredient in a combination product.[9] Since there is yet no U.S. process for "generic biologics," the impact of PREA is not an issue for potential biosimilars.

As a final note on the U.S. status, it should be recognized that, unlike PREA, which had no text in the amendment defining a legislative sunset date, BPCA had a sunset of October 2007.[3,7,8] The renewal of BPCA met with active debate, but it had consistently been expected that the program of exclusivity incentive would be renewed, although draft versions of the legislation had varied in whether, in deference to generic companies' pressures, special restrictions should be placed on the period of exclusivity for "blockbuster" products.

In point, not every law contains a sunset provision. While BPCA did, the sunset term was not used in the 2003 PREA legislation. Nonetheless, essentially, PREA was set to cease if and when BPCA did. In general, by including an end date or another indication of a predetermined termination date, Congress provides "an 'action-forcing' mechanism, carrying the ultimate threat of termination, and a framework or guidelines for the systematic review and evaluation of past performance." For this reason, it is not surprising that Congress has now linked BPCA and PREA into a common legislative sunset dating. Moreover, the concept of a sunset for BPCA, itself a renewed legislation from what was in the 1997 FDAMA, has precedence and was readily accepted. Arguably, a sunset for PREA may be less appropriate but, nonetheless, the FDA Amendments Act of 2007 (FDAAA), as signed on September 27, 2007, finalized the continuation of both aspects of regulation for pediatric development (BPCA and PREA) with a common 5-year period to sunset on October 1, 2012.[10] Notably, there was no change to the 6-month exclusivity for blockbuster products.

Even as renewed under the FDAAA of 2007, without qualifying for 6 months of exclusivity, generic drug manufacturers still have no incentive to provide pediatric data under BPCA, although the FDA working through the NIH has options to secure the data. For those generic manufacturers considering Suitability Petition type changes, including many generic drugs known to be used in children, the PREA requirements will specifically be triggered. However, it is worthy to note that the change to a generic RLD requiring a pediatric assessment under PREA would potentially qualify for 3 years of exclusivity under the Waxman–Hatch Act for that labeling. If a Written Request was also issued, 6 months of exclusivity could be added to that 3-year exclusivity period. This may well change the balance of decision-making for generic drug companies in the future.

REFERENCES

1. Beer DO. *Generic Drugs: A Guide to FDA Approval Requirements*. Upper Saddle River, NJ: Prentice Hall Law & Business; 1988.
2. Merrill RA, Hutt PB. *Food and Drug Law Cases and Materials*, 2nd ed. Mineola, NY: The Foundation Press; 1991.
3. CDER Pediatric Initiatives. 2007. Available at http://www.fda.gov/cder/pediatrics/index.htm.
4. The Content and Format for Pediatric Use Supplements. 2007.
5. FDA Moderization Act. 1997.
6. Rose K, Vanden Anker JN, eds. *Guide to Paediatric Clinical Research*. Basel, Switzerland: Karger; 2007.
7. Politis P. Transition from the carrot to the stick: the evolution of pharmaceutical regulations concerning pediatric drug testing. *Widener Law Rev*. 2005;12:271.
8. Tauzin B. *Medicines in Development for Children*. Washington, DC: Pharmaceutical Research & Manufacturers Assoc. 2007.
9. Karst KR. Is the ANDA Suitability Petition Process Dead? *Regulatory Affair Focus* 2005;10(5):35.
10. Best Pharmaceuticals for Children Act. 2007.

PART IV

PRECLINICAL SAFETY ASSESSMENT

CHAPTER 16

Introduction and Overview

TIMOTHY P. COOGAN, PhD, DABT

Senior Director, Toxicology and Investigational Pharmacology, Centocor Research & Development, Inc., 145 King of Prussia Road, Radnor, Pennsylvania 19087

The role of preclinical development (toxicology and ADME) is similar with regard to supporting drug development in pediatric populations and adults. The goals of these investigations are to characterize the absorption, distribution, metabolism and excretion (ADME) and toxicokinetics (TK) in animals, as well as to identify potential target organs for toxicity. In addition, an understanding of reversibility for any identified targets, as well as safety margins relative to human exposure is desired. With regard to pediatric drug development an additional level of complexity is involved, as evaluation of human relevance not only involves consideration of cross species applicability of findings but also the dimension of age. In comparisons across species, physiologic age is considered, not chronologic age and, as an example, a general guide for this comparison with regard to central nervous system and reproductive development is presented in Figure 16.1.[1] Finally, just as adult rats, dogs or monkeys are not fully representative of adult humans, juvenile animals do not always mirror pediatric populations.

In the past 15–20 years, interest in the use of juvenile animals has greatly increased in a hope to provide adequate models for safety evaluation for clinical testing in pediatric populations. Much like the situation for adult animal toxicity testing, juvenile animals offer potential benefit to flag possible concerns or identify biomarkers for clinical evaluation. Testing in juvenile animals is thought to fill the "gap" between safety evaluations in standard reproductive/developmental toxicology programs and adult animal general toxicology assessments (Figure 16.2). In the pre- and postnatal developmental toxicity studies, animals are exposed, first *in utero* via direct dosing of the dam followed by indirect, lactational exposure. Exposure via lactation is both uncontrolled and indeterminate, thereby resulting in highly variable exposure levels or even unexposed animals. However, these studies do include a number of evaluations not routinely conducted in general toxicology programs, including assessment of developmental milestones, effects on behavior, learning and reproductive performance of off-spring. On the other hand, general toxicology programs routinely provide evaluations of clinical pathology (hematology, clinical chemistry and urinalysis), and histopathology of numerous tissues, in an attempt to identify target organ

Pediatric Drug Development: Concepts and Applications
Edited by Andrew E. Mulberg, Steven A. Silber, and John N. van den Anker

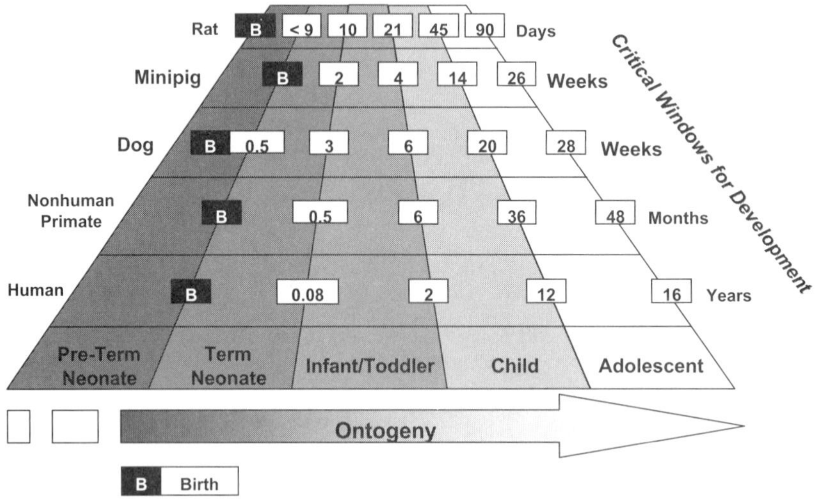

FIGURE 16.1 Comparative age categories across species based on CNS and reproductive development. Estimates are based on combined general developmental events occurring in both sexes, and represent only the overall schedule for CNS and reproductive development in these species. Note that the end age of the comparative category to human Infant/Toddler corresponds roughly to the usual age at weaning for laboratory species. (With permission- Buekle-Sam, 2001)

toxicity. Juvenile animal studies represent a hybrid of these study designs and goals, resulting in a thorough evaluation of potential drug related effects with direct dosing, thus providing information on both target organ toxicity and developmental effects. An example of the complexity involved in this hybrid study design is presented in Figure 16.3.

To understand the relevance of juvenile animal toxicity testing, a number of factors must be taken into consideration. These include, but are not limited to, the timing of development

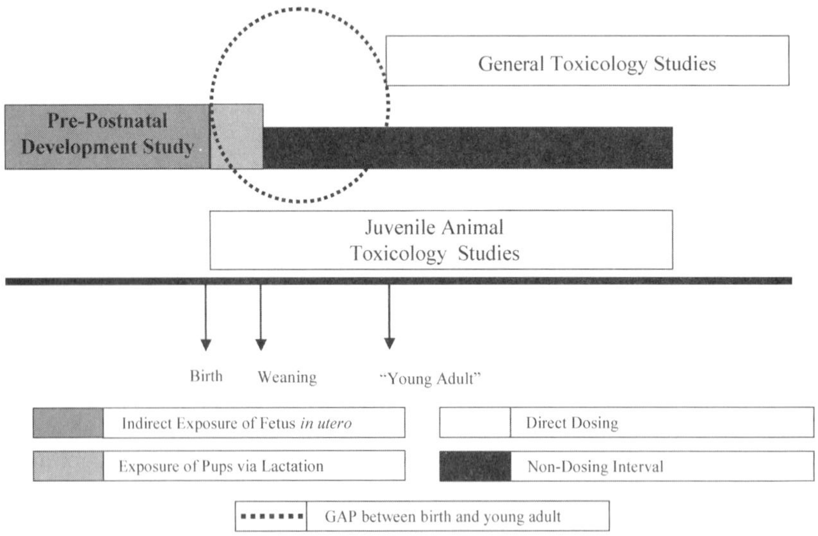

FIGURE 16.2 Juvenile animal toxicology studies fill the gap in assessment between birth and young adult.

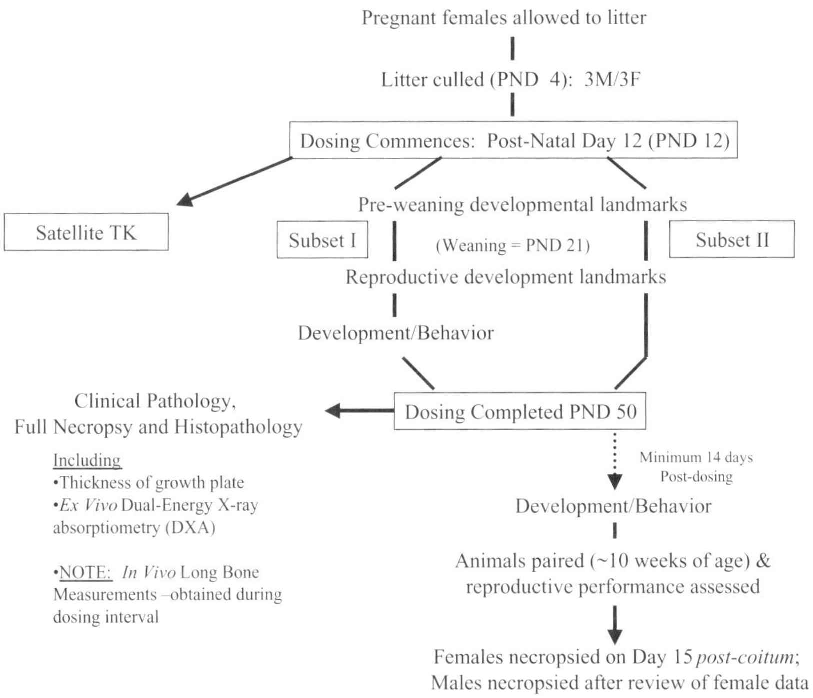

FIGURE 16.3 An example of the complexity associated with Juvenile animal studies. In this example, rats are dosed starting on post-natal day (PND) 12 and continue through PND 50 to bridge with the adult rat studies. Also included in this example are assessment of reproductive performance, development/behavior (acutely during dosing and delayed after dosing interval), TK in a satellite group of animals as well as evaluation of effects on bone (e.g., DXA).

of specific organ systems in each species relative to human. Additionally, ontogeny of metabolism enzymes and transport proteins and their corresponding development in humans has a key role in the design and interpretation of findings from toxicology studies in juvenile animals. These concepts are further expanded in subsequent chapters in this section.

The history of regulatory requirements by global health authorities for assessment in juvenile animals parallels, and in some ways has driven investigations in this area. Concerns regarding an understanding of clinical relevance of findings in juvenile animals, as well as the potential impact of findings on pediatric development programs and pipeline portfolio decisions initially slowed the acceptance of this type of testing. Regulatory requirements for these types of investigations have further supported conduct of juvenile animal studies across all therapeutic areas. The US FDA has issued a guidance[2] and the European EMEA, a draft guidance[3] regarding nonclinical safety evaluation of pediatric drug products. Both documents provide valuable information regarding study design and conduct, as well as potential timing of studies relative to clinical pediatric testing. Although similar in intent, there exists a key difference between these guidances and this involves the philosophy behind dose selection. The US FDA takes a general toxicology view regarding dose selection and advises selection of a top dosage to include evidence of identifiable toxicity, either general or developmental. In contrast, the EMEA takes a reproductive toxicology approach to dose selection, advising top dose selection to include what would be viewed as minimally toxic. The latter approach permits evaluations that are not confounded by overt, excessive toxicity.

TABLE 16.1 Key Discussion Points Impacting the Design and Conduct of Juvenile Animal Studies

- Lowest age to be included in the clinical pediatric program (e.g. neonates, or > 2 year old)
- Duration of treatment (e.g. acute vs. chronic)
- Identified toxicity in adult clinical program
- Identified target organs in adult animal toxicity assessments
- Previously identified developmental toxicity from the Reproductive Toxicology program
- Route of administration
- Unique formulation requirements with novel excipients
- PK and Metabolism in adult animals and humans
- Species selection supporting overall development (e.g. rat and dog)
- Any species specific toxicity (e.g. dog only)

Clearly, these differing views have a major impact on the design and interpretation of juvenile animal studies to support global development programs, as it is common to default to the more stringent requirements in this situation. Perhaps only greater experience with a broad range of compounds will provide insight into the more predictive approach. Additionally, these regional differences suggest that consideration of this topic by the International Conference on Harmonization (ICH) is warranted.

There are a number of factors that impact the design of toxicologic assessment in juvenile animals. A list of key questions/discussion points whose answers impact the design and conduct of preclincial juvenile animal studies is presented in Table 16.1. Information from these questions determine the age of animals to start testing, the species to include, additional assessments, as well as technical considerations for route of exposure and for obtaining biological samples for toxicokinetics and clinical pathology evaluations. It is obvious from these questions that the knowledge obtained on a compound from adult clinical and adult animal studies shapes assessments in juvenile animals.

Due to the evolving Regulatory landscape regarding the timing of pediatric drug development, pediatric clinical programs are starting earlier with a direct impact on the preclincial studies that need to precede them. Regulatory requirements to address pediatric drug development earlier in the overall drug development program has shifted the timing for pediatric drug development from a post-marketing commitment to development in parallel with the adult program. In fact, current European guidance supports submission of the pediatric investigational plan (PIP) "...no later than the completion of the relevant human pharmacokinetic studies in adults."[4] Thus, the benefit of full knowledge from both adult human clinical data and longer term adult animal data would not be available to aid in the design of the preclinical juvenile toxicity program. Regardless of the timing of the pediatric clinical program, a basic preclincial study package is required to support entry into a pediatric population. This basic study package is presented in Table 16.2 and includes those studies applicable for adult clinical use, as described in ICH M3 (R1)[5] with additional juvenile animal investigations. Although ICH guidance indicate the need for a full reproductive test program (Segment I, II and III), the need for all these studies is questionable and open for discussion. For example, a study in female pediatric subjects that have not reached child bearing age would not require embryo-fetal development or pre- and postnatal development studies since these study designs are intended to evaluate the risk of *in utero* and/or milk exposure. As previously mentioned, the juvenile toxicology program

TABLE 16.2 Basic Preclinical Study Package to Support Clinical Studies in Pediatric Populations

Toxicology

- Safety Pharmacology (Core Battery)[a]
  - Central nervous system (Irwin)
  - Cardiovascular
  - Respiratory
- Single/Repeat Dose (2 species)
  - Duration as per ICH Guidelines[b]
- Genetic Toxicology Package
  - Ames
  - Chromosomal aberration – in vitro (e.g., mouse lymphoma)
  - Chromosomal aberration – in vivo (e.g., mouse micronucleus)
- Reproductive Toxicology
  - Fertility
  - Developmental toxicology (2 species)[c]
  - Pre- and postnatal development[c]
- Juvenile Animal Toxicology[d]

Preclinical ADME

- In vitro metabolism comparison in animals and human
- Plasma protein binding in animals and human
- In vitro identification of major enzymes involved and drug–drug interaction potential in humans
- Tissue distribution in male and pregnant female rats
- Absorption, excretion, and metabolism (AEM) in animals and humans (with radiolabel)

[a] Additional assessments may be needed based on outcome from Core Battery
[b] ICH M3 (R1)[104]
[c] May not be relevant/required (see text)
[d] Needed on a case-by-case basis

incorporates most of the assessments included within the pre- and postnatal study design, with the benefit of controlled direct dosing of pups similar to direct dosing of pediatric populations. It is highly recommended to initiate early interactions with Health Authorities to seek agreement on the overall study package supporting the intended pediatric population and also to agree upon the overall study design and species selection for the juvenile animal program.

It is fitting to discuss target organ toxicity and susceptibility of juvenile populations. There is a general feeling that pediatric populations are more susceptible to toxic outcomes but this is not always the case. There are examples where susceptibility is either increased (chloramphenicol[6]) or decreased (acetaminophen[7]) in a human pediatric population. Often differences in sensitivity are related to differing pharmacokinetics as a function of age. Factors impacting pharmacokinetics and/or sensitivity include, but are not limited to, ontogeny of metabolizing enzymes and transporters, receptor expression and function, body composition and plasma protein content. Of course from a societal viewpoint, there is greater concern regarding harm to children and this may, in part, drive the thinking that children are more susceptible. Developing organisms (children, juvenile animals) may actually have greater capacity for repair and recovery from toxic insult and therefore, have a decreased susceptibility. Conversely, for a target organ with less repair capacity, the long term outcome of early insult during development could be devastating. Clearly, this is an area that requires greater understanding and research. It is likely that outcomes in this area will be target organ specific, and appropriately so, the desire will be to err on the conservative side of the discussion when making a risk-benefit analysis.

In conclusion, the preclinical science of toxicology and ADME plays an important role in pediatric drug development. Adult human data and prior, as well as, on-going development of pediatric safety databases carry great weight in the overall assessment of a therapeutics' pediatric safety profile. Animal data can provide an initial flag of potential target organs that would result in greater scrutiny during pediatric clinical programs. In addition, some possible targets are only or more easily assessed in animals from both a practical and ethical viewpoint. The following chapters within this section provide a more detailed analysis of the science of toxicity assessment in juvenile animals including an understanding of absorption, distribution, metabolism, and excretion as they shape our thinking in interpretation of outcomes in animals and in an attempt to determine the clinical relevance of any potential findings from these programs.

REFERENCES

1. Buelke-Sam J. Design considerations in juvenile toxicity testing. *Neurotoxicol Teratol.* 2003; 25:389.
2. Guidance for Industry: Nonclinical Safety Evaluation of Pediatric Drug Products, U.S. Department of Health and Human Services, Food and Drug Administration, Center for Drug Evaluation and Research (CDER), February 2006,
3. Committee for Human Medicinal Products (CHMP): Guideline on the Need for Non-Clinical Testing in Juvenile Animals on Human Pharmaceuticals for Paediatric Indications (DRAFT). Doc. Ref. EMEA/CHMP/SWP/169215/2005, September 29, 2005.
4. Frequently asked questions on regulatory aspects of Regulation (EC) No 1901/2006 (Paediatric Regulation) amended by Regulation (EC) No 1902/2006. Doc. Ref. EMEA/520085/2006, 12 January 2007.
5. ICH Harmonized Tripartite Guideline: Maintenance of the ICH Guideline on Non-Clinical Safety Studies for the Conduct of Human Clinical Trials for Pharmaceuticals M3 (R1), November 9, 2000.
6. Kapusnik-Uner JE, Sande MA, Chambers HF. Antimicrobial agents. In: Hardman JG, Limbird LE, Molinoff PB, Ruddon RW, and Gilman AG, eds. *Goodman & Gilman's The Pharmacologic Basis of Therapeutics*, 9th Ed. New York: McGraw-Hill; 1996. pp. 1124–1153.
7. Insel PA, Analgesic-antipyretic and anti-inflammatory agents. In: Hardman JG, Limbird LE, Molinoff PB, Ruddon RW, and Gilman AG, eds. *Goodman & Gilman's The Pharmacologic Basis of Therapeutics*, 9th Ed. New York: McGraw-Hill; 1996. p. 632.

CHAPTER 17

Preclinical Juvenile Toxicity Assessments and Study Designs

LUC M. DE SCHAEPDRIJVER, DVM, PhD and GRAHAM P. BAILEY, BSc
Johnson & Johnson Pharmaceutical Research & Development, 2340 Beerse, Belgium

INTRODUCTION

An increasing number of neonatal and juvenile animal toxicity studies are being conducted during the development of pharmaceutical or biotechnology products for pediatric use.[1–6] The objective of these studies is to assess whether there are toxicities that might be unique to immature animals, what stage of development might be the most sensitive, and if there are safety margins sufficient to support dosing in human pediatric populations. In addition, the relevance of the toxicological findings when extrapolating from juvenile animals to children is evaluated. The study design of these complex studies requires a case-by-case approach and careful programming and planning for each test material.

DIFFERENT STUDY TYPES

Juvenile animal studies are designed either to perform a general screening or to assess particular aspects of function or development of a particular target organ(s) such as central nervous system (CNS), skeletal, male and female reproductive, pulmonary, cardiovascular, renal, and immune systems.[1–6] With respect to the general screen or "standard design," these juvenile studies mirror the repeated dose toxicity studies in adults and as such comprise the following standard toxicological measures: in life monitoring of the animals including daily examination for signs of reaction to treatment, body weight (twice weekly preweaning) and food consumption assessment, ophthalmoscopy, laboratory investigations (hematology, biochemistry, and urinalysis), and gross and histopathological examinations of selected tissues and organ weights. In addition, the standard design comprises physical and sexual developmental milestones; effects on behavior,

Pediatric Drug Development: Concepts and Applications
Edited by Andrew E. Mulberg, Steven A. Silber, and John N. van den Anker

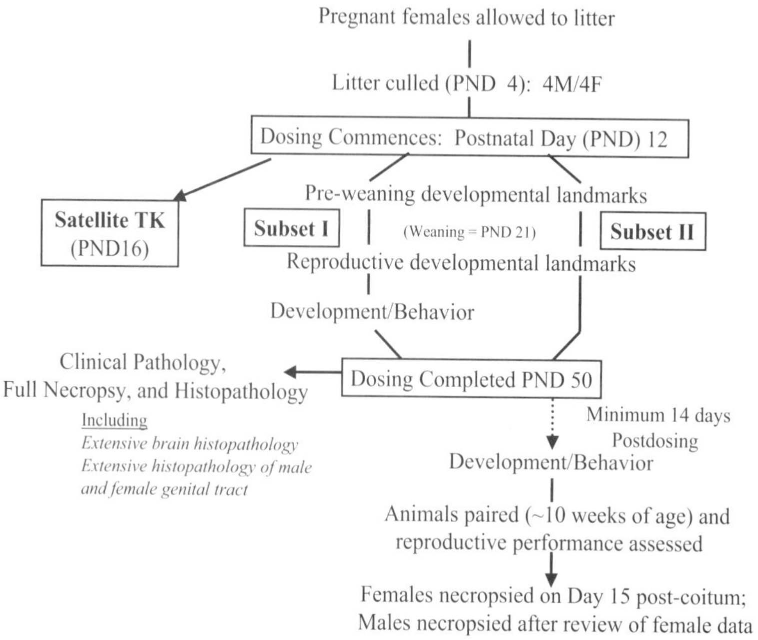

FIGURE 17.1 Main juvenile toxicity study in the rat.

learning, and memory; and reproductive performance of the offspring. An example of a study design of a main juvenile toxicity study for a CNS active compound is given in Figure 17.1.

To date, most juvenile studies have been designed on a "case-by-case" basis.[3,5,6] These complex studies should be designed purposefully and based on scientific rationale, with each endpoint carefully considered in terms of practicality and interpretability of data generated (including availability of historical control data). Numerous technical challenges exist, which need to be understood and anticipated at the outset.

Prior to the conduct of a definitive good laboratory practice (GLP) juvenile toxicity study, a range-finding study is usually performed for dose selection or to address a specific question. In addition, the inclusion of toxicokinetic/pharmacokinetic parameters in these preliminary studies is highly recommended.[3,5,6]

In our laboratory, we have developed a good alternative to conducting a stand-alone juvenile rat range-finding study. This approach involves modifying the existing pre- and postnatal developmental toxicity range-finding study by means of a simple "add-on" to the end of the study.[7] In essence, this combined range finding study—further referred to as the "combo" study design (see Figure 17.2)—not only provides justification for the selection of dose levels for the definitive pre- and postnatal study but also generates early scientific data on exposure and potential toxicity in juvenile animals. To summarize the design: during the pregnancy/lactation study phase, five groups (including two control groups) of pregnant females are dosed from Day 6 of gestation until Day 7 of lactation. The females are allowed to litter and standard littering and survival observations are made. After completion of maternal treatment on Day 7, pups are selected for a subsequent juvenile study phase. After

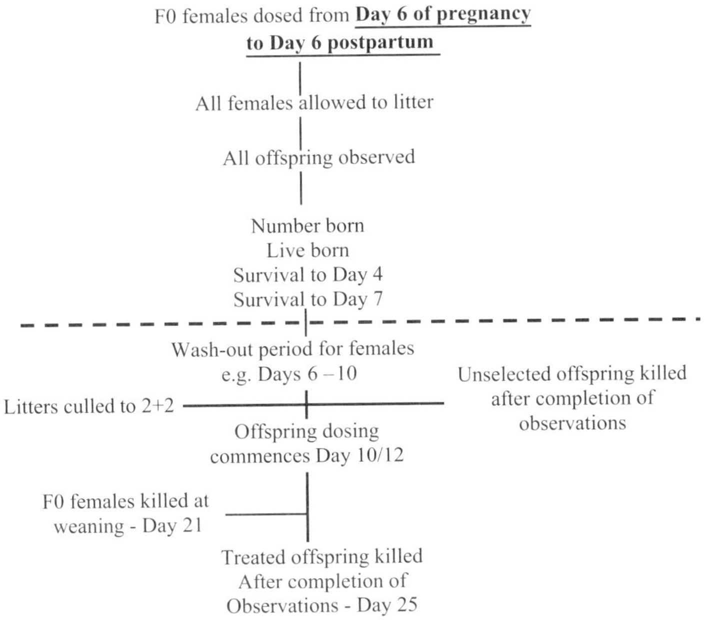

FIGURE 17.2 Combined pre- and postnatal developmental toxicity and juvenile toxicity range-finding study in the rat.

an appropriate washout period, four groups of pups, including those from one of the control groups (providing initially naive animals), are dosed directly with the test material until a suitable point postweaning. Toxicokinetic data can be collected from unselected pups (exposure via the maternal milk) and from directly dosed animals for single dose and "steady-state" assessments. This combo study can differentiate between maternal dosing mediated effects and those related to direct pup dosing and allows an early assessment of potential safety concerns. Other benefits of this approach are a decrease in drug development time and a reduction in animal usage and cost.

KEY STUDY DESIGN CONSIDERATIONS

To date, little has been published on juvenile study designs, although the number of papers on this topic will increase as more and more juvenile animal studies are being conducted.[3,5,7–12] The latter studies represent a hybrid of the pre- and postnatal developmental toxicity studies and the repeated dose toxicity studies but with direct dosing of the juvenile animals. Also, upon initial dosing, the young animals are still with their mothers and important considerations in designing any juvenile toxicity study are potential effects on mother–offspring interaction.

More importantly, study designs should take into account the following key factors: (1) matching the developmental age of the animals at the start of treatment to the age of the intended pediatric population; (2) the inclusion of specific in-life or postmortem assessments to investigate potential effects of the compound on the expected target organ system; (3) and the need to discriminate between acute effects and developmental toxicity.

Species Selection

Factors that need to be considered when choosing the appropriate species include the pharmacodynamic, pharmacokinetic, and toxicological properties of the medicinal product; comparative developmental status of the major organs of concern between juvenile animals and pediatric patients, sensitivity of the species, and the feasibility of conducting the study.[1–3]

Although traditionally both rodent (rat) and nonrodent (dog) species are used in routine nonclinical toxicology studies, current approaches for juvenile toxicity studies have been to use the most appropriate species to enable the main endpoints to be adequately assessed. Rats are usually the species of first choice since there is significant experience with the rat as an experimental model and hence a large historical control data set is available. Importantly, the appropriate developmental window in the rat[13] covers all the postnatal developing targets of interest such as CNS and skeletal, male and female reproductive, pulmonary, cardiovascular, renal, and immune systems. For example, both rats and humans undergo extensive postnatal neurological development and most brain regions/structures that develop postnatally in humans also develop postnatally in rats. In cases where, in light of the findings in the adult animal, the appropriate endpoints for investigation in the juvenile animal are behavioral and sexual development, the rat is clearly the more suitable species. In terms of reproductive development, rats mature younger and the sexual development and fertility can be easily assessed. This would be much more protracted in the dog and, while not impossible, the reproductive parameters are much less well characterized. With regard to behavioral assessments, again the routine behavioral battery used as part of the pre- and postnatal developmental studies have been developed over a considerable period and animal responses are well characterized.[3] This makes the interpretation of the data from the rat much more meaningful than in the less well characterized juvenile dog.

In some cases, however, the rat is not appropriate and other species must be considered. When drug metabolism differs substantially from humans or when a target organ of toxicity is not observed in the rat, another species such as the dog, minipig,[9] rabbit, or nonhuman primate[12] may be more appropriate. The minipig is becoming more recognized as an acceptable nonrodent alternative to dogs and primates.[14]

The use of a single species for juvenile toxicity testing is generally believed to be sufficient.[1,2,5] Therefore, if the rat has proved to be a suitable animal model for the test article, with adequate prediction of human risk, and where the toxicological responses of rats and dogs to that compound are essentially similar, there is no scientific justification for including dogs in addition to rats. However, health authorities have the option to request a second species and, therefore, early interaction with regulatory agencies is recommended.

Study Duration/Age at Dosing Start

The duration of the dosing period in juvenile animal studies, and the age of the animals at the initiation of dosing, will depend on the developing organ system(s) likely to be affected by the medicinal product and taking the age and the duration of exposure of the intended pediatric clinical population into consideration.[1–3] A good understanding of the development of the different target systems is critical. In this respect, a series of

TABLE 17.1 Developmental Events that Differ Postnatally Across Species

Organ System	Morphological and Functional Markers
Central nervous	Locomotor and fine motor development, sensory and reflexive development, cognitive development, social play, myelination, and blood–brain barrier
Male reproductive	Preputial separation, testes descent, blood–testis barrier, sex hormone production, puberty
Female reproductive	Vaginal opening, sex hormone production, puberty
Renal	Urine volume control, concentration ability, acid–base equilibrium, glomerular filtration rate, and tubular secretion
Pulmonary	Lung volume, alveolar and saccular maturation
Cardiovascular	Electrophysiology (ECG), cardiac output and hemodynamics, coronary vasculature, cardiac innervation
Skeletal	Longitudinal bone growth (at epiphyseal growth plate long bones), diametric bone growth, fusion of secondary ossification centers, vascularity
Immune	Functional immunocompetence (B and T cell, NK cell), T-dependent antibody response, IgG levels
Digestive	Gastrointestinal motility, gastric pH, gastric emptying, HCl and pepsin secretion, bacterial colonization

informative articles have been published in the journal *Birth Defects Research*, detailing the comparative development of a number of organ systems including heart,[15] immune system,[16] reproductive system,[17,18] CNS,[19,23] kidneys,[20] lung,[21] and bone.[22] More information on functional markers whose development varies across species is also given in Table 17.1.

In general, studies are conducted with duration to "bridge" to the adult animal program. When adverse reactions are expected on systems with a long development period (e.g., brain development, bone growth, immune function), investigations should ensure that the duration includes the development interval of concern. When adverse effects are expected on an organ system with a relatively short critical developmental window, such as the kidneys and lungs, then the study might be confined to that particular period of development. The importance of age versus development consideration is illustrated by the age- and species-dependent toxicity profile of angiotensin converting enzyme (ACE) inhibitors. These compounds can induce fetal renal failure. However, as nephrogenesis is complete in humans before birth and continues in rats for 4–6 weeks postnatally, rats must be tested prior to this latter period to successfully predict nephrotoxicity.[20]

Route of Administration

Ideally, the intended clinical route of administration and dosage formulation should be used unless studies in adult animals have demonstrated that an alternative route is more relevant to human use. However, when selecting the route of administration for juvenile animals, practical issues such as feasibility, dosing methodologies and equipment, dosing volume, formulation, and experience of technical staff must be taken into account. Each route of administration poses specific technical challenges depending on the species and

TABLE 17.2 Dosing Routes in Neonatal/Juvenile Animal Species

Administration Route (Earliest Day Postpartum)	Species				
	Rat	Mouse	Dog	Minipig	Rabbit
Oral gavage	From D1	From D4	From D1	From D7	From D14
Subcutaneous	From D1		From D1	From D7	
Intravenous bolus (repeated)	From D15		From D1	From D7	From D14
Intravenous infusion	From D21		From D56	From D7	
Inhalation	From D4	From D21	From D10	From D7	
Dermal[a]	From D21	From D21	From D42		From D35

[a]Not recommended in preweaned animals.

the age of the animal involved. A good overview regarding dosing methodologies for various preweaned species and routes of administration is given in an ILSI/Health and Environmental Sciences Institute (HESI) document.[24] An overview is given in Table 17.2 of the earliest possible dosing start via oral gavage; subcutaneous, intramuscular, intravenous bolus; intravenous infusion; inhalation; and dermal route for the rat, mouse, dog, minipig, and rabbit.

The feasibility of dosing via a particular route of administration such as the intravenous route is one of the first issues to consider. Switching to another systemic route is possible provided that there are only small differences in the systemic exposure profile. When using oral gavage, administering a viscous suspension through a small gauge cannula in a rat pup of 4 days of age is technically possible but not advisable; a more fluid suspension might work for gavage from midlactation. Generally, the earlier the treatment starts, the more restrictions apply. Appropriate dosing volume of 5–10 mL/kg, corresponding to 0.05–0.1 mL for a 10-g rat pup, is acceptable. Due to various factors including possible maternal rejection and maternal oral ingestion, which may result in oral exposure to the pups via the milk, the dermal route is not recommended in preweaning animals.

Dose Selection

The primary purpose of juvenile toxicity studies is to identify potential safety concerns and effects on developing organ systems in young animals. Therefore, dose levels should represent reasonable multiples of the expected exposure(s) in the pediatric therapeutic range, and where possible the high dose should achieve some identifiable toxicity (e.g., a small reduction in body weight gain). Also, in order to compare with adult data, it is interesting to include the adult no observed effect level (NOEL) or no observed adverse effect level (NOAEL) in the juvenile study. The use of a maximum tolerated dose (MTD) may result in secondary nonrepresentative effects on normal development and it is unclear what advantage demonstration of "frank toxicity" may give to any study design. In view of the differences between the U.S. Food and Drug Administration (FDA)[1] and the European Medicines Evaluation Agency (EMEA)[2] guidelines concerning the choice of the high dose, dose selection is discussed with the particular regulatory agency before study conduct.

Toxicokinetics and ADME

The inclusion of toxicokinetic (TK) assessment in juvenile studies is highly recommended to assist in explaining differences in toxicity profile between juvenile and adult animals. Neonates, in particular, have immature systems for metabolism and excretion, which can lead to marked differences in drug levels with age. This topic will be discussed in more detail later. Routinely, toxicokinetics is included in range-finding studies and TK data are most useful in designing any definitive juvenile study. Serial blood sampling is possible in larger animals, such as the dog and minipig, but not in rodents. Therefore, special attention should be given to using small amounts of blood sample in young rats since blood volumes are limited (0.1 mL/animal in young preweanlings) and sampling in preweaning animals is a terminal procedure in rodents thus requiring large numbers of offspring. On the other hand, the much higher sensitivity of the current liquid chromatography–mass spectrometry/mass spectrometry (LC-MS/MS) systems used for bioanalysis allows use of much lower blood or plasma aliquots than in the past.

C_{max}, AUC, and T_{max} are the most relevant TK parameters. Differences in plasma drug levels linked with development changes in metabolizing enzymes and transporters can be investigated by incubation of in vitro systems like isolated hepatocytes, subcellular fractions, or appropriate tissue preparations from animals at different ages, with the drug or a probe as substrates. These assessments may be valuable for elucidating (unexpected) toxicity findings and to support risk assessment.

Study Endpoints

There is a considerable discussion whether a juvenile study should be targeted to known or anticipated effects related to the mode of action observed in adult toxicity studies, or be a general study screen to evaluate all possible outcomes and to avoid missing a unique toxicity. It is generally considered that juvenile studies should be designed to determine drug effects on the overall development and function of the organ systems of specific concern. Studies should include, at a minimum, clinical signs, body weight, clinical pathology (e.g., in nonrodent species), measurement of growth (body weight and terminal tibial length), external indices of physical and sexual maturation, plus organ weights and gross and microscopic examinations of the major organ systems that develop postnatally. In practice, juvenile rat toxicity studies often include some evaluation of the CNS development with assessment of reflex ontogeny, sensory function, locomotor activity, and learning and memory test. In addition, sexual development may be assessed by measuring the onset of vaginal opening in females and balanopreputial separation in males as indicators of onset of puberty and a mating trial in animals of at least 10 weeks of age to assess potential effects on fertility.

Clinical pathology measurements may be difficult because of technical limitations with obtaining adequate samples, particularly with juvenile rats. In dogs, electrocardiograms and respiratory minute volume can also be assessed as early as Day 10 postpartum. The histopathological assessment of tissues from juvenile toxicity studies requires special expertise and a battery of historical control data. The pathologist reading the tissues from juvenile animals must be familiar with the equivalent developmental stages between laboratory animals and humans and should have access to suitable reference material to aid in interpretation.[25]

Besides this general screen, study designs should be adapted according to the therapeutic class and all available prior knowledge of the pharmacological or toxicological target.

Sample Size

Determination of the appropriate number of animals for use in a juvenile toxicity study is driven mainly by the statistical power of the design required to detect potentially adverse effects and the logistical issues involved in conducting such a complex study. Generally, it is accepted in developmental neurotoxicity (DNT) studies, where the litter is the experimental unit for biological and statistical analysis, that a minimum of 20 animals per sex per group is needed to detect statistically significant changes in neurobehavioral development. In cases where a nonrodent species such as the dog is used, animal ethics and costs, availability of test article, and other considerations may limit the sample size, increasing the likelihood of missing potential adverse effects. Toxicokinetics will usually require additional satellite animals to be allocated to each group for blood sample collection, in particular, in preweaning rodents, where collection is a terminal procedure. A full set of recovery groups may be used for functional testing or there is an option to use a high dose and control group only; in either case, this will impact the starting group size.

Allocation of Animals to Study Groups

The allocation procedure[26] is another important study design consideration when dosing commences preweaning. Different options are possible, each with advantages and disadvantages:

1. Split litter design, where one male and one female from each litter are randomly allocated to each study group. This organizational model minimizes the effects of maternal influence, gives an even distribution of littermates, and reduces animal usage. However, it increases the risk of cross-contamination between groups, and also untreated controls may dominate the litter and/or treated pups may be weak and subsequently rejected by the mother. This procedure also may increase the risk of dosing errors.
2. Whole litter design, where all pups within a litter receive the same dose. This model, which is routinely used in DNT studies, allows siblings to be assigned to separate investigations, minimizes the risk of cross-contamination, and is logistically easy to manage. However, maternal care may be a confounding factor and there could be genetic bias among groups. Also, relatively large numbers of animals are needed, which makes the study less cost effective.
3. Cross-fostering design, where pups are cross-fostered resulting in a variety of genetic backgrounds in each litter but all receive the same dose, thus eliminating the risk for cross-contamination. Theoretically, both the genetic component (littermates) and the maternal influences are randomly distributed among groups. However, if a dam rejects a cross-fostered litter—which is rare if new litters are formed within a few days after birth—a significant portion of the group could be compromised.

The selection of the allocation method should be based on knowledge of the test material, potential contamination risks, and existing toxicity and kinetic data. Obviously, the choice of litter allocation is only critical for preweaning dosing in rats; in the case of postweaning dosing, the standard is that treatments are spread across littermates.

DESIGN CONSIDERATIONS IN SPECIFIC ORGAN SYSTEMS

While a "standard design" juvenile animal study may be appropriate in a large number of cases where the compound to be tested has shown no particular toxicity in the adult animal studies or in the F1 generation of the pre- and postnatal developmental toxicity study, there will be a significant number of occasions where a more specific design is required. This will apply if the chemical/pharmacological class of compound or previous studies in humans or animals have given cause for concern for the overall growth and function of the developing system. In particular, more adequate testing should be included if an organ system of concern is considered to be at increased risk for toxicity based on the stage of development in the intended clinical population.

CNS

Study designs including an assessment of the CNS development are generally considered to be required for compounds where the CNS is a main target organ in children such as in treatments for attention deficit hyperactivity disorders (ADHD) and anticonvulsant therapy. In addition, any compound where CNS related changes in the pre- and postnatal developmental toxicity study have been observed should trigger further juvenile animal testing, including potential recovery of CNS adverse effects. CNS assessments are probably the most common additions to the standard study design either by requirement as above or at the request of a regulatory agency. In fact, this is becoming so common as to almost be considered part of "the standard" design.

Triggers that would indicate the need for neurological testing include treatment-related neurological effects in adult animals such as clinical signs of neurotoxicity, neuropathological findings, or functional and behavioral effects. In the developing animal following pre- and/or postnatal exposure, these may be nervous system malformations or functional or behavioral changes observed in the postnatal developmental testing battery.

Structural activity relationships to known neurotoxicity compounds or alterations in neurotransmitter responses should also trigger the inclusion of neurological testing.

The FDA guidelines[1] for juvenile studies state that, for developmental neurotoxicity assessments, "well-established methods should be used to monitor key functional domains of the central nervous system, including assessments of reflex ontogeny, sensorimotor function, locomotor activity, reactivity, and learning and memory."

There is no single test that assesses behavior comprehensively; therefore, it is necessary to use a battery approach. Apical testing integrates performance on a diverse set of tasks; for example, visual perception, motor function, and coordination provide an overall assessment of the animal.

An initial consideration is the choice of species. For CNS active drugs, the availability of methods for neurotoxicity assessments positions the rat as the species of first choice. In contrast to the situation in dogs, the neurobehavioral batteries in rats used as part of routine pre- and postnatal development toxicity studies for drug submission and for the developmental neurotoxicity studies in the submissions for chemicals are well established (validated). This makes the interpretation of neurobehavioral data in the juvenile rat much more meaningful than that in the poorly characterized juvenile dog. In addition, most brain regions/structures that develop postnatally in humans also develop postnatally in rats. In areas such as cognitive function and synaptogenesis, the rat may not be the most appropriate subject, in which case an alternative species such as the nonhuman primate may be considered.

In dogs, observations are generally limited to a veterinary style neurological examination, which includes spinal reflexes, postural reactions, and cranial nerve assessments and is a long way from the comprehensive testing battery available for the rat. In addition, the number of dogs used in testing is generally insufficient to pick up potential adverse effects.

Other species may be considered, such as the minipig, guinea pig, and the nonhuman primate. Behavioral testing available in the minipig is generally comparable with that conducted in the dog in its complexity. In the guinea pig, the pup is born with a mature nervous system that undergoes little postnatal maturation and as such is inappropriate for juvenile CNS developmental assessments. While the nonhuman primate may give a far more relevant assessment, ethical considerations in addition to the obvious cost implication would generally rule this species out unless absolutely necessary due to compound specificity or if further targeted investigations are envisaged following a rodent study. Therefore, in the majority of cases, the species used will be the rat.

The age at commencement of dosing should represent the anticipated age in the pediatric population. The commonly used age for commencement of dosing in juvenile rat studies is Day 4 postpartum, as this is a point at which repeated dosing can comfortably be achieved. In children, the treatment for ADHD, however, is unlikely to be aimed at an age range of less than 3 years of age, which relates to a rat pup of between 3 and 5 weeks. With respect to brain development, a 2-year-old child is approximately equivalent to a 21-day-old rat, and from this period until puberty, at 35–45 days of age in the rat, there is a broadly comparable "childlike" period of development.[23] This could therefore be considered to be comparable to human. However, rats have a very compressed developmental time window and attain adult-like motor patterns at approximately 14 days of age and reach full adult forms by 21 days of age, which in comparative terms is markedly earlier than the human child.

The development of the blood–brain barrier (BBB) is an important factor for consideration where drugs are targeted at the CNS or where effects have been seen in postnatal assessments and the patient population is less than 1 year of age. The BBB is present in all vertebrate brains and in humans is laid down within the first trimester of fetal life. The permeability of the BBB varies during development but is established in humans at around 6 months of age,[27] whereas in the rat the BBB is functional at approximately Day 10 and achieves full functionality between 33 and 40 days of age.[28] With the immaturity of the barrier in pediatric subjects less than 1 year of age, there exists the potential for damaging levels of drug in the developing brain. This problem was observed with oseltamivir, a treatment for influenza.[29] When 7-day-old rats received a single dose of 1000 mg/kg, deaths occurred which were associated with unusually high exposure to the prodrug, although a single dose of 2000 mg/kg to a 14-day-old rat resulted in no significant effects. Further investigations revealed an age-related increased brain exposure in the younger animals (1500-fold and 650-fold higher exposure levels at 7 and 14 days of age, respectively).

Therefore, the comparability of a 21-day-old rat pup with a 2-year-old child is not reflected in the development of the BBB. In the child, this is well developed but not so in the rat, and therefore results from the animal model may not be representative for the human

TABLE 17.3 Reflex Ontogeny and Behavioral Testing Battery

Parameter	Example Tests
Reflex ontogeny (assessed on an age basis)	Surface righting Air righting Negative geotaxis Auditory startle (Preyer response)
Sensorimotor function, locomotor activity, reactivity	Extensor thrust response Hopping Grip strength Swimming ontogeny Rotorod Hindlimb landing foot splay Auditory startle, or auditory startle with pre-pulse inhibition and evoked potentials
Learning and memory	Passive avoidance[30] Biel maze[31] Morris water maze[32] Radial arm maze Cincinnati T-maze[33]

situation. This does not apply only to CNS active compounds, as any xenobiotic administered at an age in rats where the BBB is not fully developed has the potential to result in brain levels of the compound in the young animal, which are not attained or observed in the adult. Therefore, it is important that this be taken into account in the age selection for the study, particularly where dosing prior to 10 days of age is proposed.

Neurotoxicity assessment needs to examine the key functions of the CNS and testing may be necessary both during the treatment period and during a posttreatment recovery period, where acute effects of treatment—a confounding factor in the interpretation of data—will no longer be apparent. The testing battery should include assessments as delineated in Table 17.3.

In addition to the in-life assessments, histopathology of both the central and peripheral nervous system may need to be evaluated, possibly including morphometry.[34] All main regions of the brain should be examined to ensure adequate assessment (i.e., olfactory bulbs, cerebral cortex, hippocampus, basal ganglia, thalamus, hypothalamus, midbrain, brainstem, and cerebellum).

Guidance on the types of assessments and procedures can be found in the U.S. EPA Health Effects Test Guidelines OPPTS 870.6300 for Developmental Neurotoxicity Studies.[35]

Reproduction

The requirements for some form of reproduction assessment in juvenile toxicity studies is becoming more common irrespective of whether a "signal" was seen in either the reproductive or repeat dose studies. Reproductive development and performance are assessed, as part of the pre- and postnatal developmental toxicity and fertility studies; however, while these studies cover the main developmental periods there exist "gaps" that the juvenile study fills. The fertility study covers the development from puberty, as the animals need to be fully sexually mature. For the mating phase, rats are generally around 6–7 weeks old at the commencement of dosing. In the pre- and postnatal study, meanwhile, the fertility of the developing offspring is assessed following exposure in utero and following exposure during lactation via the milk. It therefore assesses the impact of the test article on the developing reproductive system but it does not fully cover exposures from weaning through to puberty. Additionally, exposure via lactation may be lacking if drug is not excreted in the milk at an appreciable level. This results in the period at least from weaning until puberty where no assessment of drug exposure is made and hence there is a driver for incorporating a fertility assessment in the juvenile toxicity study.

The above evaluations are routinely performed in the rodent reproduction toxicology battery. Similarly in juvenile studies, this is predominantly assessed in the rodent species for much the same reasons as for the CNS assessment; the procedures in the rodent species are well documented, the normal performance of the commonly used strains is established, and the time frames and ease of assessment make this the most attractive option. This should not, however, preclude the use of other species where appropriate. As stated previously, this would be much more protracted in the dog and, while not impossible, the reproductive parameters are much less well characterized.

In the rat, the parameters included are for an initial screen aimed at detection, not characterization of effects or elucidation of the mechanism. Should an issue arise, then more detailed follow-up studies such as hormone level assessment (e.g., estradiol, progesterone, prolactin, testosterone, follicle stimulating hormone, luteinizing hormone) may be performed. However, these parameters and outcomes are not designed to look at subtle changes in gene/cell function and generally the cellular/molecular mechanism is unknown.

The initial assessments would be at the physical level in terms of sexual development and onset of puberty and these include the testes descent and balanopreputial separation in males and vaginal opening in females.

Testes descent, under the control of androgens and mechanical processes, occurs prenatally in the human, at 5–6 weeks old in the dog, and 15 days of age in the rat. While this is measurable, it is highly subjective and has been replaced in routine use by the much less subjective balanopreputial penis separation, separation of the prepuce from the glans penis. In rats, this occurs around 45 days of age and in the human between 9 months and 3 years and in both species androgens play a key role.

Vaginal opening/patency occurs in rats between 34 and 36 days old following the day after the first preovulatory surge of gonadotropins and is the indication of the onset of puberty. The normal estrous cycle in rats comprises four stages—proestrus, estrus, metestrus, and diestrus—and is characterized by assessment of vaginal cell cytology. The normal cycle pattern is 4–5 days in length and is controlled by ovarian steroid and pituitary profiles, which are intimately linked. Estrous cycle monitoring for abnormal cycle patterns could commence shortly after vaginal opening, but it is adequate to start monitoring for a period prior to pairing.

An overview of the further assessments of mating, sperm analysis, and terminal sacrifice are best presented in the ICH Harmonised Tripartite Guideline.[36]

It is now generally accepted that fertility in the rat is a relatively insensitive indicator of testicular problems and effects on spermatogenesis, due to the large excess of sperm produced in rodents, but it does still allow an assessment of mating behavior, precoital interval, and so on to be conducted. Testicular histopathology, however, is considered a sensitive indicator of testicular problems; sperm counts, motility, and morphological assessment (seminology) can detect problems that neither mating nor histopathological investigation pick up (e.g., late maturation in the epididymis and sperm function effects).

Immune System

Within immunology, the mouse is the most studied species, whereas the rat is the most used rodent species for toxicology including developmental immunotoxicity testing. The dog, however, closely models timing of immune development in humans, but is less used due to limited knowledge and availability of assays and reagents. For the pig, there is very little development information available and the nonhuman primate due to its extended juvenile period would require extensive follow-up over several years.

The FDA guideline[1] on safety testing for pediatric drugs does provide some information on the immune system and organ development as well as comparative tables of developmental periods but makes no specific test recommendations. The EMEA (CHMP) guidelines[2] state that immunotoxicity testing is only required if the compound class or previously conducted studies give cause for concern for the developing immune system and that testing strategies should be based on validated assays, but as with all studies in young animals, flexibility should be maintained. In addition to the functional assays, such as T-cell-dependent antibody response and cell-mediated immune assay, histopathology should also be included in line with the weight of evidence approach described in the ICH immunotoxicology guideline.[37]

Immune development involves a complex series of ontogenic mechanisms and the objective of immune testing in juvenile studies is to detect immune modulation, which may increase the susceptibility of juveniles to disease.

A single screening battery should be used to:

- Test pups during the dosing period to detect direct immune modulation.
- Evaluate immune function posttreatment (after washout) and in later life to detect developmental immunotoxicity.

The screen will include routine parameters such as hematology (differential white blood cell counts), organ weights, and histopathology (e.g., thymus, spleen, lymph nodes, Peyer's patches, bone marrow smears), but it should also include developmental milestones such as thymus development as well as the above-mentioned functional tests to assess the integrity of humoral and cellular immunity.[38]

The most widely accepted general method for immune function assessment is the determination of effects on the immune response to a T-cell-dependent immunogen (T-cell-dependent antibody response, TDAR). The plaque-forming cell assay measures primary antibody response to sheep red blood cells (SRBCs) and is achievable in rats from 2 weeks of age. The test is performed by sensitizing the animals with SRBCs either by intravenous or intraperitoneal injection with a necropsy 4 days later. The plaques formed are counted in a Cunningham slide[39] or an enzyme-linked immunoadsorbent assay (ELISA) technique[40] and may be used to measure the antibody response.

Humoral response can be assessed using keyhole limpet hemocyanin (KLH)[41], which is administered either subcutaneously or intravenously, and serum titers of anti-KLH IgM and anti-KLH IgG are assessed 7 days later. This assay is feasible from approximately 4 weeks of age. Both tests measure primary antibody response, although the TDAR involves elements of both cellular and humoral systems and is therefore less specific.

The delayed hypersensitivity response, thought to be the most sensitive test, is reported as assessable in weanling rats.[42] The animals are tested by sensitization with bovine serum albumin (BSA) mixed with Freund's complete adjuvant and challenged 6 days later with heat-aggregated BSA in one footpad and the footpad thickness is measured.

Other tests not routinely performed, including the assessment of cytokines and immunoglobulin levels using flow cytometry or ELISA techniques and host resistance assays, are well covered by Barrow and Ravel.[43]

Kidney

In humans, nephrogenesis is complete by 34 weeks of pregnancy although postnatal maturation of the nephrons and elongation of the tubules continues during the first year of life. This is a similar situation in mice and guinea pigs, where nephrogenesis is completed by birth, precluding these species from use in the assessment of postnatal developmental nephrotoxicity. In rats, nephrogenesis occurs between birth and 8 days of age and is complete by 4–6 weeks of age. In the dog, nephrogenesis commences on approximately embryonic day 29 and is completed by 2 weeks of age.

Glomerular filtration rate (GFR) is a reasonable measure of functionality in the kidney and in humans this commences prenatally and continues to increase and reaches adult levels at 1–2 years of age. In the rat, the GFR rises sharply during the first 6 weeks postnatally and in the dog this also increases between postnatal weeks 1 and 6.

In view of these differences, design and interpretation of studies in juvenile animals regarding renal development requires consideration of the variability in maturation time both anatomically and functionally between the species. This may well influence the choice of species between the rat and dog as well as the age at which dosing commences.

Assessments of the urine are conducted on routine repeat dose toxicity studies and should be considered for inclusion in the juvenile toxicity studies. These "simple" tests of kidney function can be performed on the urine: (1) volume, (2) osmolality, (3) pH, (4) sodium and potassium concentrations, (5) protein concentration, and (6) excretion of enzymes of renal origin—*N*-acetyl-β-D-glucosaminidase (NAG).

These urinary parameters together with hematological assessments for blood urea nitrogen and serum creatinine as well as histopathological examination of the kidney should provide a suitable initial screen for potential kidney damage.

While all of these assessments are quite straightforward in the adult animal, in the juvenile rat and mouse there are a number of practical issues, not least the collection of urine. It is impractical to collect urine by routine methods when the pups are very young and still reliant on their mother. The small quantities of urine produced from preweaned pups and the necessary separation from the mother for extended periods would generally preclude simple methods of collection. It is possible to collect urine directly from the bladder at necropsy but this would require very large numbers of animals on the study to enable multiple timepoints. It is generally acceptable therefore, within the context of screening for nephrotoxicity, to limit the collection of urine to shortly after weaning when standard procedures can be used successfully.

Pulmonary

The rat is considered the most acceptable model for juvenile studies because of the broad similarities between the rat and human in the stages of lung development as shown below:

Early Alveolarization

Human: birth to 1–2 years

Rat: day 4 to day 13 postnatal

Microvascular Maturation

Human: several months to 2–3 years

Rat: postnatal week 2 and 3

Late Alveolarization

Human: ends at ca. 8 years; from 2 years to adulthood, growth proportional to body weight

Rat: slow progression throughout life

While the few studies that have been conducted in the dog that address postnatal lung development have presented conflicting findings due to the biological variability at birth, it is still considered to be an appropriate species for the safety assessment of inhaled drugs for children under 2 years of age. The rabbit and nonhuman primate are not considered suitable for juvenile assessment of lung development due to their advanced stage of development at birth.

For inhaled products, dosing in rats can start as early as Day 1 postpartum and is by whole body exposure for a maximum exposure time of approximately 6 hours. This is a fairly inexact route of exposure as the animals may also be exposed to the compound dermally and orally following the onset of grooming activity and possibly through the mother's milk, as she will also have to be within the inhalation chamber. Studies can be conducted by nose only exposure in dogs from approximately 2 weeks of age when the pups can be removed from their mothers for up to 4 hours[4] and in rats generally from weaning at 4 weeks of age.

Cardiovascular

Cardiovascular development in humans, dogs, and rats varies in a number of small details in biochemical development and electrophysiology.[15] Structures and functions that are still developing in all species at or shortly after birth include the coronary blood vessels and capillaries, innervation and response to the autonomous nervous system, and the baroreceptor reflex. While most anatomic abnormalities are caused by prenatal insults, postnatal insults are more likely to cause biochemical and functional abnormalities and these may only show differences in quantity rather than quality of the effect (e.g., cardiac glycosides: immature animals are more sensitive) and neonate reactions may be completely different from adults (e.g., catecholamines cause tachycardia in adults, but bradycardia in the fetus). In the rat, functional innervation occurs after birth with the parasympathetic system maturing first. Positive ionotropic responses to sympathetic nerve response do not develop until between 2 and 3 weeks of age. In the dog, cardiac innervation is also structurally and functionally immature at birth with both the sympathetic and parasympathetic systems becoming fully functional at around 7 weeks of age. Standard repeat dose toxicology measures can still be made such as blood pressure, heart rate, and ECG (PQ, QT, and QRS intervals). If specific concerns exist about the heart as a target organ, it is recommended that a detailed literature search be undertaken before making a choice of species.

Bone

Postnatal bone growth and development pattern are similar in both laboratory species and humans. As an initial screen of bone development, a simple in-life measurement of the crown rump length or long bone growth such as the femur and tibia are routinely performed. Biochemical markers in the blood and urine and postmortem measurements such as bone mineral density (densitometry), histomorphometry, and evaluation of bone quality (biomechanics) may be included on a case-by-case basis when there is a cause for concern.

For biochemical markers of bone formation, collagen-derived products in serum can be assessed by the following techniques:

Serum P1CP: Serum type I collagen carboxy terminal propeptide—*ELISA*.

Serum Osteocalcin: Noncollagenous protein secreted by mature osteoblasts, odontoblasts, and hypertrophic chondrocytes interacts with calcium hydroxyapatite—*ELISA*.

Osteoblast-Derived Enzymes: Serum alkaline phosphatase activity (total and/or bone). In bone, ALP is synthetized by osteoblasts and participates in the initiation of mineralization—*Colorimetric*.

For markers of bone resorption the following collagen-derived products can be assessed in urine:

Hydroxyproline: Derived from collagen breakdown—*Colorimetric*.

Galactosyl Hydroxylysine: Derived from collagen breakdown—*HPLC*.

Pyridinoline: Hydroxylysyl pyridinoline and lysyl pyridinoline five fold higher in children than adults, and higher in primary hyperparathyroidism and hyperthyroidism—*ELISA*.

Acid Phosphatase: Secreted by osteoclasts into the resorptive space—*Colorimetric*.

Bone mineral density can be measured in vivo and ex vivo using a dual energy X-ray absorptiometry (DEXA or DXA) osteometer, which emits a source of X-rays at two wavelengths. The basic principle is that the larger the bone mass, the more radiation is absorbed and the less reaches the detector. This technique is a noninvasive, quantitative, reproducible, precise method, which allows for serial measurements over time including reversibility. For assessments in-life, the animal is anesthetized and either whole body (if the animal is small enough) or an area such as a limb is measured. Postmortem assessments can be made on isolated samples of either whole bone or regions such as the epiphyses.

Histomorphometry is a quantitative evaluation of bone histology in order to characterize bone architecture and gives structural, static, dynamic, and microarchitectural parameters by looking at bone architecture on undecalcified bone sections. Biomechanical testing involves the measurement of bending and torsional strength in the long bones, compression strength in the vertebral bodies, and the femoral neck strength.

As strength is a function of geometry and size, cortical thickness, density, architecture, and composition, it is therefore related to bone morphometry.

When these parameters should be assessed depends on the pharmacological activity and the existing toxicity data from previous repeat dose toxicity studies or range-finding studies in juvenile animals. As a general rule, it is advisable to use a tiered approach, where early data can be used to trigger further investigations.

CONCLUSIONS

Regulatory guidelines regarding preclincial support for pediatric drug development programs have resulted in an increasing number of protocols being developed for studies in juvenile animals. The most important factors when designing these studies are the pharmacological activity of the drug, the expected target organ or system toxicity, and specific concerns from adult animal toxicity studies and adult clinical experience.

A variety of additional considerations are to be taken into account, such as species, age at start of treatment, study duration, route of administration, study endpoints, and feasibility. Screens to assess the development of particular organ systems such as CNS and skeletal, male and female reproductive, pulmonary, cardiovascular, renal, and immune systems may be added to both rodent and nonrodent studies. Follow-up studies may be used for further investigations of findings noted during screening or specific assessments may be incorporated into initial studies when appropriate.

While most dosing, sampling, and evaluation procedures are available for the standard laboratory species, care in testing program design and planning is needed to achieve scientifically valid studies. Using the correct strategy and design, potential safety issues for drugs intended for pediatric populations can be identified. As the knowledge from juvenile animal studies grows, a better understanding of the clinical applicability/relevance of findings will be achieved.

REFERENCES

1. US FDA, Center for Drug Evaluation and Research. *Guidance for Industry, Nonclinical Safety Evaluation of Pediatic Drug Products*. Rockville, MD: US Department of Health and Human Services; February 2006.
2. European Medicines Evaluation Agency (EMEA), Committee for Human Medicinal Products (CHMP). *Guideline on the Need for Non-clinical Testing in Juvenile Animals on Human Pharmaceuticals for Pediatric Indications*. January 2008.
3. Beck MJ, Padgett EL, Bowman CJ, et al. Nonclinical juvenile toxicity testing. In: Hood RD, ed. *Developmental and Reproductive Toxicology. A Practical Approach*. Boca Raton, FL: CRC Press Taylor & Francis Group; 2006. p. 263.
4. Brent RL. Utilization of juvenile animal studies to determine the human effects and risks of environmental toxicants during postnatal developmental stages. *Birth Defects Res (Part B)*. 2004;71:303–320.
5. Hurtt ME, Daston G, Davis-Bruno K, et al. Juvenile animal studies: testing strategy and design. *Birth Defects Res (Part B)*. 2004;71:281–288.
6. Baldrick P. Developing drugs for paediatric use: a role for juvenile animal studies? *Regul Toxicol Pharmacol*. 2004;39:381–389.

7. De Schaepdrijver L, Bailey GP. An elegant study design to generate juvenile animal data early in drug development. Current experiences. *Reprod Toxicol.* 2005;20:481.
8. Maguire SR, Ridings J, Parkinson M, et al. Assessment of new drugs with potential CNS activity in juvenile rats. *Reprod Toxicol.* 2004;18:736.
9. Makin A, Christensen A, El Salanti Z. Practical aspects of juvenile animal studies in minipigs. *Reprod Toxicol.* 2003;17:498.
10. Barrow PC, Ravel G. Immune assessments in developmental and juvenile toxicology: practical considerations for the regulatory safety testing of pharmaceuticals. *Regul Toxicol and Pharmacol.* 2005;43:35–44.
11. Youssef AF, Turck P, Fort FL. Safety and pharmacokinetics of oral lansoprazole in preadolescent rats exposed from weaning through sexual maturity. *Reprod Toxicol.* 2003;17:109–116.
12. Rasmussen AD, Nelson JK, Chellman GJ, et al. Use of barusiban in a novel study design for evaluation of tocolytic agents in pregnant and neonatal monkeys, including neurobehavioural and immunological endpoints. *Reprod Toxicol.* 2007;23: 471–479.
13. Buelke-Sam J. Comparative schedules of development in rats and humans: implications for developmental neurotoxicity testing. Presented at the 2003 Annual Meeting of the Society of Toxicology, Salt Lake City.
14. Jacobs A. Use of nontraditional animals for evaluation of pharmaceutical products. *Expert Opin. Drug Metab. Toxicol.* 2006;2:345–349.
15. Hew KW, Keller KA. Postnatal anatomical and functional development of the heart: a species comparison. *Birth Defects Res (Part B).* 2003;68:309–320.
16. Holsapple MP, West LJ, Landreth KS. Species comparison of anatomical and functional immune system development. *Birth Defects Res (Part B).* 2003;68:321–334.
17. Marty MS, Chapin RE, Parks LG, et al. Development and maturation of the male reproductive system. *Birth Defects Res (Part B).* 2003;68:125–136.
18. Beckman DA, Feuston M. Landmarks in the development of the female reproductive system. *Birth Defects Res (Part B).* 2003;68:137–143.
19. Wood SL, Beyer BK, Cappon GD. Species comparison of postnatal CNS development: functional measures. *Birth Defects Res (Part B).* 2003;68:391–407.
20. Zoetis T, Hurtt ME. Species comparison of anatomical and functional renal development. *Birth Defects Res (Part B).* 2003;68:111–120.
21. Zoetis T, Hurtt ME. Species comparison of lung development. *Birth Defects Res (Part B).* 2003;68:121–124.
22. Zoetis T, Tassinari MS, Bagi C, et al. Species comparison of postnatal bone growth and development. *Birth Defects Res (Part B).* 2003;68:86–110.
23. Watson RE, DeSesso JM, Hurtt ME, et al. Postnatal growth and morphological development of the brain:a species comparison. *Birth Defects Res (Part B).* 2006;77:471–484.
24. Zoetis T, Walls I, eds. ILSI Risk Science Institute Expert Working Group on Direct Dosing of Pre-weaning Mammals in Toxicity Testing and Research. *Principles & Practices for Direct Dosing of Pre-weaning Mammals in Toxicity Testing and Research.* Washington DC: ILSI Press; 2003.
25. Kaufmann W, Gröters S. Developmental neuropathology in DNT studies—a sensitive tool for the detection and characterization of developmental neurotoxicants. *Reprod Toxicol.* 2006; 22:196–213.
26. Myers DP, Bottomley AM, Willoughby CR, et al. Juvenile toxicity studies: key issues in study design. *Reprod Toxicol.* 2005;20:475–476.
27. Costa LG, Aschner M, Vitalone A, Syversen T, Soldin OP. Developmental neuropathology of environmental agents. *Annu Rev Pharmacol Toxicol.* 2004;44:87–110.

28. Clark JB, Bates TE, Cullingford T, Land JM. Development of enzymes of energy metabolism in the neonatal mammalian brain. *Dev Neurosci* 1993;15:174–180.

29. CDER pediatric. Available at http://www.fda.gov/cder/foi/label/2006/021087s033lbl.pdf.

30. Tilson HA, Harry GJ. Neurobehavioural toxicology. In: Abou-Donia MB, ed. *Neurotoxicology*. Boca Raton, FL: CRC Press; 1999: 527.

31. Biel WC. Early age differences in maze performance in the albino rat. *J Gen Psychol.* 1940;56:439.

32. Morris RGM. Spatial localization does not require the presence of local cues. *Learn Motivat.* 1981;12:239.

33. Vorhees CV. Maze learning in rats: a comparison of performance in two water mazes in progeny parentally exposed to different doses of phenytoin. *Neurotoxicol Teratol.* 1987;9:235–241.

34. de Groot DM, Bos-Kuijpers MH, Kaufmann WS. Regulatory developmental neurotoxicity testing: a model study focussing on conventional neuropathology endpoints and other perspectives. *Environ Toxicol Pharmacol.* 2005;19:745–755.

35. US EPA. Health Effects Test Guidelines OPPTS 870.6300 for Developmental Neurotoxicity Study. August 1998.

36. International Conference on Harmonisation. *Guideline on Detection of Toxicity of Reproduction for Medicinal Products.* ICH Harmonised Tripartite Guideline Endorsed by ICH Steering Committee. Washington DC, June 1993.

37. International Conference on Harmonisation. *Immunotoxicity Studies for Human Pharmaceuticals S8.* Approval by the Steering Committee under Step 4 and recommendation for adoption to the three ICH regulatory bodies. September 2005.

38. Holsapple MP, Burns-Naas LA, Hastings KL, et al. A proposed testing framework for developmental immunotoxicology (DIT). *Toxicol Sci.* 2005;83:18–24.

39. Holsapple MP. The plaque-forming cell (PFC) response in immunotoxicology: an approach to monitoring the primary effect function of B-lymphocytes. In: Burleson GR, Dean JH, Munson AE, eds. *Methods in Immunotoxicology,* Volume 1. Hoboken, NJ:John Wiley & Sons; 1995:150–176.

40. Temple L, Kawabata TT, Munson AE, White KL Jr. Comparison of ELISA and plaque-forming cell assays for measuring the humoral immune response to SRBC in rats and mice treated with benzo[a]pyrene or cyclophosphamide. *Fundam Appl Toxicol.* 1993;21:412–419.

41. Ulrich P, Paul G, Perentes E, et al. Validation of immune function testing during a 4-week oral toxicity study with FK 506. *Toxicol. Lett.* 2004;149:123–131.

42. Bunn TL, Dietert RR, Ladics GS, Holsapple MP. Developmental immunotoxicology assessment in the rat:age gender, and strain comparisons after exposure to lead. *Toxicol Methods.* 2001; 11:1–118.

43. Barrow PC, Ravel G. Immune assessments in developmental and juvenile toxicology: practical considerations for the regulatory safety testing of pharmaceuticals. *Regul Toxicol Pharmacol.* 2005;43:35–44.

CHAPTER 18

Absorption, Distribution, Metabolism, and Excretion (ADME) and Pharmacokinetic Assessments in Juvenile Animals

LOECKIE L. DE ZWART, PhD and JOHAN G. MONBALIU, PhD

Johnson & Johnson Pharmaceutical Research & Development, 2340 Beerse, Belgium

PIETER P. ANNAERT, PhD

Department of Pharmaceutical Sciences, Katolieke Universiteit Leuven, 3000 Leuven, Belgium

INTRODUCTION

Absorption, distribution, metabolism, and excretion (ADME) are the processes governing drug disposition in animals and humans. Pharmacokinetics may be different between animals due to species-specific differences in those processes. Within one species, the disposition of a test compound may change between birth and adult age due to alteration in one or more disposition processes during development and aging.

With respect to timing, the ICH guideline (M3)[1] requires that exposure data in animals should be evaluated prior to the conduct of human clinical trials in adults. In addition, further information on ADME should be made available to compare human and animal metabolic pathways.

In contrast, specific preclinical guidelines for conducting preclinical ADME studies during pediatric drug development do not exist. However, both the European Medicines Evaluation Agency (EMEA)[2] and the U.S. Food and Drug Administration (FDA)[3] recommend collecting blood samples in juvenile toxicity studies to study the exposure of a test compound (and/or major metabolites).[4] This allows comparison of the exposure obtained in the various juvenile age groups with the exposure obtained in adult animals, which may help to understand and to explain differences in toxicities between juvenile and adult animals.

When the ADME is well characterized in adult animals (rodent and nonrodent), both above-mentioned guidelines do not require additional preclinical ADME studies in juvenile

Pediatric Drug Development: Concepts and Applications
Edited by Andrew E. Mulberg, Steven A. Silber, and John N. van den Anker

animals apart from exposure data. Nevertheless, additional knowledge of ADME in juvenile animals may be of value in the efficacy and/or safety evaluation in case the ADME characteristics in juvenile animals are comparable to the human situation at a similar stage of development.[2] Therefore, additional experiments should be considered on a case-by-case basis.

In general, in the preclinical pediatric study package, preclinical ADME will be limited to toxicokinetics in the juvenile toxicity studies. Depending on the species used and the age of the animals, individual blood samples may be used or blood samples will need to be pooled for the analysis of the test compound. Additional ADME studies, such as determination of cytochrome P450 (CYP450) enzyme activities at different ages may be performed on a case-by-case basis in order to explain the toxicokinetic findings (e.g., in case of increased exposure in neonatal rats in comparison with adults). However, alteration of exposure during aging from neonate to adult may be due to a simultaneous change of various disposition factors. Increased toxicity, for example, may be attributed to higher exposure to the test compound due to the ontogeny of the metabolizing enzymes, as well as to an increased brain penetration in neonate animals due to the immature development of the blood–brain barrier. This was the case for Prezista, a recently approved anti-HIV compound.[5] Besides in vivo evaluation, in vitro metabolism in liver microsomes and/or hepatocytes from juvenile animals of different age ranges can help to explain whether and to what extent the ontogeny of the metabolizing enzymes contributes to age-dependent changes in exposure.[5]

GENERAL CONSIDERATIONS

Differences in efficacy and toxicity in children in comparison with adults may be due to age-related differences in pharmacokinetics (ADME) and/or differences in pharmacodynamics. In humans, many examples are known. Acetaminophen is a classical example of overdosing where children are less sensitive than adults, because they possess a higher rate of glutathione turnover and more active sulfate conjugation in comparison with adults.[6]

Valproic acid[7] is more toxic in young children due to its hepatotoxicity, which is related to its Phase II conjugate metabolism. Chloramphenicol, which is well known for the "gray baby syndrome," has a higher mortality in newborns due to their inadequate glucuronidation.[8] As in humans, newborn animals are not always more sensitive or more deleteriously affected by drugs or chemicals.[5,9] As organs do develop at different rates in various animal species and humans, it is important to consider comparisons in pharmacokinetics and toxicokinetics between various age groups in children and juvenile animals to take into account the comparable age group per species from a point of view of age-related physiological development. In this respect, the particular class of enzymes (e.g., CYP450 versus flavin-containing monooxygenase (FMO)) that is driving elimination of a specific drug in animals and humans can be expected to define how relevant studies in juvenile animals are to get insight into the pharmacokinetics in children. For instance, compared to the CYP450 superfamily of drug metabolizing enzymes, the FMO class of enzymes shows remarkable sequence identity (76–86%) between orthologous proteins, indicative of substantial conservation of these genes among species.[10] On the other hand, differences in the ontogeny of the orthologous proteins in various animal species and humans may still exist. A full understanding of the clinical implications of these differences remains to be elucidated.

In humans, the age effects from birth to adult on the various aspects of pharmacokinetics have been thoroughly investigated in in vivo as well as in vitro studies.[11,12] In contrast, only few in vivo data on the pharmacokinetics in animals of various age groups are available. For theophylline, the half-life was about five times higher in newborn rabbits compared to adult rabbits,[13] which was explained by a markedly lower clearance of theophylline in newborns. In vitro data on the ontogeny of metabolizing enzymes and transporters in various animal species are somewhat more abundant.

A key question is: Can we predict differences in toxicity—due to age differences in ADME of the test compound—in children in comparison with adults, based on age differences in ADME in animal species? Further mechanistic studies in animals as a function of age certainly may provide the answer.

In this chapter an overview is given of the in vivo ADME data in various preclinical animal species for different age groups from birth to adult. Similarly, an overview is given on the ontogeny of metabolizing enzymes and transporters in animals at different ages, based on literature and in-house generated data. In addition, the ontogeny of other mechanisms, which may have an influence on the exposure in animals immediately after birth, is discussed.

Effect of Age on ADME in Laboratory Animals

Maturation of enzyme systems occurs *generally* in the first 2–3 weeks in rats as compared to the first 2–3 months in humans.[11]

Dosing of animals within the first days of life probably provides the most useful mechanistic insights with respect to sensitivities due to immature ADME-relevant proteins.[11,12]

An overview of representative literature data regarding the effect of age on ADME of a variety of drugs in animal species is provided in Table 18.1. In several cases, body-weight-normalized clearance appears to be higher in children and young animals compared to adults. This phenomenon often somewhat compensates for higher sensitivity of developing organs.[11] Nevertheless, Table 17.1 illustrates that, for most compounds, increased half-life and decreased clearance were observed. These findings can generally be attributed to limited metabolizing capacity at very young age.

The different time frames for development of ADME-specific processes between animals and humans coupled with substantial species differences in enzyme and transporter affinity profiles make interpretation of juvenile toxicity data challenging.

Absorption Limited information is available on the effect of age on drug absorption in preclinical animals. However, a few studies report on the physiological differences between immature and adult intestine and intestinal content. For instance, Ee et al.[14] observed limited reesterification of triglycerides in the intestine of suckling rats. Instead, higher levels of free fatty acids and cholesterol were observed. Clearly, altered composition of small intestinal content is expected to have substantial impact on dissolution and thus absorption of especially poorly soluble drugs. Another intraluminal condition affecting oral drug absorption and disposition in young animals is the limited β-glucuronidase activity at young age. A 15-fold increase in the activity of this hydrolytic enzyme in the intestine of rats PND (postnatal day) 7–35 has been reported.[15] Also, as clear influences of the diet on intestinal absorption characteristics of orally ingested nutrients[16] and drugs have been demonstrated, substantial age-related differences are to be expected in intestinal drug dissolution and solubility behavior, as well as intestinal barrier function, between preweanling (milk-based diet) and adult (solid food) animals. Sangild et al.,[17] who

TABLE 18.1 Effect of Age on ADME in Various Animal Species

Species	Compound	ADME Parameter	Effect of Age	References
Mouse	Cyclosporin A, digoxin	Brain/blood distribution ratio	Increased in neonatal and young animals.	25
Mouse	Digoxin	Renal clearance	Digoxin clearance peaks around weaning; clearance is very low in first weak after birth.	41
Mouse	Genistein	Exposure	Higher exposure in neonatal animals attributed to lower capacity for genistein conjugation in perinatal mice.	28
Rat	Bisphenol A	Biliary excretion	Possibly reduced at young age (no secondary peak in plasma profile as seen in adult animals)—see also effects on metabolism; limited enterohepatic circulation at young age.	42
Rat	Domperidone	AUC	Higher AUC at PND 1 and PND 6; brain/plasma ratio higher at PND 1.	26
Rat	Ethanol	Metabolism	Limited capacity at PND 5–30.	57
Rat	Kanamycin	$t_{1/2}$	2.5-fold increase at PND 12 versus PND 25.	58
Rat	Ouabain, digoxin	Hepatic elimination	Strongly reduced in newborn rats, most likely related to Oatp1a4 ontogeny.	48, 59
Rabbit	Diltiazem	V, Cl, AUC	No changes with age; only higher elimination in rabbits at PND 30.	39
Rabbit	Phenobarbital	$t_{1/2}$, V, Cl	Longer $t_{1/2}$, slower Cl (bound and unbound drug), no difference in V, larger fraction unbound (fu) in rabbit pups (PND 19–20).	60

Rabbit	Theophylline	$t_{1/2}$, V, Cl, fu (in vivo and isolated perfused liver)	Longer $t_{1/2}$, larger V, slower Cl with maximum at week 16, larger fu in newborns; limited metabolizing capacity.	13
Rabbit	Verapamil	$t_{1/2}$, V, Cl, fu	Longer $t_{1/2}$, lower Cl, no difference in V and fu.	40
Dog	Theophylline	$t_{1/2}$, V, Cl	Longer $t_{1/2}$, lower Cl in neonates, no difference in V; lower metabolizing capacity.	61
Dog	Phenytoin	$t_{1/2}$	Lowest in pups at PND 30.	62
Monkey	Stavudine	Cl, V, $t_{1/2}$, CSF/plasma ratio	Cl and V increased 2.3 times; $t_{1/2}$ decreased between 1 week and 4 months and then remained constant.	63
Pig	Cyclosporin A	$t_{1/2}$, Cl	Longer $t_{1/2}$, lower Cl, no difference in V; lower metabolizing capacity in neonates (PND 5).	64
Pig	Lidocaine	$t_{1/2}$, Cl, fu	Longer $t_{1/2}$, lower Cl, no difference in V; lower metabolizing capacity and fu lower in newborns.	24
Pig	Parathion	Clearance	Clearance after intravenous administration increased from 7 to 35 and 121 mL/min/kg in newborn, 1-week-old, and 8-week-old piglets.	65

Abbreviations: $t_{1/2}$, half-life; Cl, clearance; AUC, area under the curve; fu, fraction unbound; V, distribution volume; PND, postnatal day.

studied the ontogeny of gastric function in the pig, found that gastric acid secretion and the synthesis and secretion of gastrin matured in late fetal and early postnatal life in the pig. In the rat, however, the acid secretory capacity does not reach mature levels until after weaning.[18] In dogs, hydrochloric acid secretion was reported at postnatal days 1–2 (reviewed in Walthall et al.[19]).

Finally, protein expression data for the efflux transport protein P-glycoprotein (P-gp; encoded by Mdr1a/b in rodents; see Table 17.3) obtained in two independent studies [20,21] have demonstrated about fivefold lower levels in the small intestine of neonatal compared to adult mice. As P-gp can play a modulating role in the intestinal absorption [22] of its orally administered substrates (e.g., digoxin [23]), it is expected to contribute to age-dependent absorption kinetics of certain drugs.

Distribution Physiological factors that are known to have an influence on the distribution of compounds, and for which ontogenic differences are observed in humans, include body composition and plasma protein binding. These types of data are scarce in neonatal animals. Our own data (unpublished) indicate that, in Sprague–Dawley rats, the total plasma protein and plasma albumin concentration increased gradually (twofold) from birth to adult age (PND 42). Information on plasma concentrations of α_1-acid glycoprotein for binding of basic lipophilic compounds would be interesting. A fast decline of α_1-acid glycoprotein plasma concentration in pigs from 14 μg/mL at Day 1 to 0.70 μg/mL at 1 month of age was observed. This decrease most probably contributed to the increase in clearance of lidocaine with age.[24]

The role of drug transporters in various tissues becomes increasingly recognized to mediate drug uptake in various tissues. Thus, it is clearly of interest to study the effect of age on tissue distribution of transporter substrates. For instance, in neonatal mice [25] and rats [26] the brain/plasma or blood ratios of cyclosporin A, digoxin, and domperidone were higher in comparison with adult animals. This may be due to the immature blood–brain barrier and ontogeny of certain efflux transporters (i.e., P-glycoprotein) in the brain. [20,25,27]

Metabolism The ontogeny of drug metabolism enzymes is probably one of the major factors responsible for age-related changes in clearance for metabolically cleared drugs. The rate of drug metabolism is affected by the relative liver size and by the liver blood flow, but the immaturity of several enzyme systems at early age is also a major factor. Drug metabolizing enzymes (DMEs) are typically classified as enzymes mediating phase I (or oxidative) and phase II (or conjugative) metabolism. Isoenzymes belonging to the cytochrome P450 (CYP450) class are by far the most important phase I enzymes, although other enzymes such as those belonging to the FMO class also mediate oxidation of certain drugs (in the latter case oxidation typically occurs at heteroatoms, e.g., to form N-oxide metabolites).

The most important phase II enzymes are glucuronosyltransferases (UGTs), sulfotransferases (SULTs), and glutathione-*S*-transferases (GSTs), mediating the conjugation of phase I metabolites or drugs, forming glucuronide, sulfate, and glutathione conjugates, respectively. For some compounds, direct conjugative metabolism can also occur. For instance, glucuronidation is the main elimination mechanism for genistein in the mouse, and limited glucuronidation capacity in neonatal animals was reported to cause higher exposure levels in developing animals [28] (see Table 17.1).

Age-related differences in these metabolizing enzymes have been studied at different levels, including the mRNA level, protein level, or enzyme activity level (using probe substrates). Although the information on mRNA expression of drug metabolizing enzymes may be useful, literature data as well as our own data [29] indicate that mRNA expression levels do not always coincide with the activity of the corresponding enzymes. However, information on the enzyme activities is not always available and mRNA expression may be useful to observe a trend in the ontogeny of enzymes and can help in relevant selection of enzymes to be investigated at the activity level.

Most information is available on the ontogeny of CYPs in mouse, rat, and rabbit. In Table 18.2 the results from literature concerning the ontogeny of enzymes determined by enzyme activity or protein expression in the liver from various species are listed. This table is not meant to be complete with all studies performed but provides representative literature data in preclinical species for different metabolizing enzyme systems. Information concerning the ontogeny of metabolizing enzymes in humans is given (among others) elsewhere.[12]

With respect to the age dependency of FMO-mediated metabolism, some older in vitro studies were performed in mouse, rat, and rabbit liver microsomes with dimethylaniline as a substrate (see Table 17.2). Formation of the N-oxide metabolite of dimethylaniline was measured at various ages to determine ontogeny of FMO activity in these species. Comparison of the data illustrates that ontogeny of this FMO-mediated metabolism appears to depend on the species considered. For instance, while the most notable increase in activity is observed immediately after birth in the mouse,[30] a gradual increase is observed in the rabbit [31] between neonatal and adult age.

Alnouti and Klaassen[32] studied the ontogeny of several isoenzymes of the sulfotransferase family in different organs of C57BL/6 mice. These enzymes demonstrated different ontogenic expression patterns in different organs. In liver, expression of a group of sulfotransferases (Sult1a1, 1c2, 1d1, 2a1/2 and PAPSs2) increases gradually from birth until about 3 weeks and declined somewhat thereafter. Sulfotransferase Sult1c1 showed the highest expression before birth and declined thereafter, whereas Sult3a1 mRNA expression was very low in fetal liver and remained low in males, but increased dramatically in females after PND 30.

The organ mainly responsible for the metabolism of various drugs and compounds is, of course, the liver. However, other organs may also contribute to the metabolism of certain compounds. In relatively older literature,[33] the ontogeny of benzo[*a*]pyrene (BaP) hydroxylase, epoxide hydrolase (EH), and glutathione-*S*-transferase (GST) was studied in lung and brain of C57BL/6 mice. In lung tissue BaP hydroxylase, EH, and GST activities were high at birth and then decreased in the first days after birth. BaP hydroxylase activity showed a burst again at PND 6, EH activity increased gradually to adult value, and GST activity increased to reach adult level at weaning. In brain tissue, BaP hydroxylase activity decreased after birth and remained stable thereafter; EH activity increased only slightly after birth and GST activity decreased in the first days after birth and then continuously increased from PND 6 until weaning.

γ-Glutamyltransferase (γ-GT) is a key enzyme in the metabolism of glutathione and glutathione-substituted molecules. The ontogeny of γ-GT was studied in Sprague–Dawley rats. It was found that the alveolar type II cell is the only cell producing γ-GT in the newborn lung and that it synthesizes a form of γ-GT that appears to differ from that produced at a later time point by the Clara cell. The enzyme activity was low in the newborn rat and a sudden increase was observed from approximately PND 10 until adult level.[34]

Several CYP enzymes have been studied in the small intestine of rats. CYP2B protein expressed throughout maturation in Ao/01a Hsd rat intestine, but was low at birth and increased modestly until adult level.[35] CYP1A1 expression in Wistar rats increased sharply

TABLE 18.2 Ontogeny of Drug Metabolizing Enzymes in Liver of Several Animal Species

Species	Isoform	Protein Expression or Enzyme Activity	Ontogeny	References
Mouse	B(a)P hydroxylase	Activity	Gradual increase from birth until weaning.	33
	EH	Activity	High levels at birth; decreased first days after birth and then increased to adult value.	
	GST	Activity	Low at birth; starts increasing on PND 6; adult level not reached at weaning.	
	FMO	Dimethylaniline N-oxidase activity	Rapid increase at birth to reach 60–80% of adult levels at PND 3; slow gradual increase thereafter.	10, 30
Rat	GST α en μ	Protein	No expression in fetal hepatocytes.	66, 67
	GST π	Protein	Present in fetal hepatocytes.	
	Microsomal GSTs	Activity	Very low in fetal liver; increasing continuously after birth; adult level at PND 50–150.	68, 69
	FMO	Dimethylaniline N-oxidase activity	Activity in liver microsomes doubles between 3 and 12 weeks of age.	10, 70
Rat (Wistar)	UGT1A1	Activity	UGT1A1 activity increased gradually from birth to PND 6, followed by a sharper increase until adult level.	71
Rat (Sprague–Dawley)	Carboxylesterase	Activity	Approximately 20% at PND 3 and 50% at PND 21 of adult level.	43
	A-esterase	Activity	20–50% at PND 3 and 90% at PND 21 of adult level.	
Rat	CYP1A	Protein	(Mainly CYP1A1) Sharp increase just before weaning; then fourfold decrease to adult levels by PND 60.	36
Rat (Wistar)	CYP3A	Activity (6OHT) and protein (CYP3A2)	Gradual increase up to PND 25 then decrease in female and further increase in male.	37
Rat	CYP3A	EROD activity	Activity increased approximately sixfold in males from PND 14 to PND 60; no increase observed in females.	72
	CYP2D	Protein	Sharp increase from PND 3 to PND 14.	73

Rat (Wistar)	CYP1A	EROD activity	Slow and minor increase from PND 10 to PND 20; strong increase (ca. fivefold) from PND 20 to PND 40.	38
	CYP2B	PROD activity	Strong (fourfold) increase from PND 10 to PND 40.	
	CYP2E1	CYP2B1	Gradual increase from PND 10 to PND 40.	
Rabbit	CYP3A	Demethylase activity	Sharp increase from PND 1 to PND 16; maximum activity at PND 30, then decrease to adult level.	40
	Esterase	DTZ acetylase activity	Gradual decrease in activity after birth until adulthood.	39
	CYP3A	DTZ demethylase acitvity	Increased with age at PND 30.	39
	FMO	Dimethylaniline N-oxidase activity	Activity levels at PND 4 are ca. 20% of adult levels; gradual increase.	10, 31
Dog	P450	Protein	Low at birth; increases fivefold at PND 28–42.	74
	G6P	Activity	Fourfold increase from birth to adult level at PND 28–42.	
	UGT	Activity	Fourfold increase from birth to PND 28–42; then slight decrease to adult.	
	CYP1A	EROD activity	Marked increase during first 8 weeks, then decrease to lower level.	75
	CYP2B	Aminopyrine N-demethylase	Increase during first 8 weeks, then more gradual further increase.	
	CYP2E1	Aniline hydroxylase	Increase from birth until 3 weeks of life, then constant.	
	CYP2A	Coumarin hydroxylase	Increase during first 5 weeks, then constant.	
Pig	CYP450 content		Low at birth and increasing to maximal levels at 4–6 weeks.	76
	CYP	*p*-Nitroanisole O-demethylation	Marked increase to reach maximal activity at 4 weeks of age.	76
	UGT	*p*-Nitrobenzoic acid reduction and Phenolphthalein	Marked increase from birth to reach maximum activity at 4–6 weeks of age.	

Abbreviations: EH, epoxide hydrolase; GST, glutathione-*S*-transferase; UGT, glucuronosyltransferase; CYP, cytochrome P450 enzyme; 6OHT, 6-OH testosterone; EROD, ethoxy-resorufine O-demethylation; PROD, pentoxy-resorufine O-deethylation; G6P, glucose-6-phosphatase; PND, postnatal day.

just before weaning to levels fourfold higher than adult level at PND 60, whereas CYP1A2 expression increased sharply at weaning to plateau from PND 60 at adult levels.[36] CYP3A activity, as measured by 6OH-testosterone hydroxylation, and CYP3A2 protein expression in Wistar rats increased sharply (seven- to tenfold) from weaning to adult level.[37] CYP1A, CYP2B, and CYP2E1 activity in brain of Wistar rats was found to gradually increase from PND 10 to PND 40.[38] In rabbits, the diltiazem (DTZ) demethylase activity (CYP3A) was found to increase with age up to PND 30 in liver and gut, but no change with age was observed in blood.[39]

Also in rabbits, the esterase activity was measured by diltiazem (DTZ) deacetylase activity in several tissues. A gradual increase in activity was observed in liver, lung, brain, and blood until adulthood, but no change with age was observed in the gut.[39,40]

In our laboratory the ontogeny of several metabolizing enzymes in liver and various transporters in liver, kidney, and brain of Sprague–Dawley rats was studied.[29] In liver, both enzyme activity and mRNA expression were determined. The transporters were only studied by mRNA expression (see section on Transporters). The results of the metabolizing enzyme activities (Figure 18.1) showed that there were considerable differences between the individual isoenzymes. Moreover, the patterns observed for mRNA expression levels were not always reflected in enzyme activities. For example, maximum mRNA expression of CYP3A2 in rats was observed at PND 7, whereas enzyme activity of CYP3A1/2 showed a gradual increase toward adulthood. Another aspect is that mRNA expression is specific for

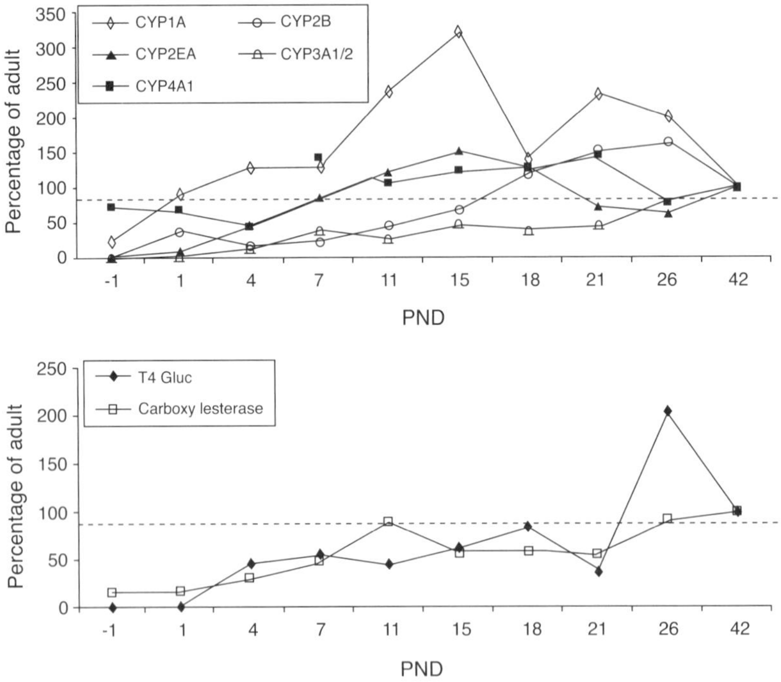

FIGURE 18.1 CYP450 isoenzyme activities (top graph), carboxylesterase and T4-GT activities (bottom graph) as function of age in male Sprague–Dawley rats (the activities are expressed as a percentage of adult values in male rats).

an isoenzyme, such as CYP1A1 or CYP1A2, whereas a substrate is not always isoenzyme specific.

Excretion The functional capacity of liver and kidney as organs mediating excretion of xenobiotics is typically low at birth. For instance, in the liver, many transport proteins (e.g., those belonging to the Oatp family) mediating the hepatic uptake of xenobiotics—the first step in hepatobiliary excretion—exhibit undetectable or low expression. Also in the kidney, the active secretion of transporter substrates such as digoxin is lower at birth, peaks during adolescence, then declines a bit toward adult levels.[41]

Studying the effect of age on biliary excretion of xenobiotics is clearly more challenging, partly due to the relatively invasive methods required to measure biliary excretion in vivo. Nevertheless, comparison of plasma pharmacokinetics in young versus adult animals often provides indirect evidence for at least reduced biliary excretion of drugs in immature animals. For instance, substantially increased levels of the endocrine disrupting agent bisphenol A and its monoglucuronide were measured in neonatal animals.[42] Also, in contrast to adult animals, neonatal rats dosed with bisphenol A do not show a secondary maximum in the plasma concentration time profile, which is consistent with reduced biliary excretion and/ or enterohepatic circulation. Reduced enterohepatic circulation may also be attributed to inefficient hydrolysis of bisphenol A glucuronide in the intestine of neonatal animals.[15]

Is There a Relation Between Ontogeny of Metabolism and Toxic/ Pharmacodynamic Effects? Juvenile rats are more susceptible to the acute toxicity of the organophosphorus (OP) insecticides, like parathion and chlorpyrifos, than adult rats. The main mechanism of action of OP insecticides or their active metabolites is the inhibition of acetylcholinesterase (AchE), but they also interact with other esterases relevant to cholinergic toxicity (i.e., carboxylesterases and A-esterases). The reactive metabolites of OP insecticides are the oxons and the increased sensitivity of juvenile rats to OP insecticides appears to be due to a greater portion of oxon metabolites reaching the brain. The oxons are formed by desulfuration by CYPs, whereas another metabolic route of OP insecticides is CYP-mediated dearylation, which is a detoxification route. In hepatic microsomes at all ages tested (PND 1, 3, 12, 21, 33, and 80), a higher dearylation rate than desulfuration rate was measured, indicating a more efficient detoxification than activation. These results suggested that other (than CYP) detoxification mechanisms are responsible for the age-related differences in toxicity of these compounds.

Another detoxification route for the oxons are esterases. Carboxylesterase activity and A-esterase activity in liver, lung, and plasma were low in neonatal rats (PND 7) but gradually increased to adult level at PND 90. The lower esterase capacity in the liver may be responsible for more oxon metabolites escaping the liver and thus causing toxicity at target sites.[43,44] Recently, Anand et al.[45] investigated the hypothesis that inefficient detoxification was the primary reason for the greater sensitivity of immature Sprague–Dawley rats to the acute neurotoxicity of deltamethrin (DLM). It was shown that the internal exposure to DLM, as determined by the blood DLM areas under the concentration versus time curve (AUC), was closely correlated with toxic signs (salivation and tremors). DLM metabolism in vitro by plasma and liver carboxylesterases and hepatic CYPs progessively increased with maturation.

Transporters As mentioned earlier, drug transport proteins (further referred to as DTP or transporters) are increasingly recognized as determinants of ADME processes in animals and humans. While the ontogeny of DTP was initially mostly studied in rats, several recent studies have been performed using mouse tissues. The mouse recently gained interest as a more common laboratory species due to availability of the mouse genome sequence as well as transgenic and knockout strains that are useful in pharmacokinetic studies.

Data extracted from representative studies on the ontogeny of transporters in various organs and various preclinical species are presented in Table 18.3. Not much data are available yet on the ontogeny of transporters in humans.

In a recent study conducted in our laboratory, age-dependent mRNA expression was studied for various transporters in liver, kidney, and brain (cerebellum) of Sprague–Dawley rats (Figure 18.2).[29] Generally, mRNA expression in liver could be divided into three groups of ontogeny profiles: (1) expression high at birth and decreasing toward adulthood (Bcrp, Mrp1, Mrp2, Mrp3, and Mrp6); (2) expression low at birth and increasing toward adulthood (Oct1, Oct2, Oat2, and Oatp1a4); and (3) expression increasing from birth to reach a maximum around PND 15–22 and then decreasing to adult level (Mdr1a, Oat3, and Oat1). In addition, the ontogeny of transporter mRNA expression was shown to be organ specific.

Notably, the ontogeny of DTPs (in mouse and rat) appears to depend on the role of these DTPs in the transport of endogenous versus exogenous substrates. For instance, Oatp1b2 expression in rat liver matures substantially earlier than the Oatp1a1 and 1a4 isoforms.[46]

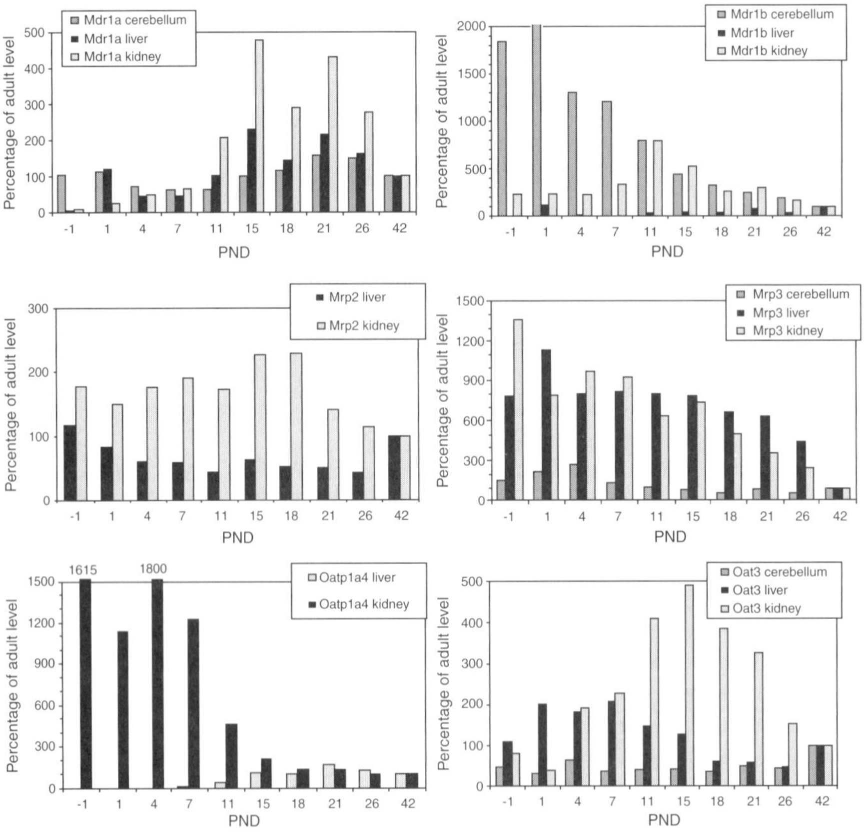

FIGURE 18.2 mRNA expression of several transporters in liver, kidney, and brain as function of age in male Sprague–Dawley rats (the data are expressed as a percentage of adult values).

TABLE 18.3 Ontogeny of Transporters in Several Organs of Various Animals at Different Levels of Effect

Species	Organ	Isoform (Gene Symbol)	mRNA, Protein, or Activity	Ontogeny	References
Mouse	Brain	Mdr1a (*Abcb1a*)	mRNA	PND 1 level is 30% (females) to 50% (males) of adult level (PND 42).	25
		Mdr1b (*Abcb1b*)		Constant between PND 1 and PND 42.	
	Kidney	Mdr1a (*Abcb1a*)		Twofold (males) to sixfold (females) increase between PND 1 and PND 42.	
		Mdr1b (*Abcb1b*)		Constant between PND 1 and PND 42, except fivefold increase in females at PND 42.	
	Liver	Mdr1a (*Abcb1a*)		For females, PND 1 level is 1/6 of adult level, reached at PND 12.	
				For males, PND 1 level is equal to adult level, with eightfold peak at PND 12.	
		Mdr1b (*Abcb1b*)		At PND 1 level is approximately twofold that of adult (PND 42) and PND approximately five fold for 12–19.	
	Brain	Mdr1a (*Abcb1a*)	Protein	PND 1 levels are 20% of adult levels, reached at PND 21.	20
	Intestine	Mdr1a/b (*Abcb1a/b*)	Protein	PND 1 levels are 20% of adult levels, gradual increase.	20, 21
	Liver			Expressed at adult levels from birth onwards.	
	Kidney			Expressed at adult levels from birth onwards.	
	Liver	Mrp2/4 (*Abcc2/4*)	mRNA	PND 1 level approximately equal to adult levels, limited age-dependency in between.	77
		Mrp3 (*Abcc3*)		PND 1 level is 25% adult level, reached at PND 30.	
		Mrp6 (*Abcc6*)		Not detected before PND 10, 50% reduction between PND 10 and PND 15.	
	Kidney	Mrp1 (*Abcc1*)	mRNA	PND 1 levels equal to adult level in females or twice the adult level in males.	77
		Mrp2 (*Abcc2*)		PND 1 level is 30% of PND 45 level; rapid increase at PND 15.	
		Mrp3 (*Abcc3*)		Sharp increase at PND 15, back to perinatal levels in males at PND 45.	

(*continued*)

TABLE 18.3 *(Continued)*

Species	Organ	Isoform (Gene Symbol)	mRNA, Protein, or Activity	Ontogeny	References
Mouse	Kidney	Mrp4 (*Abcc4*)		Constant in males, gradual increase by threefold in females (PND 1–PND 4, 5).	
		Mrp5 (*Abcc5*)		Highest levels at birth, approximately twofold reduction by PND 45.	
		Mrp6 (*Abcc6*)		Constant levels for PND 1–45.	
	Liver	Oct1 (*Slc22a1*)	mRNA	Level at birth is 10% of adult (PND 45) level, reached by PND 22.	78
	Kidney	Oct1 (*Slc22a1*)		Gradual increase between PND 0 and PND 45.	
		Oct2 (*Slc22a2*)		Constant levels in female mice for PND 0–45, 50% of adult levels in male mice for PND 0–22.	
		Octn1/2 (*Slc22a4/5*)		10–30% of adult levels PND 0–10, adult levels by PND 15.	
	Liver	Oatp1a1 (*Slco1a1*)	mRNA	Absent before PND 30; female = 30% of male.	46
		Oatp1a4 (*Slco1a4*)		Absent before PND 10; male = 30% of female.	
		Oatp1b2 (*Slco1b2*)		Level at birth is 30% of adult level, reached at PND 23.	
	Kidney	Oatp2b1 (*Slco2b1*)		Very low until PND 15, adult levels reached at PND 23.	
		Oatp1a1 (*Slco1a1*)		Absent until PND 30, then only expressed in males (PND 45).	
		Oatp1a4 (*Slco1a4*)		Low expression from before birth until adulthood.	
		Oatp1a6 (*Slco1a6*)		Level at birth is 25% of adult level, reached at PND 15–22.	
		Oatp2b1 (*Slco2b1*)		Level at birth corresponds to adult level.	
		Oatp3a1 (*Slco3a1*)		Constant in female PND 0–45, doubles in males PND 30–45.	
Rat	Brain	Mdr1a/b (*Abcc1a/b*)	Protein	Increased with postnatal development starting on PND 7.	27
	Intestine	Octn2 (*Slc22a5*)	mRNA	Perinatal levels are twofold higher than adult.	79
			Activity	Na^+-dependent L-carnitine uptake significantly higher in newborn compared to suckling rats; activity disappears after weaning.	

Rat	Kidney	Oat1 (*Slc22a6*)	mRNA	Large increase in expression directly after birth.	80
			Activity	Probenecid-sensitive PAH accumulation in renal cortical slices of neonatal rats is 75% of adult level.	
	Kidney	Oat1 (*Slc22a6*)	mRNA	Level at birth is 20% of adult level, gradual increase.	81
		Oat2 (*Slc22a7*)		Very low until PND 35, then rapid increase in female rats only.	
		Oat3 (*Slc22a8*)		Level at birth is 30% of adult level, gradual increase.	
	Liver	Ntcp (*Slc10a1*)	mRNA	PND 1 expression is 35% of adult level, reached after 1 week.	55
			Protein	PND 1 expression is 50% of adult level, reached after 1 week.	
	Liver	Bsep (*Abcb11*)	Immunodetectable protein at membrane surface	First detected at GD 19, adult levels reached by PND 12.	50
		Mrp2 (*Abcc2*)		First detected before GD 15, adult levels by PND 10.	
		Mrp6 (*Abcc6*)		First detected before GD 15, adult levels by PND 10.	
		Ntcp (*Slc10a1*)		First detected at GD 19, adult levels reached by PND 4.	
		Oatp1a1 (*Slco1a1*)		First detected at PND 19, adult levels reached by PND 29.	
		Oatp1a4 (*Slco1a4*)		First detected at PND10, adult levels reached by PND 29.	
		Oatp1b2 (*Slco1b2*)		First detected at PND 4, adult levels reached by PND 29.	
	Liver	Oatp1a4 (*Slco1a4*)	mRNA	In males, PND 1 level ~30% and ~75% of PND 30 and adult (PND 45) levels; in females, PND 1 level ~30% of adult level, reached at PND 25.	49
			Protein	PND 0–20: very low expression; expression peaks in males at PND 35, then 50% decline by PND 45; adult level in females reached at PND 30.	
	Liver	Oatp1b2 (*Slco1b2*)	mRNA	PND 1 level is 20% of adult (PND 45) level, gradual increase with age.	47
	Liver	Ntcp and/or Oatp(s)	Activity	±Twofold lower V_{max} for taurocholate uptake in hepatic basolateral membrane vesicles from suckling rat compared to adult rats.	51
	Liver	Ntcp and/or Oatp(s)	Activity	Progressively increased V_{max} for taurocholate uptake in freshly isolated hepatocyte suspensions for rats from PND 7 to PND 54.	51

Abbreviation: GD, gestation day.

Furthermore, Oatp1b2 expression was not altered upon treatment of rats with prototypical inducers.[47] These observations have been attributed to the more important involvement of Oatp1b2 in hepatic uptake of endogenous compounds (such as bilirubin, and bile salts).

Oatp1a4 is the only hepatic uptake transporter with affinity for cardiac glycosides. The absence of hepatic Oatp1a4 in rats before PND 10 (see also data shown in Figure 17.2) explains the 100-fold lower LD_{50} for ouabain in neonatal versus adult rats.[48] In addition, induction of hepatic Oatp1a4 in neonatal rats by the prototypical inducer PCN (pregnenolone-16α-carbonitrile) has been observed.[49] Treatment with PCN accelerates the maturation of Oatp1a4 during development and protects neonatal rats against increased sensitivity to cardiac glycoside toxicity. The increase in Oatp1a4 expression around weaning (see Table 17.3) has also been related to the dramatic increase in bile acid concentrations in serum at this age. Therefore, it seems that the relatively late onset of Oatp1a4 expression is one way the liver copes with higher capacity needs for bile acid uptake during development. Furthermore, the expression of Mrp2 in fetal rat liver has been related to the initiation of bile-salt-independent bile flow before birth.[50] The rapid increase in Ntcp and Bsep levels in the perinatal period is in line with the onset of enterohepatic circulation around birth, and clearly precedes the increase in Oatp expression levels. These findings generally suggest that transporter maturation processes are tailored to support acute changes in hepatobiliary disposition of endogenous and exogenous compounds around suckling, weaning, and nutrition.[50] Interestingly, these recent findings at the molecular level are in line with the few transporter activity measurements that were conducted about two decades ago with taurocholic acid as a probe. Indeed, two independent studies[51,52] (see Table 18.3), conducted in suspended hepatocytes and in hepatic basolateral membrane vescicles, clearly indicate age-dependent increase in the maximum velocity at which the rat liver can accumulate bile salts.

Ontogenic profiles of many transporters contribute to a better understanding of pharmacokinetic differences that have long been known. For instance, the limited ability of the neonatal kidney in humans, dogs, rats, rabbits, and sheep to eliminate the model organic anion *para*-aminohippurate (PAH) has been reported by several groups.[53,54] Recently, this could be attributed to the gradually increasing organic anion transporter expression and activity in kidney after birth.[53] It is noteworthy that particularly young rats also appear susceptible to induction of organic anion transport by thyroid hormones and dexamethasone.[53]

With respect to transporter expression levels measured at the protein level, it should be noted that intracellular presence of transport proteins often significantly precedes active transport protein at plasma membranes. This illustrates the need for further studies investigating the ontogeny of transporter activity. Even immunological detection of protein at the relevant membrane domain does not necessarily correspond to transport activity due to intracellular protein maturation processes such as glycosylation (e.g., Ntcp).[55]

Other Factors Responsible for Age-Related ADME Differences in Animals

Apart from the ontogeny of enzyme/transporter expression, other factors could affect differences in pharmacokinetics between developing and adult animals. For instance, in situ brain perfusion experiments with the paracellular leakage marker ^{14}C-mannitol were conducted in 1-, 2-, 3-week-old and adult rats.[56] Results from these ex vivo experiments revealed a significantly increased initial volume of distribution (reflecting higher association with endothelial cells) of mannitol in 1-week-old rats compared to older rats. The data did not support different blood–brain barrier permeability values for mannitol among the age groups.

CONCLUSION

Differences in ADME between various age groups of juvenile animals may be due to one of the factors determining the processes of absorption, distribution, metabolism, or elimination separately or to a combination of more than one disposition factor. Data for the relationship of preclinical ADME to pediatric drug development guidelines are limited. Both the EMEA and the FDA require evaluation of the systemic exposure in different age groups in juvenile toxicity studies. Additional knowledge of ADME characteristics of a compound in various age groups can be considered on a case-by-case basis. However, it is important for the efficacy/safety evaluation that the ADME characteristics in juvenile animals are comparable to the human situation at a similar stage of development. Very limited information is available on the effect of age on drug absorption in juvenile animals. Most data are available on metabolism. The ontogeny of the activity of various enzymes—especially in the rat—is well documented and is, in combination with the relative liver size and liver blood flow, a major factor in age differences in the elimination of drugs in young animals.

Besides the ontogeny of drug metabolizing enzymes, knowledge on the ontogeny of drug transport proteins is increasing and apart from rat and humans is also being studied in the mouse. The ontogeny of drug transporters is related with the ontogeny of transport of endogenous versus exogenous substrates.

Differences in toxicity between adult and neonate and young animals may be due to differences in ADME of the test compound under development. The different time frame for the ontogeny of ADME-relevant processes between animals and humans is an additional key consideration and potential hindrance in the extrapolation of toxicological observations in juvenile animals toward children.

REFERENCES

1. Maintenance of the ICH Guideline on non-clinical safety studies for the conduct of human clinical trials for pharmaceuticals M3 (R1).
2. EMEA/CHMP/SWP/169215/2005. Draft guideline on the need for non-clinical testing in juvenile animals on human pharmaceuticals for paediatric indications. September 2005.
3. *Guidance for Industry: Nonclinical Safety Evaluation of Pediatric Drug Products*. US Department of Health and Human Services, Food and Drug Administration Center for Drug Evaluation and Research (CDER).
4. ICH Harmonised Tripartite Guideline (S3A). Note for guidance on toxicokinetics. The assessment of systemic exposure in toxicity studies. October 1994.
5. Bailey GP, Verbeeck J, Raoof AA, et al. Challenging toxicity and pharmacokinetic findings in juvenile rat. Presented at the Annual Meeting of the European Teratology Society, Padua (Italy). 2006.
6. Insel PA. Analgesic-antipyretic and anti-inflammatory agents and drugs employed in the treatment of gout. In: Hardman JG, Limbird LE, Molinoff PB, et al, eds. *Goodman & Gilman's Pharmacological Basis of Therapeutics*. New York: McGraw-Hill; 1996:617–657.
7. Dreifuss FE, Santilli N, Langer DH, et al. Valproic acid hepatic fatalities: a retrospective review. *Neurology*. 1987;37(3):379–385.
8. Kapusnik-Uner JE, Sande MA, Chambers HF. Antimicrobial agents: tetracycline, chloramphenicol, erythromycin and miscellaneous antibacterial agents. In: Hardman JG, Limbird LE, Molinoff PB, et al, eds. *Goodman & Gilman's Pharmacological Basis of Therapeutics*. New York: McGraw-Hill; 1996:1124–1154.

9. Brent RL. Utilization of animal studies to determine the effects and human risks of environmental toxicants (drugs, chemicals, and physical agents). *Pediatrics*. 2004;113(4 Suppl):984–995.
10. Hines RN. Developmental and tissue-specific expression of human flavin-containing monooxygenases 1 and 3. *Expert Opin Drug Metab Toxicol*. 2006;2(1):41–49.
11. Ginsberg G, Hattis D, Sonawane B. Incorporating pharmacokinetic differences between children and adults in assessing children's risks to environmental toxicants. *Toxicol Appl Pharmacol*. 2004;198(2):164–183.
12. de Zwart LL, Haenen HE, Versantvoort CH, et al. Role of biokinetics in risk assessment of drugs and chemicals in children. *Regul Toxicol Pharmacol*. 2004;39(3):282–309.
13. Bortolotti A, Corada M, Barzago MM, et al. Pharmacokinetics of theophylline in the newborn and adult rabbit. In vivo and isolated perfused liver approaches. *Drug Metab Dispos*. 1991;19 (2):430–435.
14. Ee LC, Zheng S, Yao L, et al. Lymphatic absorption of fatty acids and cholesterol in the neonatal rat. *Am J Physiol Gastrointest Liver Physiol*. 2000;279(2):G325–G331.
15. Chang J, Chadwick RW, Allison JC, et al. Microbial succession and intestinal enzyme activities in the developing rat. *J Appl Bacteriol*. 1994;77(6):709–718.
16. Pacha J. Development of intestinal transport function in mammals. *Physiol Rev*. 2000;80 (4):1633–1667.
17. Sangild PT, Cranwell PD, Hilsted L. Ontogeny of gastric function in the pig: acid secretion and the synthesis and secretion of gastrin. *Biol Neonate*. 1992;62(5):363–372.
18. Hervatin F, Moreau E, Ducroc R, et al. Development of acid secretory function in the rat stomach: sensitivity to secretagogues and corticosterone. *J Pediatr Gastroenterol Nutr*. 1989;9 (1):82–88.
19. Walthall K, Cappon GD, Hurtt ME, et al. Postnatal development of the gastrointestinal system: a species comparison. *Birth Defects Res B Dev Reprod Toxicol*. 2005;74(2):132–156.
20. Watchko JF, Daood MJ, Mahmood B, et al. P-glycoprotein and bilirubin disposition. *J Perinatol*. 2001;21(Suppl 1):S43–S47; discussion S59–S62.
21. Mahmood B, Daood MJ, Hart C, et al. Ontogeny of P-glycoprotein in mouse intestine, liver, and kidney. *J Investig Med*. 2001;49(3):250–257.
22. Fromm MF. P-glycoprotein: a defense mechanism limiting oral bioavailability and CNS accumulation of drugs. *Int J Clin Pharmacol Ther*. 2000;38(2):69–74.
23. Stephens RH, Tanianis-Hughes J, Higgs NB, et al. Region-dependent modulation of intestinal permeability by drug efflux transporters: in vitro studies in mdr1a(−/−) mouse intestine. *J Pharmacol Exp Ther*. 2002;303(3):1095–1101.
24. Satas S, Johannessen SI, Hoem NO, et al. Lidocaine pharmacokinetics and toxicity in newborn pigs. *Anesth Analg*. 1997;85(2):306–312.
25. Goralski KB, Acott PD, Fraser AD, et al. Brain cyclosporin A levels are determined by ontogenic regulation of mdr1a expression. *Drug Metab Dispos*. 2006;34(2):288–295.
26. Heykants J, Knaeps A, Meuldermans W, et al. On the pharmacokinetics of domperidone in animals and man. I. Plasma levels of domperidone in rats and dogs. Age related absorption and passage through the blood–brain barrier in rats. *Eur J Drug Metab Pharmacokinet*. 1981;6 (1):27–36.
27. Matsuoka Y, Okazaki M, Kitamura Y, et al. Developmental expression of P-glycoprotein (multidrug resistance gene product) in the rat brain. *J Neurobiol*. 1999;39(3):383–392.
28. Doerge DR, Twaddle NC, Banks EP, et al. Pharmacokinetic analysis in serum of genistein administered subcutaneously to neonatal mice. *Cancer Lett*. 2002;184(1):21–27.
29. de Zwart LL, Scholten M, Monbaliu J, et al. The ontogeny of drug metabolizing enzymes and transporters in the rat. *Drug Metab Rev*. 2006;38(Sup.1):107–108.

30. Wirth PJ, Thorgeirsson SS. Amine oxidase in mice—sex differences and developmental aspects. *Biochem Pharmacol.* 1978;27(4):601–603.

31. Devereux TR, Fouts JR. N-oxidation and demethylation of *N*, *N*-dimethylaniline by rabbit liver and lung microsomes. Effects of age and metals. *Chem Biol Interact.* 1974;8(2):91–105.

32. Alnouti Y, Klaassen CD. Tissue distribution and ontogeny of sulfotransferase enzymes in mice. *Toxicol Sci.* 2006;93(2):242–255.

33. Rouet P, Dansette P, Frayssinet C. Ontogeny of benzo[*a*]pyrene hydroxylase, epoxide hydrolase and glutathione-*S*-transferase in the brain, lung and liver of C57Bl/6 mice. *Dev Pharmacol Ther.* 1984;7(4):245–258.

34. Oakes SM, Takahashi Y, Williams MC, et al. Ontogeny of gamma-glutamyltransferase in the rat lung. *Am J Physiol.* 1997;272(4 Pt 1):L739–L744.

35. Patel HR, Hewer A, Hayes JD, et al. Age-dependent change of metabolic capacity and genotoxic injury in rat intestine. *Chem Biol Interact.* 1998;113(1):27–37.

36. Johnson TN, Tanner MS, Tucker GT. Developmental changes in the expression of enterocytic and hepatic cytochromes P4501A in rat. *Xenobiotica.* 2002;32(7):595–604.

37. Johnson TN, Tanner MS, Tucker GT. A comparison of the ontogeny of enterocytic and hepatic cytochromes P450 3A in the rat. *Biochem Pharmacol.* 2000;60(11):1601–1610.

38. Johri A, Dhawan A, Lakhan Singh R, et al. Effect of prenatal exposure of deltamethrin on the ontogeny of xenobiotic metabolizing cytochrome P450s in the brain and liver of offsprings. *Toxicol Appl Pharmacol.* 2006;214(3):279–289.

39. Fraile LJ, Bregante MA, Garcia MA, et al. Development of diltiazem deacetylase and demethylase activities during ontogeny in rabbit. *Xenobiotica.* 2001;31(7):409–422.

40. Solans C, Bregante MA, Aramayona JJ, et al. Pharmacokinetics of verapamil in New Zealand white rabbits during ontogeny. *Biol Neonate.* 2000;78(4):321–326.

41. Pinto N, Halachmi N, Verjee Z, et al. Ontogeny of renal P-glycoprotein expression in mice: correlation with digoxin renal clearance. *Pediatr Res.* 2005;58(6):1284–1289.

42. Domoradzki JY, Thornton CM, Pottenger LH, et al. Age and dose dependency of the pharmacokinetics and metabolism of bisphenol A in neonatal Sprague–Dawley rats following oral administration. *Toxicol Sci.* 2004;77(2):230–242.

43. Karanth S, Pope C. Carboxylesterase and A-esterase activities during maturation and aging: relationship to the toxicity of chlorpyrifos and parathion in rats. *Toxicol Sci.* 2000;58 (2):282–289.

44. Atterberry TT, Burnett WT, Chambers JE. Age-related differences in parathion and chlorpyrifos toxicity in male rats: target and nontarget esterase sensitivity and cytochrome P450-mediated metabolism. *Toxicol Appl Pharmacol.* 1997;147(2):411–418.

45. Anand SS, Kim KB, Padilla S, et al. Ontogeny of hepatic and plasma metabolism of deltamethrin in vitro: role in age-dependent acute neurotoxicity. *Drug Metab Dispos.* 2006;34(3):389–397.

46. Cheng X, Maher J, Chen C, et al. Tissue distribution and ontogeny of mouse organic anion transporting polypeptides (Oatps). *Drug Metab Dispos.* 2005;33(7):1062–1073.

47. Li N, Hartley DP, Cherrington NJ, et al. Tissue expression, ontogeny, and inducibility of rat organic anion transporting polypeptide 4. *J Pharmacol Exp Ther.* 2002;301(2):551–560.

48. Klaassen CD. Immaturity of the newborn rat's hepatic excretory function for ouabain. *J Pharmacol Exp Ther.* 1972;183(3):520–526.

49. Guo GL, Johnson DR, Klaassen CD. Postnatal expression and induction by pregnenolone-16alpha-carbonitrile of the organic anion-transporting polypeptide 2 in rat liver. *Drug Metab Dispos.* 2002;30(3):283–288.

50. Gao B, St Pierre MV, Stieger B, et al. Differential expression of bile salt and organic anion transporters in developing rat liver. *J Hepatol.* 2004;41(2):201–208.

51. Suchy FJ, Courchene SM, Blitzer BL. Taurocholate transport by basolateral plasma membrane vesicles isolated from developing rat liver. *Am J Physiol.* 1985;248(6 Pt 1):G648–G654.
52. Suchy FJ, Balistreri WF. Uptake of taurocholate by hepatocytes isolated from developing rats. *Pediatr Res.* 1982;16(4 Pt 1):282–285.
53. Sweet DH, Bush KT, Nigam SK. The organic anion transporter family: from physiology to ontogeny and the clinic. *Am J Physiol Renal Physiol.* 2001;281(2):F197–F205.
54. Stopp M, Braunlich H. In vitro analysis of postnatal maturation of tubular *p*-aminohippurate transport in rat kidney. *Acta Biol Med Ger.* 1980;39(7):825–832.
55. Hardikar W, Ananthanarayanan M, Suchy FJ. Differential ontogenic regulation of basolateral and canalicular bile acid transport proteins in rat liver. *J Biol Chem.* 1995;270(35):20841–20846.
56. Preston JE, al-Sarraf H, Segal MB. Permeability of the developing blood–brain barrier to ^{14}C-mannitol using the rat in situ brain perfusion technique. *Brain Res Dev Brain Res.* 1995;87 (1):69–76.
57. Zorzano A, Herrera E. Decreased in vivo rate of ethanol metabolism in the suckling rat. *Alcohol Clin Exp Res.* 1989;13(4):527–532.
58. Henley CM, Weatherly RA, Ou CN, et al. Pharmacokinetics of kanamycin in the developing rat. *Hear Res.* 1996;99(1–2):85–90.
59. Cagen SZ, Gibson JE. Characteristics of hepatic excretory function during development. *J Pharmacol Exp Ther.* 1979;210(1):15–21.
60. Yoo SD, Burgio DE, McNamara PJ. Phenobarbital disposition in adult and neonatal rabbits. *Pharm Res.* 1994;11(8):1204–1206.
61. Alberola J, Perez Y, Puigdemont A, et al. Effect of age on theophylline pharmacokinetics in dogs. *Am J Vet Res.* 1993;54(7):1112–1115.
62. Sanders JE, Yeary RA, Powers JD, et al. Relationship between serum and brain concentrations of phenytoin in the dog. *Am J Vet Res.* 1979;40(4):473–476.
63. Keller RD, Nosbisch C, Unadkat JD. Pharmacokinetics of stavudine (2′, 3′-didehydro-3′-deoxythymidine) in the neonatal macaque (*Macaca nemestrina*). *Antimicrob Agents Chemother.* 1995;39(12):2829–2831.
64. Tsao PW, Ito S, Wong PY, et al. Pharmacodynamics and pharmacokinetics of cyclosporin A in the newborn pig. *Dev Pharmacol Ther.* 1992;18(1–2):20–25.
65. Nielsen P, Friis C, Gyrd-Hansen N, et al. Disposition of parathion in neonatal and young pigs. *Pharmacol Toxicol.* 1991;69(4):233–237.
66. Tee LB, Gilmore KS, Meyer DJ, et al. Expression of glutathione-*S*-transferase during rat liver development. *Biochem J.* 1992;282(Pt 1):209–218.
67. Tee LB, Smith PG, Yeoh GC. Expression of alpha, mu and pi class glutathione-*S*-transferases in oval and ductal cells in liver of rats placed on a choline-deficient, ethionine-supplemented diet. *Carcinogenesis.* 1992;13(10):1879–1885.
68. Borlakoglu JT, Scott A, Henderson CJ, et al. Expression of P450 isoenzymes during rat liver organogenesis. *Int J Biochem.* 1993;25(11):1659–1668.
69. Lundqvist G, Morgenstern R. Ontogenesis of rat liver microsomal glutathione transferase. *Biochem Pharmacol.* 1995;50(3):421–423.
70. Das ML, Ziegler DM. Rat liver oxidative N-dealkylase and N-oxidase activities as a function of animal age. *Arch Biochem Biophys.* 1970;140(1):300–306.
71. Bustamante N, Cantarino MH, Arahuetes RM, et al. Evolution of the activity of UGT1A1 throughout the development and adult life in a rat. *Life Sci.* 2006;78(15):1688–1695.
72. Murakami T, Sato A, Inatani M, et al. Effect of neonatal exposure of 17beta-estradiol and tamoxifen on hepatic CYP3A activity at developmental periods in rats. *Drug Metab Pharmacokinet.* 2004;19(2):96–102.

73. Chow T, Imaoka S, Hiroi T, et al. Developmental changes in the catalytic activity and expression of CYP2D isoforms in the rat liver. *Drug Metab Dispos*. 1999;27(2):188–192.

74. Tavoloni N. Postnatal changes in hepatic microsomal enzyme activities in the puppy. *Biol Neonate*. 1985;47(5):305–316.

75. Kawalek JC, el Said KR. Maturational development of drug-metabolizing enzymes in dogs. *Am J Vet Res*. 1990;51(11):1742–1745.

76. Short CR, Maines MD, Westfall BA. Postnatal development of drug-metabolizing enzyme activity in liver and extrahepatic tissues of swine. *Biol Neonate*. 1972;21(1):54–68.

77. Maher JM, Slitt AL, Cherrington NJ, et al. Tissue distribution and hepatic and renal ontogeny of the multidrug resistance-associated protein (Mrp) family in mice. *Drug Metab Dispos*. 2005;33 (7):947–955.

78. Alnouti Y, Petrick JS, Klaassen CD. Tissue distribution and ontogeny of organic cation transporters in mice. *Drug Metab Dispos*. 2006;34(3):477–482.

79. Garcia-Miranda P, Duran JM, Peral MJ, et al. Developmental maturation and segmental distribution of rat small intestinal L-carnitine uptake. *J Membr Biol*. 2005;206(1):9–16.

80. Nakajima N, Sekine T, Cha SH, et al. Developmental changes in multispecific organic anion transporter 1 expression in the rat kidney. *Kidney Int*. 2000;57(4):1608–1616.

81. Buist SC, Cherrington NJ, Choudhuri S, et al. Gender-specific and developmental influences on the expression of rat organic anion transporters. *J Pharmacol Exp Ther*. 2002;301(1):145–151.

PART V

PHARMACOLOGICAL PRINCIPLES IN PEDIATRIC DRUG DEVELOPMENT

CHAPTER 19

Pediatric Clinical Pharmacology: Why, Where, How, When?

JOHN N. VAN DEN ANKER, MD, PhD
Division of Pediatric Clinical Pharmacology, Children's National Medical Center, Washington DC, 20010

INTRODUCTION

The majority of today's medicines have been tested predominantly in adults. Testing for safety has become mandatory in the United States only after 1934 and testing for efficacy only in the 1960s.[1,2] Until 1990, children remained the "therapeutic orphans"—as described by Harry Shirkey in 1963 to emphasize the plight of children being excluded from new pharmaceutical treatments.[3,4] Since 1962, pediatric disclaimers resulted in medicines prescribed to children "off-label," transferring responsibility from the manufacturer to the prescribing physician, who could either not prescribe or prescribe with the risk of under- or overdosing. Numerous publications describe the extent of existing off-label use of medicines.[5–10] As a consequence, governments in the United States, the European Union, Japan, and other regions are taking action to improve the availability of medicines suited for children.[11–15] Multiple contributors discuss these issues throughout this book and the reader will find various chapters addressing these issues.

U.S. PEDIATRIC LEGISLATION

Since 1960, pediatricians have criticized off-label use of medicines in children. Due predominantly to pediatricians' lobbying of the Food and Drug Administration (FDA) and the National Institutes of Health (NIH), the U.S. government implemented a two-pronged approach in the late 1990s. FDAMA (FDA Modernization Act)[16] 1997 offered pharmaceutical companies a voluntary incentive of 6 months of added market exclusivity (pediatric exclusivity (PE)) in return for pediatric research.[17] In 2002, FDAMA was reauthorized under the name of Best Pharmaceuticals for Children Act (BPCA).[18] Furthermore, in 2003 President Bush signed the Pediatric Research Equity Act (PREA),[19] which

Pediatric Drug Development: Concepts and Applications
Edited by Andrew E. Mulberg, Steven A. Silber, and John N. van den Anker

made it mandatory for pharmaceutical companies to submit a pediatric assessment for each age group in medicines filed for regulatory approval in the United States, unless a deferral or a waiver had been granted.[19,20] A pediatric assessment contains the data of pediatric studies for which the assessment is required. A pediatric plan, that is, a statement of intent that outlines pediatric studies the company plans to conduct, should be discussed with the FDA at preinvestigational new drug (pre-IND) and end-of-phase 1 meetings for products intended for life-threatening or severely debilitating illnesses; for products intended for all other diseases, the pediatric plan should be submitted and discussed no later than the end-of-phase 2 meeting. The FDA had issued a similar requirement already in 1998 as the Pediatric Rule,[21] which was, however, suspended by court order after objections that the FDA had overstepped its authority.[11,12] As PREA replaced the original Pediatric Rule, it applies retrospectively to its original time frame; that is all submissions done on or after April 1, 1999.[21] Both legislations (i.e., BPCA and PREA) have recently been reauthorized. Maldonado extensively discusses these issues in Chapter 12 in this book.

DEVELOPMENT OF THE PEDIATRIC PHARMACOLOGY RESEARCH UNIT

A crucial approach to the solution of the therapeutic orphan dilemma in the United States occurred at the National Institutes of Health with the development of the Pediatric Pharmacology Research Unit (PPRU) network. This network was established in 1994 following a conference of the Forum of Drug Development of the Institute of Medicine of the National Academy of Sciences. This conference strongly recommended that steps be taken to eliminate the therapeutic orphan situation. In response to the need for appropriate therapy for pediatric patients, the National Institute of Child Health and Development (NICHD) established the PPRU network with the mission of facilitating and promoting pediatric labeling of new drugs or drugs already on the market. In this process, the network strives to foster cooperative and complementary research efforts among the academy, industry, and health professionals. The overall goal of the network is to provide data for the safe and effective use of drugs in children.

The network has three functions: (1) to conduct studies on the pharmacokinetics and pharmacodynamics of drugs in infants and children, (2) to provide a focus for pre- and postmarketing clinical trials in children conducted by clinical pediatric pharmacologists in collaboration with the pharmaceutical industry and contract research organizations, and (3) to serve as an advisory body to the pharmaceutical industry, regulatory agencies, health professionals, and the public on the appropriate use of drugs in children.

The principal investigators in the PPRU network are pioneers in the field of pediatric pharmacology. In January 1999, the network of 7 sites was expanded to 13 sites. With this expansion in the number of sites, the network has access to a large, all-inclusive pediatric population with more than 2 million outpatient visits per year and more than 100,000 pediatric in-patient admissions. The network serves as a major resource for the training of health professionals in pediatric pharmacology and clinical trial methodology as well as the study of drugs in infants and children.

The involvement of children in clinical trials requires study designs that minimize discomfort in patients and do not disturb family life. The PPRU pediatric pharmacologists use their combined experience and skills and access large numbers of children to shorten the length of the study period. The PPRU network strives to develop child-friendly protocols with minimal risk for all pediatric patients regardless of their condition. Pediatric patients involved in drug studies include those with common disorders, such as allergies, asthma, and upper respiratory infections, as well as those with less common disorders such as cystic fibrosis, severe infections, HIV infection and AIDS, sickle cell anemia, cancer, and childhood depression. The combined patient capabilities of the network are available to ensure appropriate and objective evaluation of drugs including, but not limited to, antipyretics, analgesics, antibiotics, decongestants, antihypertensives, diuretics, and bronchodilators. The network is also committed to developing safe and effective drug therapy for children in intensive care and in life-threatening situations.

An overriding consideration for the PPRU pediatric pharmacologists who are in the network is to delineate the effects of development during infancy and childhood on the pharmacokinetics of drugs, the influence of age-specific changes in drug disposition and pharmacodynamics, and the interplay among disease states, stage of development, and response to drugs. In addition, studies of translational research involving molecular mechanisms of action and pharmacogenetics and pharmacogenomics are encouraged in the network.

EU PEDIATRIC REGULATION

In Europe, the debate about off-label use of medicines in children and possible consequences started in parallel to a similar discussion in the United States. After publication of a consultation paper[22] and the submission of a draft regulation to the European parliament[23] by the EU Commission in 2004, the EU pediatric regulation passed a first reading in the EU parliament and the EU Health Council in 2005.[24] It passed second reading in the EU parliament and was put in force in January 2007 after final approval by the EU Health Council and EU Commission. Please see Chapter 13 in this book by Rose for further discussion of the EU legislation.

The key institution that will evolve out of the new EU regulation will be the Pediatric Committee (PC), whose main tasks will be to assess submitted Pediatric Investigation Plans (PIPs), to grant waivers and deferrals, and to evaluate compliance with the original PIP once clinical pediatric data are generated. Waivers to the law are granted if a disease does not exist in children or there are other reasons that make pediatric development impossible; deferrals postpone pediatric development. The new law also stipulates that future EU Marketing Authorization Applications (MAAs) must include the data outlined in the PIP, unless a waiver or a deferral has been issued. PIP results will have to be submitted for new indications, new pharmaceutical forms, and new routes of administration. A PIP is to be submitted not later than upon completion of the human pharmacokinetic studies in adults. PIP compliance will be rewarded with a 6 months' extension of a Supplementary Protection Certificate (SPC). The SPC is a European regulatory tool designed to make up for the considerable part of the patent life that is lost during drug development. SPC prolongation corresponds to the 6 months' PE granted in the United States and reflects that in the EU no common patent system exists.

For products no longer patent protected, the pediatric regulation plans a special incentive for pediatric formulations of off-patent medicines in the form of Pediatric Use Marketing Authorization (PUMA). Formulations authorized under PUMA will carry a special symbol and will receive 10 years of data protection.

Other measures of the draft EU regulation include plans to increase the robustness of pharmacovigilance of medicines marketed for children; an EU inventory of therapeutic needs of children to focus research, development, and authorization of medicines; an EU network of investigators and clinical trial centers to conduct research and development on medicines for children; a system of scientific advice for the pharmaceutical companies free of charge by the EMEA; and a database of pediatric clinical trials.

CONCLUSION

The U.S. and EU pediatric legislation shows the preparedness of the U.S. government, the FDA, the EU Commission, and EMEA to facilitate pediatric research in the United States and Europe. This alone has given momentum to the process and continues to help remove mental barriers that still exist against pediatric research. It will increasingly lend energy and support and enthusiasm for companies to address pediatric trials as part of the drug development strategy for new chemical entities and products.

To optimize these opportunities it will be essential to establish an intensified dialogue between the partners in healthcare—regulators, health authorities, patients and parents, and the pharmaceutical industry.

ACKNOWLEDGMENTS

This work was supported by grant 1K24RR19729, National Center for Research Resources, and grant 1U10HD45993, National Institute of Child Health and Development, Bethesda, Maryland.

REFERENCES

1. Hilts J. *Protecting America's Health.* New York: Alfred A Knopf; 2003.
2. Hawthorne F. *Inside the FDA.* Hoboken, NJ: John Wiley & Sons; 2005.
3. Murphy D. Pediatric trials: the impact of US legislative and regulatory efforts. *Appl Clin Trials.* 2005;14(1):38–42.
4. Shirkey H. Therapeutic orphans. *Pediatrics.* 1968;72:119–120.
5. McIntyre J, Conroy S, Avery A, et al. Unlicensed and off label prescribing of drugs in general practice. *Arch Dis Child.* 2000;83:498–501.
6. Turner S, Longworth A, Nunn AJ, et al. Unlicensed and off label drug use in paediatric wards: prospective study. *BMJ* 1998;316:343–345.
7. Ekins-Daukes S, Helms PS, Taylor MW, et al. Off-label prescribing to children: attitudes and experience of general practitioners. *Br J Clin Pharmacol.* 2005;60 (2):145.
8. Conroy S, Choonaral I, Impicciatore P, et al. Survey of unlicensed and off-label drug use in paediatric wards in five European countries. *BMJ* 2000;320:79–82.
9. 't Jong GW, Vulto AG, de Hoog M, Schimmel KJM, Tibboel D, van den Anker JN. A survey of the use of off-label and unlicensed drugs in a Dutch children's hospital. *Pediatrics.* 2001;108:1089–1093.
10. 't Jong GW, Eland IA, Sturkenboom MC, van den Anker JN, Stricker BH. Unlicensed and off-label prescription of drugs to children in the general population. *BMJ* 2002;324:1313–1314.
11. Chesney RW, Christensen ML. Changing requirements for evaluation of pharmacologic agents. *Pediatrics.* 2004;113:1128–1132.
12. Caldwell PHY, Murphy SB, Butow PN, Craig JC. Clinical trials in children. *Lancet* 2004;264:803–811.

13. FDA. The Pediatric Exclusivity Provision January 2001—Status Report to Congress. Available at http://www.fda.gov/cder/pediatric/reportcong01.pdf.
14. Rose K. Pediatric drug development. Implementation of pediatric aspects into the general drug development process. *Appl Clin Trials*. 2005;14(1):50–53.
15. Uchiyama A. Pediatric clinical studies in Japan: regulations and current status. *Appl Clin Trials*. 2002;11(7):57–59.
16. FDA homepage. FDAMA. Available at http://www.fda.gov/cder/guidance/105-115.htm#SEC.%20111.
17. FDA homepage. Guidance for Industry: Qualifying for Pediatric Exclusivity. Available at http://www.fda.gov/cder/guidance/2891fnl.htm.
18. FDA homepage. BPCA. Available at http://www.fda.gov/cder/pediatric/PL107-109.pdf.
19. FDA homepage. PREA. Available at http://www.fda.gov/cder/pediatric/S-650-PREA.pdf.
20. FDA homepage. Guidance for Industry: How to Comply with the Pediatric Research Equity Act. Available at http://www.fda.gov/cder/guidance/6215dft.pdf.
21. FDA homepage. Pediatric Rule. Available at http://www.fda.gov/ohrms/dockets/98fr/120298c.pdf.
22. EMEA homepage. Commission consultation on a draft proposal for a European Parliament and Council Regulation (EC) on medicinal products for paediatric use. Available at http://pharmacos.eudra.org/F2/pharmacos/docs/Doc2004/mar/Paediatric%20consultation%20document%20final%208%20March%2004.pdf.
23. EMEA homepage. Proposal for a regulation of the European Parliament and of the Council on Medicinal Products for paediatric use. Available at http://pharmacos.eudra.org/F2/Paediatrics/docs/Paeds%20reg%20adopted%2029%20September%202004%20English.pdf.
24. EMEA homepage. Amended proposal for a regulation of the European Parliament and of the Council on Medicinal Products for paediatric use and amending Regulation (EEC) No 1768/92, Directive 2001/83/EC and Regulation (EC) No 726/2004. Available at http://pharmacos.eudra.org/F2/Paediatrics/docs/COM_2005_0577_EN.PDF.

CHAPTER 20

Developmental Pharmacology Issues: Neonates, Infants, and Children

NATELLA Y. RAKHMANINA, MD

Divisions of Pediatric Clinical Pharmacology and Infectious Diseases, Children's National Medical Center, Washington, DC 20010

JOHN N. VAN DEN ANKER, MD, PhD

Division of Pediatric Clinical Pharmacology, Children's National Medical Center, Washington, DC 20010

INTRODUCTION

Clinical pharmacology intends to predict drug-specific effects and side effects based on pharmacokinetics (i.e., absorption, distribution, metabolism, and elimination) and pharmacodynamics (i.e., dose–effect relationship). Developmental pharmacology focuses on the maturational aspects of these phenomena during perinatal life and later stages of infancy. Important alterations in renal clearance and hepatic metabolism occur in perinatal life, resulting in maturational trends in drug metabolism and elimination in preterm and term infants.[1–3]

The history of pediatric pharmacotherapy is replete with examples of adverse reactions to drugs in neonates, infants, children, and adolescents. In 1937, 107 people—primarily children—died after taking elixir of sulfanilamide to treat streptococcal infection. Sulfanilamide was not very water soluble, but a chemist at Massengill Co. found that it dissolved well in diethylene glycol (more commonly known as antifreeze), which is now known to be highly toxic. In 1956, Silverman and colleagues[4] at Columbia reported an excessive mortality rate and an increased incidence of kernicterus among premature babies receiving a sulfonamide antibiotic compared with those receiving chlortetracycline. Then, in 1959, Sutherland[5] described a syndrome of cardiovascular collapse in three newborns receiving high doses of chloramphenicol for presumed infections. More recently, the therapeutic misadventures experienced by low birth weight infants exposed to a parenteral vitamin E formulation[6] and the "gasping syndrome" by infants who received excessive amount of benzyl alcohol[7] all serve to underscore the generally held perception that newborn infants are more likely to experience adverse reactions to drugs. More recently, all therapeutic issues surrounding the retinoic acid embryopathy and maternal antidepressant drug use have refocused attention on the effects of drugs on the fetus and newborn.[8]

Pediatric Drug Development: Concepts and Applications
Edited by Andrew E. Mulberg, Steven A. Silber, and John N. van den Anker

During the past two decades, tremendous strides have been made to tailor therapies for the needs of children. As our knowledge of normal growth and development has increased, so has our recognition that developmental changes profoundly affect the responses to medications and produce a need for age-dependent dose requirements. Rovner and Zemel address the importance of growth assessment and differences in pediatrics in Chapter 28 in this book.

Prior to the clinical integration of developmental pharmacology into therapeutic decision-making, numerous approaches (e.g., Young's Rule, Clark's Rule) for determining pediatric drug doses were recommended. These approaches vary with some using discrete age points and others using allometric principles that generally assume predictable, linear relationships between mass (e.g., cell mass, body weight) and/or body surface area between infants, children, adolescents, and adults. However, as human growth is not a linear process and age-associated changes in body composition and organ function are dynamic and can be discordant during the first decade of life, simplified dosing approaches are not adequate for individualizing drug doses across the span of childhood.[9] As a result, these old "dosing equations" have been abandoned and, in most instances, replaced by simple "normalization" of drug dose as a function of either body weight (mg/kg) or body surface area (mg/m^2). While such guidelines are generally adequate for the initiation of therapy, they may not be sufficient for age-based individualization of continued (e.g., chronic) treatment, where dramatic developmental differences in pharmacokinetics and/or pharmacodynamics (i.e., the determinants of appropriate dosing regimen) may occur. Thus, the provision of safe and effective drug therapy for pediatric patients requires a fundamental understanding and integration of the role of development on drug disposition and action. Additionally, in this postgenome era, the impact of pharmacogenetics on these important developmental changes in pharmacokinetics and pharmacodynamics needs to be investigated in a rigorous but feasible way to elucidate the link between genetics and developmental biology for the different phases along the pediatric age continuum.

This chapter primarily discusses the age-associated changes in absorption, distribution, metabolism, and elimination and, where appropriate, touches on the impact of pharmacogenetics on drug clearance. Further discussion of pharmacogenomics is provided by the insights of Cohen and Ness in Chapter 21 on this topic in this book.

ABSORPTION

For therapeutic agents administered by extravascular routes, the process of absorption is reflected by the ability of a drug to overcome chemical, physical, mechanical, and biological barriers. Developmental differences in the physiologic composition and function of these barriers can alter the rate and/or extent of drug absorption. While factors influencing drug absorption are multifactorial in nature, developmental changes in the absorptive surfaces (e.g., gastrointestinal tract, skin, pulmonary tree) can be determinants of bioavailability.

The skin represents an often overlooked, but important organ for systemic drug absorption. Chemical agents applied to the skin of a premature infant may result in inadvertent poisoning. There are numerous reports in the literature of neonatal toxicity related to the cutaneous exposure to drugs and chemicals. They include hexachlorophene,[10] pentachlorophenol-containing laundry detergents,[11] hydrocortisone,[12] and aniline-containing disinfectant solution.[13] Therefore, extreme caution should be exercised in using topical therapy in newborn infants.

The morphologic and functional development of the skin as well as the factors that influence penetration of drugs into and through the skin have been reviewed.[14–16] Basically, the percutaneous absorption of a compound is directly related to the degree of skin hydration and relative absorptive surface area and inversely related to the thickness of the stratum corneum.[15] The integument of the full-term neonate possesses intact barrier function and is similar to that of an older child or adult. However, the ratio of surface area to body weight of the full-term neonate is much higher than that of an adult.[16] Thus, the infant will be exposed to a relatively greater amount of drug topically than will older infants, children, or adults. Theoretically, if a newborn receives the same percutaneous dose of a compound as an adult, the systemic availability per kilogram of body weight will be approximately 2.7 times greater in the neonate.

In contrast, studies in premature infants suggest the existence of an immature barrier to percutaneous absorption.[17] Nachman and Esterly[17] studied the blanching response to topical 10% phenylephrine in preterm and term infants. Newborn infants of 28–34 weeks' gestational age had a rapid response lasting from 30 minutes to as long as 6–8 hours. No response was apparent under the same study conditions at 21 days of postnatal age. Newborns of gestational age 35–37 weeks had a less dramatic response with a longer latency period, and term infants failed to demonstrate a blanching response.

Finally, if the integrity of the integument is compromised (e.g., denuded, burned, or inflamed skin), then percutaneous translocation of compounds into the blood will be enhanced.

The oral route is the principal means for drug administration to pediatric patients. Changes in intraluminal pH can directly impact both drug stability and degree of ionization, thus influencing the relative amount of drug available for absorption. Additionally, the ability to solubilize and subsequently absorb lipophilic drugs can be influenced by age-dependent changes in biliary function.[18,19] Complete information regarding these processes and the effect of development remains unexplored in most cases.

Gastric emptying and intestinal motility are primary determinants of the rate at which drugs are presented to and dispersed along the mucosal surface of the small intestine. Generally, the rate at which most drugs are absorbed is generally slower, and thus, the time to achieve maximum plasma concentrations is prolonged in neonates and young infants relative to older infants and children.

Despite their incomplete characterization,[20] developmental differences in the activity of intestinal drug metabolizing enzymes and efflux transporters have the potential to markedly alter drug bioavailability. Notably, data on developmental expression of the efflux transporter P-glycoprotein (MDR1) in humans are absent.

DISTRIBUTION

Age-dependent changes in body composition[21] alter the physiologic "spaces" into which a drug may distribute. Larger extracellular and total body water spaces in neonates and young infants result in lower plasma concentrations for drugs that distribute into this compartment when administered in a weight-based fashion. Changes in the composition and amount of circulating plasma proteins (e.g., albumin, α_1 acid-glycoprotein) can also influence the distribution of highly bound drugs. A reduction in the quantity of total plasma proteins (including albumin) in the neonate and young infant increases the free fraction of drug, thereby influencing the availability of the active moiety. Other factors associated with development and/or disease such as variability in regional blood flow, organ perfusion, permeability of cell membranes, changes in acid–base balance, and cardiac output can also influence drug binding and/or distribution.

DRUG METABOLISM

Cardiovascular collapse associated with the "gray baby syndrome" in newborns treated with chloramphenicol is often cited as a clinically significant consequence of developmental deficiencies in drug metabolizing enzyme activities.[22,23] Multiple examples exist of clinically important developmental changes in drug biotransformation sufficient to produce the need for age-appropriate dose regimen selection in neonates and young infants (e.g., methylxanthines, captopril, midazolam, zidovudine, morphine). As reflected by recent reviews, distinct patterns of isoform-specific developmental changes in drug biotransformation are apparent for many Phase I (primarily oxidation) and Phase II (conjugation) drug metabolizing enzymes.[24,25] Please see Chapter 20 by Behm in this book for further description of the hepatic development of Phase I and II enzymes in the infant and child.

Phase I Enzymes

Development has a profound effect on the expression of Phase I enzymes such as the cytochromes P450 (CYPs). CYP3A7 is the predominant CYP isoform expressed in fetal liver, where it may play a fetoprotective role by detoxifying dehydroepiandrosterone sulfate and potentially teratogenic retinoic acid derivatives.[26] CYP3A7 expression peaks shortly after birth and then declines rapidly to levels that are undetectable in most adults.[27] Distinct patterns of isoform-specific developmental CYP expression have been observed postnatally. Within hours of birth, CYP2E1 activity surges,[28] followed closely by the onset of CYP2D6 expression.[29] CYP3A4 and CYP2C (CYP2C9 and 2C19) activities appear during the first week of life,[27,30] whereas CYP1A2 is the last hepatic CYP to be acquired with significant expression being delayed until 1–3 months of life.[31]

Insight into the ontogeny of drug metabolism also can be derived from pharmacokinetic studies of drugs metabolized by specific CYP isoforms. Midazolam plasma clearance, which primarily reflects hepatic CYP3A4/5 activity after intravenous administration,[32] increases approximately fivefold (1.2–9 mL/min/kg) over the first 3 months of life.[33] Carbamazepine plasma clearance, also largely dependent on CYP3A4,[34] is greater in children relative to adults,[35,36] thereby necessitating higher weight-adjusted (i.e., mg/kg) doses of the drug to produce therapeutic plasma concentrations.

CYP2C9 and, to a lesser extent, CYP2C19 are primarily responsible for phenytoin biotransformation.[37] Phenytoin apparent half-life is prolonged ($\sim$75 h) in preterm infants but decreases to $\sim$20 h in term infants less than 1 week postnatal age and to $\sim$8 h after 2 weeks of age.[38] Saturable phenytoin metabolism does not appear until approximately 10 days of postnatal age, demonstrating the developmental acquisition of CYP2C9 activity.

Caffeine and theophylline are the most common CYP1A2 substrates used in pediatrics. Caffeine elimination in vivo mirrors that observed in vitro with full 3-demethylation activity (mediated by CYP1A2) observed by approximately 4 months of age.[39] Formation of CYP1A2-dependent theophylline metabolites reaches adult levels by approximately 4–5 months of postnatal age[40] and in older infants and young children, theophylline plasma clearance generally exceeds adult values.[41] Furthermore, caffeine 3-demethylation in adolescent females appears to decline to adult levels at Tanner stage II relative to males, where it occurs at stages IV/V,[42] thus demonstrating an apparent sex difference in the ontogeny of CYP1A2.

Omeprazole, a proton pump inhibitor and substrate for the polymorphically expressed enzyme CYP2C19, is routinely used in infants, children, and adolescents all over the world to treat gastroesophageal reflux disease and a variety of other conditions, where control of intragastric pH is necessary. A pharmacokinetic study of this compound in children and adolescents that incorporated CYP2C19 genotyping did not show a gene–dose effect in these pediatric patients.[43]

Tramadol, a nonopiate analgesic agent that is substrate for the polymorphically expressed enzyme CYP2D6, was studied in the same way and a gene–dose effect for CYP2D6-mediated drug biotransformation was evident in this case.[44] These latter illustrations emphasize the need to incorporate relevant pharmacogenetics in pediatric clinical trials to further improve pediatric pharmacotherapy.

Phase II Enzymes

Phase II reactions generally result in pharmacological inactivation or detoxification by conjugating xenobiotics with small molecules such as UDP-glucuronic acid, glutathione, or acetyl coenzyme A. These reactions are catalyzed by a variety of enzymes, the activity of which appears to be associated with development. While the impact of ontogeny on Phase II enzymes has not been investigated to the same extent as for Phase I enzymes, a conceptual understanding of their known developmental profiles is important to understanding the acquisition of metabolic competence in the neonate and its potential therapeutic implications.[25]

The mammalian UGTs are responsible for the glucuronidation of hundreds of hydrophobic endogenous (e.g., bilirubin, bile acids, thyroxine, and steroids) and exogenous (e.g., morphine, acetaminophen, and NSAIDs) xenobiotics. Additionally, UGTs detoxify an extensive group of potentially carcinogenic or teratogenic xenobiotics that enter the body as components of the diet or as airborne pollutants.[45]

The UGTs comprise a superfamily of enzymes that are subdivided into families based on sequence homology. In neonatology, serious clinical consequences of allelic variants in UGT isoforms are well known. Over 30 different perturbations of the *UGT1A* gene results in absent or reduced enzyme activity that can lead to lethal hyperbilirubinemia (Crigler–Najjar syndrome).[46] Mutations of the promoter region of the *UGT1* gene have been associated with a milder form of congenital unconjugated hyperbilirubinemia (Gilbert syndrome).[47] From a pharmacologic perspective, failure to recognize the impact of development on the glucuronidation of chloramphenicol and its implications for age-associated individualization of therapy led to the historical catastrophe of the gray baby syndrome.[5]

Low levels of immunoreactive UGT protein are found early in gestation in liver, spleen, and kidney. Relatively greater reactivity has been observed in red blood cells as early as 32 days postconception.[48,49] Functional UGT activity, as assessed by bilirubin conjugation, is nearly undetectable in fetal liver with activity increasing immediately after birth in parallel with an increase in protein. The increase in catalytic activity is not dependent on gestational age,[50–52] thereby suggesting that postnatal events are essential for the expression and/or activation of the *UGT* gene.

The most complete information on the development of UGT activity as it relates to drug metabolism comes from studies of morphine glucuronidation by UGT2B7. Morphine is metabolized by UGT2B7 to morphine-6-glucuronide and morphine-3-glucuronide.[53] In vitro studies have shown that liver microsomes from fetuses aged 15–27 weeks glucuronidate morphine at a rate that is only 10–20% of that seen in adult microsomes.[54,55] No correlation was seen between gestational age and the rate of glucuronidation, again suggesting that birth-related events play a role in the activation of this enzyme. In vivo studies demonstrate that the mean plasma clearance of morphine was fivefold lower in premature infants (gestational age 24–37 weeks) when compared to children 1–16 years of age. Generally, morphine clearance reaches adult levels between 2 and 6 months but may take as long as 30 months.[56] Since glucuronidation is the primary metabolic pathway for morphine metabolism, the impact of development on clearance appears to accurately reflect the ontogeny of UGT2B7.

The impact of development on the activity of other UGT isoforms can also be indirectly assessed using pharmacologic substrates for this enzyme. Acetaminophen is metabolized by UGT1A6 and, to a lesser extent, by UGT1A9.[57] Acetaminophen glucuronidation, as reflected by urinary metabolite data, appears neglible in the fetus, is low at birth, and appears to approach full competence by 9–12 months of postnatal life.[58] Zidovudine plasma clearance in neonates less than 2 weeks of age is only half that of infants 2 weeks and older.[59] Collectively, the pharmacokinetic data for the aforementioned UGT substrates illustrate that the developmental profile for acquisition of enzyme activity is isoform and substrate specific.

RENAL ELIMINATION

Maturation of renal function is a dynamic process that begins early during fetal organogenesis and is complete by early childhood. The developmental increase in glomerular filtration rate (GFR) involves active nephrogenesis, a process that begins at 9 weeks and is complete by 36 weeks of gestation, followed by postnatal changes in renal and intrarenal blood flow.[60] Following birth, the GFR is approximately 2–4 mL/min/1.73 m^2 in term neonates and as low as 0.6–0.8 mL/min/1.73 m^2 in preterm neonates. GFR increases rapidly during the first 2 weeks of life followed by a steady rise until adult values are reached by 8–12 months.[61,62] Similarly, tubular secretory pathways are immature at birth and gain adult capacity during the first year of life. Collectively, these changes dramatically alter the plasma clearance of compounds with extensive renal elimination and thus provide a major determinant for age-appropriate dose regimen selection. Pharmacokinetic studies of drugs primarily excreted by glomerular filtration such as ceftazidime[62] and famotidine[63] have demonstrated significant correlations between plasma drug clearance and normal, expected maturational changes in renal function. For example, tobramycin is eliminated predominantly by glomerular filtration, necessitating dosing intervals of 36–48 hours in preterm and 24 hours in term newborns.[64] Failure to account for the ontogeny of renal function and adjust aminoglycoside dosing regimens accordingly can result in exposure to potentially toxic serum concentrations.[65] Also, concomitant medications (e.g., betamethasone, indomethacin) may alter the normal pattern of renal maturation in the neonate.[66] Thus, for drugs with extensive renal elimination, both maturational and treatment associated changes in kidney function must be considered and used to individualize treatment regimens in an age-appropriate fashion. Further discussion on the development of renal function in the infant and child by Boven is provided in Chapter 28 in this book. (See Table 20.1 for summary of developmental changes affecting drug disposition.)

TABLE 20.1 Summary of Developmental Changes in Drug Disposition

Physiological System	Age-Related Trends	Pharmacokinetic Implications	Clinical Implications
Gastrointestinal tract	Neonates and young infants: reduced and irregular peristalsis with prolonged gastric emptying time.	Slower rate of drug absorption (e.g., increased T_{max}) without compensatory compromise in the extent of bioavailability; impaired retention of suppository formulations.	Potential delay in the onset of drug action following oral ingestion; potential for reduced extent of bioavailability from rectally administered drugs.
	Neonates: greater intragastric pH (>4) relative to infants.		
	Infants: enhanced lower GI motility.		
Integument	Neonates and young infants: thinner stratum corneum (neonates only), greater cutaneous perfusion, enhanced hydration, and greater ratio of total BSA to body mass.	Enhanced rate and extent of percutaneous drug absorption; greater relative exposure of topically applied drugs as compared to adults.	Enhanced percutaneous bioavailability and potential for toxicity; need to reduce amount of drugs applied to skin.
Body compartments	Neonates and infants: decreased fat, decreased muscle mass, increased extracellular and total body water spaces.	Increased apparent volume of distribution for drugs distributed to body water spaces and reduced apparent volume of distribution for drugs that bind to muscle and/or fat.	Requirement of higher weight-normalized (i.e., mg/kg) drug doses to achieve therapeutic plasma drug concentrations.
Plasma protein binding	Neonates: decreased concentrations of albumin and α_1-acid glycoprotein with reduced binding affinity for albumin bound weak acids.	Increased unbound concentrations for highly protein-bound drugs with increased apparent volume of distribution and potential for toxicity if the amount of free drug increases in the body.	For highly bound (i.e., >70%) drugs, need to adjust dose to maintain plasma levels near the low end of the recommended "therapeutic range."

(continued)

TABLE 20.1 *(Continued)*

Physiological System	Age-Related Trends	Pharmacokinetic Implications	Clinical Implications
Drug metabolizing enzyme (DME) activity	Neonates and young infants: immature isoforms of cytochrome P450 and phase II enzymes with discordant patterns of developmental expression.	Neonates and young infants: decreased plasma drug clearance early in life with an increase in apparent elimination half life.	Neonates and young infants: increased drug dosing intervals and/or reduced maintenance doses.
	Children 1–6 years: apparent increased activity for selected drug metabolizing enzymes over adult normal values.	Children 1–6 years: increased plasma drug clearance (i.e., reduced elimination half life) for specific pharmacologic substrates of drug metabolizing enzymes.	Children 1–6 years: for selected drugs, need to increase dose and/or shorten dose interval in comparison to usual adult dose.
	Adolescents: attainment of adult activity after puberty.		
Renal drug excretion	Neonates and young infants: decreased glomerular filtration rates (first 6 months) and active tubular secretion (first 12 months) with adult values attained by 24 months.	Neonates and young infants: accumulation of renally excreted drugs and/or active metabolites with reduced plasma clearance and increased elimination half-life, greatest during first 3 months of life.	Neonates and young infants: increased drug dosing intervals and/or reduced maintenance doses during first 3 months of life.

CONCLUSION

The advances in pediatric clinical pharmacology during the past decade reside with an enhanced understanding of the influence of growth and development on drug disposition and action. However, significant information gaps remain with respect to our ability to completely and, in many cases, accurately characterize the impact of ontogeny on the activity of important drug metabolizing enzymes and transporters. As this moves forward, it is essential that the ultimate goal be kept clearly in sight: specifically, providing infants and children with safe and effective drug therapy made possible by including them in the process of development of medications essential to ensure their health.

This postgenome era presents unprecedented opportunities to make significant contributions to pediatric clinical pharmacology and drug development. The goal of rational drug therapy in neonates, infants, children, and adolescents resides with the ability to individualize it based on known developmental differences, which impact drug disposition and action. The clinical challenge in this is accounting for the variability in all of the contravening factors that influence pharmacokinetics (e.g., polymorphic expression of drug metabolizing enzymes and efflux transporters, effect of disease and/or concomitant therapy on enzyme/transporter activity, altered drug delivery associated with drug formulations that may not be suited for accurate drug dosing) and pharmacodynamics (e.g., polymorphic expression of cellular transporters and/or receptors) between patients of a given age and developmental stage. Despite the fact that most therapeutic drugs are polyfunctional substrates for drug metabolizing enzymes (both Phase I and II) and that their activity (as well as transporters and, potentially, drug receptors) is polygenically determined, knowledge of substrate specificity for a given drug and the patterns of expression for activity, which are associated with age (development), can afford the clinician with an element of prediction, be it applied to projecting either the dose of a drug or its pharmacologic consequences in the pediatric patient.

ACKNOWLEDGMENTS

This work was supported by grants 1K12RR17613 (NR) and 1K24RR19729 (JNA), National Center for Research Resources, and grant 1U10HD45993 (JNA), National Institute of Child Health and Development, Bethesda, Maryland.

REFERENCES

1. Kearns GL, Abdel-Rahman SM, Alander SW, Blowey DL, Leeder JS, Kaufman RE. Developmental pharmacology: drug disposition, action, and therapy in infants and children. *N Engl J Med.* 2003;349:1157–1167.
2. Van den Anker JN. Pharmacokinetics and renal function in the preterm infant. *Acta Paediatr.* 1996;85:1393–1399.
3. Alcorn J, McNamara PJ. Ontogeny of hepatic and renal systemic clearance pathways in infants: part 1. *Clin Pharmacokinet.* 2002;41:959–998.
4. Silverman WA, Anderson DH, Blanc WA, et al. A difference in mortality rate and incidence of kernicterus among premature infants allotted to two prophylactic antibacterial regimens. *Pediatrics.* 1956;18:614–624.

5. Sutherland JM. Fatal cardiovascular collapse of infants receiving large amount of chloramphenicol. *Am J Dis Child.* 1959;97:761–767.
6. Lorch V, Murphy D, Hoersten LR, Harris E, Fitzgerald J, Sinha SN. Unusual syndrome among premature infants: association with a new intravenous vitamin E product. *Pediatrics.* 1985;75: 598–602.
7. Lovejoy FH. Benzyl alcohol poisoning in neonatal intensive care units. A new concern for the pediatrician. *Am J Dis Child.* 1982;136:974–975.
8. Einarson A, Portnoi G, Koren G. Update on motherisk updates: seven years of questions and answers. *Can Fam Physician.* 2002;48:1301–1304.
9. Kearns GL. Impact of developmental pharmacology on pediatric study design: overcoming the challenges. *J Allergy Clin Immunol.* 2000;106:S128–S138.
10. Tyrala EE, Hillman LS, Hillman RE, Dodson WE. Clinical pharmacology of kerochlorophene in newborn infants. *J Pediatr.* 1977;91:481–486.
11. Armstrong RW, Eichner ER, Klein DE, et al. Pentachlorophenol poisoning in a nursery for newborn infants. II. Epidemiologic and toxicologic studies. *J Pediatr.* 1969;75:317–325.
12. Feinblatt BI, Aceto T, Beckhorn G, Bruck E. Percutaneous absorption of hydrocortisone in children. *Am J Dis Child.* 1966;112:218–224.
13. Fisch RO, Berglund EB, Bridge AG, et al. Methemoglobinemia in a hospital nursery. *JAMA.* 1963;185:760–763.
14. Choonara I. Pecutaneous drug absorption and administration. *Arch Dis Child.* 1994;71:F73–F74.
15. Radde IC, McKercher HG. Transport through membranes and development of membrane transport. In MacLeod SM, Radde IC, eds. *Textbook of Pediatric Clinical Pharmacology* Littleton, MA: PSG Publishing Company; 1985: 1–16.
16. Lester RS. Topical formulary for the pediatrician. *Pediatr Clin North Am.* 1983;30:749–765.
17. Nachman RL, Esterly NB. Increased skin permeability in preterm infants. *J Pediatr.* 1980;96:99–103.
18. Suchy FJ, Balistreri WF, Heubi JE, Searcy JE, Levin RS. Physiologic cholestasis: elevation of the primary serum bile acid concentrations in normal infants. *Gastroenterology.* 1981;80: 1037–1041.
19. Poley JR, Dower JC, Owen CA, Stickler GB. Bile acids in infants and children. *J Lab Clin Med.* 1964;63:838–846.
20. Hall SD, Thummel KE, Watkins PB, et al. Molecular and physical mechanisms of first-pass extraction. *Drug Metab Dispos.* 1999;27:161–166.
21. Friis-Hansen B. Water distribution in the foetus and newborn infant. *Acta Paediatr Scand.* 1983;305:7–11.
22. Weiss CF, Glazko AJ, Weston JK. Chloramphenicol in the newborn infant. A physiologic explanation of its toxicity when given in excessive doses. *N Engl J Med.* 1960;262:787–794.
23. Young WS III, Lietman PS. Chloramphenicol glucuronyl transferase: assay, ontogeny and inducibility. *J Pharmacol Exp Ther.* 1978;204:203–211.
24. Hines RN, McCarver DG. The ontogeny of human drug-metabolizing enzymes: phase I oxidative enzymes. *J Pharmacol Exp Ther.* 2002;300:355–360.
25. McCarver DG, Hines RN. The ontogeny of human drug metabolizing enzymes: phase II conjugation enzymes and regulatory mechanisms. *J Pharmacol Exp Ther.* 2002;300:361–366.
26. Chen H, Fantel AG, Juchau MR. Catalysis of the 4-hydroxylation of retinoic acids by CYP3A7 in human fetal hepatic tissues. *Drug Metab Dispos.* 2000;28:1051–1057.
27. Lacroix D, Sonnier M, Moncion A, Cheron G, Cresteil T. Expression of CYP3A in the human liver. Evidence that the shift between CYP3A7 and CYP3A4 occurs immediately after birth. *Eur J Biochem.* 1997;247:625–634.

28. Vieira I, Sonnier M, Cresteil T. Developmental expression of CYP2E1 in the human liver. Hypermethylation control of gene expression during the neonatal period. *Eur J Biochem*. 1996;238:476–483.

29. Treluyer J-M, Jacqz-Aigrain E, Alvarez F, Cresteil T. Expression of CYP2D6 in developing human liver. *Eur J Biochem*. 1991;202:583–588.

30. Treluyer J-M, Gueret G, Cheron G, Sonnier M, Cresteil T. Developmental expression of CYP2C and CYP2C-dependent activities in the human liver: in-vivo/in-vitro correlation and inducibility. *Pharmacogenetics*. 1997;7:441–452.

31. Sonnier M, Cresteil T. Delayed ontogenesis of CYP1A2 in the human liver. *Eur J Biochem*. 1998;251:893–898.

32. Kinirons MT, O'Shea D, Kim RB, et al. Failure of erythromycin breath test to correlate with midazolam clearance as a probe of cytochrome P4503A. *Clin Pharmacol Ther*. 1999;66: 224–231.

33. Payne K, Mattheyse FJ, Liedenberg D, Dawes T. The pharmacokinetics of midazolam in paediatric patients. *Eur J Clin Pharmacol*. 1989;37:267–272.

34. Kerr BM, Thummel KE, Wurden CJ, et al. Human liver carbamazepine metabolism. Role of CYP3A4 and CYP2C8 in 10,11-epoxide formation. *Biochem Pharmacol*. 1994;47:1969–1979.

35. Pynnönen S, Sillanpäa M, Frey H, Iisalo E. Carbamazepine and its 10, 11-epoxide in children and adults with epilepsy. *Eur J Clin Pharmacol*. 1977;11:129–133.

36. Riva R, Contin M, Albani F, Perucca E, Procaccianti G, Baruzzi A. Free concentration of carbamazepine and carbamazepine-10, 11-epoxide in children and adults. Influence of age and phenobarbitone co-medication. *Clin Pharmacokinet*. 1985;10:524–531.

37. Bajpai M, Roskos LK, Shen DD, Levy RH. Roles of cytochrome P4502C9 and cytochrome P4502C19 in the stereoselective metabolism of phenytoin to its major metabolite. *Drug Metab Dispos*. 1996;24:1401–1403.

38. Loughnan PM, Greenwald A, Purton WW, Aranda JV, Watters G, Neims AH. Pharmacokinetic observations of phenytoin disposition in the newborn and young infant. *Arch Dis Child*. 1977;52:302–309.

39. Aranda JV, Collinge JM, Zinman R, Watters G. Maturation of caffeine elimination in infancy. *Arch Dis Child*. 1979;54:946–949.

40. Kraus DM, Fischer JH, Reitz SJ, et al. Alterations in theophylline metabolism during the first year of life. *Clin Pharmacol Ther*. 1993;54:351–359.

41. Milavetz G, Vaughan LM, Weinberger MM, Hendeles L. Evaluation of a scheme for establishing and maintaining dosage of theophylline in ambulatory patients with chronic asthma. *J Pediatr*. 1986;109:351–354.

42. Lambert GH, Schoeller DA, Kotake AN, Flores C, Hay D. The effect of age, gender, and sexual maturation on the caffeine breath test. *Dev Pharmacol Ther*. 1986;9:375–388.

43. Kearns GL, Andersson T, James LP, et al. Omeprazole disposition in infants and children following single-dose administration. *J Clin Pharmacol*. 2003;43:840–848.

44. Abdel-Rahman SM, Leeder JS, Wilson JT, et al. Concordance between tramadol and dextromethorphan parent/metabolite ratios: the influence of CYP2D6 and non-CYP2D6 pathways on biotransformation. *J Clin Pharmacol*. 2002;42:24–29.

45. De Wildt SN, Kearns GL, Leeder JS, van den Anker JN. Glucuronidation in humans: pharmacogenetic and developmental aspects. *Clin Pharmacokinet*. 1999;36: 439–452.

46. Clarke DJ, Moghrabi N, Monaghan G, et al. Genetic defects of the UDP-glucuronosyltransferase-1 (UGT1) gene that cause familial non-haemolytic unconjugated hyperbilirubinaemias. *Clin Chim Acta*. 1997;266:63–74.

47. Bosma PJ, Chowdhury JR, Bakker C, et al. The genetic basis of the expression of bilirubin UDP-glucuronosyltransferase 1 in Gilbert's syndrome. *N Engl J Med*. 1995;333:1171–1175.
48. Hume R, Coughtrie MW, Burchell B. Differential localization of UDP-glucuronosyltransferase in kidney during human embryonic and fetal development. *Arch Toxicol*. 1995;69(4):242–247.
49. Hume R, Burchell A, Allan BB, et al. The ontogeny of key endoplasmic reticulum proteins in human embryonic and fetal red blood cells. *Blood*. 1996;87(2):762–770.
50. Burchell B, Coughtrie M, Jackson M, et al. Development of human liver UDP-glucuronosyltransferases. *Dev Pharmacol Ther*. 1989;13:70–77.
51. Leakey JEA, Hume R, Burchell B. Development of multiple activities of UDP-glucuronosyltransferase in human liver. *Biochem J*. 1987;243:859–861.
52. Kawade N, Onishi S. The prenatal and postnatal development of UDP-glucuronosyltransferase activity towards bilirubin and the effect of premature birth on this activity in the human liver. *Biochem J*. 1981;196(1):257–260.
53. Coffman BL, Rios GR, King CD, et al. Human UGT2B7 catalyzes morphine glucuronidation. *Drug Metab Dispos*. 1997;25(1):1–4.
54. Pacifici GM, Franchi M, Biuliani L, et al. Development of the glucuronyltransferase and sulphotransferase towards 2-naphthol in human fetus. *Dev Pharmacol Ther*. 1989;14(2):108–114.
55. Pacifici GM, Sawe J, Kager L, et al. Morphine glucuronidation in human fetal and adult liver. *Eur J Clin Pharmacol*. 1982;22(6):553–558.
56. Choonara IA, McKay P, Hane R, et al. Morphine metabolism in children. *Br J Clin Pharmacol*. 1989;28(5):599–604.
57. Bock KW, Forster A, Gsaidmeier H, et al. Paracetamol glucuronidation by recombinant rat and human phenol UDP-glucuronosyltransferases. *Biochem Pharmacol*. 1993;45(9):1809–1814.
58. Behm MO, Abdel-Rahman SM, Leeder JS, Kearns GL. Ontogeny of phase II enzymes: UGT and SULT. *Clin Pharmacol Ther*. 2003;73:P29.
59. Bouche FD, Modlin JF, Weller S, et al. Phase I evaluation of zidovudine administered to infants exposed at birth to the human immunodeficiency virus. *J Pediatr*. 1993;122(1):137–144.
60. Robillard J, Guillery E, Petershack J. Renal function during fetal life. In: Barratt TM, Avner ED, Harmon WE, eds. *Pediatric Nephrology*. Baltimore, MD: Lippincott Williams & Wilkins; 1999: 21–37.
61. Arant BS Jr. Developmental patterns of renal functional maturation in the human neonate. *J Pediatr*. 1978;92:705–712.
62. Van den Anker JN, Schoemaker R, Hop W, et al. Ceftazidime pharmacokinetics in preterm infants: effects of renal function and gestational age. *Clin Pharmacol Ther*. 1995;58:650–659.
63. James LP, Marotti T, Stowe C, Farrar HC, Taylor B, Kearns GL. Pharmacokinetics and pharmacodynamics of famotidine in infants. *J Clin Pharmacol*. 1998;38:1089–1095.
64. Brion LP, Fleischman AR, Schwartz GJ. Gentamicin interval in newborn infants as determined by renal function and postconceptional age. *Pediatr Nephrol*. 1991;5:675–678.
65. Szefler S, Wynn R, Clarke D, Buckwald S, Shen D, Schentag J. Relationship of gentamicin serum concentrations to gestational age in preterm and term infants. *J Pediatr*. 1980;97:312–315.
66. Van den Anker JN, Hop W, de Groot R, et al. Effects of prenatal exposure to betamethasone and indomethacin on the glomerular filtration rate in the preterm infant. *Pediatr Res*. 1994;36:578–581.

CHAPTER 21

Developmental Hepatic Pharmacology in Pediatrics

MARTIN OTTO BEHM, MD

Merck and Co., Inc., Clinical Pharmacology, North Wales, Pennsylvania 19454

INTRODUCTION

Pediatric pharmaceutical research must take into account normal growth and development pathways. Understanding these growth and development patterns can help in designing rational research studies, which will maximize the information gained from the studies. Hepatic metabolism is an important step in determining the ultimate exposure of many pharmaceutical agents including new chemical entities. This metabolism has been divided conventionally into two phases. Phase I hepatic metabolism usually results in modifying the therapeutic agent or xenobiotic (e.g., through oxidation) in order to make the molecule more polar. Phase II hepatic metabolism usually results in addition of a small molecule (e.g., glucuronide) to the therapeutic agent in order to make it more polar. Understanding the ontogeny of hepatic drug metabolizing enzymes in these two metabolic phases is extremely useful in designing rational pediatric pharmaceutical research trials.

In November 1998, the United States Food and Drug Administration (FDA) published their *Guidance for Industry: General Considerations for Pediatric Pharmacokinetic Studies for Drugs and Biological Products.*[1] In this guidance, it is acknowledged that children in general "will form the same metabolites as adults via pathways such as oxidation, reduction, hydrolysis and conjugation, but rates of metabolite formation can be different." These various "rates of metabolite formation" are governed by predictable hepatic growth and development patterns. Furthermore, this guidance uses the definition of pediatric populations from the 1994 Pediatric Rule:

1. Neonate: Birth to 1 month.
2. Infant: 1 month to 2 years.
3. Children: 2 years to 12 years.
4. Adolescents: 12 years to <16 years.

Pediatric Drug Development: Concepts and Applications
Edited by Andrew E. Mulberg, Steven A. Silber, and John N. van den Anker

While these various age groups are useful in general for designing pediatric studies, they do not necessarily take into account rapid predictable changes in the developmental patterns due to the maturation of hepatic drug metabolism pathways. Therefore, situations may arise when it would be more rational for a given pediatric development program to utilize other age divisions than provided in the FDA guidance.[1] Specific pediatric development programs need to justify the margins of ages for the specific trials based on scientific support.

Over the past two decades, there has been an explosion in the amount of information generated about metabolism of therapeutic agents in children.[2] In vivo data has been generated largely through two means. One is through dedicated "ontogeny" studies in which a probe drug (e.g., dextromethorphan[3,4] or acetaminophen[5] is given to children of various age groups or to the same children over a period of time. The other manner in which this in vivo data has been developed is serendipitously over the course of industry-sponsored or investigator-initiated pediatric clinical trials, which utilize the traditional age groups, and both anticipated as well as unexpected results reveal new data about the drug metabolizing enzymes involved.

With the recent increase in pediatric clinical trials, there is a resulting increase in review articles in the literature, which captures what is known about normal growth and development pathways with respect to absorption, distribution, metabolism, and elimination of drugs. The scope of this chapter is not to recapitulate what is already available in these review articles,[2,6–9] but rather to define what is known and not known in the four age groups with respect to predictable maturation of hepatic drug metabolism in order to provide rationally designed clinical trials.

This chapter is largely divided into two parts. The first part concerns the developmental patterns of hepatic drug metabolizing enzymes in the first 2 years of life. This roughly correlates to the first two age groups provided in the FDA guidance document previously presented. The second part concerns the maturation of hepatic drug metabolizing enzymes between the ages of 2 and 16 years of life. This roughly corresponds to the second two age groups described in the FDA guidance.

NEONATES AND INFANTS

No other age group defines such a period of rapid growth and development. Indeed, as the FDA guidance does not define these age groups based on postconceptual age but rather "birth" to 2 years of age. It is well established that infants who are barely into their second trimester of gestational life born as small as a few hundred grams can survive. On the other extreme, by the end of the first month of postnatal life, large for gestational age infants may weigh upwards of several kilograms. Indeed, the 95th weight percentile is approximately 15 kilograms. No other age group can be defined in differences measured logarithmically! As one might expect, there are similar tremendous developmental changes in hepatic drug metabolizing enzymes during this time frame. Understanding these implications is important for individualized clinical development programs.

Phase I Enzymes

CYP3A The CYP3A subfamily represents the majority of CYP total content in the liver. Indeed, it is estimated that over one-half of all drugs prescribed are metabolized by CYP3A.

The CYP3A subfamily consists of CYP3A4, 3A5, 3A7, and 3A43. CYP3A43 is not known to play a significant role in hepatic metabolism. It has been established that CYP3A4 is the predominant CYP3A enzyme in adults, whereas CYP3A7 is the predominant CYP3A enzyme in the fetus and infants by Lacroix et al.[10] Moreover, there is a great deal of overlap of specificity of ability for CYP3A4 and CYP3A7 to metabolize therapeutic agents. In 2003, Stevens et al.[11] published the results of examining the largest collection of fetal and pediatric liver samples to date. The study included 212 samples. Stevens and colleagues demonstrated that CYP3A7 is highest between 94 and 168 postconceptual days on a pmol/mg basis of total hepatic protein. The level at birth is less than half that of the high prenatal value. However, it remains higher than that of even adult CYP3A4 levels. Furthermore, these hepatic samples demonstrated that there is minimal CYP3A4 activity prenatal that continues to increase after birth. Nevertheless, CYP3A7 content remains higher than CYP3A4 content until at least 6 months of age.

To date, two probe drugs have been researched extensively, which have demonstrated the lower activity of CYP3A4 at birth and in neonates. In 2001, de Wildt et al.[12] published the results of midazolam metabolism given to 24 preterm infants. Only 19 of 24 preterm infants produced detectable levels of 1-OH-midazolam. Furthermore, these results firmly established that premature infants had lower CYP3A4 activity than full-term infants, than did children and adults historically. Oral cisapride has also been demonstrated to be a suitable substrate for CYP3A4 activity.[13,14] Cisapride has demonstrated a similarly low activity for CYP3A4 in the neonatal period, as did midazolam.[15–17]

In conclusion, CYP3A7 activity is very high before birth and continues to have high activity after birth and is even present into adulthood. CYP3A4 possesses very low activity at birth and very slowly increases in the neonatal period. Thus, when designing studies with substrates for CYP3A4 in young infants and children, great care needs to be taken to adjust for this low activity in order to achieve the goal of the FDA guidance that in children exposure and C_{max} are not higher than that in adults.

CYP1A2 One of the first CYP enzymes to be studied utilizing a probe drug in the first year of life is CYP1A2. Two methylxanthines (caffeine and theophylline) have been utilized extensively to evaluate CYP1A2 in vivo in young children.[18–24] Theophylline and caffeine are two commonly utilized medications in neonates for the treatment of apnea. These medications are frequently continued from the neonatal period during the first year of life. At birth, caffeine 3-demethylation, a measure of CYP1A2 activity, is very low. Consequently, Erenberg et al.[19] published in 2000 that the efficacious dose of caffeine is 10 mg/kg every day. The half-life of caffeine is 72–96 hours in infants compared to approximately 5 hours in older children and adults. Similarly, 8-hydroxylation of theophylline is reduced at birth.[24–27] Nevertheless, longitudinal data indicates a rapid maturation process for CYP1A2, as it appears to reach adult levels within the first year of life, often within the first 6 months of life. Finally, it is important to note that caffeine activity is highly inducible by drugs, diet, and exogenous toxins, such as cigarette smoke. In adult literature, variability up to 100-fold has been reported.[28] Moreover, Blake et al.[29] reported that caffeine elimination half-life in neonates who are breast-fed is longer than that of bottle-fed infants.

In conclusion, it is evident that CYP1A2 activity is highly reduced in young infants. Additionally, activity of the enzyme is highly inducible. Finally, maturation of CYP1A2 activity is rapid in the first year of life. Therefore, when designing clinical studies, which include neonates, great care must be taken to assure that this variability in drug response is properly assessed, especially within the first 6–12 months of life.

CYP2D6 CYP2D6 is one of the most polymorphically expressed enzymes in humans.[30–33] Some estimates indicate that fewer than 90% of individuals are homozygous for the wild-type allele.

In 1991, Treluyer et al.[33] published the results of liver samples from fetuses aged 17–40 weeks postconception. These results demonstrated that the concentration of hepatic CYP2D6 protein was very low or undetectable in these fetuses. This lack of CYP2D6 activity at birth led to the hypothesis that birth-related events may trigger maturation of the enzyme.

In 2007, Blake et al.[3] published in vivo results that provided further understanding of CYP2D6 activity in the first year of life. These results came from dosing infants with dextromethorphan at 0.5, 1, 2, 4, 6, and 12 months of age and measuring the metabolites in urine. These dextromethorphan results demonstrate indeed that there is low level of activity at birth, but that there is a rapid acquisition of CYP2D6 activity in the first year of life. Verily, within the first 2 weeks of life, there is measurable acquisition of CYP2D6 activity.

Taken together, these results demonstrate the need for careful pharmacokinetic studies in infants who are provided a pharmacologic agent, which is primarily metabolized by CYP2D6. Not only does one need to be cognizant of potential infants who are predestined by their genome to be poor metabolizers, but potential studies need to realize the implications of low levels of CYP2D6 at birth and also the rapid maturation process that occurs within the first year of life.

CYP2C9/CYP2C19 Lee et al.[34] and Koukouritaki et al.[35] have pushed the most extensive reviews to date of CYP2C activity in humans. They demonstrate that the two main representatives of the CYP2C subfamily of enzymes (CYP2C9 and CYP2C19) conveniently follow the CYP2C rule of 20%. Namely, approximately 20% of hepatic CYP content of adult livers is CYP2C; approximately, CYP2C enzymes metabolize 20% of pharmaceuticals developed to date.

Although not to the same extent as CYP2D6, the two main CYP2C representatives are polymorphically expressed. To date, over 10 genetic polymorphisms of CYP2C9 have been identified[34] and at least 15 genetic polymorphisms of CYP2C19 have been reported in the literature.[35] Just as with CYP2D6, these polymorphisms may result in poor metabolizer status, which may confound studies in infants and young children.

The ontogeny of CYP2C9 is much better established than CYP2C19. Indeed, hepatic liver samples have shown that CYP2C9 activity is functionally very low just prior to birth. However, much like CYP2D6, this activity increases quickly in the first year of life. The classic example of the effects of this very low level of CYP2C9 activity at birth can be seen with phenytoin.[36] Indeed, the recommended daily dose for newborns is 5 mg/kg/day, but by 6 months to 3 years of age this increases to 8–10 mg/kg/day consequent to increased CYP2C9 activity.

Two major pharmaceutical classes of drugs (i.e., benzodiazepines and proton pump inhibitors) have major representative therapeutic agents that are metabolized by CYP2C19.[37,38] Indeed, characteristic representatives from these classes are used in the literature to indirectly ascertain the ontogeny of CYP2C19 activity. Hydroxylation of diazepam is attributed to CYP2C19 activity[39] and is a classic example of the effects of the maturation process of CYP2C19. In neonates, the half-life of diazepam is reported to be 50–90 hours. Within the first year of life, that half-life of 40–50 hours is much closer to the adult value, which is reported as 20–50 hours.[40]

More recently, the effects of ontogeny on proton pump inhibitor metabolism have been reviewed. To date, all proton pump inhibitors other than rabeprazole are metabolized by CYP2C19. As expected, exposures of the CYP2C19 metabolism-dependent proton pump inhibitors are universally increased in the youngest infants when genetic polymorphisms of CYP2C19 are fully accounted.[37,38]

Taken together, these results demonstrate an important trend when designing pharmaceutical studies that depend on hepatic metabolism through the two major CYP2C enzyme pathways. It is extremely important to be cognizant of the limited activity of these enzymes in early childhood. Moreover, much like with CYP2D6, it is important to recognize the impact of genetic polymorphisms when studying individuals who intake substrates of these enzymes.[41] Finally, the first month of life represents a dramatic maturation time for these enzymatic pathways. Therefore, one can assume there will be great variability of exposure in studies with infants in this age group.

CYP2E1 CYP2E1 is being increasingly recognized for its importance in the oxidative metabolism of a wide variety of pharmaceuticals (e.g., acetaminophen, halothane, and ethanol).[42] However, only in the last 5 years has the developmental patterns of this important enzyme been well understood.[43] Nevertheless, human hepatic CYP2E1 developmental expression is difficult to appreciate due to the multiple levels of regulation in its activity. For example, CYP2E1 is known to be elevated in individuals who have high levels of ethanol consumption, in individuals who are obese, and finally in individuals who have type 2 diabetes.[44] Finally, an increasing number of genetic polymorphisms, which lead to lower CYP2E1 protein concentration, have been demonstrated in the literature.[45]

To date, Johnsrud et al.[43] have published the largest study of the activity of fetal ($N = 73$ and pediatric ($N = 165$) liver samples to determine the ontogeny of CYP2E1. Measurable CYP2E1 activity was demonstrated in 18 of 49 second trimester livers and 12 of 15 third trimester samples. Moreover, measurements of mean concentration of CYP2E1 protein as part of total milligrams of microsomal protein found that second trimester infants averaged 0.35 pmol/mg, third trimester infants averaged 6.7 pmol/mg, newborns averaged 8.8 pmol/mg, older infants aged 30–90 days averaged 23.8 pmol/mg, and finally children aged 90 days to 18 years averaged 41.4 pmol/mg. Thus, this implies a rapid maturation starting in late fetal life and continuing through early infancy in the activity of CYP2E1. It would appear that this data demonstrates that careful attention would be required in studies of new CYP2E1 substrates in infants under the age of 90 days.

FMO The flavin-containing monooxygenases (FMOs) are important for the metabolism of several common pediatric medications, such as chlorpromazine,[46] promethazine,[47] and brompheniramine.[48] Unlike the CYP450 enzymes, there are only six known FMO enzymes known to date in mammals. In the fetus and young children, FMO1 is the most common FMO enzyme, while in adults, FMO3 is known to be more common.[49] In 2002, Koukouritaki et al.[50] examined 240 liver samples from 8 weeks gestation through 18 years of age. FMO1 expression on a picomole per milligram of liver protein basis is the highest at 8–15 weeks gestation with suppression occurring within 3 days postpartum. Thus, it appears that this process is more tightly tied to birth rather than gestational age. Additionally, the appearance of FMO3 is highly variable during infancy, with most infants having detectable levels around 1–2 years of life. However, these levels appear to increase slowly through adolescence. Additionally, this hepatic study demonstrated twofold to 20-fold variability within each age group.

While it has been known since the early 1990s that chlorpromazine, promethazine, and brompheniramine are metabolized by FMOs and since 2002 that the development pathway of FMO3 is very slow, the link between this ontogeny and possible dose adjustment for pure FMO metabolized therapeutic agents has yet to be established. Thus, it is important to proceed in an extremely conservative manner when developing drugs that are metabolized mainly through FMO metabolism in young infants.

Phase II Enzymes

While Phase I hepatic enzymatic processes are characterized by the removal of part of the molecule to make it more polar, Phase II enzymatic processes generally add a small molecular weight organic donor molecule to make a molecule more polar. Indeed, the ontogeny of Phase II enzymatic pathways is approximately 5–10 years behind Phase I pediatric and adult research. Nevertheless, great progress has been made over the last decade in understanding the developmental pathways of these enzymatic processes.

Glucuronidation

UGT1A Just as there is great variety in the patterns of development for CYP enzymatic pathways, so too is there great variety in the development of glucuronidation in infancy. To date, over 15 different UGT human enzymes have been identified to date. At least two-thirds of these are encoded at the *UGT1* locus.

Perhaps the best understood of the UGT1 enzymes is UGT1A1, which is responsible for glucuronidation of bilirubin.[51] In fetal hepatic samples, the activity of UGT1A1 is extremely limited. However, in the first year of life, this activity greatly increases. Indeed, this may be seen with the metabolism of bilirubin. In young infants, benign neonatal jaundice due to immature UGT1A1 expression is ubiquitous. Nevertheless, some children go beyond benign neonatal hyperbilirubinemia of infancy and develop pathological jaundice. Some of these cases are due to the polymorphic expression of UGT1A1. The majority of these polymorphisms are the result of TATA repeats in the promoter region of UGT1A1. Indeed, Gilbert syndrome is now known to result from a polymorphism of UGT1A1. Finally, the activity of UGT1A1 even in infancy is highly inducible by pharmacologic substrates (e.g., rifampin and some benzyodiazepines).[52]

The other UGT1 enzymes for which there is understanding for infants include UGT1A3, UGT1A6, and/or UGT1A9. In contrast to many other UGT isoforms, in 1989, Burchell et al.[51] discovered that UGT1A3 activity in fetal livers is 30% of that in adult livers. However, the clinical consequences of a substrate metabolized primarily by UGT1A3, as studied in infants, have yet to be determined.

Alternatively, much is known about the maturation of the glucuronidation of acetaminophen.[53] Fortunately, acetaminophen is metabolized by multiple metabolic pathways including glucuronidation and sulfation. Therefore, since the 1970s, the ratio of acetaminophen glucuronide to other acetaminophen metabolites in the urine has been used as a surrogate for measuring the maturation of glucuronidation. Unfortunately, this process actually predates the current genetic understanding of the genetics of glucuronidation, and thus, there are contradictions in the literature concerning the relative contributions of UGT1A6 and UGT1A9 to the glucuronidation of acetaminophen.[54]

The ontogeny of acetaminophen glucuronidation has been understood for over three decades. Miller et al[53], in 1976, published the first results of acetaminophen usage in various age groups (i.e., 0–2 days old, 3–9 years old, 12 year olds, and adults). As expected, acetaminophen glucuronidation in the youngest infants was almost nonexistent.

A study by Behm et al.[5] in which acetaminophen was given to over 60 infants at each of their seven well baby visits (i.e., 2 weeks, 1 month, 2 months, 4 months, 6 months, 9 months, and 12 months) in the first year of life has been reported. In the first month of life at the first two well baby visits, acetaminophen glucuronide levels were very low. Some of the infants had levels that were undetectable. However, unlike the ontogeny of some of the CYP enzyme activity described earlier for the Phase I data, acetaminophen activity did not suddenly increase at a certain point in the first year of life to near adult levels, but rather gradually and consistently increased as was initially demonstrated by Miller et al.[53]

UGT2B Historically, one of the first well-documented cases of a failure of glucuronidation occurred with the use of chloramphenicol in what has since become known as "gray baby syndrome," which was first reported by Sutherland in 1959.[55] Although it is currently not known what UGT isoform is responsible for the metabolism of chloramphenicol, McCarver and Hines[56] hypothesized in 2001 that "it appears to be a member of the UGT2B subfamily." In gray baby syndrome, young infants who were given chloramphenicol for several days turned grey, suffered cardiovascular collapse, and subsequently died. From this point forward, it was recognized that lack of understanding of the ontogeny of hepatic glucuronidation could result in catastrophic consequences.

In contrast to the disastrous gray baby syndrome, morphine is another glucuronidated medication used successfully in young infants. Morphine is metabolized by UGT2B7. In 1982, Pacifici et al.[57] published that morphine is glucuronidated in fetal hepatic samples as young as 15–27 weeks postconception. This rate corresponded to 10–20% of adult levels. Moreover, Choonara et al.[58,59] published that morphine glucuronidation frequently approached adult levels within the first year of life but occasionally required up to 30 months to reach adult levels. Moreover, with respect to tramadol, it was demonstrated in 2007 that both postnatal and postmenstrual age contribute to the interindividual variability of its glucuronidation, further demonstrating the variability in early glucuronidation.[60]

Taken together, these data suggest that UGT2B isoform ontogeny is grossly similar to that of the aforementioned UGT1A isoforms in that care is needed when designing studies in young infants, especially in the first year of life.

Sulfation Of the Phase II hepatic drug metabolizing enzymes, none perhaps is as well developed as in sulfation in prenatal individuals and young infants. The developing human fetus depends on many complex hormonal and neurochemical molecular signals for proper growth. In order to regulate these signals, the fetus is dependent on metabolizing these hormones and endogenous small molecules. Indeed, this is thought to play a role in the relatively mature ability to sulfate in the developing fetus. SULT1A1 and SULT1A3 are essential to the metabolism of iodothyronines and catecholamines in fetal life.[61] Thus, their relatively mature conjugating ability in fetal life is essential for fetal survival. Indeed, in the studies by Miller et al.[53] and Behm et al.,[5] which analyzed the acetaminophen metabolites in the first year of life, acetaminophen sulfate was by far the major metabolite in the urine of young infants in contrast to acetaminophen glucuronide, which was the major metabolite in the urine of adults. Infants depend on this highly developed sulfation ability for biotransformation of doses of 10 mg/kg of acetaminophen, a dose comparable on a milligram per kilogram basis to that consumed by adults.

In conclusion, therapeutic agents studied in children, which are highly metabolized by hepatic sulfation, can be extrapolated to be metabolized relatively easily in infants and young children on a milligram per kilogram basis.

NAT2 As described previously, the ontogeny of metabolism of caffeine in the first year of life is well established. Caffeine is metabolized by both CYP1A2 as well as NAT2.[62] While the ontogeny of NAT1 is not well established, the work of Pariente-Khayat et al.[23] in 1991 established the maturation process for caffeine acetylation. This studied the ratio of caffeine metabolites from 3-day-old to 630-day-old children. While NAT2 is a polymorphically expressed enzyme, this study established that the majority of infants reached their genetic potential for acetylation between birth and 15 months of age. Similar maturation progress was seen by Pariente-Khayat et al.[63] in 1997 when examining the maturation of the ability of children to acetylate isoniazid.

Therefore, it appears that when developing therapeutic agents that are metabolized by NAT2, two important considerations must be included. First, one must determine if the young child has the genetic predisposition to be a fast or slow acetylator. Second, one must carefully design the study to determine metabolic ability before and after the age of 15 months.

OLDER INFANTS AND CHILDREN 2–16 YEARS

The FDA guidance has divided pediatric studies into four major groups. The third of these age groups includes children 2–6 years of age and the fourth includes children 6–16 years of age.[1] As has hopefully been demonstrated, great maturation occurs in the first 2 years of life for many hepatic enzyme pathways.

There are many studies that demonstrate through in vitro hepatic samples and in vivo ontogeny studies the activity in young infants under 2 years of age. Indeed, Blake et al.[3] have only pushed dextramethophan results to 1 year of age and Behm et al.[5] have only published acetaminophen results to 12 months of age. As Miller et al.[53] demonstrated in 1976, acetaminophen glucuronidation continues to mature throughout childhood. Therefore, in this section we review what is known and for the most part what is still to be determined in the developmental patterns of hepatic enzymes in this older age group with respect to designing rational drug studies.

Phase I Enzymes

CYP3A There are no definitive ontogeny studies to measure the development of CYP3A activity in children 2–16 years of age. However, in the literature there are two definitive pharmacokinetic (PK) papers of classic CYP3A4 substrates utilized in children which provide insight into the development patterns of CYP3A4 activity in children.

In 2001, Reed et al.[17] published the results of dosing midazolam at three doses: 0.25 mg/kg, 0.5 mg/kg, and 1.0 mg/kg, in three age groups, 0.5–2 years, 2–12 years, and 12–16 years. This study showed that the disposition of midazolam appeared linear over the dose range studied and was independent of patient age. There was an apparent trend toward increasing $t_{1/2}$ with increasing age. Overall, the disposition of midazolam as compared to its CYP3A4 metabolite, α-hydroxymidazolam, also was linear over the dose range and was independent of age. This study demonstrated significant stability of CYP3A4 activity over these age ranges.

According to the label for itraconazole, the major metabolite of itraconazole, hydroxylitraconazole, is formed through CYP3A4.[64] Pharmacokinetic results were reported for a single 2.5-mg/kg dose of itraconazole to 33 children aged 7 months to 17 years. The total body exposures of itraconazole and hydroxylitraconazole demonstrated no significant difference with respect to age. Additionally, there was no difference with respect to age and total body clearance. This study further demonstrated a lack of clinically significant changes in CYP3A4 activity in these ages with respect to dosing itraconazole.[65,66]

Earlier in this chapter, a difference was demonstrated with respect to very young infants and CYP3A4 activity. To date, a clinically meaningful difference in the activity of two important CYP3A4 substrates and their metabolites has yet to be demonstrated beyond infancy. While action is always prudent, it would not be unrealistic to anticipate this trend to continue with other CYP3A4 substrates.

CYP1A2 It appears that beyond the first year of life, caffeine, when dosed in a weight-adjusted manner, results in exposures that are close to those in adults. Similarly, other pharmacokinetic parameters, such as terminal half-life, are much closer to adult values by the end of the first year of life. In conformation, beyond the first year of life, the mean half-life of theophylline is 3.4 hours in children aged 1–4 years and 3.7 in children 16–17 years of age. Thus, there appears to be stability in CYP1A2 activity beyond 2 years of age. Nevertheless, in 1986, Lambert et al.[22] published the results of a [^{13}C] caffeine breath test in children and adolescents, which appeared to show an effect of gender, age, and puberty on the 2-hour results. However, the clinical significance of these results has yet to be determined in the pharmacokinetics of any pure CYP1A2 substrates. Unfortunately, unlike for CYP3A4, there are no newer drugs that are predominantly metabolized by CYP1A2, which have had detailed modern pharmacokinetic studies in children. Therefore, there is a large gap in the knowledge base with respect to changes that occur between the ages of 2 and 16 with respect to CYP1A2 activity. Thus, if one is investigating the pharmacokinetics of a pure CYP1A2 substrate in this age group, caution needs to be taken before presuming heterogeneous activity of CYP1A2 in this group.

CYP2D6 As is the theme for the ontogeny of drug metabolizing enzymes in children between 2 and 16 years of age, there are no dedicated studies definitively demonstrating enzyme activity in this age range. The work of Blake and colleagues showed that CYP2D6 activity is well established in the first year of life. This corroborates the work of Evans et al.,[20] who suggest mature activity in early childhood.

In 2002, Abdel-Rahman et al.[67] demonstrated a concordance between tramadol formation of the M1 metabolite by CYP2D6 and the gold standard dextromethorphan CYP2D6 mediated formation of dextorphan in children ages 7–16. Future detailed tramadol pharmacokinetic studies in children ages 2–16 could shed light on the ontogeny of CYP2D6 in this age group.

To date, the most detailed study of the ontogeny of CYP2D6 activity can be found in the 2005 abstract by Park et al.,[68] which showed that the urinary ratio of dextramethorphan to dextorphan in children aged 3–10, adolescents aged 10–15, young adults aged 20–25, and older adults (over 60) remains constant.

In conclusion, while detailed ontogeny studies are not completed in the 2–16-year-old age group, it appears that CYP2D6 activity remains relatively constant and comparable to mature adult values. Nevertheless, the vast number of CYP2D6 polymorphisms needs to be appreciated in the design of studies in children and teens in addition to infants and young children.

CYP2C9/CYP2C19 As stated previously, in 2004, Koukouitaki et al.[35] evaluated 237 liver samples from 8 weeks gestation to 18 years of age to evaluate the activity of CYP2C9 and CYP2C19. This study demonstrated a great stability of CYP2C9 activity by 5 months of age through the end of childhood. Both CYP2C9 and CYP2C19 are responsible for the metabolism of phenytoin, but CYP2C9 appears to be the major metabolic pathway for phenytoin in children as well as adults. As demonstrated previously, beyond the age of 2 years, the pharmacokinetics of phenytoin are stable in children who have a wild-type genotype. However, as with adults, those children with one or more CYP2C9*3 alleles have reduced ability to metabolize and excrete phenytoin.[36] Although the data with warfarin is not as well established in children, similar trends are seen. Beyond infancy, allelic variance plays a larger role in metabolic ability than age.[69]

With CYP2C19, the initial in vitro work by Koukouitaki et al.[35] is confirmed by the in vitro data in the study of modern proton pump inhibitors.[37,38]

CYP2E1 The work of Johnsrud et al.[43] provides the most comprehensive analysis of the maturation of CYP2E1 from the prenatal stage through the age of 18. Previously, we showed three important factors beyond the neonatal range that must be considered when developing therapeutic agents for children. First, there are nearly adult levels of CYP2E1 in young children and adolescents. Second, CYP2E1 is highly inducible and demonstrates a high degree of interindividual variability.[70] Finally, a number of clinically significant CYP2E1 genetic polymorphisms have recently been identified. Therefore, when developing therapeutic agents for children beyond the neonatal period, while CYP2E1 hepatic ability may be relatively mature, there are additional therapeutic concerns.

FMO In 2002, Koukouritaki et al.[50] published the most comprehensive analysis of pediatric hepatic samples from fetal life through age 18 with respect to concentrations of FMO1 and FMO3. This analysis demonstrated that FMO1 appears to be most common before birth with FMO3 becomes the predominant flavin containing monooxygenase in childhood and adulthood. Of all the hepatic drug metabolizing enzymes discussed in this chapter, the work of Koukouritaki et al.[50] demonstrates that FMO3 may be the slowest to deslope. Indeed, it appears that on a picomole per milligram of hepatic protein basis, the amount of FMO3 present even at age 11 years is only half that of later teens. Nevertheless, an in vivo probe study has yet to be conducted to demonstrate the clinical consequences of reduced FMO3 throughout much of childhood. Thus, caution when designing pediatric studies with novel FMO3 metabolized therapeutic agents must be considered.

Phase II Enzymes

UGT1A The work of Miller et al.,[53] which examines the ontogeny of acetaminophen glucuronidation, is important beyond childhood. Indeed, this study appears to show that it takes many years into early childhood for glucuronidation to fully mature and develop. In contrast, it appears that bilirubin levels in non-Gilbert syndrome individuals appear stable beyond the first year of life. This could imply that UGT1A1 matures more quickly than other

UGT1A isoforms due to the evolutionary need to metabolize this ubiquitous endogenous substance.

The exact isoform of UGT1A responsible for acetaminophen metabolism remains to be elucidated. Therefore, much work needs to be completed to understand fully the various maturation patterns of UGT1A isoforms beyond early childhood. Until more information is gathered, when faced with a substance that is metabolized almost exclusively by a UGT1A isoform (other than UGT1A1 in children) beyond early childhood, it is possible that adult levels of the particular UGT1A isoform may not be present until beyond early childhood.

UGT2B Two of the more common therapeutic agents metabolized by UGT2B isoforms are morphine and chloramphenicol. While it has been demonstrated that dose adjustments should be utilized in infants beyond a milligram per kilogram basis when utilizing these medications, thankfully, it does not appear that such adjustments are needed in older children and teens.[71,72] Nevertheless, greater research into the ontogeny of specific UGT2B isoforms (e.g., UGT2B7) is needed to confirm these clinical conclusions.

Sulfotransferases The sulfotransferases evolved to metabolize a wide range of endogenous chemicals and hormones. SULT1A1 and SULT1A3 are present at very high levels even at birth, thus allowing infants to metabolize successfully certain medications such as acetaminophen at very young ages at levels close to adults. Indeed, the recommended dose of acetaminophen at birth is very close to that recommended in adults on a milligram per kilogram basis. Thus, while detailed ontogeny studies are lacking which show how therapeutic agents are metabolized through sulfation in older children and teens, it is a fair assumption that there is no clinically significant regression of sulfation ability in this age group.

NAT The landmark work of Pariente-Khayat et al.[63] in determining the ontogeny of caffeine and isoniazid acetylator ability is well established in young children. Indeed, while NAT1 ontogeny is not well established, this work demonstrates that hepatic NAT2 ability is very near genetically predisposed adult levels by the age of 2 years. Thus, with respect to development of rational clinical trials in children and teens, the ontogeny of NAT2 is far less important than determining the clinical trial subject's genetic potential to acetylate in order to properly appreciate the study results.

CONCLUSION

This chapter has focused on the developmental patterns of hepatic drug metabolizing enzymes in order to provide further guidance to design rational studies of new and existing therapeutic agents in children. To date, the U.S. government has provided draft guidance on the design of pharmacologic studies in children. In this guidance, the age groups have been somewhat artificially divided into four groups.

The goal of this chapter was to divide children from birth to 16 years into two large groups and then describe what is known and what remains to be known about the predictable developmental patterns of several of the most important Phase I and Phase II enzymes. This chapter has covered a wide range of ontogeny patterns of hepatic drug metabolizing enzymes. Some hepatic drug metabolizing enzymes are present at relatively high levels at birth (e.g., SULT and CYP3A7). Of these, SULT appears to remain at high levels throughout childhood due to the importance of metabolizing endogenous substances, while CYP3A7 activity decreases through childhood. Some hepatic enzymes, which are present at very low levels at birth, have activity that appears very rapidly, such as the activity of CYP1A2 and CYP2D6 in the first year of life. Finally, some hepatic enzymes appear to take much longer to approach adult levels over the course of years such as UGT enzymes.

By examining these ontogeny patterns of hepatic drug metabolizing enzymes, this chapter has strived to provide insight into dividing the age groups of future studies on therapeutic agents in children into more rational age divisions based on what is known about the ontogeny of the hepatic drug metabolizing enzymes responsible for the biotransformation of these agents. In this way, the greatest knowledge may be gained from the children in the pharmaceutical research so that their study participation is maximized.

FUTURE DIRECTIONS

As much has been learned about the ontogeny of hepatic drug metabolizing enzymes, much has still to be discovered. There are significant gaps in the understanding of the changes in activity over childhood.

Much has been discovered about the changes in early childhood: several enzymes that are present at low levels at birth approach adult levels in the first year or two of life. However, many clinicians have anecdotal experiences in which it appears there are times in childhood (such as during the toddler period as well as adolescent growth spurts) when hepatic drug metabolic activity appears to surpass adult activity when adjusted for weight. Indeed, there are a handful of studies that have looked in detail at ontogeny beyond the first several years of life into later childhood. In the meantime, most novel therapeutic agents as per the FDA guidance rely on small groups (e.g., $N = 12$ per age group) to define the pharmacokinetic parameters in older children. More work remains to be completed on the ontogeny of drug metabolizing enzymes in older children.

Additionally, a handful of studies have examined how hepatic drug metabolizing enzymes change during the course of chronic childhood illnesses, such as asthma or cystic fibrosis. Nevertheless, these innovative studies have been conducted in small numbers of individuals, examining very small numbers of hepatic drug metabolizing enzymes. To date, these results have been inconclusive and somewhat discouraging. However, more definitive studies remain to be conducted.

Finally, Phase II drug metabolizing enzyme studies in adults have frequently taken second stage to studies of Phase I enzymes. However, as was demonstrated with gray baby syndrome, infants are very vulnerable to the ontogeny of Phase II drug metabolizing enzymes. Additionally, many of the studies to date with novel therapeutic agents have not taken into account the UGT or SULT isoforms metabolizing these substances. Significant work remains to be completed in both adults and young children to further understand the implications of Phase II hepatic drug metabolism.

REFERENCES

1. USFDA. *Guidance for Industry: General Considerations for Pediatric Pharmacokinetic Studies for Drugs and Biological Products*. US Department of Health and Human Services, Food and

Drug Administration Center for Drug Evaluation and Research (CDER), Center for Biologics Evaluation and Research (CBER), 1998.

2. Ginsberg G, Hattis D, Sonawane B, et al. Evaluation of child/adult pharmacokinetic differences from a database derived from the therapeutic drug literature. *Toxicol Sci.* 2002; 66(2):185–200.
3. Blake MJ, Gaedigk A, Pearce RE, et al. Ontogeny of dextromethorphan O- and N-demethylation in the first year of life. *Clin Pharmacol Ther.* 2007;81(4):510–516.
4. Blake MJ, Castro L, Leeder JS, Kearns GL. Ontogeny of drug metabolizing enzymes in the neonate. *Semin Fetal Neonatal Med.* 2005;10(2):123–138.
5. Behm MO, Abdel-Rahman SM, Leeder SJ, Kearns GL. Ontogeny of Phase II enzymes: UGT and SULT. *Clin Pharmacol Ther.* 2003;73:P29.
6. Hattis D, Ginsberg G, Sonawane B, et al. Differences in pharmacokinetics between children and adults—II. Children's variability in drug elimination half-lives and in some parameters needed for physiologically based pharmacokinetic modelling. *Risk Anal.* 2003;23(1):117–142.
7. Kearns GL, Abdel-Rahman SM, Alander SW, et al. Developmental pharmacology—drug disposition, action, and therapy in infants and children. *N Engl J Med.* 2003;349(12):1157–1167.
8. Leeder JS, Kearns GL. Pharmacogenetics in pediatrics. Implications for practice. *Pediatr Clin North Am.* 1997;44(1):55–77.
9. Leeder JS, Kearns GL. The challenges of delivering pharmacogenomics into clinical pediatrics. *Pharmacogenomics J.* 2002;2(3):141–143.
10. Lacroix D, Sonnier M, Moncion A, et al. Expression of CYP3A in the human liver—evidence that the shift between CYP3A7 and CYP3A4 occurs immediately after birth. *Eur J Biochem.* 1997;247(2):625–634.
11. Stevens JC, Hines RN, Gu C, et al. Developmental expression of the major human hepatic CYP3A enzymes. *J Pharmacol Exp Ther.* 2003;307(2):573–582.
12. de Wildt SN, Kearns GL, Hop WC, et al. Pharmacokinetics and metabolism of intravenous midazolam in preterm infants. *Clin Pharmacol Ther.* 2001;70(6):525–531.
13. Kearns GL, Robinson PK, Wilson JT, et al. Cisapride disposition in neonates and infants: in vivo reflection of cytochrome P450 3A4 ontogeny. *Clin Pharmacol Ther.* 2003;74(4):312–325.
14. Lowry JA, Kearns GL, Abdel-Rahman SM, et al. Cisapride: a potential model substrate to assess cytochrome P4503A4 activity in vivo. *Clin Pharmacol Ther.* 2003;73(3):209–222.
15. de Wildt SN, Kearns GL, Hop WC, et al. Pharmacokinetics and metabolism of oral midazolam in preterm infants. *Br J Clin Pharmacol.* 2002;53(4):390–392.
16. de Wildt SN, de Hoog M, Vinks AA, et al. Population pharmacokinetics and metabolism of midazolam in pediatric intensive care patients. *Crit Care Med.* 2003;31(7):1952–1958.
17. Reed MD, Rodarte A, Blumer JL, et al. The single-dose pharmacokinetics of midazolam and its primary metabolite in pediatric patients after oral and intravenous administration. *J Clin Pharmacol.* 2001;41(12):1359–1369.
18. Bosso JA, Liu Q, Evans WE, Relling MV. CYP2D6, N-acetylation, and xanthine oxidase activity in cystic fibrosis. *Pharmacotherapy.* 1996;16(5):749–753.
19. Erenberg A, Leff RD, Haack DG, et al. Caffeine citrate for the treatment of apnea of prematurity: a double-blind, placebo-controlled study. *Pharmacotherapy.* 2000;20(6):644–652.
20. Evans WE, Relling MV, Petros WP, et al. Dextromethorphan and caffeine as probes for simultaneous determination of debrisoquin-oxidation and N-acetylation phenotypes in children. *Clin Pharmacol Ther.* 1989;45(5):568–573.
21. Kennedy MJ, Scripture CD, Kashuba AD, et al. Activities of cytochrome P450 1A2, *N*-acetyltransferase 2, xanthine oxidase, and cytochrome P450 2D6 are unaltered in children with cystic fibrosis. *Clin Pharmacol Ther.* 2004;75(3):163–171.

22. Lambert GH, Schoeller DA, Kotake AN, et al. The effect of age, gender, and sexual maturation on the caffeine breath test. *Dev Pharmacol Ther*. 1986;9(6):375–388.
23. Pariente-Khayat A, Pons G, Rey E, et al. Caffeine acetylator phenotyping during maturation in infants. *Pediatr Res*. 1991;29(5):492–495.
24. Tateishi T, Asoh M, Yamaguchi A, et al. Developmental changes in urinary elimination of theophylline and its metabolites in pediatric patients. *Pediatr Res*. 1999;45(1):66–70.
25. Nassif EG, Weinberger MM, Shannon D, et al. Theophylline disposition in infancy. *J Pediatr*. 1981;98(1):158–161.
26. Kraus DM, Hatzopoulos FK, Reitz SJ, Fischer JH. Pharmacokinetic evaluation of two theophylline dosing methods for infants. *Ther Drug Monit*. 1994;16(3):270–276.
27. Kraus DM, Fischer JH, Reitz SJ, et al. Alterations in theophylline metabolism during the first year of life. *Clin Pharmacol Ther*. 1993;54(4):351–359.
28. Dorne JL, Walton K, Renwick AG. Uncertainty factors for chemical risk assessment. Human variability in the pharmacokinetics of CYP1A2 probe substrates. *Food Chem Toxicol*. 2001;39 (7):681–696.
29. Blake MJ, Abdel-Rahman SM, Pearce RE, et al. Effect of diet on the development of drug metabolism by cytochrome P-450 enzymes in healthy infants. *Pediatr Res*. 2006;60 (6):717–723.
30. Dorne JL, Walton K, Slob W, Renwick AG. Human variability in polymorphic CYP2D6 metabolism: is the kinetic default uncertainty factor adequate? *Food Chem Toxicol*. 2002;40 (11):1633–1656.
31. Cascorbi I. Pharmacogenetics of cytochrome p4502D6: genetic background and clinical implication. *Eur J Clin Invest*. 2003;33 (Suppl 2):17–22.
32. Gaedigk A, Simon SD, Pearce RE, et al. The CYP2D6 activity score: translating genotype information into a qualitative measure of phenotype. *Clin Pharmacol Ther*. 2008;83(2):234–242.
33. Treluyer JM, Jacqz-Aigrain E, Alvarez F, Cresteil T, et al. Expression of CYP2D6 in developing human liver. *Eur J Biochem*. 1991;202(2):583–588.
34. Lee CR, Goldstein JA, Pieper JA. Cytochrome P450 2C9 polymorphisms: a comprehensive review of the in vitro and human data. *Pharmacogenetics*. 2002;12(3):251–263.
35. Koukouritaki SB, Manro JR, Marsh SA, et al. Developmental expression of human hepatic CYP2C9 and CYP2C19. *J Pharmacol Exp Ther*. 2004;308(3):965–974.
36. Suzuki Y, Mimaki T, Cox S, et al. Phenytoin age–dose–concentration relationship in children. *Ther Drug Monit*. 1994;16(2):145–150.
37. Kearns GL, Winter HS. Proton pump inhibitors in pediatrics: relevant pharmacokinetics and pharmacodynamics. *J Pediatr Gastroenterol Nutr*. 2003;37 (Suppl 1):S52–S59.
38. Kearns GL, Andersson T, James LP, et al. Omeprazole disposition in children following single-dose administration. *J Clin Pharmacol*. 2003;43(8):840–848.
39. Jung F, Richardson TH, Raucy JL, Johnson EF. Diazepam metabolism by cDNA-expressed human 2C P450s: identification of P4502C18 and P4502C19 as low K(M) diazepam *N*-demethylases. *Drug Metab Dispos*. 1997;25(2):133–139.
40. Klotz U. The role of pharmacogenetics in the metabolism of antiepileptic drugs: pharmacokinetic and therapeutic implications. *Clin Pharmacokinet*. 2007;46(4):271–279.
41. Brandolese R, Scordo MG, Spina E, et al. Severe phenytoin intoxication in a subject homozygous for CYP2C9*3. *Clin Pharmacol Ther*. 2001;70(4):391–394.
42. Jimenez-Lopez JM, Cederbaum AI. CYP2E1-dependent oxidative stress and toxicity: role in ethanol-induced liver injury. *Expert Opin Drug Metab Toxicol*. 2005;1(4):671–685.
43. Johnsrud EK, Koukouritaki SB, Divakaran K, et al. Human hepatic CYP2E1 expression during development. *J Pharmacol Exp Ther*. 2003;307(1):402–407.

44. Caro AA, Cederbaum AI. Oxidative stress, toxicology, and pharmacology of CYP2E1. *Annu Rev Pharmacol Toxicol.* 2004;44:27–42.

45. Hanioka N, Tanaka-Kagawa T, Miyata Y, et al. Functional characterization of three human cytochrome P450 2E1 variants with amino acid substitutions. *Xenobiotica.* 2003;33(6): 575–586.

46. Bhamre S, Bhagwat SV, Shankar SK, et al. Flavin-containing monooxygenase mediated metabolism of psychoactive drugs by human brain microsomes. *Brain Res.* 1995;672(1–2): 276–280.

47. Clement B, Lustig KL, Ziegler DM. Oxidation of desmethylpromethazine catalyzed by pig liver flavin-containing monooxygenase. Number and nature of metabolites. *Drug Metab Dispos.* 1993; 21(1):24–29.

48. Cashman JR, Celestial JR, Leach A, et al. Tertiary amines related to brompheniramine: preferred conformations for N-oxygenation by the hog liver flavin-containing monooxygenase. *Pharm Res.* 1993;10(8):1097–1105.

49. Krueger SK, Vandyke JE, Williams DE, Hines RN. The role of flavin-containing monooxygenase (FMO) in the metabolism of tamoxifen and other tertiary amines. *Drug Metab Rev.* 2006;38(1–2):139–147.

50. Koukouritaki SB, Simpson P, Yeung CK, et al. Human hepatic flavin-containing monooxygenases 1 (FMO1) and 3 (FMO3) developmental expression. *Pediatr Res.* 2002;51(2):236–243.

51. Burchell B, Coughtrie M, Jackson M, et al. Development of human liver UDP-glucuronosyltransferases. *Dev Pharmacol Ther.* 1989;13 (2–4):70–77.

52. de Wildt SN, Kearns GL, Leeder JS, et al. Glucuronidation in humans. Pharmacogenetic and developmental aspects. *Clin Pharmacokinet.* 1999;36(6):439–452.

53. Miller RP, Roberts RJ, Fischer LJ. Acetaminophen elimination kinetics in neonates, children, and adults. *Clin Pharmacol Ther.* 1976;19(3):284–294.

54. Court MH, Duan SX, von Moltke LL, et al. Interindividual variability in acetaminophen glucuronidation by human liver microsomes: identification of relevant acetaminophen UDP-glucuronosyltransferase isoforms. *J Pharmacol Exp Ther.* 2001;299(3):998–1006.

55. Sutherland JM. Fatal cardiovascular collapse of infants receiving large amounts of chloramphenicol. *AMA J Dis Child.* 1959;97(6):761–767.

56. McCarver DG, Hines RN. The ontogeny of human drug-metabolizing enzymes: phase II conjugation enzymes and regulatory mechanisms. *J Pharmacol Exp Ther.* 2002;300(2):361–366.

57. Pacifici GM, Säwe J, Kager L, Rane A. Morphine glucuronidation in human fetal and adult liver. *Eur J Clin Pharmacol.* 1982;22(6):553–558.

58. Choonara I, Lawrence A, Michalkiewicz A, et al. Morphine metabolism in neonates and infants. *Br J Clin Pharmacol.* 1992;34(5):434–437.

59. Choonara IA, McKay P, Hain R, Rane A. Morphine metabolism in children. *Br J Clin Pharmacol.* 1989;28(5):599–604.

60. Allegaert K, Vanhole C, Vermeersch S, et al. Both postnatal and postmenstrual age contribute to the interindividual variability in tramadol glucuronidation in neonates. *J. Early Hum Dev.* 2008; 84(5):325–330.

61. Richard K, Hume R, Kaptein E, et al. Sulfation of thyroid hormone and dopamine during human development: ontogeny of phenol sulfotransferases and arylsulfatase in liver, lung, and brain. *J Clin Endocrinol Metab.* 2001;86(6):2734–2742.

62. Butler MA, Lang NP, Young JF, et al. Determination of CYP1A2 and NAT2 phenotypes in human populations by analysis of caffeine urinary metabolites. *Pharmacogenetics.* 1992;2(3):116–127.

63. Pariente-Khayat A, Rey E, Gendrel D, et al. Isoniazid acetylation metabolic ratio during maturation in children. *Clin Pharmacol Ther.* 1997;62(4):377–383.

64. Gubbins PO, Gurley BJ, Williams DK, et al. Examining sex-related differences in enteric itraconazole metabolism in healthy adults using grapefruit juice. *Eur J Clin Pharmacol.* 2008;64 (3):293–301.

65. Abdel-Rahman SM, Johnson FK, Connor JD, et al. Developmental pharmacokinetics and pharmacodynamics of nizatidine. *J Pediatr Gastroenterol Nutr.* 2004;38(4):442–451.

66. Abdel-Rahman SM, Johnson FK, Manowitz N, et al. Single-dose pharmacokinetics of nizatidine (Axid) in children. *J Clin Pharmacol.* 2002;42(10):1089–1096.

67. Abdel-Rahman SM, Leeder JS, Wilson JT, et al. Concordance between tramadol and dextromethorphan parent/metabolite ratios: the influence of CYP2D6 and non-CYP2D6 pathways on biotransformation. *J Clin Pharmacol.* 2002;42(1):24–29.

68. Park J, Kim K, Park P, et al. Effect of age on the activity of cytochrome P4502D6 using dextromethorphan phenotyping in humans. *Clin Pharmacol Ther.* 2005;77(2):P24.

69. Desai H, Farrington E. Anticoagulation with warfarin in pediatrics. *Pediatr Nurs.* 2000;26(2): 199–203.

70. Aleynik MK, Leo MA, Aleynik SI, Lieber CS. Polyenylphosphatidylcholine opposes the increase of cytochrome P-4502E1 by ethanol and corrects its iron-induced decrease. *Alcohol Clin Exp Res.* 1999;23(1):96–100.

71. Sawe J, Pacifici GM, Kager L, et al. Glucuronidation of morphine in human liver and interaction with oxazepam. *Acta Anaesthesiol Scand Suppl.* 1982;74:47–51.

72. Nahata MC, Powell DA. Bioavailability and clearance of chloramphenicol after intravenous chloramphenicol succinate. *Clin Pharmacol Ther.* 1981;30(3):368–372.

CHAPTER 22

Applications of Pharmacogenomics to Study Design in Pediatrics

NADINE COHEN, PhD and SETH NESS, MD, PhD

Johnson & Johnson Pharmaceutical Research and Development, Raritan, New Jersey 08869

INTRODUCTION

Pharmacogenomics is the study of the genetic basis of variability among individuals in response to drugs. It is a very active area of research, with the pharmaceutical industry gaining experience applying it and integrating it into the drug development process. It is being used as an innovative tool to develop safer and more efficacious drugs and for therapeutic differentiation. Personalized medicine based on the genetic make-up of patients has the potential to improve the therapeutic response, but also to lead a clinically important reduction in adverse drug reactions for the pediatric population.[1–3]

The goals of pharmacogenomics research are to (1) improve identification of patients who will benefit from genetically defined therapy and avoid futile therapeutic attempts, (2) improve proof of principle for efficacy trials and salvage drugs, (3) identify optimal dosing, (4) improve drug safety and understand adverse events in development and postapproval, and (5) improve discovery of drugs targeted to human disease (Figure 22.1). Pharmacogenomics is the investigation of the relationship between genetic variability in genes encoding proteins involved in drug absorption/distribution/metabolism/elimination (so-called ADME including drug metabolizing enzymes and transporters), in drug mode of action (i.e., intended targets), in off-target mechanisms (i.e., unintended targets) or in disease susceptibility/etiology, and in variability in clinical parameters relevant to pharmacokinetics, efficacy, or safety, following drug exposure.[1–4]

Currently, a number of pharmaceutical and biotech companies are working toward the goal of producing medicines that will be correctly administered to the patients expected to respond to treatment. The most classic example of personalized medicine is the development of Herceptin (trastuzumab). Herceptin is a targeted therapy that inhibits the proliferation of human tumor cells that overexpress HER2. Labeling of Herceptin requires that patients be screened for HER2 overexpression prior to treatment with the drug.[5] This example has encouraged other investigators and the pharmaceutical industry to identify similar

Pediatric Drug Development: Concepts and Applications
Edited by Andrew E. Mulberg, Steven A. Silber, and John N. van den Anker

- Drive Innovation and Pipeline Growth
- Provide mechanism for knowledge transfer to Discovery
- Enrich efficacy trials/reduce costs and timelines
- Improve drug safety, and understand adverse events
- Potentially salvage compounds by improving risk/beneafits
- Find optimal dose
- Improve identifications of patients who will benefit from stratified (genetically-defined) therapy, and avoid futile therapeutic attempts
- Therapeutic differentiation

FIGURE 22.1 Pharmacogenomics opportunities in the pharmaceutical life cycle.

opportunities. Recent discoveries have unveiled the roles of variants that are pertinent for other drugs demonstrating variable efficacy or adverse effect profiles and are being used to either identify patients who will respond to treatment (such as Gleevec and Tarceva)[6,7] or to identify patients who will develop severe toxicities (6-mercaptopurin, irinotecan, and Strattera).[8–10] The label of these drugs, in contrast to Herceptin, either includes pharmacogenomics information or a recommendation about pharmacogenomics testing, but no mandate (Table 22.1). Warfarin is a widely used anticoagulant where incorrect dosage carries a high risk of either severe bleeding or failure to prevent thromboembolism. Recently, Warfarin's drug label has also been modified to point out that "genetic variations" in the CYP2C9 and VKORC1 enzymes are one of the factors that "may" influence the response of the patient to Warfarin.[11] These approved labeling changes are further steps showing the commitment of the U.S. Food and Drug Administration (FDA) to personalized medicine, but as of yet, no specific guidelines on pharmacogenomics in pediatric studies have been released.

Two terms are used to describe this type of research: pharmacogenomics and pharmacogenetics. Several definitions have been proposed for each of them.[1,12] In this chapter we have

TABLE 22.1 Examples of "Personalized Medicine:" Information, Recommendation, or Required Test in Drug Label

Drug	Indication	Gene and Label	Category
6-MP	ALL	*TPMT*—Recommendation	Toxicity
Irinotecan	Colorectal cancer	*UGT1A1*—Recommendation	Toxicity
Warfarin	Anticoagulant	*CYP2C9* and *VKORC1*—Recommendation	Toxicity
Strattera	ADHD	*CYP2D6*—Information	Toxicity
Erbitux	Colorectal cancer	*EGFR*—Required	Efficacy
Gleevec	Gastrointestinal stromal tumors	*c-kit*—Information	Efficacy
Tarceva	NSCLC	*EGFR*—Information	Efficacy
Herceptin	Metastatic breast cancer	Overexpression of HER2 receptor—Required	Efficacy/salvage

decided to use only the term of pharmacogenomics and to focus on the most mature DNA technologies, while acknowledging that great progress is also being made in other complementary approaches such as RNA expression, proteomics, and metabolomics technologies.

Pediatric pharmacogenomics interventions have the potential to improve the quality of healthcare for children.[13] Some diseases that affect children such as acute lymphocytic leukemia (ALL) are rarely seen in adults, while several complex disorders that manifest early in childhood, such as asthma, autism, attention deficit hyperactivity disorder (ADHD), epilepsy, and juvenile rheumatoid arthritis, continue into adulthood and thus require chronic treatment. The focus of pharmacogenomics applications in the pharmaceutical industry is generally on common disorders, resulting from mutations on several genes in combination with environmental factors, such as schizophrenia, diabetes, obesity, cardiovascular disorders, and various types of cancers. Little research is being done on many of the rare and monogenic disorders resulting from single genetic lesion, such as Huntington disease or cystic fibrosis, even though they are quite serious diseases and represent, in our view, good models for the application of pharmacogenomic therapeutic intervention.

Presently, pediatricians use a combination of knowledge of clinical pharmacology, developmental processes, and trial and error to best treat children and to identify the appropriate dose. The great variability in drug response observed in children creates huge opportunities to improve drug dosing and to identify genetic markers of efficacy or toxicity in those patients. Applications of pharmacogenomics to the treatment of pediatric disorders need to consider the added challenge of age-specific disease processes and age-related variability in drug response.[14–16] The molecules involved in disease states and in determining gene expression at various ontological periods are key targets for pharmacotherapeutic intervention and for the design of pharmacogenomics studies in children. Thus, pharmacogenomics studies for a pediatric disorder should consider the characterization of the variability of genes involved in a pathway throughout growth and development.

The present chapter describes the strategies and some of the challenges and opportunities in applying pharmacogenomics in children. It also provides a few selective examples of pharmacogenomics applications in several common disorders in pediatric patients and gives a brief overview of the rationale for pharmacogenomics studies in rare monogenic disorders affecting children.

STRATEGIES FOR GENETIC ASSOCIATION STUDIES OF DISEASE AND DRUG RESPONSE

Pharmacogenomics seeks to examine the genetic basis for individual variation in response to a given therapeutic.[17] All genes harbor variants termed single nucleotide polymorphisms (SNPs); however, identification of those that are most relevant with respect to their influence on disease susceptibility or drug response traits remains a challenge. While our knowledge of the human genome and how it varies among people has increased dramatically over the past decade,[18,19] there is still relatively little known about how these genetic differences impact individual response to drugs.[20] Appropriately designed and analyzed pharmacogenomics association studies have the potential to improve the discovery, development, and ultimately the prescription and usage of drugs by identifying the most meaningful DNA sequence variation and hence identifying subgroups of patients who are more likely to respond to a drug or to develop unacceptable toxicity. In 2004, initial guidelines for the submission of genomic data were published by the FDA in order to encourage pharmacogenomics research during clinical drug development.[21] In 2006, a draft consensus guideline regarding terminology in pharmacogenomics (E15), agreed upon by the regulatory authorities of the European Union, Japan, and the United States, was published by the ICH.[12] Genetic association studies measure the correlation between variations in human genes ("genetic polymorphisms") and phenotypes of interest. Traditionally, linkage and association studies have led to the identification of the genetic lesions that underlie monogenic disorders. The sequencing of the human genome and the completion of the International Hap Map project have enabled the generation of new large scale and powerful genomics tools, such as the 500k or 1 million DNA single nucleotide polymorphism (SNP) chips from Affymetrix[22] or Illumina.[23] These tools are being used routinely to implement genetic association analyses of drug response and disease predisposition. This progress has led to several major publications this year on genome-wide screen association studies in several common disorders affecting children and/or adults.

The choice of pharmacogenomics study design and genotyping strategy depend on many factors. Some are practical, such as economical cost and DNA sample availability. Others are related to the current state of knowledge and the overall purpose of the study, for example, whether the study is designed to test a specific hypothesis (e.g., a drug mechanism of action) or rather to generate a new hypothesis (e.g., to identify novel drug targets for a disease of interest).

Study Designs

Candidate Polymorphism Candidate polymorphism studies involve the genotyping of one or a few polymorphisms in a limited number of genes. This approach may apply in cases where there is a strong hypothesis for the role of the polymorphism in the cause of the disease, or strong evidence for a functional genetic variant that is relevant to the disease or the phenotype of interest. In these cases, the hypothesis that such variant is causal to the disease phenotype can be tested directly by comparing the frequency of the genetic variant in a group of patients and controls. Candidate polymorphism studies in pharmacogenomics are being performed to study the effect of functional alleles in genes affecting the drug pharmacokinetics, in particular, drug metabolizing enzymes and transporters (ADME), but also in genes affecting drug pharmacodynamics (targets and disease genes).[1–4,24]

Candidate Gene The candidate gene approach has been widely used to study the genetic basis of pharmacogenomic traits and is being used if there is sufficient scientific knowledge to formulate testable hypotheses. A candidate gene study may involve studying one to thousands of genes in a single experiment. Small numbers of SNPs (5–50) selected within coding sequence, untranslated sequence, and flanking regions, and possibly splice or regulatory regions, are generally genotyped within each gene. A gene may be either a positional candidate that results from a prior linkage study or a functional candidate identified based on its known or presumed role in the phenotype of interest.

Candidate gene approaches represent the bulk of the genetic association studies published to date and have led to the identification of a number of clinically important genetic variants. However, many of the identified associations have not been replicated, most probably because initial studies were not properly designed, and the number of samples might have been insufficient. The discovery of genetic variants in *VKORC1* and *GGCX* (together with the drug metabolism enzyme CYP2C9) is a finding of significant clinical relevance explaining one-third of the variance in dosing of Warfarin.[25,26]

Genome-Wide Genome-wide association studies aim to identify genetic associations between a disease or drug response phenotype and genotypes by testing 100,000–1,000,000 well-selected polymorphisms in a single experiment. An advantage of genome-wide over candidate-gene studies is that they do not rely on a priori hypotheses for the role of a specific gene or biological function. This raises the possibility of identifying novel disease genes, thus improving knowledge of the underlying causative mechanisms. However, as a large number of SNPs covering the genome are being genotyped in a genome-wide association study, this requires a large number of statistical tests that greatly raises the threshold for statistical significance. Consequently, interpretable results can only be achieved with large sample sizes, unless the impact of the associated variant on the phenotype is very large.

Recently, several genome-wide association studies have been published,[27–36] resulting from collaborative efforts aimed at identifying genetic associations for major diseases in thousands of cases and controls. To date, however, no genome-wide scan of drug response has been published. This might result from the relative difficulty of evaluating drug response retrospectively and collecting sufficiently large patient populations with homogeneous drug response phenotypes. The genome-wide association studies published were found to be highly robust in identifying variants that associate with and predispose to complex disease, such as age-related macular degeneration, type 2 diabetes, and coronary artery disease, disorders that predominantly affect adults. More recent studies have identified significant association of novel genes predisposing to type 1 diabetes and autism, and replicated associations to irritable bowel disease (IBD) and obesity genes in children. These findings stirred new hope for the mapping of genes that regulate drug response related to pediatric conditions. Furthermore, these studies indicate that modern high-throughput SNP genotyping technologies, when applied to large and comprehensively phenotyped patient cohorts might capture the most clinically relevant disease-modifying and drug response genes.

Fine Mapping Once a chromosomal region containing a signal of positive association with a disease trait or a genetic linkage peak is identified, a fine mapping study can be carried out to obtain an in-depth assessment of this region. Fine-mapping studies generally involve the genotyping of several hundred polymorphisms over large genomic regions (up to 10 Mb in length) that may span several genes. In-depth characterization of all genetic variants present in the chromosome region can be obtained by resequencing.

Resequencing Resequencing of genomic regions, and ultimately of entire genomes, might become in the near term the strategy of choice for comparing patient groups given the fact that the cost of DNA sequencing is decreasing rapidly. This approach will allow a comprehensive evaluation of the contribution of both rare and common genetic variants, and will ultimately be preferable. Current efforts are seeking to ultimately reduce the cost of sequencing an entire human genome to less than US $1000.

CHALLENGES OF PHARMACOGENOMICS APPLICATIONS IN CHILDREN

Pharmacogenomics is one of the newest disciplines in medicine to be applied in the context of drug development. The pharmaceutical industry, having gained experience in this field and also having a greater appreciation of the hurdles in applying it, is starting to set more realistic expectations for the implications of this research. However, there are yet several challenges to meet regarding the appropriate implementation of this technology to deliver greatest value for the pharmaceutical industry overall and, more importantly, to benefit the patients. These challenges have been extensively discussed in several reviews.[37,38] They include strategic, commercial, and regulatory issues, the complexity of translating pharmacogenomics findings into clinical practice, the lack of education of heathcare professionals, and the difficulty in interpreting the pharmacogenomics information for a drug label. These challenges will hopefully gradually be addressed as more successes are described. The pharmacogenomics science itself is also extremely complex and this is by far the most critical challenge. Pharmacogenomics studies require a large number of samples to be available for identifying and replicating valid genetic associations. Large sample collections are not easy to obtain, especially in pediatric clinical trials. It is thus important to routinely collect DNA samples with an appropriate informed consent form in all pediatric clinical trials to be able to test or generate hypotheses. Such studies can be done prospectively or retrospectively. In order to increase the chances of success, it is our recommendation to apply pharmacogenomics in clinical trials as a balancing act, including routinely both experiments where hypotheses are being tested, and exploratory analyses to generate hypotheses (Figure 22.2). The former might include the analysis of valid known biomarkers and generate submissible data as part of a New Drug Application (NDA); the latter might generate data that could be part of a voluntary genomic data submission per the recent guideline developed by the FDA.[21] This guideline provides definitions for the so-called valid known and probable valid biomarkers. It also provides guidance on when and under which format to submit pharmacogenomics data. This guideline is part of a series of guidelines being currently developed by the FDA and also by other regulatory agencies such as the European Medicines Evaluation Agency (EMEA) or the Japanese agency, MHLW, and are aimed at encouraging the pharmaceutical industry to integrate pharmacogenomics within their drug development process. A table of valid genomic biomarkers has been created by the FDA, providing a reference for genomic biomarkers in labels of FDA approved drug products.[39] So far, there are only few examples of drugs used in the treatment of pediatric disorders (Strattera, 6-mercaptopurine), where pharmacogenomics data are part of the drug label. Overall, in spite of recent changes in the labels of a few drugs, there is still a lack of dosing instructions associated with label information. Many questions about the standards

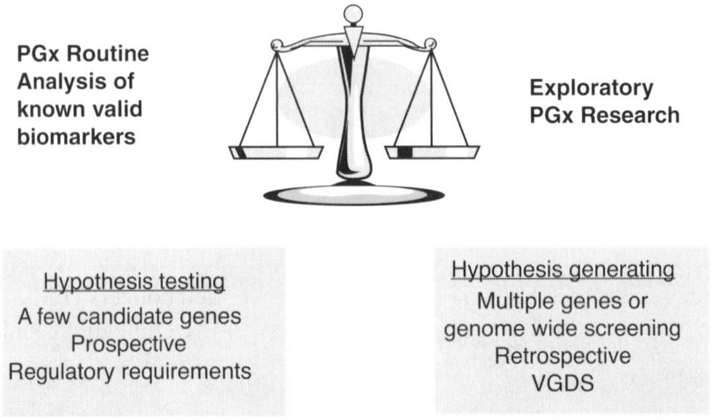

FIGURE 22.2 Pharmacogenomics in clinical trials is a balancing act.

required to validate genotype–phenotype associations to determine optimal dosing in clinical practice or for regulatory policies regarding label updates are undergoing intense debate among the key stakeholders.

Pharmacogenomics research also raises several ethical issues that have been discussed extensively in several reviews and reports.[40,41] Obtaining, storing, and analyzing DNA samples have always been perceived to be more problematic than other types of samples and do receive particular attention by ethic committees and regulatory agencies. Privacy and confidentiality issues remain as a challenge to gaining trust of all consumers. There currently are expectations for a higher level of stringency for maintaining confidentiality and preventing unintended access to or release of genomic samples and data. The pharmaceutical industry in general has developed stringent and appropriate procedures to routinely collect samples for pharmacogenomic applications in clinical trials.[42,43] Another topic of debate within the pharmacogenomics community has been whether and how individual genetic research results should be returned to study participants. A paper from the industry Pharmacogenomics Working Group (http://www.I.pwg.org/cms/) has summarized the key points to be considered when making those decisions.[44] When pharmacogenomics research involves children, additional issues must also be considered. For example, consent to pharmacogenomics research and children's assent can become more complicated due to the complexity of information that needs to be transmitted to the research participants and family. The questions of how and to whom (child, legal representative, doctor, someone else) the research results should be returned are even more complicated, because of the potential impact the information might have on a child's clinical care, among several issues.

PHARMACOGENOMICS APPLICATIONS AND OPPORTUNITIES IN CHILDREN

We provide here a brief overview of conclusive pharmacogenomics discoveries made in pediatric disorders such as attention deficit hyperactivity disorder (ADHD), diabetes, obesity, asthma, and cancer, and some considerations on the study of rare genetic disorders that often affect children.

Attention Deficit Hyperactivity Disorder

ADHD is a chronic disease affecting about 5% of the world's population. It typically presents during childhood and is characterized by a persistent pattern of inattention and/or hyperactivity, as well as impulsivity and distractibility.

Treatment generally consists of stimulant therapy combined with behavioral modifications and/or counseling. Recently, atomoxetine, a nonstimulant therapy whose mechanism of action is thought to involve selective inhibition of the presynaptic norepinephrine transporter, has been approved. Atomoxetine is significantly metabolized by CYP2D6, a liver enzyme that plays a role in the breakdown of many drugs. CYP2D6 has been known to have many genetic variants that affect its activity and these variants are resposnsible for individuals being poor metabolizers (PMs) or extensive metabolizers (EMs).[45]

In the case of atomoxetine, this distinction was found to be clinically meaningful.[46] The following adverse reactions occurred in at least 2% of PM patients and were either twice as frequent or statistically significantly more frequent in PM patients compared with EM patients: decreased appetite (23% of PMs, 16% of EMs); insomnia (13% of PMs, 7% of

EMs); sedation (4% of PMs, 2% of EMs); depression (6% of PMs, 2% of EMs); tremor (4% of PMs, 1% of EMs); early morning awakening (3% of PMs, 1% of EMs); pruritus (2% of PMs, 1% of EMs); and mydriasis (2% of PMs, 1% of EMs). Thus, the FDA required the following text in the prescribing information: poor metabolizers (PMs) of CYP2D6 have a tenfold higher AUC and a five-fold higher peak concentration to a given dose of atomoxetine compared with extensive metabolizers (EMs).[9] Approximately 7% of Caucasian populations are PMs. Laboratory tests are available to identify CYP2D6 PMs. The blood levels in PMs are similar to those attained by taking strong inhibitors of CYP2D6. The higher blood levels in PMs lead to a higher rate of some adverse effects of atomoxetine.

Although genotyping of CYP2D6 is not mandatory, atomoxetine is one of the first examples of pharmacogenetics being mentioned in the prescribing information for a drug.

Asthma

Asthma is a chronic disease of the respiratory system affecting as many as 1 in 4 urban children.[47,48] Treatment can combine acute therapy, most commonly with beta-agonist bronchodilators during an attack and chronic therapy, commonly with corticosteroids.

The study of disease genetics has elucidated many of the genes and pathways influencing the disease.[49–51] Genetic variations responsible for differences in efficacy and side effects of drugs are just beginning to be found.[52,53]

Beta-2 agonists act as bronchodilators through their action on the beta-2 adrenergic receptor and are considered the first line treatment for immediate relief of acute symptoms. There can be significant variability of response to beta agonist therapy up to three-quarters of which is genetically based.[54] Several point mutations in the receptor have been found to affect response to beta agonists.[55] One group found bronchodilator desensitization was much higher with homozygosity for glycine at the 16th amino acid than with homozygosity for arginine as measured by maximal forced expiratory volume in 1-s (FEV1) response.[56]

Another group reported that asthmatic patients who were homozygous for Arg-16 had a more than fivefold greater bronchodilator response to albuterol than homozygous Gly-16 individuals.[57] However, replication of these findings has been inconsistent.[58–60]

Other mutations and polymorphisms have been implicated in responsiveness and although further work and replication are needed, genotyping may be useful in tailoring treatment, especially in children resistant to therapy.

Diabetes

Diabetes affects more than 18 million people in the United States, with approximately 90% of those affected having type 2 diabetes (T2D). Type 1 diabetes (T1D) risk has been shown to be strongly influenced by multiple genetic loci and environmental factors. Several loci have already been identified as being associated with a significant proportion of the familial clustering of T1D: major histocompatibility complex loci on 6p (MHC),[61–63] the insulin locus (INS),[64–66] the protein tyrosine phosphatase-22 (*PTPN22*) gene on 1p13,[67,68] and the gene encoding the cytotoxic T-lymphocyte-associated protein 4 (CTLA4) on 2q31,[69–71] *KIAA0350*, which is a new locus of unknown function.[72–74]

Type 2 diabetes (T2D) is a late onset disease (>40 years). Due to an ageing population and also to increasing obesity, its prevalence is increasing. Furthermore and most concerning, an increasing number of children and young adults are now developing T2D.[75] The chronic complications of diabetes include development of cardiovascular and microvascular disease. Maturity onset diabetes of the young (MODY) includes any of several rare hereditary forms of diabetes mellitus caused by autosomal dominant defects of insulin secretion and beta cell dysfunction. Generally, MODY presents as a mild version of T1D, with partial insulin production and normal insulin sensitivity. It is typically diagnosed in patients under 25 years old.[76,77] Because each of the types of MODY may respond differently to specific therapies, it is likely that tailored drug therapy will be possible based on pharmacogenetics testing. MODY due to glucokinase mutations has only mild hyperglycemia and does not require treatment; however, it is often misdiagnosed as T1D or T2D and inappropriately treated.[78] Ineffective treatment and exposure to the risks of therapy could be avoided with genetic diagnosis. On the other hand, the greater hyperglycemia in MODY due to mutations in hepatic nuclear factor 1 alpha is very responsive to sulfonylureas.[79–82] Appropriate testing would allow these patients to avoid insulin therapy.

Obesity

Obesity is a complex disorder that involves interactions between environmental and genetic factors. It is a global health problem for the adult, but childhood obesity rates are soaring as well, and it is an important risk factor for T2D, cardiovascular disease, and overall mortality.[83,84]

Several drugs have been developed to facilitate weight reduction in obese individuals. A significant difference in response to the centrally acting noradrenaline and serotonine reuptake inhibitor, sibutramine, in programs aimed to reduce weight, was attributed to the C825T polymorphism in the guanine nucleotide-binding protein beta-3 (*GNB3*) gene.[85] Thus, genotyping of the *GNB3* C825T polymorphism may help identify obese individuals who may benefit from sibutramine therapy. There are many efforts currently devoted to studying obesity and potential therapeutic interventions. Those studies will hopefully deliver new targets and pharmacogenomic approaches to this huge medical problem in the short to medium term. Of particular interest is the search of genetic factors that might help to predict which obese patients are at risk to develop T2D or cardiovascular diseases.

Childhood Leukemia

Acute lymphoblastic leukemia (ALL) is the most common pediatric cancer. Great progress has been made, with event-free survival rate currently of almost 80% for most patients.[86,87] Nevertheless, many challenges still exist and some children experience disease relapse or drug toxicity. There is a need for more efficacious and safer treatment options that would enable one to predict more accurately which patients would require less toxic therapies and which patients would benefit from therapy with dose intensification. Increasingly, patients who do not respond to standard therapy are being treated with novel treatment options.[13]

The study of genetic polymorphisms in various metabolic pathways has led to the identification of predisposing factors to the development of ALL or to risk for relapse or adverse effects from standard treatment options. Among the cytochrome P450 enzymes that typically mediate phase I drug metabolism, several studies were published reporting an association between CYP1A1 and 2A variants with worse therapeutic outcome in some,[88,89] but not other studies.[90]

Among the phase II drug metabolism enzymes, studies have focused on the family of glutathione-*S*-transferase (GST) enzymes, particularly GSTM1, GSTT1, and GSTP1.

Polymorphisms in those enzymes have been associated with the risk of de novo cancer and therapy-related cancers following chemotherapy and with the efficacy and toxicity of cancer chemotherapy.[91–93] In childhood ALL, reported studies have modest sample sizes and thus limited statistical power. Results are contradictory and would require prospective genotyping and appropriately designed studies. Initial studies did not show any correlation between GST gene deletions and risk of relapse in childhood ALL.[94,95] A subsequent study reported on the correlation of the GSTT1-null genotype with improved response to prednisone in childhood ALL. GSTT1 or GSTM1 deletion polymorphisms were protective against relapse.[96]

Thiopurine methyltransferase (TPMT) catalyzes the S-methylation of the 6-mercaptopurine (6MP), which is a key medication used for the treatment of ALL, into the inactive metabolite *S*-methyl-thioinosine 5′-monophosphate. Genetic variations in the TPMT gene have been investigated in numerous studies. It was found that patients with TPMT deficiency are at risk for severe hematological toxicities when treated with conventional doses of 6-MP because these polymorphisms lead to a decrease in the rate of 6-MP metabolism.[97,98] The molecular basis for polymorphic TPMT activity has been well defined. TPMT*2, TPMT*3A, and TPMT*3C account for nearly 95% of TPMT deficiency.[99] Each of these alleles encodes TPMT proteins that undergo rapid degradation, leading to enzyme deficiency. The type and frequencies of TPMT alleles have also been reported to vary among various ethnic groups.[100–104] Seventy-one percent of patients with bone marrow intolerance to 6-MP were found to be TPMT heterozygotes or homozygous deficient, and to be most likely hospitalized.[13] The TPMT heterozygotes required a mean dose reduction of 35%, whereas the TPMT-deficient patients required a mean dose reduction of 90%. Appropriate 6-MP dose reductions for TPMT-deficient patients have allowed for similar toxicity and survival outcomes as patients with normal TPMT levels.[105–107] Genotyping methods have been established for the molecular diagnosis of TPMT and can now assist with determining a safe starting dose for 6-MP therapy in children with ALL.[108] Recently, the label of the immunosuppressant 6-MP has been updated, recommending TPMT genotyping or phenotyping in order to identify patients who are homozygous deficient or have low or intermediate TPMT activity.[10] So far the uptake of TPMT testing to guide dosing has been minimal, possibly because the physicians were used to dosing patients empirically and monitoring their absolute neutrophile count. It is very possible that uptake in clinical practice will be stimulated if the regulatory agencies would "require" the testing prior to prescribing the drug instead of "recommending" it.

Pharmacogenomics and Rare Genetic Disorders

The progress in the human genome project has also led to major scientific breakthrough in the field of rare monogenic disorders that mostly affect children. The rare monogenic diseases provide the most clearly evident examples of pathology resulting from a single genetic lesion. These are quite serious diseases, often leading to death in childhood or early adulthood, and still not fully understood. There are at present an estimated 6000 characterized monogenic diseases. Those disorders are observed with higher frequency in founder populations, such as the Amish community,[109] or ethnic populations with increased frequency of consanguineous families, such as the Arab-Israeli population.[110] The genes for an increasing number of rare genetic disorders are being identified faster and faster, provided these diseases have been clinically defined and samples from affected children are available (Online Mendelian Inheritance Catalog).[111]

Most genomics research is currently being done in complex multifactorial disorders and we are in fact almost overwhelmed by the massive amount of genomics data. With advances in genomics, we have also begun to recognize an astonishing complexity not only in how a gene may interact with others within its own genome, but also how these genes interact in complex ways with other organisms and the environment. While drug development and research for rare disorders is generally less appealing to the large pharmaceutical companies, it does make sense to do more research in these disorders as they are simpler to study, might provide some clues for our understanding of genomics diseases, and might open the way to new drug development. By studying the rare monogenic disorder familial hypercholesterolemia, for example, researchers gained a fundamental understanding of cholesterol metabolism and its relation to heart disease. This research of a rare, monogenic disease also led directly to the development of the statin drugs, the most widely prescribed class of medications in the United States. Furthermore, the rare monogenic diseases might provide an easier model for the application of pharmacogenomic therapeutic approaches since the cause-and-effect relationship between genetic lesion and pathology may be more clearly delineated. In light of the seriousness of so many rare monogenic diseases, such patients are also in great need of therapeutic advances. Remarkable efforts are currently ongoing in several academic research facilities to elucidate the genetic basis of rare genetic disorders, and to translate the findings into clinical practice using prenatal diagnosis, genetic testing at birth, and/or implementing prevention therapies, such as the clinic for special children (www.clinicforspecialchildren.org).

THE WAY FORWARD

During the past decade, technologies to identify genetic variations have evolved from labor-intensive, time-consuming chores to being highly automated, efficient, and less expensive, thus enabling high-throughput genome-wide screen associations and large-scale candidate gene studies. Pediatric pharmacogenetics studies have the potential to improve the quality of medical care for children. The pediatric population presents a unique pharmacogenetic challenge as children not only demonstrate the same interindividual genetic variability seen in adults, but also present further differences arising from the various stages of development. It is expected that within the next few years, we will see an increase of investment in pharmacogenomics research in pediatric disorders by the pharmaceutical industry and academia coupled with more evidence of clinical relevance for pharmacogenomics data. This will lead to more examples of pediatric drugs with genetic information in their prescribing drug labels. A noteworthy effort is the creation at the Children's Hospital of Philadelphia (CHOP) of a biorepository and high-throughput pediatric genomics center aiming at genotyping over 100,000 children in 3–4 years. The facility is coupled to electronic medical records within the healthcare network at CHOP for those patients who volunteer to participate. The diseases that are being examined include some of the most common complex pediatric disorders, such as asthma, obesity, T1D, ADHD, autism, atopic dermatitis, and neuroblastoma. Such efforts combined with those of the pharmaceutical industry and biotech companies are likely to bear fruit, given the recent advances in the technology platforms, both in the area of disease genetics and pharmacogenomics, and will deliver a new generation of drugs and diagnostics and lead to a major paradigm shift from conventional medicine to efficient predictive medicine. Pharmacogenomics research has already greatly enhanced the safety of treating children diagnosed with ALL, by identifying TPMT-deficient individuals using genetic testing. Additional benefits of pharmacogenetics research in pediatric populations is expected in drug treatments for other common diseases in children such as asthma, autism, and ADHD, and will allow pediatricians to gain confidence regarding which patients are best suited for a particular therapeutic agent and which patients may be at risk for serious potentially life-threatening complications from standard treatment regimens. Study of rare genetic disorders might provide appropriate and additional models to identify meaningful pharmacogenomics findings and to translate genetic findings into the clinic.

REFERENCES

1. Weinshilboum RM, Wang L. Pharmacogenetics and pharmacogenomics: development, science and translation. *Annu Rev*. 2006;7; 223–245.
2. Roses AD. Pharmacogenetics' place in modern medical science and practice. *Life Sci*. 2002;70 (13):1471–1480.
3. Roses AD. Pharmacogenetics and the practice of medicine. *Nature*. 2000;405(6788):857–865.
4. Tomalik-Scarte D, Lazar A, Fuhret AL. The clinical role of genetic polymorphisms in drug-metabolizing enzymes. *Pharmacogenomics J*. 2008;8:4–15.
5. Herceptin Product Label. Available at http://www.fda.gov/cder/biologics/products/trasgen092598.htm. (Accessed July 25, 2007.)
6. Gleevec Revised Product Label. Available at http://www.fda.gov/CDER/drug/infopage/gleevec/default.htm. (Accessed October 30, 2006.)
7. Tarceva Product Label. Available at http://www.fda.gov/cder/foi/label/2004/021743lbl.pdf. (Accessed July 25, 2007.)
8. Camptosar Product Label. Available at http://www.fda.gov/cder/foi/label/2006/020571s030lbl.pdf. (Accessed July 25, 2007.)
9. Strattera Product Label. Available at http://www.fda.gov/cder/foi/label/2007/021411s004s012s013s015s021lbl.pdf. (Accessed July 25, 2007.)
10. Purinethol Product Label. Available at http://www.fda.gov/cder/foi/label/2004/09053s024lbl.pdf. (Accessed July 25, 2007.)
11. Coumadin Product Label. Available at http://www.fda.gov/cder/foi/label/2006/009218s102lbl.pdf. (Accessed July 25, 2007.)
12. ICH Terminology in Pharmacogenomics (E15). Available at http://www.fda.gov/cder/guidance/7619dft.pd.
13. Husain A, Loehle JA, Hein DW. Clinical pharmacogenetics in pediatric patients. *Pharmacogenomics*. 2007;8(10):1403–1410.
14. Leeder JS. Developmental and pediatric pharmacogenomics. *Pharmacogenomics*. 2003;4: 331–341.
15. Hines RN, McCarver DG. The ontogeny of human drug-metabolizing enzymes: phase I oxidative enzymes. *J Pharmacol Exp Ther*. 2002;300:355–360.
16. McCarver DG, Hines RN. The ontogeny of human drug-metabolizing enzymes: phase II conjugation enzymes and regulatory mechanisms. *J Pharmacol Exp Ther*. 2002;300(2):361–366.
17. Goldstein DB, Tate SK, Sisodiya SM. Pharmacogenetics goes genomic. *Nat Rev Genet*. 2003;4 (12):937–947.
18. Venter JC, Adams MD, Myers EW, et al. The sequence of the human genome. *Science*. 2001;291(5507):1304–1351.
19. Lander ES, Linton LM, Birren B, et al. Initial sequencing and analysis of the human genome. *Nature*. 2001;409(6822):860–921.

20. A haplotype map of the human genome. *Nature*. 2005;437 (7063): 1299–1320.

21. US FDA. Guidance for Industry: Pharmacogenomic Data Submissions. Available at http://www.fda.gov/cder/guidance/6400fnl.pdf. (Accessed July 25, 2007.)

22. Ragoussis J, Elvidge G. Affymetrix GeneChip system: moving from research to the clinic. *Expert Rev Mol Diagn*. 2006;6(2):145–152.

23. Steemers FJ, Gunderson KL. Whole genome genotyping technologies on the BeadArray platform. *Biotechnol J*. 2007;2(1):41–49.

24. Evans WE, Relling MV. Pharmacogenomics: translating functional genomics into rational therapies. *Science*. 1999;206:487–491.

25. Wadelius M, Chen LY, Downes K, et al. Common VKORC1 and GGCX polymorphisms associated with Warfarin dose. *Pharmacogenomics J*. 2005;5(4):262–270.

26. Rieder MJ, Reiner AP, Gage BF, et al. Effect of VKORC1 haplotypes on transcriptional regulation and Warfarin dose. *N Engl J Med*. 2005;352(22):2285–2293.

27. Sladek R, Rocheleau G, Rung J, et al. A genome-wide association study identifies novel risk loci for type 2 diabetes. *Nature*. 2007;445(7130):881–885.

28. Edwards AO, Ritter R 3rd, Abel KJ, Manning A, Panhuysen C, Farrer LA. Complement factor H polymorphism and age-related macular degeneration. *Science*. 2005;308 (5720):421–424.

29. Kiessling A, Ehrhart-Bornstein M. Transcription factor 7-like 2 (TCFL2)—a novel factor involved in pathogenesis of type 2 diabetes. Comment on: Grant et al., *Nat Genet*. 2006, Published online 15 January 2006. *Horm Metab Res*. 2006;38(2):137–138.

30. Herbert A, Gerry NP, McQueen MB, et al. A common genetic variant is associated with adult and childhood obesity. *Science*. 2006;312(5771):279–283.

31. Amundadottir LT, Sulem P, Gudmundsson J, et al. A common variant associated with prostate cancer in European and African populations. *Nat Genet*. 2006;38(6):652–658.

32. Gudmundsson J, Sulem P, Manolescu A, et al. Genome-wide association study identifies a second prostate cancer susceptibility variant at 8q24 *Nat Genet*. 2007;39(5):631–637.

33. Horvath A, Boikos S, Giatzakis C, et al. A genome-wide scan identifies mutations in the gene encoding phosphodiesterase 11A4 (PDE11A) in individuals with adrenocortical hyperplasia. *Nat Genet*. 2006;38(7):794–800.

34. Duerr RH, Taylor KD, Brant SR, et al. A genome-wide association study identifies IL23R as an inflammatory bowel disease gene. *Science*. 2006;314(5804):1461–1463.

35. Helgason A, Palsson S, Thorleifsson G, et al. Refining the impact of TCF7L2 gene variants on type 2 diabetes and adaptive evolution. *Nat Genet*. 2007;39(2):218–225.

36. Helgadottir A, Manolescu A, Helgason A, et al. A variant of the gene encoding leukotriene A4 hydrolase confers ethnicity-specific risk of myocardial infarction. *Nat Genet*. 2006;38 (1):68–74.

37. Roden DM, Altman RB, Benowitz NL, et al. Pharmacogenomics: challenges and opportunities. *Ann Intern Med*. 2006;145:749–757.

38. Lesko LJ. Personalized medicine: elusive dream or imminent reality? *Clin Pharmacol Ther*. 2007;81:807–816.

39. US FDA. Genomics at FDA, Table of Valid Genomic Biomarkers in the Context of Approved Drug Labels. Available at http://www.fda.gov/cder/genomics/genomic_biomarkers_table.htm. (Accessed July 25, 2007.)

40. Nuffield Council report on bioethics of pharmacogenetics. Available at http://www.nuffieldbioethics.org/filelibrary/paf1/pharmacogenetics_report.pdf.

41. Sciences CfIOoM. *International Ethical Guidelines for Biomedical Research Involving Human Subjects*. Geneva; 2002.

42. Anderson C, Gomez-Mancilla B, Spear B, et al. Elements of informed consent for pharmacogenetic research; perspective of the Pharmacogenetics Working Group. *Pharmacogenomics J.* 2002;2(5):284–292.

43. Spear B, Heath-Chiozzi M, Barnes D, Cheeseman K, Shaw P, Campbell D. Terminology for sample collection in clinical genetic studies. *Pharmacogenomics J.* 2001;1:101–103.

44. Renegar G, Webster C, Stuerzebecher S, et al. Returning genetic research results to individuals: points to consider. *Bioethics.* 2006;20(1):24–36.

45. Ring BJ, Gillepsie JS, Eckstein JA, et al. Identification of the human cytochromes P450 responsible for atomoxetine metabolism. *Drug Metab Dispos.* 2002;30(3):319–323.

46. Michelson D, Read HA, Ruff DD, et al. CYP2D6 and clinical response to atomoxetine in children and adolescents with ADHD. *Jam Acad Child Adolesc Psychiatry.* 2007;46(2):242–251.

47. Worldwide variations in the prevalence of asthma symptoms: the International Study of Asthma and Allergies in Childhood (ISAAC) *Eur Respir J.* 1998;12(2): 315–335.

48. Eder W, Ege MJ, von Mutius E. The asthma epidemic. *N Engl J Med.* 2006;355(21):2226–2235.

49. Van Eerdewegh P, Little RD, Dupuis J, et al. Association of the ADAM33 gene with asthma and bronchial hyperresponsiveness. *Nature.* 2002;418(6896):426–430.

50. Zhang Y, Leaves NI, Anderson GG, et al. Positional cloning of a quantitative trait locus on chromosome 13q14 that influences immunoglobulin E levels and asthma. *Nat Genet.* 2003;34 (2):181–186.

51. Laitinen T, Polvi A, Rydman P, et al. Characterization of a common susceptibility locus for asthma-related traits. *Science.* 2004;304(5668):300–304.

52. Choudhry S, Ung N, Avila PC, et al. Pharmacogenetic differences in response to albuterol between Puerto Ricans and Mexicans with asthma. *Am J Respir Crit Care Med.* 2005;171(6):563–570.

53. Fenech A, Hall IP. Pharmacogenetics of asthma. *Br J Clin Pharmacol.* 2002;53(1):3–15.

54. Reihsaus E, Innis M, MacIntyre N, Liggett SB. Mutations in the gene encoding for the beta 2-adrenergic receptor in normal and asthmatic subjects. *Am J Respir Cell Mol Biol.* 1993;8 (3):334–339.

55. Lima JJ, Thomason DB, Mohamed MH, et al. Impact of genetic polymorphisms of the beta 2-adrenergic receptor on albuterol bronchodilator pharmacodynamics. *Clin Pharmacol Ther.* 1999;65(5):519–525.

56. Tan S, Hall IP, Dewar J, Dow E, Lipworth B. Association between beta 2-adrenoceptor polymorphism and susceptibility to bronchodilator desensitisation in moderately severe stable asthmatics. *Lancet.* 1997;350(9083):995–999.

57. Martinez FD, Graves PE, Baldini M, Solomon S, Erickson R. Association between genetic polymorphisms of the beta2-adrenoceptor and response to albuterol in children with and without a history of wheezing. *J Clin Invest.* 1997;100(12):3184–3188.

58. Kukreti R, Bhatnagar P, Rao B, et al. Beta(2)-adrenergic receptor polymorphisms and response to salbutamol among Indian asthmatics. *Pharmacogenomics.* 2005;6(4):399–410.

59. Hancox RJ, Sears MR, Taylor DR. Polymorphism of the beta2-adrenoceptor and the response to long-term beta2-agonist therapy in asthma. *Eur Respir J.* 1998;11(3):589–593.

60. Silverman EK, Kwiatkowski DJ, Sylvia JS, et al. Family-based association analysis of beta2-adrenergic receptor polymorphisms in the childhood asthma management program. *J Allergy Clin Immunol.* 2003;112(5):870–876.

61. Cucca F, Lampis R, Congia M, et al. A correlation between the relative predisposition of MHC class II alleles to type 1 diabetes and the structure of their proteins. *Hum Mol Genet.* 2001;10 (19):2025–2037.

62. Nerup J, Platz P, Andersen OO, et al. HL-A antigens and diabetes mellitus. *Lancet.* 1974;2 (7885):864–866.

63. Noble JA, Valdes AM, Cook M, et al. The role of HLA class II genes in insulin-dependent diabetes mellitus: molecular analysis of 180 Caucasian, multiplex families. *Am J Hum Genet.* 1996;59(5):1134–1148.

64. Bell GI, Horita S, Karam JH. A polymorphic locus near the human insulin gene is associated with insulin-dependent diabetes mellitus. *Diabetes.* 1984;33(2):176–183.

65. Bennett ST, Lucassen AM, Gough SC, et al. Susceptibility to human type 1 diabetes at IDDM2 is determined by tandem repeat variation at the insulin gene minisatellite locus. *Nat Genet.* 1995;9(3):284–292.

66. Vafiadis P, Bennett ST, Todd JA, et al. Insulin expression in human thymus is modulated by INS VNTR alleles at the IDDM2 locus. *Nat Genet.* 1997;15(3):289–292.

67. Bottini N, Musumeci L, Alonso A, et al. A functional variant of lymphoid tyrosine phosphatase is associated with type I diabetes. *Nat Genet.* 2004;36(4):337–338.

68. Smyth D, Cooper JD, Collins JE, et al. Replication of an association between the lymphoid tyrosine phosphatase locus (LYP/PTPN22) with type 1 diabetes, and evidence for its role as a general autoimmunity locus. *Diabetes.* 2004;53(11):3020–3023.

69. Kristiansen OP, Larsen ZM, Pociot F. CTLA-4 in autoimmune diseases—a general susceptibility gene to autoimmunity? *Genes Immunity.* 2000;1(3):170–184.

70. Ueda H, Howson JM, Esposito L, et al. Association of the T-cell regulatory gene *CTLA4* with susceptibility to autoimmune disease. *Nature.* 2003;423(6939):506–511.

71. Anjos SM, Tessier MC, Polychronakos C. Association of the cytotoxic T lymphocyte-associated antigen 4 gene with type 1 diabetes: evidence for independent effects of two polymorphisms on the same haplotype block. *J Clin Endocrinol Metabol.* 2004;89 (12):6257–6265.

72. Wellcome Trust Case Control Consortium. Genome-wide association study of 14,000 cases of seven common diseases and 3,000 shared controls *Nature.* 2007;447(7145): 661–678.

73. Todd JA, Walker NM, Cooper JD, et al. Robust associations of four new chromosome regions from genome-wide analyses of type 1 diabetes. *Nat Genet.* 2007;39:857–864.

74. Hakonarson H, Grant SF, Bradfield JP, et al. A genome-wide association study identifies *KIAA0350* as a type 1 diabetes gene. *Nature.* 2007;448:591–594.

75. Zimmet P, Alberti KG, Shaw J. Global and societal implications of the diabetes epidemic. *Nature.* 2001;414(6865):782–787.

76. Tattersall RB. Mild familial diabetes with dominant inheritance. *Q J Med.* 1974;43(170):339–357.

77. Tattersal RB, FajansSS. Prevalence of diabetes and glucose intolerance in 199 offspring of thirty-seven conjugal diabetic parents. *Diabetes.* 1975;24(5):452–462.

78. Froguel P, Zouali H, Vionnet N, et al. Familial hyperglycemia due to mutations in glucokinase. Definition of a subtype of diabetes mellitus. *N Engl J Med.* 1993;328(10):697–702.

79. Frayling TM, Bulamn MP, Ellard S, et al. Mutations in the hepatocyte nuclear factor-1alpha gene are a common cause of maturity-onset diabetes of the young in the UK. *Diabetes.* 1997;46 (4):720–725.

80. Heiervang E, Folling I, Sovik O, et al. Maturity-onset diabetes of the young. Studies in a Norwegian family. *Acta Paediatr Scand.* 1989;78(1):74–80.

81. Sovik O, Njolstad P, Folling I, et al. Hyperexcitability to sulphonylurea in MODY3 *Diabetologia.* 1998;41(5):607–608.

82. Pearson ER, Liddell WG, Shepherd M, Corrall RJ, Hattersley AT. Sensitivity to sulphonylureas in patients with hepatocyte nuclear factor-1alpha gene mutations: evidence for pharmacogenetics in diabetes. *Diabet Med.* 2000;17(7):543–545.

83. Rennert NJ, Charney P. Preventing cardiovascular disease in diabetes and glucose intolerance: evidence and implications for care. *Prim Care.* 2003;30(3):569–592.

84. Dominiczak MH. Obesity, glucose intolerance and diabetes and their links to cardiovascular disease. Implications for laboratory medicine. *Clin Chem Lab Med.* 2003;41 (9):1266–1278.

85. Hauner H, Meier M, Jockel KH, Frey UH, Siffert W. Prediction of successful weight reduction under sibutramine therapy through genotyping of the G-protein beta3 subunit gene (*GNB3*) C825T polymorphism. *Pharmacogenetics.* 2003;13(8):453–459.

86. Carroll WL, Bhojwani D, Min DJ, et al. Pediatric acute lymphoblastic leukemia. *Hematol Am. Soc. Hematol. Educ. Program.* 2003;102–131.

87. Silverman LB, Sallan SE. Newly diagnosed childhood acute lymphoblastic leukemia: update on prognostic factors and treatment. *Curr Opin Hematol.* 2003;10:290–296.

88. Krajinovic MD, Labuda D, Richer C, Karimi S, Sinnett D. Susceptibility to childhood acute lymphoblastic leukemia: influence of *CYP1A1, CYP2D6, GSTM1,* and *GSTT1* genetic polymorphisms. *Blood.* 1999;93:1496–1501.

89. Krajinovic MD, Labuda D, Mathonnet G, et al. Polymorphisms in genes encoding drugs and xenobiotic metabolizing enzymes, DNA repair enzymes, and response to treatment of childhood acute lymphoblastic leukemia. *Clin. Cancer Res.* 2002;8:802–810.

90. Balta G, Yusek N, Ozyurek E, et al. Characterization of *MTHFR*, *GSTM1*, *GSTT1*, *GSTP1*, and *CYP1A1* genotypes in childhood acute leukemia. *Am J Hematol.* 2003;73:154–160.

91. Meissner B, Stanulla M, Ludwig WD, et al. The *GSTT1* deletion polymorphism is associated with initial response to glucocorticoids in childhood acute lymphoblastic leukemia. *Leukemia.* 2004;18:1920–1923.

92. Stanulla M, Schaffeier E, Arens S, et al. *GSTP1* and *MDR1* genotypes and central nervous system relapse in childhood acute lymphoblastic leukemia. *Int. J. Hematol.* 2005;81:39–44.

93. Rocha JC, Cheng C, Liu W, et al. Pharmacogenetics of outcome in children with acute lymphoblastic leukemia. *Blood.* 2005;105(12):4752–4758.

94. Chen CL, Liu Q, Pui CH, et al. Higher frequency of glutathione *S*-transferase deletions in black children with acute lymphoblastic leukemia. *Blood.* 1997;89(5):1701–1707.

95. Davies SM, Bhatia S, Ross JA, et al. Glutathione *S*-transferase genotypes, genetic susceptibility, and outcome of therapy in childhood acute lymphoblastic leukemia. *Blood.* 2002;100(1):67–71.

96. Stanulla M, Schrappe M, Brechlin AM, et al. Polymorphisms within glutathione *S*-transferase genes (*GSTM1, GSTT1, GSTP1*) and risk of relapse in childhood B-cell precursor acute lymphoblastic leukemia: a case–control study. *Blood.* 2000;95(4):1222–1228.

97. Lennard L, Lilleyman JS, van Loon J, Weinshilboum RM. Genetic variation in response to 6-mercaptopurine for childhood acute lymphoblastic leukaemia. *Lancet.* 1990;336:225–229.

98. Evans WE, Horner M, Chu YQ, Kalwinski D, Roberts WM. Altered mercaptopurine metabolism, toxic effects, and dosage requirement in a thiopurine methyltransferase-deficient child with acute lymphocytic leukemia. *J. Pediatr.* 1991;119:985–989.

99. Otterness D, Szumlanski C, Lennard JT, et al. Human thiopurine methyltransferase pharmacogenetics: gene sequence polymorphisms. *Clin Pharmacol Ther.* 1997;62(1):60–73.

100. Larovere LE, Kremer RD, Lambooy LH, et al. Genetic polymorphism of thiopurine *S*-methyltransferase in Argentina. *Ann Clin Biochem.* 2003;40 (Part 4):388–393.

101. Indjova D, Atanasova S, Shipkova MN, et al. Phenotypic and genotypic analysis of thiopurine *S*-methyltransferase polymorphism in the bulgarian population. *Ther Drug Monit.* 2003;25 (5):631–636.

102. Hon YY, Fesing MY, Pui CH, et al. Polymorphism of the thiopurine *S*-methyltransferase gene in African-Americans. *Hum Mol Genet.* 1999;8(2):371–376.

103. Hiratsuka M, Inoue T, Omori F, et al. Genetic analysis of thiopurine methyltransferase polymorphism in a Japanese population. *Mutat Res.* 2000;448(1):91–95.

104. Zhang JP, Guan YY, Wu JH, et al. Genetic polymorphism of the thiopurine *S*-methyltransferase of healthy Han Chinese. *Ai Zheng*. 2003;22(4):385–388.

105. Andersen JB, Szumlanski C, Weinshilboum RM, et al. Pharmacokinetics, dose adjustments, and 6-mercaptopurine/methotrexate drug interactions in two patients with thiopurine methyltransferase deficiency. *Acta Paediatr*. 1998;87:108–111.

106. Evans WE. Pharmacogenetics of thiopurine *S*-methyltransferase and thiopurine therapy. *Ther Drug Monit*. 2004;26:186–191.

107. Evans WE, Bomgaars L, Coutre S, et al. Preponderance of thiopurine *S*-methyltransferase deficiency and hetrozygosity among patients intolerant to mercaptopurine or azathioprine. *J Clin Oncol*. 2001;19:2293–2301.

108. Ezzeldin HH, Diasio R. Genetic testing in cancer therapeutics. *Clin Cancer Res*. 2006;12: 4137–4141.

109. Patton MA. Genetic studies in the Amish community. *Ann Hum Biol*. 2005;32(2):163–167.

110. Labay V, Raz T, Baron D, et al. Mutations in *SLC19A2* cause thiamine-responsive megaloblastic anaemia associated with diabetes mellitus and deafness. *Nat Genet*. 1999;22(3):300–304.

111. McKusick VA. Online Mendelian Inheritance in Man, OMIM (TM). McKusick–Nathans Institute of Genetic Medicine, Johns Hopkins University, Baltimore, MD, and National Center for Biotechnology Information, National Library of Medicine, Bethesda, MD.

CHAPTER 23

General Principles of Population Pharmacokinetics in Pediatrics

MAHESH SAMTANI, PhD and HUI KIMKO, PhD

Advanced Modeling & Simulation Department, Johnson & Johnson Pharmaceutical Research & Development, Raritan, New Jersey 08869

INTRODUCTION

Physicians treating infants often prescribe drugs that have not been approved by the Food and Drug Administration (FDA) for use in this population. The unapproved use of approved drugs (for adults) occurs unfortunately because drugs are often approved without labeling guidelines or with labeling disclaimers about pediatric use. This has led to past therapeutic catastrophes such as the thalidomide incident. In the late 1960s, thousands of children were born with stunted arms and legs because their mothers were treated with thalidomide for treating morning sickness during pregnancy. For many years thereafter, studies in children were considered unethical and this led to off-label use of drugs in this population, which in turn again led to significant morbidity.[1] This caused regulatory agencies to start various initiatives to facilitate examination of drugs in this sensitive population.

With the advent of population pharmacokinetics, pediatric data analysis became the poster child for nonlinear mixed effect modeling of data. This technique could allow better understanding of pediatric pharmacology because it could deal with the constraints associated with pediatric studies such as limited and sparse availability of pharmacokinetic and pharmacodynamic data. Grasela and Donn[2] demonstrated the usefulness of population analysis approach in pediatrics for phenobarbital. Traditional analysis with fewer patients was subject to biased and imprecise description of data, whereas the population analysis approach allowed not only the use of sparse data from routine clinical care, but also the identification of influential covariates such as APGAR scores on the pharmacokinetics of the drug. An appreciation of these types of analysis has led to an inclusion of population-based techniques as one of the preferred methods for data analysis with special populations in a number of regulatory documents by the FDA, International Conference on Harmonisation (ICH), and CPMP.[3]

Pediatric Drug Development: Concepts and Applications
Edited by Andrew E. Mulberg, Steven A. Silber, and John N. van den Anker

REGULATORY PERSPECTIVE ON PEDIATRIC DRUG LABELING

Children are considered therapeutic orphans because, for both safety and efficacy, two-thirds of drugs have information that is either missing or insufficient.[4] This has led to off-label use of medications in pediatrics. This practice can lead to underdosing that can cause lack of efficacy or overdosing with possible developmental harm in this special population. The potential for developmental harm often prevents prescribers from using potentially useful medications in this population. These problems have led to initiatives sponsored by regulatory agencies that encourage the pharmaceutical industry to gather information regarding the safe and efficacious use of medications in the pediatric population. These issues are fully represented in other chapters by Maldonado (Chapter 12), Rose (Chapter 13), and Nakamura and Ono (Chapter 14) in this book; the reader is urged to review these chapters for further information and background.

NECESSITY OF SPARSE SAMPLING AND POPULATION PHARMACOKINETICS IN PEDIATRIC STUDIES

Ethical and logistical issues that make it difficult to collect pediatric samples include the following:

- Patient autonomy wherein children and/or their parents refuse to provide blood samples for nontherapeutic studies.
- Limited volume of blood available for sampling especially from neonates.
- Limited flexibility available with pediatric patients due to their fear of pain associated with venipuncture.

Given these logistical issues, the benefit of collecting samples from children getting the drug therapeutically becomes quite obvious. This strategy poses less of a concern than nontherapeutic studies in a pediatric population. Collection of 2–4 samples per individual with samples collected during routine clinical visits is acceptable for population pharmacokinetic analysis. The methodology involves collecting microvolume samples that can be analyzed using modern and highly sensitive analytical methods. In order to get more information from a limited number of subjects and samples, it is preferred that subjects are not sampled at the same time; instead, they are divided into groups and samples from different groups are obtained at various sequences of time. Thus, a complete pharmacokinetic profile is available from the entire population without obtaining a full profile from any single individual, allowing treatment of the population (rather than the individual) to be the unit for the population analysis. This method therefore permits analysis of data from individuals with truncated profiles and missing samples.

Analysis and interpretation of sparsely sampled routine clinical data from therapeutic drug monitoring can be utilized to provide dosing recommendations that can be used for other patients. This methodology has an additional advantage that data from different studies can be pooled to obtain a global characterization of data rather than performing analysis on a study-by-study basis. Traditionally, the number of subjects used in this type of analysis is about 50 or more[1] and the number of subjects has to increase if a large number of subject-specific characteristics need to be evaluated for their influence on the pharmacokinetics of the drug. Moreover, the characterized individual pharmacokinetic profiles can also help with predicting the clinical outcome for individuals if there is a known relationship between drug concentrations and response.

MODEL-BASED ANALYSIS AND SIMULATION

The use of modeling and simulation is not new and has been used by other industries such as aerospace engineering and weather forecasting to understand and optimize dynamic systems.[5] The models of drug absorption, drug disposition, drug action, and disease progression have experienced an exponential growth in the last few decades due to the availability of better computers and assay methods. These models have evolved from simple empirical functions to more mechanistic models that capture the underlying physiology and pharmacology of the system.[6]

The systems model captures the time course of drug concentration and response at the population level using mathematical and stochastic equations that can be solved using a mixed-effects modeling approach. The parameters of the systems model represent the theoretical typical values in an entire population and are referred to as fixed effects. Random effects are used to explain within-subject, between-subject, between-occasion, between-trial random variability. Modeling software also provides estimates of uncertainty in these model parameters, which can be taken into account when prospective scenarios are evaluated in the simulation mode. Very often in pediatric modeling, since data can be very sparse, complex systems models may be difficult to accommodate.

Some patient-specific factors, such as body weight, age, gender, and renal function, can cause the pharmacokinetic parameters to deviate from the typical population value, and their influence on pharmacokinetics can be identified by plotting an individual-specific parameter against patient factors. This evaluation can also suggest bimodal distribution of parameters, which can be indicative of pharmacogenomic differences between subjects. A good example is the drug isoniazid used for treating tuberculosis. Bimodal distribution for this drug's clearance has been observed in pediatric subjects and this is suggestive of slow and fast metabolizer subgroups for the acetylation pathway of isoniazid.[7] These types of trends related to patient factors are tested statistically to decide whether a patient factor is a covariate in the systems model and the likelihood ratio test is commonly used for this purpose.

Models developed from these exercises should allow generalization and must not be more complex than needed. The complexity of the model should be based on the quantity of the available data and the goal of the modeling exercise. The model performance should be examined, but commonly used techniques such as data splitting for model evaluation are generally not feasible in pediatrics due to the limited quantity of pediatric data that is available. However, the model should at least be capable of recreating the data from which the model was built. This is generally referred to as the visual predictive check,[8] wherein simulated data from the final pharmacokinetic model are compared to the original data and 90% of the observations are expected to lie within the 90% prediction intervals of the simulated profiles.

Once a qualified systems model is available, it can be used for simulation exercises that can be used to scrutinize the pharmacokinetic properties of the system.[9] These simulations allow the generation of hypotheses and the exploration of numerous what-if scenarios, which would otherwise be difficult due to the necessity of enrolling large numbers of pediatric subjects for investigation. Therefore, simulation provides an objective basis for selecting optimal dosing regimens and study designs that will be most cost effective in terms of demonstrating efficacy and safety in pediatric drug development.

The virtual population generated during simulations with the systems model should display reasonable patient covariate distributions expected in real life. These distributions should reflect the heterogeneity in the patient population and their joint distributions (e.g., correlation between gender and weight) to avoid the generation of unrealistic representations of virtual patients. These simulation exercises can be applied to (but are not restricted to) the following types of investigations in pediatric populations: characterization and bridging of exposure–response relationships from adults to children; dealing with censored and sparse data due to assay limitations; decision-making for selection of dose and dosing regimen; optimization of future clinical trial designs; prediction of pharmacokinetics from single dose to multiple dose scenarios; and formulation optimization.

DEALING WITH COVARIATES IN PEDIATRIC ASSESSMENTS: SPECIAL CONSIDERATIONS

A substantial change in body size, as a function of age, is an important temporal phenomenon not seen in adults. This complexity makes evaluation of covariates, such as body size and age, a difficult problem in pediatric population analysis due to the strong collinearity between them. Thus, for a particular drug, clearance may change as a function of age, weight, and/or glomerular filtration rate, which are all related with the maturational stage of the child. Bonate[10] has investigated this problem of collinearity and shown that when the degree of correlation is greater than 0.5, then the model may indicate that only some of the correlated covariates are important when in fact all factors could be controlling the pharmacokinetics. Thus, collinearity can potentially introduce errors when identifying the importance of covariates in pediatric modeling.

A solution that has been suggested to circumvent this problem is to fix certain parameters of the systems model a priori using a well-described physiologic framework (i.e., allometric size adjustment based on body weight). Pharmacokinetic parameters can be related to body weight through a power function and the power can be fixed to (1) 0.75 for clearance, (2) 1 for volume of distribution, and (3) 0.25 for half-lives.[11] Once the size factor is fixed, the effect of other covariates such as maturational age and organ function can be separated out.

OPTIMAL SAMPLING STRATEGIES

An important aspect about pediatric studies is the sparse nature of study design so as to accommodate ethical concerns associated with performing studies in this special population. It is important to optimize the sampling strategy to ensure that the data generated would be informative and amenable to population modeling. A good set of prior information (e.g., pharmacokinetic information including covariates from adults and the distributions of the covariates of pediatric patients) can be used for developing and evaluating a sparse sampling strategy. The D-optimality sampling method allows selection of a sequence of sampling time windows that can potentially provide the maximum information about parameter estimates of a given model.[12] The objective of the method is to find a set of sampling times that could yield the smallest uncertainty (i.e., standard errors) in the estimates of the parameters by maximizing the Fisher's information matrix with respect to the sampling times.

The performance of these sampling windows can be evaluated through Monte Carlo simulations, where drug concentrations at sampling times obtained using the D-optimality method are simulated based on the prior knowledge of pharmacokinetic parameters. The simulated concentrations are put through the estimation procedure, and the estimated parameter values can be compared to those used for simulations. This comparison provides bias and precision that can be expected in the parameter estimates if an actual study were to be conducted with a sampling strategy identified using D-optimality sampling method.

CASE STUDIES

Etoposide is used to treat childhood acute lymphocytic leukemia and its pharmacokinetics has been related to efficacy and toxicity endpoints. Limited and optimal sampling strategies for etoposide have facilitated the assessment of etoposide pharmacokinetics in pediatrics. Limited sampling times have been estimated and only two samples per subjects were found to be optimal for this drug.[13] Using the empirical Bayesian estimation technique, pharmacokinetics of this drug could be estimated with good precision and accuracy (mean bias and precision <10%).

Therapeutic drug monitoring of neonatal subjects in intensive care units generates sparsely sampled data, which has greatly facilitated our understanding of the disposition of aminoglycoside antibiotics in this population.[14] It has greatly helped with understanding the influence of age and body weight on the pharmacokinetics of these drugs. This knowledge has been used for dosage optimization of aminoglycosides in neonates. Similarly, the influence of nonsteroidal anti-inflammatory drugs such as aspirin and ibuprofen on the clearance of these drugs has also been delineated. It is now understood that these nonsteroidal anti-inflammatory drugs can have an impact on renal function, which may reduce the clearance of aminoglycosides such as amikacin by 22% during the first day of neonatal life.

Busulfan is used before hematopoietic stem cell transplantation for adults suffering from chronic myelogenous leukemia to achieve bone marrow ablation. The drug has a narrow therapeutic index (AUC: 900–1500 μmol/L/min) below which there exists the possibility of graft failure with cancer relapse and above which there is the potential for fatal hepatic venoocclusions. To understand the appropriate dosing regimen for the use of busulfan in pediatric subjects needing transplantation, a population pharmacokinetic model was developed.[15,16] The model contained data from 24 children within the age range of 4–17 years. Model-based simulations indicated that different regimens with 2–7 steps and combinations of different body weights would give at most 60% chance of being in the therapeutic window. This finding was mainly attributed to the narrow therapeutic index and the large between-subject variability for the drug. Based on this analysis, a dosing regimen based on weight (1.1 mg/kg for infants less than or equal to 12 kg and 0.8 mg/kg for children greater than 12 kg) was recommended along with therapeutic drug monitoring in the label for this drug to ensure that concentrations are maintained within the therapeutic window. It is noteworthy to mention that the proposed dosing regimen was not studied directly in actual trials and the recommendation is based on the modeling results.

Sotalol is used in the treatment of tachycardias and two clinical trials have been conducted using biomarkers (e.g., heart rate) to obtain dosing guidelines in pediatric subjects in the age range of 1 month to 12 years.[15, 17] Based on the trial information, 30 mg/m^2 starting dose three times a day (with upward titration up to 60 kg/m^2) could be prescribed for children older than 2 years of age. To obtain dosing recommendations for children less than 2 years of age, an exposure–response model was developed keeping in mind that, for a given exposure, drug effects are similar between adults and children for this drug. The only correction needed in this modeling was the incorporation of the physiologic factor that kidneys undergo a maturation phenomenon up to 2 years of age. Thus, up to an age of 2 years, the clearance of sotalol increases and beyond this age clearance mainly depends on body size.[15, 17] For ages less than 2 years, an age factor correction is needed for sotalol. As an example, it can be recommended that for a 1-month-old infant the starting dose (30 mg/m^2) should be modified by a factor of 0.68 (20 mg/m^2).[15]

Oxcarbazepine is used for adjunct therapy in adults and pediatrics for treating epilepsy. It is also approved for monotherapy in adults with epileptic seizures. The placebo–response and exposure–response for this drug are similar between adults and children based on an exposure–response model using the data from the three indications above.[15] Using the exposure–response model developed with a population analysis approach, dosing recommendation for monotherapy with oxcarbazepine in pediatrics could be derived by matching the simulated exposure obtained from proposed pediatric regimens with the exposures seen with adult monotherapy. Thus, monotherapy trials for oxcarbazepine in children could be circumvented according to the *Federal Code of Regulations* that allows for such a provision as long as scientific criteria (such as similar exposure–response relationship between pediatrics and adults) are met.

CONCLUSION

Modeling techniques can be used efficiently to integrate and quantify the information collected during routine clinical care of pediatric patients. This knowledge can be combined with other relevant information such as exposure–response relationship and disease behavior from the adult population to bridge the gap between adults and children. Mixed-effect models provide a way of quantifying the relationship between an outcome of interest and underlying explanatory variables while at the same time accounting for the inherent between-subject variability. The proper management of the knowledge generated with the modeling techniques allows the exploration of different scenarios of interest through simulations without conducting extensive testing in the pediatric population.

Modeling and simulation can guide pediatric drug development by serving as a decision-making tool that can track the benefit/risk ratio of using pharmacological agents in this vulnerable population. Thus, simulation-based approaches will facilitate the development of better and safer dosing regimens for new medicines and medications already approved for adult indications. This will ultimately help maximize the benefit of the drug therapy while improving the quality of life for precious pediatric patients.

REFERENCES

1. Anderson BJ, Allegaert K, Holford NH. Population clinical pharmacology of children: general principles. *Eur J Pediatr*. 2006;165:741–746.

2. Grasela T, Donn S. Neonatal population pharmacokinetics of phenobarbital derived from routine clinical data. *Dev Pharmacol Ther*. 1985;8:374–383.
3. Hsochhaus G, Barrett JS, Derendorf H. Evolution of pharmacokinetics and pharmacokinetic/dynamic correlations during the 20th century. *J Clin Pharmacol*. 2000;40:908–917.
4. Roberts R, Rodriguez W, Murphy D, Crescenzi T. Pediatric drug labeling: improving the safety and efficacy of pediatric therapies. *JAMA*. 2003;290:905–911.
5. Kimko H, Duffull S, eds., *Simulation for Designing Clinical Trials: A Pharmacokinetic–Pharmacodynamic Modelling Perspective*, New York: Marcel Dekker; 2003.
6. Sheiner LB, Steimer JL. Pharmacokinetic/pharmacodynamic modeling in drug development. *Annu Rev Pharmacol Toxicol*. 2000;40:67–95.
7. Pariente-Khayat A, Rey E, Gendrel D, et al. Isoniazid acetylation metabolic ratio during maturation in children. *Clin Pharmacol Ther*. 1997;62:377–383.
8. Duffull SB, Aarons L. Development of a sequential linked pharmacokinetic and pharmacodynamic simulation model for ivabradine in healthy volunteers. *Eur J Pharm Sci*. 2000;10:275–284.
9. Holford N, Kimko H, Monteleone J, Peck C. Simulation of clinical trials. *Annu Rev Pharmacol Toxicol*. 2000;40:209–234.
10. Bonate PL. The effect of collinearity on parameter estimates in nonlinear mixed effect models. *Pharm Res*. 1999;16:709–717.
11. Meibohm B, Laer S, Panetta JC, Barrett JS. Population pharmacokinetic studies in pediatrics: issues in design and analysis. *AAPS J*. 2005;7:E475–E487.
12. D'Argenio DZ. Optimal sampling times for pharmacokinetic experiments. *J Pharmacokinet Biopharm*. 1981;9:739–756.
13. Panetta JC, Wilkinson M, Pui CH, Relling MV. Limited and optimal sampling strategies for etoposide and etoposide catechol in children with leukemia. *J Pharmacokinet Pharmacodyn*. 2002;29:171–188.
14. Allegaert K, Anderson BJ, Cossey V, Holford NH. Limited predictability of amikacin clearance in extreme premature neonates at birth. *Br J Clin Pharmacol*. 2006;61:39–48.
15. Bhattaram VA, Booth BP, Ramchandani RP, et al. Impact of pharmacometrics on drug approval and labeling decisions: a survey of 42 new drug applications. *AAPS J*. 2005;7:E503–E512.
16. Busulfan. Center for Drug Evaluation and Research, United States Food and Drug Administration. Available at http://www.fda.gov/cder/foi/nda/2002/20-954S004_Busulfex.htm.
17. Betapace (Sotalol Hydrochloride). Center for Drug Evaluation and Research, United States Food and Drug Administration. Available at http://www.fda.gov/cder/foi/nda/2001/19-865s10_Betapace.htm.

CHAPTER 24

Development of Oncology Drugs for Children

ROBIN NORRIS

Clinical Instructor, Division of Oncology, Clinical Pharmacology & Therapeutics, Children's Hospital of Philadelphia, Philadelphia, Pennsylvania 19146

WAYNE RACKOFF, MD

Vice President, Clinical Oncology, Ortho Biotech Oncology Research & Development, Johnson & Johnson Pharmaceutical Research and Development, Raritan, New Jersey 08869

STEVEN HIRSCHFELD, MD, PhD

CAPT USPHS, Associate Director for Clinical Research, National Institute of Child Health and Human Development, 31 Center Drive, Bethesda, Maryland 20814

PETER C. ADAMSON, MD

Chief, Division of Clinical Pharmacology & Therapeutics, Director, Office of Clinical and Translational Research, The Children's Hospital of Philadelphia, University of Pennsylvania School of Medicine, Philadelphia, Pennsylvania 19104

INTRODUCTION

In 1948, Dr. Sidney Farber and colleagues published the first account of inducing remission in children with acute leukemia using folic acid antagonists.[1] Since that time, research has transformed childhood cancer from an almost uniformly fatal disease to one having a 5-year overall survival rate approaching 80%.[2] A cornerstone of these advances has been collaborative clinical research, which is an integral component of the mission of pediatric oncologists worldwide (Figure 24.1). The first national pediatric oncology cooperative group was formed by the National Institutes of Health (NIH) in 1955.[3] Today, the Children's Oncology Group (COG) is the largest pediatric cancer cooperative group in the world, with over 90% of children with cancer in the United States cared for at member institutions and nearly half of children entered into at least one trial.[3,4] In contrast, only 2% of adult cancer patients are entered into cooperative group trials.[3] Participation in clinical trials is a routine component of pediatric oncology care, and pediatric oncologists are trained to align clinical research with clinical care in their everyday practice.[5]

Pediatric Drug Development: Concepts and Applications
Edited by Andrew E. Mulberg, Steven A. Silber, and John N. van den Anker

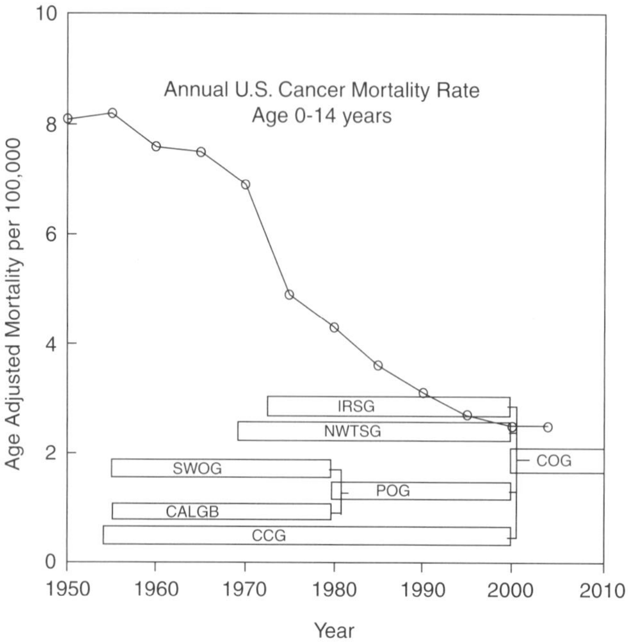

FIGURE 24.1 The pediatric cooperative groups and national decline in rate of cancer deaths in children 0–14 years of age. (Adapted from Bleyer.[3] Mortality data from SEER Cancer Statistics Review.[2]) CALGB, Cancer and Acute Leukemia Group B; CCG, Children's Cancer Group; COG, Children's Oncology Group; IRSG, Intergroup Rhabdomyosarcoma Group; NWTSG, National Wilms Tumor Study Group; POG, Pediatric Oncology Group; SWOG, Southwest Oncology Group.

Despite these advances, several childhood cancers still have unacceptably low cure rates,[6] and even when treatment is successful, the acute and long-term morbidity of current therapy can be substantial.[7,8] One potential reason for this is few new agents have been developed for the treatment of childhood cancer. Since 1980, only three novel agents have been labeled for use in pediatric cancers, compared to greater than 100 drugs approved for the treatment of adult malignancies.[9,10] In the United States, legislative initiatives, including the Best Pharmaceuticals for Children Act (BPCA) and the Pediatric Research Equity Act (PREA), have impacted upon this and begun to change the pediatric oncology drug development landscape. Similar legislative initiatives under the European Medicines Evaluation Agency (EMEA) have recently been enacted in Europe. Nonetheless, significant gaps in pediatric oncology drug development remain.[11]

HISTORICAL OVERVIEW

Early Oncology Drug Discovery

The pediatric oncology drug development process has evolved significantly since Farber and colleagues published their results of testing the folate antagonist, aminopterin, in 16 children with acute leukemia.[1] While this study is renowned because it is the first published account of a treatment leading to remission, it should be noted that the investigators also stressed the toxic effects of aminopterin and suggested that toxicity could prevent the drug from widespread use. Subsequently, methotrexate (MTX), an antifolate with similar efficacy but a more predictable toxicity profile than aminopterin,[7] was approved for use by the FDA in 1953.[8]

6-Mercaptopurine (6-MP) was developed and synthesized by Gertrude Elion in 1951. This rationally developed purine antagonist was found to cause toxicity in both *Lactobacillus casei* and in the frog embryo.[12] The compound, noted to have an effect on rodent tumors, was quickly tested in children with acute leukemia and found to induce remissions in patients with disease resistant to MTX and cortisone.[13] By the end of 1953, only 2 years after its discovery, 6-MP was approved for use in leukemia patients.

Even though MTX and 6-MP remain the cornerstone of maintenance chemotherapy for children with acute lymphoblastic leukemia (ALL), the process and timelines of oncology drug development have changed dramatically since their approval more than 50 years ago. Important safeguards and oversight now exist for all participants in clinical research, with additional protections afforded children.[14] Relative to cancer in adults, childhood cancer is rare, and thus there has been virtually no economic incentive for industry to pursue pediatric oncology drug development. There are now more than 400 new agents under development for adult cancers, posing another challenge for pediatric oncology drug developers, as only a very small proportion of these agents can realistically be tested in children.[15] However, increasing experience with early phase pediatric trials by clinical investigators, the pharmaceutical industry, and the U.S. Food and Drug Administration (FDA), coupled with legislative changes that provide incentives, have begun to lower the barriers to pediatric oncology drug development.

Role of National Cancer Institute's Cooperative Groups in Drug Development

The formation of the Children's Cancer Group (CCG) in 1955 and the Cancer and Leukemia Group B (CALGB) in 1956 were critical steps contributing to the dramatic improvements in outcome for children with cancer.[3,16] The first studies from the CALGB were aimed at determining the efficacy of combination chemotherapy, the impact of sequence of drug administration, and the role of chemotherapy in maintaining leukemic remissions.[17–20] Investigators noted that by collaborating with several institutions, the time to complete a study with large enough numbers to produce statistically meaningful results was drastically shortened.[19] With nationwide collaboration, results were also more efficiently disseminated, leading to a rapid and dramatic improvement in survival.[21] Additional benefits of the collaborative group paradigm are that a unified data base was created to study the history of incremental changes in therapy leading to significant outcome differences without the confounding effects of changing institutional participants.[22] Studies have also shown that children treated on standardized protocols have improved outcomes, providing evidence that current patients in addition to future patients benefit from cooperative group research.[23,24] With the established success of the pediatric oncology cooperative group, the Pediatric Oncology Group was formed from the pediatric divisions of the Southwest Oncology Group and the CALGB in 1980.[22] By the 1990s CCG and POG were the largest pediatric oncology cooperative groups in the world. In 2000, the Children's Oncology Group (COG) was formed from the merger of four pediatric cancer cooperative groups–CCG, POG, the National Wilms Tumor Study Group, and the International Rhabdomyosarcoma Study Group.[25] It is estimated that more than 90% of children with cancer in the United States are seen at a COG institution, allowing access to state-of-the-art treatment protocols for the vast majority of the country's pediatric oncology patients.

The formation of the Cancer Therapy Evaluation Program (CTEP) has also played a significant role in the impact of the cooperative group paradigm on drug development in the United States. As part of the National Cancer Institute's (NCI) Division of Cancer Treatment and Diagnosis, CTEP oversees all aspects of extramural clinical cancer trials and promotes collaborations among scientists, experts, and clinicians in both the academic research community and the pharmaceutical/biotechnology industry. In addition, CTEP funds an extensive national program of cancer research and by sponsoring clinical trials to evaluate new anticancer agents, with a particular emphasis on translational research of molecular targets and mechanisms of drug effects.[26]

Impact of Legislative Changes

Federal interest in influencing the development of pediatric therapeutics began in 1979 with regulation calling for a Pediatric Use subsection in the Product Label of approved drugs. In 1983, the Orphan Drug Act was passed establishing the precedent of federal financial incentives to support product development in areas of public health need. In 1994, the FDA issued the Pediatric Rule, permitting the use of efficacy data from adult studies to be extrapolated to a pediatric population, if the disease under study exists in both the pediatric and adult populations. Unfortunately, these programs did not increase the proportion of new products labeled for pediatric use but their failure did lead to the formation of both voluntary and mandatory programs aimed at improving the development and product labeling of therapeutics for pediatric use.

Today, the primary voluntary incentive program in the United States is the Best Pharmaceuticals for Children Act (BPCA) and the primary mandatory program is the Pediatric Research Equity Act (PREA) (Table 24.1). These issues are more fully discussed in chapters by Maldonado (Chapter 12) and Rose (Chapter 13) in this book. Approved in 2002, the BPCA was based on the 1997 Food and Drug Modernization Act (FDAMA), which contained a provision to provide financial incentive for including pediatric data for certain product categories, primarily small molecule drugs. The BPCA has additional provisions

TABLE 24.1 Comparison of the Pediatric Research Equity Act and the Best Pharmaceuticals for Children Act

Pediatric Research Equity Act (former 1998 Pediatric Rule)	Best Pharmaceuticals for Children Act (former 1997 FDA Modernization Act)
Mandatory program	Incentive program
Applies to all drugs and biologics	Includes Orphan Designation
Excludes Orphan Designation	
Applies only to drug product and indication under review	Applies to all products with same active moiety
Only applies if an approved or pending application occurs in adults and children	Eligible indications for study must occur in pediatric populations
Only applies if there is a therapeutic advantage or widespread use	Only applies when there is underlying patent or exclusivity protection
May be used as often as public health need arises	Limited use in a product lifetime

and programs including a government funded program to develop new information and product labeling for off-patent drugs, a report to Congress on access to investigational agents for pediatric oncology, the establishment of an FDA pediatric oncology advisory subcommittee, and a directive to the National Cancer Institute to establish a pediatric preclinical testing program. The PREA, based on the 1998 Pediatric Rule, allows the FDA to require studies in relevant populations, including children, when a therapeutic advance occurs or widespread use is anticipated. Both the BPCA and the PREA have shown to be successful, resulting in label changes in about 175 products, and have stimulated hundreds of clinical studies in the pediatric population.

PEDIATRIC ONCOLOGY DRUG DEVELOPMENT TIMELINE

The historic lack of attention to pediatric oncology drug development has often led to protracted development timelines for new cancer agents for children. For example, at one end of the timeline spectrum is the development of irinotecan (CPT-11), a water-soluble drug analog of camptothecin that undergoes deesterification to form the more potent topoisomerase I inhibitor, SN-38. Irinotecan was first synthesized in the early 1980s and found to have *in vivo* activity against a broad spectrum of murine tumor cell lines.[27] The first Phase I study of irinotecan in adults, conducted in Japan, was published in 1990 and Phase I trials began soon thereafter in Europe and the United States.[28] By 1995, there were already more than 10 published Phase I studies of irinotecan in solid tumors in adults.[29,30] Based on the results from three multicenter open-label, Phase II single-agent studies involving a total of 304 patients in 59 centers, irinotecan was granted accelerated approval in 1996 for the treatment of patients with metastatic carcinoma of the colon or rectum whose disease had progressed or recurred after treatment with 5-fluorouracil (5FU).[9] Full approval for this indication was granted in 1998 and by April 2000, irinotecan was given approval for use as a first-line treatment.

The potential for camptothecins did not escape the attention of pediatric investigators. Irinotecan was found to have significant activity in a range of pediatric tumor models, most notably rhabdomyosarcoma and neuroblastoma.[31,32] However, pediatric Phase I trials were only begun in 1996, the same year as irinotecan was granted accelerated approval in adults.[33] The first pediatric Phase II results were reported in 2002, with an additional two trials reported in 2007.[34–36] Thus, for irinotecan, pediatric drug development began at the tail end of the adult development process and resulted in more than a decade lag in development.

Even prior to the legislative initiatives of the past decade, the infrastructure for pediatric oncology drug development has existed, as evidenced most recently by the purine analog, nelarabine. Nelarabine was initially developed in 1983 as an antiviral therapy but was found to be preferentially cytotoxic to human malignant T cells compared to human malignant B cells.[37] Thus, clinical evaluation in acute leukemia, with a focus on the less common patients with T-cell ALL, was initiated. The first Phase I trial of nelarabine in patients with leukemia and lymphoma began in 1994 and included pediatric patients at each dose level.[38] A remarkable overall response rate of 31% was observed, with the most significant responses occurring in patients with T-cell malignancies. Separate Phase II studies of nelarabine in pediatric and adult patients began in 1997 and 1998, respectively. The pediatric Phase II study reported a 55% response rate in patients in first relapse and a 27% response rate in patients in the second or greater relapse.[39] The adult Phase II study reported a complete remission rate of 31% and an overall response rate of 41%.[40] Based on these two Phase II studies, nelarabine was granted accelerated approval in October 2005 for the treatment of both pediatric and adult patients with T-cell acute lymphoblastic leukemia and T-cell lymphoblastic lymphoma whose disease has not responded to or has relapsed following treatment with at least two chemotherapy regimens.[9]

The case of nelarabine is an example of how important it is to have influential supporters of pediatric drug development within the pharmaceutical industry. In this case, (1) the passion of Gertrude Elion, a Nobel prize winner who revolutionized the process of drug development and synthesized nelarabine and 6-MP, (2) a department head of early clinical development who was a pediatric oncologist, and (3) team members with clinical experience in pediatric oncology—all led to the development of a drug for a pediatric indication. Interestingly, the simultaneous development in adults and children led to the recommendation of two different efficacious dose schedules for adolescents. This is not problematic for physicians and patients, but can provide challenges to regulatory agencies and manufacturers who must agree on product information details.

EARLY PHASE TRIALS: SPECIAL CONSIDERATIONS FOR CHILDREN

The first step in clinical evaluation of a novel anticancer agent is a Phase I study. The primary aims of a Phase I study are to determine a drug's toxicity and pharmacokinetic parameters, including absorption, distribution, metabolism, and elimination (ADME), with the goal of defining a safe dose and schedule for continued clinical investigation in Phase II trials.[41] In other areas of medicine, the participants of a Phase I study are typically healthy subjects who are offered compensation for their participation in a Phase I study.[42] By contrast, participants in Phase I oncology trials are most often patients with cancer who have relapsed or have refractory disease with neither curative therapeutic options nor an accepted standard of therapy. Standard eligibility criteria for pediatric Phase I studies typically include: age less than 22 years, histologic proof of malignancy from archival tissue, measurable or evaluable disease, failure of conventional treatment or absence of conventional therapeutic options, life expectancy of at least 8 weeks, adequate organ function, as well as a performance status of Karnofsky $\geq$50 or the equivalent.[43] A second feature of Phase I oncology trials is the secondary aim of initial observation of antitumor activity, as responses observed on a Phase I trial can further guide the development of the new agent by helping to define potential disease targets to pursue in Phase II.[42,44] There is thus implicit therapeutic intent when administering a novel agent to patients enrolled in oncology Phase I trials.[42]

The historic paradigm for cytotoxic chemotherapy is that more is better, and thus Phase I oncology trials have attempted to define the maximum tolerated dose (MTD) in patients with refractory cancer. The starting dose for Phase I trials in adult oncology patients is most often based on animal toxicology and often begins at a dose equivalent to 10% of the dose found to be lethal to 10% of mice.[45] One dose level is typically studied in 3–6 subjects and toxicity is graded by the NCI Common Toxicity Criteria. If no dose limiting toxicities (DLTs) are observed, subjects are then entered on a higher dosage level. The MTD is defined as the highest dose at which no more than 1 in 6 patients experience a DLT. As the initial dose is quite low in adult Phase I trials, often more than 10 dose levels are tested before a MTD is identified. As responses may not be apparent in a Phase I study, the presence of toxicity is indication that if a response is not achieved, at least a biologically active dose was given.[46] With an increasing number of molecularly targeted agents in development, an important focus of current studies is on defining biomarkers to determine appropriate doses, doses that are not necessarily the MTD.

Most anticancer drugs are considered to carry a greater than minimal risk to participants in a clinical trial. Thus, the additional protections afforded children require that there must be potential for direct benefit for participating in such trials.[14] Furthermore, the observation that for cytotoxic agents the MTD in children is often higher than that in adults,[47,48] coupled with a limited number of pediatric patients who are candidates for Phase I trials, historically led to starting pediatric Phase I trials at approximately 80% the adult MTD.

Pediatric Phase I studies have been performed safely for more than three decades. In a review of studies conducted by St. Jude Children's Research Center or the Pediatric Oncology Group since 1967, a total of 577 children with leukemia or solid tumors were enrolled in 27 Phase I trials. The mortality due to direct toxicity of the investigational agent was 2.4% and the overall response rate was 5.9%. Progressive disease was responsible for the majority of deaths of patients enrolled in these trials (78%).[49] A more recent report examined pediatric Phase I clinical trials of antineoplastic agents to identify trends in response and toxicity over time. A total of 1606 children with cancer were enrolled in 56 single-agent Phase I trials between 1978 and 1996. The overall response rate was 7.9%. Death occurred in 7% of all enrolled patients, and only 0.7% of patients experienced a death related to drug toxicity.[50] In a subsequent review of Phase I pediatric oncology trials published between 1990 and 2004, the treatment-related mortality was less than 0.5%.[51] This supports the fact that well-designed Phase I clinical trials do not expose children to unacceptable risk of toxicity and offer hope for a therapeutic effect.

Once the MTD is identified, an agent can be studied in a Phase II trial. The primary objective of Phase II trials is to define the spectrum of antitumor activity for a new agent administered at the optimal dose and schedule determined in the Phase I trial.[52] Phase II trials are restricted to patients with specific histologic types of cancer, which are based on activity of the drug in preclinical cancer models, the mechanism of action of the drug, and activity observed in Phase I trials. For conventional cytotoxic drugs, the endpoint in a Phase II trial is response, which is measured as the percent decrease in size of the tumor compared to the pretreatment tumor size. Therefore, patients enrolled in conventional Phase II trials must have measurable tumor that is refractory to standard therapy.

To improve efficiency, pediatric Phase II trials often define a number of different strata based on tumor histology and conduct multiple disease efficacy evaluations in the context of a single trial. Furthermore, pediatric Phase II trials usually follow a two-stage design, which allows for early closure of a disease stratum if no responses are observed in the initial cohort of patients. Trials with a two-stage design usually require suspension of enrollment after accrual to the first stage is complete until response data have been collected. For cytotoxic agents, response can be evaluated over a relatively short time (weeks to months).

PEDIATRIC PRECLINICAL TRIALS PROGRAM

The establishment of the Pediatric Preclinical Trials Program (PPTP) marked a significant advancement in addressing the challenges inherent to the development of new therapies for children with cancer.[53] Because of the low incidence of childhood cancers as well as the effective first-line therapies for childhood malignancies, only a very small proportion of children are eligible to evaluate experimental therapies currently in the developmental pipeline. Over the past 50 years, systematic screening programs using both in vitro and in vivo models have been a successful tool to help identify drugs with the highest probability of clinical benefit.[54–56] Retrospective analysis has shown that when differences in drug exposure between tested animals and humans are taken into account, childhood cancer xenograft models can identify clinically active agents and effective drug combinations.[57]

Based on the validity of systematic preclinical investigation using xenograft models as well as the successes of cooperative group paradigm in pediatric clinical drug evaluation, the PPTP, a multi-institutional program for preclinical testing of new agents already in clinical development for their applicability to childhood cancers, was funded by the NCI in 2004.[58] The PPTP has the ability to test approximately 12 new agents annually in over 30 xenograft models of childhood cancers (Figure 24.2). In the first stage of testing, novel agents are tested at their MTD in the entire xenograft panel of the PPTP. Those agents that demonstrate significant activity at their MTD will be further evaluated in the sensitive tumor models in order to establish a dose–response relationship. Later stages of testing in the PPTP include performing pharmacokinetic and pharmacodynamic studies, evaluating the novel agent in combination with conventional anticancer agents, and exploring the agent in other relevant tumor models. Using these strategies, the PPTP will prioritize agents for further clinical investigation in children with the primary goal of improving outcome for all children with cancer. Results from the initial agents evaluated, including molecularly targeted agents, by the PPTP have now been published.[59–61]

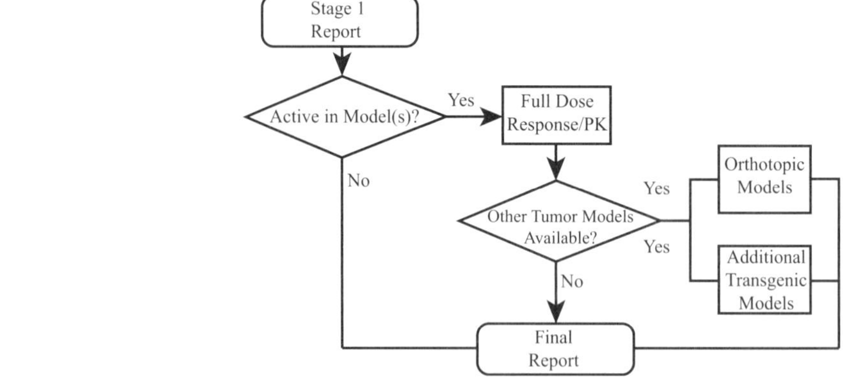

FIGURE 24.2 Paradigm for the preclinical testing of cancer drugs in a panel of human xenograft models. (Adapted from Houghton et al.[53])

INDUSTRY'S ROLE IN PEDIATRIC CANCER DRUG DEVELOPMENT

Despite the infrastructure that has developed around cooperative group clinical studies, and the fact that some academic institutions have made drugs that are being tested for their effect on childhood cancers,[62] the majority of new agents for the treatment of children with cancer emerge from the pharmaceutical industry. Early in the history of oncology drug development, a number of factors contributed to a dearth of new agents being introduced into clinical studies of children with cancer: (1) drug screening was conducted in cell-based systems, almost all of which did not include pediatric tumor types; (2) a concern among drug development departments that the approval of drugs for adults might be hampered if a child experienced a toxic effect not seen in adults; (3) there were few pediatric oncologists working in the pharmaceutical industry who could serve as champions for pediatric drug development; and (4) the current government programs and incentives that promote pediatric drug development were not in place.

The commonality of many neoplastic processes in nature may turn out to be a fortunate occurrence for children with cancer. Most new antineoplastic drugs now are developed by high-throughput screening of chemical libraries against specific molecular targets, as opposed to low-throughput screening for activity against panels of cell lines predominantly derived from adult malignancies. While the complete set of genetic aberrations that exist in childhood cancers are different from adult cancers, many of the individual growth pathways and genes that contribute to the pathogenesis of childhood cancers appear to be the same as those that contribute to malignancy in adults. Drugs that affect targets such as IGF1, AKT, cMET, p53, and BCR-ABL may be effective in pediatric tumor cells that are addicted to the protein product of the aberrant gene, just as are tumor cells from adult malignancies. Therefore, as the industry delves deeper into the pool of possible targets, the higher the likelihood that drugs will be useful for children as well as adults.

In this regard, the PPTP serves an important role by setting priorities for both the industry and the pediatric oncology cooperative groups. Given the limited number of pediatric oncology patients, negative PPTP screens are as important as positive screens. Especially for recently approved drugs, companies want to take advantage of the additional patent life offered as an incentive for pediatric studies. Using data from the PPTP, the cooperative groups can set priorities and the companies can respond appropriately to health authorities. The more common ground that exists, coupled with government programs like the PPTP, incentives, and the increased presence of pediatric oncologists in the pharmaceutical industry, the more progress that will be made toward earlier access to and the development of new drugs for children with cancer. The result is likely to be that more of the drugs discovered to affect these targets will be relevant to childhood cancers.

Still, impediments remain. Basic scientists working in discovery departments of drug companies may be unaware of resources such as the PPTP, even though a number of pharmaceutical industry representatives participated in the development of the model material transfer agreement that simplifies the process by which a company gains access to an excellent and wide array of preclinical models of childhood cancers. Such models are not always readily available in pharmaceutical companies, and the ability to obtain test results from multiple institutions under one agreement simplifies the process. However, the public nature of the PPTP report can impede early access to the models: the publication of the report must be timed to coincide with the public disclosure of the company's program so that patent rights are protected.

During the 1990s, and continuing to the present day, pediatric oncologists are taking positions in many large and small pharmaceutical companies. In 2007, the discovery and early development department of one of the largest pharmaceutical companies in the world is led by a pediatric oncologist; in many other companies, pediatric oncologists now have the tenure, stature, and numbers to affect decisions about the direction of drug development. The presence of knowledgeable pediatricians in drug companies, as well as data that demonstrate that MTDs are generally higher in children than adults, can calm the fear of those who still worry about the early introduction of a drug into children.

The impact on the pharmaceutical industry of legislation in the United States and in the European Union that provides incentives for pediatric drug development cannot be overstated. Within pharmaceutical companies, consideration of a pediatric development plan is now a routine part of the development process. European legislation, which mandates that a pediatric plan be presented to the EMEA, as well as the patent extension incentives offered in the United States and EU for the completion of pediatric studies, has changed how companies approach pediatric oncology. In the past, there were few people promoting development for pediatric indications. With the current incentives and requirements comes the support of regulatory affairs and commercial staff in addition to the voices of pediatric oncologists.

The relative rarity of cancer in childhood is a fortunate circumstance for children, but stands as an impediment to drug development. The current government programs, the presence of pediatric oncologists in companies, and the altruistic tendencies of some companies combine to advance the cause of children with cancer. However, without the government incentives, pediatric oncologists and others within companies who support pediatric oncology studies would have a hard time competing for company resources against studies of adult malignancies that offer the potential for substantial sales and return on investment.

FUTURE DIRECTIONS

The exponential growth in our understanding of the biology of pediatric malignancies is providing an extraordinary opportunity to improve the care for children with cancer. Constant attention must be paid, however, to increasing the profile of pediatric oncology drug development within the pharmaceutical and biotechnology industries, with a goal of initiating pediatric Phase I trials of select new agents upon completion of adult Phase I investigation. Continuing to forge collaborations among academic centers, government agencies, and industry will be critical to successful pediatric drug development efforts, including the development of a formal public–private partnership to spearhead development of new agents that may be uniquely targeted to pediatric malignancies.[63] The success of the latter half of the last century will soon fade if we do not identify and develop targeted new agents that improve cure rates and diminish the acute and long-term toxicities of today's therapy.

REFERENCES

1. Farber S, Diamond LK, Mercer RD, Sylvester RF, Wolff JA. Temporary remissions in acute leukemia in children produced by folic acid antagonist, 4-aminopteroyl-glutamic acid (aminopterin). *N Engl J Med.* 1948;238(23):787–793.

2. SEER Cancer Statisitics Review 1975–2003. National Cancer Institute. 2006. Available at http://seer.cancer.gov/csr/1975_2003/.
3. Bleyer WA. The US pediatric cancer clinical trials programmes: international implications and the way forward. *Eur J Cancer*. Aug 1997;33(9):1439–1447.
4. Ross JA, Severson RK, Pollock BH, Robison LL. Childhood cancer in the United States. A geographical analysis of cases from the Pediatric Cooperative Clinical Trials groups. *Cancer*. Jan 1996;77:1:201–207.
5. Hirschfeld S, Shapiro A, Dagher R, Pazdur R. Pediatric oncology: regulatory initiatives. *Oncologist*. 2000;5(6):441–444.
6. Adamson PC, Blaney SM. New approaches to drug development in pediatric oncology. *Cancer J*. Jul–Aug 2005;11(4):324–330.
7. Jonsson OG, Kamen BA. Methotrexate and childhood leukemia. *Cancer Invest*. 1991;9(1):53–60.
8. Burchenal JH, Karnofsky DA, Kingsley-Pillers EM, et al. The effects of the folic acid antagonists and 2,6-diaminopurine on neoplastic disease, with special reference to acute leukemia. *Cancer*. May 1951;4(3):549–569.
9. United States Food and Drug Administration. Listing of approved oncology drugs with approved indications [Web page]. Available at http://www.fda.gov/cder/cancer/druglistframe.htm. (Accessed May 24, 2007.).
10. Hirschfeld S, Ho PT, Smith M, Pazdur R. Regulatory approvals of pediatric oncology drugs: previous experience and new initiatives. *J Clin Oncol*. Mar 2003;21(6):1066–1073.
11. Hirschfeld S. Pediatric patients and drug safety. *J Pediatr Hematol Oncol*. Mar 2005;27 (3):122–124.
12. Elion GB. Historical background of 6-mercaptopurine. *Toxicol Ind Health*. Sep 1986;2 (2):1–9.
13. Burchenal JH, Murphy ML, Ellison RR, et al. Clinical evaluation of a new antimetabolite, 6-mercaptopurine, in the treatment of leukemia and allied diseases. *Blood*. Nov 1953;8 (11):965–999.
14. Code of Federal Regulations Title 45 Part 46 [Web page]. Available at http://www.hhs.gov/ohrp/humansubjects/guidance/45cfr46.htm. (Accessed September 22, 2007.).
15. 2006 Survey: Medicines in development for cancer. Available at www.phrma.org. (Accessed May 26, 2007.).
16. Zubrod CG. Chapter 7, In: Lazlo JM (ed.). *The Cure of Childhood Leukemia: Into the Age of Miracles*. New Brunswick, NJ: Rutgers University Press; 1996: 93–107.
17. Frei E III, Karon M, Levin RH, et al. The effectiveness of combinations of antileukemic agents in inducing and maintaining remission in children with acute leukemia. *Blood*. Nov 1965;26 (5):642–656.
18. Holland JF, Glidewell O. Chemotherapy of acute lymphocytic leukemia of childhood. *Cancer*. Dec 1972;30(6):1480–1487.
19. Frei E III, Freireich EJ, Gehan E, et al. Studies of sequential and combination antimetabolite therapy in acute leukemia: 6-mercaptopurine and methotrexate. *Blood*. 1961;18 (4):431–454.
20. Freireich E, Gehan E, Frei E III, et al. The effect of 6-mercaptopurine on the duration of steroid-induced remissions in acute leukemia: a model for evaluation of other potentially useful therapy. *Blood*. 1963;21(6):699–716.
21. Kersey JH. Fifty years of studies of the biology and therapy of childhood leukemia. *Blood*. 1997;90(11):4243–4251.
22. Pediatric Oncology Group. Progress against childhood cancer: the Pediatric Oncology Group experience. *Pediatrics*. Apr 1992;89(4 Pt 1):597–600.

23. Meadows AT, Kramer S, Hopson R, Lustbader E, Jarrett P, Evans AE. Survival in childhood acute lymphocytic leukemia: effect of protocol and place of treatment. *Cancer Invest.* 1983;1 (1):49–55.
24. Lennox EL, Stiller CA, Morris Jones PH, Kinner Wilson LM. Nephroblastoma: treatment during 1970–3 and the effect on survival of inclusion in the first MRC trial. *Br Med J.* 1979;2:567–569.
25. Ross JA, Olshan AF. Pediatric cancer in the United States: the Children's Oncology Group Epidemiology Research Program. *Cancer Epidemiol Biomarkers Prev.* 2004;13(10):1552–1554.
26. The Mission of the Cancer Therapy Evaluation Program (CTEP). Available at http://ctep.cancer.gov/. (Accessed September 22, 2007.).
27. Kunimoto T, Nitta K, Tanaka T, et al. Antitumor activity of 7-ethyl-10-[4-(1-piperidino)-1-piperidino]carbonyloxycamptothecin, a novel water-soluble derivative of camptothecin, against murine tumors. *Cancer Res.* 1987;47(22):5944–5947.
28. Taguchi T, Wakui A, Hasegawa K, et al. [Phase I clinical study of CPT-11. Research group of CPT-11.] *Gan To Kagaku Ryoho.* Jan 1990;17(1):115–120.
29. O'Reilly S, Rowinsky EK. The clinical status of irinotecan (CPT-11), a novel water soluble camptothecin analogue: 1996. *Crit Rev Oncol Hematol.* Sep 1996;24(1):47–70.
30. Abigerges D, Chabot G, Armand J, Herait P, Gouyette A, Gandia D. Phase I and pharmacologic studies of the camptothecin analog irinotecan administered every 3 weeks in cancer patients. *J Clin Oncol.* 1995;13(1):210–221.
31. Thompson J, Zamboni W, Cheshire P, et al. Efficacy of systemic administration of irinotecan against neuroblastoma xenografts. *Clin Cancer Res.* 1997;3(3):423–431.
32. Houghton PJ, Cheshire PJ, Hallman JC, Bissery MC, Mathieu-Boue A, Houghton JA. Therapeutic efficacy of the topoisomerase I inhibitor 7-ethyl-10-(4-[1-piperidino]-1-piperidino)-carbonyloxy-camptothecin against human tumor xenografts: lack of cross-resistance in vivo in tumors with acquired resistance to the topoisomerase I inhibitor 9-dimethylaminomethyl-10-hydroxycamptothecin. *Cancer Res.* 1993;53(12):2823–2829.
33. Furman WL, Stewart CF, Poquette CA, et al. Direct translation of a protracted irinotecan schedule from a xenograft model to a Phase I trial in children. *J Clin Oncol.* 1999;17(6):1815–1824.
34. Turner CD, Gururangan S, Eastwood J, et al. Phase II study of irinotecan (CPT-11) in children with high-risk malignant brain tumors: the Duke experience. *Neuro-Oncology.* 2002;4:102–108.
35. Pappo AS, Lyden E, Breitfeld P, et al. Two consecutive Phase II window trials of irinotecan alone or in combination with vincristine for the treatment of metastatic rhabdomyosarcoma: the Children's Oncology Group. *J Clin Oncol.* Feb 2007;25(4):362–369.
36. Vassal G, Couanet D, Stockdale E, et al. Phase II trial of irinotecan in children with relapsed or refractory rhabdomyosarcoma: a joint study of the French Society of Pediatric Oncology and the United Kingdom Children's Cancer Study Group. *J Clin Oncol.* Feb 2007;25 (4):356–361.
37. Cohen A, Lee J, Gelfand E. Selective toxicity of deoxyguanosine and arabinosyl guanine for T-leukemic cells. *Blood.* 1983;61(4):660–666.
38. Kurtzberg J, Ernst TJ, Keating MJ, et al. Phase I study of 506U78 administered on a consecutive 5-day schedule in children and adults with refractory hematologic malignancies. *J Clin Oncol.* May 2005;23(15):3396–3403.
39. Berg SL, Blaney SM, Devidas M, et al. Phase II study of nelarabine (compound 506U78) in children and young adults with refractory T-cell malignancies: a report from the Children's Oncology Group. *J Clin Oncol.* May 2005;23(15):3376–3382.
40. DeAngelo DJ, Yu D, Johnson JL, et al. Nelarabine induces complete remissions in adults with relapsed or refractory T-lineage acute lymphoblastic leukemia or lymphoblastic lymphoma: Cancer and Leukemia Group B study 19801. *Blood.* Jun 2007;109(12):5136–5142.

41. Pratt CB. The conduct of Phase I-II clinical trials in children with cancer. *Med Pediatr Oncol.* 1991;19(4):304–309.
42. Roberts TG Jr, Goulart BH, Squitieri L, et al. Trends in the risks and benefits to patients with cancer participating in Phase 1 clinical trials. *JAMA.* Nov 2004;292(17):2130–2140.
43. Bernstein ML, Reaman GH, Hirschfeld S. Developmental therapeutics in childhood cancer. A perspective from the Children's Oncology Group and the US Food and Drug Administration. *Hematol Oncol Clin North Am.* Aug 2001;15(4):631–655.
44. Sekine I, Yamamoto N, Kunitoh H, et al. Relationship between objective responses in Phase I trials and potential efficacy of non-specific cytotoxic investigational new drugs. *Ann Oncol.* Aug 2002;13(8):1300–1306.
45. Smith M, Bernstein M, Bleyer W, et al. Conduct of Phase I trials in children with cancer. *J Clin Oncol.* 1998;16(3):966–978.
46. Marsoni S, Wittes R. Clinical development of anticancer agents—a National Cancer Institute perspective. *Cancer Treat Rep.* Jan 1984;68(1):77–85.
47. Marsoni S, Ungerleider RS, Hurson SB, Simon RM, Hammershaimb LD. Tolerance to antineoplastic agents in children and adults. *Cancer Treat Rep.* Nov 1985;69(11):1263–1269.
48. Carlson L, Ho P, Smith M, Reisch J, Weitman S. Pediatric Phase I drug tolerance: a review and comparison of recent adult and pediatric Phase I trials. *J Pediatr Hematol Oncol.* 1996;18 (3):250–256.
49. Furman W, Pratt C, Rivera GK, Krischer JP, Kamen BA, Vietti TJ. Mortality in pediatric Phase I clinical trials. *J Natl Cancer Inst.* 1989;81:1193–1194.
50. Shah S, Weitman S, Langevin AM, Bernstein M, Furman W, Pratt C. Phase I therapy trials in children with cancer. *J Pediatr Hematol Oncol.* Sep–Oct 1998;20(5):431–438.
51. Lee DP, Skolnik JM, Adamson PC. Pediatric Phase I trials in oncology: an analysis of study conduct efficiency. *J Clin Oncol.* Nov 2005;23(33):8431–8441.
52. Fox E, Adamson PC. Phase I trial design: considerations in the pediatric population. In: Budman D, Calvert H, Rowinski E, eds. *Handbook of Anti-Cancer Drug Development*: Baltimore, MD: Lippincott/Williams & Wilkins; 2003:297–308.
53. Houghton PJ, Adamson PC, Blaney S, et al. Testing of new agents in childhood cancer preclinical models: meeting summary. *Clin Cancer Res.* Dec 2002;8(12):3646–3657.
54. Henderson ES. Treatment of acute leukemia. *Ann Intern Med.* Sep 1968;69(3):628–632.
55. Monga M, Sausville EA. Developmental therapeutics program at the NCI: molecular target and drug discovery process. *Leukemia.* Apr 2002;16(4):520–526.
56. Kung AL. Practices and pitfalls of mouse cancer models in drug discovery. *Adv Cancer Res.* 2007;96:191–212.
57. Voskoglou-Nomikos T, Pater JL, Seymour L. Clinical predictive value of the in vitro cell line, human xenograft, and mouse allograft preclinical cancer models. *Clin Cancer Res.* Sep 2003;9 (11):4227–4239.
58. Houghton PJ, Morton CL, Tucker C, et al. The pediatric preclinical testing program: description of models and early testing results. *Pediatr Blood Cancer.* Dec 2007;49(7):928–940.
59. Maris JM, Courtright J, Houghton PJ, et al. Initial testing of the VEGFR inhibitor AZD2171 by the pediatric preclinical testing program. *Pediatr Blood Cancer.* Mar 2008;50(3):581–587.
60. Houghton PJ, Morton CL, Kolb EA, et al. Initial testing (stage 1) of the proteasome inhibitor bortezomib by the pediatric preclinical testing program. *Pediatr Blood Cancer.* Jan 2008;50 (1):37–45.
61. Houghton PJ, Morton CL, Kolb EA, et al. Initial testing (stage 1) of the mTOR inhibitor rapamycin by the pediatric preclinical testing program. *Pediatr Blood Cancer.* Apr 2008;50(4):799–805.

62. Kushner BH, Kramer K, Cheung N-KV. Phase II trial of the anti-GD2 monoclonal antibody 3F8 and granulocyte-macrophage colony-stimulating factor for neuroblastoma. *J Clin Oncol*. Nov 2001;19(22):4189–4194.

63. National Cancer Policy Board (US) Committee on Shortening the Time Line for New Cancer Treatments. Adamson P, Weiner S, Simone J, Gelband H, eds. *Making Better Drugs for Children with Cancer*. Washington DC: The National Academies Press; 2005.

PART VI

CLINICAL TRIAL OPERATIONS: UNDERSTANDING DIFFERENCES BETWEEN PEDIATRIC AND ADULT STUDY SUBJECTS—DEVELOPMENT ISSUES RELATED TO ORGAN DEVELOPMENT AND ENDPOINT CHOICES

CHAPTER 25

Brain and Central Nervous System Development: Physiological Considerations for Assessment of Long-Term Safety

EUGENE SCHNEIDER, MD

Amicus Therapeutics, 6 Cedar Brook Drive, Cranbury, New Jersey 08512

INTRODUCTION

A firm understanding of what makes the embryonic, fetal, and postnatal nervous system different from that of an adult, and how xenobiotics may alter neurodevelopment, is a prerequisite for adequate assessment of pharmaceutical safety. Human neurogenesis spans a protracted period of time, beginning with the formation of the neural tube during the first month of gestation, and progressing through a sequence of complex, well-organized processes of cellular proliferation, migration, differentiation, axono-/dendritogenesis, synaptogenesis, gliogenesis, and myelination. No less critical for the formation of a functional cytoarchitecture of a mature brain are important regressive phenomena of axonal retraction, synaptic elimination, and apoptosis. Orchestration of these complex and overlapping ontogenetic events is conducted via a great number of chemically diverse molecules—trophic factors, neuromodulators, and neurotransmitters—which, along with neuronal activity itself, are the powerful forces shaping the developing central nervous system (CNS). Although neurogenesis is not fully completed until well into the third decade of life, it is thought that the majority of critical developmental milestones occur between the first trimester of pregnancy and late adolescence, making this period critically important in terms of heightened susceptibility to adverse environmental influences. Therefore, when designing pediatric clinical trials of CNS-active medications, it is of the utmost importance to have a comprehensive, proactive safety assessment plan that takes into account the specific mechanism of action of the tested product, and employs multiple structural and functional endpoints that are relevant to the timing of exposure and its duration. Such a plan should also make use of informative animal models and in vitro data, particularly when assessment of delayed CNS toxicity is necessary.

Pediatric Drug Development: Concepts and Applications
Edited by Andrew E. Mulberg, Steven A. Silber, and John N. van den Anker

The concept of developmental susceptibility as a unique feature of immature CNS has only been at the forefront of neuroscience for the last few decades, despite having been first introduced by developmental biologists of the early 1920s.[1] A key premise that a xenobiotic is more likely to cause adverse effects if exposure to it coincides with the development of an underlying structure, rather than occurring before or after its genesis, has been substantiated by experimental evidence.[2,3] Follow-up studies of the effects of ionizing radiation from the atomic bombs exploded over Hiroshima and Nagasaki had revealed that the highest risk of mental retardation was in children exposed to radiation between 10 and 17 weeks gestation—the period of robust neuronal proliferation—while exposure prior to that period did not raise the risk for retardation, and a later exposure only produced mild cognitive impairment later in life.[2] In general, toxic insults during the protracted period of neurogenesis can result in a number of outcomes, ranging from the gross structural abnormalities that become apparent during early embryogenesis and are incompatible with fetal survival, to the ultrastructural or molecular abnormalities that are manifested as functional deficits much later in adult life. It therefore behooves researchers involved in the design of clinical trials in pediatric subjects to have a keen understanding of the timelines of normal neural development in humans as well as other species used in toxicological testing. This chapter provides a brief overview of human neurogenesis, focusing primarily on the brain, and discusses the implications of neurodevelopmental processes for conducting safety assessments in pediatric clinical trials.

NEURODEVELOPMENTAL PROCESSES

Development of the human nervous system commences with the induction of the *neural plate* and continues well into postnatal period. Cellular precursors of the future CNS begin to develop early in embryogenesis when, around the second week of gestation, the *notochord*—a cellular rod defining a primitive axis of the developing embryo—induces the overlying ectodermal tissue to form the neural plate. This plate then invaginates, forming the midline groove with two folds, which by the end of the third week of gestation fuse together in a zipper-like fashion forming the *neural tube*. The process of formation of the neural tube from primitive ectoderm, called *neurulation*, is of paramount importance since the entire human central nervous system subsequently develops from the walls of the neural tube. Although neurulation is completed in humans by approximately 26–28 days of gestation, the story of neurogenesis is only in its opening chapter. From this point on, specific areas of the brain begin to organize with neurogenesis and migration of cells forming three main regions: forebrain, midbrain, and hindbrain. This regional organization takes place through a sequence of major histogenetic events: cell proliferation, migration, differentiation, axono- and dendritogenesis, synaptogenesis, synaptic remodeling, apoptosis, and myelination. Each one of these processes is conducted to the tune of hundreds, perhaps even thousands, of positive and negative molecular and environmental influences, which modern scientists are only beginning to decipher. While generally occurring sequentially, these processes frequently overlap with one another and, depending on the region, progress with different cadence (hence an important concept of *spatiotemporal development*). These regional ontogenetic differences explain why certain brain structures in humans are fully formed at birth (brainstem and basal ganglia), while others do not mature until well into adolescence (prefrontal cortex). In general, regional development of the mammalian brain follows a clear spatial gradient, with more *caudal* (from the Latin for "tail") structures like the hindbrain developing earlier than more *rostral* (from the Latin for "beak") ones like the forebrain, and more *medial* (from the Latin for "close to the midline") aspects of these structures maturing earlier than more *lateral* (from the Latin for "toward the side") ones.* Evidence suggests that these ontogenetic gradients are established by diffusible morphogens, like fibroblast growth factor-8, sonic hedgehog, and Wnt, which are released from distinct locations.[4, 5]

Shortly after the neural tube formation, specific proliferation areas are formed in the ventricular and, later on, the subventricular zone—the two areas histologically known as the *germinal matrix*. This structure is the birthplace of all the neurons and the majority of the glial cells that will form the human brain. Generally, the majority of neurons are generated between weeks 4 and 25 of gestation (with notable exceptions of the granular neurons of the cerebellum and hippocampus, whose genesis is thought to occur postnatally). On the other hand, the bulk of neuroglia are produced between weeks 20 and 40 of gestation, with one important exception. It has long been recognized that radial glia appear to play a critical role in neuronal migration, which explains why the two cell types are generated concurrently in the ventricular zone from the same multipotent stem cells.

As neurons are generated, they begin migrating to their final destination. This occurs either via passive displacement of the earlier generated cells by the newly formed ones, or, more commonly, through an orchestrated, genetically programmed active process. The active migration is particularly critical for the formation of the human neocortex, where the sites of final neuronal destination can lie at substantial distances from the sites of cell genesis. Although the exact mechanisms of active neuronal migration remain poorly understood, it appears that cortical neurons reach their final destination through a highly organized process of migration in either radial (vast majority) or tangential (some interneurons) direction. Most radially migrating neurons move along the processes of radial glia, using them as a scaffold. This process is regulated by complex molecular interactions between the two cell types. As each wave of migrating cells travels past their predecessors, distinct cellular layers are formed in an inside-out manner (i.e., younger neurons are closer to the cortical surface than the older ones). Neuronal migration in the human cerebral cortex peaks between the 3rd and 5th months of gestation and appears to be completed at around 30 weeks. The first recognizable cortical layer is formed when, around week 7 of gestation, postmitotic cells migrate in a radial fashion from the germinal matrix, forming an important structure called the *preplate*. This transient structure will remain active for about 3–4 weeks, eventually differentiating into the outer layer called the *marginal zone* and the inner layer called the *subplate*. The latter will serve as the "waiting room" for the afferent fibers from the subcortical structures like the thalamus, basal ganglia, and the brainstem nuclei, which are destined to reach cortical neurons once those complete their migration and final differentiation.

Although neuronal differentiation on a molecular level begins soon after the cells have been born, a major portion of axonal and dendritic sprouting occurs after the cells have reached their target location. Akin to active migration, the guidance for axonal sprouting is provided by a multitude of molecular and cellular cues—cell surface molecules, elements of the extracellular matrix, and trophic factors produced by the target cells—acting in concert either as chemoattractants or chemorepellents. In contrast to the robustness of axonogenesis,

---

*One notable exception to this rule is the development of the caudally located *cerebellum* (Latin for "little brain"), which takes place after the more rostral regions like the midbrain, thalamus, and hypothalamus. Cerebellar maturation in humans, particularly of its cortex, takes place postnatally during the first year of life.

dendritic development of cortical neurons proceeds relatively slowly during the first two trimesters of gestation. However, it accelerates significantly from the third trimester onwards and remains extremely active during the first postnatal year, staying apace until about age 5. During this period of several years, the most striking development of dendritic arborization takes place, making it one of the most critical periods of increased vulnerability of the developing CNS to exogenous influences. Abnormalities of dendritic development have been linked to undernutrition and ingestion of neurotoxins, and described in conjunction with a number of pathologies, including Angelman and fragile X syndromes, autism, and Duchenne muscular dystrophy.[6] The strongest evidence for abnormal dendritogenesis exists in mental retardation, where disturbances in the number, length, and spatial arrangement of dendritic branching are consistently described.[7–10] Also, studies looking at the effects of ventilatory support in infants born before term had demonstrated abnormally thin dendrites with decreased number of dendritic spines, especially in the medulla, implicating oxygen toxicity as the potential etiology of disrupted dendritogenesis.[11]

In concert with axonal and dendritic growth, another critical neurogenetic event is taking place. After the first cortical synapses appear at around the 9th week of gestation, the rate of formation of new synapses is steadily increasing by about 5% a week until 24–26 weeks of gestation, with the more or less even distribution across all cortical regions.[12] Some time around gestational week 28, this previously steady rate suddenly explodes, leading to a six-fold increase in synaptic density by the time the peak has been reached, which occurs in some areas, like the primary sensory cortex, as early as 3 months after birth, and in others, like frontal cortex and hippocampus, much later.[13] In the frontal cortex, synapse formation does not reach its maximal density until after 15 postnatal months. In the middle frontal gyrus (MFG), for example, which is implicated in abstract thinking and reasoning, synaptic density reaches its maximum at 3.5 years.[13] It was previously believed that, following this peak, the rate of new synapse formation gradually decreases until adult levels are reached around the time of puberty. However, the more recent data in primates suggest that after reaching its peak, the rate of synaptogenesis plateaus until puberty, and then begins to decline.[14] Scientists believe that this occurs because the process of synaptic elimination becomes more dominant. This is in line with modern understanding of synaptogenesis as a dynamic process in which synaptic formation is constantly being counterbalanced by synaptic elimination. Both processes are thought to occur throughout our life span, resulting in continuous reorganization of our synaptic architecture.

Formation of specific neuronal populations that interconnect with one another in functional networks is often referred to as an activity-independent process because, in general, neuronal activity is not required for this process to take place. Instead, the genetically specified cues are responsible for guiding neuronal differentiation and the formation of basic cytoarchitecture.[15] The next crucial step in forming precise functional neural connections involves the refinement of the crude early pathways to form specific patterns of connectivity that characterize the mature brain. This process of "fine-tuning" is highly activity dependent, meaning that the electrochemical neuronal activity is required for strengthening or adding some neural connections while eliminating others. In fact, one of the leading theories of the neuropathology of schizophrenia is that it is a developmental disorder of neural disconnectivity. Experimental evidence implicates decreased activity of the subplate neurons leading to abnormalities in neural connectivity between the thalamus and the frontal cortex that ultimately manifest in the schizophrenic phenotype.[16]

In addition to the synaptic elimination, other regressive phenomena critical to the success of neurogenesis have been identified. Apoptosis, or programmed cell death, is one of them.

Apoptosis is a highly phylogenetically conserved process that is crucial for controlling the final number of neurons and glial cells in the CNS. In humans, two apoptotic waves can be distinguished. The first one takes place prior to synaptic formation, having been observed in the proliferative zones of the human telencephalon around the 7th gestational week.[17] The second apoptotic wave occurs much later in development and appears to be linked to synaptogenesis and network formation.[18] This process of programmed cell death is quite sensitive to chemical influences and, therefore, could easily be disturbed. Both ethanol and methylmercury have been shown to disrupt normal apoptotic processes in vitro.[19, 20]

CLASSICAL NEUROTRANSMITTERS

The human brain is as functionally elegant as it is complex. This is made possible through development via a series of overlapping cascading events, which are highly coordinated by a number of diverse chemicals, collectively known as neurotrophins and neuromodulators. The most well-studied trophic factors to date—nerve growth factor (NGF), brain-derived neurotrophic factor (BDNF), and neurotrophin-3 (NT-3)—act via second or third messengers to either alter gene expression or affect the cellular signaling cascades.[21, 22] The molecules that will later become neurotransmitters are produced by the developing embryo during the first weeks of its gestation, even before the formation of the nervous system. The transition from prenervous to the neuronal stages of ontogenesis is accompanied by changes in the physiologic role of neurotransmitters. Experimental data suggest that many classical neurotransmitters initially function as intracellular regulators, then become local hormones, and finally acquire the role of synaptic transmitters.[23] More than 40 substances are now known or strongly suspected to have neurotransmitter function. Although this group includes molecules of extremely diverse chemical nature, from peptides to nitric oxide and CO, this discussion will be limited to a few substances known as "classical neurotransmitters." These include acetylcholine, the biogenic monoamines serotonin, norepinephrine, and dopamine, as well as the amino acids glutamate and γ-aminobutyric acid (GABA). It is difficult to overestimate the importance of neurotransmitters and neuromodulatory substances for development of the human nervous system. These substances play a pivotal role in nearly every ontogenic process, from neuronal proliferation and migration (acetylcholine, glutamate), to the formation of functional networks of cells via dendritic arborization and synaptic pruning (glutamate, GABA, acetylcholine). When discussing the developmental trajectories of specific neurotransmitter systems, it is important to keep in mind that these systems extend beyond a single receptor–single ligand structure and often include several different receptor subtypes (some with important regional gradients, as well as temporal patterns of expression of their protein subunits), various transporter molecules, degradation enzymes, ion-gated channels, and many other molecular elements that coordinate the neurotransmitter activity. Moreover, since most of the available data on neurotransmitter ontogeny had examined the expression of specific receptors in experimental animals, our ability to extrapolate these data to the functional development of transmitter systems in humans is somewhat limited. With these caveats in mind, let's briefly examine what is known about the ontogeny of the classical neurotransmitter systems and their role in human neurogenesis.

Nicotinic acetylcholine receptors have been detected as early as week 6 of gestation in a number of CNS regions including the spinal cord, brainstem, cerebellum, and mesencephalon.[24] Studies looking at the expression of acetylcholinesterase (AChE) in the nucleus basalis complex—a major source of cholinergic innervation of the neocortex—show reactivity for the enzyme from gestational week 11 onwards.[25] Between weeks 20 and 24, the AChE-reactive fibers reach the subplate on their way to the frontal, temporal, parietal, and occipital cortices.[26] Normal cholinergic innervation appears to be critical for cortical development, plasticity, and sexual differentiation.[27] In addition, activation of nicotinic ACh receptors in the hippocampal neurons has been shown to promote synaptogenesis during postnatal development.[28] Disturbances in cholinergic neurotransmission have been implicated in the development of Down syndrome, lead and ethanol toxicity, and perinatal asphyxia.[29]

Catecholaminergic systems emerge early in neurogenesis. Noradrenergic cells can be found in the medulla, pons, and locus caeruleus, and the dopaminergic cells can be found in the brainstem, midbrain, and hypothalamus, beginning at 6 weeks' gestation.[30,31] At around week 13, many catecholaminergic fibers reach the subplate, and beginning at week 15, these projections enter the developing cortex.[31,32] Between weeks 22 and 26 of gestation, at the time when cortical lamination occurs, noradrenergic and dopaminergic innervation of the cortex begins to acquire adult-like regional patterns, with noradrenergic fibers concentrated in the primary motor and sensory cortical regions, with dopaminergic innervation extending more globally.[33] Noradrenaline and dopamine are both essential for normal brain development. Regulation of the development of one of the most important early neuronal cell types, the Cajal–Retzius cells, by noradrenergic input appears to be a crucial initial step in cortical migration and laminar formation.[34] Since dopaminergic transmission is paramount for higher cognitive functioning including abstract reasoning, problem solving, and executive planning, disturbances of this axis have long been postulated to be associated with attention deficit hyperactivity disorder (ADHD). Indirect evidence in support of this relationship comes from the data on low-level exposure to methylmercury during development affecting the dopaminergic transmission (D_2 receptors, specifically), which results in long-lasting alterations in brain cytoarchitecture and function, manifesting in learning deficits, language difficulties, and attentional problems.[35]

Serotonergic cells, which appear between the 5th and 12th weeks of human gestation, send their axons to the forebrain, reaching all cortical layers by the time of birth. Serotonin (5-HT) appears to play a critical role in coordination of complex sensory–motor patterns during different behavioral states. Although its concentration varies widely during neurogenesis, serotonin is an important influence during the critical period of synaptogenesis and the formation of regional brain networks. Rodent and human data indicate that transient expression of the high affinity serotonin transporter (SERT) resulting in increased uptake and storage of 5-HT in the developing thalamic neurons occurs during the critical period of formation of the cortical somatosensory maps.[36,37] A connection between abnormal levels of serotonin during neurogenesis and the development of autism has been proposed.[38] It is known that serotonin concentrations in young children are much higher than in adults, and normally decline during the first few years of life. This decline seems to be absent in autistic children in whom the supernormal levels of 5-HT persist.[38] Additionally, a point mutation of the gene encoding for the enzyme monoamine oxidase, which is responsible for inactivating several monoamine neurotransmitters, including 5-HT, has been found to be strongly associated with antisocial behavior—one of the hallmarks of autism.[37]

Amino acid neurotransmitters are among the most abundant in the human CNS. They play a crucial role in the wiring of networks and establishment of the proper cytoarchitecture. Glutamate is the predominant excitatory neurotransmitter in the human brain. Its receptors are found in the fetal cortex as early as 10 weeks' gestation. There are two periods of overexpression of glutamate receptors. One occurs between weeks 13 and 21 of gestation, when neuronal migration is at its peak, and the other around the time of birth, which may be related to the high rate of synaptogenesis and synaptic refinement taking place.[39] Also of importance are the ontogenic differences in glutamate receptor types, with AMPA and kainate receptors being dominant during the fetal life, while NMDA becomes more active postnatally. Evidence suggests that early NMDA receptors are qualitatively different: they express the "fetal" isoform of its subunit, NR2B, rather than the "adult" NR2A, which makes the entire receptor channel more permeable to Ca^{2+} inflow and effectively increases receptor sensitivity.[40] Given the critical role of NMDA receptors in learning and memory storage, it is thought that this "primes" the newborn's brain for the novel sensory input. Unfortunately, having hypersensitive NMDA receptors creates a liability if perinatal complications such as hypoxia-ischemia or brain injury should occur, since excessive influx on intracellular Ca^{2+} results in excitotoxicity and neuronal death. Exposure to NMDA antagonists during neuronal development is highly detrimental, as is illustrated by the unequivocal damage produced by ethanol to the fetal brain.[41]

GABA is the predominant inhibitory neurotransmitter in the local cortical circuits. Early in development GABA functions as an excitatory transmitter and becomes inhibitory after birth due to a switch from high to low chloride content of neurons.[42] This developmental switch, along with the presence of GABA-ergic receptors in the specific subplate population of neurons not found in adults, strongly suggests the role of GABA in regulation of neurogenesis. GABA is an essential neurotransmitter for the developing brain. Erroneous exclusion of vitamin B_6—a cofactor in GABA synthesis—from infant formula in the 1960s resulted in several deaths from intractable seizures caused by GABA deficiency.[43]

When reviewing human neurogenesis, it is important not to focus exclusively on the neuron, and examine other cells and processes that are of critical importance for the development of a fully functional mature brain. Chief among them are gliogenesis (i.e., generation of glial cells that support proper functioning of neurons), myelination, and angiogenesis (i.e., growth of the blood vessels that supply neurons and glia with nutrients). Only a cursory description of these processes can be included in this chapter.

Two basic types of glial cells exist: *microglia* and *macroglia*. Microglia are thought to be derived from the hematopoietic system and enter the developing CNS during the early stages of neurogenesis. These cells essentially function as brain macrophages, cleaning up the debris that is left following cell death. Several types of macroglia can be distinguished, but the main ones are *oligodendrocytes* and *astrocytes*. Oligodendrocytes are the cells that provide axonal myelination in the CNS. Astrocytes are responsible for regulating the synaptic environment and maintaining appropriate levels of neurotransmitters and growth factors.[44,45] In addition, they are involved in neuronal signaling and regulation of synaptic and nonsynaptic transmission.[46,47] Oligodendrocytes and astrocytes are generated from the same precursor cells as neurons; however, they are born well after the initial waves of neurogenesis, and their differentiation typically lags behind that of neurons in a given structure. As mentioned earlier, radial glia—members of the macroglia family—are generated concurrently with neurons and provide a general scaffold for neuronal migration. Production of myelin—a fatty sheath that surrounds axons and increases the conduction speed and efficiency—typically begins after the oligodendrocytes have matured and reached their final destination, which in humans occurs during the last trimester of pregnancy and continues well past adolescence and young adulthood.[48,49] Myelination is a process that is exquisitely sensitive to exogenous disruption, which can occur from malnutrition, hormonal dysfunction, or exposure to xenobiotics like ethanol and lead.[50,51]

BRAIN PLASTICITY

A discussion of human neurogenesis and the vulnerability of nascent brain structures to external insults would be incomplete without touching upon a phenomenon of brain plasticity. Research conducted during the last decade has greatly enhanced our understanding of brain plasticity, with functional neuroimaging becoming an invaluable tool. Today, it is clear that our brains retain the capacity of modifying neuronal circuits well into advanced age, both in response to cognitive and motor experience and in response to mechanical or toxic injury. Plasticity may produce both positive and negative effects during the critical developmental periods, and these effects are usually long-lasting or even permanent. An example of such is the cross-modal plasticity leading to cortical reorganization in blind and deaf children. Several imaging experiments have demonstrated that speech processing and auditory localizations activate the visual cortex in congenitally blind people.[52,53] Also, cerebral organization for language seems to differ in deaf and hearing subjects, with sign language activating separate regions in deaf and hearing signers.[54] Lack of exposure to language during the critical period of brain development leading to permanent inability to fully acquire language later in life is a good example of evolutionary plasticity resulting in devastating, irreparable consequences.[55] Many types of plasticity have been identified thus far, and a thorough discussion of these is well beyond the scope of this chapter. Fortunately, several excellent reviews have recently been published, and the reader would be well served to refer to these for further information.[56,57]

IMPLICATIONS FOR ASSESSMENT OF SAFETY

At this point, hopefully, you can appreciate the enormous complexity of human neurogenesis—the exceedingly high number of different cell types and an intricate organization of synaptic connections that must be made and pruned during critical periods of development. It is clear that inherent in this complexity are both an opportunity and a risk. An opportunity stems from an intrinsic plasticity, with which thousands of years of evolution have endowed the developing human brain, allowing it to weather the many challenges encountered during its protracted maturation. Yet, the developing brain remains highly vulnerable to the influences from its environment and, in particular, to xenobiotic substances that may disrupt some of the critical ontogenetic processes, for instance, by interacting with trophic and neuromodulator molecules. The results of such influences might be subtle changes in cytoarchitecture and molecular interactions manifesting as behavioral dysfunction, rather than gross malformations. Therein lies one of the greatest challenges the field of developmental neurotoxicology is facing today: an ability to reliably predict and detect cognitive and behavioral disturbances caused by exposure to chemicals or drugs during the protracted period of neurogenesis. With respect to pediatric neuropharmacology, this dilemma translates into finding the appropriate exposure times and levels for the CNS-active drugs that would minimize or, ideally, eliminate both short- and long-term structural and functional consequences of such exposure. This objective should be at the heart of any prospective risk assessment program.

REFERENCES

1. Child CM. *The Origin and Development of the Nervous System*. Chicago, IL: University of Chicago Press; 1921.
2. Otake M, Schull WJ. In utero exposure to A-bomb radiation and mental retardation. *Br J Radiol*. 1984;57:409–414.
3. Olney JW. New insights and new issues in developmental neurotoxicology. *Neurotoxicology*. 2002;23:659–668.
4. Shimogori T, Banuchi V, Ng HY, et al. Embryonic signaling centers expressing BMP, WNT and FGF proteins interact to pattern the cerebral cortex. *Development*. 2004;131:5639–5647.
5. Grove EA, Cole S, Simon J, et al. The hem of the embryonic cerebral cortex is defined by the expression of multiple Wnt genes and is compromised in Gli3-deficient mice. *Development*. 1998;125:2315–2325.
6. Volpe J. *Neurology of the Newborn*. Philadelphia: Saunders; 1995: 43–94.
7. Huttenlocher P. Synaptic and dendritic development and mental defect. *UCLA Forum Med Sci*. 1975;18:123–140.
8. Huttenlocher P. *Brain Mechanisms in Mental Retardation*. New York: Academic Press; 1979.
9. Purpura D. Dendritic differentiation in human cerebral cortex: normal and aberrant developmental patterns. *Adv Neurol*. 1975;12:91–134.
10. Purpura D. Normal and abnormal development of cerebral cortex in man. *Neurosci Res Program Bull*. 1982;20:569–577.
11. Tasashima S, Mito T. Neuronal development in the medullary reticular formation in sudden infant death syndrome and premature infants. *Neuropediatrics*. 1985;16:76.
12. Zecevic N. Synaptogenesis in layer I of the human cerebral cortex in the first half of gestation. *Cereb Cortex*. 1998;8:245–252.
13. Huttenlocher PR, Dabholkar AS. Regional differences in synaptogenesis in human cerebral cortex. *J Comp Neurol*. 1997;387:167–178.
14. Bourgeois JP, Rakic P. Changes of synaptic density in the primary visual cortex of the macaque monkey from foetal to adult stage. *J Neurosci*. 1993;13:2801–2820.
15. Goodman CS, Shatz CJ. Developmental mechanisms that generate precise patterns of neuronal connectivity. *Cell*. 1993;72 (Suppl):77–98.
16. Bunney WE, Bunney BG. Evidence for a compromised dorsolateral prefrontal cortical parallel circuit in schizophrenia. *Brain Res Rev*. 2000;31:138–146.
17. Rakic S, Zecevic N. Programmed cell death in the developing human telencephalon. *Eur J Neurosci*. 2000;12:2721–2734.
18. Lossi L, Merighi A. In vivo cellular and molecular mechanisms of neuronal apoptosis in the mammalian CNS. *Prog Neurobiol*. 2003;69:287–312.
19. Castoldi AF, Barni S, Randine G, et al. Ethanol selectively interferes with the trophic action of NMDA and carbachol on cultured cerebellar granule neurons undergoing apoptosis. *Dev Brain Res*. 1998;111:279–289.
20. Bulleit RF, Cui H. Methylmercury antagonizes the survival-promoting activity of insulin-like growth factor on developing cerebellar granule neurons. *Toxicol Appl Pharmacol*. 1998; 153:161–168.
21. Klein R. Role of neurotrophins in mouse neuronal development. *FASEB J*. 1994;8:738–744.
22. Zhou J, Bradford HF. Nerve growth factors and the control of neurotransmitter phenotype selection in the mammalian central nervous system. *Prog Neurobiol*. 1997;53:27–43.
23. Buznikov GA, Shmukler YB, Lauder JM, et al. Changes in the physiological roles of neurotransmitters during individual development. *Neurosci Behav Physiol*. 1999;29:11–21.

24. Hellstrom-Lindahl E, Gorbounova D, Seiger A, et al. Regional distribution of nicotinic receptors during prenatal development of human brain and spinal cord. *Brain Res Dev Brain Res.* 1998;108:147–160.
25. Kostovic I. Prenatal development of nucleus basalis complex and related fiber systems in man—a histochemical study. *Neuroscience.* 1986;17:1047–1077.
26. Kostovic I, Judas M. Correlation between the sequential ingrowth of afferents and transient patterns of cortical lamination in preterm infants. *Anat Rec.* 2002;267:1–6.
27. Hohmann CF. A morphogenetic role for acetylcholine in mouse cerebral neocortex. *Neurosci Biobehav Rev.* 2003;27:351–363.
28. Maggi L, Le Magueresse X, Changeux JP, et al. Nicotine activates immature "silent" connections in the developing hippocampus. *Proc Natl Acad Sci.* 2003;18:2059–2064.
29. Herlenius E, Lagercrantz H. Development of neurotransmitter systems during critical periods. *Exp Neurol.* 2004;190:S8–21.
30. Sundstrom E, Kölare S, Souverbie F, et al. Neurochemical differentiation of human bulbospinal monoaminergic neurons during the first trimester. *Brain Res Dev Brain Res.* 1993;75:1–12.
31. Zecevic N, Verney C. Development of the catecholamine neurons in human embryos and fetuses, with special emphasis on the innervation of the cerebral cortex. *J Comp Neurol.* 1995;351:509–535.
32. Verney C, LeBrand C, Gaspar P, et al. Changing distribution of monoaminergic markers in the developing human cerebral cortex with special emphasis on the serotonin transporter. *Anat Rec.* 2002;267:87–93.
33. Verney C, Milosevic A, Alvarez C, et al. Immunocytochemical evidence of well-developed dopaminergic and noradrenergic innervations in the frontal cerebral cortex of human fetuses at midgestation. *J Comp Neurol.* 1993;336:331–344.
34. Naqui SZ, Harris BS, Thomaidou D, et al. The noradrenergic system influences in fate of Cajal–Retzius cells in the developing cerebral cortex. *Dev Brain Res.* 1999;113:75–82.
35. Dare E, Fetissov S, Hokfelt T, et al. Effects of prenatal exposure to methylmercury on dopamine-mediated locomotor activity and dopamine D2 receptor binding. *Naunyn Schmiedebergs Arch Pharmacol.* 2003;367:500–508.
36. Lebrand C, Cases O, Adelbrecht C, et al. Transient uptake and storage of serotonin in developing thalamic neurons. *Neuron.* 1996;17:823–835.
37. Gaspar P, Cases O, Maroteaux L, et al. The developmental role of serotonin: news from mouse molecular genetics. *Nat Rev Neurosci.* 2003;4:1002–1012.
38. Chugani DC. Role of altered brain serotonin mechanisms in autism. *Mol Psychiatry.* 2002;7 (Suppl 2):S16–S17.
39. Ritter LM, Unis AS, Meador-Woodruff JH, et al. Ontogeny of ionotropic glutamate receptor expression in human foetal brain. *Dev Brain Res.* 2001;127:123–133.
40. Tang YP, Shimizu E, Dube GR, et al. Genetic enhancement of learning and memory in mice. *Nature.* 1999;401:63–69.
41. Cohen G, Han ZY, Grailhe R, et al. Beta 2 nicotinic acetylcholine receptor subunit modulates protective responses to stress: a receptor basis for sleep-disordered breathing after nicotine exposure. *Proc Natl Acad Sci.* 2002;99:13272–13277.
42. Herlenius E, Lagercrantz H. Neurotransmitters and neuromodulators during early human development. *Early Hum Dev.* 2001;65:21–37.
43. Frimpter GW, Andelman RJ, George WF, et al. Vitamin B_6-dependency syndromes. New horizons in nutrition. *Am J Clin Nutr.* 1969;22:794–805.
44. Vernadakis A. Glia–neuron intercommunications and synaptic plasticity. *Prog Neurobiol.* 1996;49:185–214.

45. Chvatal A, Sykova E. Glial influence on neuronal signaling. *Prog Brain Res.* 2000;125:199–216.
46. Araque A, Carmignoto G, Haydon PG, et al. Dynamic signaling between astrocytes and neurons. *Annu Rev Physiol.* 2001;63:795–813.
47. Ullian EM, Sapperstein SK, Christopherson KS, et al. Control of synapse number by glia. *Science.* 2001;291:657–661.
48. Paus T, Gijdenbos A, Worsley K, et al. Structural maturation of neural pathways in children and adolescents: in vivo study. *Science.* 1999;283:1908–1911.
49. Sowell ER, Trauner DA, Garnst A, et al. Development of cortical and subcortical brain structures in childhood and adolescence: a structural MRI study. *Dev Med Child Neurol.* 2002;44:4–16.
50. Zoeller RT, Butnariu OV, Fletcher DL, et al. Limited postnatal ethanol exposure permanently alters the expression of mRNAs encoding myelin basic protein in myelin-associated glycoprotein in cerebellum. *Alcohol Clin Exp Res.* 1994;18:909–916.
51. Rothenberg SJ, Poblano A, Garza Morales S, et al. Prenatal and perinatal low level lead exposure alters brainstem auditory evoked responses in infants. *Neurotoxicology.* 1994;15:695–699.
52. Weeks R, Horwitz B, Aziz-Sultan A, et al. A positron emission tomographic study of auditory localization in the congenitally blind. *J Neurosci.* 2000;20:2664–2672.
53. Roder B, Stock O, Bien S, et al. Speech processing activates visual cortex in congenitally blind humans. *Eur J Neurosci.* 2002;16:930–936.
54. Bavelier D, Brozinsky C, Tomann A, et al. Tmpact of early deafness and early exposure to sign language on the cerebral organization for motion processing. *J Neurosci.* 2001;21:8931–8942.
55. Robinson K. Implications of developmental plasticity for the language acquisition of deaf children with cochlear implants. *Int J Pediatr Otorhinolaryngol.* 1998;46(1):71–80.
56. Trojan S, Pokorny J. Theoretical aspects of neuroplasticity. *Physiol Res.* 1999;48:87–97.
57. Johansson BB. Brain plasticity in health and disease. *Keio J Med.* 2004;53(4):231–246.

CHAPTER 26

Cognitive Development Considerations for Long-Term Safety Exposures in Children

MARY PIPAN, MD

Children's Hospital of Philadelphia, Division of Child Rehabilitation and Developmental Medicine, Philadelphia, Pennsylvania, 19104

INTRODUCTION

Children are in a constant state of growth and development. As reflected in Chapter 24 by Schneider on central nervous system (CNS) development, brain development occurs in tandem. We know that brain development can be susceptible to outside influences, which can either improve or harm current and future development for the individual child at risk. In utero, there are clearly certain medications and illicit substances that can impair brain development. However, the influence of medications on brain development after delivery has been less well defined. Of particular concern to healthcare providers and families is the effect that medication might have on growth and development. When parents ask about long-term side effects, they are asking if this medication will adversely affect their child's potential to function optimally throughout their school age into adulthood. Conversely, when children have a disorder that adversely affects growth and development, parents may seek medication that could enhance future growth and development. Parents of children with disabilities can be particularly vulnerable to these claims and may not have access to the resources that can help them make a reasoned decision. In either case, healthcare professionals, government regulators, and the pharmaceutical industry have a responsibility to assure that medications do not adversely affect cognition, in terms of both immediate cognitive processes and future cognitive development. Their responsibility also extends to guarding against medication that falsely claims to enhance that future.

Any medication that affects the brain and behavior needs to be studied for cognitive effects. These include (1) medications designed to directly ameliorate neurobehavioral abnormalities, such as medications for seizures, attention deficit disorders, autism, schizophrenia, and bipolar disorders; (2) medications developed to influence behavior that can be

Pediatric Drug Development: Concepts and Applications
Edited by Andrew E. Mulberg, Steven A. Silber, and John N. van den Anker

detrimental to optimal behavioral functioning or to learning, such as those to treat disruptive behaviors, anxiety, or depression; (3) medications designed to enhance neurobehavioral development, such as the cholinesterase inhibitors; and (4) medications that treat medical conditions but have neurocognitive effects such as thyroid medication, chemotherapy, or antihistamines.

The physiology of the brain of a child changes over time. Thus, medications that are used across the life span of childhood need to be studied in children within different age groups. Children also need to be studied as a separate group from adults. When medications have adverse cognitive effects in the adult studies, there is good reason to use extra caution in a child. However, medications deemed "safe" in adults cannot be assumed to have the same safety profile in children. Children are not adults in small packages. Since the child's brain development changes over time and continues into young adulthood, the effect of medication on cognition needs to be studied over years or even decades. Although the scope of such assessments seems daunting, the goal of establishing medication safety in children is paramount. The establishment of set protocols for following patients over time for cognitive effects of neuroactive agents, and the provision of financial incentive for doing so, could make such assessment a routine part of the U.S. Food and Drug Administration's (FDA's) medication approval process.

This chapter summarizes the issues involved in determining the effects of medication on cognition and development in children. It discusses what is known about the cognitive effects of the various classes of psychotropic medications used in pediatrics, considering adult studies if there are no data in children. The complexities involved when considering cognitive effects in different neurobehavioral diagnoses are also addressed. Lastly, it explores the complications involved in the assessment of change of cognition in children and briefly describes the tools that have been used to assess change.

STIMULANT MEDICATION: SHORT- AND LONG-TERM EXPOSURES AND EFFECTS ON COGNITIVE FUNCTION

The short-term cognitive effects of immediate release methylphenidate have been studied extensively in children with attention deficit disorder (ADD) and attention deficit hyperactivity disorder (ADHD). Pietrzak et al.[1] provided an excellent summary of the scientific literature up to 2006. They summarize 40 studies that investigated the effects of immediate-release methylphenidate on cognitive function in children with ADHD. Inclusion requirements were that studies utilized a placebo control, used only methylphenidate, included only children that met formal criteria for ADD or ADHD, used cognitive or neuropsychological tasks for assessment, had a sample size of at least 10, and excluded studies where all participants had a comorbid disorder. The review found that 63.5% of the studies found improvement in at least one area of functioning. These areas included eye movement control, planning/cognitive flexibility, attention/vigilance, and inhibitory control. Less consistent were improvements in working memory and divided attention, areas that are also commonly impaired in children with ADHD.

Mollica et al.[2] studied the short-term cognitive and behavioral effects of dexedrine at two different doses, using a quick-to-administer battery of seven tests (the Cog State) assessing psychomotor function, visual attention, executive function, and memory in 14 children with ADHD compared to 14 age- and gender-matched controls, with equivalent IQ ranges. The children with ADHD did significantly better than controls on the higher dose of dexedrine, both behaviorally and cognitively. The trial was over a 3-day period examining the effect of a single dose of dexedrine on day 1 and 3.

Pearson et al.[3] studied the short-term cognitive effects of methylphenidate on children with mental retardation and ADHD. They studied 24 children with mild to moderate mental retardation (mean IQ 56.5) in a within-subject, crossover, placebo-controlled design and varying doses of methylphenidate. Cognitive tests included tests of sustained attention, selective attention, impulsivity/inhibition, and immediate memory and were performed at the end of each week of the 4 week drug trial. Improvements were seen in performance of sustained attention, selective attention, and inhibition/impulsivity, more so at higher doses.[3] Previous studies showed mixed results, with smaller sample sizes.[4–7] However, in a reevaluation of their data, Pearson et al.[8] also noted cognitive decline in a substantial percentage of children (35% on the low dose and 23% on the high dose).

ADHD is increasingly being diagnosed in preschool age children. However, only 48% of symptomatic children retain the diagnoses of ADHD into school age years.[9] Despite the often transient nature of ADHD symptoms, use of stimulant medication in this age group has greatly increased. Connor[10] reviewed the studies on stimulant medication in preschoolers in 2002. Nine double-blind placebo-controlled studies were included in his review. Only one study utilized laboratory psychological tests of cognition.[11] All studies lasted from 3 to 9 weeks, with only one longer at 20 weeks. There are no long-term data on preschool age children treated with stimulant medication in the literature.[12]

The long-term cognitive effects of stimulants have been far less studied. Gimpel et al.[13] studied 31 children with ADHD, with baseline (before-medication) WISC-III and ADHD behavioral scales, and performed the same assessment at least 1 year later comparing the group who were taking stimulant medication at that time ($n = 24$) to those not taking medication ($n = 7$). They found significant albeit mild improvements in the verbal and performance scale IQ scores in subjects taking medication. For children not taking medication, there were no significant differences in scores across the year.[13]

Limitations of the studies of the cognitive effects of the stimulants in children with ADHD include the variability of performance measures utilized, making comparisons among studies difficult. Measures that are given repeatedly may show practice effects,[14,15] and thus a comparable control group is needed in order to assess whether change across time can truly be attributed to the medication itself. Children also improve on task measures as they develop, so improvements of repeated measures over a long period may simply be developmental improvements.[16] The instruments are often selected to assess functions known to be improved by increased dopamine and norepinephrine, namely, vigilance, inhibitory control, and working memory.[17] Whether improvements in these functions relate to functional improvements in school performance or other areas of complex thinking has not been shown. The response to medication seems to vary with task complexity. The more complex tasks did not show as great an improvement with stimulant therapy as more simple tasks.[18]

The relationship between improvement on cognitive tasks and behavior also needs to be better defined. Parents and teachers are often looking for behavioral improvement, not improvement on cognitive tasks. In studies that looked at behavioral outcome and cognitive outcome, the behavioral improvement as rated by parents and teachers did not consistently correlate with the improvement in cognitive tasks, with cognitive tasks showing more improvement than behavioral ratings.[19,20]

Another limitation of most studies was the lack of discussion regarding adverse cognitive effects. Pearson's study[21] in children with mental retardation (MR) showed that even though there was overall statistical improvement in cognitive performance, one-quarter of subjects with MR had adverse cognitive effects. Similar reviews of the cognitive data on typical children are also needed.

Long-term cognitive effects of the stimulants have not been adequately studied, and need to be, especially in younger children. These studies should include comparing the cognitive and academic performance of children with ADHD every 1–2 years, based on medication treatment status. It should also examine whether certain children have adverse cognitive outcomes and examine if there is a link to stimulant treatment. Such information would be a valuable adjunct to counseling families on the benefits and potential costs of stimulants in young children.

ANTIPSYCHOTIC MEDICATION

Atypical neuroleptic medications are increasingly prescribed to children and adolescents for a variety of diagnoses including schizophrenia, bipolar disorder, and autism and a variety of symptoms including aggression, depression, delirium, and psychosis.[22,23] First generation antipsychotic medications clearly carry high risk of significant side effects in children and adolescents, including sedation and extrapyramidal symptoms,[24,25] which have by and large precluded their use since the introduction of the atypical antipsychotic medications. Most trials of atypical antipsychotics in children and adolescents in the literature are limited to open-label studies, case reports, retrospective case series, and chart reviews.[26] Very few consider the cognitive effects of these medications.

The atypical antipsychotics are the most widely used class of psychotropic medication used to treat aggression and other maladaptive behaviors in children and adolescents.[27–29] Risperidone is the only atypical antipsychotic studied in a double-blind placebo-controlled randomized trial in children with autism spectrum disorders.[30,31] Multiple open-label studies have also examined the effects of atypical antipsychotics on aggression and disruptive behavior in the autistic spectrum disorders and most found positive results.[32] None looked at cognitive effects.

In children with conduct disorder and disruptive behavior disorders, one randomized double-blind, placebo-controlled study has been done and showed efficacy of risperidone in patients ages 5–15 years, over 10 weeks.[33] Another open-label safety study of aripiprazole also showed efficacy.[34] Cognitive effects were not examined.

In children with cognitive disability, short-term efficacy of risperidone has been shown to be effective for aggression and disruptive behavior in four randomized, double-blind, placebo-controlled studies. Medication was only used for 4–6 weeks, and risperidone was found to be superior to placebo in all studies.[35–38] Cognitive effects were not examined.

Antipsychotic medication is being increasingly used in the preschool population. Biederman et al.[39] conducted an open-label study of risperidone and olanzepine in preschool age children with bipolar disorder, and Masi et al.[40] examined the open-label use of risperidone in preschool children with pervasive developmental disorders (PDDs). Luby et al.[41] did so in a randomized placebo-controlled study of risperidone in preschool children with pervasive developmental disorder–not otherwise specified (PDD-NOS). Behavioral and physiologic measures were utilized to assess effects, but none of these studies used any measures to assess cognitive effects.

Very few of the studies of atypical antipsychotic medication in children measure cognitive effects of the medication over time. Given the heterogeneity of subjects, even within a diagnostic or symptomatic category, assessment of cognitive effects involves taking into account baseline cognitive deficits and carefully selected control groups. In adults, atypical antipsychotic medications have shown mixed results when examining the cognitive effects of atypical antipsychotics. Short-term studies of cognitive effects of atypical antipsychotics have shown improvements in visual memory, delayed recall, and executive function.[42–45] However, Hori et al.[46] examined adult patients taking nonstandard high doses of medications and showed poorer performance than those on standard medication on tasks of visual memory, delayed recall, performance IQ, and executive function. There is very little written about the long-term cognitive effects of these medications. Croonenberghs et al.[47] evaluated the long-term cognitive effects of risperidone in children with intellectual impairment after 1 year. This study was an open-label study, 504 children with IQs ranging from 36 to 84, ages 5–14 years, excluding PDD-NOS, conducted across 32 sites in 12 countries. The California Verbal Learning Test and the Continuous Performance Test were performed at baseline and at 6 and 12 months. There were no control groups. Children showed slight improvement in the tests done.[47] Turgay et al.[48] did a similar open-label study of 77 children at 9 sites, using the same cognitive measures, and found statistically significant improvement in both tasks, and overall no cognitive deterioration. There was no control group. Kravariti et al.,[49] in a cross-sectional sample of adolescent onset schizophrenic patients, showed that the longer the patient was exposed to antipsychotic medication, the poorer the performance was on Attention/Concentration, Within-Search Errors, Between Search Errors, Trails A, and Motor Speed/Reaction Time.

The Hori et al.[46] and Kravariti et al.[49] studies demonstrate the importance of using valid cognitive testing as part of medication trials, at different doses of medication, and following these cognitive tests over time. If the findings of these two studies are valid, then studying the cognitive effects of these medications in children is imperative. Children may effectively be exposed to larger doses of medication because of their body mass and different neurological substrate. Also, children on antipsychotic medication are likely to remain on the medication for a long time, sometimes years. To date, no study has adequately and specifically examined the cognitive effects of any atypical antipsychotic using valid neuropsychological instruments assessing a range of abilities over the short or long term in children. Because of children's development over time and the natural improvement in testing with repeated measures, control groups will also need to be part of these studies. The heterogeneity of diagnoses and symptoms for which these medications are used will also make careful subject selection criteria and control group selection very important. Subjects would need to be matched based on the specific neurodevelopmental disorder as well as the degree of cognitive ability. Data bases of children with certain diagnoses should be established, so that specific populations can be studied and matched for cognitive and academic performance. In addition, they would need to be assessed annually and followed for medication and dosing status. Children would need to be evaluated every 1–2 years, with testing to include behavioral questionnaires and observations as well as cognitive and academic performance measures.

ANTIEPILEPTIC MEDICATION

Assessment of the cognitive effects of antiepileptic medication in children has particular challenges. Approximately one-third of children with epilepsy have neurocognitive impairment and an even higher percentage have behavioral impairments.[50] Certain epilepsy syndromes are associated with cognitive arrest or decline, but even "benign" epilepsy may be associated with cognitive problems.[51] Thus, treatment of the seizure disorder may result in improved cognition because the seizures are reduced, not as a direct effect of the medication itself. Approximately 20% of children with mental retardation have a seizure disorder.[52] Children with mental retardation may be more susceptible to side effects of medication, including cognitive effects.[53] Antiepileptic medication can also cause behavioral side effects (e.g., sedation), which may have an effect on the assessment of cognition and on the child's ability to effectively learn. A few studies and case reports show that these medications may affect certain individuals quite differently, causing more cognitive and behavioral effects than would be expected in the general population.[54,55]

The goal in treating children with antiepileptic medication is to reduce the frequency of seizures. Therefore, most studies focus on seizure reduction, rather than on cognitive effects, and adverse cognitive effects may be accepted if the agent used is effective at reducing seizure frequency. However, study of cognitive and behavioral effects and side effects of these medications has become more important as the number of medications from which to choose continues to grow. When there is a choice of medication, it is imperative to choose that drug that may have the potential to do the least harm.

Phenobarbital has long been associated with behavioral changes thought to affect cognitive development.[56] Vining et al.[57] compared valproic acid to phenobarbital in 21 school age children with mild seizure disorder and average intelligence in a double-blind, crossover trial with each child receiving 6 months of each medication. On four tests of neuropsychological function, children performed significantly less well while receiving phenobarbital. Four children dropped out of the study because of severe behavioral problems while receiving phenobarbital. Vining cautioned in his study that although there were some behavioral differences between the two groups, these were not so great in most cases that one would have suspected cognitive problems without the formal testing. Farwell et al.[58] followed 217 children with febrile seizures between 8 and 36 months who were given either phenobarbital or placebo over 2 years. In the phenobarbital group, the IQ score was 7 points lower than in the placebo group following the 2-year trial, and 5 points lower 6 months following the medication taper. Calandre et al.[59] compared phenobarbital and valproic acid in a open parallel study of 64 children on phenobarbital, 64 on valproic acid (VPA) and 60 controls and found lower performance IQ at baseline in children on phenobarbital, and an improvement in performance IQ from baseline in the control and the VPA groups, but not in the phenobarbital group.

Ideally, in order to assess whether medication affects children adversely, groups of normal children would be given medication and cognitive tests followed. However, this study design would be considered unethical, so one can also study developing animals in experimental conditions and postulate similar effects in the human brain. Bolanos et al.[60] studied the long-term effects of valproate and phenobarbital in young rats after kainic acid induced epilepsy. Rats received daily injections of phenobarbital, valproate, or saline. After tapering the drugs, the rats were tested in a water maze and handling test. In the control and phenobarbital groups, water maze performance was impaired. Valproate showed no impaired learning.

The cognitive effects of phenytoin have not been studied extensively in children. Forsythe et al.[61] randomly assigned 64 children with new cases of epilepsy to treatment with carbamazepine, phenytoin, and valproate and performed cognitive tests at baseline and every 3–4 months over the course of the next year. They found that carbamazepine adversely affected memory but not valproate and phenytoin. Two adult studies show negative cognitive effects in patients with epilepsy treated with phenytoin. Pulliainen and Jokelainen[62] examined the cognitive functions of 43 newly diagnosed epileptic patients and randomly assigned them to carbamazepine or phenytoin, and retested patients after 6 months. A control group was also tested 6 months apart. Both anticonvulsants decreased the practice effect seen in normal controls, more so in the phenytoin group. Subjects on phenytoin also became slower, and visual memory decreased. In a study by Aldenkamp et al.,[63] two groups of 25 patients were compared in an open, parallel group, nonrandomized investigation of the cognitive effects of carbamazepine and phenytoin monotherapy. Neuropsychological battery assessing speed factors, memory, and attention was administered. The phenytoin group showed lower performance on all tests measuring motor speed, and data also suggested slower speed of information processing.

Valproate and carbamazepine have been shown to have fewer cognitive side effects than either phenobarbital or phenytoin except in the Forsythe study,[64] where carbamazepine adversely affected memory. Stores et al.[65] studied 63 school-age children with seizure disorders randomly assigned either to valproate or carbamazepine, and compared these groups with matched controls. Lower performance on visual motor coordination and attention and mild behavioral differences were found at baseline in the children with seizure disorder. At 12 months, tests of intelligence and behavioral measures revealed no difference between patients and controls.

Seidel and Mitchell[66] examined the cognitive and behavioral effects of carbamazepine in ten children with benign Rolandic epilepsy, ages 6–12 years old, and performed the same neuropsychological assessment on 14 unmedicated controls with migraine headaches, but no epilepsy. Children with epilepsy were quicker on a visual search task and recalled stories better when off medicine, than when treated with carbamazepine. Practice effects were observed in the control group, but not in the carbamazepine group.

Valproic acid has been associated with mild impairment of mental and psychomotor speed. No evidence has linked the effects of valproic acid to dose or drug levels.[67]

Stefan and Feuerstein[68] in 2007 and Aldenkamp et al.[69] in 2003 reviewed the effects and side effects of newer anticonvulsant drugs. Gabapentin in three studies showed no impairment in cognitive functioning. Somnolence and fatigue were side effects that could affect cognition. No studies prospectively evaluating the effects of gabapentin on cognitive effects in children have been done.

Lamotrigine in 13 studies demonstrated either improved or no effect on cognition in 11 studies and two with negative effects (decreased cerebral efficiency and hyperactivity in MR population). Somnolence was the only side effect that could affect cognition. No studies examining lamotrigine and cognitive effects in children have been done.

Topiramate in six studies shows negative effects on cognition in all studies (language, attention, verbal memory, verbal fluency, and verbal intelligence). Speech and language effects, somnolence, psychomotor slowing, nervousness, concentration difficulty, memory difficulty, and confusion were adverse effects that could affect cognition. No studies looking at topiramate and cognitive effects in children have been done.

Tiagabine in three studies shows no cognitive impairment. Somnolence, nervousness, concentration difficulty, and confusion were side effects that could affect cognition. No studies looking at tiagabine and cognitive effects in children have been done.

Levetiracetam in one study shows no impairment in cognitive functions. No studies looking at levetiracetam and cognitive effects in children have been done.

Oxcarbazepine in five studies showed no cognitive impairment in four studies and improvement in cognition in one study. Tzitiridou et al.[70] studied the cognitive effects of oxcarbazepine in children. Seventy patients with benign childhood epilepsy with centrotemporal spikes (BECTS) were treated with oxcarbazepine in an open-label study. No children had been on an antiepileptic drug (AED) in the past or at the start of the study. Cognitive testing was administered at baseline to all subjects and 45 controls. There were no significant differences in cognitive measures at baseline except on the information subscale of the WISC III. Neuropsychological evaluation was repeated after 18 months of medication in the treatment group only, and slight improvements were seen in two subtests of the WISC-III.

Pulsifer et al.[71] examined the cognitive effects of the ketogenic diet at 1 year follow-up and found significant increases in the developmental quotients of subjects. However, the developmental quotient was calculated from parental report, and no controls were included in the study.

SELECTIVE SEROTONIN REUPTAKE INHIBITORS (SSRIs)

SSRIs are used to treat depression, anxiety, and obsessive–compulsive symptoms in children. Recent reviews of the literature by Courtney[72] and Whittington et al.[73] call into question the evidence to support the efficacy and safety profile of SSRIs except for fluoxetine when used in children and adolescents in the treatment of depression. Whittington et al.[73] accessed unpublished data, which showed higher risks for adverse events and suicidal ideation than had been reported in published studies. Cognitive effects of SSRIs have been examined in adults, but not children and adolescents. Wadsworth et al.[74] compared cognitive performance on a number of neuropsychological tests in adults on SSRIs and controls and found decreased recall and recognition memory, when subjects were matched to controls for levels of depressive symptoms. Mowla et al.[119] found that fluoxetine improved immediate and delayed logical memory scores on the Wechsler Memory Scale III and improved functioning on the Mini Mental Status Examination in individuals with mild cognitive impairment.

LITHIUM

Several studies have shown cognitive impairment in short-term memory, long-term memory, and psychomotor speed in bipolar adult patients taking lithium.[75] A study looking at normal subjects taking lithium for 3 weeks, by Stip et al.,[76] showed deficits in short-term memory tasks and long-term memory. Bipolar disorder itself is associated with neuropsychological abnormalities including deficits in executive function, memory, attention, speed of information processing, visual spatial perception, and psychomotor speed, confirmed in a study by Basso et al.,[77] comparing bipolar patients to controls.

BENZODIAZEPINES

Benzodiazepines clearly produce transient cognitive effects in adults following single doses. These include increased sedation, impairments in reaction times, impairment in psychomotor skills that involve focused attention and visual motor coordination, working memory impairments, and impaired learning of new information.[78–80] Patients were thought to develop tolerance to these effects with long-term use but Curran[80] found that memory impairments last throughout 2 months of treatment and persist following discontinuation up to several weeks after withdrawal.

The cognitive effects of long-term benzodiazepine use were investigated as part of a meta-analysis of the literature by Barker et al.[81] All 13 studies in the meta-analysis involved adult patients who had been on benzodiazepines for 1–34 years. They concluded that for all cognitive categories neuropsychological assessment showed declines in cognition, which were more pronounced as related to the longer benzodiazepines were used. In a follow-up meta-analysis to assess whether these cognitive declines persisted after medication withdrawal, they found that 3–6 months after medication discontinuation, there was some improvement in cognitive scores, but scores did not approach the level of the controls.[82] Limitations of the meta-analyses included the wide range of doses and duration of medication in the studies, the variability of diagnoses of subjects, the lack of control for coexisting drug and alcohol use, and the length of time from benzodiazepine dose to cognitive testing.

CHOLINESTERASE INHIBITORS

Cholinesterase inhibitors, galantamine, rivastigmine, donepezil, and tacrine have been studied with some success as cognitive enhancers in Alzheimer's disease.[83] Heller et al.[120] conducted a 22-week trial of donepezil in seven children with Down syndrome. They administered two broad language instruments at intervals between baseline and 22 weeks, and showed gains in language scores on one of the tests. No control group was utilized. Chez et al.[84] administered rivastigmine to children with autism in an open-label, double-blind study (no control group) showing improvements in the Expressive One-Word Picture Vocabulary Test given at 6 and 12 weeks over baseline. A trial with donepezil showed similar results.[85] Lee et al.[86] and Choinard et al.[87] studied cholinesterase inhibitors given as an add-on to antipsychotic medications in adults in an effort to improve the cognitive deficits of schizophrenia, and showed no significant cognitive improvement. Wilens et al.[88] used donepezil as an adjunctive medication to stimulants to improve executive function in a 12-week open-label trial of seven children and six adults with ADHD. Six subjects dropped out due to adverse events. There were no significant improvements on the executive function checklist in the seven who completed the trial.

OTHER MEDICATIONS

Memantine is an *N*-methyl D-aspartate (NMDA) antagonist, used to slow decline in Alzheimer's disease.[89] Owley et al.[90] studied its effects in 14 children, ages 3–12 years, with pervasive developmental disorders in an 8-week open-label trial. No control group was utilized and measured outcomes were differences from baseline. Significant improvement from baseline was seen in the memory test, but no differences were seen on measures of language or nonverbal IQ.

L-Carnosine is a dipeptide that is thought to enhance frontal lobe function and be neuroprotective, and possibly have anticonvulsant properties. Chez et al.[121] studied the cognitive and behavioral effects of L-carnosine on 31 children, ages 3–12 years, with pervasive developmental disorders in an 8-week double-blind, placebo-controlled study. Children given carnosine showed significant improvement in the Receptive One-Word Picture Vocabulary Test scores, as well as autism rating scales compared to their baseline, while no significant improvements were seen in the placebo group.

Exogenous steroids have been shown in two studies to affect behavior, but no measure of cognition was done.[91,92]

Methylphenidate, amantadine, and donepezil have been used in children and adults with traumatic brain injury. Donepezil showed improvement in memory functioning in a case report of three adolescents.[93] Amantadine was studied in a retrospective, case-controlled study of 54 children. Subjective improvement was noted and an increase in the Rancho scale better than control group, who were placed on no psychostimulant medication. No formal neuropsychological testing was done. In a double-blind, placebo-controlled study of Ritalin for traumatic brain injury (TBI), in 18 adult patients, there were significant improvements in response accuracy for working memory and visual spatial attention tasks and decrease in the response time on the working memory tasks in the methylphenidate compared to placebo group.[94] Jin and Schachar[95] reviewed the evidence for efficacy of Ritalin on ADHD symptoms and cognition in subjects with both TBI and ADHD. The studies reviewed showed that methylphenidate showed positive effects on hyperactivity and impulsivity, but effects on cognition were less apparent, and that furthermore, more rigorous treatment outcome research was needed.

Van Trotsenburg et al.[96] recently studied the effects of thyroxine started in the neonatal period on development and growth of children with Down syndrome in the first 2 years of life. Infants who met the defined inclusion criteria were randomized in the neonatal period, to receive either thyroxine or placebo, and thyroid function tests, growth parameters, and development using the Bayley Scales of Infant Development were measured at regular intervals. At age 2 years, they found a lower developmental age delay on the mental subtest test results ($p = .032$) and on the motor scales ($p = 0.015$) in the thyroxine-treated children at age 2.

ASSESSMENT OF CHANGE IN COGNITION

Much needs to be done to assess the cognitive effects of the medications that can affect children's neurological status and behavior. The science of detecting short-term cognitive change has been developing in adult populations, looking at cognitive effects of medications and medical conditions,[97,98] as well as the effects of acute events, such as coronary surgery and concussions.[99,100] In children, formal study of the assessment of change has the added challenge of looking at change in the context of a changing neurological system. Studies thus need to determine whether medication causes slower or more rapid change than would already be expected.

The challenges of designing valid studies in order to determine the cognitive effects of these medications in both the short term and long term are listed and described next.

Subject Selection

Because of the heterogeneity within the diagnoses and symptoms for which psychotropic medications are being used, rigorous, careful definition of inclusion and exclusion criteria are needed. Inclusion criteria should include definitions of age and cognitive ability, as well as strict diagnostic criteria. The experimental group needs to be as homogeneous as possible, particularly when looking at changes in cognitive performance as an outcome measure. If wide ranges of age and ability are included, then outcome data should be examined within more specific groupings (e.g., preschool, school age, adolescence).

The Chez et al.[84] study is an example of wide heterogeneity in subject selection. In this study of rivastigmine in children with pervasive developmental disorders, the 32 subjects ranged in age from 3 to 12 years, with baseline autism scores on the childhood autism rating scale within one standard deviation of the mean ranging from 28 (nonautistic) to 40 (severely autistic)[101] and the mean raw score at baseline of the Expressive One-Word Picture Vocabulary Test, Revised (the significant outcome measure) of 34.91 (age equivalent 3 years 4 months) with a standard deviation of 33.61, corresponding to age equivalents within one standard deviation of the mean ranging from less than 1 year to 6 years 7 months.[102] In addition, 40% of the subjects had abnormal EEGs and 60% were on antiepileptic medications.[103] These wide variations in subject characteristics make any "significant" results difficult to interpret.

More homogeneity of subjects was achieved in the Van Trotsenberg et al.[104] study of thyroxine in children with Down syndrome; the group was well defined by diagnoses, age, and inclusion criteria. The baseline scores for all parameters studied had relatively narrow standard deviations within the groups, also indicating a relatively uniform study population.

Control Group Selection

Control groups are essential when studying complex disorders over time, as there are generally no "norm" of the test itself in that population, much less norms for what constitutes expected changes over time. Controls need to be matched to the subjects selected by diagnostic criteria, age, and cognitive ability. A normal control group matched by age and ability is also useful in many studies. The importance of a normal control group in addition to a diagnosis-specific control group is seen in the Calendre et al.[105] study of phenobarbital and valproic acid, in which both the control and the valproic acid groups made improvements in the test over time, and the phenobarbital group remained stable. If the study was performed only with the phenobarbital treatment group, it could have been concluded that phenobarbital had no effect on cognition, as there was no change in scores across the study. If the study was just with the phenobarbital group and the valproic acid group, one could have concluded that valproic acid acted to enhance cognition. With the control group included, the improvement in the valproic acid group is seen as part of the normal course of development and the study thus concludes that phenobarbital actually has a negative cognitive effect.

Measures to Assess Change

Detecting a change across time requires test instruments that show stability across time, or predictable changes. Thus, instruments must be normalized for repeated assessments in the population being studied. Particularly when using repeated instruments over short intervals, the instruments need to be relatively quick to administer and have alternative forms when possible. This refers to tests with different versions that measure the same property, but use different test items to minimize the learning effect. For example, if measuring verbal memory with word lists, the actual words on the list should be different with each administration.

Measures should also show stability in test performance (test–retest reliability), because the sensitivity to true change will decrease as the measurement error increases.[106] The instrument also needs to be sensitive to cognitive change. Measures also need to encompass the range of cognitive abilities of subjects studied, within the time measured, to avoid floor and ceiling effects. The measures should show a correlation with performance on accepted cognitive measures and, ideally, be correlated to adaptive and behavioral functioning. Thus, several considerations need to be made in selecting outcome measures and in administration of those measures, especially in children with varied cognitive abilities.

Reliability The reliability of a measure can be defined as how closely that measure represents the "true" score of what is being measured.[107] For example, if a child achieves a standard score of 75 on a reading assessment tool, how likely is it that the result of that measure reflects the child's actual reading ability? The actual test score thus reflects a "true" score and a certain amount of measurement error, either random, that is, error introduced by the vagaries of the individual situation, or systematic, that is, error caused by factors that affect the measurement across the sample.

Random error in a large sample should not affect the results across the sample, as it increases the variability of the test results but does not affect the mean of the sample (random errors will randomly have positive and negative effects on the scores and cancel each other out). Thus, the standard error associated with the reliability coefficient will be lower with a larger sample size. However, in smaller samples and certainly when evaluating the reliability of an individual's test performance, random errors can greatly affect the reliability of the test results, that is, how likely the test result reflects "true" ability.

Systematic errors (or bias) are those errors caused by factors that affect measurement across the sample. For example, an examiner who systematically administers test items incorrectly, causing higher than "true" scores, or testing that is done routinely late in the afternoon, when most of the sample is tired and hungry, causing lower than "true" scores.

Factors especially contributing to error in test administration to children include the following:[108]

- Factors attributable to the subject being tested, including attitude, affect, health, level of alertness, distractibility, comprehension, attention to test items, familiarity with testing situation and materials, impulsivity, and frustration tolerance.
- Factors attributable to the test, including test length (in time) and number of items assessing a particular factor, homogeneity of test items, how often the subject guesses, test–retest interval, the chance that certain test items will be more familiar to some examinees than others, test items that are misinterpreted, and scoring guidelines that are overly restrictive or, in contrast, not specific enough.
- Factors attributable to the situation in which the testing takes place (e.g., the setting in which testing takes place (home, office, play room), noise and distractions in the location, waiting time).
- Factors attributable to the examiner doing the testing, including rapport, attitude, affect, health, level of alertness, level of understanding of subject, experience with that class of subjects, preconceived perceptions, errors in test administration, data recording, scoring, and interpretation.

In order to estimate reliability, tests are evaluated based on their consistency and repeatability by considering the following:[109]

- The consistency of performance on items within the test that measure the same thing (internal consistency reliability).
- The consistency of a test from one administration to another (test–retest reliability).
- The consistency of the results of two forms of the same test (parallel forms reliability).
- The consistency of test results between examiners or raters (inter-rater reliability).

Most assessment tools are standardized in what is considered to be a normal population. Thus, if the experimental group is significantly different from the population from which a test was normed, caution on the reliability of that test must be taken. Also, very few assessment tools are normed (referring to testing in the "normal' population to assess how the scores are distributed in that population, in order to better assess when a subject's results are abnormal) to assess changes across time. Thus, a study assessing the effects of a medication across time needs to include a control group in order to be sure the changes found are attributable to the medication being used.

Validity In addition to reliability, one must consider validity of the test in the population of children studied. Types of validity to consider include content validity, concurrent validity, predictive validity, and construct validity.

- Content validity refers to how well the test measures what it claims to be measuring. The items need to cover the area of interest adequately, be pertinent to the area of interest, and cover the level of skills in the population being tested. For example, a math test for school age children needs to include calculation and word problems, and the level of difficulty needs to encompass the levels of abilities being tested. Many tests for children are designed for certain age groups and level of abilities. The performance of children tested at either end of the age range designated will not fully describe their abilities, either because they were able to complete the test without being challenged to their capability, or because they were unable to answer enough questions to establish whether abilities are actually lower than the test measures. In either case, content of the test would not be a true measure of the child's actual ability.
- Concurrent validity refers to how well a particular measure compares to another standard assessment tool measuring the same skill. For example, to establish the validity of Expressive One-Word Picture Vocabulary Test in assessing overall expressive language, it would need to be compared to a standard expressive speech and language tool.
- Predictive validity refers to how well the current test results correlate with future test scores. For example, does a higher score on the Bayley Test of Infant Development at age 2 predict the performance academic achievement test at age 6? If there is a significant correlation, then the Bayley Test can be said to have good predictive validity and would lead one to take much more seriously an intervention that leads to improvement in Bayley scores at age 2.
- Construct validity refers to how well a certain test measures the construct it purports to represent. In assessing the "cognitive" effects of medication, one first must define cognition as a construct. If, for example, cognition is defined as intelligence, then any measure utilized would need to be compared to an accepted measure of IQ (assuming IQ is an acceptable representation of the construct of intelligence). Of course, intelligence itself has many definitions and theoretical constructs.[110] Intelligence could be defined as academic achievement, in which case the measures needed to validate the test used would be a standard educational assessment. Intelligence could also be defined as "how" one thinks, that is, information processing, in which case one might compare test measures to tasks of executive function. In developmentally disabled populations, cognition is often measured as functionality of skills, that is, by adaptive functioning usually measured by caregiver or parent report.

Both reliability and validity of cognitive testing may vary significantly depending on the age group and the degree of disability. For example, when studying children with cognitive disability, one needs to choose instruments that will be sensitive to small variations. For most standardized cognition tests, children with cognitive disability are in the "floor" range; that is, they cannot answer enough questions on the test to establish a reliable true measure of their ability, and a small change in cognition over time may be missed because that instrument is insensitive to performance change.[111] Some studies will use the raw score to assess change instead of the standard score, but the variance of the raw score result in that population then needs to be studied in order to establish reliability of the raw score result.

Thus, critical analysis of any study on cognitive effects of medications must include consideration of reliability and validity of the measures utilized. Time and money constraints may push investigators to use short and easy-to-administer instruments to represent cognition, and to investigate these effects without a control group, resulting in a compromise of reliability and validity that may compromise the significance of the study itself.

Experimental Design

After taking into account selection of the experimental and control groups, and the reliability and validity of the test measures, the experimental design needs to be able to differentiate the medication effects of changes in cognition from other causes, for example, normal development and educational or therapeutic intervention affected by a disease process, environmental deprivation, or medication side effects. In adults, this assessment is somewhat simplified, as the cognitive functioning of adults is assumed to remain stable across a number of years. Adults' scores on certain tests may improve with repeated measures because of practice effects or learning, but this can be taken into account with assessing the normative practice effects over time.

In children, however, in addition to practice effects, there are developmental effects; that is, children improve from one testing to another because their brain development allows them to acquire new skills. This brain development coupled with educational intervention complicates the determination that a particular medication affects cognition. For example, a child identified with ADHD and learning disability is placed on medication, at the same time behavioral and educational interventions are put in place. One year later, cognitive and achievement tests have improved. If this child was part of a medication trial, it would appear that the medication had a large effect, when, in fact, it may account for only a small part of the change. Standard scores have been used to obviate the developmental factor, as a standard score measures performance relative to same age norms. Standard scores could be compared across time if they were relatively stable across time. Unfortunately, children's standard scores can change significantly over time, especially in the preschool period. After age 5, IQ scores become more consistent as children get older, but still can vary up to 20 points in some children.[112]

Therefore, if the study aims to show long-term cognitive outcome of a particular medication, the subject group needs to be well defined and matched to a control group as similar to the subjects as possible, except not on the medication. Matching needs to take into account not only diagnosis, comorbidities, and cognitive ability, but also the specific socioeconomic situation and educational opportunities. Studies should be designed to follow children over several years. Each group would need to be large enough to be able to detect subtle changes after taking into account the intrasubject variability of test scores.

Many studies aim to show short-term cognitive effects of medication, which require frequent measures of "cognition" over a short period of time. These tests can be quite useful in testing a medication's short-term safety and could potentially be used to assess the efficacy of medication. In adults, there are a few batteries of measures that have been used including the EpiTrack (six subtests assessing attention, cognitive tracking, and working memory),[113] the CANTAB battery (three subtests assessing visual memory, attention, and working memory and planning),[114] and the Cog State Battery (seven subtests assessing visual attention, executive function, and memory).[115] These batteries are easy to administer and quick and the CANTAB and the Cog State are computer based. Adults with developmental disabilities were included in the Lutz and Helmstaedter[113] study of EpiTrack. The CANTAB has been used extensively in populations with neurodegenerative disorders. Mollica et al.[116] studied the Cog State in 87 children, ages 8–12, with average cognitive abilities, and showed that children's scores improved from the first to second administration, but then remained stable from the second to fourth administration, and showed reasonable within-subject standard deviation, to be able to show modest changes of an applied treatment. In normal children, that study showed no relationship between behavioral rating scales and performance on the tests.

Mollica et al.[117] also studied the Cog State battery in 14 children with ADHD, at baseline and with two different doses of dexedrine. Surprisingly, they found the within-subject standard deviation to be similar in the ADHD group compared to the control group. The larger dose of dexedrine resulted in improvements in the Cog State battery on 3 of the 7 tests, and in 10 of the 14 children using a z-score composite of all the tests. The study also found a positive relationship between changes in performance on the Cog State battery and changes in behavioral rating scales in children with ADHD on medication. The Mollica et al.[117] study also outlined a statistical approach to classifying treatment response in children with ADHD.

The predictive utility of these batteries will also have to be established, perhaps by adding a longitudinal component to a short-term study, looking at broader cognitive testing as well as academic and functional performance. The same design could also be applied to other disorders, using cognitively and diagnostically matched controls. Pearson et al.[118] did use a similar battery in children with mental retardation and ADHD, but within-subject variability was not established in the group prior to the study, and there was no control group.

CONCLUSION

The long-term cognitive effects of medication in children have not been adequately studied for most classes of medication that affect the nervous system. Such study is vitally important, as the major task we assign to our children is to learn. Therefore, understanding the role of medications on cognitive development should be considered in the drug development paradigm.

Studies of stimulants show short-term improvement in certain cognitive tests involving eye movement control, planning/cognitive flexibility, attention/vigilance, and inhibitory control, but lack of improvement in more complex cognitive tasks. A single study of long-term effect of overall cognition showed a slight improvement in WISC III scores over a year. In children with cognitive disability, short-term improvement was also seen, but 25–35% showed cognitive deterioration.

There have been no adequate studies of cognitive effects of the atypical antipsychotic medications in children. Generally, studies in adults are mixed although three studies suggest adverse cognitive effects.

Studies of antiepileptic medication show that phenobarbital has definite adverse cognitive effects in children with long-term use. Topiramate has adverse cognitive effects in adults and has not been studied in children. The data on phenytoin and carbamazepine show mixed results, with a few studies in adults raising concerns for adverse cognitive effects. The studies on oxcarbazepine and valproic acid in children show no significant cognitive impairment. Many of the newer antiepileptic medications show no significant cognitive impairment in adults but have not been studied in children.

More research is needed to further investigate the cognitive side effects of the newer antiepileptic medications in children. Because of the nature of neurodevelopment, and the fact that children are on these medications for many years, studies should follow long-term cognitive effects on a regular basis through young adulthood. Short-term studies will miss effects on more complex thinking that may not be testable when a child is younger. Antiepileptic medications are increasingly being used in the pediatric population to treat behavioral disorder. The cognitive effects of these medications when used for that purpose also need to be studied.

There are no studies of cognitive effects of SSRIs in children. The few studies in adults show mixed results.

Adult studies of lithium and benzodiazepines show adverse cognitive effects. No studies have been done in children.

The anticholinesterase inhibitors and memantine have shown positive effects in Alzheimer's disease in adults. The few studies of these medications in children are inadequate to develop any conclusions. The potential use of these types of drugs in Down syndrome or other disorders in children needs to be elucidated.

The study of cognitive effects of medication in children is more complicated than in adults because of changing neurodevelopment and the variability of that development, both within and between subjects and across time. This variability especially needs to be considered in diagnostic groups containing heterogeneity with respect to age, cognitive abilities, severity of symptoms, and comorbidities. Careful definition of inclusion and exclusion criteria in subject selection will help make the subject group less heterogeneous. In order to isolate a medication as the cause of cognitive effects from other factors that also affect cognition, the careful selection of a control group is crucial. Matching criteria should take into account age, cognitive level, diagnosis, comorbidities, and educational and social environments.

Measures selected need to have good reliability and validity in the population that is being studied. If the subject group is significantly different from the group on which the measure was normed, then norms for the study group may need to be established separately, especially in terms of the reliability of the measure. Tests also have to encompass the breadth of skills being assessed, so as to avoid floor and ceiling effects. Children present many more challenges to reliability than adults, including characteristics of the children, the test, the testing situation, and the examiner.

The validity of the tests being used also needs to be carefully considered. The instruments used need to be measuring what they purport to measure (*content validity*), and that they compare favorably to other accepted tests that measure the same content (*concurrent validity*). Test results need to indicate future performance on a similar test (*predictive validity*). Test results also need to have a significant correlation with the construct which that test represents (*construct validity*).

Comparison among studies will be much easier if there are uniform standards for psychometric instruments to use. These instruments will need to be standardized for the study population examined if the study population differs significantly from the norm.

Experimental design needs to take into account this heterogeneity of children in order to assure that the effects found experimentally are truly due to the medication and not other factors. Studies need to be large enough to take into account both the within-subject and among-subject variability, to assure that results are significant. The statistics and methodology for the assessment of cognitive change in children need to be established, so that similar methodology can be applied across a range of ages and levels of development.

Testing for the cognitive effects of any medication that affects the neurological system should be required as part of the field testing of a new chemical entity (NCE) or medication and, particularly for children, should be part of the required ongoing monitoring after the medication or NCE is approved for use. The batteries to assess short-term cognitive effects are relatively easy to administer in adults. These need to be adapted and further tested for their use in children and could be a valuable indicator of both positive and negative cognitive effects.

There is a clear increasing trend in the use of psychotropic medications that have been tested and approved in adults to treat children, and at younger and younger ages. Many of the newer medications mentioned earlier and studied in children have not been approved for use in children. Federal regulations need to require that medications likely to be used in children are field tested in children as well as adults. Technically, the use of medications not approved for children should require an Investigational New Drug (IND) application, when being studied in children. Physicians also need to be made aware of the lack of safety and efficacy of these medications in children, and of the potential adverse cognitive effects when known.

REFERENCES

1. Pietrzak RH, Mollica CM, Maruff P, Snyder PJ. Cognitive effects of immediate-release methylphenidate in children with attention-deficit/hyperactivity disorder. *Neurosci Biobehav Rev*. 2006;30:1225–1245.
2. Mollica CM, Maruff P, Vance A. Development of a statistical approach to classifying treatment response in individual children with ADHD. *Hum Psychopharmacol Clin Exp*. 2004;19:445–456.
3. Pearson DA, Santos CW, Casat CD, et al. Treatment effects of methylphenidate on cognitive functioning in children with mental retardation and ADHD. *J Am Acad Child Adolesc Psychiatry*. 2004;43:677–685.
4. Aman MG, Marks RE, Turbott SH, Wilsher CP, Merry SN. Methylphenidate and thioridazine in the treatment of sub-average children: effects on cognitive-motor performance. *J Am Acad Child Adolesc Psychiatry*. 1991;30:816–824.
5. Hagerman RJ, Murphy MA, Wittenberger MD. A controlled trial of stimulant medication in children with the fragile X syndrome. *Am J Med Genet*. 1988;30:377–393.
6. Handen BL, Breaux AM, Gosling A, Ploof DL, Feldman H. Efficacy of methylphenidate among mentally retarded children with attention deficit hyperactivity disorder. *Pediatrics*. 1990;86:922–930.
7. Handen BL, Breaux AM, Janosky J, McAuliffe S, Feldman H, Gosling A. Effects and noneffects of methylphenidate in children with mental retardation and ADHD. *J Am Acad Child Adolesc Psychiatry*. 1992;31:455–461.
8. Pearson DA, Lane DM, Santos CW, et al. Effects of methylphenidate treatment in children with mental retardation and ADHD: individual variation in medication response. *J Am Acad Child Adolesc Psychiatry*. 2004;43:686–698.

9. Barkley RA, *Attention Deficit Hyperactivity Disorder: A Handbook for Diagnosis and Treatment*, 2nd ed. New York: Guilford Press; 1998.
10. Connor DF. Preschool attention deficit hyperactivity disorder: a review of prevalence, diagnosis, neurobiology, and stimulant treatment. *J Dev Behav Pediatr*. 2002;23:S1–S9.
11. Byrne JM, Bawden HN, Dewolfe NA, et al. Clinical assessment of psychopharmacological treatment of preschoolers with ADHD. *J Clin Exp Neuropsychol*. 1998;20:613–627.
12. Kratochvil CJ, Greenhill LL, March JS, Burke WJ, Vaughan BS. The role of stimulants in the treatment of preschool children with attention-deficit hyperactivity disorder. *CNS Drugs*. 2004;18:957–966.
13. Gimpel HA, Collett BR, Veeder MA, et al. Effects of stimulant medication on cognitive performance of children with ADHD. *Clin Pediatr*. 2005;44:405–411.
14. McCaffrey RJ, Ortega A, Orsillo SM, Nelles WB, Haase RF. Practice effects in repeated neuropsychological assessments. *Clin Neuropsychologist*. 1992;6:32–42.
15. Mollica CM, Maruff P, Collie A, Vance A. Repeated assessment of cognition in children and the measurement of performance change. *Child Neuropsychol*. 2005;11:303–310.
16. Berman T, Douglas VI, Barr RG. Effects of methylphenidate on complex cognitive processing in attention deficit hyperactivity disorder. *J. Abnormal Psychol*. 1999;108:90–105.
17. Pietrzak RH, Mollica CM, Maruff P, Snyder PJ. Cognitive effects of immediate-release methylphenidate in children with attention-deficit/hyperactivity disorder. *Neurosci Biobehav Rev*. 2006;30:1225–1245.
18. Berman T, Douglas VI, Barr RG. Effects of methylphenidate on complex cognitive processing in attention deficit hyperactivity disorder. *J. Abnormal Psychol*. 1999;108:90–105.
19. Pearson DA, Santos CW, Casat CD, et al. Treatment effects of methylphenidate on cognitive functioning in children with mental retardation and ADHD. *J Am Acad Child Adolesc Psychiatry*. 2004;43:677–685.
20. Gimpel HA, Collett BR, Veeder MA, et al. Effects of stimulant medication on cognitive performance of children with ADHD. *Clin Pediatr*. 2005;44:405–411.
21. Pearson DA, Lane DM, Santos CW, et al. Effects of methylphenidate treatment in children with mental retardation and ADHD: individual variation in medication response. *J Am Acad Child Adolesc Psychiatry*. 2004;43:686–698.
22. Findling RL. Dosing of atypical antipsychotics in children and adolescents. *J Clin Psychiatry*. 2003;5 (Suppl 6):10–13.
23. Aparasu RR, Bhatara V. Patterns and determinants of antipsychotic prescribing in children and adolescents, 2003–2004. *Curr Med Res Opin*. 2007;23:49–56.
24. Pool D, Bloom W, Mielke DG, et al. A controlled evaluation of loxitane in seventy-five adolescent schizophrenic patients. *Curr Ther Res Clin Exp*. 1976;19:99–104.
25. Realmuto GM, Erickson WD, Yellin AM, et al. Clinical comparison of thiothixene and thioridazine in schizophrenic adolescents. *Am J Psychiatry*. 1984;141:440–442.
26. Findling RL, Steiner H, Weller EB. Use of antipsychotics in children and adolescents. *J Clin Psychiatry*. 2005;66 (Suppl 7):29–40.
27. Heyneman EK. The aggressive child. *Child Adolesc Psychiatr Clin N Am*. 2003;12: 667–677.
28. Pappadapulos E, MacIntyre JC II, Crismon ML, et al, Treatment recommendations for the use of antipsychotics for aggressive youth (TRAAY). Part 2. *J Am Acad Child Adolesc Psychiatry*. 2003;42:145–161.
29. Pappadopulos E, Jensen PS, Schur SB, et al. "Real world" atypical antipsychotic prescribing practices in public child and adolescent inpatient settings. *Schizophr Bull*. 2002;28: 111–121.

30. McCracken JT, McGough J, Shah B, et al. Risperidone in children with autism and serious behavioral problems. *N Engl J Med.* 2002;347:314–321.
31. Shea S, Turgay A, Carroll A, et al. Risperidone in the treatment of disruptive behavioral symptoms in children with autistic and other pervasive developmental disorders. *Pediatrics.* 2004;114:e634–e641.
32. Findling RL, Steiner H, Weller EB. Use of antipsychotics in children and adolescents. *J Clin Psychiatry.* 2005;66(Suppl 7):29–40.
33. Findling RL, McNamara NK, Branicky LA, et al. A double–blind pilot study of risperidone in the treatment of conduct disorder. *J Am Acad Child Adolesc Psychiatry.* 2000;39:509–516.
34. Findling RL, Blumer JL, Kauffman R, et al. Pharmacokinetic effects of aripiprazole in children and adolescents with conduct disorder. (Poster, presented at the 24th Collegium Internationale Neuro-Psychopharmacologicum Congress; June 20–24, 2004.
35. Aman MG, DeSmedt G, Derivan A, et al. Double-blind placebo-controlled study of risperidone for the treatment of disruptive behaviors in children with subaverage intelligence. *Am J Psychiatry.* 2002;159:1337–1346.
36. Buitelaar JK, van der Gaag RJ, Cohen-Kettenis P, et al. A randomized controlled trial of risperidone in the treatment of aggression in hospitalized adolescents with subaverage cognitive agilities. *J Clin Psychiatry.* 2001;62:239–248.
37. Snyder R, Turgay A, Aman M, et al. Effects of risperidone on conduct and disruptive behavior disorders in children with subaverage IQs. *J Am Acad Child Adolesc Psychiary.* 2002;41:1026–1036.
38. Van Bellinghen M, De Troch C. Risperidone in the treatment of behavioral disturbances in children and adolescents with borderline intellectual functioning: a double-blind placebo-controlled pilot trial. *J Child Adolesc Psychopharmacol.* 2001;11:5–13.
39. Biederman J, Mick E, Hammerness P, et al. Open-label, 8-233k trial of olanzapine and risperidone for the treatment of bipolar disorder in preschool-age children. *Biol Psychiatry.* 2005;58:589–594.
40. Masi G, Cosenza A, Mucci M, Brovedani P. A 3-year naturalistic study of 53 preschool children with pervasive developmental disorders treated with risperidone. *J Clin Psychiatry.* 2003;64:1039–1047.
41. Luby J, Mrakotsky C, Stalets MM, et al. Risperidone in preschool children with autistic spectrum disorders: an investigation of safety and efficacy. *J Child Adolesc Psychopharmacol.* 2006;16:575–587.
42. Bender S, Dittmann-Balcar A, Schall U, et al. Influence of atypical neuroleptics on executive functioning in patients with schizophrenia: a randomized double blind comparison of olanzapine vs. clozapine. *Int J Neuropsychopharmacol.* 2006;135–145.
43. Bilder RM, Goldman RS, Volavka J, et al. Neurocognitive effects of clozapine, olanzapine, risperidone, and haloperidol in patients with chronic schizophrenia or schizoaffective disorder. *Am J Psychiatry.* 159:2002;1018–1028.
44. Kern RS, Green MF, Marshall BD, et al, Risperidone versus haloperidol on secondary memory: Do newer medications aid learning? *Schizophr Bull.* 1999;25:223–232.
45. McGurk SR, Carter C, Goldman R, et al. The effects of clozapine and risperidone on spatial working memory in schizophrenia. *Am J Psychiatry.* 2005;162:1013–1016.
46. Hori H, Noguchi H, Hashimoto R, et al, Antipsychotic medication and cognitive function in schizophrenia. *Schizophrenia Research* 2006;86:138–146.
47. Crooenberghs J, Fegert JM, Findling RL, deSmedt G, VanDongen S, Risperidone Disruptive Behavior Study Group. Risperidone in children with disruptive behavior disorder and sub-average intelligence: a 1-year open-label study of 504 patients. *J Am Acad Child Adolesc Psychiatry.* 2005;44:64–72.

48. Turgay A, Binder C, Snyder R, Fisman S. Long-term safety and efficacy of risperidone for the treatment of disruptive behavior disorders in children with subaverage IQs. *Pediatrics*. 2002; 110:e34.

49. Kravariti E, Morris RG, Rabe-Hesketh S, Murray RM, Frangou S. The Maudsley Early-Onset Schizophrenia Study: cognitive function in adolescent-onset schizophrenia. *Schizophr Res*. 2003;65:95–103.

50. Besag FMC. Childhood epilepsy in relation to mental handicap and behavioral disorders. *J Child Psychol Psyhiatry*. 43:2002;103–131.

51. Besag FMC. Cognitive and behavioral outcomes of epileptic syndromes: implications for education and clinical practice. *Epilepsia*. 2006;47 (Suppl 2):119–125.

52. Bowley C, Kerr M. Epilepsy and intellectual disability. *J Intellect Disabil Res*. 2000;44: 529–543.

53. Working Group of the International Association of the Scientific Study of Intellectual Disability. Clinical guidelines for the management of epilepsy in adults with an intellectual disability. *Seizure*. 2001;10:401–409.

54. Vaquerizo J, Gomez MH, Gonzalez IE, Cardesa JJ. Reversible neuropsychological deterioration associated with valproate. *Rev Neurol. (Pairs)* 1995;23:148–150.

55. Seidel WR, , Mitchell WG. Cognitive and behavioral effects of carbamazepine in children: data from the benign Rolandic epilepsy. *J Child Neurol*. 1998;14:716–723.

56. deSilva M, MacArdle B, McGown M, et al. Randomised comparative monotherapy trial of phenobarbitone, phenytoin, carbamazepine, or sodium valproate for newly diagnosed childhood epilepsy. *Lancet*. 1996;347:709–713.

57. Vining EPG, Mellits ED, Dorsen MM, et al. Psychologic and behavioral effects of antiepileptic drugs in children: a double blind comparison between phenobarbital and valproic acid. *Pediatrics*. 1987;80:165–174.

58. Farwell JR, Lee YJ, Hirtz DG, Sulzbacher SI, Ellenberg JH, Nelson KB. Phenobarbital for febrile seizures—effects on intelligence and on seizure recurrence. *N Engl J Med*. 1990;322:364–369.

59. Calandre EP, Dominguez-Granado R, Gomez-Rubio M, MolinaFont JA. Cognitive effects of long-term treatment with phenobarbital and valproic acid in school children. *Acta Neurol Scand*. 1990;81:504–506.

60. Bolanos AR, Sarkisian M, Yang Y, et al. Comparison of valproate and phenobarbital treatment after status epilepticus in rates. *Neurology*. 1998;51:41–48.

61. Forsythe I, Butler R, Berg I, McGuire R. Cognitive impairment in new cases of epilepsy randomly assigned to carbamazepine, phenytoin, and sodium valproate. *Dev Med Child Neurol*. 1991;33:524–534.

62. Pulliainen V, Jokelainen M. Effects of phenytoin and carbamazepine on cognitive functions in newly diagnosed epileptic patients. *Acta Neurol Scand*. 1994;89:81–86.

63. Aldenkamp AP, Alpherts WCJ, Diepman L, Van't Slot B, Overweg J, Vermeulen J. Cognitive side-effects of phenytoin compared with carbamazepine in patients with localization-related epilepsy. *Epilepsy Res*. 1994;19:1–87.

64. Forsythe I, Butler R, Berg I, McGuire R. Cognitive impairment in new cases of epilepsy randomly assigned to carbamazepine, phenytoin, and sodium valproate. *Dev Med Child Neurol*. 1991;33:524–534.

65. Stores G, Williams PL, Styles E, Zaiwalla Z. Psychological effects of sodium valproate and carbamazepine in epilepsy. *Arch Dis Child*. 1992;67:1330–1337.

66. Seidel WR, Mitchell WG. Cognitive and behavioral effects of carbamazepine in children: data from the benign Rolandic epilepsy. *J Child Neurol*. 1998;14:716–723.

67. Aldenkamp A, Vigevano F, Arzimanoglou A, Covanis A. Role of valproate across the ages. Treatment of epilepsy in children. *Acta Neurol Scand.* 2006;114(Suppl 184):1–13.
68. Stefan H, Feuerstein TJ. Novel anticonvulsant drugs. *Pharmacol Ther.* 113:2007;165–183.
69. Aldencamp AP, DeKrom M, Reijs R. Newer antiepileptic drugs and cognitive issues. *Epilepsia.* 2003;44 (Suppl 4):21–29.
70. Tzitiridou M, Panou T, Ramantani G, Kambas A, Spyroglou K, Panteliadis C. Oxycarbazepine monotherapy in benign childhood epilepsy with centrotemporal spikes: a clinical and cognitive evaluation. *Epilepsy Behav.* 2005;7:458–467.
71. Pulsifer MB, Gordon JM, Brandt J, Vining EPG, Freeman JM. Effects of ketogenic diet on development and behavior: preliminary report of a prospective study. *Dev Med Child Neurol.* 2001;43:301–306.
72. Courtney DB. Selective serotonin reuptake inhibitor and venlafaxine use in children and adolescents with major depressive disorder: a systematic review of published randomized controlled trials. *Can J Psychiatry.* 2004;49:577–563.
73. Whittington CJ, Kendall T, Fonagy P, Cottrell D, Cotgrove A, Boddington E. Selective serotonin reuptake inhibitors in childhood depression: systematic review of published versus unpublished data. *Lancet.* 2004;363:1341–1345.
74. Wadsworth JK, Moss SC, Simpson SA, Smith AP. SSRIs and cognitive performance in a working sample. *Hum Psychopharmacol Clin Exp.* 2005;20:561–572.
75. Pachet AK, Wisniewski AM. The effects of lithium on cognition: an updated review. *Psychopharmacology.* 2003;170:225–234.
76. Stip E, Dufresne J, Lussier I, Yatham L. A double-blind, placebo controlled study of the effects of lithium on cognition in healthy subjects: mild and selective effects on learning. *J Affect Disord.* 2000;60:147–157.
77. Basso MR, Lowery N, Neel J, Purdie R, Bornstein RA. Neuropsychological impairment among manic, depressed, and mixed-episode inpatients with bipolar disorder. *Neuropsychology.* 2002; 16:84–91.
78. Block RI, Berchou R. Alprazolam and lorazepam effects on memory acquisition and retrieval processes. *Pharmacol Biochem Behav.* 1984;20:233–241.
79. File SE, Lister RG. Do lorazepam-induced deficits in learning result from impaired rehearsal, reduced motivation or increased sedation? *Br J Clin Pharmacol.* 1982;14:545–550.
80. Curran HV. Tranquilizing memories: a review of the effects of benzodiazepines on human memory. *Biol Psychol.* 1986;23:179–213.
81. Barker MJ, Greenwood KM, Jackson M, Crowe SF. Cognitive effects of long-term benzodiazepine use, a meta-analysis. *CNS Drugs.* 2004;18:37–48.
82. Barker MJ, Greenwood KM, Jackson M, Crowe SF. Persistence of cognitive effects after withdrawal from long-term benzodiazepine use: a meta-analysis. *Arch Clin Neuropsychol.* 2004;19:437–454.
83. Behl P, Lanctot KL, Streiner DL, Guimont I, Black SE. Cholinesterase inhibitors slow decline in executive functions, rather than memory in Alzheimer's disease: a 1-year observational study in the Sunnybrook dementia cohort. *Curr Alzheimer Res.* 2006;3:147–156.
84. Chez MG, Buchanan CP, Mrazek S, Tremb R. Treating autistic spectrum disorders in children: utility of the cholinesterase inhibitor rivastigmine tartrate. *J Child Neurol.* 2004;19: 165–169.
85. Chez MG, Nowinski CV, Buchanan CP. Donepezil hydrochloride use in children with autistic spectrum disorders. *Ann Neurol.* 200;48:541 (abstract).
86. Lee S, Lee J, Lee B, Kim YH. A 12-week, double-blind, placebo-controlled trial of galantamine adjunctive treatment to conventional antipsychotics for the cognitive impairments in chronic schizophrenia. *Int Clin Psychopharmacol.* 2007;22:63–68.

87. Choinard S, Stip E, Poulin J, et al. Rivastigmine treatment as an add-on to antipsychotics in patients with schizophrenia and cognitive deficits. *Curr Med Res Opin*. 2007;23:575–583.

88. Wilens TE, Waxmonsky J, Scott M, et al. An open trial of adjunctive donepezil in attention-deficit/hyperactivity disorder. *J Child Adolesc Psychopharmacol*. 2005;15:947–955.

89. Cosman KM, Boyle LL, Porsteinsson AP. Memantine in the treatment of mild to moderate Alzheimer's disease. *Expert Opin Pharmacotherapy*. 2007;8:203–214.

90. Owley T, Salt J, Guter S, et al. A prospective, open-label trial of memantine in the treatment of cognitive behavioral and memory dysfunction in pervasive developmental disorders. *J Child Adolesc Psychopharmacol*. 2006;16:517–524.

91. Su TP, Pagliaro M, Schmidt PJ, Pickar D, Wolkowitz O, Rubinow DR. Neuropsychiatric effects of anabolic steroids in male normal volunteers. *JAMA*. 1993;269:2760–2764.

92. Wolkowitz OM. Prospective controlled studies of the behavioral and biological effects of exogenous corticosteroids. *Psychoneuroendocrinology*. 1994;19:233–255.

93. Trovato M, Slomine B, Pidcock F, Christensen J. The efficacy of donepezil hydrochloride on memory functioning in three adolescents with severe traumatic brain injury. *Brain Inj*. 2006;20:339–343.

94. Kim Y, Ko M, Na S, Park S, Kim K. Effects of single-dose methylphenidate on cognitive performance in patients with traumatic brain injury: a double-blind placebo-controlled study. *Clin Rehab*. 2006;20:24–30.

95. Jin C, Schachar R. Methylphenidate treatment of attention-deficit/hyperactivity disorder secondary to traumatic brain injury: a critical appraisal of treatment studies. *CNS Spectrums*. 2004;9:217–226.

96. Van Trotsenburg ASP, Vulsma SL, Rutgers Van Rozenburg-Marres SL, et al. The effect of thyroxine treatment started in the neonatal period on development and growth of two-year-old Down syndrome children: a randomized clinical trial. *J Clin Endocrinol Metab*. 2005;90: 3304–3311.

97. Fray PJ, Robbins TW. CANTAB battery: proposed utility in neurotoxicology. *Neurotoxicol Teratol*. 1996;18:499–504.

98. Lutz MT, Helmstaedter C. EpiTrack: tracking cognitive side effects of medication on attention and executive functions in patients with epilepsy. *Epilepsy Behav*. 2005;7:708–714.

99. Colle A, Darby DG, Falleti MG, Silbert BS, Maruff P. Determining the extent of cognitive change after coronary surgery: a review of statistical procedures. *Ann Thorac Surg*. 2002;73:2005–2011.

100. Collie A, Maruff P, Makdissi M, McCrory P, McStephen M, Darby D. Cog sport: reliability and correlation with conventional cognitive tests used in post-concussion medical evaluations. *Clin J Sport Med*. 2003;13:28–32.

101. Schopler E, Reichler RJ, Renner BR, *The Childhood Autism Rating Scale*. Western Psychological Services, 1988.

102. Brownell R. ed. *Expressive One Word Picture Vocabulary Test*, 3rd ed. Los Angeles CA: Western Psychological Services.

103. Chez MG, Buchanan CP, Mrazek S, Tremb R. Treating autistic spectrum disorders in children: utility of the cholinesterase inhibitor rivastigmine tartrate. *J Child Neurol*. 2004;19:165–169.

104. Van Trotsenburg ASP, Vulsma SL, Rutgers Van Rozenburg-Marres SL, et al. The effect of thyroxine treatment started in the neonatal period on development and growth of two-year-old Down syndrome children: a randomized clinical trial. *J Clin Endocrinol Metab*. 2005;90:3304–3311.

105. Calandre EP, Dominguez-Granado R, Gomez-Rubio M, MolinaFont JA. Cognitive effects of long-term treatment with phenobarbital and valproic acid in school children. *Acta Neurol Scand*. 1990;81:504–506.

106. Mollica CM, Maruff P, Collie A, Vance A. Repeated assessment of cognition in children and the measurement of performance change. *J Child Neuropsychol.* 2005;11:303–310.

107. Trochim W. *The Research Methods Knowledge Base*, 2nd ed. Cincinnati, OH: Atomic Dog Publishing; 2000.

108. Sattler JM. Useful statistical and measurement concepts. In: Sattler JM. *Assessment of Children: Cognitive Applications*, 4th ed. San Diego, CA: Jerome M Sattler, Publisher, Inc; 2001: Chap 4.

109. Sattler JM. Useful statistical and measurement concepts. In: Sattler JM. *Assessment of Children: Cognitive Applications*, 4th ed. San Diego, CA: Jerome M Sattler, Publisher, Inc; 2001: Chap 4.

110. Sattler JM. Historical survey and theories of intelligence. In: Sattler JM. *Assessment of Children: Cognitive Applications*, 4th ed. San Diego, CA: Jerome M Sattler, Publisher, Inc; 2001: Chap 5.

111. Heller JH, Spiridigloizze GA, Crissman BG, Sullivan-Saarela JA, Li JS, Kishnani PS. Clinical trials in children with Down syndrome: issues from a cognitive research perspective. *Am J Med Genet Part C Semin Med Genet.* 2006;42C:187–195.

112. Satter JM. Issues related to the measurement and change of intelligence. In: Sattler JM. *Assessment of Children: Cognitive Applications*, 4th ed. San Diego, CA: Jerome M Sattler, Publisher, Inc; 2001.

113. Lutz MT, Helmstaedter C. EpiTrack: tracking cognitive side effects of medication on attention and executive functions in patients with epilepsy. *Epilepsy Behav.* 2005;7:708–714.

114. Fray PJ, Robbins TW. CANTAB battery: proposed utility in neurotoxicology. *Neurotoxicol Teratol.* 1996;18:499–504.

115. Falleti MG, Maruff P, Collie A, Darby DG. Practice effects associated with the repeated assessment of cognitive function using the CogState battery at 10-minute, one week and one month test–retest intervals. *J Clin Exp Neuropsychol.* 2006;28(7):1095–1112.

116. Mollica CM, Maruff P, Collie A, Vance A. Repeated assessment of cognition in children and the measurement of performance change. *J Child Neuropsychol.* 2005;11:303–310.

117. Mollica CM, Maruff P, Vance A. Development of a statistical approach to classifying treatment response in individual children with ADHD. *Hum Psychopharmacol Clin Exp.* 2004;19:445–456.

118. Pearson DA, Santos CW, Casat CD, et al. Treatment effects of methylphenidate on cognitive functioning in children with mental retardation and ADHD. *J Am Acad Child Adolesc Psychiatry.* 2004;43:677–685.

119. Mowla A, Mosavinasab M, Pani A. Does fluoxetine have any effect on the cognition of patients with mild cognitive impairment? A double blind, placebo-controlled, clinical trial. *J Clin Psychopharmacol.* 2007;27:67–70.

120. Heller JH, Spiridigliozzi GA, Doraiswamy PM, Sullivan JA, Crissman BG, Kishanani PS. Donepezil effects on language in children with Down syndrome: Results of the first 22-week pilot clinical trial. *Am J Med Genet Part A.* 2004;130A;325–326.

121. Chez MG, Buchanan CP, Aimonovitch MC, Becker M, Schaefer K, Black C, Komen J. Double-blind, placebo-controlled study of L-carnosine supplementation in children with autistic spectrum disorders. *J. Child Neurology.* 2002;17:833–837.

CHAPTER 27

Cardiovascular and QTc Issues

BERT SUYS, MD, PhD

Congenital and Pediatric Cardiology, University Hospital Antwerp, Wilrijkstraat 10, 2650 Edegem (Antwerp), Belgium

LUC DEKIE, PhD

Director of Pharmaceutical Operations, Biomedical Systems, Chaussée de Wavre 1945, 1160 Brussels, Belgium

CARDIOVASCULAR DEVELOPMENT CHANGES AND IMPLICATIONS FOR TREATMENT WITH CARDIOVASCULAR DRUGS: DEFICIENT DATA SUPPORTING DOSING FOR EFFICACY AND SAFETY

Comparable to other medicinal classes, many cardiovascular products used currently in children have not been tested adequately but are given on an "off-label" or "unlicensed" basis, in doses extrapolated from adult data. However developmental changes in the cardiovascular system (Table 27.1) could have an important impact on the goals of efficacy and safety, including the development of adverse events related to the differences in the pharmacokinetic (PK) and pharmacodynamic (PD) properties of individual drugs. Besides differences from adults with respect to many aspects of metabolism, clearance, and protein binding, neonates and children have important age-related physiological changes in cardiac output and portal vein flow, but also in vascular space and volume distribution in general, with age-specific adaptive mechanisms.[1,2] These critically important developmental changes in hemodynamic factors in neonates, infants, and children will affect the response to inotropic (enhanced myocardial contractility) or pressor (elevated systemic vascular resistance) therapy. Developmentally based pharmacodynamic differences, in part due to changes in myocardial structure, cardiac innervation, and function of the adrenergic receptor, also affect the responses to both therapeutic and toxic effects of inotropes. For example, the immature myocardium has fewer contractile elements and therefore a decreased ability to increase contractility; it also responds poorly to standard techniques of manipulating preload.[3] Despite increasing investigation in the area of cardiovascular instability, large and extensive gaps in knowledge remain, especially in neonates and surely

Pediatric Drug Development: Concepts and Applications
Edited by Andrew E. Mulberg, Steven A. Silber, and John N. van den Anker

TABLE 27.1 Cardiac Developmental Changes

Anatomy/Morphology

- Differences in position and size; absolute weight of the heart triples in the first year.
- The left ventricular wall grows over proportionally; the left ventricular wall becomes thicker, the lumen smaller than the right.

Histology/Biochemical

- Intense growth of cardiomyocytes—first by proliferation, later by enlargement.
- Myocyte volume increases 25 × from birth to adulthood.
- With age, cells acquire more contractile elements.
- Increase in sarcoplasmic reticulum makes electromuscular coupling more efficient.
- Switch from glucose to lipids as energy source occurs before birth; change from α to β MHC shortly after birth.

Electrophysiology

- Cardiac innervation is functionally and structurally immature at birth.
- In the early postnatal period, hearts have generally higher stimulation thresholds and favor excitation over inhibition.
- In the neonatal period, the sympathetic system is dominant.
- α_1-adrenergic, β-adrenergic, and muscarinic receptors and their signaling cascades change from the postnatal period to childhood.

Hemodynamics

- Blood flow rapidly redistributes to the lung after birth with decreasing resistance.
- Blood pressure, heart rate, and cardiac output rise in the neonatal period.
- Neonates especially respond to higher demands mainly with increased heart rate, rather than stroke volume; cardiac output is low in the neonatal and infant period.
- Steep increase in blood pressure postnatally and in the first months.
- Rapid increase in heart rate the first days (until Day 10), then decreases up to Day 100 postnatally.
- Baroreceptor reflex is immature in all neonates and infants.

in the preterm population because none of the current treatments have been well studied, and data regarding safety and efficacy are lacking.

Therefore, in the United States, a specific neonatal Cardiology Group was established in 2003, as part of the Newborn Drug Development Initiative.[4] Studies have revealed that, with age, left ventricular volume and mass increase, mass/volume ratio remains almost constant, and the afterload increases. Apparent modulations in echocardiographic diastolic flow pattern and flow duration have also been observed from fetal life throughout infancy and childhood.[5,6] Despite this knowledge on developmental changes in systolic and diastolic function, strikingly little research exists in children with ventricular dysfunction, in terms of well-designed large-scale studies of the epidemiology or multicenter-controlled clinical therapeutic trials. The implications of this lack of knowledge directly impact on intervention in this vulnerable population. For example, despite the fact that β-blockers have been shown to significantly improve morbidity and mortality in adults with chronic congestive heart failure, there is little reported experience with their use in children; a comparable but slightly better situation exists for the use of angiotensin-converting enzyme (ACE) inhibitors in pediatric patients with heart failure.[7,8] Antiarrhythmic drugs have been administered to children following data extrapolated from adults. Sotalol, for example, is generally accepted as a good first-line treatment for supraventricular tachycardias in children. Until recently, however, there was no large trial giving correct PK/PD data in this population, and thus no "adapted" dosage and no correct formulation. In a recent trial, PK data showed that sotalol adheres to the physiological maturation process of renal function in the developing child, explaining the higher drug exposure in neonates and young infants. There was also evidence from animal data that developmental changes include not only kidney maturation but also maturation of the myocardium potassium channels, which influences the action potential duration.[9] Due to interference with these channels, sotalol is known to prolong the QT interval. The higher drug exposure in young children consequently leads to higher QT intervals and thus a higher proarrhythmic potential (risk of torsades de pointes, leading to often lethal arrhythmias). This study provided an example of rational drug dosage in children, addressing interpatient variability, and led to individually guided therapy based on effective levels, and thus different dosing recommendations for subgroups of different ages.[10]

Children may also be more vulnerable to adverse cardiac events due to immaturity in autonomic control of the heart; these changes are actually incompletely understood, but recent findings suggest that the sympathetic nervous system is more sensitive to developmental changes than the parasympathetic system. This can be important for heart rate variability adaptation and blood pressure regulation.[11] The paucity of knowledge in addressing hemodynamic-specific developmental changes as it relates to appropriate dosing of cardiovascular active drugs needs to be addressed on an individual basis for proper safety and efficacy monitoring.

DEVELOPMENTAL CHANGES IN INFANTS AND CHILDREN AND IMPLICATIONS FOR UNDERSTANDING CARDIOTOXICITY OF NONCARDIOVASCULAR DRUGS

Noncardiovascular drugs can have important cardiovascular (adverse) side effects. The best known example is the proarrhythmic effect of certain pharmaceutical products due to QT interval lengthening, leading to potential actions including withdrawal from the market, severe prescribing restrictions, or the implementation of the thorough QT trial to evaluate the effect of the drug on the cardiac repolarization (QTc interval) before gaining regulatory approval. The important implications for pediatric drug development (differences in electrocardiogram (ECG) reading, QT interval measurements) are discussed in detail further in this chapter.

Drug-Induced Disorders of the Myocardium: Influence on Cardiac Function

Although heart failure is predominantly caused by cardiovascular conditions such as hypertension, coronary heart disease, and valvular heart disease (conditions less frequent in the pediatric population), it can also be an adverse reaction induced by drug therapy.[12] Cardiomyopathy is a well-recorded and well-known toxic sign of some antineoplastic agents, such as the anthracyclines (doxorubicin, daunorubicin), mitoxantrone, and cyclophosphamide. These drugs are processed by myocardial cells, destroying the cardiac muscle, and so decrease the ventricular wall and cardiac mass. These effects are cumulative dose and dose intensity dependent; data suggest that children are particularly vulnerable but may develop symptoms of insidious onset several years after therapy.[13–16] Other drugs that may adversely affect the heart are some antidiabetic drugs, amphetamines, antifungal drugs, and nonsteroidal anti-inflammatory drugs (NSAIDs).

Drug-Induced Blood Pressure Changes

Most drugs that cause a rise in blood pressure do so by causing retention of salt and water; the best examples are the class of corticosteroids. More commonly used medications in the pediatric practice (with often over-the-counter availability) that can lead to changes in blood pressure are the nasal decongestants and several cough/cold medications, especially when incorrect dosages are used.[17,18]

Recent warnings for the stimulant–drug therapy for attention deficit disorder (with or without hyperactivity) and potential inherent cardiovascular risk have raised concerns related to the indications of these drugs in children and adults.[19] This therapeutic class of drugs is known to increase blood pressure and heart rate, which are well-documented risk factors for cardiovascular events.[20] The implications for children may be understood in light of the developmental age-related immaturity of the autonomic regulation. Changes in blood pressure or heart rate may make children more vulnerable.[21] An increase in heart rate can potentially provoke life-threatening arrhythmias when the QT interval does not compensate for the increase.

Drug-Induced Electrocardiographic Changes and Arrhythmias

Many drugs can lead to electrocardiographic changes and potentially life-threatening arrhythmias, often caused by a delayed cardiac repolarization and reflected on the ECG by QT prolongation. The issue of QT prolongation is controversial in children due to difficulty in obtaining quality ECGs and due to important electrophysiological changes. Compared to adults, there are important differences in interpretation of ECGs in infants and children, with important differences from those in adults.

Reading Pediatric ECGs Due to differences in position and size, but also to changes in physiology, pediatric ECGs do not have the same morphology as those recorded on adults, especially during the first 12 months of life when the electrical activity of the heart changes significantly. This change impacts both amplitude and time interval measurements.[22,23] In a pediatric setting, it is therefore of the utmost importance that physicians interpreting pediatric electrocardiograms are aware of the age-specific normal limits (Table 27.2) to prevent incorrect interpretation and resulting care.[24]

Several studies have been published where the same batch of ECGs was interpreted by the emergency room pediatrician, a regular pediatrician, and a pediatric cardiologist.[25,26] Overall, accordance was pretty high; however, when interpreting the clinically significant ECGs, discordance was very significant. As a result, ECGs are best interpreted by a trained pediatric cardiologist, to assure that the proper care can be given to the subject's clinical status. In clinical trials, all ECGs of the subject should be read by the same pediatric cardiologist to avoid inter-reader variability.

TABLE 27.2 Normal ECG Limits[a]

Lead	0–1 month	1–3 months	3–6 months	6–12 months	1–3 years	3–5 years	5–8 years	8–12 years	12–16 years
Heart rate (beats/min)	160 (129, 192)	152 (126, 187)	134 (**112**, **165**)	128 (106, 194)	**119** (97, **155**)	98 (73, 123)	88 (62, 113)	78 (55, **101**)	73 (**48**, **99**)
	155 (136, 216)	154 (126, 200)	139 (**122**, **191**)	134 (106, 187)	**128** (95, **178**)	101 (78, 124)	89 (68, 115)	80 (58, **110**)	76 (**54**, **107**)
P axis	56 (13, 99)	52 (10, 73)	49 (-5, 70)	49 (9, 87)	48 (-12, 78)	43 (-13, 69)	41 (-54, 72)	39 (-17, 76)	40 (-24, 76)
	52 (24, 80)	48 (20, 77)	51 (16, 80)	50 (14, 69)	47 (1, 90)	44 (-6, 90)	42 (-13, 77)	42 (-15, 82)	45 (-18, 77)
P duration (ms)	78 (64, 85)	79 (65, 98)	81 (64, 103)	80 (66, 96)	80 (63, 113)	87 (67, 102)	92 (73, 108)	98 (78, 117)	100 (82, 118)
	79 (69, 106)	78 (62, 105)	78 (63, 106)	80 (64, 07)	83 (62, 104)	84 (66, 101)	89 (71, 107)	94 (75, 114)	98 (78, 122)
PR interval (ms)	99 (77, 120)	98 (85, 120)	106 (87, 134)	114 (82, 141)	118 (86, 151)	121 (98, 152)	129 (99, 160)	134 (105, 174)	139 (107, 178)
	101 (91, 121)	99 (78, 133)	106 (84, 127)	109 (88, 133)	113 (78, 147)	123 (99, 153)	124 (92, 156)	129 (103, 163)	135 (106, 176)
QRS axis	97 (75, 140)	87 (37, 138)	66 (-6, 107)	68 (14, 122)	64 (-4, 118)	70 (7, 112)	70 (-10, 112)	70 (-21, 114)	65 (-9, 112)
	110 (63, 155)	80 (39, 121)	70 (17, 108)	67 (1, 102)	69 (2, 121)	69 (3, 106)	74 (27, 117)	66 (5, 117)	66 (5, 101)
QRS duration (ms)	67 (50, 85)	64 (52, 77)	66 (54, **85**)	**69** (52, **86**)	**71** (54, 88)	**75** (58, 92)	**80** (63, 98)	**85** (67, 103)	**91** (**78**, 111)
	67 (54, 79)	63 (48, 77)	64 (50, **78**)	**64** (52, **80**)	**68** (54, 85)	**71** (58, 88)	**77** (59, 95)	**82** (66, 99)	**87** (**72**, 106)
QTc interval (ms)[b]	413 (378, 448)	419 (**396**, 458)	422 (391, 453)	411 (379, 449)	412 (383, 455)	412 (377, 448)	411 (371, 443)	411 (373, 440)	407 (362, 449)
	420 (379, 462)	424 (**381**, 454)	418 (386, 448)	414 (381, 446)	417 (381, 447)	415 (388, 442)	409 (375, 449)	410 (365, 447)	414 (370, 457)

[a] Bold values indicate that the 95% confidence intervals of the percentile estimates for boys and girls do not overlap.

[b] Corrected QT interval, according to Bazett's formula: $QTc = QT \cdot \sqrt{\text{heart rate}/60}$.

Source: From Rijnbeek et al.,[22] with permission.

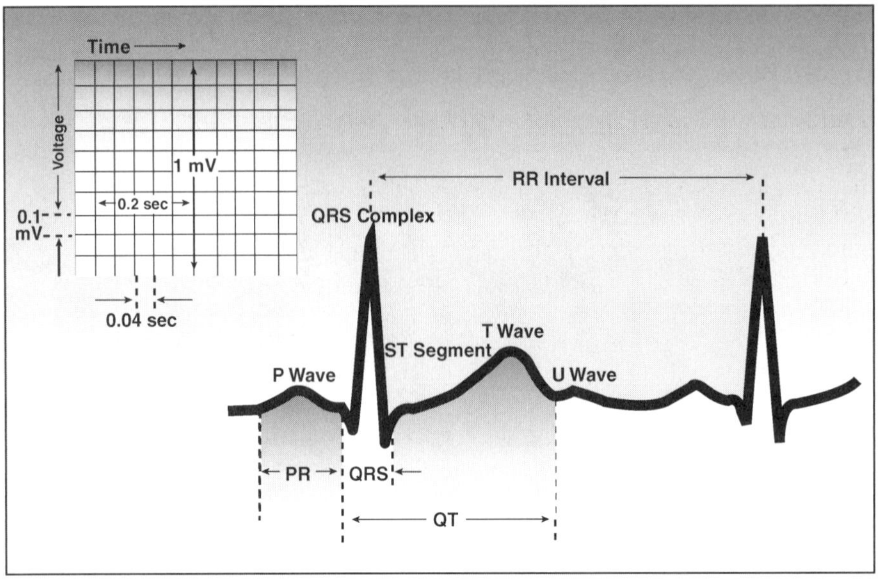

FIGURE 27.1 ECG intervals.

QT Interval The QT interval represents the ventricular depolarization and repolarization of the heart. The depolarization phase is represented on the electrocardiogram (ECG) by the QRS interval. The ventricular repolarization is expressed by the JT interval (Figure 27.1).[27]

The QT interval is affected by a multitude of factors: autonomic tone, time point, meals, age, gender, and drug administration, just to name a few.[28]

Since the late 1980s and early 1990s, the QT interval has received growing attention by the U.S. and EU regulatory agencies due to reported instances of torsades de pointes after terfenadine intake.[29] Since then, many other antiarrhythmic and nonantiarrhythmic drugs have been linked to this polymorphic ventricular arrhythmia.[30–33] There is a relationship between the extent of QT prolongation by a drug or its metabolites and torsades de pointes.[28,34] However, there is no consensus on how much prolongation is too much and the QT is generally accepted to be a biomarker for torsades de pointes.[27] There are also confounding subject-specific risk factors, which should be avoided to predict the possible proarrhythmic risk of a drug just on the QT interval prolongation.[31] QT prolongations of over 500 ms and changes compared to baseline of more than 60 ms are generally accepted as substrate or important risk factors for ventricular tachyarrhythmia; when these adverse events occur, special attention should be given to the subject and investigation should be done to exclude other factors than drug administration that could cause this prolongation (e.g., genetic mutations in the ion channels, metabolic inhibition or drug interactions).[33,35–38]

One drug that was extensively administered to children for many years before it was withdrawn from the market was cisapride. In two studies with healthy adult volunteers, cisapride administration was found to prolong the QTc interval (QT interval corrected for heart rate) by 6 and 18 ms.[39,40] Concomitant use of clarithromycin, which inhibits CYP3A4, led to a higher plasma concentration of cisapride and resulted in a QTc interval increase of 25 ms.[39] Addition of a CYP450 inhibitor is also known to prolong the QTc interval after cisapride intake. It is important to point out that neonates also have a limited CYP450 function which results in higher blood plasma levels and corresponding QTc prolongation compared to other pediatric age groups.[41,42] Other reports, however, support the practical experience that, with appropriate dosing, cisapride can be effective and safe.[43]

Measurement of the QT Interval and Correction to Heart Rate The QT interval should be measured from the beginning of the QRS interval (i.e., either first negative or positive deflection depending on the presence of the Q wave) to the end of the T wave. There are different reasons and conditions that prevent the correct assessment and hamper the comparison of data across studies. This chapter does not have the intention to discuss all the causes and limits itself to problems related to recording quality, measurement methodology, T wave morphology and U wave presence, HR variability, and diurnal change. A debate also exists on the correction formulas, especially in a pediatric population with extremely varying heart rates.

ECG Quality It is obvious that a clinical database needs only source data of high quality. In case of an ECG database, this implies that the electrocardiograms are recorded at a sufficiently high sampling frequency and are free of high- and low-frequency artifacts. Additionally, the subject's skin needs to be well prepared and electrodes correctly placed.

Rijnbeek et al.[22,44] and MacFarlane et al.[45] reported that a sampling frequency of at least 500 Hz is necessary for pediatric ECG recordings. Lower sampling rates might affect the amplitude measurements, leading to lower values. Low- and high-frequency noise will in many cases lead to prolonged QT interval measurements. Therefore, the person recording the ECGs should make certain that the subject is calm and resting and that the electrodes are correctly placed. Many of the commercial available ECG recording machines have a screen, allowing a view of the tracing prior to collection. Proper use of this "preview" should prevent ECGs from being recorded with flat lines in one or several leads, significant sinus arrhythmia, and low- or high-frequency noise.

Exact lead placement and good skin preparation are crucial for the correct interpretation of the ECG.[46–48] Only adequately trained personnel should be allowed to prepare the subject for electrode placement and a refresher course on a periodic basis is advisable.

Manual Measurement Methodologies In many hospital settings, the physician or cardiologist will only use automatic measurements and interpretation provided by the ECG machine. In clinical trials, however, it is generally accepted that measurements and statements provided by an automatic algorithm should be used with caution.[48] In case of a clear tracing without baseline artifact or other noise, a nicely shaped T wave, and in the absence of merging U waves, a well validated automatic algorithm will probably give a very good overreading.[49] Unfortunately, ECGs are not always of such high quality leading to algorithms giving false measurements that result in either under- or overtreatment (Figure 27.2). In a multiple site study or a large hospital with several ECG machines, different ECG machine models might be in use. The algorithms can change significantly from manufacturer to manufacturer and within a brand also from version to version; so caution should be taken when comparing ECGs of a subject recorded on different machines.[50] As a result, it is better to have ECGs interpreted by a person who specializes in cardiology[24,27] To reduce reader variability, all ECGs should be read by the same expert or a small group of experts, where one reader will analyze all the ECGs of a given subject.

Unfortunately, there is no consensus on how the QT interval should be measured and there are also no "official" guidelines.[51] The methodologies most used in clinical trials are: (1) average of the QT measurements performed on 3–5 individual beats of a

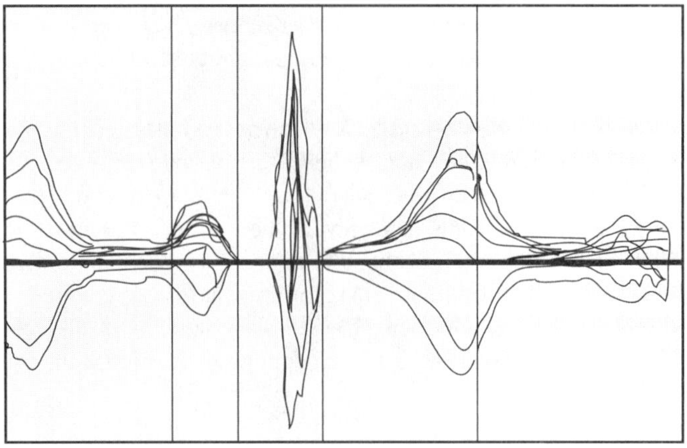

FIGURE 27.2 Examples of faulty T offset caliper placement by computer algorithm.

given lead; (2) measurements on the median beat (i.e., representative beat) of a given lead; and (3) measurement on a global superimposed beat, which consists of a superposition of the median beats of all leads. Most of the time, the end of the T wave is determined as the point where the end of the T wave returns to the isoelectric line or alternatively as the intersection of the isoelectric line and the tangent to the steepest slope of the descending limb of the T wave.[52] Usually the isoelectric line is either defined as the PQ segment or the TP segment.

Because measurements will change depending on the methodology used, the same method should be used consistently within a study. As a result, measurements should not be compared across studies where different methodologies were used.[52–54]

Another very important finding is that the QT interval is not the same in every lead. The difference is considered to be due to repolarization differences between the different areas of the heart. This interlead difference between the longest QT and the shortest QT provides the QT dispersion.[55] Comparing QT measurements performed on different leads in serial ECGs can lead to the wrong assumption of a drug-induced QTc prolongation or might mask the rightful QTc prolongation.

T Wave Morphology and U Wave Presence It is generally accepted that the beginning of the Q wave is mostly easy to find; correct placement of the T wave caliper, however, can be problematic due to noise (as discussed before), the shape of the T wave, and the presence of U waves. In the case of flat T waves, it is not easy or may be even impossible to find the exact end of the repolarization. Use of the tangent methodology can then result in very significant overestimation of the QT interval compared to the threshold method.[53] In the case of notched T waves, the entire complex should be included. In the case of U waves, discrete U waves should be excluded from the QT measurement. Correct measurements, however, can be hampered when T waves merge with U waves. In some cases, it might be difficult to differentiate between a biphasic/notched T wave and a merged U wave. People have suggested performing interval measurements on lead II since U waves are less prominent in this lead.[56]

In 2000, an expert group proposed that large U waves merging with the T wave should be included in the QT interval.[34] They also stated that a QT-U measurement might be a more correct formulation.

Heart Rate Variability and Correction Formulas Heart rate and QT length are inversely related. At higher heart rates the QT interval will be shorter and the converse is also true. To be able to compare QT intervals at different heart rates, people have been trying for many decades to come up with correction formulas that should allow calculating a QTc interval value that is independent of the heart rate.[57] Note that most of these correction formulas hold limitations since they do not take any other influences on the QT interval but heart rate into consideration (e.g., hysteresis, diurnal changes).[27]

Four commonly used correction formulae are the following:

$$\text{QTc Bazett: QTcB} = \text{QT}/\sqrt{\text{RR}};\quad \text{RR in seconds}$$
$$\text{QTc Fridericia: QTcF} = \text{QT}/\sqrt[3]{\text{RR}};\quad \text{RR in seconds}$$
$$\text{QTc Framingham: QTcS} = \text{QT} + 0.154(1000 - \text{RR});\quad \text{RR in milliseconds}$$
$$\text{QTc Hodges: QTcH} = \text{QT} + 1.75(\text{rate} - 60)$$

QTc Bazett (QTcB) is probably the most used, although everyone seems to be in agreement that it is also one of the correction formulas that will overcorrect at high heart rate frequencies and undercorrect at lower heart rates.[57–59] QTcB is routinely used by pediatricians and pediatric cardiologists since it is automatically printed by most electrocardiographs or easy to calculate.[60] Caution is thus necessary when drawing medical conclusions from the Bazett corrected QT interval in this study population since small children have higher heart rates than adults.[61]

Two other correction formulas that are widely used are QTc Fridericia and QTc Framingham.[62,63] Both are considered "better" but are still far from optimal. Most of the correction formulas have been derived from an adult population. Pediatric studies have been published where the common QTc formulas were tested for their utility and others where an optimum pediatric QT correction formula was proposed. In one study, it was shown that none of the previously described correction formulas, including QTcHodges, were appropriate during pediatric exercise testing.[64] In a second pediatric study, where the four correction formulas were tested during sleep, QTcHodges gave the best but not optimum results.[65]

Wernicke et al. tried to derive an age and gender specific correct formula based on a study population of 2288 mostly male children and adolescents diagnosed with ADHD (age between 6 and 17 years). The optimum formula found in this population was $\text{QT}/\text{RR}^{0.38}$. Within the different age and gender groups, different optimal correction factors were found but these need further investigation using different datasets.

Another study performed on 12,543 children, however, concluded that 0.31 is the best correction factor. Based on these results, we can conclude that it is not possible to develop an optimum correction formula that will work for every dataset.

The best correction formula is a subject-specific correction (QTcS) since the RR and QT interval seem to have a subject specific relationship.[67,68] In clinical trials, where the effect of a drug on the QT/QTc interval is investigated, a drug-free baseline day can be used to record the QT/HR relationship. In normal clinical trials, there are usually not enough "drug-free ECGs" recorded to calculate a subject-specific correction formula. In a pediatric population, it might also not be ethical to record many discrete ECGs during a day. A solution might be to hook the subjects onto a 12-lead Holter monitor for 24 hours at baseline and post-baseline days of interest, and extract ECGs to calculate a QTcS and look at drug induced QT/QTc changes.[69,70] Important benefits of 12-lead Holter over standard rest ECG recordings is that retrospective analysis of other time points is always possible and staff will not have to spend a lot of time recording discrete ECGs and therefore have less cost implications for the sponsor of the study. In 2006, a study was published in which 24 children (average age 12 years) were monitored with a 12-lead Holter for 24 hours without any toleration problems.[71] Holter has the additional advantage that in patients having QT prolongation, T wave changes and presence of pathological T waves might help to diagnose long QT syndrome.

Diurnal Changes The QT interval changes throughout the day with the longest QT occurring during sleep and the peak just after awakening, predominantly caused by differences in autonomic stimulation.[72] It is therefore very important that serial ECGs are recorded at the same time of day whenever possible to avoid false positive or negative conclusions with regard to QT prolongation.

Does All QTc Prolongation Lead to Polymorphic Ventricular Tachycardia?

As previously discussed, the QT/QTc interval certainly is a substrate for the risk of torsades de pointes but is, on the other hand, an imperfect marker. It is widely accepted that a QT/QTc prolongation of more than 500 ms and an increase of more than 60 ms compared to baseline pose a significant risk for the development of arrhythmias. However, not every QT prolongation leads to dangerous or life-threatening tachycardias. According to the ICH E14 document, a drug that prolongs the QTc by 5 ms is considered safe.[38] However, having a drug that prolongs the QT/QTc more than this threshold should not prevent its further development if the added value for society is high due to the lack of a safer, better alternative.

Amiodarone has been shown to prolong the QT interval to over 500 ms but yet seems to be safe. The exact reason is unclear but might be due to the fact that it interacts on different levels of ion channel activity.[73] On the contrary, levofloxacin intake does not prolong the QT interval but is associated with cases of polymorphic ventricular tachycardia.[74,75]

Coinciding circumstances (electrolyte disturbances, comedication, and changes in glycemia) and genetic factors play an important role in the fact that some subjects develop a life-threatening arrhythmia having a QTc of 500 ms, and others do not, having a QTc even up to 600 ms.[76] Therefore, in all cases, care to weigh the risk/benefit of a certain compound should be made. Discontinuation of a potentially very effective drug on the basis of a surrogate marker for torsades de pointes should be avoided in these authors' opinion.

CONCLUSION

Due to important developmental cardiovascular changes (morphology, hemodynamics, and electrophysiology), care should be taken to not simply extrapolate data from adults in both cardiovascular and noncardiovascular drug studies. This can lead to additional difficulties/needs in the development design of a drug for pediatric use, especially when QT prolonging effects are known or expected. Extra care should be addressed to good quality ECGs, and the ECGs should be evaluated by an expert in pediatric ECG reading, taking into account the age-specific interval limits.

REFERENCES

1. Ginsberg G, Hattis D, Sonawane B, et al. Evaluation of child/adult pharmokinetic differences from a database derived from the therapeutic drug literature. *Toxicol Sci.* 2002;66:185–200.
2. Edginton AN, Schmitt W, Willmann S. Development and evaluation of a generic physiologically based pharmacokinetic model for children. *Clin Pharmacokinet.* 2006;45:1013–1034.

3. Steinberg C, Notterman DA. Pharmacokinetics of cardiovascular drugs in children. Inotropes and vasopressors. *Clin Pharmacokinet*. 1994;27:345–367.
4. Evans JR, Lou Short B, Van Meurs K, et al. Cardiovascular support in preterm infants. *Clin Ther*. 2006;28:1366–1384.
5. Schmitz L, Koch H, Bein G, et al. Left ventricular diastolic function in infants, children, and adolescents. Reference values and analysis of morphologic and physiologic determinants of echocardiographic Doppler flow signals during growth and maturation. *J Am Coll Cardiol*. 1998;32:1441–1448.
6. Schmitz L, Xanthopoulos A, Koch H, et al. Doppler flow parameters of left ventricular filling in infants: How long does it take for the maturation of the diastolic function in a normal left ventricle to occur? *Pediatr Cardiol*. 2004;482–491.
7. Bruns LA, Canter CE. Should beta-blockers be used for the treatment of pediatric patients with chronic heart failure? *Paediatr Drugs*. 2002;4:771–778.
8. Momma K. ACE inhibitors in pediatric patients with heart failure. *Paediatr Drugs*. 2006;8:55–69.
9. Obreztchikova MN, Sosunov EA, Plotnikov A, et al. Developmental changes in *I* Kr and *I* Ks contribute to age-related expression of dofetilide effects on repolarization and proarrhythmia. *Cardiovasc Res*. 2003;59:339–350.
10. Läer S, Elshoff JP, Meibohm B, et al. Development of a safe and effective pediatric dosing regimen for sotalol based on population pharmacokinetics and pharmacodynamics in children with supraventricular tachycardia. *J Am Coll Cardiol*. 2005;46:1322–1330.
11. Laitinen T, Niskanen L, Geelen G, et al. Age dependency of cardiovascular autonomic responses to head-up tilt in healthy subjects. *J Appl Physiol*. 2004;96:2333–2340.
12. Slordal L, Spigset O. Heart failure induced by non-cardiac drugs. *Drug Safety*. 2006;29: 567–586.
13. Lipshultz SE. Exposure to anthracyclines during childhood causes cardiac injury. *Semin Oncol*. 2006;33:S8–S14.
14. De Wolf D, Suys B, Matthys D, et al. Stress echocardiography in the evaluation of late cardiac toxicity after moderate dose of anthracycline therapy in childhood. *Int J Pediatr Hematol Oncol*. 1994;1:399–404.
15. De Wolf D, Suys B, Maurus R, et al. Dobutamine stress echocardiography in the evaluation of late anthracycline cardiotoxicity in childhood cancer survivors. *Pediatr Res*. 1996;39:504–512.
16. Simbre VC, Duffy SA, Dadlani GH, et al. Cardiotoxicity of cancer chemotherapy: implications for children. *Paediatr Drugs*. 2005;7:187–202.
17. Hatton RC, Winterstein AG, McKelvey RP, et al. Efficacy and safety of oral phenylephrine: systematic review and meta-analysis. *Ann Pharmacother*. 2007;41:381–390.
18. Schroeder K, Fahey T. Over-the-counter medications for acute cough in children and adults in ambulatory settings. *Cochrane Database Syst Rev*. 2004;18:CD 001831.
19. Knight M. Stimulant-drug therapy for attention-deficit disorder (with or without hyperactivity) and sudden cardiac death. *Pediatrics*. 2007;119:154–155.
20. Nissen SE. ADHD drugs and cardiovascular risk. *N Engl J Med*. 2006;354:1445–1448.
21. Galanter CA, Wasserman G, Sloan RP, et al. Changes in autonomic regulation with age: implications for psychopharmacologic treatments in children and adolescents. *J Child. Adolesc Psychopharmacol*. 1999;9:257–265.
22. Rijnbeek PR, Witsenburg M, Schrama E, et al. New normal limits for the paedriatic electrocardiogram. *Eur Heart J*. 2001;22:702–711.
23. Davignon A, Rautaharju P, Boisselle E. Normal ECG standards for infants and children. *Pediatr Cardiol*. 1979/80;1:123–131.
24. Drew BJ, Califf RM, Funk M, et al. Practice standards for electrographic monitoring in hospital settings: an American Heart Association scientific statement from the Councils on Cardiovascular

Nursing, Clinical Cardiology, and Cardiovascular Disease in the Young: endorsed by the International Society of Computerized Electrocardiology and the American Association of Critical-Care Nurses. *Circulation*. 2007;1110:2721–2746.

25. Snyder CS, Bricker JT, Fenrich AL, et al. Can pediatric residents interpret electrocardiograms? *Pediatr Cardiol*. 2005;26:396–399.
26. Wathen JE, Rewers AB, Yetman AT, et al. Accuracy of ECG interpretation in the pediatric emergency department. *Ann Emerg Med*. 2005;46(6):507–511.
27. Bednar MM, Harrigan EP, Anziano RJ, et al. The QT interval. *Progr Cardiovasc Dis*. 2001; 43(5):1–45.
28. Malik M, Camm AJ. Evaluation of drug-induced QT interval prolongation. *Drug Safety*. 2001;24 (5):323–351.
29. Monahan BP, Ferguson CL, Killeavy ES. Torsade de pointes occurring in association with terfenadine use. *JAMA*. 1990;264(21):2788–2790.
30. Darpö B. Spectrum of drugs prolonging QT interval and the incidence of torsades de pointes. *Eur Heart J Suppl*. 2001;3 (Suppl K):K70–K80.
31. De Ponti F, Poluzzi E, Cavalli A, et al. Safety of non-antiarrhythmic drugs that prolong the QT interval or induce torsade de pointes. *Drug Safety*. 2002;25(4):263–286.
32. Haverkamp W, Breithardt G, Camm AJ, et al The potential for QT prolongation and pro-arrhythmia by non-anti-arrhythmic drugs: clinical and regulatory implications report on a policy conference of the European Society of Cardiology. *Cardiovasc Res*. 2000;47:219–233.
33. Roden DM. Drug-induced prolongation of the QT interval. *N Engl J Med*. 2004;350(10): 1013–1022.
34. Anderson ME, Al-Khatib SM, Roden DM, et al. Cardiac repolarization: current knowledge, critical gaps, and new approaches to drug development and patient management. *Am Heart J*. 2002;144(5):769–781.
35. Russell MW. The long QT syndromes. *Progr Pediatr Cardiol*. 1996;6:43–51.
36. Zareba W, Moss AJ. Long QT syndrome in children. *J Electrocardiol*. 2001;34 (Suppl): 167–171.
37. Moss AJ. Long QT syndrome. *JAMA*. 2003;289(16):2041–2044.
38. ICH Topic E14. The Clinical Evaluation of QT/QTc Interval Prolongation and Proarrhythmic Potential for Non-Antiarrhythmic Drugs. Available at http://www.emea.eu.int/pdfs/human/ich/000204en.pdf.
39. van Haarst AD, van 't Klooster GAE, van Gerven JMA, et al. The influence of cisapride and clarithromycin on the QT intervals in healthy volunteers. *Clin Pharmacol Ther*. 1998;64: 542–546.
40. Kivistö KT, Lilja JJ, Backman JT, et al. Repeated consumption of grapefruit juice considerably increases plasma concentrations of cisapride. *Clin Pharmacol Ther*. 1999;66(5):448–453.
41. Benatar A, Feenstra A, Decraene T, et al. Cisapride and proarrhythmia in childhood. *Pediatrics*. 1999;103:856–857.
42. Bernardini S, Semama DS, Huet F, et al. Effects of cisapride on QTc interval in neonates. *Arch Dis Child Fetal Neonatal Ed*. 1997;77:F241–F243.
43. Chhina S, Peverini RL, Deming DD, et al. QTc interval in infants receiving cisapride. *J Perinatol*. 2002;22:144–148.
44. Rijnbeek PR, Witsenburg M, Szatmari AS, et al. PEDMEANS: a computer program for the interpretation of pediatric electrocardiograms. *J Electrocardiol*. 2001;34 (Suppl):85–91.
45. MacFarlane PW, Coleman EN, Houston A, et al. A new 12-lead pediatric ECG interpretation program. *J Electrocardiol*. 1990;23 (Suppl):76–81.
46. Drew BJ, Ide B, Sparacino PS. Accuracy of bedside electrocardiographic monitoring: a report on current practices of critical care nurses. *Heart Lung*. 1991;20:597–609.

47. Medina V, Clochesy JM, Omery A. Comparison of electrode site preparation techniques. *Heart Lung*. 1989;18:456–460.

48. Kligfield P, Gettes LS, Bailey JJ, et al. Recommendations for the standardization and interpretation of the electrocardiogram. Part II: Electrocardiography diagnostic statement list. A scientific statement from the American Heart Association Electrocardiography and Arrhythmias Committee, Council on Clinical Cardiology; the American College of Cardiology Foundation; and the Heart Rhythm Society. Endorsed by the International Society for Computerized Electrocardiology. *Circulation*. 2007;115(10):1306–1324.

49. Darpö B, Agin M, Kazierad DJ, et al. Man versus machine: Is there an optimal method for QT measurements in thorough QT studies? *J Clin Pharmacol*. 2006;46:598–612.

50. Kligfield P, Hancock EW, Helfenbein ED, et al. Relation of QT interval measurements to evolving automated algorithms from different manufacturers of electrocardiographs. *Am J Cardiol*. 2006;98:88–92.

51. Moss AJ. Drugs that prolong the QT interval: regulatory and QT measurement issues from the United States and European perspective. *Ann Noninvasive Electrocardiol*. 1999;4 (3):255–256.

52. Lepeschkin E, Surawicz B. The measurement of the Q-T interval of the electrocardiogram. *Circulation*. 1952;6:378–388.

53. Goldenberg I, Moss AJ, Zareba W. QT interval: how to measure it and what is "normal." *J Cardiovasc Electrophysiol*. 2006;17(3):333–336.

54. Azie NE, Adams G, Darpö B, et al. Comparing methods of measurement for detecting drug-induced changes in the QT interval: implications for thoroughly conducted ECG studies. *Ann Noninvasive Electrocardiol*. 2004;9(2):166–174.

55. Antzelevitch C, Shimizu W, Yan GX, et al. Cellular basis for QT dispersion. *J Electrocardiol*. 1998;30 (Suppl):168–175.

56. Garson A. How to measure the QT interval—What is normal? *Am J Cardiol*. 1993;72(6): 14B–16B.

57. Luo S, Michler K, Johnston P, et al. A comparison of commonly used QT correction formulae: the effect of heart rate on the QTc of normal ECGs. *J Electrocardiol*. 2004;37(Suppl):81–90.

58. Bazett JC. An analysis of time relations of electrocardiograms. *Heart*. 1920;7:353–370.

59. Indik JH, Pearson EC, Fried K, et al. Bazett and Fridericia QT correction formulas interfere with measurement of drug-induced changes in QT interval. *Heart Rhythm*. 2006;3(9):1003–1007.

60. Schwartz PJ, Garson A, Paul T, et al. Guidelines for the interpretation of the neonatal electrocardiogram. *Eur Heart J*. 2002;23:1329–1344.

61. Pearl W. Effects of gender, age, and heart rate on QT interval in children. *Pediatr Cardiol*. 1996;17:135–136.

62. Sagie A, Larson MG, Goldberg RJ, et al. An improved method for adjusting the QT interval for heart rate (the Framingham Heart Study). *Am J Cardiol*. 1992;79:797–801.

63. Fridericia LS. Die Systolendauer im Elektrokardiogramm bei normalen Menschen und bei Herzkranken. *Acta Med Scand*. 1920;53:469–486.

64. Benatar A, Decraene T. Comparison of formulae for heart rate correction of QT interval in exercise ECGs from healthy children. *Heart*. 2001;86:199–202.

65. Benatar A, Ramet J, Decraene T, et al. QT interval in normal infants during sleep with concurrent evaluation of QT correction formulae. *Med Sci Monitor*. 2002;8(5):CR351–CR356.

66. Wernicke JF, Faries D, Breitung R. QT correction methods in children and adolescents. *J Cardiovasc Electrophysiol*. 2005;16(1):76–81.

67. Hnatkova K, Malik M. "Optimum" formulae for heart rate correction of the QT interval. *PACE*. 1999;22(11):1683–1687.

68. Malik M, Färbom P, Batchvarov V, et al. Relation between QT and RR intervals is highly individual among healthy subjects: implications for the heart rate correction of the QT interval. *Heart*. 2002;87:220–228.
69. Sarapa N, Morganroth J, Couderc JP, et al. Electrographic identification of drug-induced QT prolongation: assessment by different recording and measurement methods. *Ann Noninvasive Electrocardiol*. 2004;9(1):48–57.
70. Morganroth J. Cardiac repolarization and the safety of new drugs defined by electrocardiography. *Clin Pharmacol Ther*. 2007;81(1):108–113.
71. Emmel M, Sreeram N, Schickendantz S, et al. Experience with an ambulatory 12-lead Holter recording system for evaluation of pediatric dysrhythmia. *J Electrocardiol*. 2006;39:188–193.
72. Pater C. Methodological considerations in the design of the trials for safety assessment of new drugs and chemical entities. *Curr Controlled Trials Cardiovasc Med*. 2005;6:1.
73. Hohnloser SH. Proarrhythmia with class III antiarrhythmic drugs: types, risks, and management. *Am J Cardiol*.1997;80(8A):82G–89G.
74. Paltoo B, O'Donoghue S, Mousavi MS. Levofloxacin induced polymorphic ventricular tachycardia with normal QT interval. *Pacing Clin Electrophysiol*. 2001;24:895–897.
75. Makaryus AN, Byrns K, Makaryus MN, et al. Effect of ciprofloxacin and levofloxacin on the QT interval: Is this a significant "clinical" event? *Southern Med J*. 2006;99(1):52–56.
76. Suys B, Heuten S, De Wolf D, et al. Glycemia and corrected QT interval prolongation in young type 1 diabetic patients: What is the relation? *Diabetes Care*. 2006;29:427–429.

CHAPTER 28

Renal Function Issues

KATIA BOVEN, MD

Global Clinical Development, Tibotec Inc., Yardley, Pennsylvania 19067

INTRODUCTION

Excretion of drugs by the kidney is dependent on three processes (Figure 28.1). First, glomerular filtration is a passive process whereby a drug not bound to plasma proteins is able to enter the glomerular filtrate. Second, drugs may be excreted into the proximal tubular lumen by an "active" or energy-dependent process. Finally, drugs may be reabsorbed from the tubule by a passive process determined by the physicochemical characteristics of the drug (e.g., lipid solubility and degree of ionization in the proximal tubular fluid).

The most important mechanisms for renal handling of drugs are glomerular filtration and tubular secretion. These mechanisms undergo significant maturational changes in infants and children. These changes are responsible for the differences in renal excretion of drugs in the pediatric population as compared to the adult population. The contribution of each of these mechanisms to the renal clearance of a certain drug is dependent on its chemical structure and physicochemical characteristics. Relatively limited information exists about the maturation of renal transporters, but it is known that significant maturation occurs postnatally, especially in infants less than 34 weeks gestation. Rapid maturation of proximal tubular cells, the site of tubular handling of organic acids, occurs between gestational week 32 and 35, but depending on the specific transport function and factors that regulate it, complete maturation may occur at various ages.[1]

Renal blood flow and renal plasma flow increase with age as a result of increase in cardiac output and a reduction in peripheral vascular resistance.[2] Renal plasma flow averages 12 mL/min at birth and increases to 140 mL/min by 1 year of age.[3] Using the clearance of *para*-aminohippurate (PAH) to estimate renal plasma flow, it has been demonstrated that renal plasma flow reaches adult rates by 5 months of age. Renal blood flow appears to increase in proportion to the development of renal tubules.[3] In the mature kidney, about 20% of renal plasma flow is filtered; this percentage of a drug, if present as free drug in plasma, is extracted continuously. The remainder or a fraction of it may or may not be extracted by the process of tubular secretion as blood flows through the peritubular capillaries.

Pediatric Drug Development: Concepts and Applications
Edited by Andrew E. Mulberg, Steven A. Silber, and John N. van den Anker

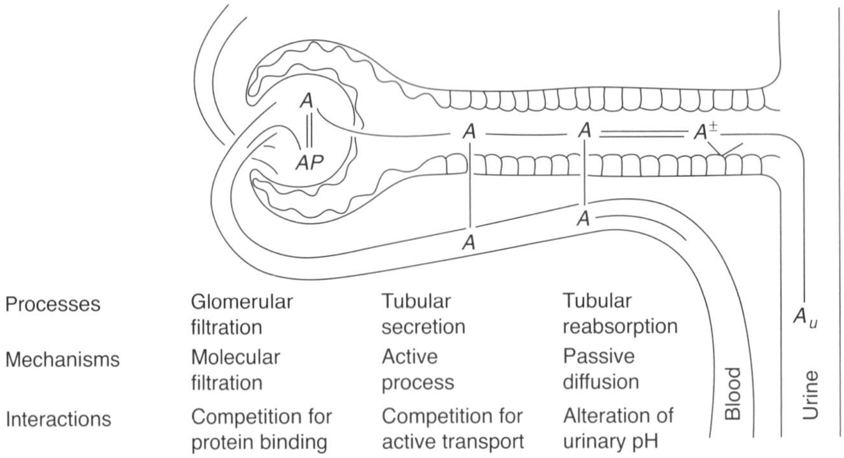

FIGURE 28.1 Mechanisms of renal excretion of drugs and potential sites for drug interactions. *A*, unbound drug; *AP*, drug bound to plasma proteins; A_u, unbound drug in urine.

MATURATION OF RENAL FUNCTIONS

Glomerular Filtration

Glomerular filtration rate (GFR) is reduced in the newborn compared to adults;[4] it is ~15 mL/min/1.73 m^2 at 1 week of postnatal age in preterm infants (gestational age 29–34 weeks), ~40 mL/min/1.73 m^2 at 1 week of postnatal age in full-term infants (gestational age 38–42 weeks), and rises to adult values of 100 mL/min/1.73 m^2 at ~3–6 months of age.[5–8]

To estimate GFR accurately in a clinical setting, the substance to be used must be filterable by the glomerular capillaries to the same extent as water, must be neither secreted nor reabsorbed by the tubule, and must not be metabolized or synthesized by the kidney. Substances that fulfill these criteria include chromium-51 EDTA (^{51}Cr-EDTA), iothalamate, and inulin, but all of these need to be administered intravenously. Inulin clearance is regarded to be the gold standard and can either be measured by a constant intravenous infusion technique, with or without urine collection, or by a single-injection technique. This method is cumbersome for the patient and allergic reactions to inulin have been reported. Iothalamate, because of tubular secretion, approximates creatinine clearance rather than inulin clearance. ^{51}Cr-EDTA provides results similar to inulin; however, it is not licensed for use in humans in several countries. The above approaches for estimation of GFR are recommended when an accurate prediction of GFR is required, particularly in patients with reduced renal function; however, they are not easily applicable in a clinical trial because they require injection of an exogenous compound, radiation exposure, withdrawal of multiple blood samples, and special specimen handling. Glomerular filtration rate can more easily be estimated from serum creatinine concentration or calculated creatinine clearance, the latter based on both serum and urinary creatinine measurements. These methods offer a less precise estimation of renal function but can generally be performed with minimum patient inconvenience and at low cost. Several mathematical equations and nomograms have been proposed in order to predict creatinine clearance on the basis of readily available patient characteristics, such as weight, height, age, and plasma creatinine. The most widely used equation in pediatrics is the Schwartz formula, which gives an approximate GFR, based on body length (L in cm), and plasma creatinine concentration (P_{cr} in mg/dL):[9,10]

$$\text{eGFR}\ (\text{mL/min}/1.73\ \text{m}^2) = \frac{k \cdot L(\text{cm})}{P_{cr}\ \text{mg/dL}}$$

where k, a constant of proportionality, is a function of urinary creatinine excretion per unit of body size. The value of k is 0.33 in low birth weight infants and 0.45 in other infants aged ≤ 1 year, 0.55 in children aged 1–12 years and in adolescent girls aged 13–21 years, and 0.70 in adolescent boys aged 13–21 years. P_{cr} is ~0.8–0.9 mg/dL at birth, decreases to reach ~0.4 mg/dL by the middle of the second postnatal week, and then remains relatively stable during the first 2 years of life. Beyond the age of 2 years it increases, with male children attaining values progressively higher than those in female children, such that mean P_{cr} is 0.9 mg/dL in males and 0.7 mg/dL in females by 18–20 years of age.[10,11]

Validation of the Schwartz equation in pediatric populations has revealed inaccuracies, especially for infants and for children with an age around those proposed for change of constant (i.e., 1 and 12 years old). Recently, approaches based on nonlinear mixed effects models, often referred to as population models, have been developed, which involved the use of patient-specific ^{51}Cr-EDTA plasma data together with supplementary covariate information, such as plasma creatinine, age, body weight, and height. However, these methods were not developed in healthy children but in specific patient populations, for example, the formula described by Léger in children with kidney diseases or the one described by Cole in cancer patients, and therefore need further validation before they can be applied more widely.[12,13]

Several pitfalls exist when using plasma creatinine to estimate GFR. Endogenous creatinine production is related to muscle mass, which explains the rising P_{cr} with age and the higher values in males. Serum or plasma creatinine is dependent on age, gender, and muscle mass, and its tubular secretion, which accounts for about 12–15% of its total renal clearance, increases in chronic renal failure and therefore it is not an ideal marker of GFR, especially in children. Creatinine is secreted by the proximal tubules by both the anionic and cationic secretory pathways.[14] Cimetidine, trimethoprim, pyrimethamine and salicylates can inhibit secretion of creatinine by the proximal tubule.[15–18] Corticosteroids and vitamin D metabolites probably modify the production rate and the release of creatinine.[19, 20] Drugs can also erroneously cause higher plasma creatinine concentrations by interfering with the analytical methods used for creatinine determination. Some cephalosporins, acetohexamide, high doses of furosemide, parenteral methyldopa, and phenacemide were reported to interfere with the Jaffe reaction, while flucytosine and lidocaine influence the enzymatic assay systems. Modification of the colorimetric methodology (e.g., rate-blanked Jaffe reaction) circumvents these interferences but might have other issues. All of these factors make creatinine a less than optimal marker of GFR.[21]

Besides creatinine, several endogenous low-molecular-weight proteins have been studied as candidate markers of GFR, of which cystatin C seems to be the most reliable. Cystatin C is a small protein that is produced by all nucleated cells and its production rate is unaltered in inflammatory conditions. Early methods for cystatin C quantification were slow, impractical for single-sample analysis, and did not allow automation. Newer methodologies, such as particle-enhanced nephelometry and a particle-enhanced turbidimetric assay, are automated but are still not widely available.[20]

Serum cystatin C concentrations are independent of age beyond 1 year of age, height, weight, muscle mass, and gender.[21] The molecule is freely filtered through the glomeruli and almost completely reabsorbed by the proximal tubule, where almost all of it is metabolized. Several formulas to estimate GFR (eGFR) from serum cystatin C are available, of which the most widely used are the following:

Hoek formula[22]	$\text{eGFR} = -4.32 + 80.35 \times 1/\text{CysC}$
Larsson formula[23]	$\text{eGFR} = 77.239 \times \text{CysC}^{-1.2623}$

Values of eGFR for infants and children of various ages using classic clearance techniques are shown in Table 28.1.[24]

The clinical implications of the maturation of GFR become apparent in the case of drugs that are eliminated primarily by glomerular filtration, for example, aminoglycosides (e.g., gentamicin, tobramycin, amikacin) and glycopeptides (e.g., vancomycin). Several studies have investigated the pharmacokinetics of aminoglycosides in preterm and term infants. With increasing gestational age in neonates >7 days of age, the plasma elimination half-life for gentamycin decreases.[25] In preterm children, there is a much stronger correlation between gentamycin half-life and postconceptual age compared with postnatal age.[26] Similar results have been obtained with tobramycin.[27] Intrauterine growth restriction also has an impact on the normalized weight of the kidney, on the number of nephrons, on GFR, and on tubular function. Renal drug clearance of both amikacin and vancomycin is significantly lower in preterm neonates born small for gestational age than in neonates appropriate for gestational age.[28] This reduced clearance was not only observed at birth, but also up to 4 weeks postnatal age.

A factor that influences glomerular filtration of a drug, but is not directly linked to GFR, is its protein binding, which can be different in children compared to adults. Plasma protein binding of a compound depends on the amount of available binding proteins, the affinity constant of the compound for the protein(s), the number of available binding sites, and the presence of pathophysiological conditions or circulating molecules that may alter the drug–protein

TABLE 28.1 Values of Mean (Range) eGFR for Infants and Children of Various Ages Using Classic Clearance Techniques

Age	eGFR ($mL/min/1.73\,m^2$)
Preterm neonates at birth[a]	
28 weeks gestation	0.35
32 weeks gestation	0.50
36 weeks gestation	1.21
Full-term neonates at birth[a]	2.24
2–8 days	39 (17–60)
4–28 days	47 (26–78)
37–95 days	58 (30–86)
1–6 months	77 (39–114)
6–12 months	103 (49–157)
12–19 months	127 (62–191)
2–12 years	127 (89–165)

[a]mL per minute uncorrected for body surface area.

binding interaction. In general, acidic drugs bind to albumin, whereas basic drugs bind to globulins, α_1-acid glycoprotein (AAG), and lipoproteins.[29,30] Although there are well-established data on the levels of plasma proteins in the pediatric population and how they compare to adult values, data on the binding of drugs to plasma proteins in this age group are scarce. Protein binding in neonates and infants tends to be reduced as compared to adults and likely for several reasons. The concentration of plasma proteins is lower in neonates (5.9 g/dL) than in adults (7.2 g/dL) and the proteins are qualitatively different and generally have lower binding capacities in neonates.[31,32] Physiological and pathophysiological increases in bilirubin and free fatty acid plasma concentrations often occur in the neonatal period. Increased concentrations of nonesterified fatty acids, as well as the increased levels of bilirubin and other endogenous molecules that competitively bind to albumin, may reduce drug binding.[33,34] The concentration of AAG, also low at birth, increases over the first zyear to reach adult levels.[35]

Tubular Secretion

A large number of drugs are actively transported across the proximal tubule (Table 28.2).[36] Certain substances are both actively secreted and reabsorbed. With the exception of creatinine and a few other neutral substances, secreted drugs and endogenous compounds are bases or weak organic acids. Organic anions and cations are secreted into the proximal tubular lumen by specific transport processes.

TABLE 28.2 Examples of Substances Secreted or Reabsorbed by Active Mechanisms in the Proximal Tubule

Acids	Bases
Secreted	
Penicillins	Morphine
Cephalosporins	Tetraethylammonium
Acetazolamide	Choline
Thiazides	*N*-Methylnicotinamide
Furosemide	Triamterene
Salicylate	Thiamine
Phenylbutazone	Tolazoline
Uric acid[a]	Aminoglycosides
p-Aminohippuric acid[a]	Histamine
Taurocholic acid[a]	Catecholamines
Glucuronide and sulfate conjugates of metabolized drugs	
Reabsorbed	
Glucose	
Amino acids	
Urate	
Bile acids	
p-Aminohippuric acid	
Taurocholic acid	
Lithium	
Bromide	

[a]Secretion of acids and bases may not be demonstrable when large amounts are subsequently reabsorbed.

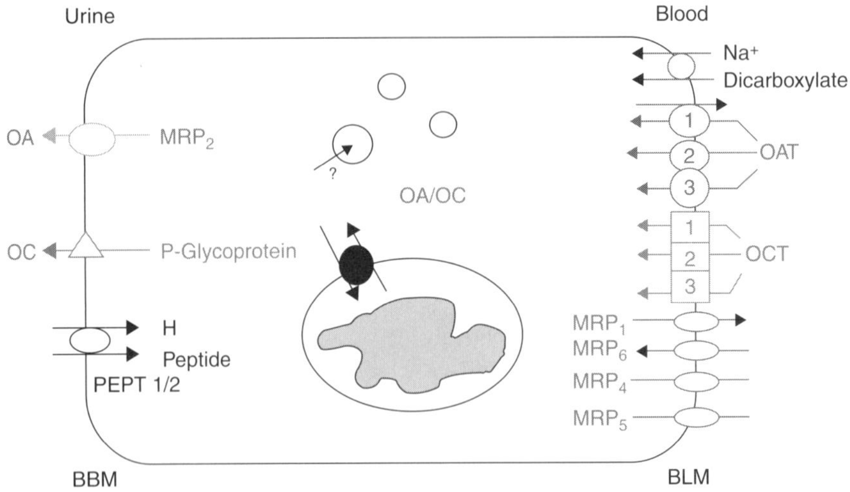

FIGURE 28.2 Schematic model of organic anion and cation transporters in renal proximal tubule. Uptake of organic anions (OAs) across the basolateral membrane (BLM) is mediated by the classic sodium-dependent OAT system, which includes α-ketoglutarate (α-KG2)/OA exchange via the OAT1 and sodium-ketoglutarate cotransport via the Na^+/dicarboxylate cotransporter. The apical (brush border) membrane (BBM) contains various transport systems for efflux of OA into the lumen or reabsorption from the lumen into the cell. The multidrug resistance transporter, MRP_2, mediates primary active luminal secretion. Cellular uptake of organic cation (OC) across the BLM is mediated by organic cation transporters (OCTs) such as OCT1, OCT2, and OCT3. Exit of cellular OC across the BBM is mediated by P-glycoprotein. PEPT1 and PEPT2 mediate luminal uptake of peptide drugs.

Organic anion transporters (OATs) and organic cation transporters (OCTs) play a critical role in protecting against the toxic effects of such substances, whether of endogenous or exogenous origin, by removing them from the blood via a transport mechanism located in the basolateral membrane of renal epithelial cells. Many of these transporter proteins have been identified and characterized and a recent overview is presented in Figure 28.2.[37] The OAT family handles a wide variety of clinically important compounds (e.g., antibiotics, non-steroidal anti-inflammatory drugs, antiviral agents) and toxins. This system plays a critical role in protecting against the toxic effects of anionic substances.[38] OAT-gene expression has been studied in adults, but not in infants and children. Studies in rats and mice showed that mRNA expression increased through birth, with the highest levels detectable at 1 day postpartum, followed by a decrease to adult levels.[39,40] This same pattern was found for OCT1.[41]

Plasma protein binding generally does not prevent tubular extraction, although it may delay it, because the active secretory transport system shuttles free drug across the tubular cell rapidly enough to permit the equilibrium between protein-bound drug and free drug to be reestablished.

Maturation of Tubular Secretion of Anionic Drugs The renal tubular secretion capacity, as measured by PAH secretion, increases over the first months of life and then declines to reach the adult level (per unit of body area) at ~7 months to 1 year of age.[6,8] This increase in organic anion secretion was disproportionate to the increase in renal mass and is thought to reflect the specific maturation of the OAT system. Thus, active tubular secretion takes somewhat longer to reach adult levels than glomerular filtration. As a result, in neonates GFR may be relatively more important than tubular secretion in the excretion of some drugs (e.g., penicillins and cephalosporins) for which active secretion is quantitatively most important in the adult.[5,7] In children and adolescents, the tubular secretion capacity can even exceed that in adults. This was observed with imipenem in children aged 3–12 years receiving imipenem-cilastatin. In comparison to available adult data, the imipenem renal clearance in children was 1.95-fold greater than the estimated creatinine clearance, thus suggesting significant tubular secretion of imipenem in children.[42] In the case of digoxin, the renal tubular secretion also plays a more important role in the excretion of the drug in children and in adolescents than it does in adults.[43] Thus, the inhibition of renal tubular secretion by compounds such as amiodarone may cause a steeper increase in serum digoxin concentration in children.[44] Penicillin is actively secreted by the PAH pathway. In animals, it has been demonstrated that the capacity of secretory pathways responsible for penicillin secretion may undergo substrate stimulation. Although substrate stimulation has not been formally studied in human neonates, there is evidence that it does occur.[45]

Maturation of Tubular Secretion of Cationic Drugs Similar maturation studies do not appear to have been conducted with cationic drugs.

Tubular Reabsorption

Passive transport of drug molecules into renal cells from the tubular lumen is the major mechanism by which the elimination of drugs is slowed and renal accumulation is achieved. Tubular reabsorption becomes more important the more lipid soluble a drug is and the less it is metabolized.[29,37] Most drugs are weak electrolytes. At physiologic pH, they are present in ionized and nonionized forms. Cell membranes are relatively impermeable to the lipid-insoluble ionized form; they delay the passage of lipid-soluble nonionized forms far less. Since the range of urinary pH is much broader than plasma pH, the concentration of a nonionized drug present in tubular urine, and thus the amount that can be reabsorbed passively, depends on urinary pH. The theoretical equilibrium distribution of weak acids between urine and peritubular plasma can be calculated from the Henderson–Hasselbalch equation:

$$\mathrm{pH} = \mathrm{p}K_\mathrm{a} + \log(\mathrm{A}/\mathrm{HA}) \quad \text{or} \quad \frac{\mathrm{A\ (ionized)}}{\mathrm{HA\ (nonionized)}} = 10^{(\mathrm{pH}-\mathrm{p}K_\mathrm{a})}$$

Procedures that alkalinize the urine (e.g., administration of lactate or bicarbonate) will slow the rate of tubular reabsorption and thus increase the excretion of lipid-soluble acids with $\mathrm{p}K_\mathrm{a}$ values in the appropriate range. The urinary excretion of phenobarbital ($\mathrm{p}K_\mathrm{a} = 7.2$) can be accelerated considerably by procedures that increase urine pH. Maintenance of a high rate of urine flow will further facilitate the excretion of these drugs. In a similar manner, excretion of an organic base is enhanced by acidification of the urine.

The extent to which drugs accumulate in renal cells is not dependent merely on the entry mechanisms discussed earlier, but also on the presence in renal cells of both specific and nonspecific receptors with which the drug can react.

Although most reabsorption for drugs is predominantly passive, many endogenous substances undergo active tubular reabsorption, including glucose, urate, and amino acids.

Endocytosis is an important process for the uptake of larger molecules such as polypepetides into the proximal tubular cell. Examples include insulin, growth hormone, and β_2-microglobulin. Endocytosis is also involved in the uptake of aminoglycoside antibiotics and radiocontrast agents, leading to accumulation and toxicity to the proximal tubular cell. In the example of aminoglycosides, this is possible through the binding to acid phospholipids in the brush border of proximal renal tubules.[46]

The development of the renal tubular reabsorbing capacity appears to be largely unknown. The concentrations of retinol-binding protein and microalbumin in the urine of healthy subjects aged from birth to 60 years have been measured as a marker of renal tubular and glomerular development and maturation, respectively.[47] Data suggest that the development and maturation of the glomerular permeability functions and the renal tubular reabsorption are gradual and continuous processes from birth to adolescence, but the key stage of their maturation may be at ~1 and 3 years, respectively. Determination of low-molecular-weight (LMW) proteins in the urine is also a way of detecting early damage in the renal proximal tubules, whereby tubular reabsorption of these LMW proteins is decreased.

GUIDELINES ON PEDIATRIC DOSING FOR RENALLY CLEARED DRUGS

Dosing regimens are devised to maintain the drug plasma concentrations within the therapeutic window. The dosing regimen is commonly based on the product of the total body clearance and the desired plasma concentration. Although the total body clearances of most drugs are available for adults, there is little quantitative information about clearance values in infants and children. Consequently, when drug clearance values are unknown in the pediatric patient, the suggested dosing rate is the product of the adult dosing rate and the fraction of the adult body surface area of the child. Implicit in this calculation is that the total body clearance of the drug is directly proportional to the body surface area in adults and children; that is, proportional to body weight to the two-thirds power.

For renally eliminated drugs, clearance values in children follow the body surface area relationship reasonably well, although caution is recommended for very young children because clearance mechanisms are immature at birth and, depending on the mechanism, achieve adult clearance capacity at different ages. Both maturation and growth are involved in the age-associated increase in renal clearance capacity and thus it is important to use a model that accounts for both maturation and growth influences.

The development of a method to normalize renal function while including all mechanisms that contribute to renal drug clearance over the preadult period has important practical implications for the development of dosing regimens for renally cleared drugs.

Renal plasma flows (RPFs) were examined in relation to body weight, body height, body surface area, average kidney weight, and average basal metabolic rate. None of these factors individually accounted for the changes observed in infants and children.[48] Creatinine clearances from 5146 normal subjects (from 68 publications) ranging in age over the entire life span were compiled, but normalization to body weight, lean body mass, and body surface area did not produce a unifying value for this parameter.[49] More recently, linear regression techniques were used to develop an empirical mathematical model that related GFR to age and weight.[12,13,50] Although these models predicted GFR, they failed to provide insight into the kinetics of maturation as being separate from the kinetics of growth.[51] Recently, a model was developed that characterized the maturation and growth of the renal function parameters: glomerular filtration rate, active tubular secretion, and renal plasma flow.[52–54]

HEALTH AUTHORITY GUIDELINES FOR THE USE OF RENALLY CLEARED DRUGS

Pharmacokinetics in Patients with Impaired Renal Function

A pharmacokinetic (PK) study in patients with impaired renal function is recommended when the drug is likely to be used in these patients and renal impairment is likely to significantly alter the pharmacokinetics of a drug and/or its active/toxic metabolites and a dosage adjustment is likely to be necessary for safe and effective use in such patients. In particular, a study in patients with impaired renal function is recommended when the drug or its active metabolites exhibit a narrow therapeutic index and when excretion and/or metabolism occurs primarily via renal mechanisms (excretion or metabolism). A study also should be considered when a drug or an active metabolite exhibits a combination of high hepatic clearance (relative to hepatic blood flow) and significant protein binding. In this setting, renal impairment could induce a significant increase in the unbound concentrations after parenteral administration due to a decreased plasma protein binding coupled with little or no change in the total clearance (decrease in unbound clearance). If a study in renally impaired subjects is indicated, such a study will typically be performed in adults and different design options are discussed in the specific guidances from the U.S. Food and Drug Administration[55] and from the European Medicines Agency.[56]

Impact of Renal Immaturity When Investigating Drugs Intended for Pediatric Use

A draft guideline on the investigation of medicinal products in the term and preterm neonate and the impact of immaturity of different organ systems on drug handling was issued by the European Medicines Agency and is available at http://www.emea.europa.eu/pdfs/human/paediatrics/26748407en.pdf.

REFERENCES

1. Jones DP, Chesney RW. Development of tubular function. *Clin Perinatol.* 1992;19:33–57.
2. Hook, JB, Baillie MD. Perinatal renal pharmacology. *Annu-Rev Pharmacol Toxicol.* 1979;19: 491–509.
3. Calcagno PL, Rubin MI. Renal excretion of *para*-amino hippurate in infants and children. *J Clin Invest.* 1963;42:1632–1639.
4. Barnett HL, McNamara H, Schultz S, Thompsett R. Renal clearance of sodium penicillin G, procaine penicillin G and insulin in infants and children. *Pediatrics.* 1949;3:418–422.
5. Rane A, Wilson JT. Clinical pharmacokinetics in infants and children. *Clin Pharmacokinet.* 1976;2:2–24.
6. Brown RD, Campoli-Richards M. Antimicrobial therapy in neonates, infants and children. *Clin Phamacokinet.* 1989;17:105–115.
7. Routledge PA. Pharmacokinetics in children. *J Antimicrob Chemother* 1994;34:19–24.
8. West JR, Smith HW, Chasis H. Glomerular filtration rate, effective renal blood flow, and maximal tubular excretory capacity in infancy. *J Pediatr.* 1948;32:10–18.
9. Schwartz GJ, Haycock GB, Edelman CM, Spitzer A. A simple estimate of glomerular filtration rate in children derived from body length and plasma creatinine. *Pediatrics.* 1976;58:259–263.

10. Schwartz GJ, Brion LP, Spitzer A. The use of plasma creatinine concentration for estimating glomerular filtration rate in infants, children and adolescents. *Pediatr Clin North Am*. 1987; 34:571–590.

11. Schwartz GJ, Haycock GB, Chir B, Spitzer A. Plasma creatinine and urea concentration in children: normal values for age and sex. *J Pediatr*. 1976;5:828–830.

12. Léger F, Bouissou F, Coulais Y, Tafani M, Chatelut E. Estimation of glomerular filtration rate in children. *Pediatr Nephrol*. 2002;17:903–907.

13. Cole M, Price L, Parry A, et al. Estimation of glomerular filtration rate in pediatric cancer patients using ^{51}Cr-EDTA population pharmacokinetics. *Br J Cancer*. 2004;90:60–64.

14. Sica DA, Schoolwerth AC. Renal handling of organic anions and cations and renal excretion of uric acid. In: Brenner BM, Rector FC, eds. *The Kidney*. Phialdelphia, PA: WB Saunders; 1996; 607–626.

15. van Acker BAC, Koomen GCM, Koopman MG, de Waart DR, Arisz L. Creatinine clearance during cimetidine administration for measurement of glomerular filtration rate. *Lancet*. 1992; 340:1326–1329.

16. Berg KJ, Gjellestad A, Nordby G, Rootwelt K, Djoseland O, Fauchald P. Renal effects of trimethoprim in cyclosporine- and azathioprine-treated kidney-allografted patients. *Nephron*. 1989;53:218–222.

17. Opravil M, Keusch G, Luthy R. Pyrimethamine inhibits renal secretion of creatinine. *Antimicrob Agents Chemother*. 1993;37:1056–1060.

18. Burry HC, Dieppe PA. Apparent reduction of endogenous creatinine clearance by salicylate treatment. *Br Med J*. 1976;2:16–17.

19. George CRP. Non-specific enhancement of glomerular filtration by corticosteroids. *Lancet*. 1974; ii: 728–729.

20. Swan SK. The search continues—an ideal marker of GFR. *Clin Chem*. 1997;43:913–914.

21. Bertoli M, Luisetto G, Ruffatti A, Urso M, Romagnoli G. Renal function during calcitriol therapy in chronic renal failure. *Clin Nephrol*. 1990;33:98–102.

22. Andreev E, Koopman M, Arisz L. A rise in plasma creatinine that is not a sign of renal failure: Which drugs can be responsible? *J Intern Med*. 1999;246:247–252.

23. Pondraka L, Feber J, Lepage N, Filler G. Intra-individual variation of cystatin C and creatinine in pediatric solid organ transplant recipients. *Pediatr Transplant*. 2005;9:28–32.

24. Hoek FJ, Kemperman FAW, Krediet R. A comparison between cystatin C, plasma creatinine and the Cockcroft–Gault formula for the estimation of glomerular filtration rate. *Nephrol Dial Transplant*. 2003;18:2024–2031.

25. Larsson A, Malm J, Grubb A, Hansson LO. Calculation of glomerular filtration rate expressed in mL/min from plasma cystatin C values in mg/dL. *Scand J Clin Lab Invest*. 2004;64:25–30.

26. Goldsmith DI, Novello AC. Clinical and laboratory evaluation of renal function. In: Edelman CM, ed. *Pediatric Kidney Disease*. Boston: Little, Brown and Company; 1992;461–473.

27. Szefler SJ, Wynn RJ, Clarke DF, Buckwald S, Shen D. Relationship of gentamicin serum concentrations to gestational age in preterm and term neonates. *J Pediatr*. 1980;97:312–315.

28. Kasich JW, Jenkins S, Leuschen P, Nelson RM. Postconceptual age and gentamicin elimination half life. *J Pediatr*. 1985;106:502–505.

29. Arbeter AM, Saccar CL, Eisner S, Sarni E, Yaffe SJ. Tobramycin, sulfate elimination in premature infants. *J. Pediatr*. 1983;103:131–135.

30. Allegaert K, Anderson BJ, van den Anker JN, Vanhaesbrouck S, de Zegher F. Renal drug clearance in preterm neonates: relation to prenatal growth. *Ther Drug Monit*. 2007;29:284–291.

31. Olive G. Pharmacocinetique et biotransformation des medicaments chez l'enfant. *Louvain Med*. 1991;110:565–569.

32. Anderson BJ, McKee AD, Holford HG. Size, myths and the clinical pharmacokinetics of analgesia in pediatric patients. *Clin Pharmacokinet*. 1977;33:313–327.

33. Ehrnebo M, Agurell S, Jalling B, Boreus LO. Age differences in drug binding by plasma proteins: studies in human fetuses, neonates and adults. *J Clin Pharmacol*. 1971;3:189–193.

34. Kurz H, Michels H, Stickel HH. Differences in the binding of drugs to plasma proteins from newborn to adult man. *Eur J Clin Pharmacol*. 1977;11:469–472.

35. Morselli PL. Clinical pharmacokinetics in neonates. *Clin Pharmacokinet*. 1976;1:81–98.

36. Rane A, Lunde PKM, Jalling B, Yaffe J, Sjökvist F. Plasma protein binding of diphenylhydantoin in normal and hyperbilirubinemic infants. *Pediatr Pharmacol Ther*. 1971; 78:877–882.

37. Routledge PA. Pharmacokinetics in children. *J Antimicrobial Chemother*. 1994;34:19–24.

38. Launay-Vacher V, Izzedine H, Karie S, Hulot JS, Baumelou A, Deray G. Renal tubular drug transporters. *Nephron Physiol*. 2006;103:97–106.

39. Cafruny EJ, Feinfeld DA, Schwartz GJ, Spitzer A. Effects of drugs, toxins, and heavy metals on the kidney. In: Edelman CM, ed. *Pediatric Kidney Disease*. Boston: Little, Brown and Company; 1992: 1707–1726.

40. Sweet DH, Bush KT, Nigam SK. The organic anion transporter family: from physiology to ontogeny and the clinic. *Am J Physiol Renal Physiol*. 2001;281:F197–F205.

41. Lopez-Nieto CE, You G, Bush KT, Barros EJ, Beir DR. NKT, a gene product related to the organic cation transporter family that is almost exclusively expressed in the kidney. *J Biol Chem*. 1997;272:6471–6478.

42. Nakajima N, Sekine T, Cha SH. Developmental changes in multispecific organic anion transporter 1 expression in the rat kidney. *Kidney Int*. 2000;57:1608–1616.

43. Pavloova A, Sakurai H, Leclercq B, Beier DR, Yu AS. Developmentally regulated expression of organic ion transporters NKT (OAT1), OCT1, NLT (OAT2) and Roct. *Am J Physiol*. 2000;278: F635–F643.

44. Jacobs RF, Kearns GL, Brown AL, Trang JM, Kluza RB. Renal clearance of imipenem in children. *Eur J Microbiol*. 1984;3:471–474.

45. Lindlay L, Engle MA, Reidenberg MM. Maturation and renal digoxin clearance. *Clin Pharmacol Ther*. 1981;30:735.

46. Koren G, Hesslein PS, MacLeod SM. Digoxin toxicity associated with amiodarone therapy in children. *J Pediatr*. 1984;104:467–470.

47. De Zwart LL, Haenen Hemg Versantwoort CHM, Wolterink G, Van Engelen JGM, Sips Ajam. Role of biokinetics in risk assessment of drugs and chemicals in children. *Regul Toxicol Pharmacol*. 2004;39:282–309.

48. Sastrasinh M, Knauss TC, Weinberg JM. Identification of the aminoglycoside binding site in rat renal brush border membranes. *J Pharm Exp Ther*. 1982;222:350–358.

49. Hua MJ, Kun HY, Jie CS, Yun NZ, De WQ, Yang Z. Urinary microalbumin and retinol-binding protein assay for verifying children's nephron development and maturation. *Clin Chim Acta*. 1997;264:127–132.

50. Rubin M, Bruck E, Rapoport M. Maturation of renal function in childhood: clearance studies. *J Clin Invest*. 1949;28:1144–1162.

51. Hallynck T, Soep H, Thomis J, Boelaert J, Daneels R, Dettli L. Should clearance be normalized to body surface or to lean body mass? *Br J Pharmacol*. 1981;11:523–526.

52. Heilbron DC, Holliday MA, Al-Dahwi A, Kogan BA. Expressing glomerular filtration rate in children. *Pediatr Nephrol*. 1991;5:5–11.

53. Anderson B, McKee A, Holford N. Size, myths and the clinical pharmacokinetics of analgesia in pediatric patients. *Clin Pharmacokinet*. 1997;33:313–327.

54. Hayton WL. Maturation and growth of renal function: dosing renally cleared drugs in children. *AAPS Pharm Sci*. 2002;2:1–7.

55. US Department of Health and Human Services, Food and Drug Administration. Guidance for Industry (FDA). Pharmacokinetics in Patients with Impaired Renal Function—Study Design, Data Analysis, and Impact on Dosing and Labelling. Available at http://www.fda.gov/CDER/GUIDANCE/1449fnl.pdf.

56. European Medicines Agency, Committee for Medicinal Products for Human Use (CHMP). Note for Guidance on the Evaluation of the Pharmacokinetics of Medicinal Products in Patients with Impaired Renal Function. Available at http://www.emea.europa.eu/pdfs/human/ewp/022502en.pdf.

CHAPTER 29

Growth and Physical Maturation

ALISHA J. ROVNER, PhD

Division of Epidemiology, Statistics and Prevention, Eunice Kennedy Shriver National Institute of Child Health and Human Development, National Institutes of Health, Rockville, Maryland 20852

BABETTE S. ZEMEL, PhD

Division of Gastroenterology, Hepatology and Nutrition, The Children's Hospital of Philadelphia, Philadelphia, Pennsylvania 19104

INTRODUCTION

Growth Versus Development

Although "growth and development" are often considered a single biological phenomenon, these processes are contemporaneous but distinct. *Growth* is an increase in cell size and number, while *development* (or maturation) represents a progression through the stages leading to maturity. From a biological standpoint, the endpoint of human growth is the size attained by adulthood, whereas the endpoint of pubertal development is mature reproductive function. Similarly, the endpoint of skeletal development is the closure of the epiphyses.

The processes of human growth and development are incredibly complex and controlled by both genetic and environmental factors. A well-accepted concept of genetic potential for growth holds that at conception a child obtains a genetic blueprint for the potential to achieve a particular adult size and shape. Environmental factors alter this potential: in a neutral or favorable environment a person's genetic potential for growth may be fully realized. The ability of environmental influences to alter genetic potential depends on several factors, including the time at which they occur; the strength, duration, and frequency of their occurrence; and the age and gender of the child.

Growth During Infancy, Childhood, and Puberty

Postnatal growth can be classified broadly into three distinct phases: infancy, childhood, and puberty. Infancy is a high velocity, rapidly decelerating phase of growth. A plateau in growth velocity occurs between 5 and 10 years of age in both boys and girls and is interrupted by a

Pediatric Drug Development: Concepts and Applications
Edited by Andrew E. Mulberg, Steven A. Silber, and John N. van den Anker

mid-childhood (or juvenile) growth spurt, which occurs between ages 6 and 8 years.[1] The adolescent growth spurt, which occurs during puberty, ends with the closure of the epiphyses and the attainment of final adult height.[2] In actuality, growth in all stages proceeds through a series of mini growth spurts interspersed with periods of stasis.[3] Interference during any of the three phases of growth can jeopardize the realization of genetic potential for adult body size.

Growth Status Versus Growth Velocity

Growth status is the interpretation of a growth measurement such as height or weight, in relation to an appropriate growth reference (i.e., Centers for Disease Control or World Health Organization growth charts). For example, knowing that a child has a height measurement of 130 centimeters provides little information until height status is determined by comparing the measurement to a growth chart, at which point one can determine that the child's height is at the 3rd percentile for age and gender. In addition to percentiles, growth status can also be characterized with z-scores, which indicate the number of standard deviations above or below the mean of a reference population. For example, a child whose height is classified at the 50th percentile for her age has a height-for-age z-score of zero. Height status must be interpreted in light of the genetic potential for growth as assessed by factors such as mid-parental height. Additional factors that affect growth status include ethnicity, skeletal and sexual maturation, underlying chronic diseases, and nutritional and other environmental factors.

Growth velocity is the change in height (or weight) divided by the time between the measurements. It is an excellent indicator of current nutritional status and overall well-being for children whose growth is not yet complete. Growth velocity varies with age, gender, maturational status, and season, so these factors must be considered in the interpretation of velocity measures. Growth velocity measurements should be used with caution since the accuracy of the growth velocity measurement is sensitive to the accuracy and precision of the measurements on which it is based. Growth velocity is derived from two measurements, each with their own measurement error; therefore, the measurement error associated with the velocity is greater than that of a single measurement. Growth velocity should be calculated over a 6- or 12-month interval depending on the velocity reference standards used. For infants, shorter intervals are used to assess growth velocity because of the rapid deceleration in the rate of infant growth. Assessment of growth velocity over a longer or shorter interval than that used in the standards may overestimate or underestimate the true growth velocity.

Body Composition

Body composition refers to the amount and relative proportions of fat, muscle, and bone and their chemical components. Two-compartment models of body composition whereby the body is divided into the compartments of fat mass (FM) and fat-free mass (FFM) are most commonly used. Three- and four-compartment models can also be used and are based on methods or combinations of methods that can measure FM along with two or more components of FFM.

Significant changes in body composition occur during growth and development, especially during infancy and puberty. Analysis of body composition provides an in-depth understanding of growth processes by describing changes in the size of body compartments and the chemical composition of the body. Body composition assessment for drug trials is important since dosage regimens for medical therapies often are based on estimated lean body mass. Age- or disease-related factors may cause deviations from assumptions inherent in standard formulas to estimate lean body mass, and thus can affect the delivery of therapy.

Pubertal Development

Puberty is a milestone in human development and involves a rapid transformation of anatomy, physiology, and behavior.[2] The central feature of puberty is the maturation of the primary reproductive endocrine axis. The secondary features of puberty include the development of secondary sexual characteristics, the development of sexually dimorphic anatomical features, and the acceleration and completion of linear growth. Information on the onset and progression of pubertal development is important in interpreting the endocrine and growth status of a child. The effects of pharmaceutical agents on growth in children may differ depending on what stage of pubertal development the child is in when the drug is introduced. Therefore, it is important to assess pubertal development in children participating in drug trials.

METHODS TO ASSESS GROWTH, BODY COMPOSITION, AND PUBERTAL DEVELOPMENT

Anthropometry

Anthropometry provides the most technologically simple approach to assessing growth and body composition in children. This method uses tools that are simple, precise, portable, and inexpensive, and the exam itself is rapid and noninvasive. Moreover, anthropometry is feasible for children of all ages and has the distinct advantage of being the only body composition method that has widely available reference data. Weight and height (or length in infants) are the most fundamental measures of growth. Additional measures, including circumferences, skinfold thicknesses, and lengths and breadths of bony tissue, provide further information for estimating body size and composition.

In order to obtain accurate growth measurements, a properly trained anthropometrist is necessary. Even among well-trained anthropometrists, stylistic differences can develop over time, so periodic checks on intra- and interobserver reliability are recommended. Intra- and interobserver reliability for anthropometric techniques has been described elsewhere.[4] There are several comprehensive publications describing anthropometric techniques.[5–7] In addition, the Centers for Disease Control's (CDC's) website offers training modules on accurate measuring, equipment, and techniques and the proper use of growth and body mass index (BMI) charts.[8] The World Health Organization's (WHO's) website also provides training modules on measuring weight, length, and height and interpreting growth indicators.[9] The WHO website further has a link to an anthropometry training video (available in English, French, and Spanish) that describes measurement procedures and how to calibrate equipment.

Growth

Weight Weight is an essential measure in the assessment of both growth and nutritional status. As a measure of overall body mass, it is an indicator of growth. Additionally, it is an outstanding short-term measure of nutritional status because it can change rapidly. The weight status of a child can be determined by comparing the child's weight with age- and gender-specific weight charts. However, in order to determine underweight and overweight status, a measurement of height is required since children of the same weight who differ in height can have vastly different "relative weights." BMI is the preferred measure of relative weight for children 2–20 years of age. For children younger than 2 years of age, the weight-for-length chart is used to assess relative weight.

Proper equipment and measurement techniques are crucial for obtaining accurate weight measurements. Weight should be measured on a digital electronic or beam balance scale to the nearest 0.1 kg and the scale should be checked weekly with a known calibration weight to assure proper functioning. Prior to each measurement, the scale should be set to zero to make sure that there is no drift in the functioning of the equipment. Shoes and heavy clothing (including diapers for infants) should be removed, and pockets emptied prior to weight measurements. In research studies, weight measurements should be taken at the same time of day after the bladder has been emptied, so that children are always evaluated in the same physiological state.

Height (or Length) Height is a cumulative measure of a child's nutritional history. Deficits in height can be caused by either previous nutritional insults or current growth failure. Consecutive height measurements can be used to assess current growth failure or catch-up growth. Unlike weight, a longer period of observation is required to detect changes in height status (usually 6 months in children).

Proper equipment and techniques for measuring height are essential. A wall-mounted stadiometer with a smoothly gliding headboard that is firmly perpendicular to the wall should be used to measure height. Calibration of the stadiometer should be performed daily with a fixed calibration rod to assure proper functioning. Alternatively, a carefully positioned wall-mounted tape measure can be used, provided the markings are readable and a head paddle is available that will fit at a 90-degree angle to the wall. Height measurements should be accurate to 0.1 cm. Stadiometers and wall-mounted tape measures should be placed in a room where the floor is level and the child can be positioned with the heels parallel to the back.

In order to obtain accurate height measurements, shoes and interfering hair adornments must be removed. If the hair is plaited and cannot be flattened to the skull, then the distance from the skull to the top of the plait should be measured and the measurement should be subtracted from the height measurement. The child should be positioned with the heels, buttocks, and back of the head against the stadiometer.[7] The arms should be extended by the side and relaxed, and the heels positioned as close together as is comfortable for the child. The head should be positioned with the Frankfort plane parallel to the floor. The Frankfort plane is an imaginary line extending from the lower margin of the orbit to the upper margin of the auditory meatus (Figure 29.1). In order to significantly reduce measurement error, consistent positioning of the head is required, especially for longitudinal growth measurements and calculation of growth velocity. Poor posture also contributes to measurement error. The child should be instructed to stand as straight as possible and to inhale deeply and hold his or her breath briefly as the height measurement is taken. For older children, an alternate technique involves a "stretched" height measurement, whereby the anthropometrist glides the palm of his or her hand along the child's spine in an upward sweeping motion and applies upward pressure to the mastoid processes to encourage a fully erect posture. To reduce measurement error, height measurement should be taken in triplicate and the mean used.

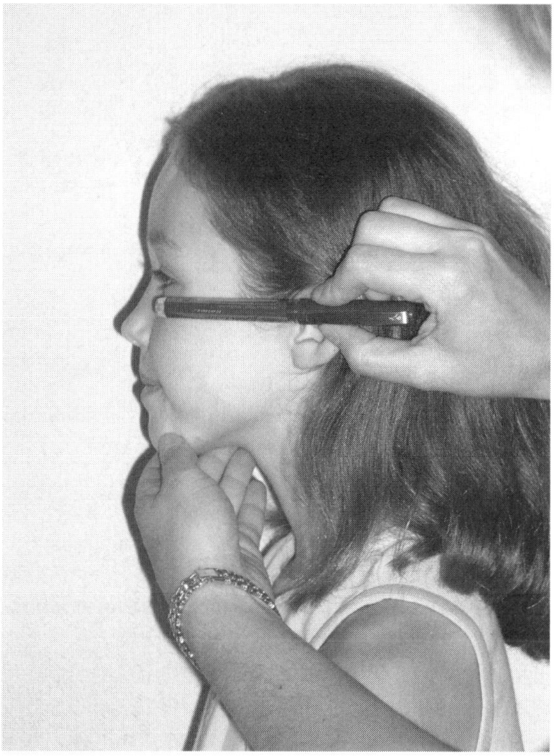

FIGURE 29.1 The Frankfurt plane extends from the auditory meatus to the lower border of the orbit. For proper measurement of height, the Frankfurt plane should be parallel to the floor as shown in the figure.

For infants younger than 2 years of age, supine length should be measured with an infantometer or inflexible length board with a fixed headboard and moveable footboard. Two anthropometrists are required to take the infant's length measurement—one holds the infant's head in position while the other positions the torso and legs (Figure 29.2). The infant's head should be positioned with the Frankfort plane parallel to the headboard. The knees should be flat with the hips parallel to the footboard and the feet flat against the footboard. Measurements should be taken in triplicate and the average used.

Genetic potential for growth can be estimated from the heights of the biological parents. The parents' heights should be recorded on the child's growth chart, and the mid-parent height calculated (the average height of the mother and father). Tables with the values to adjust a child's height based on mid-parent height are age and gender specific, so periodic reevaluation of genetic potential for growth is suggested.[10]

Alternate Length or Stature Measurements Length and stature measures are not appropriate for certain children, including those with spinal curvature, contractures, or any muscoskeletal deformities, because these conditions may interfere with measurement positioning. For those children, upper arm length and lower leg length are alternative measurements that can be used. For infants, sliding calipers can be used, and for older infants and children, an anthropometer is used. The right side should be measured unless a physical

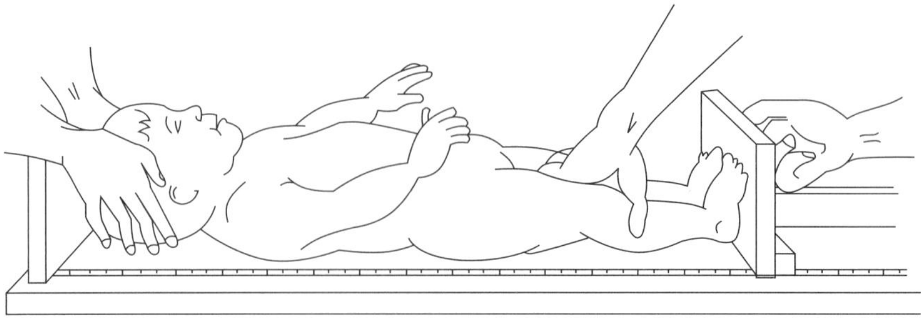

FIGURE 29.2 Proper technique for measuring infant length. Two anthropometrists are required to properly position the infant. (From Frisancho[50] with permission.)

deformity exists with unilateral involvement affecting the right side. The least affected side should be measured and the side that is measured should be recorded so that all subsequent measurements are performed on the same side.

For young children (less than 24 months), upper arm legnth is measured as shoulder–elbow length, from the superior lateral surface of the acromion to the inferior surface of the elbow with the arm flexed at a 90-degree angle. For children older than 2 years of age, upper arm length is measured from the superior lateral surface of the acromion to the radial with the arm relaxed.

In children younger than 24 months, lower leg length is measured as knee–heel length from the heel to the superior surface of the knee. The infant should lie on his/her back with the leg flexed to 90 degrees at the hip, the knee, and the ankle. For children older than 2 years of age, the lower leg measurement is taken from the lower border of the medial malleolus to the medial tip of the tibia while the child is sitting in a relaxed position.

Reference percentiles for upper arm length and lower leg length are available.[11] Children whose growth is outside these ranges may have an unusual growth pattern.

Body Mass Index A desirable relative measure of weight is one that is independent of height. In adults, BMI is a simple relative weight measure that has a zero correlation with height and does not differ significantly by gender. Therefore, among adults, a nutritional classification scheme is based on ranges of BMI as shown in Table 29.1. For children, BMI varies with both age and gender, so it is crucial to compare BMI values to the CDC charts in order to determine age- and gender-specific BMI percentiles. The weight status category for BMI-for-age percentile for children are presented in Table 29.2.

TABLE 29.1 CDC Body Mass Index (BMI) Categories for Adults[a]

BMI	Weight Status
Below 18.5	Underweight
18.5–24.9	Normal
25.0–29.9	Overweight
30.0 and above	Obese

[a] For adults 20 years old and older, BMI is interpreted using standard weight status categories that are the same for all ages and for both men and women.

Source: Kuczmarski et al.[52]

TABLE 29.2 CDC Body Mass Index Categories for Children[a]

BMI Percentile	Weight Status
Less than the 5th percentile	Underweight
5th to less than the 85th percentile	Normal
85th to less than the 95th percentile	At risk overweight
Equal to or greater than the 95th percentile	Overweight

[a]For children, the interpretation of BMI is both age and gender specific.
Source: Kuczmarski et al.[52]

Head Circumference Head circumference represents brain growth and is an important aspect of nutritional assessment in young children. Because of the rapid growth of the brain in infancy, head circumference increases at a faster rate than weight and height early in life. Head circumference is typically measured in infants and children until age 3 years. For children with medical conditions resulting in macrocephaly or microcephaly, head circumference is not an accurate measure of nutritional status.

An accurate head circumference measure is obtained with a flexible nonstretchable measuring tape. The infant should be positioned comfortably in the arm or lap of a parent, and any braid or hair decorations should be removed. Head circumference, or occipital to frontal circumference, is measured from the most prominent part of the back of the head (occiput) to just above the eyebrows (supraorbital ridges), which is the largest circumference of the head. The tape is pulled snugly to compress the hair and underlying soft tissues. The measurement is read to the nearest 0.1 cm. The tape should be repositioned, and the head circumference measurements should be taken in triplicate and the mean used.

Body Composition and Nutritional Status

In addition to growth, anthropometric measures can be used to estimate body composition and nutritional status.

Upper Arm Anthropometry and Skinfold Measurements Upper arm anthropometry is frequently used as an indicator of nutritional status. Mid–upper arm circumference is a good measure of short-term nutritional status and is a composite measure of muscle, fat, and bone in the arm. It is an easily accessible measurement site that requires simple equipment.

A nonstretchable, flexible measuring tape should be used with measurements accurate to 0.1 cm. The midpoint of the upper arm is identified with the arm (usually the right to keep measurements standarized) flexed to 90 degrees and the palm facing upward. The midpoint of the upper arm is located midway between the tip of the acromion process and the olecranon.[7] The midpoint is marked with a washable ink pen. Prior to the arm circumference measurement, the arm should be extended to a fully relaxed position (gently shaking the arm usually assures that it is relaxed). The circumference measurement should be taken over the marked midpoint with the tape perpendicular to the long axis of the arm. Proper measurement technique involves checking to ensure there is no pinching or gaping of the tape. Measurements should be taken in triplicate and the average used.

Skinfold measurement can be used to further estimate body composition.[12,13] This technique is based on prediction equations established from comparisons of skinfold measures with a criterion method, such as hydrodensitometry. The use of skinfold measurements assumes that the prediction equations used are generalizable from the samples from which they were derived and that body density is the same across varying age and gender groups. Despite these assumptions, the fat-free mass and fat mass estimates from skinfold measurements do correlate with other independently derived estimates such as dual energy X-ray absorptiometry.[14]

The triceps skinfold thickness is taken at the same place as the mid-upper arm circumference over the triceps muscle on the back of the upper arm.[7] It is a measure of subcutaneous fat stores and serves as a good indicator of energy stores. Spring-loaded skinfold calipers are necessary to accurately measure skinfold thickness. Holtain skinfold calipers are scaled to 0.2 mm, and Lange calipers are scaled to 0.5 mm.

To measure the triceps skinfold thickness, the child should stand with his/her arms hanging down in a relaxed position. The fold of fat and skin is lifted away from the underlying triceps muscle at the level marked previously for the mid-upper arm circumference measurement. While continuously holding the skinfold in position, the calipers are placed on the skin next to the examiner's fingers lifting up the fold. The calipers are then released so that they exert a constant pressure on the subcutaneous fat fold. The reading should be taken 4s after releasing the caliper's handles.

The mid-upper arm circumference and triceps skinfold thickness can be combined to calculate upper arm fat area and upper arm muscle area. These measures correlate well with total body stores of fat and muscle. Reference data from the National Health and Nutrition Examination Survey are available for upper arm fat area and muscle area.[15] The formulas for calculating upper arm fat area and upper arm muscle area are presented in Table 29.3.

In addition, Table 29.4 summarizes some of the common prediction equations used to estimate body composition from skinfold measurements. Details on techniques for other skinfold measurements have been described by Lohman et al.[7]

Dual Energy X-Ray Absorptiometry Dual energy X-ray absorptiometry (DXA) has now become a fairly widely used technique to assess body composition. DXA is a rapid, safe, accurate, and reproducible method that measures bone mass, lean body mass (LBM), and fat mass (FM). The density of these tissue compartments varies so they will differentially attenuate the energy beams. Using two low-energy X-ray beams, the computer software is able to determine the mass of the three tissue compartments.[16] The radiation exposure from DXA scans is very low (see Table 29.5) and scan times are rapid, so it is quite feasible to use in children. However, the use of DXA to assess body composition in multicenter studies can be problematic because of differences between DXA manufacturers, and between different software and hardware versions from the same manufacturer in body composition determination, especially for small children. In addition, metal implants and movement artifacts among very young children, or children with seizures or cerebral palsy result in loss of usable data in the research setting.

TABLE 29.3 Formulas for Calculating Upper Arm Fat Area and Upper Arm Muscle Area

Area	Formula
Upper arm area	$\text{Armcirc}^2/4\pi$
Arm muscle area (mm^2)	$[\text{Armcirc} - (\text{triceps} \times \pi)]^2/4\pi$
Arm fat area (mm^2)	Arm area − Arm muscle area

Source: Frisancho.[15]

TABLE 29.4 Equations for Predicting Body Composition from Anthropometry

Two-Skinfold Method for Prediction of Percent Body Fat[a]	
Prepubescent white males:	% Body fat = $1.21(T + S) - 0.008(T + S)^2 - 1.7$
Prepubescent black males:	% Body fat = $1.21(T + S) - 0.008(T + S)^2 - 3.2$
Pubescent white males:	% Body fat = $1.21(T + S) - 0.008(T + S)^2 - 3.4$
Pubescent black males:	% Body fat = $1.21(T + S) - 0.008(T + S)^2 - 5.2$
Postpubescent white males:	% Body fat = $1.21(T + S) - 0.008(T + S)^2 - 5.5$
Postpubescent black males:	% Body fat = $1.21(T + S) - 0.008(T + S)^2 - 6.8$
All females:	% Body fat = $1.33(T + S) - 0.013(T + S)^2 - 2.5$
When sum of triceps and subscapular is greater than 35 mm, use the following:	
All males:	$0.783(T + S) + 1.6$
All females:	$0.546(T + S) + 9.7$
Four-Skinfold Method for Prediction of Percent Body Fat[b]	
% Body fat = [(4.95/body density) − 4.5] × 100, where body density is defined as follows:	
Prepubertal children (1–11 years)	
Males:	Density = 1.1690 − 0.0788 log sum of 4 skinfolds
Females:	Density = 1.2063 − 0.0999 log sum of 4 skinfolds
Adolescent children (12–16 years)	
Males:	Density = 1.1533 − 0.0643 log sum of 4 skinfolds
Females:	Density = 1.1369 − 0.0598 log sum of 4 skinfolds
Ages 17–19 years	
Males:	Density = 1.1620 − 0.0630 log sum of 4 skinfolds
Females:	Density = 1.1549 − 0.0678 log sum of 4 skinfolds
Ages 20–29 years	
Males:	Density = 1.1631 − 0.0632 log sum of 4 skinfolds
Females:	Density = 1.1599 − 0.0717 log sum of 4 skinfolds

[a]T = triceps; S = subscapular.
[b]Sum of four skinfolds equals triceps plus biceps plus subscapular plus suprailiac.
Source: Adapted from Zemel et al.[51]

TABLE 29.5 Radiation Exposure of Methods Used to Assess Bone Density in Children

Source	Effective Dose Equivalent (μSv)
Natural radiation sources	
Natural background radiation at sea level	3,000 per year
Roundtrip transcontinental airplane flight	60
DXA (Hologic QDR-4500)	
Lumbar spine	3.8
Lateral spine	1.4
Whole body	2.6
Hip	1.3
Hand–wrist X-ray	1

Infant body composition by DXA can be obtained for research purposes using some Hologic DXA devices.[17–19] Measurement of body composition among older infants and toddlers is more problematic since they are more likely to require sedation and validation studies in this age range are lacking. Attention to the software version and the assumptions involved are important considerations among children of all ages. For example, the pediatric whole body analysis algorithm used by Hologic, Inc. in the Version 11 software was intended for use in children under age 10 to 12 years old. There were no evidence-based guidelines regarding the exact age or body size range that was best suited for this software mode, and the scan analysis results were sufficiently different that data derived from children analyzed in pediatric mode and adolescents analyzed in adult mode could not be combined.[20] The latest generation of Hologic software (Version 12) uses a weight-based adjustment to the body composition algorithm to improve the detection of bone in small subjects.[21,22] This new algorithm also changes the body composition results, and the new algorithm has not been validated for children.

A trained technician with expertise should perform DXA measurements. The International Society for Clinical Densitometry provides bone densitometry courses and offers certification for those who perform DXA scans.[23] Prior to having a DXA measurement, children should remove any articles of clothing that contain metal (snaps, jewelry, glasses, bras, and hairpieces) and change into hospital scrubs. For a whole body DXA scan, the child lays on the table with arms at the side and fingers flat on the table. A whole body scan takes less than 5 minutes to complete. If there is significant movement, the scan should be repeated.

Bioelectrical Impedance Analysis Bioelectrical impedance analysis (BIA) is a simple, portable, and inexpensive method used to estimate body composition. It is based on the principle that the impedance of a small electrical current as it passes through the body is proportional to the amount of water and electrolytes. Initially, results from BIA devices were quite variable and not generally accepted as accurate measures of body composition. However, in recent years, technological improvements have made BIA a more reliable and therefore a more acceptable way of assessing body composition.

Two source and two detector electrodes are used on the wrist and ipsilateral ankle to estimate the total body water (TBW) of the entire body. Other BIA models measure leg impedance using a stand-on scale with electrodes positioned under the heel and foot. Positioning the electrodes over smaller segments of the body provides regional estimates of body composition. One of the major limitations of single-frequency BIA systems is that 28% of the whole body resistance occurs in the arm, and 33% in the lower leg, even though they represent only 1.5–3% of body weight compared to the trunk.[24] Furthermore, sources of measurement error can arise from the skin surface where the electrodes are placed, and leakage can occur along the surface between terminals.[25] For young children with small hands and feet, this is particularly concerning. Another concern is the estimation of TBW, FFM, FM, and percent body fat derived from BIA measurements. Prediction equations are required to estimate FFM and FM from BIA measurements.[26] Several equations have been developed for use with children.[27–31] Differences in the prediction equations may be attributable to the characteristics of the reference sample in terms of age and body proportions, or the reference method used to generate the prediction equation. Compared to a four-compartment model of body composition, BIA measures of FFM and fatness do not perform much better than using height and weight to estimate body composition.[32]

Skeletal and Pubertal Development

Age is an imperfect indicator of maturation because individuals differ in their rate of biological maturation. Skeletal maturation and pubertal development are the most commonly used measures of biological maturation due to the ease of measurement and standardized methods of assessment. Skeletal maturation is a particularly useful index of biological maturation in research studies because (1) it can be used to characterize children prior to the onset of puberty, and (2) it is a continuous measure of maturation until fusion of the epiphyses occurs, and therefore has certain advantages for statistical analyses in research studies.

Skeletal maturation, or "bone age," is determined from a hand–wrist radiograph (usually the left for standardization) (Figure 29.3). Epiphyseal maturation of the hand and wrist is compared to an atlas or standard based on the development of healthy children. The Greulich and Pyle atlas[33] is most commonly used in the United States. The atlas was based on a longitudinal study of well-off Caucasian children from Cleveland, Ohio, who participated in the Brush Foundation study between 1931 and 1942. The atlas contains a series of hand–wrist radiographs that illustrate typical bone development from birth to 18 years in girls, and from birth to 19 years in boys. For each bone age, the standard deviation is given. Two standard deviations above or below the bone age are considered to be clinically significant.

Outside the United States, the Tanner–Whitehouse III (TWIII) method[34] is most commonly used. Rather than evaluating the entire hand and wrist, the TWIII technique evaluates 13 individual bones (the distal radius; the distal ulna; the first, third, and fifth metacarpals; and the proximal, middle, and distal phalanx) and assigns a score to each bone.

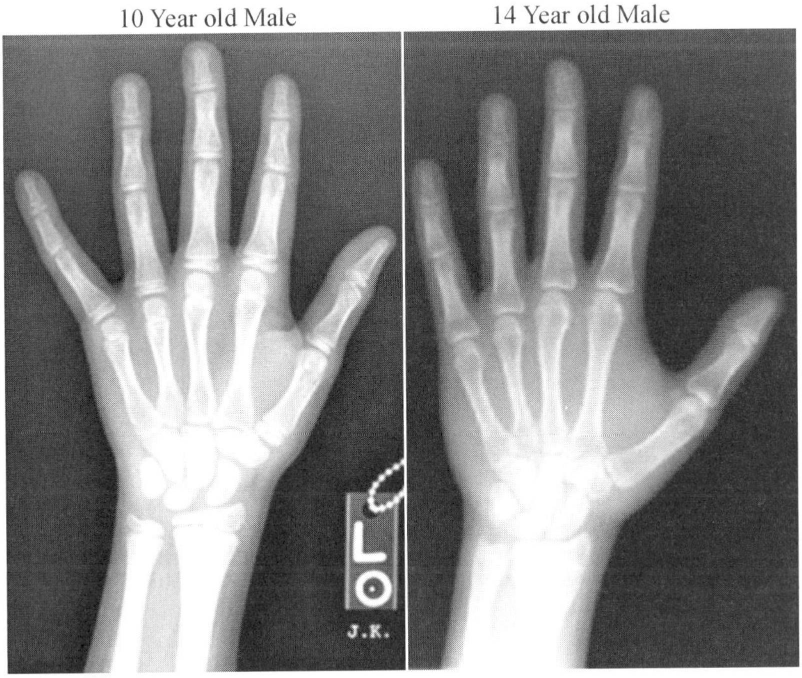

FIGURE 29.3 Hand–wrist X-ray used to determine "bone age".

A summary score is calculated and used to determine bone age. Because the TWIII method has been used so widely, the reference range is based on large international studies conducted from the 1950s to the 1990s, and a computer program is available to calculate z-scores and percentiles based on these reference data. The time required to evaluate a hand–wrist X-ray is longer using the TWIII method compared to the Greulich and Pyle method. However, the TWIII method is somewhat more flexible than the Greulich and Pyle atlas for evaluating children with mosaic maturation of the bones in the hand and wrist.

Delayed or advanced sexual maturation is not uncommon among children with chronic illnesses. Delayed sexual maturation often occurs in combination with growth failure and may manifest as a delay in the onset or progression through stages of sexual development. Classification of sexual maturity is based on breast development in girls, genital development in boys, and pubic hair development in both boys and girls. These are categorized into the five stages described by Tanner (often referred to as Tanner stages).[35] Sexual maturity staging can be done by a trained clinician or by a self-assessment questionnaire that contains line drawings and written descriptions (Figure 29.4).[36] To optimize results using the self-assessment method, the questionnaire should be completed in a private area, with clear instructions to use both the written descriptions and the pictorial representations and to view themselves in a mirror when selecting their stage of development. Parents may assist with the self-assessment. In boys, sexual maturation can also be assessed by physical examination of the testes using an orchidometer to determine testes size and the corresponding stages of genital development (see Table 29.6).

Another important component of sexual maturation is the timing of menarche, the onset of menses in girls, and by spermarche, the onset of nocturnal emissions in boys. Retrospective information regarding age at menarche can also be obtained. Table 29.7 shows the median age at entry into Tanner stages for girls and boys from the U.S. National Health and Nutrition Examination Survey. Ethnic differences in the timing of onset and progression through the stages of sexual maturation have been reported.[37,38]

Use of Reference Data

Reference data is central to the use and interpretation of anthropometric measures of growth and nutritional status. Body dimensions and composition vary as a function of age and gender, and the variability is also influenced by maturation. Therefore, age- and gender-specific reference ranges are needed to determine growth and nutritional status.

As outlined previously, the quality of anthropometric data is dependent on proper equipment and measurement technique. The interpretation of the measurements is equally dependent on the quality of the reference data. The key characteristics of ideal reference data are: (1) they should be based on a large, representative sample of children; (2) the children should be a well-nourished population; and (3) the variability in the measures should be appropriately characterized in the reference curves. Table 29.8 lists the recommended reference data for anthropometric measures of growth and nutritional status.

Growth as an Outcome Measure in Studies of Drug Treatment Effects

Growth assessment is part of the standard pediatric examination because it is a sensitive indicator of the overall well-being of the child. Drug treatments that accommodate normal growth are preferred and, in some cases, may have growth-promoting effects. As increasing numbers of children with chronic diseases are surviving to adulthood, there is growing

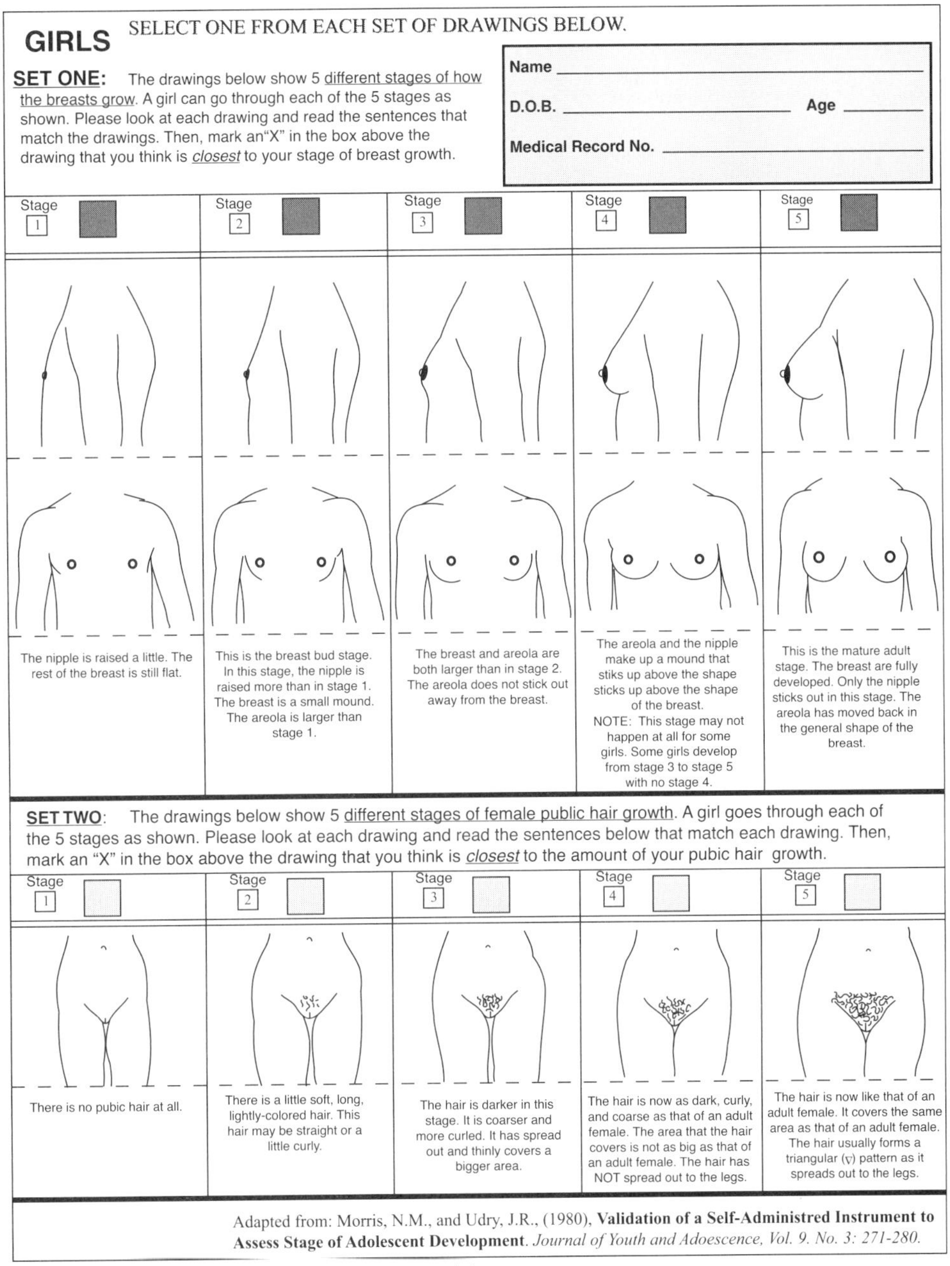

GIRLS SELECT ONE FROM EACH SET OF DRAWINGS BELOW.

Name ______

D.O.B. ______ **Age** ______

Medical Record No. ______

SET ONE: The drawings below show 5 different stages of how the breasts grow. A girl can go through each of the 5 stages as shown. Please look at each drawing and read the sentences that match the drawings. Then, mark an"X" in the box above the drawing that you think is *closest* to your stage of breast growth.

Stage 1	Stage 2	Stage 3	Stage 4	Stage 5
The nipple is raised a little. The rest of the breast is still flat.	This is the breast bud stage. In this stage, the nipple is raised more than in stage 1. The breast is a small mound. The areola is larger than stage 1.	The breast and areola are both larger than in stage 2. The areola does not stick out away from the breast.	The areola and the nipple make up a mound that stiks up above the shape sticks up above the shape of the breast. NOTE: This stage may not happen at all for some girls. Some girls develop from stage 3 to stage 5 with no stage 4.	This is the mature adult stage. The breast are fully developed. Only the nipple sticks out in this stage. The areola has moved back in the general shape of the breast.

SET TWO: The drawings below show 5 different stages of female public hair growth. A girl goes through each of the 5 stages as shown. Please look at each drawing and read the sentences below that match each drawing. Then, mark an "X" in the box above the drawing that you think is *closest* to the amount of your pubic hair growth.

Stage 1	Stage 2	Stage 3	Stage 4	Stage 5
There is no pubic hair at all.	There is a little soft, long, lightly-colored hair. This hair may be straight or a little curly.	The hair is darker in this stage. It is coarser and more curled. It has spread out and thinly covers a bigger area.	The hair is now as dark, curly, and coarse as that of an adult female. The area that the hair covers is not as big as that of an adult female. The hair has NOT spread out to the legs.	The hair is now like that of an adult female. It covers the same area as that of an adult female. The hair usually forms a triangular (▽) pattern as it spreads out to the legs.

Adapted from: Morris, N.M., and Udry, J.R., (1980), **Validation of a Self-Administred Instrument to Assess Stage of Adolescent Development**. *Journal of Youth and Adoescence, Vol. 9. No. 3: 271-280.*

(a)

FIGURE 29.4 Self-assessment questionnaire for assessment of the Tanner stages of sexual maturation: (a) girls and (b) boys.

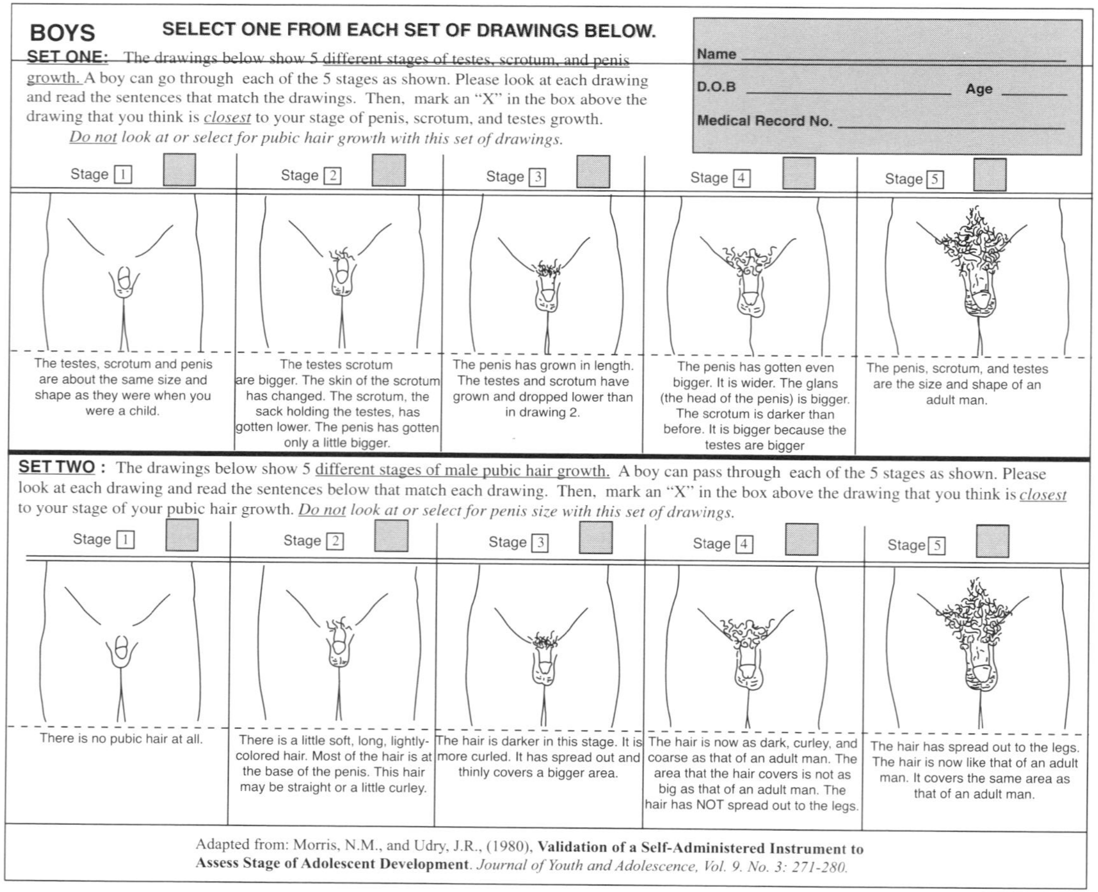

BOYS **SELECT ONE FROM EACH SET OF DRAWINGS BELOW.**

Name ______ D.O.B ______ Age ______ Medical Record No. ______

SET ONE: The drawings below show 5 different stages of testes, scrotum, and penis growth. A boy can go through each of the 5 stages as shown. Please look at each drawing and read the sentences that match the drawings. Then, mark an "X" in the box above the drawing that you think is *closest* to your stage of penis, scrotum, and testes growth. *Do not look at or select for pubic hair growth with this set of drawings.*

Stage 1	Stage 2	Stage 3	Stage 4	Stage 5
The testes, scrotum and penis are about the same size and shape as they were when you were a child.	The testes scrotum are bigger. The skin of the scrotum has changed. The scrotum, the sack holding the testes, has gotten lower. The penis has gotten only a little bigger.	The penis has grown in length. The testes and scrotum have grown and dropped lower than in drawing 2.	The penis has gotten even bigger. It is wider. The glans (the head of the penis) is bigger. The scrotum is darker than before. It is bigger because the testes are bigger	The penis, scrotum, and testes are the size and shape of an adult man.

SET TWO: The drawings below show 5 different stages of male pubic hair growth. A boy can pass through each of the 5 stages as shown. Please look at each drawing and read the sentences below that match each drawing. Then, mark an "X" in the box above the drawing that you think is *closest* to your stage of your pubic hair growth. *Do not look at or select for penis size with this set of drawings.*

Stage 1	Stage 2	Stage 3	Stage 4	Stage 5
There is no pubic hair at all.	There is a little soft, long, lightly-colored hair. Most of the hair is at the base of the penis. This hair may be straight or a little curley.	The hair is darker in this stage. It is more curled. It has spread out and thinly covers a bigger area.	The hair is now as dark, curley, and coarse as that of an adult man. The area that the hair covers is not as big as that of an adult man. The hair has NOT spread out to the legs.	The hair has spread out to the legs. The hair is now like that of an adult man. It covers the same area as that of an adult man.

Adapted from: Morris, N.M., and Udry, J.R., (1980), **Validation of a Self-Administered Instrument to Assess Stage of Adolescent Development**. *Journal of Youth and Adolescence, Vol. 9. No. 3: 271-280.*

(b)

FIGURE 29.4 *(Continued).*

TABLE 29.6 Ranges of Testes Size Corresponding to Tanners Stages of Genital Development

Tanner Stage	Testes Size
1	$\leq$3 cc
2	4–6 cc
3	8–10 cc
4	12–15 cc
5	>15 cc

recognition of the importance of maintaining normal growth and achieving normal adult size. The following scenarios offer examples of the importance of growth as an outcome in clinical trials.

Effects of Corticosteroids on Growth in Children with Cystic Fibrosis Airway obstruction and recurrent respiratory infections lead to inflammation, lung damage, and mortality in people with cystic fibrosis (CF). Anti-inflammatory agents, including oral

TABLE 29.7 Median Ages in Years at Entry into Tanner Stages

	Non-Hispanic White	Non-Hispanic Black	Mexican-American
		Girls	
Breast			
Stage 2	10.4	9.5	9.8
Stage 3	11.8	10.8	11.4
Stage 4	13.3	12.2	13.1
Stage 5	15.5	13.9	14.7
Pubic Hair			
Stage 2	10.6	9.4	10.4
Stage 3	11.8	10.6	11.7
Stage 4	13.0	11.9	13.2
Stage 5	16.3	14.7	16.3
		Boys	
Genitalia			
Stage 2	10.0	9.2	10.3
Stage 3	12.3	11.8	12.5
Stage 4	13.5	13.4	13.8
Stage 5	16.0	15.0	15.8
Pubic Hair			
Stage 2	12.0	11.2	12.3
Stage 3	12.7	12.5	13.1
Stage 4	13.6	13.7	14.1
Stage 5	15.7	15.3	15.8

Source: Sunn et al.[37]

TABLE 29.8 Recommended Reference Data for the Assessment of Growth and Nutritional Status

Measure	Reference Data Source
Height	CDC growth charts [52]
	WHO growth charts [53]
Weight	CDC growth charts [52]
	WHO growth charts [53]
Body mass index (BMI)	CDC growth charts [52]
	WHO growth charts [53]
Growth velocity	Incremental growth Tables [54] and height velocity standards[55]
Upper arm anthropometry	Norms for upper limb fat and muscle[15]
Skeletal maturation	Greulich and Pyle atlas[33]
	Tanner–Whitehouse III method[34]
Pubertal maturation	National Health and Nutrition Examination Survey[37]

corticosteroids, are prescribed to children with CF since complications due to inflammation can occur early in life. Oral prednisone-equivalent doses of 1–2 mg/kg on alternate days appear to slow the progression of lung disease in children with CF.[39] However, there are several known adverse effects of this treatment, including impaired linear growth. Three studies have examined the effect of steroids on growth in children with CF and are discussed below.

In 1985, Auerbach et al.[40] published the first randomized study of alternate-day prednisone in children with CF. Forty-five children with CF (ages 1–12 years) from a single center were randomly assigned to either placebo or prednisone (2 mg/kg/day). All of the children had mild to moderate lung disease, the only inclusion criteria used in this study. Although Auerbach and co-workers reported that there were no adverse effects on growth after 4 years, the study has several limitations. The mean heights and weights were reported without any comment on height or weight status: z-scores or percentiles were not reported. In addition, the number of males and females and the age distributions in each group were not reported. Although the goal of randomization is to increase the likelihood that groups will be similar, it does not guarantee it. Therefore, if more males or older children were in the prednisone group, then the mean heights and weights of that group would most likely be higher, thus potentially masking a deleterious effect of the drug on growth. This study underscores the importance of comparing growth z-scores or percentiles in studies assessing growth since such measures take age and gender into consideration while weight or height do not.

Ten years after the Auerbach et al.[40] study was published, Eigen et al.[41] published a larger randomized controlled study of alternate-day prednisone in 285 children (ages 6–14 years) from 15 North American CF centers. Children were assigned to low-dose prednisone (1 mg/kg), high-dose prednisone (2 mg/kg), or placebo and were followed every 6 months for 48 months. Beginning 6 months onward, height z-scores fell in the 2-mg/kg groups compared with the placebo group. For the 1-mg/kg group, height z-scores were lower 24 months after treatment and onward. Both the low-dose and the high-dose treatments had a beneficial effect on lung function. The authors concluded that low-dose prednisone should be considered for children with CF since improvements in lung function outweighed the potential for adverse effects when the treatment period was less than 24 months. The study by Eigen and co-workers illustrates another important concept in pediatric drug trials: some adverse events from drugs may be immediate whereas others, such as growth failure, can take longer to observe. Therefore, it is important to ensure that children involved in drug trials are followed for a sufficient time period to determine if there is an adverse effect on growth or development. This is an example of a critical issue related to adverse drug events and relatedness, which is discussed by Hetherington (Chapter 40) and Kaplan (Chapter 41) in this book.

In 1999, Lai et al.[42] published a paper that evaluated the growth of 224 participants from the study by Eigen and co-workers 6–7 years after the steroid treatment had been discontinued. In both boys and girls, height z-scores declined for those who received low-dose and high-dose prednisone but began to increase about 2 years after the drug was discontinued (Figure 29.5). Among the boys in that study who were 18 years or older,

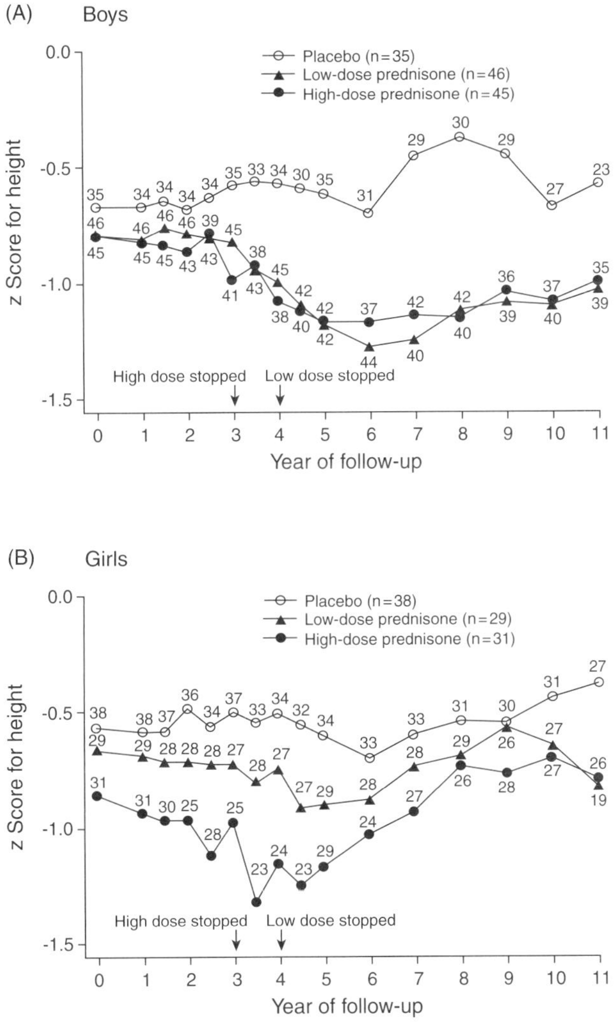

FIGURE 29.5 These graphs show the relationship of height z-scores to years of follow-up for (a) boys and (b) girls who received placebo, low-dose prednisone (1 mg/kg), and high-dose prednisone (2 mg/kg). For boys who received prednisone, the height z-scores remained lower compared to boys who received placebo ($p = 0.03$). For girls, after 6 years, there were no statistically significant differences among treatment groups. (From Lai[42] with permission.)

those who received prednisone were shorter by approximately 13 percentile points (4 cm) than those who received placebo. In addition, boys who were younger when they started treatment had lower final heights, suggesting that the timing of prednisone exposure is important. A similar persistent adverse affect on growth was not seen in girls. This study raises an important consideration: a drug's potential effects on growth and development may differ by gender and age. Therefore, drug trials should have an adequate sample size so that there is sufficient power to detect growth differences by gender and age.

Effect of Zinc Supplementation on Growth in Children with Sickle Cell Disease It has been recognized for several decades that children with homozygous SS sickle cell disease (SCD-SS) have poor growth and delayed pubertal maturation.[43] Despite normal weight and length at birth, reduced weight-for-age and height-for-age z-scores are common in children with SCD and become more pronounced with age.[44,45] Zinc deficiency has been the most widely studied micronutrient in SCD and has been suggested as one of the causes of growth failure in this population.[46,47] Zinc deficiency in SCD was initially described in the 1970s; however, it was not until a decade later that a zinc supplementation study was conducted in this population. That study examined the effect of zinc supplementation on growth in 16 males (ages 14–18 years) with SCD.[48] Although the authors report a beneficial effect of zinc supplementation on growth, this study has the limitation of not reporting growth status.

In 2002, Zemel et al.[49] published a 12-month randomized, placebo-controlled study of zinc supplementation in 42 prepubertal children (ages 4–10 years) with SCD-SS. Randomization was stratified by age group (4.0–6.9 years and 7.0–10.9 years) and by initial height status (height-for-age z-scores ≥ -0.15 and < -0.15) to ensure comparability of the two groups. Children were followed at 3, 6, and 12 months after initiation of zinc supplementation. After 12 months, the children who received the placebo demonstrated significant declines in height-for-age (HAZ) and weight-for-age (WAZ) z-scores. In contrast, the children in the zinc group did not have a significant decline in HAZ or WAZ, indicating a protective effect of zinc (Figure 29.6). These results demonstrate the importance of using a

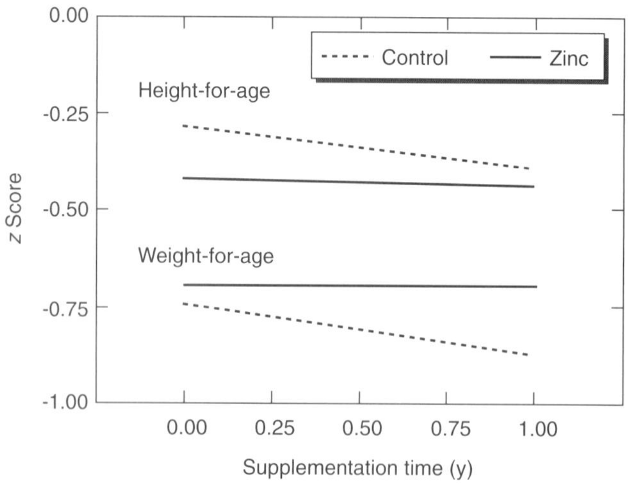

FIGURE 29.6 This graph illustrates the changes in growth status in the zinc and control groups. Height and weight z-scores decreased significantly over time in the control group but not in the zinc group. (From Zemel et al.[49] with permission.)

control group in growth studies: if this study had lacked a control group then the conclusion that zinc is not beneficial in treating growth failure in children with SCD-SS could have been made erroneously. Preventing growth decline is an important treatment effect for many pediatric illnesses.

CONCLUSION

The anthropometric exam provides important information about short- and long-term growth but requires attention to detail to obtain accurate and precise measures. Comparison of measures with well-designed reference data is necessary to determine growth and nutritional status and monitoring changes over time. Percentiles or z-scores are the preferred way to express growth status in children since they take age and gender into account. For some studies, the growth of specific body compartments, such as muscle, fat, and bone, are important outcomes. Delayed skeletal and pubertal maturation and assessment of genetic potential for growth provide a context for interpreting growth and nutritional status.

REFERENCES

1. Lejarraja H. Growth in infancy and childhood: a pediatric approach. In: Cameron N, ed. *Human Growth and Development*. Oxford, UK: Academic Press; 2002: 21–44.
2. Ellison PT. Puberty. In: Cameron N, ed. *Human Growth and Development*. Oxford, UK: Academic Press; 2002: 65–84.
3. Lampl M, Veldhuis JD, Johnson ML. Saltation and stasis—a model of human growth. *Science* 1992;258:801–803.
4. Ulijaszek S, Lourie J. Intra- and inter-observer error in anthropometric measurement. In: *Anthropometry*. Cambridge, UK: Cambridge University Press; 1994: 30–54.
5. Cameron N. The methods of auxological anthropometry. In: Falkner F, Tanner JM, ed. *Human Growth: A Comprehensive Treatise*. New York: Plenum Press; 1986: 3–46.
6. Gibson R. *Principles of Nutritional Assessment*. New York: Oxford University Press; 1990.
7. Lohman T, Roche AF, Martorell R. *Anthropometric Standardization Reference Manual*. Champaign, IL: Human Kinetics Books; 1988.
8. Center for Disease Control. Growth Chart Training. 2007. Available at http://www.cdc.gov/nccdphp/dnpa/growthcharts/training/index.htm.
9. World Health Organization. 2007. Child Growth Standards Training Course. Available at http://www.who.int/childgrowth/training/en.
10. Himes JH, Roche AF, Thissen D, et al. Parent-specific adjustments for evaluation of recumbent length and stature of children. *Pediatrics*. 1985;75:304–313.
11. Stallings VA, Zemel BS. Nutrition assessment of the disabled child. In: Sullivan PB, ed. *Clinics in Developmental Medicine: Feeding the Disabled Child*. London: MacKeith Press; 1996.
12. Brook CG. Determination of body composition of children from skinfold measurements. *Arch Dis Child*. 1971;46:182–184.
13. Durnin J, Rahaman M. The assessment of the amount of fat in the human body from measurements of skinfold thickness. *Br J Nutr*. 1967;21:681–689.
14. Buison AM, Ittenbach RF, Stallings VA, Zemel BS. Methodological agreement between two-compartment body-composition methods in children. *Am J Hum Biol*. 2006;18:470–480.
15. Frisancho A. New norms of upper limb fat and muscle areas for assessment of nutritional status. *Am J Clin Nutr*. 1981;34:2450–2445.

16. Crabtree N, Leonard M, Zemel B. Dual energy X-ray absorptiometry. In: Sawyer A, Bachrach L, Fung E, ed. *Bone Densitometry in Growing Patients*. Totowa, NJ: Humana Press; 2007: 41–57.
17. Brunton JA, Weiler H, Atkinson SA. Improvement in the accuracy of dual energy X-ray absorptiometry for whole body and regional analysis of body composition: validation using piglets and methodologic considerations in infants. *Pediatr Res*. 1997;41:590–596.
18. Picaud JC, Nyamugabo K, Braillon P, et al. Dual-energy X-ray absorptiometry in small subjects: influence of dual-energy X-ray equipment on assessment of mineralization and body composition in newborn piglets. *Pediatr Res*. 1999;46:772–777.
19. Koo WWK, Hammami M, Hockman EM. Interchangeability of pencil-beam and fan-beam dual-energy X-ray absorptiometry measurements in piglets and infants. *Am J Clin Nutr*. 2003; 78:236–240.
20. Zemel BS, Leonard MB, Stallings VA. Evaluation of the Hologic experimental pediatric whole body analysis software in healthy children and children with chronic diseases. *J Bone Miner Res*. 2000;15:400.
21. Kelly TL. Pediatric whole body measurements. *J Bone Miner Res*. 2002;17:S297.
22. Shypailo RJ, Ellis KJ. Bone assessment in children: comparison of fan-beam DXA analysis. *J Clin Densitom*. 2005;8:445–453.
23. International Society for Clinical Densitometry. Education and Certification. 2007. Available at http://www.iscd.org/Visitors/certification/index.cfm.
24. Foster KR, Lukaski HC.Whole-body impedance—what does it measure? *Am J Clin Nutr*. 2007; 64:388S–396S.
25. Oldham NM. Overview of bioelectrical impedance analyzers. *Am J Clin Nutr*. 1996; 64: 405S–412S.
26. Chumlea WC, Guo SS. Bioelectrical impedance and body composition: present status and future directions. *Nutr Rev*. 1994;52:123–131.
27. Danford LC, Schoeller D, Kushner RF. Comparison of two bioelectrical impedance analysis models for total body water measurement in children. *Ann Hum Biol*. 1992;19:603–607.
28. Davies PS, Preece M, Hicks CJ, Halliday D. The prediction of total body water using bioelectrical impedance in children and adolescents. *Ann Hum Biol*. 1988;15:237–240.
29. Deurenberg P, Van Der Kooy K, Paling A, Withagen P. Assessment of body composition in 8–11 year old children by bioelectrical impedance. *Eur J Clin Nutr*. 1989;43:623–629.
30. Houtkooper LB, Lohman TG, Going SB, Hall MC. Validity of bioelectric impedance for body composition assessment in children. *J Appl Physiol*. 1989;66:814–821.
31. Kushner RF, Schoeller D, Fjeld CR, Danford L. Is the impedance index (ht^2/R) significant in predicting total body water? *Am J Clin Nutr*. 1992;56:835–839.
32. Wells JC, Fuller NJ, Dewitt O, Fewtrell MS, Elia M, Cole TJ. Four-component model of body composition in children: density and hydration of fat-free mass and comparison with simpler models. *Am J Clin Nutr*. 1999;69:904–912.
33. Greulich W, Pyle S. *Radiographic Atlas of Skeletal Development of the Hand and Wrist*. Stanford, CA: Stanford University Press; 1950.
34. Tanner J, Healey M, Goldstein H, et al. *Assessment of Skeletal Maturity and Prediction of Adult Height (TW3) Method*. London: WB Saunders; 2001.
35. Tanner J. *Growth at Adolescense*. Oxford, UK: Academic Press; 1962.
36. Morris NM, Udry JR. Validation of a self-administered instrument to assess stage of adolescent development. *J Youth Adolesc*. 1980;9:271–280.
37. Sunn S, Schubert CM, Chumlea WC, et al. National estimates of the timing of sexual maturation and racial differences among US children. *Pediatrics*. 2002;110:911–919.

38. Sunn S, Schubert CM, Liang R, et al. Is sexual maturity occurring earlier among US children? *J Adolesc Health.* 2005;37:345–355.

39. Cheng K, Ashby D, Smyth R. Oral steriods for cystic fibrosis. *Cochrane Database Sytematic Rev.* 1999;1–18.

40. Auerbach HS, Williams M, Kirkpatrick JA, Colten HR. Alternate-day prednisone reduces morbidity and improves pulmonary function in cystic fibrosis. *Lancet* 1985;2:686–688.

41. Eigen H, Rosenstein BJ, FitzSimmons S, Schidlow DV. A multicenter study of alternate-day prednisone therapy in patients with cystic fibrosis. Cystic Fibrosis Foundation Prednisone Trial Group. *J Pediatr.* 1995;126:515–523.

42. Lai HC, FitzSimmons S, Allen DB, et al. Risk of persistent growth impairment after alternate-day prednisone treatment in children with cystic fibrosis. *N Engl J Med* 2000;342:851–859.

43. Winsor T, Burch G. The habitus of patients with sickle cell anemia. *Hum Biol.*1944;16: 99–114.

44. Stevens MC, Maude GH, Cupidore L, Jackson H, Hayes RJ, Serjeant GR. Prepubertal growth and skeletal maturation in children with sickle cell disease. *Pediatrics.* 1986;78:124–132.

45. Zemel BS, Kawchak DA, Ohene-Frempong K Schall JI, Stallings VA. Effects of delayed pubertal development, nutritional status, and disease severity on longitudinal patterns of growth failure in children with sickle cell disease. *Pediatr Res.* 2007;61.

46. Phebus CK, Maciak BJ, Gloninger MF, Paul HS. Zinc status of children with sickle cell disease: relationship to poor growth. *Am J Hematol.* 1988;29:67–73.

47. Leonard MB, Zemel BS, Kawchak DA, Ohene-Frempong K, Stallings VA. Plasma zinc status, growth, and maturation in children with sickle cell disease. *J Pediatr.* 1998;132: 467–471.

48. Prasad AS, Cossack ZT. Zinc supplementation and growth in sickle cell disease. *Ann Intern Med.* 1984;100:367–371.

49. Zemel BS, Kawchak DA, Fung EB, Ohene-Frempong K, Stallings VA. Effect of zinc supplementation on growth and body composition in children with sickle cell disease. *Am J Clin Nutr.* 2002;75:300–307.

50. Frisancho A. *Anthropometric Standards for the Assessment of Growth and Nutritional Status.* Ann Arbor, MI: University of Michigan Press; 1990.

51. Zemel BS, Riley ER, Stallings VA. Evaluation of methodology for nutritional assessment in children: anthropometry, body composition, and energy expenditure. *Annu Rev Nutr.* 1997;17: 211–235.

52. Kuczmarski RJ, Ogden CL, Grummer-Strawn LM, et al. CDC growth charts: United States. *Adv Data.* 2000;1–27.

53. WHO Multicentre Growth Reference Study Group. *WHO Child Growth Standards: Length/Height-for-Age, Weight-for-Age, Weight-for-Length, Weight-for-Height and Body Mass Index-for-Age: Methods and Development.* Geneva: World Health Organization; 2006.

54. Baumgartner RN, Roche AF, Himes JH. Incremental growth tables—supplementary to previously published charts. *Am J Clin Nutr.* 1986;43:711–722.

55. Tanner J, Davies P. Clinical longitudinal standards for height and height velocity for North American children. *J Pediatr.* 1985;107:317–329.

PART VII

CLINICAL TRIAL OPERATIONS AND GOOD CLINICAL TRIALS

CHAPTER 30

The Consent Process in Pediatric Clinical Trials

M. RENEE SIMAR, PhD

INC Research, Inc., Austin, Texas 78746

INTRODUCTION

The sea of regulations and guidance on informed consent can be daunting when the moral compass is turned toward pediatric trials. Protections for pediatric consent represent overlapping perspectives of ethicists, jurists, regulators, and child advocates. The perspectives, in turn, are shaped by ethics committees into operational requirements for pediatric investigators. The perspectives generally converge on the central questions regarding consent, while divergence or silence is often the case on pragmatic issues for the process. The divergence reflects the limited research on factors that impact parental decisions to enroll their child and the extent to which the child should participate in the decision. These unresolved dilemmas along with others related to an ethically valid process for pediatric participants are unfolding with the emergence of unprecedented numbers of pediatric pharmaceutical trials in recent years.

Fortunately, intermingled findings from the fields of pediatric bioethics and therapeutics reveal a few answers. Other findings have illuminated aspects of the process that warrant standardization or further analysis. Successful management of the unresolved dilemmas entails thoughtful consideration on findings to date and a thorough appreciation of current regulatory policy. This chapter identifies many of the challenges for valid pediatric consent and solutions for the overall process. The chapter begins with a review of regulatory guidance and implications for stakeholders. The next section discusses empirical research and perspectives of investigators and families on the process. Taken together, an enhanced scheme to guide elements of the consent process in pediatric trials is presented. The chapter presumes familiarity with the fundamentals of human research protections and Good Clinical Practice (GCP).

Pediatric Drug Development: Concepts and Applications
Edited by Andrew E. Mulberg, Steven A. Silber, and John N. van den Anker

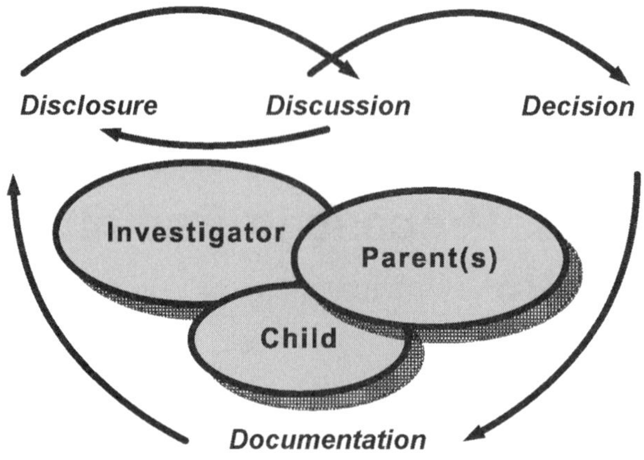

FIGURE 30.1 Process of consent and assent in pediatric trials. An interactive and iterative process among investigator, parent(s), and child satisfies regulatory policies and ethical principles for consent in pediatric clinical trials.

GUIDANCE AND OBLIGATIONS

Standards for research consent are derived from doctrines on respect and voluntary participation.[1,2] Consent refers to choices for oneself based on competency or capacity* to understand.[4] An informed decision following review of consent documents and discussion with an investigator are the core features of the consent process. While similar to the constructs of consent for clinical care, consent in pediatric research is rigorously regulated to protect the rights of both parents and children.[5–7] Regulations mandate elements of consent to ensure that each participant understands the purpose, procedures, risks, benefits, alternatives, compensation, and right to withdrawal. In simple terms, the elements of the process consist of disclosure, discussion, decision-making, and documentation (Figure 30.1). Implicit to pediatric trials is that each element of the process is amplified to accommodate both parent and child. ("Parent" in this context equates to legal guardian and "child" or "children" refers to nonemancipated children or adolescents.)

Rights of Parents

Valid consent assumes an ability to understand the disclosed information and appreciate the premise of medical research apart from clinical care.

Current guidance limits consent to agreement for oneself, thereby precluding consent by a parent for a child. Parents retain their decisional right regarding their child's participation but the terminology to describe their role has changed. "Proxy consent" is infrequently used and generally reserved for decisions on behalf of others with impaired cognition. "Parental consent" is commonly cited, although a more precise description of the parental role is denoted by the terms "informed permission" or "parental permission." In any case, the

*Competency and capacity are often used interchangeably in the consent literature. In stricter terms, competency is a legal term to describe the ability for rational decision-making. Capacity denotes ability to decide and is often used in healthcare settings to describe a task-specific ability.[3]

decision-making right of the parent for research applies until the child reaches the legal age of majority as defined by local authorities.

The requirement for permission from one or both parents is assigned by national regulations.[8,9] The U.S. Food and Drug Administration (FDA) links the requirement to the level of risk. Permission from two parents is necessary for trials involving greater than minimal risk without potential for direct benefit. (See Wendler and Varma[10] for a recent description on levels of risk under FDA.) Otherwise, one parent is sufficient. The exception to the two-parent requirement is for cases in which a parent is not reasonably available (e.g., deceased, unknown, incompetent) or only one parent has legal custody.[8] The U.S. exception is without a definition of "reasonably" and ignores a role for an involved, noncustodial parent who desires to be included in the decision.[6] Regulations for one- or two-parent consent among EU member states generally agree with U.S. regulations. One exception is Denmark's requirement for two parents regardless of risk.[9] In the absence of committee or institutional policies to define reasonable availability, the investigator is charged with assessing parental availability. A conservative approach is to involve both parents whenever possible, regardless of regulation.

Rights of Children

The complexities of assimilating information, then making a free choice about participation, preclude individuals without adequate cognitive and psychological development, as well as those who are implicitly at risk for undue influence or coercion by authority figures. These latter individuals, including children, are classified as members of a vulnerable population and are afforded special protections for research participation.[11] Standards for research competency are lacking, but children are presumed to be without adequate decisional capacity. In the absence of consent rights, the proviso of assent was established to ensure that children as well as parents are informed about participation.

Assent is described as knowing agreement, whereby the child is presented with age-appropriate information and given the opportunity to voice an opinion regarding participation.[2] The Convention on Rights of the Child further implies a promise that their views will be given weight by adults.[12] U.S. regulations advise that affirmative agreement be solicited from the child with a final decision by the parent, whereas EU regulations denote that parental consent must represent the child's presumed will.[8,9,13]

Refusal to participate or dissent may be displayed as a "deliberate objection" or refusal to undergo a procedure.[16] Although dissent is described as a right within the scope of pediatric research protections, the right is limited by some national regulations and local requirements described later.[14,15] In other words, the right to a voice in the process does not equal "unfettered freedom" for the decision-making by a child.[6,17]

Waivers

Waivers of parental permission are not allowed under current U.S. regulations, even for treatments equivocal to clinical care that would otherwise be permitted under state laws.[8] (Parental waivers are allowed in the United States under a different regulation for publicly funded research that is exempt from FDA purview.[18]) Many advocates have argued that the regulation ignores the right to consent and privacy afforded adolescents when seeking medical attention for mental health, sexual activites, and substance abuse.[19–21] Parental permission may be mandated for research, yet aspects of adolescent rights to confidentiality may still apply in some regions and should be appropriately incorporated into consent and assent documents.[9] In practice, investigators should clearly delineate to pediatric participants any matters to be discussed with parents.

Assent waivers are allowed under U.S. regulations. Ethics committees may waive assent for a group of subjects or individual subjects when board members determine that participation is in the best interest of the child. Cases for waiver often include for conditions where alternative proven treatments are unavailable, other treatments have failed, or available treatments are deemed as unacceptably high risk *and* the research offers the prospect of direct benefit.[8,9]

Legal Guardians or Representatives

In general, the transfer of parental consent rights to legal guardians or representatives is straightforward.[8,13] Less clear are protections for foster children. U.S. regulations require assignment of an advocate independent of the investigator and guardian organization prior to enrolling wards of the state.[8] Ethical issues regarding research participation of foster children led to major reforms on subject protections in the 1960s, yet the problem was brought to media attention again in the 1990s for HIV studies.[22,23] Public hearings in 2005 revealed that state policies to guide research in foster children are generally lacking in the United States.[24] Concerns have been raised in other regions, especially in South Africa, where thousands of children in need of HIV drugs are without parents. Stakeholders should be well aware of additional regional protections for these children prior to study initiation.[25]

THE ROLE OF INSTITUTIONAL ETHICS COMMITTEES

Ethics committees are responsible for ensuring that consent and assent documents comply with applicable national statues and local standards. (Note that some institutions also require collateral approval by other entities. Academic centers may impose reviews by a scientific board, pediatric departments, specialized research teams, or other committees. To enhance management of document approvals, industry sponsors should be aware of tiered review processes prior to engaging the institution.) The committees are charged with approving an age-appropriate process and informational items directed to children and their parents. Materials may include consent and assent forms, printed or visual media to accompany forms and recruitment items. Widespread differences in the local interpretation of national regulations on consent and assent were observed among U.S. committees as early as 1983.[26] Disparities were again reported by an expert panel charged with reviewing U.S. regulations on research protections for children.[20] As described later, the panel's concerns have received further scrutiny in the United States and EU, especially for age, assent documentation, and compensation. (Compensation issues are discussed in Chapter 30.)

Age for Assent

The maximum age for assent is generally the age of majority, usually between 18 and 21 years, however consent may be granted to emancipated individuals less than the age of majority. In the U.S. individuals may reach emancipation through marriage, military service, court order, or other reasons as allowed by local jurisdiction.[28] On the other hand minimum age requirements vary considerably.[9,27] Recommendations for the age of assent were originally derived from considerations that age 7 was the earliest that a child could form specific intent.[29,30] The International Conference on Harmonisation (ICH) recommends age 6 years but defers to ethics committees for interpreting local statutes.[31] In the United States, committees are given latitude to specify age requirements for individual children apart from the overall study population.[8] In practice, many U.S. committees allow investigator discretion on age limits, whereas others follow the common practice of age 7 years or greater.[32] Member states in the EU similarly require assent at 7 years in most cases, although some defer the requirement until age 16.[9]

Assent Documentation

Regulations defer to ethics committees for the documentation of assent. The ICH recommendation for signatures from children with appropriate intellectual maturity is generally interpreted as documentation by school-age children with the ability to read and write.[31] However, signature by a child is not necessarily in his/her best interest given that a child's perspective on the intent, or lack thereof, of a signature is expectedly different from adults. The requirement may be perceived as intimidating or empowering, depending on the individual child.[15,33] Children may incorrectly assume that their signature secures their choice, irrespective of parental decisions. Adolescents may understand its limitations yet perceive the signature is perfunctory and refuse. Others may appreciate the opportunity and willingly sign the form. Regardless of signature, the tenets of assent should be upheld by appropriately engaging children during the process.[33] The "school rule" preference for a signature is a *recommendation*, whereas solicitation of assent, the child's agreement and willingness to participate, is *required* unless waiver criteria are imposed by ethics committees. Therefore, an invitation to sign the form, rather than requirement, is a reasonable solution.[34]

Placement of the signature lines for assent also varies across ethics committees. A retrospective review among 55 U.S. ethics committees revealed that 83% included a means to document assent using a signature line either on the child assent form or on the parent form.[35] Until consistency is established for signature requirements, sponsors can minimize stalemates in multisite trials by inquiring about individual committees' preferences prior to submitting forms for approval.

Documentation of dissent appears to be largely ignored by ethics committees. One study revealed that only 2 of 67 committee submissions included dissent documentation.[35] Nonetheless, committees that deem assent signatures as necessary should give prospective participants an option to similarly signify dissent. In any case, the investigator should document either assent or dissent in the study records.

THE CONTEXT OF THE INVESTIGATOR IN THE PROCESS

The paramount distinction between investigator obligations in an adult study compared to a pediatric trial is the paradigm for assent. GCP protects vulnerable subjects by obligating the investigator to disclose information "to the extent compatible with the subject's understanding."[11] Guidance further requires that the presumed will of the child must be valued and weighed by the investigator.[9] Therefore, an investigator serves as informant and advocate and is responsible for fostering a process that allows for consent, assent or dissent.

Pediatric investigators often fulfill dual responsibilities for the child in the roles of investigator and physician. Children with serious or rare conditions typically receive much of their medical care from research specialists. The principle of research equipoise† limits the physician in the role of investigator to protocol disclosure, not treatment recommendations. The distinction may be difficult for parents and children to perceive when disagreements arise about participation. Either the parent or child may attempt to solicit recommendations from the physician to support their respective decisions.

Parent–child disagreements may occur when young children dissent through direct refusal or display resistance to study procedures. Older children with chronic illness or end-of-life choices may prefer continued intervention when parents perceive that the research will impose unnecessary burden or risk. Even more challenging is the situation of older adolescents with rights for medical care that are preempted in research.[36] Regardless of the circumstances, the investigator must consider the best interests of the child along with parental choice. Obviously, best interests cannot be defined or regulated and must be evaluated on an individual basis for each child by the investigator or ethics committees. (In some cases, U.S. ethics committees may defer to a federal panel to make decisions about allowable research and best interests for the child.[37]) Institutional policies for responding to dissent are often lacking; thus, the investigator is obligated to bridge the child's right to dissent and the parent to override it.

In cases of disagreement between the investigator and families or parent and child, a child research advocate independent of the investigator and family is recommended.[38] The investigator should consider seeking approval of the review committee before agreeing to overrule the child.[16] A practical response to procedural objections would be to "stop, assess and address" concerns through comfort measures or minor adjustments as allowed by the protocol.[28] If the investigator is also the child's physician, responsiveness to routine medical treatment may be a useful gauge for predicting dissent to study procedures. In all, the singular role of investigators demands scrutiny of the willingness of the child to endure research discomfort and parental perspectives on best interests for their child.

IMPLICATIONS OF POLICIES ON THE SPONSOR

GCP sponsor obligations for pediatric consent primarily include approval of consent materials by an ethics committee and documented evidence that voluntary willingness was obtained prior to a child's participation.[11] Debatable is the means for a sponsor to ensure the *process* of consent and assent. In the absence of witnessing parent or child interviews, sponsors indirectly monitor the process by reviewing letters of approval by the committee, signed forms, and source notes about the process.

In most cases, sponsor procedures for monitoring the consent process are considerably fewer than for data review and other site activites. Less attention to the consent process presents several risks to overall trial integrity. First, inadequate consent may reduce evaluable data due to noncompliance and drop-outs. Second, child safety may be jeopardized by incorrect dosing or adverse event oversight by uninformed parents. Finally, inattention to nuances of the protocol during investigator disclosure can have a direct impact on recruitment and retention when parents and children withdraw consent following unexpected events or requirements. A case example is adolescent female drop-out due to expected, temporary weight gain that is insufficiently addressed during assent. Another case is

†Research equipoise denotes genuine uncertainty among medical experts regarding the treatment arms of a clinical trial.

TABLE 30.1 Operational Considerations for Consent in Pediatric Trials

Sponsor Checklist for the Process of Consent and Assent

Document Preparation

- General considerations for content and format presentation
  - □ Demographics of child and parent
  - □ Age of child (chronological and developmental)
  - □ Requirements for translations
  - □ Seriousness of disease or condition
  - □ Duration of child's experience with the disease or condition
  - □ Prior research experience of the child
  - □ Protocol limitations on timing of the process
  - □ Number of assent forms needed for age subgroups
- Strategic consultation on content
  - □ Obtain input of children and parents
  - □ Request preapproval consultation from ethics committees
  - □ Solicit investigator perspective on consent-assent in the targeted sample

Site Training

- ■ Anticipate protocol requirements that can impact commitment, treatment compliance, and safety
- ■ Disclose with a view toward minimizing the risk of therapeutic and procedural misconceptions
- ■ Evaluate staging and iteration for the process
- ■ Utilize consent and assessment tools to promote individualized understanding
- ■ Identify practices that will foster enrollment and retention without coercion

Monitoring

- ■ Document approvals by ethics committees or other entities required by local practices
- ■ Signatures on correct versions
- ■ Documentation of the process in source notes listing each participant, including interpreters and child advocates

nonparticipation due to parental concern about risks when the investigator neglects emphasis of safety features during explanations of the protocol.

Understandable documents aid in overcoming these risks but effective use of the materials by site staff is equally important for valid consent and trial integrity. Investigator training on the use of the documents should follow the precedents for rigorous training on data collection. Risk can be further mitigated by close attention to enrollment metrics that signal problems with consent documents. Table 30.1 lists operational considerations for managing many of the obstacles presented in the next section.

THE PROCESS OF CONSENT AND ASSENT

The notion that informed permission and assent is a process and not a form or singular event is well established. Overemphasis on the form should never supersede attention to the process, otherwise full disclosure and understanding may be undermined. Research on factors that impact the validity of pediatric consent has evaluated therapeutic condition, research settings, age, and investigator practices. Trends have been identified, but definitive findings are scant due to differences in interview methods, probes for understanding, and scoring.[39,40] Although the inconsistencies have hampered the development of standards, recommendations for policy changes and further study have been promulgated.[5,20] This section reviews more recent studies and commentary on parent–child vulnerabilities, perceptions of families, timing for the process, and forms. Many of the findings have unmasked additional layers of complexities about the process, but the indisputable prerequisite for a meaningful process of consent and assent is sufficient time and adequate resources.

FACTORS THAT IMPACT PARENTAL PERMISSION

Stakeholder concerns and public sentiment over the decisional rights of parents will continue to evolve in parallel to the ever-increasing number of trials in children. Many ethicists hold that the duty of adults is to guide the child to make the best choice and, as such, the ultimate decision for a child's participation resides with the parent, not the child.[17] The counter-perspective of free choice by the child for some types of research has been tempered to emphasize a role for both parent and child in all types of research.[6,41]

Parental permission is predicated upon the ability to understand and assimilate all elements of consent. Of particular concern to the validity of informed permission are misconceptions by parents about benefit for their child. Many studies, including a recent one by Kodish et al.[42] have found that parents may not fully comprehend randomization and assume their child with a serious condition will receive the best treatment. The false assumption, termed "therapeutic misconception," undermines consent validity and is commonly found in oncology research where dose-finding studies may be construed as potentially beneficial.[43,44] Fisher[45] suggests that "procedural misconception" is another obstacle to valid consent. The term describes a lack of understanding of the research enterprise—its experimental nature, merits and risks, and financial or other reward for the investigator and study sponsor. The impact of both types of misconception is that context rather than information about the protocol and research may preferentially guide voluntariness. Given the medical setting for the consent process, a parent may falsely assume clinical care instead of experimentation (i.e., procedural misconception). The invitation to enroll their child may be perceived as their physician's recommended treatment (i.e., therapeutic misconception). Parental trust toward healthcare providers is a key factor in either type of misconception. The trust may lead some parents to forfeit their decision-making right to their child's physician.[46,47] Findings that other parents value their right to a final decision, regardless of circumstance, suggest the importance of individualizing the process for every participant.[48,49]

Despite investigator explanations to the contrary, the innate trust and contextual appearance of clinical treatment may lead parents to perceive enrollment as their only avenue of treatment for their child.[50–52] Overcoming misconceptions involves a thorough explanation of clinical research, the transformed role of physician to investigator, and details specific to the trial. Following disclosure of information, assessment of understanding is expected by ethics committees. In some cases, the committee may include a written assessment with the consent or assent form. Ethics committees generally agree that assessment is a core feature of valid permission, but even parents disagree about its value. Eder et al.[53] found that parents desired assessment for understanding through open-ended rather than closed questions and active solicitation of their questions by the investigator. Other parents reportedly perceive assessments as unnecessary or offensive, suggesting that the process should be sensitive to parental preferences.[20]

Parent Vulnerability

The precept that children are a vulnerable population applies to their parents when making treatment decisions about a sick child. Faced with an option to permit enrollment of their child into a trial, the baseline distress of the parent is expectedly magnified. Vulnerability of the parent may be exacerbated by ethnic or cultural differences between an investigator and the family.[54] The duress associated with parental decisions is not limited to the time point of signature. Phipps et al.[55] found patterns of stress up to 3 weeks following treatment agreement by parents of children undergoing stem cell transplants. The findings, although not surprising, support the requisite for ongoing discussions with parents.

Evidence that parental duress impacts recall has been reported in many studies. For example, two studies in the EU and United States found that parents of children with cancer could not recall discussing enrollment.[46,53] Another U.S. study for newborn pain similarly reported memory deficits of parents.[56] The lack of recall across these three distinct studies suggests a universal challenge shared by parents facing medical treatment or research options for their child. Other studies have also shown that parental duress is negatively associated with specific elements of consent, including lack of overall understanding, withdrawal, and right to alternative treatment.[46,57,58]

The impact of educational level, minority position, or socioeconomic status on consent reviewed elsewhere has been further investigated in recent studies. Negative correlations between low socioeconomic status and concepts for voluntariness, withdrawal options, and clinical versus research therapy were found in two studies in parents of children with leukemia.[59,60] Positive associations between level of education and capacity of understanding were shown in a systematic review of adult studies and individual pediatric trials.[39,58,61,62] Regardless of reading abilities or duress, a process to ensure valid permission is expected of investigators and sponsors. Among individuals whose reading ability or cognitive status may negatively impact understanding, additional time and breaks during the process have been shown to improve consent capacity. Information processing is also improved when the amount of information per encounter is reduced and free-recall of information is solicited.[63]

Temporal Elements of Consent

Universal acceptance of consent as a process gives weight to the importance of practices that promote continuous consent.[16] Studies on the process among parents have identified factors that improve perceptions about the process and, more importantly, enhance understanding. Among factors of importance to parents are investigator involvement, privacy, and sufficient time to review documents.[62] Eder et al.[53] found that time for decision-making was the suggestion most often cited by parents for improving the process of consent. The same group of parents valued time to review trial options with extended family or other supportive individuals.

Undue pressure imposed by the investigator or research team for rapid decisions resulted in negative perceptions among parents with children in 25 different studies.[64] The same report also identified a negative association between form readability and time allowed to review and ask questions. Investigators have an obligation to ensure adequate time and resources to support the initial interview and continuing process.[1,16] Implicit to the investigator's responsibilities are reasonable enrollment time lines by the trial sponsor.

Staged or continuous consent has been shown to improve recall by parents about study details at a time point subsequent to initial enrollment and, by extension, to optimize overall informed decision-making.[46,65] Informed permission and assent as a continuous process may take the form of ad hoc discussions at every visit or discrete events with documentation of re-consent. Trials for an acute condition, where exclusionary pretreatment would render a patient ineligible, may warrant re-consent, especially where a delay in treatment would pose risk or unnecessary discomfort to the child. For example, a study for acute otitis media may involve permission for immediate treatment and tympanocentesis in an emergency department, then re-consent during a follow-up visit for a secondary tympanocentesis.

Neonatal trials are another area where re-consent may be appropriate due to the effects of maternal medication and sleep deprivation on cognition and other stressors unique to the perinatal period.[66] Parents of neonates presented with hypothetical scenarios voiced concern that duress at the time of consent would be likely to impair an informed decision.[48] Improved understanding using staged consent has been reported in a recent neonatal trial. Mason and Allmark[67,68] attributed improvement to a three-stage process compared to a single interview and to investigator training on continuous consent.

A staged process has also been proposed for long-term trials and studies where parents are confronted with immediacy of decisions in pediatric oncology. Staging may be especially beneficial for trials involving multiple phases of treatment. A pediatric oncology study found that a two-stage process versus a single interview not only improved understanding about randomization and differences between the trial and standard treatment, but it also fostered greater parental trust in the process.[69] The value to parents of sequential discussions has also been found in other studies involving newborns and pediatric oncology.[48,53,66]

Conversely, at least one study identified dissatisfaction with staged consent when parents perceived that risks and information were increasingly detailed during later discussions.[68] Thus, a potential downside for staged consent could be a perception of coercion through partial rather than full disclosure at each stage. An alternative solution is full disclosure at repeated intervals prior to enrollment whenever the study criteria allows for days, not hours, for a decision.

FACTORS THAT IMPACT THE ASSENT PROCESS

Effective assent recognizes the asynchronous elements of psychological, cognitive, and physiological development.[70] Interindividual differences add to the complexity of assessing the ability of each individual child for decision-making. Like their parents, the ability of children to assent is impacted by severity of condition, emotional duress, and educational level. The impact of each is further compounded by family dynamics.[5,71]

Exploration by Snethen et al.[72] of decision-making among parents and participants ages 8–20 years with life-threatening conditions revealed that family patterns for decision-making could be distinguished based on parental goals, parental roles, and child involvement. One pattern described empowerment among 17–20-year-old adolescents when engaged by their parents in a collaborative process. Other parents involved their young children but made the enrollment decision independent of their child. A third group of "gatekeeper" parents perceived that exclusion of their children served the best interests of the child. Decision by delegation described a fourth group of parents who screened information, then supported the child's decision. Forethought by the investigator on family patterns is an obvious means for assessing validity of the process.

Age is another obvious factor impacting assent, although several decades of theoretical arguments have failed to resolve the dilemma on minimum age.[6] Children display increasing capacity for self-determination, yet the setpoint during development whereby an individual is competent to make informed decisions is arbitrary at best. Despite the absence of definitive answers, current thinking recognizes that comparably less maturity than adults does not necessarily preclude the ability of a child to participate in healthcare and other life decisions.[70,73] Disagreement on the setpoint is partially a reflection of different operational definitions. The limited decision-making right of assent is described as affirmative or knowing agreement. Some commentators interpret knowing agreement to mean that the child understands current and future ramifications of his/her decisions, while others follow a simpler interpretation of willingness or preference.[35]

Despite inconclusive findings, a majority of ethics committees appear to set the age range for assent between 6 or 7 years and the age of majority. Some pediatric experts contend that valid assent is not achievable until age 14 and others favor consent rather than assent from adolescents starting at age 14.[28,74] Disagreements over the degree to which assent should mirror consent suggested to one commentator that the term should be replaced with a descriptive phrase, "developmentally appropriate decision-making."[75] If assent is limited to children with capacity for prospective decision-making, then younger children will be excluded.[76] However, recent evidence that even preschoolers can demonstrate a basic understanding of illness management suggests that assent of very young children merits consideration.[73,77] Consistency in the interpretation of assent awaits further investigation using validated assessments yet to be developed.[28]

An association between prior exposure to medical treatment or research and meaningful assent is well accepted among pediatric trialists,[20, 73] although Miller's systematic review of studies on experience was inconclusive.[5] Children with chronic illness and previous trial experience may be well informed about their condition and poised for research decisions.[14] In such cases, considerable weight should be given to their choice even if they are chronologically young.

Dissent

Dissent is generally interpreted as the functional opposite of assent—lack of affirmative agreement. Many children will not be able to gauge whether or not procedures will be distressing until they are actively enrolled; therefore, dissent prior to enrollment is unlikely unless the child has previous exposure to medical or research settings.[28] Once enrolled, the investigator is not required to discontinue the child's participation unless the parent withdraws consent. Ethical precepts, nonetheless, compel the investigator to advocate for the child's choice about participation. Investigator advocacy is particularly important to older adolescents with end-stage disease. Many adolescents have adult level competencies for research decisions and "veto power" is recommended for adolescents at end of life.[15,71]

Signs of distress in children should be considered by the investigator during the consent discussion with parents. Subtle distressful behavior is not unusual in younger or less articulate children and should be interpreted within the context of the child's development and circumstances. Overly stressful reactions should preclude enrolling the child unless the study meets ethics committee criteria for waiver of assent.

Assent for Long-Term Studies

Long-term follow-up pediatric studies greater than 6 months were found to double in the United States between 2000 and 2006 and similar trends for long-term data are expected to continue.[78] Securing a valid assent process during long-term data collection demands attention to the trajectory of psychological and physical development. For example, re-assent is important when the study duration spans the periadolescence period. More specifically to females, treatment risk may differently impact their willingness to join a study if menarche occurs during the study. One proposed solution is to conduct interactive re-assent using two separate assent forms in which the importance of menses disclosure is emphasized in the initial form and teratogenic risks in the second form.[21]

Long-term assent is similarly of concern in genetic studies. Genetic materials or history may be solicited from children for population studies, research on markers of disease, or as a substudy unrelated to the primary study objective. The psychological impact of disease prediction for adolescents should be recognized and managed by fully engaging them in the process.[79] Regardless of the underlying research question, blanket consent for unlimited use of the data is unacceptable.[80] A second stage of consent is necessary when the storage period for the repository extends beyond the age requirement for consent. In such cases, provisions for long-term follow-up with participants should be detailed in the protocol to accommodate time gaps between initial assent and age of majority.

Consent and Assent Forms

Securing a child's right to voice his/her opinion presumes the availability of developmentally appropriate assent documents. The preparation of understandable documents necessitates an assessment of the quantity, content, and means for delivery to each age group involved in the study. Admittedly, the overall challenge is the difference in capacity even for children of the same chronological age. Multiple versions of forms may be needed to satisfy the needs of developmentally diverse participants, although the practice seems to be largely ignored. The study of consent/assent form submissions by Kimberly et al.[35] demonstrated that only 10% of submissions included multiple assent forms with language differentially targeted to age subgroups.

Notwithstanding decades of observations on overly complex consent forms, documents continue to resemble legal contracts such that lengthy narrative details are likely to impair rather than promote informed decisions.[81] Two decades ago, Tarnowski et al.[82] reported that increasing form length did not improve readability scores for pediatric consent documents, yet the use of complex forms has continued unabated. In addition to form length, reading levels of grade 12 and above are not unusual and ignore fundamental guides that have consistently suggested forms should target upper middle school reading levels. As mentioned earlier, reading abilities may be further compromised in psychiatric or other conditions that adversely affect comprehension and cognition.[83,84] These findings suggest the importance for consideration of developmental stage, as well as disease or condition, when designing forms.

Investigations on the content and format of consent and assent forms have revealed distinctions between readability (i.e., layout and simple phrasing) and understandability (i.e., comprehension). Davis et al.[85] reported that reducing the word count in a standard adult form and reformatting with graphics improved readability and perceptions about the form, but not comprehension. Conversely, two-column layouts and pictures improved understanding and were preferred by parents reviewing a hypothetical form.[61] A positive finding that pictograms enhanced understanding for patients with limited reading skills may have application for assent, especially in young children.[86]

Adult presumptions about type and format for printed materials to satisfy the needs of pediatric participants may not always be correct. One study found that pediatricians chose a comic book style for an adolescent patient information leaflet whereas adolescents preferred factual information and high quality illustrations.[87] The widely differing perspectives suggest that pediatric participants may be the best resource for assent document development. Ford et al.[88] recently reported the outcome of assent forms developed by 6–12-year-old children engaged in revising typical assent documents with the assistance of a facilitator. The document produced by the children used much simpler terms and was assessed at a grade 4 reading level. Supporting documents, revised formats, and/or electronic materials have been proposed to simplify generally complex consent documents, although the value of different modalities remains uncertain.[39] The most effective solution appears to involve a combination of written, oral, and technology-driven communication. Strikingly clear from individual studies and systematic reviews on consent and assent is the essential element of direct engagement involving the investigator, parents, and child, followed by sufficient time to consider the information prior to a decision.

CONCLUSION

The evidence presented in this chapter parallels earlier findings regarding the impact of demographics and situational context on permission and assent. Many of the studies have been limited in scope and require additional elucidation. Nonetheless, it is clear that a valid process demands attention to an infinite array of demographics, family dynamics, experience, and developmental stages. Understandable materials based on input from children and parents, coupled with active investigator engagement, are fundamental to the process.

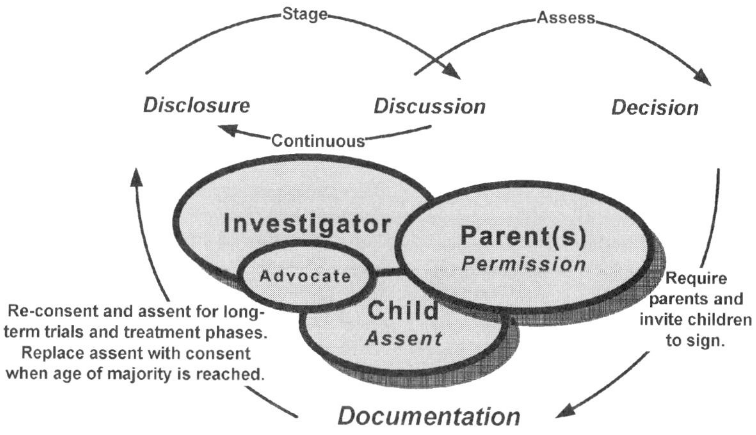

FIGURE 30.2 Enhanced process for pediatric consent and assent. The process enhancements emphasize engagement using staged and/or continuous discussions, assessment of understanding prior to decision-making, and cases for re-consent. The advocate role for the child is fulfilled by the investigator and, in some cases, by an individual independent of the study team.

Investment into training and tools that promote individualized understanding can be expected to equally improve the process and perceptions about pediatric research. The operational challenges imposed by the array can be reduced by translating feedback from participants into an enhanced process as shown in Figure 30.2.

Regulatory standardization is inadequate for many aspects of permission and assent, but the overarching consideration consistently voiced by parents and children is unlikely to be the subject of regulation. Pediatric research volunteers have asked for ample time and active engagement with the investigator. Pediatric trialists and stakeholders need to abide by their request and in doing so will satisfy the essence of human experimentation rights—respect through disclosure and an invitation to participate.

REFERENCES

1. National Commission for the Protection of Human Subjects of Biomedical and Behavioral Research. *The Belmont Report: Ethical Principles and Guidelines for the Protection of Human Subjects of Research*. Washington DC: US Government Printing Office; 1979.
2. World Medical Association. Declaration of Helsinki: Ethical Principles for Medical Research Involving Human Subjects. 52nd General Assembly; October 2000.
3. Wolpe PR, Moreno J, Caplan AL. Ethical principles and history. In: Pincus H, Lieberman J, Ferris S, eds. *Ethics in Psychiatric Research: A Resource Manual for Human Subjects Protection*. Washington DC: American Psychiatric Association; 1999:1–21.
4. Beauchamp TL, Childress JF. *Principles of Biomedical Ethics*, 5th ed. New York: Oxford University Press; 2001.
5. Miller VA, Drotar D, Kodish E. Children's competence for assent and consent: a review of empirical findings. *Ethics Behav*. 2004;14(3):255–295.
6. Ross LF. Informed consent in pediatric research. *Cambridge Q Healthcare Ethics*. 2004;13 (4):346–358.
7. Silverman WA. The myth of informed consent: in daily practice and in clinical trials. *J Med Ethics*. 1989;15(1):6–11.
8. Food and Drug Administration. 21 CFR Subpart D: Additional Safeguards for Children in Clinical Investigations of FDA-Regulated Products: Interim Rule. *Fed Regis*. 2001;66: 20589.
9. European Commission. Ethical Considerations for Clinical Trials Performed in Children. 2006.
10. Wendler D, Varma S. Minimal risk in pediatric research. *J Pediatr*. 2006;149:855–861.
11. International Conference on Harmonisation. *Consolidated Guideline for Good Clinical Practice: E6*. Geneva; 1997.
12. General Assembly of the United Nations. Convention on the Rights of the Child; 1989.
13. European Parliament. Directive 2001/20/EC of the European Parliament and of the Council: Approximation of the laws, regulations and administrative provisions of the Member States relating to the implementation of good clinical practice in the conduct of clinical trials on medicinal products for human use. 2001.
14. De Lourdes Levy M, Larcher V, Kurz R. Informed consent/assent in children. Statement of the Ethics Working Group of the Confederation of European Specialists in Paediatrics (CESP). *Eur J Pediatr*. 2003;162(9):629–633.
15. Joffe S, Fernandez CV, Pentz RD, et al. Involving children with cancer in decision-making about research participation. *J Pediatr*. 2006;149:862–868.
16. Council for International Organizations of Medical Sciences. International Ethical Guidelines for Biomedical Research Involving Human Subjects. Geneva; 2002.

17. Ackerman TF. Fooling ourselves with child autonomy and assent in nontherapeutic clinical research. *Clin Res.* 1979;27(5):345–348.

18. US Department of Health and Human Services (DHHS). Protection of Human Subjects, 45 CFR 46; 1991.

19. Levine RJ. Adolescents as research subjects without permission of their parents or guardians: ethical considerations. *J Adolesc Health.* 1995;17:287–297.

20. Institute of Medicine Committee on Clinical Research Involving Children. Understanding and Agreeing to Children's Participation in Clinical Research. In: Field J, Behrman RE, eds. *The Ethical Conduct of Clinical Research Involving Children.* Washington DC: National Academy of Sciences; 2004: 146–210.

21. O'Lonergan T, Zodrow JJ. Pediatric assent: subject protection issues among adolescent females enrolled in research. *J Law Med Ethics.* 2006;34(2):451–459.

22. Beecher H. Ethics and clinical research. *N Engl J Med.* 1966;274:1354–1360.

23. Solomon J. Researchers tested AIDS drugs on children. *San Francisco Chronicle.* May 4, 2005.

24. Protections for foster children enrolled in clinical trials. In: US House of Representatives Subcommittee on Human Resources of the Committee on Ways and Means. One Hundredth Ninth Congress, First Session; 2005.

25. Slack C, Strode A, Fleischer T, et al. Enrolling adolescents in HIV vaccine trials: reflections on legal complexities from South Africa. *BMC Med Ethics.* 2007;8(1):5.

26. Kapp MB. Children's assent for participation in pediatric research protocols: assessing national practice. *Clini Pediatr.* 1983;22(4):275–278.

27. Wendler D. Assent in paediatric research: theoretical and practical considerations. *J Med Ethics.* 2006;32 229–234.

28. Campbell, AT. State regulation of medical research with children adolescents: An overview and analysis. In: Field J, Behman RE, eds. *The Ethical Conduct of Clinical Research Involving Children.* Washington DC: National Academy of Sciences; 2004:320–387.

29. American Academy of Pediatrics: Task Force on Pediatric Research and Medical Ethics. Consent. *Pediatrics.* 1976;57(3): 414–416.

30. American Academy of Pediatrics. Guidelines for the ethical conduct of studies to evaluate drugs in pediatric populations. *Pediatrics.* 1995;286–294.

31. International Conference on Harmonisation. *Clinical Investigation of Medicinal Products in the Pediatric Population*: E11. 2000.

32. Whittle A, Shah S, Wilfond B, et al. Institutional review board practices regarding assent in pediatric research. *Pediatrics.* 2004;113:1747–1752.

33. Ungar D, Joffe S, Kodish E. Children are not small adults: documentation of assent for research involving children. *J Pediatr.* 2006;149 (1 Suppl):S31–S33.

34. Gill D, Crawley FP, LoGiudice M, et al. Guidelines for informed consent in biomedical research involving paediatric populations as research participants; the Ethics Working Group of the Confederation of European Specialists in Paediatrics (CESP). *Eur J Pediatr.* 2003;162 (7–8):455–458.

35. Kimberly MB, Hoehn KS, Feudtner C, et al. Variation in standards of research compensation and child assent practices: comparison of 69 Institutional Review Board-approved informed permission and assent forms for 3 multicenter pediatric clinical trials. *Pediatrics.* 2006;117:1706–1711.

36. Collogan LK, Fleischman AR. Adolescent research and parental permission. In: Kodish E, ed. *Ethics and Research with Children: A Case-Based Approach.* New York: Oxford University Press; 2005: 77–99.

37. Food and Drug Administration. *Process for Handling Referrals to FDA Under 21 CFR 50.54 (Draft Guidance).* Rockville, MD; 2006.

38. US Department of Health and Human Services. *Protections for Children in Research A Report to Congress in Accord with Section 1003 of P.L. 106–310, Children's Health Act of 2000*. Office of Human Research Protections; 2001.
39. Flory J, Emanuel E. Interventions to improve research participants'understanding in informed consent for research: a systematic review. *JAMA* 2004;292(13):1593–1601.
40. Kon AA, Klug M. Methods and practices of investigators for determining participants' decisional capacity and comprehension of protocols. *J Empirical Res Hum Res Ethics*. 2006;1(4):61–68.
41. Miller VA, Nelson RM. A developmental approach to child assent for nontherapeutic research. *J Pediatr*. 2006;149 (1 Suppl):S25–S30.
42. Kodish E, Eder M, Noll RB, et al. Communication of randomization in childhood leukemia trials. *JAMA* 2004;291(4):470–475.
43. Appelbaum PS, Roth LH, Lidz CW, et al. False hopes and best data: consent to research and the therapeutic misconception. *Hastings Cent Rep*. 1987;17(2):20–24.
44. Glannon W. Phase I oncology trials: why the therapeutic misconception will not go away. *J Med Ethics*. 2006;32(5):252–255.
45. Fisher JA. Procedural misconceptions and informed consent: insights from empirical research on the clinical trials industry. *Kennedy Inst Ethics J*. 2006;16(3):251–268.
46. Chappuy H, Doz F, Blanche S, et al. Parental consent in paediatric clinical research. *Arch Dis Child*. 2006;91(2):112–116.
47. Zupancic JAF, Gillie P, Streiner DL, et al. Determinants of parental authorization for involvement of newborn infants in clinical trials. *Pediatrics*. 1997;99(1):1–6.
48. Culbert A, Davis DJ. Parental preferences for neonatal resuscitation research consent: a pilot study. *J Med Ethics*. 2005;31:721–726.
49. Morley CJ, Lau R, Davis PG, et al. What do parents think about enrolling their premature babies in several research studies? *Arch Dis Child Fetal Neonatal Ed*. 2005;90(3):F225–F228.
50. Pace C, Talisuna A, Wendler D, et al. Quality of parental consent in a Ugandan malaria study. *Am J Public Health*. 2005;95(7):1184–1189.
51. Benedict JM, Simpson C, Fernandez CV. Validity and consequence of informed consent in pediatric bone marrow transplantation: the parental experience. *Pediatr Blood Cancer*. 2007;49(6):846–851.
52. Jaffe A, Prasad SA, Larcher V, et al. Gene therapy for children with cystic fibrosis—Who has the right to choose? *J Med Ethics*. 2006;32(6):361–364.
53. Eder ML, Yamokoski AD, Wittmann PW, et al. Improving informed consent: suggestions from parents of children with leukemia. *Pediatrics*. 2007;119(4):e849–e859.
54. Fisher C. Commentary: SES, ethnicity and goodness-of-fit in clinical–parent communication during pediatric cancer trials. *J Pediatr Psychol*. 2005;30(3):231–234.
55. Phipps S, Dunavant M, Lensing S, et al. Patterns of distress in parents of children undergoing stem cell transplantation. *Pediatr Blood Cancer*. 2004;43(3):267–274.
56. Ballard HO, Shook LA, Desai NS, et al. Neonatal research and the validity of informed consent obtained in the perinatal period. *J Perinatol*. 2004;24:409–415.
57. Vitiello B, Aman MG, Scahill L, et al. Research knowledge among parents of children participating in a randomized clinical trial. *J Am Acad Child Adolesc Psychiatry*. 2005;44 (2):145–149.
58. Simon CM, Siminoff LA, Kodish ED, et al. Comparison of the informed consent process for randomized clinical trials in pediatric and adult oncology. *J Clin Oncol*. 2004;22(13):2708–2717.
59. Hazen R, Drotar D, Kodish E. The role of the consent document in informed consent for pediatric leukemia trials. *Contemp Clin Trials*. 2006;28(4):401–408.
60. Miller VA, Drotar D, Burant C, et al. Clinician–parent communication during informed consent for pediatric leukemia trials. *J Pediatr Psychol*. 2005;30(3):219–229.

61. Tait AR, Voepel-Lewis T, Malviya S, et al. Improving the readability and processability of a pediatric informed consent document. *Arch Pediatr Adolesc Med.* 2005;159:347–352.

62. Tait AR, Voepel-Lewis T, Malviya S. Factors that influence parents' assessments of the risks and benefits of research involving their children. *Pediatrics.* 2004;113(4):727–732.

63. McEvoy JP, Keefe RS. Informing subjects of risks and benefits. In: Pincus H, Lieberman J, Ferris S, eds. *Ethics in Psychiatric Research: A Resource Manual for Human Subjects' Protection.* Washington DC: American Psychiatric Association; 1999: 129–154.

64. Franck LS, Winter I, Oulton K. The quality of parental consent for research with children: a prospective repeated measure self-report survey. *Int J Nurs Stud.* 2007;44(4):525–533.

65. Kupst MJ, Patenaude AF, Walco GA, et al. Clinical trials in pediatric cancer: parental perspectives on informed consent. *J Pediatr Hematol Oncol.* 2003;25(10):787–790.

66. Burgess E, Singhal N, Amin H, et al. Consent for clinical research in the neonatal intensive care unit: a retrospective survey and a prospective study. *Arch Dis Child Fetal Neonatal Ed.* 2003;88:280–286.

67. Mason SA, Allmark PJ. Obtaining informed consent to neonatal randomised controlled trials: interviews with parents and clinicians in the Euricon study. *Lancet.* 2000;356(9247):2045–2051.

68. Allmark P, Mason S. Improving the quality of consent to randomised controlled trials by using continuous consent and clinician training in the consent process. *J Med Ethics.* 2006;32:439–443.

69. Angiolillo A, Simon C, Kodish E, et al. Staged informed consent for a randomized clinical trial in childhood leukemia: impact on the consent process. *Pediatr Blood Cancer.* 2004;42:433–437.

70. Weithorn LA, Scherer DG. Children's involvement in research participation decisions: psychological considerations. In: Grodin MA, Glantz LH, eds. *Children as Research Subjects: Science, Ethics and Law.* New York: Oxford University Press; 1994: 133–179.

71. Freyer DR. Care of the dying adolescent: special considerations. *Pediatrics.* 2004;113(2):381–388.

72. Snethen JA, Broome ME, Knafl K, et al. Family patterns of decision-making in pediatric clinical trials. *Res Nurs Health.* 2006;29(3):223–232.

73. Alderson P, Sutcliffe K, Curtis K. Children's competence to consent to medical treatment. *Hastings Cent Rep.* 2006;36(6):25–34.

74. Toner K, Schwartz R. Why a teenager over age 14 should be able to consent, rather than merely assent, to participation as a human subject of research. *Am J Bioethics.* 2003;3(4):38–40.

75. Joffe S. Rethink "affirmative agreement," but abandon "assent."*Am J Bioethics.* 2003;3(4):9–11.

76. Rossi WC, Reynolds W, Nelson RM. Child assent and parental permission in pediatric research. *Theor Med Bioethics.* 2003;24(2):131–148.

77. Diekema DS. Taking children seriously: what's so important about assent? *Am J Bioethics.* 2003; 3(4):25–26.

78. Milne C. Pediatric study costs increased 8-fold since 2000 as complexity level grew. *Impact Report.* 2007;9(2):1–4.

79. Geller G. The ethics of predictive genetic testing in prevention trials involving adolescents. In: Kodish E, ed. *Ethics and Research with Children: A Case-Based Approach.* New York: Oxford University Press; 2005: 194–220.

80. Burke W, Diekema DS. Ethical issues arising from the participation of children in genetic research. *J Pediatr.* 2006;149 (1 Suppl):S34–S38.

81. Beardsley E, Jefford M, Mileshkin L. Longer consent forms for clinical trials compromise patient understanding: so why are they lengthening?*J Clin Oncol.* 2007;25(9):e13–e14.

82. Tarnowski KJ, Allen DM, Mayhall C, et al. Readability of pediatric biomedical research informed consent forms. *Pediatrics.* 1990;85(1):58–62.

83. Christopher PP, Foti ME, Roy-Bujnowski K, et al. Consent form readability and educational levels of potential participants in mental health research. *Psychiatr Services.* 2007;58(2):227–232.

84. Revheim N, Butler PD, Schechter I, et al. Reading impairment and visual processing deficits in schizophrenia. *Schizophr Res*. 2006;87(1–3):238–245.

85. Davis TC, Holcombe RF, Berkel HJ, et al. Informed consent for clinical trials: a comparative study of standard versus simplified forms. *J Nat Cancer Inst*. 1998;90(9):668–674.

86. Mwingira B, Dowse R. Development of written information for antiretroviral therapy: comprehension in a Tanzanian population. *Pharm World Sci*. 2007;29(3):173–182.

87. Jones R, Finlay F, Crouch V, et al. Drug information leaflets: adolescent and professional perspectives. *Child Care Health Dev*. 2000;26(1):41–48.

88. Ford K, Sankey J, Crisp J. Development of children's assent documents using a child-centred approach. *J Child Health Care*. 2007;11(1):19–28.

CHAPTER 31

Recruitment and Retention in Pediatric Clinical Trials

M. RENEE SIMAR, PhD
INC Research, Inc., Austin, Texas 78746

INTRODUCTION

Undue pressure for investigators to participate and, in turn, for pediatric patients to volunteer has been a long-term concern of pediatric trialists.[1] Well recognized among stakeholders is the lack of pediatric participants and qualified sites to meet the increasing demand. The number of children involved in pharmaceutical trials increased by 1.5-fold between the years 2000 and 2006 and costs per subject increased by over sixfold.[2] Barriers to pediatric recruitment and retention arise from complex study requirements, parental concerns, and site deficiencies, as well as overly restrictive regulations. Managing the barriers involves assigning as much as 30% of a study time line to patient enrollment and 11% to site recruitment.[3]

The question, "What are the most effective recruitment strategies?" posed to clinical trial leaders generally elicits pragmatic ideas on targeted advertising campaigns, study branding, and Web-based recruitment. A veteran pediatric trialist is more likely to offer a broader view and respond with "child/parent-friendly" study designs, reasonable time lines, and additional staff for recruitment. Successful pediatric programs are the outcome of merging the two perspectives, with the latter serving as foundation for the first. In this chapter, the broad view will be explored to identify strategies to support ethical recruitment. Policies that guide recruitment practices will be reviewed followed by discussion on the unique considerations for pediatric investigator recruitment and perspectives of parents and children on trial participation. Taken together, a paradigm for pediatric recruitment and retention will be presented.

POLICIES AND GUIDANCE

Regulations on recruitment and subject selection are mostly absent. Much of the framework for recruitment practices in pediatric trials is derived from the underlying principle for

Pediatric Drug Development: Concepts and Applications
Edited by Andrew E. Mulberg, Steven A. Silber, and John N. van den Anker

voluntariness in vulnerable populations.[4] Undue influence must be avoided and disclosure on study requirements and risks accurately presented. Both of these tenets must be reflected in the recruitment materials and overall practices of an investigator. Guidance to date has mostly focused on advertising and compensation.

The International Conference on Harmonisation Good Clinical Practice Guideline (ICH GCP) specifies that recruitment plans and advertising be monitored along with other core study documents to ensure that recruitment is "appropriate and not coercive."[4] The American Academy of Pediatrics (AAP) emphasizes that advertisements for pediatric trials should not misconstrue the risks and benefits.[5] The U.S. Office of Human Research Protections (OHRP) further cautions against "advertising that emphasizes payment for participation specifically directed to children, payment to referring physicians, and other circumstances that could be viewed as creating undue influence or coercive environments."[6] Children are often subject to vulnerability on multiple accounts, including economic status and scant access to medical resources; therefore, the risk of coercive payment is of special concern in pediatric research.[7]

Even if unintentional or indirect, coercion may arise from recruitment among families with limited healthcare options. The Confederation of European Specialists in Paediatrics (CESP) advises additional protections for children recruited from economically disadvantaged communities, refugee camps, or similarly vulnerable situations.[8] Guidance cautions against selectively recruiting these individuals "simply because they can be more easily induced to participate in exchange for small payments."[9] AAP emphasizes that pediatric recruitment should follow principles of distributive justice: "the study should not rely exclusively or heavily on one socioeconomic, racial or ethnic group when this type of selection is not a necessary part of the investigation."[10,11] The risk of marginalization also applies to affluent areas, where access to healthcare is disproportionate. Unintentional selective enrollment frequently occurs among pediatric experts affiliated with academic institutions among inner-city impoverished families.[12] Nonetheless, the benefits to receiving expert care not otherwise available has considerable merit for these children with few resources and excluding them would be untenable. The responsible approach is to ensure robust standards for consent are met by fully disclosing the nature of the trial and obtaining valid permission from the parent and assent to the degree the child is able. (See Chapter 30 in this book.)

Payment to Research Participants

Allowable payments for research participation differ considerably across communities. Most regions permit compensation for injury and research-related expenses but practices vary for other payments and nonmonetary remuneration. EU regulations limit payment to reasonable compensation: "there must not be financial inducement to enroll the child in the trial; no financial incentive should be offered (other than compensation of expenses and time spent)."[13] On the other hand, noncoercive payments are sanctioned in the United States under limitations set by individual ethics committees.[14]

Methods to determine equitable compensation have been proposed but consensus is lacking.[6,15,16] The margin between fair payment and coercion is difficult to standardize due to individual diversity at the time of consent. Therefore, standardized methods to determine equitable compensation have been proposed. The added difficulty in payment for the vulnerable population of children is obvious, especially when adolescents are involved.[17] Wendler et al.[18] identified four categories of payments for pediatric research: reimbursement for direct expenses, compensation for indirect expenses (e.g., lost wages), appreciation payments or token gifts, and incentives above the other three categories. The incentive category is interpreted by most commentators as inducement above the costs of participation and carries the most risk for coercion.

Ethics committees generally agree that the schedule of payments should be an even distribution with final remuneration similar or equal to other payments. However, views on the timing of payment disclosure and recipient of the payments vary. Disclosure during the consent and assent process has been recommended.[19,20] Arguments to the contrary recommend disclosure at the end of participation to minimize the risk for undue influence.[5] Weise's[21] survey of 128 institutions found the majority required disclosure to the child and parent during the consent process. Other studies interviewed institutional review board (IRB) chairpersons on subject versus parent payments. Whittle et al.[19] found that almost 50% of 188 interviewees agreed in principle on payments to children and 35% thought payments to parents were acceptable. Although Kimberly et al.[22] reported that some ethics committees were vague about the recipient, others allowed compensation with nearly equal frequency to parents, children, or both.

Kimberly et al.[22] also identified remarkable disparity among institutions for compensation in the same pediatric trials. For example, less than half of IRBs in a 14-site pain study allowed compensation, although parents incurred expenses for hospitalization, travel, and lodging. Among 19 sites in a hypertension study, Kimberly and co-workers calculated a range of US $7.50–60/hour of participation. Scherer et al.[23] similarly reported variation between US $2.65–100/hour of participation in asthma trials among 37 committees. The methods to determine payment structure in these studies were not described although other pediatric investigators report that payments are generally a function of visit length and number of procedures.[24] IRBs endorse payment amounts based on level of risk for some studies, yet the practice carries a concern of coercion and is especially questionable in pediatrics.[14] Payments linked to duration of a child's participation and not risk may be less likely to distort the process.

The report on ethical research in children by the Institute of Medicine (IOM) recommended that institutions should standardize allowable compensation and disclosure about payments.[25] Nonetheless, disparities among institutions are still being reported and one survey found that many are without policies.[24] Uncertainties in the absence of policies can lead to disagreement among sponsors, investigators, and ethics boards and subsequent study delays. Delays are less likely when forethought about institutional practices and planning occurs prior to implementation.

Indirect Compensation to Participants

Among potentially coercive inducements are the indirect benefits of trial participation.[26] Examples are reduced waiting time for clinic visits, door-to-door transportation, no-cost drug during or following the trial, or supportive devices for routine care of the child's condition. These and similar services should be implemented with caution to reduce the risk of coercion.[7] Indirect benefits are acceptable when presented in a responsible manner without overemphasis during recruitment. Moreover, investigators have an obligation to ensure that parents do not perceive opting out as a choice for substandard care.

SITE RECRUITMENT AND RETENTION

Effective patient recruitment is predicated upon sites with pediatric expertise, access to patients, and adequate resources. Beyond these fundamental GCP requirements, successful pediatric investigators are attuned to the unique challenges of finding qualified children and willing parents. Finally, creativity and initiative of site staff are integral to maintaining patient compliance with visits and treatments. Two critical components of site assessment and recruitment are cross-functional staff and enrollment estimates.

The assessment should consider the ability of the investigator to effectively manage multidisciplinary collaboration often needed in pediatric trials. Synergistic relationships within the institution or community are especially important when long-term safety assessments require expertise beyond that of the investigator. The situation increasingly occurs when regulatory authorities request monitoring of treatment effects on neurodevelopment or growth. For example, a pediatric cardiologist faced with strictly defined noncardiovascular, developmental assessments may require a developmental pediatrician, neuropsychologist, or endocrinologist. In practical terms, success in scheduling and shuttling children among the disciplines is the outcome of effective direction by the study coordinator. Productive interdepartmental relationships are also essential when recruiting for pediatric conditions that often arrive after-hours to an acute care clinic or emergency department. Successful study coordinators learn to function unobtrusively among the staff in these areas when recruiting for conditions such as otitis, seizures, and wheezing.[27]

Estimates for enrollment and site staffing should consider actual start-up timing. The estimates may change in a short span due to significant competition for pediatric sites. A postponed study initiation may skew estimates when there is a significant delay between the assessment and actual start-up. Such a case commonly arises when companies initiate pediatric site activities in advance of final protocol approval from regulatory authorities. Sites may commit to a different study during the delay, leading to competition for staff resources and patient pools. Thus, anticipatory competition should be factored into feasibility considerations whenever possible. Delays may also distort enrollment estimates unless school holidays, vacation periods, and seasonal variation in illness are included in the estimates. The solution is to conduct the assessment or reassessment of site resources in alignment with the timing of actual start-up.

Higher Costs for Pediatric Recruitment

Recruiting and supporting productive pediatric sites is not only higher than adult trials but can vary by 20-fold depending on therapeutic area and regulatory requirements.[28] Fair compensation per enrolled child may be relatively greater due to the additional effort to find and manage patients and parents. In the past, the effort was often ignored and resulted in greater susceptibility of pediatric investigators to the "hidden" costs of pharmaceutical studies.[29] Pediatric site managers have increasingly acquired expertise in projected versus actual costs. Awareness of pediatric site operations can help sponsors gauge fair compensation and avoid the need for lengthy budget negotiations.

A common oversight by project leaders familiar with adult programs is increased coordinator time per patient in pediatric trials. Procedures and assessments often take longer, especially in young children. The classic example is the effort involved in a routine pretreatment electrocardiogram in an active and frightened 3 year old. Moreover, nearly every aspect of a pediatric visit is uniquely impacted by the need to interact with the child, one or both parents, and, in many cases, siblings or other family members. The extra effort is well recognized by pediatric investigators and adequate compensation is warranted. Staffing costs for administrative tasks may also rise when sponsors attempt to follow aggressive timelines by submitting a (mostly) final protocol to ethics committees while awaiting final approval from regulators. In most cases, the revisions require another round of committee review and considerable effort by site staff to manage the process. The misdirected effort to reduce program length ultimately leads to higher costs and increased timelines.

Strategies to encourage faster recruitment should focus on realistic assessment of barriers and adequate funding, not bonus payments to investigators or coordinators. Similar to the guidance on payments to research subjects, investigator compensation policies are not specific but generally caution against payments beyond actual costs. Thus, remuneration for higher enrollment numbers or referral fees is no longer permitted at many institutions. However, increased payments are allowed for actual costs of accelerated recruitment or enhanced retention efforts.

Partnerships to Optimize Recruitment

The paucity of pediatric investigators generates reenlistment of the same sites across multiple protocols. Therefore, successful programs, especially in rare conditions, are the outcome of productive, long-term relationships with a few qualified experts. Site motivation is engendered by fair compensation, reasonable timelines, and a choice of recruitment strategies. Overly zealous enrollment reminders and competitive metrics are generally met with disdain or simply ignored unless sponsors are actively engaged with supportive measures. Conversely, investigators should respect their contractual obligations and follow due diligence for protocol compliance and enrollment. Requiring sites to propose a detailed plan for recruitment and retention is a reasonable request at study initiation to aid the sponsor in developing an overall, cost-effective enrollment strategy. The effectiveness of the plans can be tracked to identify trends and enable corrective actions across all sites. In sum, successful recruitment and retention in pediatric trials necessitates a mutual commitment between the site and the sponsor.

PATIENT RECRUITMENT BARRIERS AND MOTIVATION

The few studies on motivating factors for pediatric trials have relied on feedback mostly from parents at the time of consent, retrospective surveys, and hypothetical vignettes. The first step for incorporating the findings into ethical strategies is to recognize that recruitment is the first step in the consent process.[30] Therefore, understanding motivations for willingness to enroll and remain in the study serves two purposes—to foster consent without misconceptions and to guide effective, yet ethical, recruitment practices.

Numerous studies found that altruism toward other families and contributing to science is a universal theme among parents and children who have either considered or actually participated in medical research. Most striking is feedback from parents who recognized that their premature infant may not benefit from research but were willing to enroll into more than one study.[31] Among children, the concept of altruism may not be fully realized until maturity but even young children appreciate the importance of altruistic acts.[32] Children ostensibly want to help "doctors learn" and "kids get better."[33,34] One study found that nearly all of the participants aged 6–16 years who donated biological samples in a nontreatment study cited a desire to help others as their reason for participation.[35] Many of the children also indicated that the opportunity to learn about research was a factor. Parents in several studies have similarly placed value on giving their children an opportunity to learn about their condition through research while also enhancing their own understanding. Increased knowledge for the participants is easy to engender through media readily available for many therapeutic conditions. A clinical trial setting offers the perfect opportunity to guide reliable information gathering for parents and children. Teaching them about their disease is likely to mutually benefit both participant and investigator since educated families are more likely to be engaged, attuned to adverse events, and compliant with treatment.

The largely positive perceptions and altruistic motivations expressed by parents and children beg the question: Why do more families not enroll and why do they leave prior to completion? Factors impacting enrollment may readily be apparent when studying rare pediatric illnesses, yet difficult to manage when regulatory reviewers proscribe strict entry criteria and sample sizes. It is no surprise that enrollment challenges have given rise to a service industry devoted to finding and retaining compliant subjects.[36] Several large surveys with adult participants have found that refusal to participate and drop-out is associated with inconvenience, demands on time, additional procedures, travel costs, and negative perceptions about study staff.[37,38] Although studies are considerably smaller, similar factors reportedly impact participation in pediatric trials as shown in Table 31.1. Overcoming the barriers involves investigator attention to site logistics, accurate information in advertisements and consent documents, and satisfying individual family needs throughout the study. In parallel, sponsors need to identify reasons for recruitment and screen failure at the initiation of enrollment, well in advance of delayed enrollment milestones, by providing simple tracking tools for coordinators.

Another frequent barrier in pediatric trials is the understandable resistance of parents to enroll their child into a study of a product available from their physician through off-label prescribing. Companies and investigators have accepted their charge to obtain pediatric data, but parents may not agree, especially in placebo-controlled studies. The risk of coercion is high if an investigator or coordinator overemphasizes staff attentiveness and no-cost drug as benefits. Alternatively, the educational aspects of trial participation noted earlier are important to parents and children and should be described during the consent process.

TABLE 31.1 Barriers to Participation

Factors	Example
Consent process	Lengthy and complex forms [40,54,55]
	Lack of engagement by investigator [49]
	Insufficient or too much information [49,55]
	Insufficient time to decide[49,56]
	Timing of consent in relation to child's condition[55,57,58]
Confidentiality	Sexual health assessments[59,60]
	General confidentiality[45]
Perception of risk	Multiple procedures[61–63]
	One treatment better than the other[45]
	Overall risk[63]
Perception of burden to child	Interviews during serious illness[64]
	Additional procedures during serious illness[40,59]
Painful procedures	Phlebotomy[45,65,66]
	Intravenous treatments[59]
Lengthy interviews	Psychiatric assessments[39]
Logistics	Transportation to clinic[43]
	Duration of study and/or visits[43,67,68]

Although neglected to date, education should also include trial results. Many experts have argued that study results should be reviewed with parents but the mechanism to support the timing and delivery of results have not been established.[39] Presenting an understandable interpretation to parents who request their child's results may offer meaningful benefit in return for participation and possibly reverse negative perceptions about trial secrecy. When parents and children are fully engaged and not simply subjects of the investigation, families are more likely to enroll even for placebo studies and, in turn, promote research opportunities to other parents.

Aside from altruism, personal benefit and hope for improvement are important to parents. Parents are often compelled to participate because they perceive no other alternative for their child.[40–42] Personal benefit is perceived by parents as access to comprehensive care by experts, attentiveness by research staff and no cost medication.[33,43–45]

Personal benefit linked to no-cost medication is germane to the risk of coercion mentioned previously, especially among low income parents. Rothmier et al.[44] found that a perceived benefit of medication by parents was negatively correlated with household income. Payments have not been identified as a primary motivator for parents or adult study subjects, yet at least some parents appear to value indirect payment via no-cost treatment more than direct remuneration.[33,38] Such benefits are ethically sound when presented within the broader context of experimental risk, inconvenience, and alternatives.

The impact of direct or indirect compensation on a child's decision to participate is unknown but is likely to differ by age and experience. Disparities found in several studies signify the need for systematic research on payment and assent. Differences between children and parents were reported in a global survey in which 51% pediatric participants versus 27% parents cited payment as a reason to participate.[33] However, a targeted analysis of adolescent recruitment using hypothetical vignettes found that compensation would not affect decisions to participate.[23] Obviously, payment related to actual participation in a trial may have yielded a different outcome in the adolescent study. The impetus for further study is further highlighted by a survey of investigator perspectives on payment and assent. The investigators responded that financial reward was unlikely to impact recruitment of children and adolescents.[24] The survey did not parse the level of investigator involvement during assent, nor did it link their perceptions to specific studies. If payment is offered to children, one suggestion based on perceptions about wages among children aged 5–15 years is to offer tokens of appreciation to children <9 years and apply a wage-payment model for older children.[46] Robust investigations are clearly needed to determine the motivational impact of reward on children and the factors to guide reasonable amounts. In the meantime, precautions for managing the risk of coercion are warranted. Wendler et al.[18] suggested that ethics boards defer review of inducement payments to a child advocate. Where large payments are offered in particularly time-consuming studies with numerous procedures, risk can be further minimized by including an individual independent of the study team in the assent process.

A PARADIGM FOR PEDIATRIC RECRUITMENT

A shift in recent years for a volunteer-centric approach to recruitment and retention has been recognized for adult studies, but recruitment in pediatrics is often still subject to a top–down approach based on regulatory authority requirements. The typical process begins with a trial design that answers scientific questions and satisfies regulations without fully considering

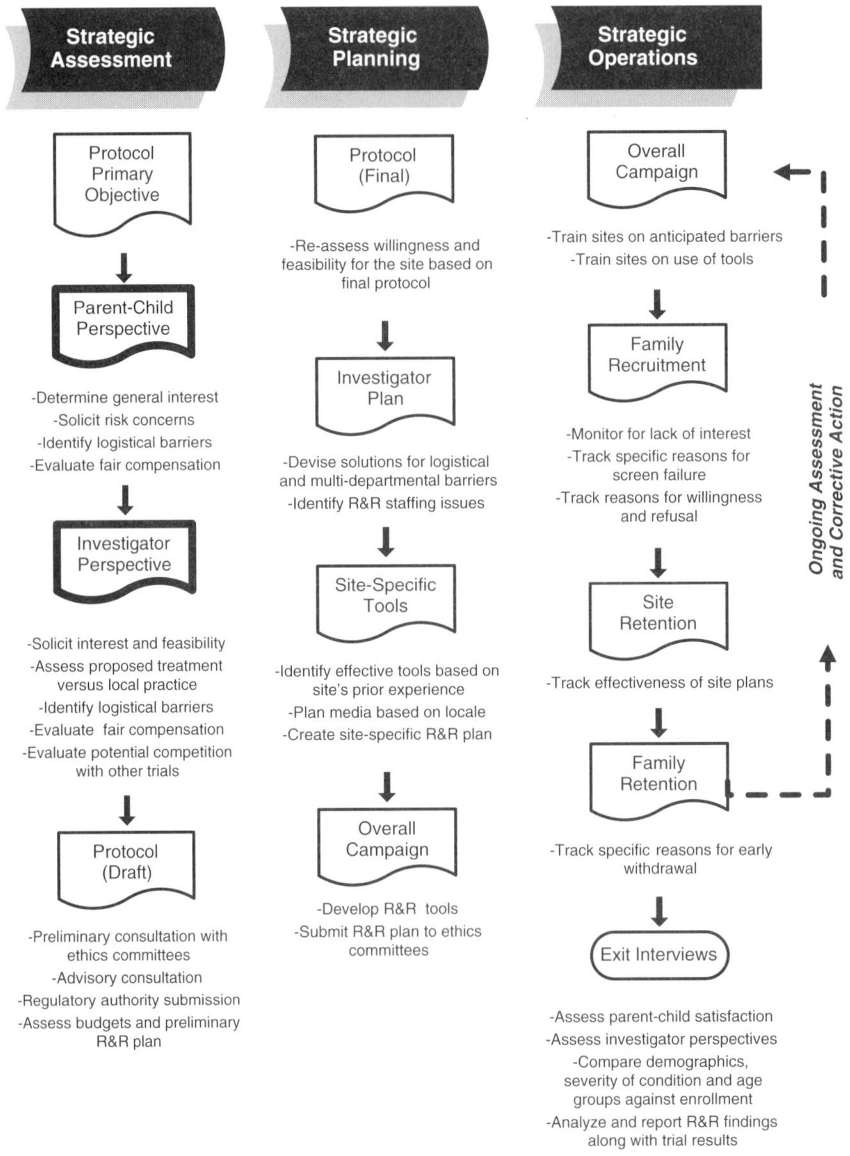

FIGURE 31.1 Paradigm for recruitment and retention of pediatric sites and patients.

the needs of families and children. The scenario to date is to negotiate a pediatric program, begin enrollment, then return to the agency for amendments when enrollment fails.

Based on the evidence presented in this chapter, a more rational starting point for recruitment success is a paradigm that integrates strategic assessment, planning, and operations (Figure 31.1). Interviewing investigators and potential families to discover perspectives on key study objectives and participant requirements is the first step. Strategic assessment of protocol objectives at the outset of protocol development is not unlike market research on a drug product and strategies involving consumer input for adult trials.[47]

Focus groups with parents and children in the targeted indication can identify questions to guide consent/assent materials and concerns about study procedures.[48,49] Researchers have already identified the value of input from children for assent materials prior to study start, as well as exit interviews to guide future protocols.[50–52] The findings can be incorporated into a realistic study design with appropriate justification at the outset of discussions with regulators.

The paradigm implies risk management with corrective actions involving reassessment and revised plans.[53] The systematic approach allows for open discourse early in program development to identify factors that could be unintentionally coercive to families or investigators. Overall, the paradigm builds upon findings from empirical studies discussed throughout this chapter, classic risk management processes, and Good Clinical Practice to optimize recruitment and ensure retention.

CONCLUSION

Ethical and effective recruitment will always present operational challenges in pediatric clinical trials for the myriad of reasons presented in this chapter. Lessons to date have revealed the complexities and now await clarification through further study using robust methodologies. The pediatric community has not only an opportunity to improve the process but also an inherent obligation to ensure valid and reliable data. Quality outcomes are more likely when ethical principles guide strategic assessment with a view toward well-designed and customized recruitment programs. Attention to the perceptions of families and investigators is a critical first step to determine supportive measures that will ensure their participation and commitment. In the final analysis, successful recruitment in pediatric trials is measured by the time, thoughtful consideration, and flexibility applied to the process.

REFERENCES

1. Walson PD. Patient recruitment: US perspective. *Pediatrics*. 1999;104(3):619–622.
2. Milne C. Pediatric study costs increased 8-fold since 2000 as complexity level grew. *Impact Report*. 2007;9(2):1–4.
3. Cutting Edge Information. *Accelerating Clinical Trials: Budgets, Patient Recruitment and Productivity*; 2004.
4. International Conference on Harmonisation. *Consolidated Guideline: Good Clinical Practice. Consensus Guideline*. 1997.
5. American Academy of Pediatrics. Guidelines for the ethical conduct of studies to evaluate drugs in pediatric populations. *Pediatrics*. 1995;95:286–294.
6. US Department of Health and Human Services, Office of Human Research Protections. *Protections for Children in Research: A Report to Congress*. 2001.
7. Church C, Santana VM, Hinds P, et al. Near the boundary of research: roles, responsibilities and resource allocation. In: Kodish E, ed. *Ethics and Research with Children: A Case-Based Approach*. New York: Oxford University Press; 2005:274–322.
8. Gill D, Crawley FP, LoGiudice M, et al. Guidelines for informed consent in biomedical research involving paediatric populations as research participants; the Ethics Working Group of the Confederation of European Specialists in Paediatrics (CESP). *Eur J Pediatr*. 2003;162 (7–8):455–458.

9. Council for International Organizations of Medical Sciences. *International Ethical Guidelines for Biomedical Research Involving Human Subjects*. 2002.
10. National Commission for the Protection of Human Subjects of Biomedical and Behavioral Research. *Ethical and Policy Issues in International Research: Clinical Trials in Developing Countries*. Bethesda, MD: Department of Health and Human Services; 2001.
11. American Academy of Pediatrics. Informed consent, parental permission, and assent in pediatric practice. *Pediatrics*. 1995;95(2):314–317.
12. Cooke RE. Ethical issues in exposing children to risks in research. In: Grodin MA, Glantz LH, eds. *Children as Research Subjects: Science, Ethics and Law*. New York: Oxford University Press; 1994:192–214.
13. European Commission. *Ethical Considerations for Clinical Trials Performed in Children*. 2006.
14. US Department of Health and Human Services, Office for Protection from Research Risks. Protecting Human Research Subjects: *Institutional Review Board Guidebook*. 1993.
15. Dunn LB, Gordon NE. Improving informed consent and enhancing recruitment for research by understanding economic behavior. *JAMA* 2005;293(5):609–612.
16. Grady C. Payment of clinical research subjects. *J Clin Invest*. 2005;115(7):1681–1687.
17. Ramsey BW. Appropriate compensation of pediatric research participants: thoughts from an Institute of Medicine committee report. *J Pediatr*. 2006;149(1 Suppl):S15–S19.
18. Wendler D, Rackoff JE, Emanuel EJ, et al. The ethics of paying for children's participation in research. *J Pediatr*. 2002;141(2):166–171.
19. Whittle A, Shah S, Wilfond B, et al. Institutional review board practices regarding assent in pediatric research. *Pediatrics*. 2004;113:1747–1752.
20. Diekema DS. Payments for participation of children in research. In: Kodish E, ed. *Ethics and Research with Children: A Case-Based Approach*. New York: Oxford University Press; 2005: 274–322.
21. Weise K. National practices regarding payment to research subjects for participating in pediatric research. *Pediatrics*. 2002;110:577–582.
22. Kimberly MB, Hoehn KS, Feudtner C, et al. Variation in standards of research compensation and child assent practices: comparison of 69 Institutional Review Board-approved informed permission and assent forms for 3 multicenter pediatric clinical trials. *Pediatrics*. 2006;117:1706–1711.
23. Scherer DG, Brody JL, Annett RD, et al. Financial compensation to adolescents for participation in biomedical research: adolescent and parent perspectives in seven studies. *J Pediatr*. 2005;146 (4):552–558.
24. Iltis AS, DeVader S, Matsuo H. Payments to children and adolescents enrolled in research: a pilot study. *Pediatrics*. 2006;118(4):1546–1552.
25. Institute of Medicine I. Chapter 6: Payments Related to Research. In: Field J, Behrman RE, eds. *The Ethical Conduct of Clinical Research Involving Children*. Washington DC: National Academy of Sciences; 2004:211–228.
26. International Conference on Harmonisation. *Clinical Investigation of Medicinal Products in the Pediatric Population*. 2000.
27. Smith S, Jaffe D, Petty M, et al. Recruitment into a long-term pediatric asthma study during emergency department visits. *J Asthma*. 2004;41(4):477–484.
28. Milne C. US pediatric studies incentive led to new labeling for nearly 100 drugs. *Impact Report*. 2005;7(4):1–4.
29. Centerwatch. Unveiling "hidden costs" in clinical trials. *Centerwatch*. 2001;10(9):1–4.
30. US Department of Health and Human Services Office of Inspector General. *Recruiting Human Subjects: Pressures in Industry Sponsored Trials*. 2000.

31. Morley CJ, Lau R, Davis PG, et al. What do parents think about enrolling their premature babies in several research studies? *Arch Dis Child Fetal Neonatal Ed.* 2005;90(3):F225–F228.
32. Weithorn LA, Scherer DG. Children's involvement in research participation decisions: psychological considerations. In: Grodin MA, Glantz LH, eds. *Children as Research Subjects: Science, Ethics and Law.* New York: Oxford University Press; 1994:133–179.
33. Niles JP. Pediatric subjects and their parents respond to a survey. *Appl Clin Trials.* 2003;12(4): 42–44.
34. Kassam-Adams N, Newman E. Child and parent reactions to participation in clinical research. *Gen Hosp Psychiatry.* 2005;27(1):29–35.
35. Wolthers OD. A questionnaire on factors influencing children's assent and dissent to non-therapeutic research. *J Med Ethics.* 2006;32(5):292–297.
36. Anderson DL. The patient recruitment market. In: Anderson DL, ed. *A Guide to Patient Recruitment and Retention.* Boston: Thomson Centerwatch; 2004:3–22.
37. Ross LF. Informed consent in pediatric research. *Camb Q Healthcare Ethics* 2004;13(4): 346–358.
38. Getz KA. Benchmarking patient recruitment and retention in clinical trials. In: Anderson DL, ed. *A Guide to Patient Recruitment and Retention.* Boston: Thomson Centerwatch; 2004:25–44.
39. Hinshaw SP, Hoagwood K, Jensen PS, et al. AACAP 2001 research forum: challenges and recommendations regarding recruitment and retention of participants in research investigations. *J Am Acad Child Adolesc Psychiatry.* 2004;43(8):1037–1045.
40. Chappuy H, Doz F, Blanche S, et al. Parental consent in paediatric clinical research. *Arch Dis Child.* 2006;91(2):112–116.
41. Pace C, Talisuna A, Wendler D, et al. Quality of parental consent in a Ugandan malaria study. *Am J Public Health.* 2005;95(7):1184–1189.
42. Benedict JM, Simpson C, Fernandez CV. Validity and consequence of informed consent in pediatric bone marrow transplantation: the parental experience. *Pediatr Blood Cancer.* 2007;49(6): 846–851.
43. Gammelgaard A, Knudsen LE, Bisgaard H. Perceptions of parents on the participation of their infants in clinical research. *Arch Dis Child.* 2006;91(12):977–980.
44. Rothmier JD, Lasley MV, Shapiro GG. Factors influencing parental consent in pediatric clinical research. *Pediatrics.* 2003;111:1037–1041.
45. Sammons HM, Atkinson M, Choonara I, et al. What motivates British parents to consent for research? A questionnaire study. *BMC Pediatr.* 2007;7:12.
46. Bagley SJ, Reynolds WW, Nelson RM. Is a "wage-payment" model for research participation appropriate for children? *Pediatrics.* 2007;119(1):46–51.
47. Aman MG, Wolford PL. Consumer satisfaction with involvement in drug research: a social validity study. *J Am Acad Child Adolesc Psychiatry.* 1995;34(7):940–945.
48. Ford K, Sankey J, Crisp J. Development of children's assent documents using a child-centred approach. *J Child Health Care.* 2007;11(1):19–28.
49. Eder ML, Yamokoski AD, Wittmann PW, et al. Improving informed consent: suggestions from parents of children with leukemia. *Pediatrics.* 2007;119(4):e849–e859.
50. Dawson A, Spencer SA. Informing children and parents about research. *Arch Dis Child.* 2005;90:235–237.
51. Gleason C. *Statement on Behalf of the Society of Pediatric Research and the American Pediatric Society before the Committee on Clinical Research Involving Children.* Institute of Medicine; 2003.
52. May DE, Hallin M, Kratchovil C, et al. Factors associated with recruitment and screening in the treatment for adolescents with depression study (TADS). *J Am Acad Child Adoles Psychiatry.* 2007;46(7):801–810.

53. Simar M, Johnson V. Play it safe: risk management for investigators in pediatric trials. *The Monitor*. 2003;17(2):45–49.
54. Culbert A, Davis DJ. Parental preferences for neonatal resuscitation research consent: a pilot study. *J Med Ethics*. 2005;31:721–726.
55. Franck LS, Winter I, Oulton K. The quality of parental consent for research with children: a prospective repeated measure self-report survey. *Int J Nurs Stud*. 2007;44(4):525–533.
56. Hazen R, Drotar D, Kodish E. The role of the consent document in informed consent for pediatric leukemia trials. *Contemp Clin Trials*. 2006;10(4):401–408.
57. Angiolillo A, Simon C, Kodish E, et al. Staged informed consent for a randomized clinical trial in childhood leukemia: impact on the consent process. *Pediatr Blood Cancer*. 2004;42:433–437.
58. Ballard HO, Shook LA, Desai NS, et al. Neonatal research and the validity of informed consent obtained in the perinatal period. *J Perinatol*. 2004;24:409–415.
59. Massicotte MP, Sofronas M, deVeber G. Difficulties in performing clinical trials of antithrombotic therapy in neonates and children. *Thrombosis Res*. 2006;118(1):153–163.
60. Slack C, Strode A, Fleischer T, et al. Enrolling adolescents in HIV vaccine trials: reflections on legal complexities from South Africa. *BMC Med Ethics*. 2007;8(1):5.
61. Truog RD. Increasing the participation of children in clinical research. *Intensive Care Med*. 2005;31(6):760–761.
62. Brody JL, Annett RD, Scherer DG, et al. Comparisons of adolescent and parent willingness to participate in minimal and above-minimal risk pediatric asthma research protocols. *J Adolesc Health* 2005;37(3):229–235.
63. Hoehn K, Wernovsky G, Rychik J, et al. What factors are important to parents making decions about neonatal research. *Arch Dis Child Fetal Neonatal Ed*. 2005;90:F267–F269.
64. Hulst JM, Peters JW, van den Bos A, et al. Illness severity and parental permission for clinical research in a pediatric ICU population. *Intensive Care Med*. 2005;31(6):880–884.
65. Dlugos DJ, Scattergood TM, Ferraro TN, et al. Recruitment rates and fear of phlebotomy in pediatric patients in a genetic study of epilepsy. *Epilepsy Behav*. 2005;6(3):444–446.
66. Langley J, Halperin S, Mills E. Parental willingness to enter a child in a controlled vaccine trial. *Clin Invest Med*. 1998;21:12–16.
67. Harth S, Thong Y. Sociodemographic and motivational characteristics of parents who volunteer their children for clinical research: a controlled study. *Br Med J*. 1990;300:1372–1375.
68. Tierney E, Aman M, Stout D, et al. Parent satisfaction in a multi-site acute trial of risperidone in children with autism: a social validity study. *Psychopharmacology (Berl)*. 2007;191(1):149–157.

CHAPTER 32

Certification of Pediatric Clinical Investigators

A. PROCACCINO

Senior Director, Technical Training, Johnson & Johnson Pharmaceutical Research and Development LLC, 1125 Trenton-Harbourton Road, Titusuille, New Jersey 08560

INTRODUCTION

Certification is the confirmation of certain characteristics of an object, person, or organization. It is a sign of mastery of knowledge. It is often provided by an external review board, or granted after the receipt of some type of specialized education, or by the passing of a test or assessment. Many professions require certification as a means to show a basic level of competency and knowledge in that particular field of interest or study. And we as the public have a certain level of confidence in a particular profession just in knowing that they are "certified." When we think of certification we think of public accountants, use of cardiopulmonary resuscitation (CPR), lifeguards, nursing assistants, and even project managers in business. These roles and responsibilities all require certification. But only in this current decade has the word certification been associated with clinical investigators who conduct human subject research. Public perception of clinical trials and those who conduct them is at an all-time low.

CERTIFYING BODIES

Those involved in biomedical research are working hard to educate the public as to the entire clinical trial process and safeguards inherit in it. Certification has been a way to let the public know that an investigator has met certain standards of knowledge and has the appropriate foundation from which to run clinical trials. Does it mean that all problems disappear because someone is certified? No, it does not. But it does give some reassurances to the subject or his/her guardian as to the investigator's competence. As clinical research evolves, certification of clinical research professionals will continue to grow in importance.

Pediatric Drug Development: Concepts and Applications
Edited by Andrew E. Mulberg, Steven A. Silber, and John N. van den Anker

The Association of Clinical Research Professionals (ACRP) and the Academy of Pharmaceutical Physicians and Investigators (APPI) are the largest certifying bodies for investigators. The ACRP delivers the Certified Trial Investigator (CTI) exam. This exam is for doctorate level or equivalent clinical research professionals, for example, PhD, PharmD, doctorate in nursing, psychologists, licensed physician assistants, and nurse practitioners who serve in an investigator role. This includes doctorate level person working in or for the pharmaceutical industry who supervise clinical investigators for clinical trial sponsors. The eligible candidate has experience, in the last two (2) years, performing investigator functions as evidenced by copies of investigator agreements, letters from supervisors, or other equivalent documentation. Physicians trained within or outside the United States who do not hold a current license are evaluated on a case-by-case basis to ensure that they meet the certification eligibility. The APPI, which is part of the overall parent ACRP organization, delivers the CPI exam. This exam is for individuals who hold a medical physician degree (MD or equivalent degree) and who are in good standing with local, state, and national licensing and regulatory authorities and certifying bodies. These individuals had to have served as a primary, sub-, or coinvestigator in one or more clinical trials within the last two (2) years or served as a monitor, supervisor, or designer of one or more clinical trials within the last two (2) years. They can also have successfully completed a clinical research degree or clinical fellowship program of at least one (1) year's duration. Those applying for the CPI certification must also have a current license to practice medicine, who are active and in good standing.

These certification programs are garnering more attention of late as we see pharmaceutical companies and even the governments of some countries demanding this notation of excellence of their employees. We are restarting to see more legislative action around this topic in the United States as well. In early 2007, the Secretary's Advisory Committee on Human Research Protections (SACHRP) recommended that the Department of Health and Human Services Office of Human Research Protection (DHHS OHRP) require that institutions require initial and continuing training of investigators, institutional review board members and staff, institutional signatory officials, and human protection administrators. The recommendation states that investigators should complete training before being allowed to conduct human subject research and that it should ensure the competence of the investigators.

Voluntary certification by investigators, study coordinator, and clinical research associates is one achievement that each member of the clinical team can use as proof of that individual's commitment to the highest standards for the conduct of research and for the well-being of each human subject. With increased public and regulatory scrutiny of clinical trials, those involved in clinical research must become increasingly more transparent in how they conduct trials and protect human subjects. Expectations are evolving and the bar for standards of excellence is continually being raised. The industry is responding by holding itself accountable in helping to create those standards. Certification in good clinical practices is one such standard that the industry is moving toward at lightening speed. The need for addressing these issues for pediatric investigations should not be far behind this process.

CHAPTER 33

Considerations and Barriers for Pediatric Patient Recruitment Strategy for Clinical Trials

JENNIFER NIESZ and ANDREW E. MULBERG, MD

Johnson & Johnson Pharmaceutical Research & Development LLC, Titusville, New Jersey 08560

CHALLENGES TO PATIENT RECRUITMENT

Patient recruitment for clinical trials in general can be very challenging. Even studies of healthy adult volunteers may face difficulties in recruiting enough patients for the study. Clinical trials involving children and adolescents face even greater challenges in terms of patient recruitment. Logistically, it is sometimes difficult to find pediatric patients for a study because there may be a relatively small number of children affected by a particular disease or illness. This in turn can limit the number of potential child subjects to be considered for recruitment for the specific trial. Technically, there may be procedures or tests involved in a study that may seem daunting to a parent, family, or potential child participant. For example, there may be a need for frequent blood sampling to determine the pharmacokinetics, and this may appeal neither to the family nor to the child.

Practically speaking, families may have a wide range of concerns that deter them from enrolling their child in a study. For example, the parents may have very demanding careers, or they may need to focus a large amount of attention on the competing needs of siblings. They may feel that they would have difficulty devoting their time and attention, as a family, to a clinical study.

Ethically, there is a wide range of concerns about studies performed in children, and these concerns can have an impact on patient recruitment. There is a perception that children are vulnerable and need to be protected from the potential harm associated with participating in a clinical study. However, it is important to conduct research in children so that we know that medications are safe and effective for them; it could even be considered unethical *not* to perform research in children. These issues are discussed more fully throughout this book, including Chapter 10 by Baer and Chapter 9 by Nelson.

Pediatric Drug Development: Concepts and Applications
Edited by Andrew E. Mulberg, Steven A. Silber, and John N. van den Anker

ETHICAL BARRIERS TO PATIENT RECRUITMENT

The ethical challenges in recruiting patients for pediatric clinical trials may be the greatest barriers and need to be considered in detail. One aspect of clinical research that has a great impact on patient recruitment is the perceived risks and benefits of a trial. This is especially important in pediatric trials, where both clinicians and parents are very concerned about not hurting a child in any way. It can be a challenge for clinicians and for parents or families to assess the balance between harm and benefit, which can be physical, psychological, or economic in nature. In general, the potential for harm should be reasonable for the amount of benefit that a child may gain from participating in a study. There have been some guidelines established for assessing this risk–benefit ratio. These issues are raised comprehensively in other chapters in this book and are not covered deeply here.

However, in pediatric clinical trials, one of the most important considerations is the family members' perception of the risk involved in a study. The media and government agencies, as well as their own personal and family views, may influence their perception or community acquired beliefs. The most direct way of determining how acceptable a study is to a family, based on the perception of risk, is for the investigator to inform the family about the potential for their child to be harmed in the course of the study, and discuss the potential harm versus benefit openly and honestly.

The issues of consent, assent, and confidentiality are central to the recruitment of patients into pediatric clinical trials. Consent is the process of communication between a patient and physician that results in the patient's authorization to undergo a medical intervention. Parents generally give consent for pediatric patients who are too young to understand the information relating to the decision to participate in a clinical trial. The style in which the study information is given, and the type of information given can have a significant effect on patient recruitment. Studies looking at parent perception of the informed consent process have found that two main reasons why parents decide to enroll their children in clinical trials are to contribute to science and to benefit their child. Primary reasons for not enrolling their children include a fear of side (adverse) effects, inconvenience, and not wanting to get involved.

Another consideration is the difficulty of imparting certain study information to parents, including basic tenets of research. Many parents do not fully understand the concept of randomization. The concept of voluntary participation can also be difficult to comprehend; research has shown that up to 35% of parents do not know that they can withdraw their child from a study, and some parents do not know that they can refuse participation.[1] For these reasons, it is important for clinicians to spend time with parents and help them to fully understand the study terminology and what is really involved in their family's participation.

Children may be able to give their assent or lack of assent to a study. According to the American Academy of Pediatrics, any child with the intellectual age of 7 years or older is able to give assent to participating in a clinical trial. This assent does not take the place of parental consent; rather, it gives the child a chance to understand information about the study in which he/she may participate, and decide for himself/herself if it is something he/she feels comfortable doing. In order for the child to understand the study information clearly, the assent should be written simply, at the level of a 6-year-old child.[1]

In terms of confidentiality, some parents may be apprehensive about enrolling their child in a study, due to concerns that his/her medical information will be shared with others or will become public information. It can be helpful for clinicians to reassure parents that their child's medical record, both study related and otherwise, will be kept confidential.

Reimbursement is another complex issue related to the recruitment of children into clinical studies. In some cases, children or their parents may decide to enroll in a clinical study because they are under the impression that they will benefit financially from doing so. Since this may not be in the best interest of the child, there have been guidelines established by the American Academy of Pediatrics stating that a financial reward for child participation should not exceed "a token gesture of appreciation" and remuneration should not be discussed with the child or the family before the study is completed. In general, the family should be reimbursed for the direct and indirect costs they incur by participating in the research. In this way, financial benefits should not affect the family's decision to enroll their child in a clinical trial, nor should financial hardships prevent them from participating. Issues of coercion are a frequent consideration to balance in the recruitment of pediatric subjects since the intensity of specific disease testing varies with each different clinical study and disease.

CROSS-CULTURAL CONSIDERATIONS FOR PEDIATRIC PATIENT RECRUITMENT

Clinical research among pediatric patients in Europe is conducted much as it is in the United States and Canada.[2] Studies have looked at various aspects of European pediatric research, including key factors for success in recruiting patients.

European requirements state that children should only be enrolled in clinical research if there is a chance for them to derive some benefit from the study. As in the United States, the risk involved in a study cannot outweigh the potential benefit. A requirement for enrollment of children in clinical trials is parental consent, in writing, while also allowing that the opinion of the child should be considered, depending on his/her age and capacity for understanding the study. Again, as in the United States, European families cannot be coerced to participate in a study by financial benefits or other gifts, but reimbursement for study-related costs is allowed. Children are not supposed to participate in clinical research as healthy volunteers; rather, they should be patients or should be involved in studies of vaccinations or other preventive measures.[3]

Throughout Europe, children are usually recruited into clinical trials by physicians who see special groups of patients or by referral from primary care. In some countries such as Italy, children are primarily cared for by pediatricians; in other countries, such as the United Kingdom, children's primary care falls to general practitioners.

METHODS OF RECRUITMENT

A number of methods are used throughout Europe to recruit pediatric patients. National and local registries of patients are used to identify and contact potential families. In some cases, clinical trials may be advertised to groups of healthcare workers, such as well-baby clinics or school health personnel, or to parents.

A review of several European pediatric clinical trials showed a number of reasons for difficult patient recruitment.[3] Among them were inconvenience (requirement to stay at study site for many hours) and limited benefit to be gained from participating. The review of these studies also showed that there are a number of common factors that influence recruitment and retention: perception of benefit to the child, good doctor–patient relationship, clear understanding of what the trial is about, and ease with which child and family can adhere to the study requirements. These factors are important to pediatric studies in Europe and in the United States as well and should be considered when formulating a recruitment strategy, which is discussed later in this chapter.

FAMILY DECISION-MAKING AND THE INFORMED CONSENT

How do families decide to participate in a pediatric clinical trial? What influences parents' consent or lack thereof? While a number of factors have been found to influence parental consent, research has found that the most important motivating factor for parents is learning more about their child's illness or, in some cases, allowing their child to learn more about his/her own illness. Another very important motivation is the opportunity to contribute to the furthering of medical knowledge. Parents are also interested in allowing their children the opportunity to receive the newest available drugs or treatments. To a lesser extent, the family's relationship with the research staff, availability of free medications, encouragement of their physician, and free office visits were motivating factors in parents' consent. Some other less important factors are the type of treatment being used in the study, clinic location, duration of study, influence of family and friends, social support, and support received as a study participant. Parents cite reasons for declining consent that are specific to study protocols. In particular, parents may decline consent due to possible painful or uncomfortable procedures their children would have to experience, such as venipuncture or fingerpricks.

Researchers have also considered how certain demographic variables may be associated with the lesser and more important factors in providing consent for pediatric research. In particular, Rothmier et al.[4] examined the variables of family income and parents' education levels, in relation to motivating factors for consenting to clinical research. They found that families with lower socioeconomic status were significantly more interested in obtaining free medications for their children through pediatric studies. They also found that parents with a higher level of education were less likely to express a desire for contributing to medical knowledge. However, this finding was not statistically significant and may have been partly explained by the fact that the entire sample of parents had a relatively high level of education. Overall, the parents' education levels were not significantly correlated with any of the possible motivating factors for participating in a study. This is an interesting and important topic for future research related to parental consent for pediatric studies; there may be additional valuable information to be gained by looking further at demographic variables and their relationship to the factors that motivate parents to enroll their children in research studies.

Overall, parents are relatively unconcerned with financial or material gains and are most concerned with learning more about their child's illness and contributing to medical knowledge. Parents also want to avoid having their child experience painful or unpleasant medical procedures.

Patterns of decision-making are also an important consideration when informing a family about a clinical trial and attempting to gain the parents' consent. Informed consent is typically considered from an individual perspective, since the patient himself/herself is usually the one who grants consent. However, in the informed consent process for pediatric studies, it can be useful to consider the process and the decision-making from a family perspective.

Family dynamics, the child's health status, and the family's view of healthcare providers may all affect the decision-making process. For clinicians, it is important to understand how families approach the informed consent process. Families may approach the process in very different ways, depending on their culture, dynamics, and family structure. Research has shown that there are several different styles of decision-making among families[5]—exclusionary, informative, collaborative, and delegated decision-making. In the *exclusionary* style of decision-making, the decision to participate in the clinical trial rests solely with the parents, and the child is not informed about what will be happening to him/her. The parents' goal is to protect the child from the illness, the treatment, and the research itself. In the *informative* pattern of decision-making, the parents talk openly with their child about the trial and inform the child about what will happen during the study, but still make the final decision themselves regarding participation in the trial. In *collaborative* decision-making, the parents involve the child in the process and explore information about the trial together. They work together to reach a shared decision and ultimately give the child the final decision. In this style of decision-making, the children are older, approximately 17–20 years in age. Finally, in *delegated* decision-making, the parents assess the nature and value of the clinical study. If they determine that the study is worthwhile, they then delegate the decision about participation to their child and support the child's decision.

Overall, consent is a complex process involving the investigator, parents, and child. Patterns of decision-making vary among families, and it is important for investigators to remember this as they approach the informed consent process. Parents will assess the information given to them regarding the clinical study and will involve their children in the decision-making process to a varying extent.

PRACTICAL APPLICATIONS FOR PEDIATRIC PATIENT RECRUITMENT

Patient recruitment for pediatric clinical trials is complex and has many important facets. Some considerations are the following. How do parents or families find out about clinical research studies? Where can clinicians locate families who may be interested in participating? What kinds of outreach and materials are effective in reaching families? What messages and information are most important to give to families who may consider a clinical study? All of these questions should be considered when formulating a recruitment strategy for a pediatric clinical trial. In the following paragraphs, we discuss a number of recruitment tactics and messaging.

The avenues by which parents learn about pediatric research studies vary to some degree depending on the type of study. In many cases, parents learn about a study from their child's pediatrician or a specialist who their child may be seeing. In other cases, they may hear about a study from some type of advertisement, via the radio, newspaper, television, Internet sites, or flyers posted in the community.

In order to reach out to parents of potential pediatric research participants, and contact them through all the methods described above, it is best to use a recruitment strategy involving a number of different tactics. The most effective strategies supply investigators and other clinicians with useful tools with which they can discuss a study clearly with parents. Effective strategies also supply parents with the information they need to learn about a study and make a decision about their child's participation. In some cases, it is also useful to supply child-friendly information about the study and the illness involved to the children directly.

TABLE 33.1 Materials and Audience for Clinical Research

	Materials for Professionals
Talking points for informed consent	Booklet designed for investigators, giving concise, patient-friendly, study-specific points to share with parents as they conduct the informed consent process. This can include basic scientific information, possible study benefits, and descriptions of study treatments and can be useful as a quick reference tool for investigators.
	Materials for Parents
Recruitment flyer	Flyer geared toward parents, giving basic study information. The flyer can include key inclusion criteria, description of the study and study treatment, length of the study, and whether financial reimbursement will be available.
Clinical trial brochure for parents	Brochure for parents of potential study participants, explaining clinical research. Information can include why research is important, why research is needed in a particular disease state, pros and cons of participation, a glossary of research terms, and specific criteria and logistics for a study.
Newspaper advertisement	Newspaper ad geared toward parents should use an appealing design and include descriptive information about the study, similar to the recruitment flyer mentioned previously. Investigators place the ad in local newspapers that potential families would be most likely to read.
Radio advertisement	Radio ad geared toward parents should give basic information about a study, similar to the recruitment flyer. The most effective radio ads include a conversation between two people, whose situation may be similar to the parents themselves. Investigators can place radio ads with local stations.
Educational brochure	Brochure for parents should give detailed yet understandable information about their child's disease or illness. Such educational information is an important benefit for parents considering enrolling their child in a study.
	Materials for Children
Clinical trial brochure for children (starting at age 7)	Brochure for potential study participants age 7 and older should explain clinical research (and specifics about the study).
Educational brochure for children (starting at age 7)	Brochure for potential study participants age 7 and older should give information about their specific illness or disease, including how it occurs and possible treatments.

Providing investigators, other clinicians, parents, and children with useful educational information about clinical research requires a varied arsenal of materials. Table 33.1 describes a number of materials that can be used with each of these audiences, to inform and educate about clinical research or a disease state.

Advertisements should reach out to the appropriate audience, include pertinent information, and also should be captivating, particularly when they are advertising a pediatric trial. Figure 33.1 an example of a newspaper advertisement for a pediatric epilepsy clinical trial, for children under the age of 2. The ad reaches out to parents with babies, gives the pertinent trial information including the type of illness, the age range, and a space where site contact

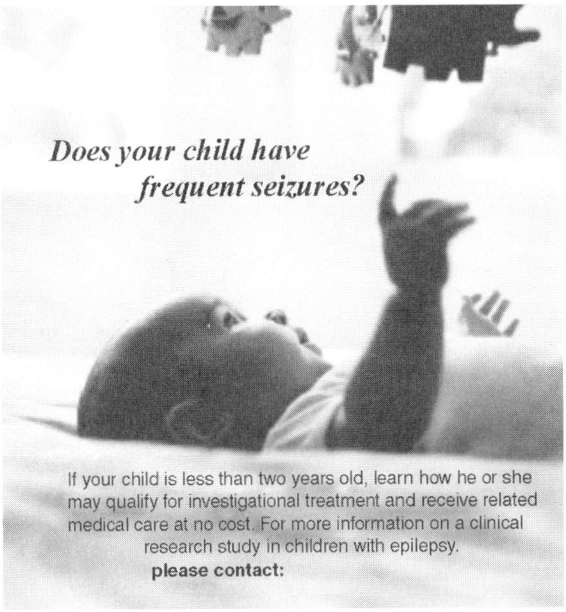

FIGURE 33.1 Newspaper ad, Johnson & Johnson Pharmaceutical Research & Development LLC.

information will be provided, and also, very importantly, uses a captivating photo to endear parents and prompt them to consider enrolling their own baby in this study.

It is crucial to create informative and educational materials that are geared toward each audience, so that maximum understanding and benefit are derived. To illustrate the importance of the way in which information is presented, Figure 33.2 is an excerpt from a clinical trial brochure geared toward parents, for a pediatric gastroesophageal reflux disease (GERD) clinical trial.

The information included in the brochure is appropriate for and can be understood by parents of potential child participants in this pediatric GERD study. The purpose of the brochure is to inform the parents about why clinical research is important, why research is needed on the subject of treatments for pediatric GERD, and why they may want to consider enrolling their child in this study.

In general, any materials given to parents regarding a study should include a fairly simple explanation of the scientific rationale, or the reason why the study is important. Materials should provide fair balance of the risks and benefits that their child may or may not experience if the child is enrolled in the study. Other important information can include background about the disease or illness, recent research leading up to the current study, other medications or treatments that are currently available, financial reimbursement that will be provided, and a discussion of logistics and procedures that will occur throughout the study. This information can serve to make parents feel more comfortable with the study, and therefore more willing to have their child participate. All of these points can be covered in educational materials, but should also be covered in the informed consent discussion.

In contrast to the parent-focused brochure, Figure 33.3 is an example of an educational booklet geared toward children, who were potential study participants for the pediatric GERD study. This booklet focused on informing children about the study, and clinical research in general, at a level they could understand and in terms to which they could relate.

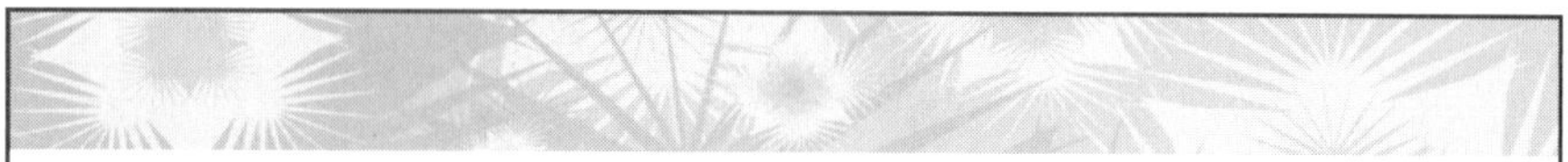

Understanding the Importance and Value of Clinical Drug Research

The treatments we have today for gastroesophageal reflux disease (GERD) are available because many people volunteered to participate in clinical research studies that were designed to evaluate those treatments. There are many treatments available for adults; however, there are many fewer treatment options for children with GERD.
By allowing your child to be part of a clinical research study, your family can contribute to improving the management of GERD in children in the future.

What is a clinical drug research study

A clinical drug research study is a scientific evaluation conducted with volunteers.
The purpose is to answer important medical questions about a potential new treatment, including: What dose and how frequently does the drug need to be given? Does it work better than medicines that are already available? Do the benifits outweigh the potential side effects? These are just some of the many questions that are answered by clinical research studies.

The process of testing new treatments can take many years. The sooner clinical research teams can recruit volunteers for studies, the sooner they can provide information to the FDA and other regulatory authorities for review and approval.

Is Clinical Research Needed for the Management of GERD in Children?

Currently, there are a number of medications approved for treating GERD in adults. One such medication is called AcipHex', and works by reducing the amount of acid produced by the stomach. In clinical research studies evaluating adults with GERD, AcipHex has been shown to be effective in the treatment of GERD, as well as other digestive system disorders. In terms of relief of symptons, AcipHex® has been shown to be even more effective than some other medications.

Medications can work differently in children than thay do in adults, and for this reason, it is important to study AcipHex® in children as well. By looking at the safety and efficacy of AcipHex® in children ages 1 to 11 years, clinical studies will help to determine age-appropriate doses of AcipHex for the treatment GERD in children.

How Can You Contribute to the Treatment of Children With GERD?

In order for doctors to determine the best treatment for children with gastroesophageal reflux disease, they need children with the condition to volunteer to participate in clinical research studies. If your child suffers from GERD, he or she may be eligible to participate in a study that is currently looking for volunteers. More information is provided at the end of this brochure.

FIGURE 33.2 Clinical trial brochure for pediatric gastroesophageal reflux disease.

This booklet and other materials written for children in this study, for the 7–11-year-old age group, were written at a fourth grade reading level. They included child-friendly photos and simple, easy-to-understand language.

Again, similar to parent-focused materials, one goal of child-focused materials is to help children understand what clinical research is and what it involves, so that they are familiar with what they may be taking part in. Another goal is to help children understand more about the disease or illness that they have (and may be participating in a study about)—in this case, pediatric GERD. This information and education, for both parents and children, is one of the potential benefits to the family participating in a pediatric clinical study.

PRACTICAL APPLICATIONS FOR PEDIATRIC PATIENT RETENTION

What keeps families who are participating in pediatric clinical trials coming back? Why do parents and children follow through to the end of the study? There are a number of factors that influence families' willingness and ability to see a trial through to its completion. Some influences are site staff attributes such as friendliness, responsiveness, and encouragement,

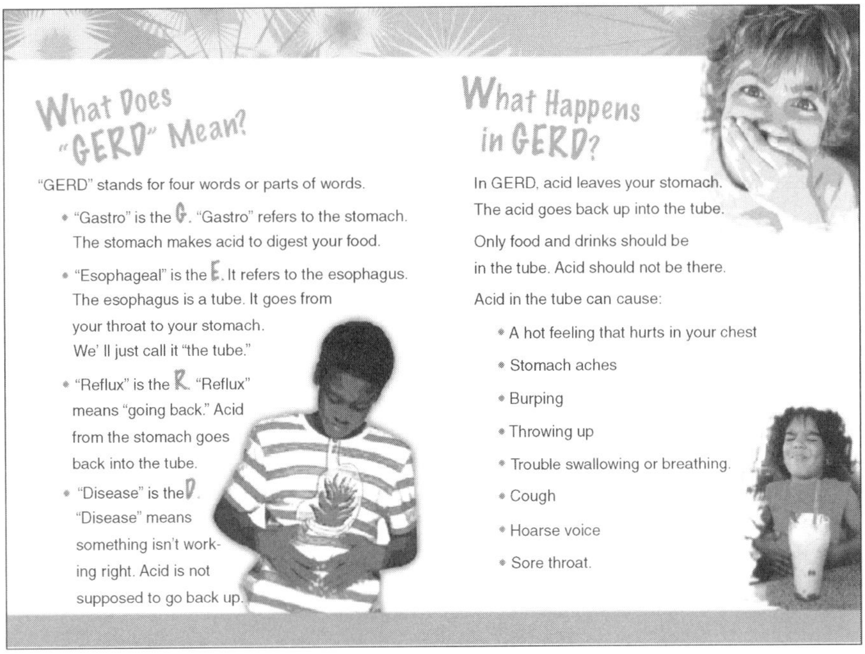

FIGURE 33.3 Educational booklet geared toward children.

while others are related to quality of care, such as quality of and completeness of exams, and free medication or medical tests/procedures. Other significant reasons for patient retention are seeing the same staff at each visit, appointment reminders, location of the site, newsletters, being part of a nationwide study, and length of study. Overall, the quality of the staff relationship with the patient and family as a whole is very important to keeping the family in the study for its entire duration.

Providing support to parents and children who are participating in clinical research is vital to retaining them in the study until completion. Support can and should be provided in a variety of ways, in order to meet all the needs of parents and children. To illustrate the importance of such support, and the way in which it may be executed, an example of a "patient support program" is described in the following paragraphs.

A patient support program was created for a pediatric epilepsy clinical trial involving patients under the age of 2. The program was called the Family Wellness Program and included a number of components. (Recruitment materials were also created for this same study but were considered separately from the Family Wellness Program.) One major component of the program was a grouping of "study assistance tools," some of which were geared toward parents, and others toward the babies. Included in the set of assistance items were a diaper bag for parents to transport their child's belongings back and forth to study visits, bibs, and a baby-friendly feeding set including a plate, spoon, fork, and sippy cup. Also included were a medication dosing card, which gave detailed, illustrated instructions to parents in how to administer the study medication to their babies, as well as a dry erase board with the study visit schedule printed on it—this was magnetized so that it could be hung on the refrigerator, and allowed parents to write in dates and notes for upcoming study visits. These materials were very useful to the study participants and made the daily dosing of medication and the task of remembering study-related procedures and visits easier for the parents.

Another even more significant aspect of the Family Wellness Program was the provision of educational handouts that were given to parents throughout the study. These handouts, 12 in total, were designed to help parents feel supported throughout the study and also to help educate parents about pediatric epilepsy. The handout topics ranged from basic information about pediatric epilepsy and child development, to dealing with family dynamics, creating a safe environment, balancing time demands, and managing emotions such as stress and anger. In addition to providing easy-to-understand information in a simple and appealing format, these handouts also provided tips and ideas for parents on ways to best care for themselves and their children with epilepsy. Figure 33.4 is a sample of the first page of the "Building a Relationship with Your Healthcare Team" (a two-page handout)—one of the many

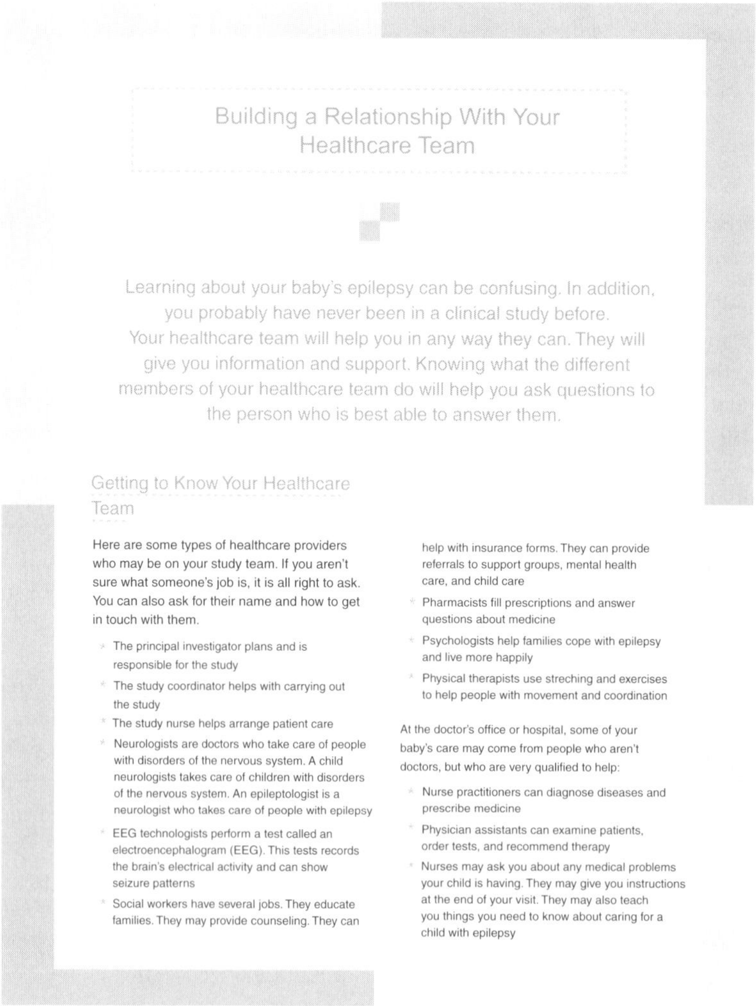

Building a Relationship With Your Healthcare Team

Learning about your baby's epilepsy can be confusing. In addition, you probably have never been in a clinical study before. Your healthcare team will help you in any way they can. They will give you information and support. Knowing what the different members of your healthcare team do will help you ask questions to the person who is best able to answer them.

Getting to Know Your Healthcare Team

Here are some types of healthcare providers who may be on your study team. If you aren't sure what someone's job is, it is all right to ask. You can also ask for their name and how to get in touch with them.

- The principal investigator plans and is responsible for the study
- The study coordinator helps with carrying out the study
- The study nurse helps arrange patient care
- Neurologists are doctors who take care of people with disorders of the nervous system. A child neurologists takes care of children with disorders of the nervous system. An epileptologist is a neurologist who takes care of people with epilepsy
- EEG technologists perform a test called an electroencephalogram (EEG). This tests records the brain's electrical activity and can show seizure patterns
- Social workers have several jobs. They educate families. They may provide counseling. They can help with insurance forms. They can provide referrals to support groups, mental health care, and child care
- Pharmacists fill prescriptions and answer questions about medicine
- Psychologists help families cope with epilepsy and live more happily
- Physical therapists use streching and exercises to help people with movement and coordination

At the doctor's office or hospital, some of your baby's care may come from people who aren't doctors, but who are very qualified to help:

- Nurse practitioners can diagnose diseases and prescribe medicine
- Physician assistants can examine patients, order tests, and recommend therapy
- Nurses may ask you about any medical problems your child is having. They may give you instructions at the end of your visit. They may also teach you things you need to know about caring for a child with epilepsy

FIGURE 33.4 First page of a two-page handout.

important topics related to pediatric epilepsy, as well as other areas of pediatric medical care and research.

In addition to providing parents with the 12 educational handouts described above, at visits throughout the course of the pediatric epilepsy study, the Family Wellness Program also provided clinician guides that corresponded to each parent handout. These clinician guides were provided because the parent educational handouts were not created in a vacuum and simply handed over to parents; rather, a study coordinator, site nurse, or psychologist visited with parents at each study visit, and this person also delivered that week's educational handout in a supportive manner. The 12 clinician guides were created to match each parent handout and to help the site professional prepare for the visit and for their supportive interaction with the parents prior to the study visit. Each clinician guide provided specific information for the clinician to use in facilitating that particular session—including objectives for the session, any preparation they needed to do beforehand, notes or questions to follow up on from the previous visit, and the main points they should cover in that visit's session. Figure 33.5 is the first page of the two-page clinician guide corresponding to the educational brochure "Building a Relationship with Your Healthcare Team."

Building a Relationship With Your Healthcare Team

STAFF GUIDE: TOPIC 2

Length of Session

* The average length of this session should be 10 minutes
* The session may last from 5 to 15 minutes
* Adjust the length of the session based on the needs of the family

Objectives

* To follow up from the last session, if needed
* To review and discuss the topic "Building a Relationship With Your Healthcare Team"
* To listen to the family's thoughts and concerns
* To provide helpful answers or suggestions where possible
* To provide appropriate referrals if the family needs additional help or information

Preparation

* Read the Family Handout: "Building a Relationship With Your Healthcare Team"
* Compare the information given in the handout to the people and roles in your particular organization
* Review the Objectives listed above
* Review any notes you may have from the previous session

Follow Up From Last Visit, If Needed

* If the family brought up any specific concerns last time, ask how things are going with these concerns (These might be specifically about living with a child with epilepsy)
* Ask how the time between sessions has gone
* If the family has continuing or new concerns, offer brief suggestions if you can, or refer them to appropriate resources
* Remember to include the child with epilepsy, if he/she is there. Look at him/her, call him/her by name, smile, and speak directly to him/her from time to time (even if the child is very young). Always use the child's name when you talk with the family about the child (Don't use impersonal terms like "the baby")

FIGURE 33.5 First page of two-page clinician guide.

Overall, the Family Wellness Program was well received at investigative sites around the world; parents were pleased to receive the education and support that were provided through the 12 educational handouts and the corresponding session with a site professional. They also found the study assistance items, including the diaper bag and medication dosing instructions, very useful in their daily compliance with the study. Clinicians at investigative sites were pleased to have the clinician guides to use in preparing for their visits with patients—they were able to be prepared for the session and present, in the conversation with the parents, just the right points for discussion. The various components of the Family Wellness Program worked together to increase the ability of clinical sites to retain families in the very challenging pediatric epilepsy clinical trial.

CONCLUSION

Many factors affect and contribute to patient recruitment for pediatric clinical trials, including ethical considerations, challenges with informed consent and family decision-making, and even variations among different cultures in regard to the way in which pediatric research is approached. The most effective way to overcome the barriers associated with pediatric patient recruitment is to use a variety of outreach strategies, including working with clinicians, advertising via radio and newspaper, and reaching out to parents directly. A number of tools and materials can be used both to recruit families for pediatric trials and also to retain them in these studies. In both cases, the materials should have the appropriate appearance and content for the audience to whom they are directed—and for many clinical studies, there should be a range of materials and tools geared to clinicians, parents, and directly to children. Retaining families in pediatric studies once they have been recruited is also a key consideration. A comprehensive patient support program should be considered, especially in more difficult studies. Parents and children alike appreciate when support and encouragement are given to them by investigative staff and, in turn, are more likely to see a study through to the end.

REFERENCES

1. Afshar K, Lodha A, Costei A, Vaneyke N. Recruitment in pediatric clinical trials: an ethical perspective. *J Urol*. September 2005;174:835–840.
2. Matsui D, Kwan C, Steer E, Rieder MJ. The trials and tribulations of doing drug research in children. *Can Med Assoc J*. November 11, 2003;169(10):1033–1034.
3. Hoppu K. Patient recruitment—European perspective. *Pediatrics*. September 1999;104(3): 623–626. (Suppl).
4. Rothmier JD, Lasley MV, Shapiro GG. Factors influencing parental consent in pediatric clinical research. *Pediatrics*. May 2003;111(5):1037–1041.
5. Snethen JA, Broome ME, Knafl K, Deatrick JA, Angst DB. Family patterns of decision-making in pediatric clinical trials. *Res Nurs Health*. DOI: 10.1002/nur.20130.

CHAPTER 34

Conducting Clinical Trials in Developing Countries: A Case Study

LEONARD R. FRIEDLAND, MD and JACQUELINE M. MILLER, MD*

GlaxoSmithKline Biologicals, King of Prussia, Pennsylvania 19406

INTRODUCTION

Fictional Pharmaceutical Company, Inc. (FPC, Inc.) is developing an investigational vaccine designed to protect against an enteric disease, known as the "GI virus." The GI virus disease occurs worldwide, and virtually every child is infected with GI virus by the age of 3 years. Infection with GI virus results in fever, abdominal pain, vomiting, and diarrhea. Symptoms typically last 5–7 days in most children. Treatment of GI virus is supportive, with use of fluids and electrolyte replacement. Most children infected with GI virus recover; however, the disease is a major cause of significant morbidity and mortality in developing countries, where children lack access to medical care. Both treatment and preventive measures have had only limited impact on the global GI virus disease burden. Therefore, FPC, Inc. is developing its investigational GI virus vaccine as vaccination against GI virus represents an important preventative strategy to control morbidity and mortality caused by this very common pediatric disease.

CLINICAL TRIALS

Phase I and Phase II clinical trials conducted by FPC, Inc. have demonstrated that the investigational GI virus vaccine when administered as a single dose to 2–3-month-old infants is efficacious, safe, and well tolerated. Phase I and Phase II clinical trials were conducted in the United States and Western Europe with appropriately designed and approved clinical trials. FPC, Inc. plans to conduct a Phase III efficacy and safety study in

*Dr. Friedland and Dr. Miller are employees of GlaxoSmithKline; information contained in this chapter does not pertain to any specific marketed or investigational products.

Pediatric Drug Development: Concepts and Applications
Edited by Andrew E. Mulberg, Steven A. Silber, and John N. van den Anker

20,000 infants in multiple countries throughout the world. FPC, Inc. plans to conduct this multinational, multicenter clinical trial in the United States, Europe, Latin America, Southeast Asia, and Africa. FPC, Inc. recognizes that, on a societal basis, it is the children in developing countries who will most benefit from the GI virus vaccine. Accordingly, the company plans to register and market GI virus vaccine in countries of the developed and developing world.

A unique feature of vaccine trials is that vaccines need to be studied when coadministered with other childhood vaccines recommended at the same age as the investigational vaccine. This is done to confirm that the investigational vaccine will not have a detrimental impact on the immunogenicity and safety of vaccines, that are already an established part of the vaccination schedule. Both the type of recommended vaccines and the ages at which they are recommended differ among countries around the world. For example, oral poliovirus vaccine and whole-cell pertussis vaccine are recommended for use in many Latin American and African countries, while inactivated poliovirus and acellular-pertussis vaccines are recommended for use in the United States and European Union. A multinational trial where GI virus vaccine is coadministered with vaccines in use in various local countries will generate clinically relevant coadministration data.

REGULATORY AUTHORITIES

In addition, plans to register GI virus vaccine in countries throughout the world will involve multiple regulatory agencies, and data generated in local markets will be useful as the dossiers are reviewed by the numerous regulatory agencies involved in the registration of the new vaccine. Demonstration of safety and efficacy of the GI virus vaccine in multiple settings will be informative and valuable to public health officials in countries around the world, and also to agencies such as the World Health Organization (WHO) and the Global Alliance Vaccine Initiative. For example, children of the developing world live in resource-poor settings where nutrition, immune status, and standard of care differ from children living in developed countries, and therefore the immune response to an oral vaccine may vary between these settings as well.

Regulatory agencies, including the U.S. Food and Drug Administration (FDA), have developed guidelines on the use of foreign data for registration files. A platform for the use of non-US data in a U.S. licensing application can be found in the Code of Federal Regulations[1–3] (CFR), 21 CFR §312.120, §314.106, and §314.126, and in the International Conference on Harmonisation (ICH) Guideline E5.[4] In review, clinical data from non-U.S. study sites should be applicable to the U.S. patient population and clinical practice; relevant non-U.S. studies should be well designed, well conducted, and carried out by qualified investigators according to appropriate ethical principles and Good Clinical Practice (GCP); and non-U.S. data must be able to be validated through on-site sponsor audits and inspection from the FDA. In addition, under a principle of reciprocity, data from U.S. study sites are applicable for licensure in other countries, including the European Union and WHO regulatory network countries. Enrolling the 20,000 infants in multiple countries will result in synergistic clinical development worldwide and thus lower development costs for FPC, Inc. Ultimately, the benefits of this development plan clearly are geared to benefit the pediatric population.

CONSIDERATIONS FOR CLINICAL TRIAL

As FPC, Inc. plans their multinational clinical trial, there are several important practical and ethical considerations to take into account.[5,6] The principles of ethics to be considered include the following: FPC, Inc. and the investigators with whom it plans to work should ensure that consistent, ethical principles are applied in the conduct of the clinical trial in the developed and the developing world. Disparities in care between developed and developing countries participating in the trial should be minimized to the greatest extent possible. Formal, signed agreements between FPC, Inc. and key parties in developing countries may be needed to include clear definitions of each party's responsibilities for the standard of care of study participants and, when necessary, for the provision of posttrial treatment. Key interested parties may include ethics committees, national governments, local health authorities, community representatives, investigators, ministries of health, and nongovernmental organizations. These agreements should be done through an informed, structured, and negotiated process in a participatory and transparent mechanism.

For specific locations of the clinical trials, FPC, Inc. should plan to run clinical trials only in countries where compliance with ICH Good Clinical Practices (GCP) guidelines can be assured. GCP training or other capacity building efforts can be put in place by FPC, Inc. to ensure that the clinical site performs to ICH standards. Clinical trials for the GI virus vaccine should be conducted only in countries where the investigational vaccine is suitable for the wider community of children. Clinical trials for GI virus vaccine should not be conducted in countries where FPC, Inc. has no intent to register the vaccine and make it available for use.

The standard of care dictated by the clinical trial study design and/or offered during the trial should, as a minimum, be consistent with local standards of care in the various countries. Capacity building efforts can be put in place by FPC, Inc. to improve standard of care, although they need to be sustainable by the local community after the trial is completed. Placebo-controlled trials should only be conducted when there are convincing and scientifically sound methodological reasons for conducting a placebo-controlled study. For the provision of locally recommended concomitant childhood vaccines during the study, FPC, Inc. may consider providing and/or funding participating clinical trial sites and the potential subjects if they are not funded through the regular healthcare infrastructure in the various countries.

Capacity Building

Capacity building refers to supportive measures that provide benefit to the community. For example, FPC, Inc. provides support for strengthening local research capacity, research-related technical or clinical equipment, and surveillance systems for adverse events that may occur in the course of conducting the study. When capacity building is provided in the context of a clinical trial, FPC, Inc. should assure that it could be sustainable by the local community posttrial. For example, if an X-ray machine is to be provided by FPC, Inc. to obtain abdominal radiographs, plans for ongoing maintenance for the machine after the conclusion of the trial should be in place prior to study initiation. FPC, Inc. should proactively address provisions for capacity building in prestudy agreements.

Selection of Investigators

FPC, Inc. should assure that investigators in developed and developing countries are qualified to accept responsibility for the medical care of subjects in the clinical trial of investigational GI virus vaccines and capable to make medical decisions regarding the enrolled subjects. Training needs with respect to the investigational vaccine, the disease the vaccine is designed to prevent, and GCP should be addressed prior to the initiation of the study.

ADDITIONAL CONSIDERATIONS

Each clinical trial presents unique operational challenges. The following are some additional considerations, depending on the safety profile of the medicinal compound to be investigated, and the potential need for posttrial medical management for the study subjects.

Data and Safety Monitoring Board (DSMB)

DSMBs are chartered groups of third-party, independent experts who review data generated in a clinical trial on an ongoing, unblinded basis. Not all clinical trials require monitoring by a DSMB. However, when there is a particular safety concern, such as the potential for extremely toxic side effects from a new anticancer pharmacologic therapy, or when there is potential efficacy against an otherwise uniformly fatal condition, a DSMB can review data on an ongoing basis and assist FPC, Inc. in making decisions about the study without harming study conduct integrity. For studies enrolling subjects in developing countries in which a DSMB will be chartered, including advisors on the DSMB with expertise in the developing world environment is an important consideration.

Posttrial Treatment

When study subjects are enrolled in the developing world, thought should be given proactively to medical management of these subjects after the clinical trial is complete. For example, if FPC, Inc. were to become involved with a clinical trial of an investigational treatment for HIV infection in children from Africa, the subjects enrolled in the trial should ethically be allowed to continue antiretroviral treatment once the study is completed. This is typically the responsibility of the national health authority of the country in question, and the ability to support continued medical treatment for study subjects is an issue that should be discussed and agreed upon with investigators, local ethics committees, health authorities, and others prior to commencement of the study. In general, FPC, Inc. should not be responsible for funding nationally licensed medicines or vaccines posttrial. However, there may be exceptional circumstances where FPC, Inc. would agree to provide access to medical treatment posttrial. In this circumstance, access to the investigational (i.e., unlicensed) compound should be accomplished in clinical trial settings through expanded access programs, and provision could continue until the product is licensed in the given country, or clinical development is terminated. Posttrial treatment should be consistent in all developing countries where the investigational product is studied.

REFERENCES

1. United States Code of Federal Regulations, Title 21, Part 50-Protection Human Subjects. Available at http://www.accessdata.fda.gov/scripts/cdrh/cfdocs/cfcfr/CFRSearch.cfm?CFRPart=50 (Accessed November 12, 2007.)
2. United States Code of Federal Regulations, Title 21, Part 312-Investigational New Drug Application. Available at http://www.accessdata.fda.gov/scripts/cdrh/cfdocs/cfcfr/CFRSearch.cfm?CFRPart=312 (Accessed November 12, 2007.)
3. United States Code of Federal Regulations, Title 21, Part 314-Applications for FDA Approval to Market a New Drug. Available at http://www.accessdata.fda.gov/scripts/cdrh/cfdocs/cfcfr/CFRSearch.cfm?CFRPart=314. (Accessed November 12, 2007.)
4. International Conference on Harmonisation Guidance. Ethnic Factors in the Acceptability of Foreign Clinical Data (ICHE5). Available at http://www.fda.gov/cber/gdlns.iche5ethnic.pdf. (Accessed November 20, 2008.)
5. World Medical Association Declaration of Helsinki. Ethical Principles for Medical Research Involving Human Subjects. Available at http://www.wma.net/e/policy/b3.htm. (Accessed November 12, 2007.)
6. World Health Organization. Ethical Considerations Arising from Vaccine Trials Conducted in Pediatric Populations with High Disease Burden in Developing Countries. Available at http://www.who.int/vaccine_research/documents/en/manu774_pdf. (Accessed November 12, 2007.)

CHAPTER 35

The Importance of Geographic Differences in Pediatric Clinical Trials

ALEXANDAR CVETKOVICH MUNTAÑOLA, MD

INC Research, Barcelona, Spain, 0811

INTRODUCTION

Due to the recent changes in the legislation mandating drug development in the United States, pharmaceutical companies are required and encouraged to perform clinical trials on the pediatric population. As a consequence, pediatric clinical trials are much more frequent than in the past and are performed globally. These global trials may involve children from different countries and continents. From the scientific and medical points of view, this achieves subject diversity but there is also the inherent potential variability of subjects. This chapter discusses the potential impact of these geographic differences in terms of their importance to the execution of international pediatric trials.

A body temperature measurement is a perfect example of obvious geographic differences in practice. Apart from using different types of devices and measurement units, the physical way that temperature measurement is taken is very different worldwide. The geographic differences can be defined from diverse geographic locations and are relevant to the conduct of a pediatric clinical trial. The geographic differences might be global or local in origin. We can speak about continent or country-related differences, but there are also differences within the same region, where we might have variations between urban and rural children's population, or variations between different socioeconomic groups. Knowing about geographic differences can help us with study plans and country selection and consequently offer answers to the following questions. *How fast* can we recruit the first patient and complete enrollment? *How many* patients can be enrolled? *How much* will the trial cost? How can we be sure about the *quality* of obtained results?

Pediatric Drug Development: Concepts and Applications
Edited by Andrew E. Mulberg, Steven A. Silber, and John N. van den Anker

REASONS FOR GEOGRAPHIC DIFFERENCES

There are several reasons for geographic differences. Among them, the most important are disease-related differences such as incidence and phenotypic variations, genetic differences, environmental (including climate) settings, lifestyle, cultural (including religious) differences, stage of industry development, types of medical care, and different legislation. Epidemiology as a discipline is not at the same development stage in all countries, and sometimes the disease-reporting procedures of the national institutes do not work appropriately. The fact is that, very often, the incidence of disease is not the same worldwide. A well-known example is the incidence variation for human immunodeficiency virus (HIV) and of the different mechanisms for transmission. Furthermore, the pattern of phenotypic expression of some diseases might differ among regions. Asthma in children, Kawasaki disease, and inflammatory bowel disease are only some examples of incidence variation and different phenotypic appearance.[1,2]

The prevalence of different genetic diseases varies worldwide. The perfect example is cystic fibrosis, the most common severe autosomal recessive disorder among Caucasians, but rare among Africans and Asians. Structural variations in the human genome are likely to make an important contribution to human diversity and disease susceptibility.[3] Differences in host immune response might explain the variety of incidence and phenotypic expression. It is possible that in the future we will have an alternative way of administrating drugs and tailoring dosages. Moreover, the *Oakland Tribune*[4] reported that pharmaceutical companies are starting to develop drugs specific to certain races and ethnic groups. These issues of genetic diversity and potential impact in pediatrics are more fully described in Chapter 21 by Cohen and Ness.

Standards of care and available medication also affect the disease expression and its severity. Diagnostic and staging procedures are not always the same and depend on country location. In less developed countries, a diagnosis is frequently made according to clinical signs and symptoms, while in developed countries additional diagnostic procedures are required. The rating scales frequently used in children's psychiatry might differ, and the truth is that in some countries they are not commonly used at all. Clinical perspectives in terms of identifying differences and similarities of disease presentation are well recognized for pediatric disorders, including gastroesophageal reflux disease (GERD), as noted by Rothman and Klemman and discussed in Chapter 39.

Finally, the patient population is not the same. In some countries, we still have a large proportion of treatment naïve patients. In other countries, we have a lot of indigenous groups and consequently the phenotypic expression of some diseases might be different. There is a high incidence of bacterial meningitis, especially among indigenous groups in industrialized countries, such as North American Eskimos, Apache Indians, and Australian Aborigines, particularly with meningitis caused by *Haemophilus influenzae* type b (Hib). Differences in host immune response to these capsular polysaccharides seem to be the most likely explanation for this observation. It has also been shown that other immunologically mediated disorders, such as Kawasaki disease and systemic lupus erythematosis, have a relatively high incidence in Sino Japanese populations, lending plausibility to inherited differences in immune response as a mechanism for these observations.[1]

The potential for adverse risk factors is not the same within diversified parts of the world. For example, when analyzing treatment of children with asthma, we should not forget patterns in their parents' smoking habits. Despite the globalization process, cultural differences are still very important and might affect the disease stage.

Different environmental and climate factors might also be responsible for incidence variations and might affect the disease's features. In the past, when there were limited treatment possibilities, people with asthma moved into hot, dry climate areas. Another example is malaria—a vector-borne infectious disease caused by protozoan parasites, widespread only in tropical and subtropical regions. The air pollution present in large urban areas is a contributing etiologic factor for many respiratory diseases.

It is important to recognize that people from different cultures have different ideals and outlooks on life. Even something so unique and essential as breastfeeding varies considerably across different racial/ethnic groups. For example, in one study it has been shown that in the United Kingdom the highest breastfeeding rates are among black and Asian mothers, which is in stark contrast to patterns in the United States, where the lowest rate is seen among non-Hispanic black mothers. The conclusion is that the contrasting racial/ethnic patterns of breastfeeding in the United Kingdom and United States necessitate very different public health approaches to reach national targets on breastfeeding, paying special attention to different social, economic, and cultural profiles of all racial/ethnic groups.[5]

The stage of industrial development has a good correlation with the level of health welfare, but also with the incidence and stages of some diseases. The high level of healthcare in Western countries has eradicated some diseases, but has brought new ones, which are still unknown or rare in developing countries. Industrialization and stressful lifestyles brought new diseases. For example, lifestyle diseases are directly or indirectly caused by unhealthy lifestyles. Lifestyle diseases of special importance for pediatric populations include obesity, diabetes, hypertension, heart disease, and diseases associated with smoking, alcohol, and drug abuse.

The type of medical care does not always have a direct correlation with the stage of industry development. In some countries, the traditional medicine has very deep roots and alternative treatment procedures and herbal medicine are sometimes more followed than classical medicine. Apart from these medical and socioeconomic reasons, there are legal reasons. Differences in local legislation are very important for the conduct of any trial but are more prominent in pediatric clinical studies. To start with, the *age of majority* as the threshold of adulthood conceptualized in law differs between countries. It varies from 16 to 21 years. In some countries, if an adolescent aged 16–18 is no longer a minor or is emancipated, then written informed consent is required, as for any adult. There are also different requirements for the informed consent that should be signed by parents. For example, in some countries, both parents should sign it and in others only one parent's signature is sufficient. The informed consent in emergency trials or informed consent of the legal representatives may also vary between countries. The minimum age of child assent is not the same in all countries, and in some of them it is not even defined in the legislation.

The next example is the country-specific obligatory vaccine calendar. The distinction is not only in the calendar but also regarding vaccine types and products used. Different medical legislation also includes different prenatal and postnatal screening tests. Finally, the submission/approval structure and procedures and timelines are different, sometimes significantly.

IMPACT OF GEOGRAPHIC DIFFERENCES ON PEDIATRIC CLINICAL TRIALS

Geographic differences might have a huge impact on several aspects of pediatric clinical trials, primarily on country and site selection, length of start-up phase, enrollment rate, study costs, and quality.

Geographic Differences and Country and Site Selection

Country selection in an international clinical trial is always a challenge and inappropriate selection might affect the whole drug development program. In order to avoid these surprises, when making country selection, several factors should carefully be analyzed. Some of the most important are disease prevalence and incidence, standard of care, experience in pediatric clinical trials, expected quality, study costs, and approval timelines. Northern versus Southern Hemisphere selection is important whenever analyzing investigational products for all diseases with seasonal variation. When selecting developing countries, we should have ethical points in mind. An ethical framework for multinational research should minimize the possibilities of exploitation. Ezekiel et al.[6] have suggested 8 principles through 31 benchmarks. These principles are collaborative partnership, social value, scientific validity, fair selection of study population, favorable risk–benefit ratio, independent review, informed consent, and respect for recruited participants and study communities.

Very often, the same feasibility approach cannot be applied for all countries. The feasibility process should respect the local healthcare infrastructure, different standards of care, and community awareness of clinical trials in some countries. Even medical specialties responsible for the indication treatment might be different. For example, in some countries there are pain clinics where pain specialists treat children suffering from cancer, and in others the pediatric oncologists are responsible for treating their pain. Anesthesiologists as well as pediatric intensive care specialists may work in pediatric intensive care units, depending on the country. Diabetic neuropathy might be treated by neurologists or by endocrinologists depending on the country or sometimes even depending on the institution. In some countries, investigators require to see the whole protocol and the Investigator's Brochure, and to be informed about economic aspects before they provide any information. Sometimes it is advisable to perform an on-site visit. In some regions, it is not unusual to have sites that are charging for every completed feasibility questionnaire.

In general, the distribution is not the same between the four types of institutions—university clinics, general hospitals, private clinics, and site management organizations (SMOs)—that participate in clinical trials. For example, in Eastern Europe almost all institutions will be university and general state hospitals, while in Western Europe, private practices and SMOs will represent a major part of the selected institutions. In some countries, the study drug should always be stored in the hospital's pharmacy, and in others drug storage is the principal investigator's responsibility and the drug will not even pass through the pharmacy. That is why it is very important that the drug storage conditions be checked in the site selection process. For trials with no central laboratory, local laboratories should be carefully checked. In some countries, laboratories have only national certificates. Apart from this, the laboratory units for some analysis could be different. Required technical equipment is sometimes absent and, when present, might vary in regard to its quality. It is also important to check the type of source data. Very few sites have part 11 CFR compliant electronic source data validated systems. Telephone, facsimile, and Internet connections are important not only for reporting serious adverse events (SAEs) or sending electronic case report forms (eCRFs), but also for electronic patient diary (PDA) transmissions. Data transmission could be a huge problem if the analog versus digital phone system was not investigated beforehand.

Clinical trials also require resources that few individual investigators possess, including non clinical time, trained research coordinators, rapid processing by institutional review boards (IRBs), and streamlined research contracting. The failure of institutions to provide necessary resources may reduce the number of available research sites and impede enrollment into trials. The evidence suggests that institutions that have developed a central trial office structure to oversee and support clinical trials have witnessed much greater growth in the number of industry sponsored trials than institutions that continue to leave this up to individuals and departments.[7]

The investigator's availability during the monitoring process varies depending on the country. In some, it is a standard that investigators will be present for a very limited and insufficient period of time. In other countries, monitors do not have such problems. Key opinion leaders are important, but not always a guarantee of expected enrollment and high quality of performed work. That is why the study team should carefully be checked in terms of their experience and availability. The study coordinator is a very important member of the study team. It can be a study nurse or a coinvestigator (physician) and in some countries a specially trained employee responsible only for clinical trials. Finally, an investigator who has strong prior performance for a particular outcome should not be involved in a study. The same applies for investigators known by their close cooperation with some pharmaceutical companies.

Finally, we can speak about different investigators' qualities. The "ideal" investigator is a well-trained and compliant physician competent with ICH GCP requirements with sufficient time and motivation to run a clinical trial; comfortable with innovation, willing to try new technologies, having access to a large patient population pool, and having high human and ethical principles. It is very hard to find such an investigator, and the percentage of candidates who fulfill these requirements differs according to the regions. The reality is that often the most important selection criterion is the investigator's patient population pool. We should insist on all requirements, because only then we will avoid compliance problems that will affect study results.

Impact of Geographic Differences on Start-up Phase

Today, the legislation of pediatric clinical trials is much more harmonized than in the past. However, regional differences still exist even within developed countries such as the United States or EU, the two bodies with the more advanced regulations on pediatric clinical research. If we add developing countries, the differences are even more prominent.

The approval structure differs among countries. Ethics committees in Europe operate much the same way that IRBs do in the United States. In some countries, the committees are created top–down: each competent authority divides its country into geographic regions and creates an ethics committee to serve each region. The large hospitals have their own ethics committees. Several countries have a two-tiered system consisting of a local, hospital-based committee and a national or central ethics committee. In those countries, studies are reviewed at both levels. The national committee is charged with reviewing the study for scientific and technical merit. Apart from that, the local committee is usually charged with reviewing the qualifications of the investigator for that particular study and with reviewing the economic aspects of the study.

For countries where ministry of health approval is required, the submission to competent authority might be done only when the ethics committee approval is obtained or in parallel with the submission to the ethics committee, depending on the country. In Russia, there is an independent pharmacology committee evaluating the Investigator's Brochure (IB) and pharmacology aspects of the investigational product.

The approval time frame mainly depends on the local legislation and the approval structure and process. The submission process might be parallel but also consecutive. The approval could be obtained within 60–70 days (EU) or 6–8 months (in some Latin American countries).

The list of documents needed for submission is not always the same and the same applies for pediatric-specific requirements. Even with respect to the most important document—the informed consent and assent form—there are country-specific differences. In Poland, for example, the signed clinical trial agreement with the institution/investigator is part of the submission package.

The time frame needed to prepare the submission package is not the same and mainly depends on the translations, contract negotiations, and time needed to obtain the local insurance. In all countries where the documents have to be translated/backtranslated into the local language, additional time for the submission preparation is required. There are countries where the whole protocol and even the IB should be translated into the local language. However, on the whole, only the study summary should be translated. The time frame for contract negotiation also varies. In some countries, it is not allowed to have a separate contract with the investigator working in the state institution, and in others it is actually required.

Trial sponsors should not overlook the fact that every country in the world besides the United States requires sponsors to carry clinical research liability insurance.[8] The requirement makes sense because most countries provide public health services; the state doesn't want to be responsible for the bill in the event subjects are injured by an investigational device or the sponsor goes out of business. In France, for example, the government has the right to sue a sponsor for the recovery of medical costs if a subject is injured during a study. Many countries even require the underwriter to be located in the host country. Trial sponsors must provide a copy of the insurance certificate to the ethics committee and, fortunately, insurance isn't hard to obtain. In some countries, a local insurance is required and should be in accordance to the local legislation, while in others the global insurance is sufficient.

The regulation for import license approval might also be different. In some countries, it is obtained together with the regulatory approval, and in others it is obtained separately and only after regulatory study approval is obtained.

It is advisable to know the country-specific vacation period. In most European countries, July and August are vacation months and some European countries do not have any meetings during this period. In the Southern Hemisphere, it is December and January. Apart from country-specific bank holidays, we should not forget that religious holidays have different calendar schedules. These facts might also affect the availability of investigators.

Impact of Geographic Differences on Patient Enrollment

The relationship between physician and patient is essential for every enrollment in a pediatric clinical trial. In pediatric trials, we have to add parents into this relationship chain. This relationship is not the same worldwide: in some countries, the role of the physician is more prominent than in others and the doctor's advice and recommendation will always be followed.

We should be aware that the family unit and structure are not the same worldwide. The social position of a woman and mother in some societies affects her decision rights in this aspect as well. Another essential factor for the enrollment's success is the investigator's motivation. Both scientific and economic interests are not the same and vary in respect to geographic location. In general, we can say that there are regions known for high enrollment. While respecting interpersonal differences, we can say that the economic factor is very important, if not the primary, and could explain this enrollment difference. These investigators not only put more energy into the study, but also more time. Countries where a direct contract with the investigator is allowed will frequently provide many more patients.

There are significant differences in respect to advertisement policy. Public announcements for clinical trials are, to this day, much more widely used in the United States than in Europe.[8] Advertisements for subjects in newspapers, on public transportation, or over the radio are less common in Europe than in the United States. Most subject recruitment is handled through the investigator's patient database or networking among investigators. In some countries, for example, in Russia, clinical trial advertising is prohibited.

Patient fees or direct payments to children or their parents/legal representatives are additional examples of geographic differences. For example, there is a different mood in Europe and the United States in this respect. European regulatory and ethics committees are worried that these payments might be coercive or unduly influential. Such payments, therefore, are frowned upon. A trial sponsor might be able to compensate a patient for travel and parking, but the amount must be disclosed to the ethics committee for its approval.[8,9]

Standard of care and availability of the drugs used for some particular indication will have a huge impact on the enrollment. Naturally, that enrollment will be higher in countries where similar medication is not available or nonaffordable for the patient.

Parents' knowledge and the available information to them are very important. Unfortunately, there are still regions where a huge proportion of parents consider injection as the only valuable way of drug administration. Although painless and comfortable for the children, inhalation therapy or needle-free devices are not accepted everywhere.

It was demonstrated in one study that what most distinguished those parents who refused randomization, from those who accepted it, was not their knowledge and information about randomized clinical trials but their beliefs, values, religion, and perception.[10]

Moreover, the perception about their children's health is not the same and might differ depending on economic factors and cultural differences. There is a study investigating parents' perception of child health, in which parents from urban, suburban, and rural areas are compared. Statistical analysis has demonstrated that parents in the predominantly low income bracket perceive their child's health to be worse than parents from the other two areas, although no significant differences in reported health conditions among the children were found.[11] Urban parents may perceive their children's health differently due to availability of social networks, proximity to polluting facilities, and level of parental education. These factors should help us when designing the patient diary or selecting the scales that should be completed by parents. Nevertheless, the investigators should be trained in how to instruct parents who are required to complete this data. Different perceptions of their child's health affect not only the enrollment rate but also the scales completion, drop-out rate, and general protocol compliance.

What about the children's perception? The general impression is that children are very homogeneous worldwide. Children as young creatures do not have time to assimilate all sociocultural environments and surrounding beliefs. They have a unique natural curiosity driving them to explore their environment and they will come to learn about the world of their own accord. Children have the same basic instincts: fear from something unknown and confidence in their parents. We are often not aware how incredibly good they are at seeing things that even adults miss, or would prefer they didn't see. That is why it is very important to give them all the necessary study-related information. We should not forget that children understand us much better then we can imagine. The critical factor is to obtain their consent/assent. It is not only a question of enrollment, but a question of compliance, cooperation, and overall study success.

Geographic Differences and Study Quality

Insufficient knowledge of ICH GCP principles and suspicious ethics are the major factors for quality issues in any trial. Unfortunately, for some investigators, participation in clinical trails does not necessarily mean that they understand why it is important to respect and follow ICH GCP principles. These problems are driven more by interpersonal variations among investigators than by differences in geographic location. Still, we expect to find better investigators in countries were ICH GCP is taught in medical school or in countries where pediatric clinical trials have a longer history. These issues of the criticality of certification are covered in Chapter 31 by Procaccino in this book.

The quality issues due to geographic differences occur whenever a study design and protocol and monitoring procedures do not respect and take into account the local specifics. These issues are more prominent in trials when the data collection required is too extensive, and frequently unnecessary for the analysis process. There are several levels where the quality might differ among the countries and selected investigators.

It is clear that the content of the informed consent form (ICF) should respect ICH GCP principles and local requirements. But the ICF is not always written in a language acceptable within the local community. The wording should respect cultural differences and should contain appropriate idioms. That is why we should not be afraid of having country-specific versions. The disclosure of the information might also vary depending on the region. The same applies for informed assent content and procedures. As previously stated, some countries have not established these requirements. As a consequence, the quality of the ICF/AF procedures might be one of the most important quality issues in multinational pediatric clinical trials.

The quality of source data might also differ between countries. Medical records, their content, and their quality are sometimes diverse. Access to source data is also not the same: in some countries, the patient's general medical files are centralized and only medical records related to specific indications are available on-site. In some cases, special permission from the hospital authorities should be requested for access to these general medical records. That is why it is important to standardize these concepts and requirements at the beginning of each study.

Similarly, there are differences in CRF completion. In some countries, it is standard to write the diagnosis in Latin. In others, it is written in the local language or English. Concomitant illnesses may be reported differently; for example, hypotension is a commonly treated disease in Germany, but it is not recognized as a disease in the United States. Drugs may be reported under different names and may not be recognized by the FDA. Herbal medications or home remedies may be used differently and not reported, and personal hygiene practices may be different. Furthermore, disease status may have progressed further in countries with long waiting lists for access to medical care. Trial sponsors should be prepared to justify to the FDA why the European clinical data are applicable to U.S. medical practice.[8]

Apart from human errors, in multinational pediatric trials we might have measurement errors. Calibration of measurement instruments is very important and should be checked carefully. Different measurement units and the lack of unique conversion tables might also affect data quality. Some American authors say that the quality of data from Europe is more variable than that from the United States.[8] "Quality" data is defined in the FDA's *Guidance for Industry: Computerized Systems Used in Clinical Trials* as data that is accurate, complete, logical, legible, contemporaneous, and attributable. European sponsors and investigators do not always share this opinion. Sometimes the query rate has significant differences between the countries. There are two possible explanatory factors: some investigators make more mistakes, or their studies are judged by a different standard. That is why training and unique monitoring standards are essential for every international pediatric trial. Even the evaluations of monitoring and auditing techniques are not the same and may vary between companies or countries. Pharmaceutical companies might have a different audit inspection approach. Control standardization should contribute to quality improvement.

GEOGRAPHIC DIFFERENCES AND STUDY COSTS

Study costs fall into three categories: regulatory and ethics committee/IRB costs; investigative site costs including investigators' grants, hospitalization, laboratory, and patient travel costs; and monitoring costs.

There is a huge difference between countries in all three categories. It is well known that it is much more expensive to perform a trial in developed countries than in developing ones and in the same way that it is more expensive to run the trial in the United States than in Europe. Even within the European Union, there are huge differences between Western European countries and recent members from Central and Eastern Europe.

Although the financial aspect is very important, it should be evaluated with respect to data quality, study results, and overall study benefit. The cost estimate should not be used as the only and most important parameter for country and site selection. It is more expensive to run a study in an inappropriately selected country/site for some specific protocol. Invested money will serve for nothing since the results might not be valid. Similarly, expensive trials will be the result when studies with a smaller *original* budget require country and site replacements, have protocol amendments, find insufficient enrollment, and suffer study prolongation.

TARGET ACTIONS FOR MULTICENTRIC PEDIATRIC CLINICAL TRIALS

There are many positive aspects of selecting different regions for a pediatric clinical trial. One of the most important is that a drug will be investigated on a more representative sample of the pediatric population. The study results and corresponding conclusions will be much more reliable and accurate. In addition, every drug is intended to be used worldwide, and consequently, there is a logic that it has to be investigated worldwide. When evaluating the benefit of an investigational product, primarily its safety and efficacy, it is essential to identify and evaluate possible different treatment responses or safety profiles. Finally, there is an important ethical aspect, since clinical trials improve healthcare systems worldwide.

On the other hand, we have to know how to minimize the possible negative impact of geographic differences on study results. This is one of the biggest challenges in all multicentric trials, especially in multinational trials.

The first step is that we should be aware that these differences exist. We should not forget them even in small national studies, where specific regional differences might occur. The second step is that we should have sufficient knowledge about these differences; we should understand them and should know their possible impact on the study results.

There are two possible scenarios: (1) to make country/site selection according to our given protocol or (2) to make the protocol and study plans according to the regions we plan to include. In both cases, the whole process could be divided into four phases: *planning*, *selection*, *training*, and *control*. The cohesive factor in all four phases is a good *communication plan*. The study organizer should know what to ask and how to ask it in order to collect necessary information, but should also have a high level of listening skills. Every clinical trial is a complex partnership between very different entities. Pharmaceutical companies and all kinds of service providers, including investigators, are part of the same team. The communication should be professional but open, friendly, and constructive, underlying mutual trust and respect.

It is necessary to have a multidisciplinary approach in the planning process and include people with medical, regulatory, data management, statistical, and quality assurance expertise and, of course, local staff familiar and experienced with the local environment. It is advisable to count on people with experience in "global studies." All study aspects should be carefully planned from the beginning. Even designing quality during the planning stages of the clinical program is recommended. In order to collect the necessary information in the study planning process, it is important to have a well-established process of generation, transmission, and reception of messages.

It is essential to involve a pediatrician, preferably a key opinion leader and expert in the indication, in the protocol design from the beginning. For large international trials, it is advisable to have consultations with key opinion leaders (KOLs) from the regions to be included in the study.

In every multinational trial, the study design should be adjusted to be feasible for all regions. Protocol design is the first and most important part of this process. The pediatric clinical trial design depends on the objective(s) of the trial and the scientific question(s) to be answered. All measures to avoid bias should be included in trials performed in children. Open and/or uncontrolled trials are subject to increased bias and should be avoided whenever possible. When unavoidable, open trials should include provisions for blinding assessment. Uncontrolled trials (refer to ICF E6) should be avoided in principle for demonstration of efficacy. They have limited value for the demonstration of safety, unless they are used prospectively for follow-up and cohorts, in predefined subgroups. Trials performed in children affected by rare diseases should follow the same methodological standards as those performed in more common diseases. Less conventional designs should be justified and it is recommended that permission be sought from the competent authorities. The size of the trial conducted in children should be as small as possible to demonstrate the appropriate efficacy with sufficient statistical power.[9,12] One of the most important parts in protocol design preparation is to define the least possible amount of collected data needed for a proper safety and efficacy evaluation and according to which the drug program decisions will be made. A data simplification strategy should be implemented to avoid quality issues and possible misinterpretation of results.

A major controversy regards the use of placebo in pediatric clinical trials. Use of placebo in children is more restricted than in adults, because children cannot consent. According to EMEA recommendations, placebo should not be used when it means withholding effective treatment, particularly for serious and life-threatening conditions. However, the use of placebo is often needed for scientific reasons, including in pediatric trials. Placebo may be warranted when evidence is lacking. As the level of evidence increases, the ethical need for placebo decreases. In any case, its use should be considered with measures to minimize exposure and avoid irreversible harm, especially in serious or rapidly evolving diseases. As appropriate, rescue treatment and escape procedures should be set up.[9]

We should make our study plans and sometimes even study design adjustments to ensure that a homogeneous patient population will be enrolled and that a standardized approach for all study procedures is followed. Data standardization is essential in study planning, primarily because of quality issues. The central readings (ECG, skin reactions, etc.) are very important for the homogeneity of the obtained results. When needed, the sites should receive conversion tables with correlation between different units. The best approach is to have a centralized study management and experienced and knowledgeable local staff.

There is no golden key for success, simply because it depends on the indication, study phase, and objectives. Every study is a separate story and, regardless of our efforts, there might arise some unpredictable problems that should be resolved during the trial. There is no unique approach for pediatric multicentric trials; it is project specific. Still, there are a couple of unique requirements that should be implemented for all multinational pediatric trials.

In addition to the standard requirements for valid research, such as adequate sample sizes and unbiased measurements, pediatric multinational research should fulfill five benchmarks.

1. *Pediatricians who are experts in a particular indication should be involved in the process of protocol design from the beginning*. For multinational trials, it is advisable to include key opinion leaders from different countries. In order to have a feasible design, it is advisable to have a preliminary consultation with parents and older children/adolescents.
2. *Study procedures should be well-designed, respecting local standards of care and standardizing all measurements and evaluations*. On the one hand, the design should not deny children the local standard care for a particular indication, and on the other hand, these procedures should be feasible in the context of the local standard of care. When needed, it should be possible to provide more extensive interventions beyond those to which participants are entitled or beyond those that are feasible and sustainable, but we have to be careful because these interventions may be unethical if they undermine the scientific objectives or make the results irrelevant to the community. Data standardization is important for quality reasons. A centralized evaluation approach is recommendable.
3. *There should be clear and meticulous protocol requirements that do not allow any type of different interpretations*. In particular, the inclusion/exclusion criteria should be very precise and should take into account the local standard diagnostic and staging procedures. Special attention should be given to concomitant diseases and adverse events reporting procedures and instructions.
4. *Data simplification is imperative*. Only data needed for the decision process should be collected. Protocols that are too complex and require a huge amount of data to be collected and analyzed are not recommended because this might affect data quality.
5. *Study design must be developed to be feasible, given the social, medical, cultural, and legal environment in which it is being conducted*. We should not forget that in some countries, for indications where there is already an available and registered drug, the studies with placebo will hardly be approved.

Once we have the final protocol, we should perform the final country and site selection. As already mentioned, it is much better to know the desired countries when developing the protocol and study design. In any case, the selection should be done very carefully, respecting all protocol requirements and local specifics. We should never forget about local differences within the same country. It is essential to perform very detailed prestudy visits, not only collecting information regarding the potential investigator, his/her experience, diagnostic and treatment procedures followed, and accessible patient population, but also checking all necessary site facilities. Frequently, it is not possible to have all sites with all the required options. It is important to recognize a valid investigator and the potential of a site. Sometimes it is necessary to provide additional technical equipment or affordable drugs.

The next phase is to focus on how to train the investigators, study coordinators, and the whole study team. Since this training is critical, it should be well designed and prepared. All investigators should receive the same instructions and information. The educational training should cover not only the basic ICH GCP principles, but also all study-specific aspects—starting with protocol objectives, study procedures, scales and forms completion, CRF data entry, AE/SAE reporting, and so on. Special attention should be paid to scales and patient diary completion, since these are areas with the most variable interpretations. Select one key opinion leader who will perform these training sessions, usually during the investigators' meeting. It is very important to have face-to-face meetings, where investigators will learn how to standardize their evaluations. Written and (if applicable) video material, translated into local languages, should be provided and used for the whole study duration. The trainer should instruct investigators on how to train patients and their parents in completing the patient diary. Study team members should also receive training in study procedures. Training should not finish when the study starts. It should be an ongoing process during the whole study duration and should be repeated in different study phases. The training should be repeated whenever needed, for example, when a site or even a country is replaced, or when some variation from the trend is identified. Equally important is the educational training of monitoring staff. As investigators, the local clinical research associates might have different understandings of the same topics. With the same centralized training approach, monitors should receive clear, protocol-specific, monitoring instructions. Special attention should be given to the educational training of monitors working for contract research organizations, simply because of the different standard operating procedures, preferences, and requirements of pharmaceutical companies they are working for.

Finally, control should be strict and present from the beginning. All study levels should be covered. The role of medical monitor is of the utmost importance. Each and every patient should be carefully checked before enrollment into the study. Frequent and early site inspections and comonitoring visits are highly recommended, as well as an early data snapshot. As with the training, control should be a permanent process. Inspections should be performed on a regular basis, in particular, when there is high enrollment (especially if it is unexpected), high staff turnover, and abnormal number of SAEs/AEs. Project management should have an action plan for every identified type and level of noncompliance. Inspection of monitoring activities has a special importance and should done through the whole study duration. Early and repeated data snapshots are recommended for early identification of any unexpected variations. Of course, this will increase the study costs but it is an absolute requirement in every multinational pediatric research.

CONCLUSION

Knowledge and awareness of geographic differences are important in study planning, country and site selection, enrollment rate calculation, data quality, and overall study success in all international pediatric clinical trials.

We should not be afraid of these differences, but we should know how to minimize their possible negative effect on study results. The evaluation and planning process should be multidisciplinary. The study design should be feasible, given the social, political, and cultural environment in which it is being conducted. The protocol should have meticulous requirements, respecting data standardization and data simplification. Country and site selection should be carefully made, matching the protocol requirements and local medical and legal specifics. The educational training and control of investigational sites and monitoring staff should be centralized and are essential and permanent requirements that must be respected. The art of communication is mandatory during the whole study duration.

REFERENCES

1. Mcintyre P. Geographic differences in bacterial meningitis: less may be as interesting as more. *J Peadiatr Child Health.* 1998;34:109–111.
2. Pinsk V, Lernberg DA, Grewal K, et al. Inflammatory bowel disease in the South Asian pediatric population of Bristish Columbia. *Am J Gastroenterol.* 2007;102:1077–1083.
3. Feuk L, Carson AR, Scherr SW, et al. Structural variation in human genome. *Nat Rev Genet.* 2006;7:85–97.
4. Vesely R. Tailoring medications for genetic differences. *Oakland Tribune* August 2006;21.
5. Kelly Y, Watt R. Racial/ethnic differences in breastfeeding initiation and continuation in the United Kingdom and comparison with findings in the United States. *Pediatrics.* 2006;118: 1428–1435.
6. Emanuel EJ, Wendler D, Killen J. What makes clinical research in developing countries ethical? *J Infect Disease.* 2004;189:930–937.
7. Schreiner M, Greeley W. Pediatric clinical trials: Shall we take a lead? *Anesth Analg* 2002; 94:1–3.
8. Stark N, Peacock J. Clinical studies: Europe or the United States. *MDDI* 2004;5:134–138. Available at www.devicelink.com/mddi/archive/04/05/003.html
9. EMEA. Note for guidance on clinical investigation of medicinal products in the paediatric population CPMP/ICH/2711/00 2001.
10. Wiley F. Parents'perceptions of randomization in pediatric clinical trials. *Cancer Pract.* 1999;7:248–256.
11. Morrone M, Crist K. Geographic differences in parental perception of child health. *J Child Health.* 2004;2:77–86.
12. Ethical considerations for clinical trials performed in children. Recommendations of the ad hoc group for the development of implementing guidelines for Directive 2001/20/EC relating to good clinical practice in the conduct of clinical trials on medicinal products for human use. Available at http://ec.europa.eu/enterprise/pharmaceuticals/paediatrics/docs/paeds_ethics_consultation20060929.pdf.

CHAPTER 36

Partnering with Industry: Academic Medical Centers and Clinical Research Centers

KATE OWEN, BS, MATTHEW HILL, JD, and SANDRA COTTRELL, PhD

Novo Nordisk, Inc., Princeton, New Jersey 08540

BRAHM GOLDSTEIN, MD, MCR

Novo Nordisk, Inc., Princeton, New Jersey, 08540 and University of Medicine and Dentistry of New Jersey – Robert Wood Johnson Medical School, New Brunswick, New Jersey 08901

INTRODUCTION

Historically, the main pathway for clinical drug development in the United States and Europe has been through close partnership of the pharmaceutical industry with academic medical centers and private clinical research centers. This pathway has resulted in improved medical care for millions of patients through the drug development process with approval of thousands of life-saving medications. While the collaboration between these groups can sometimes be described as suboptimal, time-consuming, and often inefficient, there is no question that the end result has been to the mutual benefit of patients, clinicians, researchers, medical centers, and the pharmaceutical industry. Critiquing the system and its multitude of processes is relatively easy. Suggesting constructive changes that benefit all and improve on the status quo is significantly harder to do.

The objectives of this chapter are to describe the current interactions between the pharmaceutical industry with academic medical centers and private clinical research centers including the following key areas:

- Contracting issues
- Trial budgets and subject insurance
- Trial registration

The opinions, views, and comments expressed are solely those of the authors and do not necessarily reflect the opinions or policies of Novo Nordisk, Inc., its directors, officers, employees, agents, or representatives.

Pediatric Drug Development: Concepts and Applications
Edited by Andrew E. Mulberg, Steven A. Silber, and John N. van den Anker

- Publication rights
- The relationship between investigator and sponsor

At the conclusion of this chapter, we hope the reader will have a clearer understanding of the intricacies and interrelationships of clinical drug development.

CONTRACTING ISSUES

Clinical Trial Agreements (CTAs)

In recent years, contracting has become a significant rate-limiting step to the activation of an academic site as part of an industry-sponsored clinical trial. As the negotiations of the contract becomes more complicated with greater detail being covered in the budget, as well as the increased potential for litigation over malpractice, more and more sites are investing in expansion of their infrastructure, personnel (particularly of legal and accounting divisions), and information technology (IT) systems to administer and manage clinical trials and the associated compliance functions. This lengthens the negotiations, approval, and signature processes. Hospital consolidations and a recent pattern of legal outsourcing of hospital contract services have also contributed to the escalation of the contracting procedures.

In an effort to try and reduce these critical timelines, many pharmaceutical sponsors and investigative sites are working toward master Clinical Trial Agreements (CTAs). For the most part, master CTAs eliminate the need to renegotiate contract terms for every individual clinical study. However, in some instances, the parties may wish to make exceptions to certain terms in the master CTA if, for example, intellectual property or publication rights with respect to a particular clinical trial have elevated significance to either party and different terms are desired. Nonetheless, a master CTA has benefits for both parties as it expedites the contract process and may impact the time to market.

The only aspect remaining to be negotiated under a master CTA is the Statement of Work (SOW). This reflects the particular scope of the clinical trial and the sponsored budget. Many hospitals and independent clinical research centers are becoming more familiar with master CTAs, which provide the added incentive to facilitate future work by establishing broader partnerships with investigational sites used repetitively by sponsors. The major caveat to the master CTA is that it will potentially apply to numerous clinical trials. Amendments based on legal or business policy changes will likely be more difficult than with a one-time CTA, raising the stakes for each party in the negotiation phase of a master CTA to achieve the most favorable terms possible.

Confidentiality Disclosure Agreement (CDA)

There is increasing competition to shorten the drug development cycle time and time-to-market, so more and more investigators are evaluated for participation in clinical trials. This results in more specific and "tighter" language in an attempt to assure confidentiality and nondisclosure of proprietary information. But this is often contrary to the need to expedite clinical trial execution as more cumbersome and in-depth language may result in more lengthy negotiation, thus delaying the initial stages of a clinical trial.

A CDA protects the sponsor's interest in maintaining the confidentiality of information considered to be "trade secrets," or information that in a competitor's hands could potentially be harmful to the sponsor's business. It is in the sponsor's best interests to make the definition of confidential information in the CDA as broad as possible, thereby ensuring that all information provided to the site/investigator is protected. Typically, the information disclosed during this discussion phase includes the protocol and investigator's brochure for the study.

In some cases, sites will seek to narrowly define confidential information, to include a requirement that the sponsor mark all proprietary documents as "CONFIDENTIAL," and to confirm in writing any verbal disclosures of confidential information within a specified time period. Sponsors are reluctant to agree to such conditions, as it places an administrative burden on the sponsor and creates the risk that confidential information will become public if it is disclosed without the required marking or written confirmation.

Some pharmaceutical sponsors, in an effort to reduce the time to find suitable sites, allow feasibility studies to be conducted without a CDA in place and do not supply information specific to the drug or mechanism of action being studied. The sponsor focuses their initial investigation on the potential subject population and research experience of the investigational site. If an adequate study population exists and the site appears competent, then the next step is completion of a CDA. If the site becomes an active site, the CTA also contains clauses relating to handling of confidential and proprietary information.

Finally, similar to a master CTA, a master CDA can cover all trials for a particular drug product or all clinical trials for a given period of time between the sponsor and the investigative site.

Intellectual Property Rights and the Bayh–Dole Act

Ownership of inventions arising out of a clinical trial is often a contentious issue in negotiation of a CTA between a sponsor company and an investigational site. From the sponsor's perspective, if not for the development of the drug being studied, the clinical trial site would not have the opportunity to study it and would not discover any invention. Therefore, sponsors argue that any inventions derived by a site investigator or any other site employee in connection with the clinical trial should solely be the sponsor's property.

Many sites agree that inventions developed relating to the study drug, and created in performance of the protocol, should rightfully be owned by the sponsor. However, in some cases, sites do not agree that sponsors should own every invention—specifically, inventions unrelated to the study drug and unanticipated new uses of the drug falling outside the use of the drug in accordance with the protocol. In the site's view, the sponsor is not compensating the site to develop new inventions, and any inventions or uses of the drug not anticipated by the protocol are due to the investigator's expertise and the site's contribution of subjects and resources to the trial. Therefore, any such inventions should be the property of the site.

In an effort to resolve this conflict, many sites are willing to provide sponsors with an option to negotiate an exclusive license to any site inventions that do not relate to the study drug or to a new use of the drug. Sponsors must evaluate whether such an option is acceptable, taking into account the potential for new discoveries falling outside the protocol and the value thereof. In some cases, sponsors will agree to an option on the condition that, should the sponsor choose not to exercise its option or if license negotiations between the sponsor and the site fail, the sponsor will have the option to license the invention on the same terms to a third party (i.e., a right of first refusal).

Another intellectual property issue becomes relevant when a site receives funding from the U.S. government for other research projects conducted at the site or as part of the sponsored company's research project. The federal Bayh–Dole Act, 18 U.S.C. §200,[1] was enacted in

1980 to, among other things, "promote the utilization of inventions arising from federally supported research or development." The Bayh–Dole Act gives sites receiving federal research dollars under a funding agreement the ability to retain ownership of inventions developed by the site in the performance of the government-funded project. Prior to the Bayh–Dole Act, ownership reverted to the federal agency providing funding to the site. The Bayh–Dole Act further permits sites to license those inventions to one or more third parties.

With respect to those federally funded inventions or "subject inventions," the Bayh–Dole Act gives the federal agency "march-in rights," which allow the federal agency to step in and require the site or its licensee to license the invention to a third party if, in the opinion of the federal agency, such action is necessary (the federal agency may license the subject invention itself if the site or its licensee refuses to do so). Some of the somewhat subjective reasons that a federal agency may choose to exercise its march-in rights include: (1) the site has not taken effective steps to achieve practical application of the invention; or (2) alleviating health or safety needs, which are not reasonably satisfied by the site or its licensee.[2]

For a nonfederal company sponsor, the Bayh–Dole Act march-in rights raise concerns due to the possibility that a sponsoring federal agency could claim that an invention derived in performance of the sponsor's study is a subject invention. This could happen if (1) the company-sponsored research is jointly or partially funded by a government agency at a particular site; (2) the government agency working with the site on a separate project believes that the company research diminished or distracted from the performance of the government research; or (3) the company-sponsored research creates interference with or cost to the government project.[3] If any of these factors exist, a site invention may be deemed a subject invention and a license subsequently granted by the site to the sponsor could be relicensed to a third party if the federal agency feels that the sponsor has not fulfilled its obligations with respect to the subject invention.

Precedent is on the sponsor's side. Even if a site invention is classified as a subject invention, to date the government has never exercised march-in rights under the Bayh–Dole Act. A sponsor must nonetheless take the Bayh–Dole Act issues into consideration when working with a federally affiliated site, as the government's approach to exercise of the Bayh–Dole Act march-in rights could change at any time. Ultimately, the sponsor must weigh the value of participation of said sites against the risk that a Bayh–Dole Act issue could arise.

Indemnification of Sites and Subjects

One of the areas that cause most discussion in the negotiation of a CTA is the clauses associated with indemnification. Simply stated, an agreement to indemnify another party is an agreement to accept responsibility for any claims brought by third parties against that party and pay damages that may result within certain defined parameters. In the context of a clinical trial, there are often significant differences of opinion between sponsors and sites, as to what types of claims should be indemnified by the sponsor, and the conditions upon which that indemnification will be provided.

As to the types of claims, sponsors seek to limit their indemnity obligation to the site, the investigator, and the site staff for claims and damages asserted by or on behalf of a study subject based on personal injury to the subject. The injury must be sustained as a direct result of administration of the study drug in accordance with the protocol. Furthermore, sponsors often seek to exclude indemnity (or seek reciprocal indemnity from the site) where the claim is attributable to the failure of the site, investigator, or site staff to follow the protocol or comply with FDA requirements, or any negligence or willful misconduct of the site, investigator, or site staff.

This indemnity and its limitations are intended by sponsors to create an obligation to indemnify for things that are within the sponsor's control, occurred through no fault of the party seeking indemnity, and for which the sponsor should ethically be accountable, such as drug defects, unanticipated side effects, and poor protocol design. Sites will often seek to broaden this language to include claims for injuries sustained by the subject as a result of participation in the study—not just administration of the drug in accordance with the protocol. This effort to broaden the indemnity creates tension as it exposes the sponsor to claims where the study drug was not the direct cause of the liability (e.g., injuries sustained by a subject due to a slip and fall accident on the site's property, or due to a car accident in the site parking lot).

Since the sponsor takes financial accountability for the indemnified claim, it will often require a certain level of cooperation from the indemnified party as a condition of the indemnity to protect its financial interests. For example, sponsors often require that the party seeking indemnity provide prompt notice of the claim, permit the sponsor to control the defense and disposition of the claim, cooperate with the sponsor in defending the claim, and not settle any claim without the sponsor's consent. From the sponsor's perspective, these conditions are critical to any indemnification obligation as a failure of an indemnified party to adhere to any one of them could severely prejudice the sponsor's ability to protect its interests and adequately defend the claim.

A recent audit conducted by the European Medicines Agencies (EMEA) of a major pharmaceutical clinical research organization (CRO) alluded to the fact that indemnification may not be valid in the event that investigative sites recruit ineligible subjects to the trial, even if a protocol eligibility waiver is granted by the pharmaceutical sponsor. This gray area will likely be discussed further and await future guidance.

TRIAL BUDGETS AND SUBJECT INSURANCE

Insurance coverage for subjects recruited into clinical trials is a controversial issue, especially in the United States. Historically, insurance companies have argued that costs associated with subject care in clinical trials increase their expenses dramatically. However, a number of recent studies have found that the cost of care for subjects in clinical trials is not appreciably higher than for those not enrolled.[4]

The degree of coverage varies based on whether the healthcare provider considers the studies experimental or investigational. If there is sufficient data to support that the research is both safe and efficacious, some health plans will cover some or all of the costs. Many states are reviewing and passing legislation or developing policies requiring health plans to cover certain clinical trial procedures and medications. This may potentially reduce the cost associated with drug development.[4]

Per Subject Costs

Pharmaceutical sponsors provide a budget template as part of the initial contract. It is the obligation of the investigative site to review this thoroughly and determine if the budget will support all aspects of the research. Some sponsors may provide an estimate of costs associated with required personnel time required; however, it is increasingly more common and equitable for the sponsor and investigator to agree to an itemized reimbursement on a procedural level and include additional expenses for expected administrative costs, institutional review board (IRB) fees, and an institutional overhead assessment.

In an effort to streamline the investigative site budget process, many companies look to electronic web databases and contract services to provide guidance on the estimated costs. One of the most common systems is GrantPlan® (TTC Corporate Headquarters, Philadelphia, PA). GrantPlan provides insight from 57 countries and extensive cost data for procedures and other costs, enabling the pharmaceutical sponsor to create more precise budgets regardless of the complexity of the research study. Since the system enables the investigative site to access and negotiate online and through secure e-mail, the time required to complete budget negotiations is often decreased. This method also enables the pharmaceutical sponsor to provide consistent budgets across sites and across studies for similar indications and compared to industry benchmarks.

More recently, pharmaceutical sponsors have been under much higher levels of scrutiny by the FDA as to the level of coercion sponsors may place on investigators and subjects through excessively increased budgets, or agreeing to investigator requests for additional and/or unrelated funding. Guidelines have been placed on the pharmaceutical industry to cover fair and reimbursable expenses while ensuring that financial incentives are not used to reward participation in the clinical trials. Therefore, standardized budgets are much more common and have less room for negotiation, with the exception of areas of the country that justifiably have higher cost of living and medical expenses.

More and more investigative sites are looking to ensure that the administrative costs associated with clinical trial budgets are also covered. Such items include start-up expenses, record storage costs, time for meetings, and audits. All of these items are appropriate for inclusion in a study budget but drive up the pharmaceutical sponsor's cost and are sometimes contentious.

Clinical Trial Pass-Through Expenses

Clinical trial pass-through expenses are reviewed with the pharmaceutical sponsor during the negotiation of the CTA. Generally, expenses associated with IRB review, approval, and annual renewal; processing of safety reports; and incurred for travel to attend study-specific training are all reimbursed by the pharmaceutical sponsor. There is no doubt that these necessary expenses should be covered by the pharmaceutical sponsor.

Internal Costs

Some of the costs associated with core clinical research may have to be borne by the investigative site itself since they fall under the category of "the cost of doing business." These include costs associated with the set-up and maintenance of running a clinical trial unit—general clinical trial training (non-study specific), compliance, finance, accounting, International Air Transport Association (IATA) laboratory certification for shipping biological samples, Joint Commission on Accreditation of Healthcare Organizations (JCAHO) accreditation for compliance certification, and provision of equipment to support the research conducted.[5] According to the Association of Academic Health Centers (AAHC), academic health centers are faced with a growing number of requirements and these are met with strained and limited resources.[6] Expenditure is required to create and staff systems to increase efficiency and effectiveness and to manage the complex technical aspects of the clinical trial process. In an effort to counteract some of these growing costs, many investigative sites look to the pharmaceutical sponsor and impose an indirect cost rate (overhead) to the clinical trial budget. The overhead percentages vary between investigative sites. Pharmaceutical sponsors are bearing the brunt of rising internal costs by witnessing an increase in the overhead charged and this is causing some concern. A consistent approach is being sought to cap or standardize overhead costs.

Internal Revenue Service Issues: Research Versus Unrelated Business Income

Reporting clinical trial related income is an issue for some new investigative sites. However, the definition of a clinical trial excludes it from taxation by the Internal Revenue Service (IRS). A recently proposed definition for clinical research is "any form of planned experiment which involves subjects with intent designed to elucidate the most appropriate treatment for future subjects with a given medical condition."[7] The vast majority of clinical trials center on the evaluation of a drug or a device concerned with the clinical outcome relative to the future management of subjects and not with financial interest. As such, these studies arguably have research merit and thereby fulfill tax-exemption status.

Should clinical trial research not meet the definition, then it may be considered work by a tax exempt organization. As such, the IRS will impose an Unrelated Business Income Tax (UBIT) on any income.

Financial Disclosures of Clinical Investigators

The FDA requires that for each clinical study submitted as part of a marketing application, the sponsor company must certify to the absence of certain financial interests/conflicts of its clinical investigators participating in the study in order to eliminate the potential for investigator bias. If the sponsor cannot provide such a certification for a particular investigator, it must provide detailed information regarding the arrangement and steps taken to minimize potential bias.

Financial arrangements that must be disclosed include (1) compensation paid to the investigator where the value could be affected by study outcome, (2) a proprietary interest in the tested product, (3) any equity interest in the sponsor (e.g., stock options), (4) any equity interest in a publicly held company of more than $50,000 in value during the study and for 1 year following the completion of the study, and (5) significant payments of other sorts (e.g., grants, consulting) of more than $25,000 from the company to the investigator or site, exclusive of the costs of conducting the clinical study, made during the study and for a period of 1 year following completion.[8]

This financial disclosure requirement often impacts the sponsor's timelines and resources. But full disclosure is important in order to avoid incomplete or inaccurate reporting that may impact the final marketing application.

TRIAL REGISTRATION

The U.S. National Institutes of Health (NIH), Department of Health and Human Sciences (DHHS), through its National Library of Medicine (NLM), has developed a single global website, www.clinicaltrials.gov, to provide subjects, family members, regulators, and other industry members of the public current information about clinical research studies. This website contains a summary of information about clinical trials being conducted both in the United States as well as in many countries around the world.

www.clinicaltrials.gov is a directory of federally and privately supported research trials. Historically, each entry includes a summary of the trial protocol, including the purpose, recruitment status, and criteria for subject entry. Trial location and specific contact information are also provided to assist with enrollment. Section 113 of the Food and Drug Administration Modernization Act (FDAMA) mandates registration with www.clinicaltrials.gov of all investigational new drug (IND) efficacy trials for serious diseases and conditions. In addition, registration with clinicaltrials.gov is required for journal publication. The National Committee of Medical Journal Editors (NCMJE) requires trial registration as a condition for publication of research results.

It is the responsibility of the sponsor or the clinical investigator with primary responsibility for initiating and conducting the clinical trial to register on this website. In order to initiate the registration process, the trial must be approved by an IRB and must conform to regulations of the local health authority. A clinical trial may be registered and listed as "inactive" until such time as the IRB and/or health authority approves the study. The responsibility for registration and maintenance of the data falls to the pharmaceutical sponsor, unless it is designated to the principal investigator of the clinical trial by the pharmaceutical sponsor.

Multisponsor trials are susceptible to duplicate registration and care must be taken in how these trials are registered. It is the lead sponsor in multisponsor trials that takes this responsibility.

U.S. Public Law 110-85 (Food and Drug Administration Amendments Act of 2007), Title VIII, Section 801 mandates the expansion of www.clinicaltrials.gov with new information, including safety issues, listing the responsible party for the trial (which may be different from the sponsor), and the product status with the FDA. Several data elements that were previously optional are now mandatory such as the primary and secondary outcome measures, intervention type, start date, completion date, whether the trial allows healthy volunteers, whether the trial has expanded access, IRB approval, and the target number of subjects.

There are penalties for failure to register applicable clinical trials including monetary penalties and withholding or recovery of grant funding for federally funded trials.

PUBLICATION RIGHTS

Publication of trial data is an important part of the clinical development plan as the study protocol is developed. Often, the contract and protocol includes a specific section on publication rights for those who participate in the study protocol. Two areas are covered under such documentation: publication of site data and publication of overall data. Of interest, the more controversial topic at this time is the ethical issues associated with nonpublication of trial data.

Publication of Site Data

For any sponsored study, the investigator has the right to publish data accrued from his/her own site. However, the contractual clauses typically prevent this prior to the publication of the overall data and study results. Any information to be made public from the site-specific data requires review by the pharmaceutical sponsor to remove any of the sponsor's confidential information and may also require approval from the principal investigator of the study. Review periods and selected reviewers are detailed in the individual protocol contract. In multisite studies, individual site data is not statistically powered to support any claims based on the limited results, and this is taken into consideration prior to any publication. Most clinical trials are also executed in a blinded fashion, which also limits the ability of the sponsor and investigator to publish and therefore bias the trial outcome before study completion.

Publication of Overall Study Data

The pharmaceutical sponsor or lead investigator (who is often a key opinion leader identified by the sponsor before the start of the trial) usually drafts the publication of a multicenter study. The contract specifies the details of any review period required prior to study publication. Should an individual investigator or a group of investigators wish to publish the overall study data, there is usually a requirement not only for sponsor approval but also the approval of any participating trial site.

Since the data rights are owned by the pharmaceutical sponsor, it is unlikely that any publication will be approved by the sponsor until after the study is unblinded and the sponsor's clinical trial report is submitted for a New Drug Application (NDA), for support of an existing Investigational New Drug Application (INDA), or for other publications.

Ethics and Legal Issues of Nonpublication

There is increasing debate on publication of clinical trial results regardless of trial outcome and how this information is disseminated to the public. In certain countries, the IRB (or Research Ethics Committee—REC) mandates annual reports on clinical trials including a final report at completion in order to ensure continued safety of the subjects. In the United Kingdom, based on guidance from the National Health Service, additional steps were taken for the results of the research to be reported and published prior to the Research Ethics Committee's final opinion on the conduct of the trial. This is the responsibility of the principal investigator. For multicenter trials, the names and affiliations of the members of the publication board may be required.[9]

Failure to report results of clinical trial research may be considered misconduct. There is debate whether the REC can hold the principal investigator culpable and then deny approval for any new clinical trials. However, the conditions of the CDA may make it impossible for any site results to be disseminated by a single investigator.

Pharmaceutical sponsors historically have considered the completion of the clinical trial report sufficient to meet the obligation for publication as the regulatory health authorities, the IRB, and the principal investigators receive a copy at the close of the trial. However, whether this information is sufficient and actually reaches the public is uncertain. Many studies have shown that one of the major reasons that subjects enter clinical research studies is an understanding that the research may help future subjects in a similar situation to themselves. Thus, it is reasonable to assume that dissemination of study results may increase the likelihood of consent being granted for further studies. Additionally, it can be argued that study participants are entitled to know the results of the clinical study since it was their participation that made the study possible.

It is clear that nonpublication may have harmful effects for future subjects. Failure to report study results may expose future trial subjects to duplicative, irrelevant, futile, or even dangerous research. In addition, failure to report negative results may adversely affect a subsequent meta-analysis of similar trials. In the United States, there may be legal implications for sponsors who do not communicate negative clinical trial results. In 2004, the New York Attorney General sued GlaxoSmithKline (GSK) for consumer fraud, alleging that GSK withheld the publication of critical data about the pediatric effects of one of its drugs. GSK settled the lawsuit for $2.5 million and agreed to establish a public, online database containing summaries of results of all of its clinical trials. Many other companies have followed suit in order to avoid similar lawsuits. The Pharmaceutical Research and Manufacturers of America (PhRMA) organization has created a centralized website, www.clinicalstudyresults.org, that provides public access to results of clinical trials that are voluntarily provided by company sponsors.

THE RELATIONSHIP BETWEEN INVESTIGATOR AND SPONSOR

Role of the Investigator: Collaborator or Subject Enroller?

Ideally, the academic investigator serves as a collaborator in a clinical trial. The goal is to solve a real clinical problem by applying a partnership between the sponsor and investigator. Sometimes, the partnership goes beyond serving as a site for a multicenter trial when the investigator collaborates on the study design, tracks the study throughout its life cycle and identifies problems resulting in amendments, participates in interpreting and analyzing the data, and actively assists in presenting, writing, and publishing the final scientific report.

Other times, the investigator serves more as an enroller of subjects in a trial and mostly participates by ensuring that appropriate subjects are screened and informed about the trial, assisting with the consent process, supervising the study protocol, interacting with the institutional review board (IRB) and research office, and acting on behalf of the trial should there be an audit by the sponsor or regulatory body such as the FDA. The investigator ideally will interact on a frequent basis with the clinical research coordinator to make certain the study proceeds smoothly at all levels and to address any problems. Communication with the sponsor is also a vital role. Of course, the ideal scenario is sometimes not played out, thus resulting in situations where the term PI may mean "partially involved" or "practically invisible" rather than "principal investigator."

Steering and Publication Committees

From a sponsor's perspective, a publication is defined as any publication that is intended for external use of scientific data from preclinical, clinical, chemistry, or manufacturing studies. These include papers in peer-reviewed journals, abstracts (and associated posters or slide presentations), review articles, and certain slide presentations intended for external audiences. Regulatory documents such as study protocols and clinical trial reports are not included in this definition.

Most sponsors develop a written strategy overview for all planned, ongoing, and completed studies that includes the timing for planned publications for each study. The plan reflects the sponsor's publication strategy to ensure compliance with the regulatory and marketing plans, consistency of data, selection of authors and determination of authorship, a target journal or meeting, timing and type of publications, key messages about the product, new data, unresolved issues, and a writing strategy (e.g., if done by an outside professional group or by one or more investigators). The publication committee may consist of one or more investigators plus members from various departments from within the sponsoring company such as medical development, marketing, medical affairs, medical writing, health economics, and others.

Updates to the publication strategy and plan are made at frequent intervals and after specific milestones are reached in the drug development process. Careful review and sign-off of all publications is accomplished prior to any submission.

Working with Consortiums

Identifying potential study sites for clinical trials may be a significant challenge. Knowledge of the various clinical research networks for pediatrics can lead a sponsor to choose those institutions with proven track records of successful clinical trials research as well as connect with institutions with interests and expertise that match their study. The Inventory and Evaluation of Clinical Research Networks (IECRN) (https://www.clinicalresearchnetworks.org/srchnet.asp) is funded by the NIH Roadmap for Medical Research and is a component of the Roadmap theme "Reengineering the Clinical Research Enterprise." The overall goal of this program is to improve both the efficiency and productivity of clinical research. The Inventory lists clinical research networks that meet the IECRN defi3nition of a network and has a network profile summarizing information about the network including types of studies, funding source, participating entities, and special populations. Table 36.1 lists some of the neonatal and pediatric networks that participate in Phase I–IV clinical trials.

Investigator Initiated Research

Investigator initiated research (IIR) is playing an increasingly important role in pediatric clinical drug development. From an industry perspective, IIR ideas often come about from clinical observations, new insights from basic research, or applying similar pathophysiological concepts between differing drugs and diseases. Results from IIR can be used to provide data for later phase clinical research as well as to expand potential indications for a specific drug or biologic. The associated developmental risk and costs are usually lower as the intellectual capital and time are supplied by the local investigator and his/her institution. From an investigator's perspective, IIR is useful to obtain pilot data to support future grants and apply clinical knowledge to advance medicine, and from a more practical standpoint, generate scientific publications and obtain research funds that are easier to obtain than the highly competitive and ever shrinking allocation of federally and nationally funded research grants. Finally, in terms of clinical research within U.S. academic health centers, IIR and government-funded research rank highest in terms of their desirability compared with industry-sponsored and contract research.[10]

Of note, neither all agree that IIR is uniformly a good idea nor that its goals are correct. Brown[11] suggests that the premise that the more IIR is conducted, the more benefits for society are created is fundamentally incorrect. He suggests that "trickle-down science and technology do not work any better than the famous trickle-down economics worked" and that IIR must move from remediation to prevention as a primary goal for human health.[11]

An IIR is typically developed by the investigator and an appropriate sponsor is sought to fund the proposal or a proposal may be done with a particular sponsor in mind who is

TABLE 36.1 Local and National Neonatal and Pediatric Clinical Research Networks Listed on IECRN that Participate in Phase I–IV Clinical Trials

Network or Consortium	Website
NEONATAL	www.atnonline.org
Biliary Atresia Research Consortium (BARC)	www.barcnetwork.org
Neonatal Research Network (NRN)	http://neonatal.rti.org
Vermont Oxford Network (VO)	www.vtoxford.org
PEDIATRIC	
Adolescent Trials Network for HIV/AIDS Interventions (ATN)	www.atnonline.org
American Spinal Muscular Atrophy Randomized Trials Consortium (AmSMART)	www.amsmart.org
Blood and Marrow Transplant Clinical Trials Network (BMTCTN)	www.bmtctn.net
CF[a] Foundation	
Child and Adolescent Psychiatry Trials Network (CAPTN)	www.captn.org
Childhood Arthritis & Rheumatology Research Alliance (CARRA)	www.carragroup.org
Childhood Asthma Research and Education Network (CARENET)	www.asthma-carenet.org
Children's Oncology Group (COG)	www.curesearch.org
Cincinnati Pediatric Research Group (CPRG)	www.cprg.org
Clinical Trials Network of Columbia University, Cornell University, and NY Presbyterian Hospital (CTN)	www.columbiaclinicaltrials.org/ctn
Cooperative Clinical Trials in Pediatric Transplantation (CCTPT)	Not available
Diabetes Research in Children Network (DirecNet)	http://public.direc.net
Emergency Medicine Network (EMNet)	www.emnet-usa.org
Glaser Pediatric Research Network	www.gprn.org
HIV Vaccine Trials Network (HVTN)	www.hvtn.org
International Maternal Pediatric Adolescent AIDS Clinical Trials (IMPAACT)	http://impaact.s-3.com
National Collaborative Pediatric Critical Care Research Network (CPCCRN)	http://www.cpccrn.org
Pediatric Acute Lung Injury and Sepsis Investigators (PALISI)	http://pedsccm.wustl.edu/research/palisi/palisi.html
Pediatric AIDS Clinical Trials Group (PACTG)	http://impactg.s-3.com/
Pediatric Brain Tumor Consortium (PBTC)	www.pbtc.org
Pediatric Emergency Care Applied Research Network (PECARN)	www.pecarn.org
Pediatric European Network for Treatment of AIDS (PENTA)	www.Pentatrials.org
Pediatric Eye Disease Investigator Group (PEDIG)	http://public.pedig.jaeb.org
Pediatric Heart Network (PHN)	www.pediatricheartnetwork.com
Pediatric Neuromuscular Clinical Research Network	Not available
Pediatric Pharmacology Research Units Network (PPRU)	www.ppru.org

TABLE 36.1 *(Continued)*

Network or Consortium	Website
Pediatric Rheumatology International Trials Organization (PRINTO)	www.printo.it
Project Cure SMA	www.projectcuresma.org
Research Units on Pediatric Psychopharmacology Autism Network (RUPP Autism NW)	
Southern California Permanente Clinical Trials	http://xnet.kp.org/clinicaltrials/
The National Collaborative Pediatric Critical Care Research Network (CPCCRN)	www.cpccrn.org
Therapeutics Development Network (CF TDN)	Not available
Tuberculosis Trials Consortium (TBTC)	www.TBtrialsnetwork.org/TBTC
Type 1 Diabetes TrialNet (Trial Net)	www.diabetestrialnet.org
Vaccine and Treatment Evaluation Units (VTEUs)	Not available

[a]CF—cystic fibrosis.

Source: Available at https://www.clinicalresearchnetworks.org/srchnet.asp.

known to be interested in that drug or disease state. The key is that the concept is mostly independent even though it must conform to sponsor guidelines in terms of specific disease states or drugs. The IIR protocol is submitted to the sponsor for review and, if funded, is then conducted under an Investigator's IND or uses a cross-referenced sponsor's IND. An investigation is IND exempt if the data is not used to support a new indication or change in the labeling for the drug; the data is not used to change advertising for the drug; the study does not involve a route of administration or dosage level, use in a subject population, or other factor that significantly increases the risks (or decreases the acceptability of the risks) associated with the use of the drug product; the study is done in compliance with FDA regulations on IRB review; and the study is conducted in compliance with the requirements concerning the promotion and sale of drugs.[12]

Typical types of IIR research projects include Phase I–IV clinical trials of approved drugs, clinical trials involving unapproved drugs, in vitro and in vivo animal studies, and non-interventional studies such as quality of life studies, resource utilization, and health economic studies.

Most IIR studies result in partial support of the research project to the investigator, the research team, and the investigator's institution. Typically, the institutional overhead is significantly less than charged for federally funded studies and the rate is often set by the study sponsor. Institutions rationalize this lower rate as leading sometime in the future to larger and federally funded research projects based on these preliminary or pilot data. But a common goal for both the investigator and sponsor is publication of the data within the legal framework of the IIR agreement. Recently, it is clear that even negative results need to be disseminated to the public, if not by publication in a scientific journal then by presentation at a scientific meeting and/or disclosed on an open website. For studies that hope to be published in major scientific journals, a prerequisite is often the need to register the trial prior to its start with www.clinicaltrials.gov.

There are a number of common problems with IIR proposals. In a review of deficiencies determined by nonfederal peer reviewers during the 1989 National Institute on Disability and Rehabilitation Research (NIDRR) competition, of 232 IIR applications, six (10.7%) applications were ultimately approved and funded.[13] Of potential interest to industry-sponsored IIR research, of the other 51 that merited formal review, five types of research error accounted for 76.8% of all deficiencies noted by reviewers. These included methodological errors, inadequate control of subject variables, inappropriate research design, poor conceptualization of problem/approach, and incorrect statistical analysis.[13] The top ranked singular deficiency was poor conceptualization of problems or approach while other flaws included excessive budget requests, inadequate background of investigator, and weak dissemination and utilization plans.[13] These identified areas of deficiency are very likely applicable beyond federally funded IIR grant programs and often apply to sponsored clinical trial research as well.

Additionally, problems in terms of financial overcompensation, potential fraud and abuse by the investigator, and concerns about excessive sponsor involvement or direction have occurred on occasion and must be considered by both the investigator and sponsor when an IIR is planned.

REFERENCES

1. 18 USC §200. 2007.
2. 18 USC §203. 2003.
3. 37 CFR 401.1. 2007.
4. Laws about trial costs. 2007.
5. Investing in Clinical Trial Compliance Top Academic Health Center Priority. 2007.
6. The Clinical Trials Landscape. Association of Academic Health Centers. 2007.
7. Higerd TB. Clinical trials. New policies directed at developing budgets, reporting expense and effort, and using residual funds. The Medical University of South Carolina Cost Accounting Standards and Disclosure Statement, 1999.
8. 21 CFR Part 54. 2007.
9. Mann H. Research ethics committees and public dissemination of clinical trial results. *Lancet*. 2007;360(9330):406–408.
10. Oinonen MJ, Crowley WF Jr, Moskowitz J, et al. How do academic health centers value and encourage clinical research? *Acad Med*. 2001;76(7):700–706.
11. Brown GE Jr. The freedom and the responsibility of investigator-initiated research. *Acad Med*. 1994;69(6):437–440.
12. 21 CFR Part 312. 2003.
13. Thomas JP, Lawrence TS. Common deficiencies of NIDRR research applications. *Am J Phys Med Rehabil*. 1990;69(2):73–76.

PART VIII

CLINICAL EFFICACY AND SAFETY ENDPOINTS

CHAPTER 37

Laboratory Monitoring of Efficacy and Safety Parameters in Clinical Trials for Pediatric Subjects

ANDREW E. MULBERG, MD and SAMUEL MALDONADO, MD, MPH

Johnson & Johnson Pharmaceutical Research & Development, Raritan, New Jersey 08969

INTRODUCTION

The inclusion of infants and children in pediatric clinical trials is associated with a number of different and diversified issues for the investigator to consider in the execution of clinical trials. One of the more significant issues related to assessment of efficacy and safety is the interpretation of pertinent and associated laboratory values that are considered either normal or abnormal for the pediatric subject. It is obvious to the experienced principal investigator knowledgeable about the differences between children and adults that there are critical differences related to laboratory value interpretation. There are often indeed disease-specific laboratory value differences in, for example, sickle cell disease and more recently cystic fibrosis patients.

OPERATIONAL CONSTRAINTS AND LIMIT TO BLOOD DRAWING VOLUMES FOR LABORATORY TESTING INCLUDING PHARMACOKINETICS

The absence of appropriately validated laboratory measures for clinical trials is a major unmet medical need for the pharmaceutical industry. Individual Institutional review board (IRB) limits to blood draws for age groups have clearly been differently approached depending on the individual IRB guidance and support. There are no data that are evidence based and therefore in large part are open to interpretation. According to the Guideline on the Investigation of Medicinal Products in the Term and Preterm Neonate, in the draft agreed to by the Paediatric Working Party (http://www.emea.europa.eu/pdfs/human/paediatrics/26748407en.pdf), preterm and term neonates have very limited blood volume and "are often anemic due to age and frequent sampling related to pathological conditions." To limit

Pediatric Drug Development: Concepts and Applications
Edited by Andrew E. Mulberg, Steven A. Silber, and John N. van den Anker

the need for blood samples, the use or special development of microassays, of noninvasive techniques, and of alternative methods is encouraged—like microdialysis and measuring drug levels in saliva or urine—if shown to reflect systemic exposure. However, there is clearly more lack of validation and correlation of these techniques, especially for vulnerable populations. Monitoring of actual blood loss is routinely required in preterm and term neonates. Timing of sampling should be coordinated as far as possible to avoid repeat procedures and to avoid repeat sampling during the day in order to minimize pain and distress, and the risk of iatrogenic complications. The following blood volume limits for sampling are recommended: if an investigator decides to deviate from these, this should be justified. Per individual, the trial-related blood loss (including any losses in the maneuver) should not exceed 3% of the total blood volume during a period of 4 weeks and should not exceed 1% at any single time. The actual situation of the neonate (sleep/activity, severity of anemia, and hemodynamic state) must permit such blood sampling. The total volume of blood is estimated at 80–90 mL/kg body weight; 3% corresponds to about 2.4–2.7 mL blood/kg body weight. As examples, the following table is offered for illustrative purposes. It is important to state that this is not evidence based on or supported by consensus from regulatory guidance:

PHLEBOTOMY ALLOWANCES FOR PEDIATRIC SUBJECTS*

Cohort	Age (yr)	Weight (kg)	TBV	Allowed Sample (cc)
Child	12	40	3.2 L	96
	3	15	1.2 L	36
FT neonate		4	320 mL	9.6
PT neonate		500	40 mL	1.2

CRITICALITY OF NORMAL VALUES FOR LABORATORY INTERPRETATION

The criticality of normal laboratory values cannot be underscored for its impact on clinical trial operations, including inclusion and exclusion of individual subjects for clinical trials. (See Tables 37.1–37.4) Throughout this book, individual chapters are devoted to understanding the differences in laboratory parameters as they relate to the ontogeny and developmental maturation of the individual organ system. There are two main uses of laboratory testing: defining eligibility (inclusion/exclusion) criteria, and developing a plan for identifying and monitoring adverse events. Therefore, it should be recognized that the use of any eligibility criteria that use laboratory-defined "normal" cut-offs in the definition could lead to increased rates of clinical trial exclusion that may have bearing on safety. There are clearly laboratory-specific normals generated often for the individual children's hospital or the clinical research organization that might be employed for the purposes of central laboratory monitoring. This chapter highlights the considerations of normal laboratory parameters for several diseases, including sickle cell disease, hemoglobinopathy, and cystic fibrosis, and isoforms of various laboratory parameters with diverse presentations affecting children.[1–5]

*Infants are less than 12 months old. A 3 year old is a toddler. Allowed sample should be 3% of the TBV.

TABLE 37.1 Normal Blood Values: Neonatal, Pediatric, and Adult Populations

Age	Total Leukocytes[a]		Neutrophils			Lymphocytes			Monocytes		Eosinophils	
	Mean	Range	Mean	Range	%	Mean	Range	%	Mean	%	Mean	%
Birth	18.1	9.0–30.0	11.0	6.0–26.0	61	5.5	2.0–11.0	31	1.1	6	0.4	2
12 hours	22.8	13.0–38.0	15.5	6.0–28.0	68	5.5	2.0–11.0	24	1.2	5	0.5	2
24 hours	18.9	9.4–34.0	11.5	5.0–21.0	61	5.8	2.0–11.5	31	1.1	6	0.5	2
1 week	12.2	5.0–21.0	5.5	1.5–10.0	45	5.0	2.0–17.0	41	1.1	9	0.5	4
2 weeks	11.4	5.0–20.0	4.5	1.0–9.5	40	5.5	2.0–17.0	48	1.0	9	0.4	3
1 month	10.8	5.0–19.5	3.8	1.0–9.0	35	6.0	2.5–16.5	56	0.7	7	0.3	3
6 months	11.9	6.0–17.5	3.8	1.0–8.5	32	7.3	4.0–13.5	61	0.6	5	0.3	3
1 year	11.4	6.0–17.5	3.5	1.5–8.5	31	7.0	4.0–10.5	61	0.6	5	0.3	3
2 years	10.6	6.0–17.0	3.5	1.5–8.5	33	6.3	3.0–9.5	59	0.5	5	0.3	3
4 years	9.1	5.5–15.5	3.8	1.5–8.5	42	4.5	2.0–8.0	50	0.5	5	0.3	3
6 years	8.5	5.0–14.5	4.3	1.5–8.0	51	3.5	1.5–7.0	42	0.4	5	0.2	3
8 years	8.3	4.5–13.5	4.4	1.5–8.0	53	3.3	1.5–6.8	39	0.4	4	0.2	2
10 years	8.1	4.5–13.5	4.4	1.8–8.0	54	3.1	1.5–6.5	38	0.4	4	0.2	2
16 years	7.8	4.5–13.0	4.4	1.8–8.0	57	2.8	1.2–5.2	35	0.4	5	0.2	3
21 years	7.4	4.5–11.0	4.4	1.8–7.7	59	2.5	1.0–4.8	34	0.3	4	0.2	3

[a] Numbers of leukocytes are in thousands per mm^3, ranges are estimates of 95% confidence limits, and percentages refer to differential counts.
[b] Neutrophils include band cells at all ages and a small number of metamyelocytes and myelocytes in the first few days of life.

TABLE 37.2 Hematologic Values in Children and Adults[a]

Age	Hemoglobin (g/dL)		Hematocrit (%)		Red Cell Count (10^{12}/L)		MCV (fL)		MCH (pg)		MCHC (g/dL)	
	Mean	−2 SD	Mean	−2 SD	Mean	−2 SD	Mean	−2 SD	Mean	−2 SD	Mean	−2 SD
Birth (cord blood)	16.5	13.5	51	42	4.7	3.9	108	98	34	31	33	30
1–3 days (capillary)	18.5	14.5	56	45	5.3	4.0	108	95	34	31	33	29
1 week	17.5	13.5	54	42	5.1	3.9	107	88	34	28	33	28
2 weeks	16.5	12.5	51	39	4.9	3.6	105	86	34	28	33	28
1 months	14.0	10.0	43	31	4.2	3.0	104	85	34	28	33	29
2 months	11.5	9.0	35	28	3.8	2.7	96	77	30	26	33	29
3–6 months	11.5	9.5	35	29	3.8	3.1	91	74	30	25	33	30
0.5–2 years	12.0	10.5	36	33	4.5	3.7	78	70	27	23	33	30
2–6 years	12.5	11.5	37	34	4.6	3.9	81	75	27	24	34	31
6–12 years	13.5	11.5	40	35	4.6	4.0	86	77	29	25	34	31
12–18 years												
Female	14.0	12.0	41	36	4.6	4.1	90	78	30	25	34	31
Male	14.5	13.0	43	37	4.9	4.5	88	78	30	25	34	31
18–49 years												
Female	14.0	12.0	41	36	4.6	4.0	90	80	30	26	34	31
Male	15.5	13.5	47	41	5.2	4.5	90	80	30	26	34	31

[a] The mean ± SD can be expected to include 95% of the observations in a normal population.

TABLE 37.3 Reticulocyte and Platelet Counts in Children and Adults

Age	Reticulocyte (%)	Platelets (10^3/mm^3)
Birth at term (cord blood)	3.0–7.0	84–478
1–3 days	1.8–4.6	192
1 month	0.1–1.7	
0.5–2 years	0.7–2.3	150–350
2–6 years	0.5–1.0	150–350
6–12 years	0.5–1.0	150–350
12–18 years		
Male	0.5–1.0	150–350
Female	0.5–1.0	150–350

SICKLE CELL DISEASE AND OTHER HEMATOLOGICAL DISORDERS

Morphological alterations of blood cells are observed early in most hereditary disorders. Therefore, the cytological study of blood cells is a must for the diagnosis of these disorders in neonates and children. Knowledge of the quantitative and qualitative physiological peculiarities of blood cells in neonates is mandatory for an accurate interpretation. Fenneteau et al.[6] describe the main cytological characteristics of blood cells in healthy neonates and infants and their abnormalities associated with hereditary or acquired blood disorders are reviewed. A critical study illustrating the importance of normal values of hematologic changes from birth to 5 years of age and establishing hematological reference values for infants and children with sickle cell disease has been published as a seminal study from 19 pediatric sickle cell centers. Across the United States, Brown and colleagues[7] studied 694 infants with sickle cell disease (sickle cell anemia, sickle cell-hemoglobin C disease, and sickle-beta-thalassemia) enrolled in the Cooperative Study of Sickle Cell Disease at younger than 6 months of age. These authors have provided a dataset representing longitudinal analyses of total hemoglobin concentration, percent fetal hemoglobin values, mean corpuscular volumes, total bilirubin concentration, and red blood cell (RBC), "pocked" RBC, white blood cell, platelet, and reticulocyte counts. Characteristic changes in anemia among the SS and SC populations were differentiated, leading to important conclusions that do affect the interpretation of clinical trials. The hematological profile of SC infants more closely resembled that of normal black infants, but there was mild anemia (10.5 g/dL) and slightly elevated mean values for reticulocytes (3%) and fetal hemoglobin (3%) during early childhood.[7]

In contrast, there are few reports of reference ranges for hematological values in school-age children and most studies extend over a small age range or have excluded a considerable proportion of the study population in an effort to omit those with hemoglobinopathies or anemia. Blood samples from 2135 children aged 4–19 years, from randomly selected schools, were analyzed by automated counter. Reference ranges for red blood cell, white blood cell, and platelet indices are provided from the results. Median hemoglobin and red blood cell count values for girls and boys rose together with increasing age, up to 12 years, but then diverged. Girls had a higher platelet count than boys. Mean platelet volume rose with age and was inversely related to the platelet count. Platelet count fell with age but in girls there was a peripubertal peak. Total leukocyte count fell with age. The upper limits for total leukocyte count in this study are approximately 2×10^9 lower than those quoted in modern

TABLE 37.4 Serum and Plasma Chemistry Values in Children and Adults

Test	Age	Reference Value (USA)	Reference Value (SI)
Alanine aminotransferase	0–5 days	6–50 U/L	6–50 U/L
(ALT, SGPT)	1–19 years	5–45 U/L	5–45 U/L
Albumin	Premature 1 day	1.8–3.0 g/dL	18–30 g/L
	Full term <6 days	2.5–3.4 g/dL	25–34 g/L
	<5 years	3.9–5.0 g/dL	39–50 g/L
	5–19 years	4.0–5.3 g/dL	40–53 g/L
Ammonia	<30 days	21–95 μmol/L	21–95 μmol/L
	1–12 months	18–74 μmol/L	18–74 μmol/L
	1–14 years	17–68 μmol/L	17–68 μmol/L
	>14 years	19–71 μmol/L	19–71 μmol/L
Aspartate aminotransferase	0–5 days	35–140 U/L	35–140 U/L
(AST, SGOT)	1–9 years	15–55 U/L	15–55 U/L
	10–19 years	5–45 U/L	5–45 U/L
Bicarbonate	Premature	18–26 mEq/L	18–26 mmol/L
	Full term	20–25 mEq/L	20–25 mmol/L
	>2 years	22–26 mEq/L	22–26 mmol/L
Bilirubin (conjugated)	Any age	0–0.4 mg/dL	0–8 μmol/L
Bilirubin (total)	Cord		
	Premature	<2 mg/dL	<34 μmol/L
	Term	<2 mg/dL	<34 μmol/L

	0–1 day		
	Premature	<8 mg/dL	<137 µmol/L
	Term	<6 mg/dL	<103 µmol/L
	1–2 days		
	Premature	<12 mg/dL	<205 µmol/L
	Term	<8 mg/dL	<137 µmol/L
	3–5 days		
	Premature	<16 mg/dL	<274 µmol/L
	Term	<12 mg/dL	<205 µmol/L
	>5 days		
	Premature	<2 mg/dL	<34 µmol/L
	Term	<1 mg/dL	<17 µmol/L
	Adult	0.1–1.2 mg/dL	1.7–20.5 µmol/L
Calcium (ionized)	Cord blood	5.0–6.0 mg/dL	1.25–1.50 mmol/L
	<24 hours old	4.3–5.1 mg/dL	1.07–1.27 mmol/L
	24–48 hours old	4.0–4.7 mg/dL	1.00–1.17 mmol/L
	>2 days	4.8–4.92 mg/dL	1.12–1.23 mmol/L
Calcium (total)	Cord blood	9.0–11.5 mg/dL	2.25–2.88 mmol/L
	<24 hours old	9.0–10.6 mg/dL	2.30–2.65 mmol/L
	24–48 hours	7.0–12.0 mg/dL	1.75–3.00 mmol/L
	4–7 days	9.0–10.9 mg/dL	2.25–2.73 mmol/L
	Child	8.8–10.8 mg/dL	2.20–2.70 mmol/L
	Adult	8.4–10.2 mg/dL	2.10–2.55 mmol/L
Chloride	Cord blood	96–104 mEq/L	96–104 mmol/L
	Newborn	97–110 mEq/L	97–110 mmol/L
	>Newborn	98–106 mEq/L	98–106 mmol/L

(continued)

TABLE 37.4 (*Continued*)

Test	Age	Reference Value (USA)			Reference Value (SI)		
Cholesterol (total)							
	1–3 years	45–182 mg/dL			1.15–4.70 mmol/L		
	4–6 years	109–189 mg/dL			2.80–4.80 mmol/L		
		Percentiles					
	Male	**5th**	**75th**	**95th**	**5th**	**75th**	**95th**
	6–9 years	126 mg/dL	172 mg/dL	191 mg/dL	3.26 mmol/L	4.45 mmol/L	4.94 mmol/L
	10–14 years	130 mg/dL	179 mg/dL	204 mg/dL	3.36 mmol/L	4.63 mmol/L	5.28 mmol/L
	15–19 years	114 mg/dL	167 mg/dL	198 mg/dL	2.95 mmol/L	4.32 mmol/L	5.12 mmol/L
	Female						
	6–9 years	122 mg/dL	173 mg/dL	209 mg/dL	3.16 mmol/L	4.47 mmol/L	5.41 mmol/L
	10–14 years	124 mg/dL	174 mg/dL	217 mg/dL	3.21 mmol/L	4.50 mmol/L	5.61 mmol/L
	15–19 years	125 mg/dL	175 mg/dL	212 mg/dL	3.23 mmol/L	4.53 mmol/L	5.48 mmol/L
	Adult	Desirable	Borderline	High	Desirable	Borderline	High
		<200 mg/dL	200–239 mg/dL	>240 mg/dL	<5.18 mmol/L	5.18–6.19 mmol/L	>6.2 mmol/L
Cholesterol—HDL							
	1–13 years	35–84 mg/dL			0.9–2.15 mmol/L		
	14–19 years	35–65 mg/dL			0.9–1.65 mmol/L		
	Adult (desirable)	>45 mg/dL			>1.17 mmol/L		
Cholesterol—LDL							
		Male	Female		Male	Female	
	Cord blood	10–50 mg/dL	10–50 mg/dL		0.26–1.30 mmol/L	0.26–1.30 mmol/L	
	1–9 years	60–140 mg/dL	60–150 mg/dL		1.55–3.63 mmol/L	1.55–3.89 mmol/L	
	10–19 years	50–170 mg/dL	50–170 mg/dL		1.30–4.40 mmol/L	1.30–4.40 mmol/L	
	20–29 years	60–175 mg/dL	60–160 mg/dL		1.55–4.53 mmol/L	1.55–4.14 mmol/L	
	30–39 years	80–190 mg/dL	70–170 mg/dL		2.07–4.92 mmol/L	1.81–4.40 mmol/L	
	40–49 years	90–205 mg/dL	80–190 mg/dL		2.33–5.31 mmol/L	2.07–4.92 mmol/L	
	Adult	Desirable	Borderline	High	Desirable	Borderline	High
		<130 mg/dL	130–159 mg/dL	>160 mg/dL	<3.37 mmol/L	3.37–4.13 mmol/L	>4.14 mmol/L

Creatinine		
Cord	0.6–1.2 mg/dL	53–106 µmol/L
Newborn	0.3–1.0 mg/dL	27–88 µmol/L
Infant	0.2–0.4 mg/dL	18–35 µmol/L
Child	0.3–0.7 mg/dL	27–62 µmol/L
Adolescent	0.5–1.0 mg/dL	44–88 µmol/L
Man	0.6–1.3 mg/dL	53–115 µmol/L
Woman	0.5–1.2 mg/dL	44–106 µmol/L
Glucose		
Cord blood	45–96 mg/dL	2.5–5.3 mmol/L
Premature	20–60 mg/dL	1.1–3.3 mmol/L
Full term	30–60 mg/dL	1.7–3.3 mmol/L
1 day old	40–60 mg/dL	2.2–3.3 mmol/L
>1 day old	50–90 mg/dL	2.8–5.0 mmol/L
Child	60–100 mg/dL	3.3–5.5 mmol/L
Adult	70–105 mg/dL	3.9–5.8 mmol/L
Gammaglutamyltranspeptidase (GGT, GGTP)		
Cord blood	19–270 U/L	19–270 U/L
Premature	56–233 U/L	56–233 U/L
0–3 weeks	0–130 U/L	0–130 U/L
3 weeks–3 months	4–120 U/L	4–120 U/L
>3 months, boys	5–65 U/L	5–65 U/L
>3 months, girls	5–35 U/L	5–35 U/L
1–15 years	0–23 U/L	0–23 U/L
Adult male	11–50 U/L	11–50 U/L
Adult female	7–32 U/L	7–32 U/L
Lactate dehydrogenase		
Neonate	160–1500 U/L	160–1500 U/L
Infant	150–360 U/L	150–360 U/L
Child	150–300 U/L	150–300 U/L
Adult	0–220 U/L	0–220 U/L

(continued)

TABLE 37.4 ***(Continued)***

Test	Age	Reference Value (USA)		Reference Value (SI)	
Lactate dehydrogenase isoenzymes					
LD_1 heart		24–34%		24–34%	
LD_2 heart, erythrocytes		35–45%		35–45%	
LD_3 muscle		15–25%		15–25%	
LD_4 liver, trace muscle		4–10%		4–10%	
LD_5 liver, muscle		1–9%		1–9%	
Lipase	1–18 years	3–32 U/L		3–32 U/L	
Magnesium	Any age	1.3–2.0 mEq/L		0.65–1.0 mmol/L	
Manganese	Newborn	2.4–9.6 μg/dL		0.44–1.75 μmol/L	
	2–18 years	0.8–2.1 μg/dL		0.15–0.38 μmol/L	
Phosphatase, alkaline					
	1–9 years	145–420 U/L		145–420 U/L	
	10–11 years	130–560 U/L		130–560 U/L	
		Male	Female	Male	Female
	12–13 years	200–495 U/L	105–420 U/L	200–495 U/L	105–420 U/L
	14–15 years	130–525 U/L	70–130 U/L	130–525 U/L	70–130 U/L
	16–19 years	65–260 U/L	50–130 U/L	65–260 U/L	50–130 U/L
Phosphorus, inorganic					
	0–5 days	4.8–8.2 mg/dL		1.55–2.65 mmol/L	
	1–3 years	3.8–6.5 mg/dL		1.25–2.10 mmol/L	
	4–11 years	3.7–5.6 mg/dL		1.20–1.80 mmol/L	
	12–15 years	2.9–5.4 mg/dL		0.95–1.75 mmol/L	
	16–19 years	2.7–4.7 mg/dL		0.90–1.50 mmol/L	
Plasma volume					
	Male	25–43 mL/kg		0.025–0.043 L/kg	
	Female	28–45 mL/kg		0.028–0.045 L/kg	
Potassium					
	<2 months	3.0–7.0 mEq/L		3.0–7.0 mmol/L	

	2–12 months	3.5–6.0 mEq/L		3.5–6.0 mmol/L	
	>12 months	3.5–5.0 mEq/L		3.5–5.0 mmol/L	
Protein, (L) total					
	Cord	4.8–8.0 g/dL		48–80 g/L	
	Full term	4.4–7.6 g/dL		44–76 g/L	
	1–30 days	4.4–7.6 g/dL		44–76 g/L	
	1–3 months	3.6–7.4 g/dL		36–74 g/L	
	4–6 months	4.2–7.4 g/dL		42–74 g/L	
	7–12 months	5.1–7.5 g/dL		51–75 g/L	
	13–24 months	3.7–7.5 g/dL		37–75 g/L	
	25–36 months	5.3–8.1 g/dL		53–81 g/L	
	3–5 years	4.9–8.1 g/dL		49–81 g/L	
	6–8 years	6.0–7.9 g/dL		60–79 g/L	
	9–11 years	6.0–7.9 g/dL		60–79 g/L	
	12–16 years	6.0–7.9 g/dL		60–79 g/L	
	Adult	6.0–8.0 g/dL		60–80 g/L	
Sodium					
	Newborn	134–146 mEq/L		134–146 mmol/L	
	Infant	139–146 mEq/L		139–146 mmol/L	
	Child	138–145 mEq/L		138–145 mmol/L	
	Adolescent	136–146 mEq/L		136–146 mmol/L	
Triglycerides (after >12 hours of fast)					
		Male	Female	Male	Female
	Cord blood	10–98 mg/dL	10–98 mg/dL	0.10–0.98 g/L	0.10–0.98 g/L
	0–5 years	30–86 mg/dL	32–99 mg/dL	0.30–0.86 g/L	0.32–0.99 g/L
	6–11 years	31–108 mg/dL	35–114 mg/dL	0.31–1.08 g/L	0.35–1.14 g/L
	12–15 years	36–138 mg/dL	41–138 mg/dL	0.36–1.38 g/L	0.41–1.38 g/L
	16–19 years	40–163 mg/dL	40–128 mg/dL	0.40–163 g/L	0.40–1.28 g/L
	20–29 years	44–185 mg/dL	40–128 mg/dL	0.44–1.85 g/L	0.40–1.28 g/L
	Adults	40–160 mg/dL	35–135 mg/dL	0.40–1.60 g/L	0.35–1.35 g/L

(*continued*)

TABLE 37.4 *(Continued)*

Test	Age	Reference Value (USA)	Reference Value (SI)
Urea nitrogen (BUN)			
	Cord blood	21–40 mg/dL	7.5–14.3 mmol urea/L
	Premature (1 week)	3–25 mg/dL	1.1–9.0 mmol urea/L
	Full term	3–12 mg/dL	1.1–4.3 mmol urea/L
	Infant/child	5–18 mg/dL	1.8–6.4 mmol urea/L
	Adolescent	7–18 mg/dL	2.5–6.4 mmol urea/L
Uric acid			
	0–2 years	2.4–6.4 mg/dL	0.14–0.38 mmol/L
	2–12 years	2.4–5.9 mg/dL	0.14–0.35 mmol/L
	12–14 years	2.4–6.4 mg/dL	0.14–0.38 mmol/L
	Adult male	3.5–7.2 mg/dL	0.20–0.43 mmol/L
	Adult female	2.4–6.4 mg/dL	0.14–0.38 mmol/L

hematology textbooks.[3] Lymphocyte, eosinophil, and basophil counts fell with age with little difference between the sexes. Neutrophil and monocyte counts were similar for younger girls and boys but diverged in the older children with the older girls having higher values than boys.[8]

Morphologic and laboratory aspects of diagnostic pediatric hematology are reviewed by Geaghan,[9] including developmental hematopoiesis in normal fetal blood, the principal morphologic features unique to examination of pediatric blood samples, special preanalytic considerations for the laboratory, and important analytic interferences common in the pediatric setting. National estimates of hematologic and iron-related analytes for persons 1 year of age and over, by age, sex, and race/ethnicity have been published but not cross-sectionally validated.[10] The analysis is based on data from the third National Health and Nutrition Examination Survey (NHANES III) (1988–1994), which was designed to provide information on the health and nutritional status of the civilian noninstitutionalized U.S. population. The sample used for these analyses included the 26,372 participants who had laboratory tests. This report provides mean, standard error of the mean, median, and percentile laboratory values for the U.S. population, 1988–1994, for hematological and iron-related analytes. The data on hematological and iron analytes can provide reference values for clinical and longitudinal comparisons.

CYSTIC FIBROSIS (CF)

It is important to identify clearly the appropriate laboratory parameters in CF patients participating in clinical trials, because how they compare to laboratory-specific reference ranges developed using a healthy population needs to be understood for the proper interpretation of data. Additionally, there are no data in the literature that describe the intraindividual biologic variation among CF patients participating in clinical trials and only limited data regarding this variation in clinical practice. To address this concern, cross-sectional and longitudinal analyses were performed using data from a large CF clinical trial to summarize chemistry and hematology measurements in relation to laboratory-specific reference ranges.[11] Applications of these types of data to specific development programs especially targeting the CF population are needed for validation and cross-sectional relevance and differences between subjects.

CONCLUSION

Monitoring of normal and abnormal laboratory values is critical for the operational and tactical success of pediatric clinical trials. There are clearly operational constraints for execution of clinical trial related phlebotomy for infants and children that often limits clinical trial recruitment. Availability of microassay techniques is important for the industry to develop to help protect the vulnerability of our pediatric subjects.

REFERENCES

1. Behrman RE, Kliegman RM, Arvin AM. *Nelson Texbook of Pediatrics*, 15th ed. Philadelphia: WB Saunders; 1996.

2. Siberry GK, Iannone R. *The Harriet Lane Handbook*, 15th ed. St Louis, No: Mosby; 2000.
3. Hoffman R, Benz EJ Jr, Shattal SJ, et al. *Hematology, Basic Principles and Practice*, 3rd ed. New York: Churchill Livingston; 2000.
4. Miller DR, Baehner RL, Miller LP. *Blood Diseases of Infancy and Childhood*, 6th ed. St Louis, MO: Mosby; 2000.
5. Schwartz GJ, Haycock GB, Spitzer A. Plasma creatinine and urea concentration in children: Normal values for age and sex. *Pediatr*. May 1976;88(5):828–830.
6. Fenneteau O, Hurtaud-Roux M-F, Schlegel N. Aspect cytologique normal et pathologique du sang chez le nouveau-né et le jeune enfant. *Ann Biol Clin*. 2006;64(1):17–36.
7. Brown AK, Sleeper LA, Miller ST, Pegelow CH, Gill FM, Waclawiw MA. Reference values and hematologic changes from birth to 5 years in patients with sickle cell disease. *Arch Pediatr Adolesc Med*. 1994;148:796–804.
8. Taylor, MRH, Holland CV, Spencer R, Jackson JF, O'Connor GI, O'Donnell JR. Haematological reference ranges for schoolchildren. *Clin Lab Haematol*. 1997;19(1):1–15.
9. Geaghan SM. Hematologic values and appearances in the healthy fetus, neonate and child. *Clin Lab Med*. 1999;19(1):1–32.
10. Hollowell JG, van Assendelft OW, Gunter EW, Lewis BG, Najjar M, Pfeiffer C. Hematological and iron-related analytes—reference data for persons aged 1 year and over: United States, 1988–94. *Vital Health Stat*. Mar 2005;247:1–156.
11. Christopher H, Goss CH, Mayer-Hamblett N, et al. Laboratory parameter profiles among patients with cystic fibrosis. *J Cystic Fibrosis*. 2007;6:117–123.

CHAPTER 38

Breath Testing in Pediatrics

GIGI VEEREMAN-WAUTERS, MD, PhD

Division of Pediatric Gastroenterology, Hepatology and Nutrition, Queen Paola Children's Hospital ZNA and University Hospital Antwerp, 2020 Antwerp, Belgium

INTRODUCTION

Breath tests are an attractive noninvasive tool for testing various metabolic and functional processes in children. Exhaled air contains volatile end products of metabolism, such as H_2 and CO_2. Specific metabolic or functional processes can be indirectly assessed or quantified by measuring variations of exhalation of these gases after administering a substrate challenging the function to be studied. Variations in H_2 are measurable without marker. Variations in proportions of expired CO_2 are revealed by labeling the substrate with the stable isotope ^{13}C, thus appearing as $^{13}Co_2$ in breath. The use of isotopes for most breath tests was initiated in adult medicine using ^{14}C.[1] For pediatric subjects only the stable isotope ^{13}C is acceptable due to safety concerns with ^{14}C. Administration of specific substrates must be completely safe for them to be used in pediatric investigations.

Sampling breath is obviously noninvasive, which is a great advantage for clinical studies in children. However, obtaining breath samples from very young children may be tedious for the examiner. Breath tests are subject-friendly but must be performed with great precision and adherence to methods for sampling and test meal in order to be reliable and comparable. The required time depends on the type of breath test: varying from 2 h for *Helicobacter pylori* (HP) to 10 h for orocecal transit time (OCTT).

For some tests, the samples are analyzed by easily accessible devices (H_2 measurement, HP test) but for most a specialized laboratory equipped with isotope ratio mass spectrometer (IRMS) for measuring the ratio between $^{12}CO_2$ and $^{13}Co_2$ is required.

The stable isotope breath tests that have been adapted for pediatrics and the nonmarked (H_2) breath tests are the scope of this chapter.

It is precisely their safety and noninvasiveness that make breath tests an attractive tool for studying specific clinical endpoints in pediatrics. On the downside, they are time consuming, often costly, and require experience and specialized labs. The reproducibility of breath tests may be inconsistent, probably due to methodology as well as to physiological variations.

Pediatric Drug Development: Concepts and Applications
Edited by Andrew E. Mulberg, Steven A. Silber, and John N. van den Anker

We review the breath tests that have been designed and applied in children and the clinical endpoints for which their use may be advocated.

METHODOLOGY: GENERAL PRINCIPLES ON PERFORMING BREATH TESTS IN CHILDREN

Substrates, Test Meal

Each breath test has its own test meal specific for the test. Disaccharides such as lactose and sucrose generate H_2 when they are not absorbed by the small intestine and reach the colon. They do not require a marker and are dissolved in water. Substrates that are oxidized to CO_2 need to be marked with the stable isotope ^{13}C in order to trace their fate as $^{13}CO_2$ in exhaled air. These molecules are integrated in test meals such as infant formula, chocolate, or egg yolk. The choice of test meal depends on the solubility of the substrate as well as on the acceptability and palatability for a certain age group. Test meals may not have natural ^{13}C enrichment (such as soy or corn) since that would falsify results. Reference values for a breath test are valid only for a specific test meal.

Breath Sampling

A subject needs to be fasted prior to a breath test. After obtaining a pair of basal breath samples, a substrate, specific for the function under study, is presented to the subject as a test meal. In order to maintain basal metabolism and CO_2 production constant, the subject must remain quiet during the test. For babies this implies they should not cry or be upset. Applying a mask to their face for collecting breath is therefore often not suitable. The best method for collecting breath samples in children under 3 years of age is to apply a nasal prong (Ch 6 nasopharyngeal catheter) (2) and use a two-way stopcock to slowly fill a syringe during each expiration (Figures 38.1 and 38.2). Toddlers are able to blow through a straw into a vacutainer in a similar manner as older children and adolescents (Figure 38.3). Well-sealed breath samples can be stored for a prolonged period of time: 30 days for hydrogen[2] and up to 6 months for $^{13}CO_2$.[3]

Analysis

Hydrogen (H_2) in breath is analyzed by appropriate gas chromatography followed by a detection method, such as a solid-state sensing device (e.g., Quintron®/www.quintron-usa.com). Commercially available analyzers are stationary or portable and require various volumes of expired breath. Results are expressed in parts per million (ppm) H_2. A rise above baseline of 10 ppm after substrate ingestion is generally considered a positive result.[4] Baseline is usually under 10 ppm but may be elevated due to bacterial overgrowth or fermentation of nonabsorbable dietary carbohydrates.[5] Up to 40% of healthy controls may not produce any H_2; consequently, the hydrogen test is unsuitable for them.[6] Since H_2 is produced by the host's bacterial flora, production is temporarily suppressed by oral antibiotic intake.[7]

For measuring $^{13}CO_2$ in breath, other gases (N_2, O_2, H_2) are first removed with a gas chromatograph. The purified sample can then be quantified by mass spectrometry, nuclear magnetic resonance spectroscopy, and infrared mass spectroscopy or laser resonance spectroscopy. Isotope ratio mass spectrometry (IRMS) is the best-accepted method as it

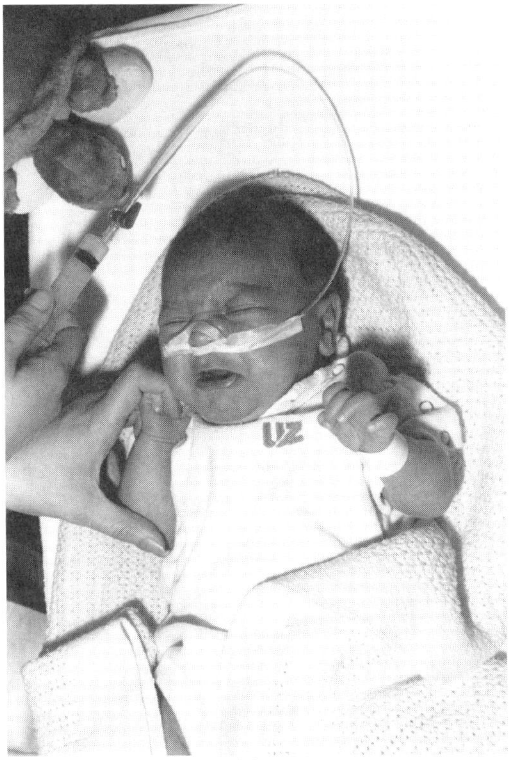

(a)

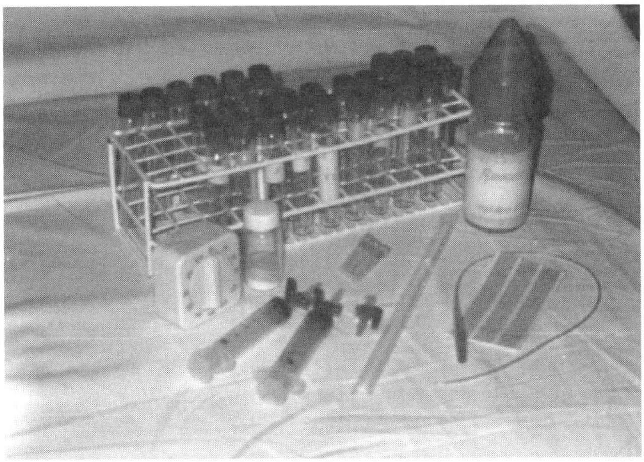

(b)

FIGURE 38.1 Sampling expired air in a newborn: method and materials.

has high accuracy for a low level of enrichment (0.001–0.01 atom percentage). The pediatric literature on stable isotope breath tests is almost completely based on IRMS analysis.

Isotope-selective nondispersive infrared spectrometry (NDIRS) offers a less expensive alternative. The validity of IRMS and of NDIRS have been compared for a number of breath tests. Barth et al.[8] reported a good correlation between these two methods for a small number

FIGURE 38.2 Toddler expiring into a vacutainer.

of samples as required for simple breath tests. Longer series, where repeated measurements are required on the NDIRS instrument lead to a decreasing correlation. The authors conclude that IRMS is superior concerning $^{13}CO_2$ kinetics over longer time periods.[8] This is confirmed by two studies in adults, the first using the [^{13}C] methacetin breath test assessing liver

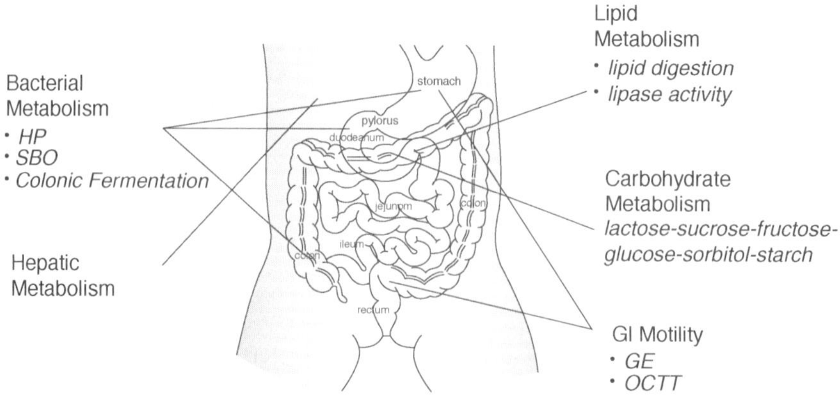

FIGURE 38.3 Breath tests in pediatrics.

function[9] and the second using the ^{13}C mixed triglyceride (MTG) breath test assessing lipase activity.[10] NDIRS is well accepted for the simple and short HP breath test and thus contributed to the wide availability of this diagnostic test.[11,12] NDIRS has been used for gastric emptying tests in adults and children.[13–15] The correlation with IRMS was shown to be reasonable in adults.[16]

^{13}C is a naturally occurring isotope consisting of appoximately 1.1% of total carbon: its abundance in nature is not constant but dependent on the origin of the carbon source, varying from around 1.06% for fossil fuels to 1.12% for carbonate stones. The natural background of ^{13}C in breath CO_2 is 1.08–1.09% in Europe. In the United States, ^{13}C in basal CO_2 excretion is higher due to the content of corn-derived components (a C_4-plant) in the diet.

To enable suitable precision, stability, and comparable data, ^{13}C abundance is always expressed against a universal reference standard, this being the carbon originating from Pee Dee Belemnite limestone (PDB). Data are expressed as per mil ($^o/_{oo}$) difference to PDB: δ (delta) $^{13}C_{PDB}(^o/_{oo})$. In practice, isotope ratios are measured by IRMS against a working CO_2 standard, which is calibrated against a certified reference standard, which in turn is calibrated against PDB limestone.

^{13}C enrichment is defined as the difference between the basal ^{13}C abundance before administration of substrate and the ^{13}C abundance at a certain time point after administration and is expressed as $\delta^{13}C_{PDB}(^o/_{oo})$. From these data, the percentage $^{13}C_{dose}$ recovered in breath can be calculated. Biological variation of baseline $\delta^{13}C_{PDB}$ values is on the order of SD 0.5 ($^o/_{oo}$).

Some dietary products are naturally richer in ^{13}C,[17] and this should be taken into account when choosing a test meal. Results are expressed numerically and graphically as percentage dose recovery (PDR) per hour of the initial administered amount and as percentage cumulative dose recovery (cPDR) after completion of the test.[18] CO_2 production is a necessary parameter for calculations of percentage of ^{13}C recovery. Basal CO_2 production is assumed to be 300 mmol/m^2•h for adults but varies with age, gender, and weight. For calculations in preterm infants, a constant CO_2 production of 20 mmol/kg h is used.[19] For infants and children, CO_2 production is calculated based on the basal metabolic rate (BMR) reported by Altman and Shofield.[18,20,21]

For some breath tests the answer is yes or no: for example lactose malabsorption and gastric HP infection. A cutoff value needs to be determined such as 10 ppm for H_2 production[4] and 5.47δ per thousand at 30 min for HP infection.[22] Other tests generate a quantitative result such as gastric half-emptying time ($T_{1/2}$) for the [^{13}C]octanoic acid breath test. Reference values are needed for clinical interpretation.[18]

SAFETY

An important advantage of breath tests with nonmarked substrates or stable isotopes is their safety, especially notable for pediatric subjects. Nonmarked substrates are for the most part dietary carbohydrates. Lactulose has a pharmacological laxative effect in children from doses of 5 g/day.

Stable isotopes such as ^{13}C are per definition free of radiation, and therefore their administration is safe.[23] The substrates are incorporated in test meals that are part of the normal diet (e.g., formula, pancakes, egg preparations). The test meals must be prepared and stored under rigorous sanitary conditions.

PHYSIOLOGICAL FUNCTIONS ACCESSIBLE FOR BREATH TESTING: CLINICAL ENDPOINTS THAT CAN BE STUDIED RELIABLY IN CHILDREN USING BREATH TESTS

In this section, specific breath tests are described for which good evidence supports their use in a clinical setting for diagnostic purposes or for assessing clinical endpoints in research (Figure 37.3). Each breath test has a proper, detailed methodology. Results and references apply only for the described methods.

Carbohydrate Digestion

Lactose The lactose hydrogen breath test is commonly accepted and used for the clinical diagnosis of lactose malabsorption.

Substrate for Breath Test This is nonmarked or [^{13}C]lactose 2 g/kg in 20% solution (maximum 50 g).

Mechanism Nonmarked lactose is not absorbed in the case of lactase deficiency; colonic fermentation yields H_2. Intestinal absorption of [^{13}C]lactose, in the presence of lactase, yields $^{13}CO_2$ exhalation.

Sampling Time Sampling is done every 30 min for 4 h.

Sensitivity and Specificity The hydrogen breath test is considered very sensitive and specific for positivity with a 10 ppm rise.[24] In predicting clinical response to dietary change, the breath hydrogen test was reported with a predictive accuracy of 96% when compared to clinical response to dietary change.[25]

A cutoff for cPDR of 14.5% was determined for the [^{13}C]lactose breath test. The $^{13}CO_2$ breath test was found to be more sensitive (0.84 vs. 0.68) and more specific (0.96 vs. 0.89) than the H_2 breath test in detecting low jejunal lactase activity.[26]

Pediatric Experience Numerous studies have been performed using the lactose hydrogen test to study primary and secondary lactase deficiency in search of explanations for chronic or recurrent abdominal pain and protracted diarrhea.[27] Although used in clinical practice, reports on the [^{13}C]lactose breath test in children are scarce.

Sucrose This functional test for sucrase activity has been proposed, as a corollary, as a marker for mucosal integrity.

Substrate for Breath Test This is nonmarked or [^{13}C]sucrose 2 g/kg in 20% solution (maximum 20 g).

Mechanism Malabsorbed sucrose is found in the case of sucrase deficiency; colonic fermentation yields H_2. Intestinal absorption of [^{13}C]sucrose in the presence of sucrase yields $^{13}CO_2$ exhalation.

Sampling Time Sampling is done every 30 min for 4 h for the hydrogen test, and every 15 min for 3 h for the [^{13}C]sucrose breath test.

Sensitivity and Specificity The hydrogen breath test yields results around 100 ppm (range 20–432 ppm) in patients with sucrase-isomaltase deficiency and around 20 ppm (range 12–51 ppm) in secondary sucrose malabsorption.[25]

For the [^{13}C]sucrose breath test, peak excretion exceeding 7% and cPDR over 20% are considered positive.[28]

Pediatric Experience The sucrose hydrogen breath test was used in children with biopsy proven sucrase-isomaltase deficiency to validate the use of a nasal prong.[2] It can be used to evaluate the effect of enzyme replacement therapy (e.g., sacrosidase).[29] Cystic fibrosis (CF) patients have a higher incidence of secondary sucrase deficiency.[30]

The [^{13}C]sucrose breath test has been applied to detect small intestinal damage associated with mucositis in pediatric cancer patients under chemotherapy. The cPDR at 90 min is significantly lower in patients with clinical mucositis compared to the unaffected chemotherapy patients and healthy controls.[31]

Fructose

Substrate for Breath Test This is fructose 0.3–0.5 g/kg in 20% solution (maximum 20 g).

Mechanism Nonabsorbed, intraluminal fructose leads to H_2 production.

Sampling Time Sampling is done every 30 min for 3 h.

Sensitivity and Specificity A rise over 20 ppm hydrogen is considered positive, but notably there are no data on sensitivity and specificity in children.

Pediatric Experience Malabsorption of fructose in fruit juices is often responsible for abdominal symptoms and diarrhea in children.[32] The [^{13}C] D-fructose test is less informative than the hydrogen breath test, because $^{13}CO_2$ in breath originates both from intestinal absorption and bacterial fermentation.[33]

Glucose

Substrate for Breath Test This is glucose 1 g/kg in 20% solution (maximum 20 g) or [^{13}C]glucose 5% in water.[34]

Mechanism Glucose malabsorption (H_2) or absorption (^{13}C) is demonstrated.

Sampling Time Sampling is done every 30 min for 4 h.

Sensitivity and Specificity Breath tests confirm proven glucose-galactose malabsorption.

Pediatric Experience Both the hydrogen breath test (with interval sampling)[35] and the stable isotope test[34] have confirmed glucose malabsorption in known glucose-galactose deficiency. Clinical application remains limited.

Sorbitol The monosaccharide sorbitol is often maldigested and thus acts as a laxative. Fruit, mostly prunes, are rich in sorbitol. It is used as low-calorie sweetener.

Substrate for Breath Test This is sorbitol 0.06 g/kg,[32] or a 2% solution.[36]

Sampling Time Sampling is done every 30 min for 4 h.

Mechanism Maldigestion leads to H_2 production.

Sensitivity and Specificity Various doses of sorbitol have been tested in healthy adults and celiac patients.[37] There are no data on sensitivity and specificity of the test in children, but a 20-ppm rise is usually considered positive.

Pediatric Experience Maldigestion of various fruit juices, due to fructose and sorbitol, has been studied.[36,38]

Starch Modeling of $^{13}CO_2$ enrichment curves after ingestion of ^{13}C-enriched wheat flour allows estimation of the contribution of the upper and lower gut to starch digestion and fermentation.[39]

Substrate for Breath Test The following can be used: [^{13}C]starch (50 g in adults)[28] and naturally enriched cornstarch 0.2–2.8 g/kg.[40]

Mechanism Digestion, not monosaccharide transport, is the rate-limiting step in the assimilation of polysaccharides.[28]

Sampling Time Sampling is done every 30 min for 6 h.

Sensitivity and Specificity There is no data.

Pediatric Experience Amarri et al.[41] studied children with CF and showed diminished starch digestion and oxidation. Pancreatic enzyme supplement restores digestion to near normal levels. A second peak in $^{13}CO_2$ enrichment, suggestive of colonic starch fermentation, is absent in healthy children, present in some children with CF, and abolished by pancreatic enzymes.[41] The stable isotope breath test can also be used to measure digestion of naturally ^{13}C enriched cornstarch in healthy children and CF,[40] as well as of weaning foods before and after partial digestion with amylase-rich flour.[42]

Lipid Digestion

Lipid Absorption The diagnosis of fat malabsorption is challenging. Precise diagnosis requires invasive duodenal intubation studies that are difficult to perform and not well tolerated by children. Laboratory determinations of nutritional parameters such as fat-soluble vitamins and fecal-fat sample measurements using the acid steatocrit method may indicate the presence of fat malabsorption. A more accurate method to quantify fat malabsorption is a 72-hour fecal-fat collection with calculation of a coefficient of fat absorption. Fat balance studies are unpleasant for regular clinical assessment.[43,44] Various lipids marked with ^{13}C have been used to assess lipid digestion. Stable isotopes can also be used to trace lipid and fatty acid metabolism and to study the influence of specific dietary supplements.[45]

Substrates for Breath Test The following can be used: [^{13}C]hiolein (mixture of long chain triglycerides 2 mg/kg),[46] [^{13}C]trioctanoin (7.5 mg/kg),[47,48] [^{13}C]triolein (17 mg/kg),[47] [^{13}C]palmitic acid (17 mg/kg),[47] or [^{13}C]cholesteryl octanoate (25 mg/kg).[49]

Sampling Time Sampling is done every 30 min for 6 h.

Mechanism Expiration of $^{13}CO_2$ reflects the end result of lipid absorption, including but not limited to pancreatic function. These tests may therefore replace fecal-fat balance. In contrast to the 72-h fecal collection, the breath test reflects fat assimilation immediately after a given test meal. Factors such as gastric emptying and fatty acid oxidation affect results.

Sensitivity and Specificity In children with known steatorrhea, the discriminative value for detection of fat malabsorption was superior for the [^{13}C]triolein breath test with 100% sensitivity and 89% specificity. Both [^{13}C]palmitic acid and [^{13}C]trioctanoin yielded may false-positive and false-negative results.[47]

In adults the sensitivity of the [^{13}C]hiolein test for detecting steatorrhoea was 91.7%, with a specificity of 85.7%.[46]

Pediatric Experience Cystic fibrosis (CF) is obviously the disease of choice in the Caucasian population for noninvasive investigations of fat absorption. Breath tests can be repeated in various circumstances, under different therapeutic regimens. Triglycerides yield different cPDR depending on the rate of hydrolysis and fat deposition.[50] So far, experience remains limited. Braden et al.[51] used the [^{13}C]hiolein breath test in children with CF. Controls had significantly higher $^{13}CO_2/^{12}CO_2$ ratios (cPDR of $39.2 \pm 18.1\%$ vs. $13.1 \pm 13.9\%$; $p < 0.001$) than CF patients, but after administration of pancreatin, CF patients increased their output to the same level as controls.[51]

McClean et al.[48] used the [^{13}C]trioctanoin breath test in infants and young children with CF. This study also showed lower (44%) cPDR in CF patients and improvement (27%) after pancreatic enzyme supplementation.

Alternatively, digestion by bile salt-stimulated lipase was studied in Gambian infants using [^{13}C]trioctanoin and [^{13}C]cholesteryl octanoate breath tests. Bile salt-stimulated lipase in human milk did not contribute to the the digestion of non-breast-milk fat in healthy, well-nourished infants.[49]

Lipase Activity Fecal fat and the previoulsy discussed breath tests reflect the result of various factors contributing to fat absorption and thus are not specific enough to evaluate exocrine pancreatic function. The MTG breath test demonstrates duodenal lipolysis due to both residual endogenous and exogenous lipase.

Substrate for Breath Test This is [^{13}C] mixed triglyceride[52] in age-adapted test meal (see Table 38.1).

Mechanism The MTG molecule consists of glycerol with stearic acid in the 1 and 3 positions and ^{13}C-labeled octanoic acid in the 2 position (Figure 38.4). The rate-limiting step

TABLE 38.1 Reference Values for the [^{13}C]MTG Breath Test in Children[52] and Adults[54]

Test Meal	
Prematures/neonates: 100 mg [^{13}C]MTG mixed with a formula with low and stable ^{13}C background activity (NAN1, Nestlé) and 1 g PEG.	
Toddlers and children (over 3 years old): a slice of white bread with 5 g butter and 15 g chocolate paste mixed with 250 mg ^{13}C-MTG and a glass of 100 ml milk	
Results	Cumulative % Dose
Preterm infants	23.9 ± 5.2%
Neonates	32.3 ± 7.5%
Children	32.5 ± 5.3%
Teenagers	28.0 ± 5.4%
Adults	33.5 ± 1.4%

for digestion of the MTG is hydrolysis of the two stearyl groups by pancreatic lipase. [^{13}C]octanoic acid is subsequently absorbed and oxidized by the liver. This process is not affected by liver disease, as occurs in CF.[53]

Sampling Time Sampling is done every 30 min for 6 h.

Sensitivity and Specificity In adults, the [^{13}C]MTG breath test was validated against duodenal intubation and fluid collection. Duodenal lipase output versus 6-h cumulative $^{13}CO_2$ excretion in breath after ingestion of [^{13}C]MTG yielded a sensitivity of 0.89 and a specificity of 0.81.[54] These findings lead to the conclusion that the [^{13}C]MTG breath test should replace the ancient invasive gold standard requiring duodenal fluid collection. For obvious reasons, it has not been possible to perform a similar validation study in children.

Pediatric Experience In children, a quantifiable measure of intraluminal lipase activity can be attained using the stable isotope [^{13}C]MTG assay.[55] Age-specific test meals and breath-sampling techniques for the MTG breath test were defined with reference values for the pediatric population[52] (Table 37.1). The study was performed in 12 premature infants (<37 weeks gestation and with body weights >2 kg), 12 full-term infants (1–6 months old), 20 children (3–10 years old), and 20 teenagers (11–17 years old); all were healthy and thriving well. Because of the noninvasive nature of the test, it was possible to obtain ethical approval and parental consent. Recovery of the tracer amounts to 20–35% in healthy

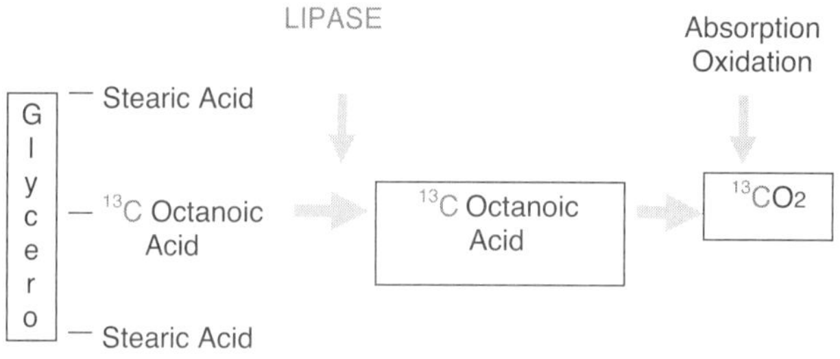

FIGURE 38.4 After ingestion of the ^{13}C mixed triglyceride (1,3 distearyl, 2(^{13}C)-octanoyl glycerol), the rate-limiting step for appearance of ^{13}C in exhaled air is the lipase activity.

subjects with normal exocrine pancreatic function and is due to sequestration of acetate in intermediary metabolism.[56] The absence of tracer in the stools has been confirmed.[57] The specificity of the test can be improved by assessing nonresting carbon dioxide production from the heart rate instead of using a constant on the calculations.[58]

In CF patients, the [^{13}C]MTG breath test yields significantly lower values than controls (cPDR 3.1% vs. 31.0%).[59] Since both endogenous and exogenous lipase activity are measured, the test can clearly demonstrate the effect of enzyme supplementation and is thereby useful as an indicator for optimal dosage since persistent steatorrhea can be due to many other factors.[59,60] Even in young infants, the most difficult population to study, optimal dosage of pancretic enzyme therapy can be assessed.[61]

Hepatic Function

The [^{13}C]methacetin breath test has so far almost exclusively been used in adults with chronic liver disease. The test indicates microsomal hepatocellular function and thus functional liver mass. The [^{13}C]ketoisocaproate test, as an indicator of mitochondrial function, is mainly used as a functional test in alcoholic liver disease.

Substrate for Breath Test The following can be used: [^{13}C]methacetin or [^{13}C] ketoisocaproate.

Mechanism Methacetin undergoes rapid O-dealkylation by hepatic microsomal enzyme systems, and the resultant CO_2 is present in the expired air.[62] Ketoisocaproate is decarboxylated in the mitochondria.[63]

Sensitivity and Specificity In adults, the methacetin breath test, expressed as 60-min cPDR, discriminates hepatic functional capacity not only between controls and liver disease patients, but also between different categories of chronic liver disease patients. The methacetin breath test is correlated with liver function tests and serum bile acids and is highly predictive of Child–Pugh score.[64]

In adult nonalcoholic steatohepatitis (NASH) patients, a sensitivity of 68% and specificity of 94% are reported for the [^{13}C]ketoisocaproate breath test.[65]

Data are lacking in the pediatric population.

Pediatric Experience Iikura et al.[66] noted significantly elevated serum glutamic–oxaloacetic transaminase levels in children with atopic dermatitis and showed lower ^{13}CO peak excretion and delayed peak time after ingestion of [^{13}C]methacetin compared with controls. Their findings suggest some relationship between atopic dermatitis and liver function.

Surprisingly, breath tests have not been used yet to explore liver function in children. Obesity and NASH are certainly an appealing area. So far, a study in adult NASH patients showed enhanced methacetin demethylation but decreased and delayed ketoisocaproate decarboxylation.[65]

Gastrointestinal Motility

Gastric Emptying Gastric emptying is a developing function is young infants. Impaired gastric emptying contributes to feeding difficulties and dyspepsia.[67]

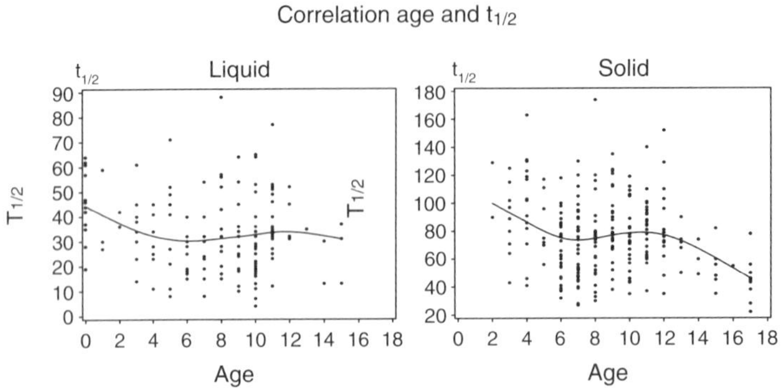

FIGURE 38.5 Correlation age and $T_{1/2}$.

Substrate for Breath Test The following can be used: [^{13}C]octanoic acid (50 microl = 45.5 mg) for solid and liquid test meals[18] or [^{13}C]acetate for liquid test meals (25–150 mg).[68]

Sampling Time Sampling is done every 15 min for 4 h.

Mechanism Gastric emptying is the rate-limiting step for appearance of $^{13}CO_2$ in the breath after duodenal absorption and oxidation of these fatty acids. Results are expressed by gastric half-emptying time ($T_{1/2}$) and gastric emptying coefficient (GEC).

Sensitivity and Specificity The [^{13}C]octanoic acid was validated against scintigraphy in adults[69] and compared to scintigraphy and dilution techniques in children.[18] Reproducibility of the [^{13}C]octanoic acid breath test in premature infants is good.[70] A large intraindividual variation was reported for both breath tests, using NDIRS.[13,14]

Pediatric Experience Pediatric reference values were established for the [^{13}C]octanoic acid breath test from birth to 17 years, with specific age-adapted test meals. These data demonstrate the maturation of gastric emptying (Figure 38.5). Gastric emptying accelerates with age, is faster in boys than girls of a similar age, and is more efficient for liquids than for solids.[18] The [^{13}C]octanoic acid breath test is used in premature infants with feeding problems[71,72] and has demonstrated faster emptying of breast milk[73] and of hydrolyzed formula.[74]

Orocecal Transit Time Orocecal transit time includes gastric emptying.

Substrate for Breath Test The following can be used:

Lactulose for hydrogen breath test (maximum 10 g) or lactose-[^{13}C]ureide ([^{13}C]LUR) 250 mg.

Priming with nonmarked LUR improves response.[75]

A study of the fate of [^{13}C]LUR suggests that glucose-[^{13}C]ureide could also be used as marker.[76]

Sampling Time Sampling is done every 15 min for 10 h.

Mechanism Lactulose, a nonabsorbable sugar, is fermented in the colon and leads to H_2 production. By stimulating colonic fermentation, lactulose has a laxative effect, which is paradoxical for a transit marker.

Lactose [^{13}C]ureide is hydrolyzed upon arrival in the cecum (probably by *Clostridium innocuum*) and then releases $^{13}CO_2$.[77]

Sensitivity and Specificity The lactulose breath test has very limited use for two reasons: 10% of the population are H_2 nonproducers and lactulose accelerates transit. In healthy children, OCTT averages 77 min with the lactulose hydrogen breath test and 244 min with the [^{13}C] LUR breath test.[18]

The stable isotope breath test was validated against scintigraphy in adults ($r = 0.94$).[78]

Pediatric Experience The [^{13}C]LUR breath test can be used in young infants after weaning.[79] Reference values for OCTT in healthy children between 3 and 17 years old are around 246 min. Children with constipation, functional gastrointestinal syptoms, and Crohn's disease were studied.[18]

The breath tests for gastric emptying and orocecal transit time can be performed simultaneously, generating a two-peaked curve. A solid test meal revealed delayed motility in children with functional complaints.[18]

Bacterial Metabolism

***Helicobacter pylori* Infection** This test undoubtedly enjoys the widest clinical application of all stable isotope breath tests.

Substrate for Breath Test This is [^{13}C]urea 50–100 mg in formula, water, or orange juice.

Mechanism As opposed to humans, *Helicobacter pylori* is a urease-producing organism.

Sampling Time Sampling is done every 15 min for 1 h; however, 30 min is sufficient.

Sensitivity and Specificity Sensitivity and specificity are very high: 95–100%.[80] cPDR after 1 hour is normally less than 1%. Values over 3.5% indicate HP infection. The use of NDIRS was validated against IRMS. NDIRS gives online results, is easier and cheaper, and thus contributes to the wide availability of the test.[11,81]

Pediatric Experience The HP breath test is used in clinical practice to prove infection and eradication after treatment.[82] Epidemiological studies have been performed in various pediatric populations.[83,84] Given the feasibility of the test, repeated testing can be performed to examine reinfection rates.[85]

Small Bowel Overgrowth (SBO) Although SBO is an important clinical problem, we still lack reliable, user-friendly methods for diagnosis.

Substrate for Breath Test Glucose is used for the H_2 test, while [^{13}C]xylose is used for $^{13}CO_2$ detection—50 mg.[86]

Mechanism In the case of SBO, nonmarked glucose is digested by bacteria and releases H_2. Xylose is a nonabsorbable sugar, fermented by intraluminal bacteria.

Sampling Time Sampling is done every 30 min for 4 h.

Sensitivity and Specificity A positive H_2 breath test is accepted as a fasting hydrogen over 15 ppm or a rise over 10 ppm over baseline during the test.[87] Dellert et al.[86] performed a validation study against culture of duodenal aspirates in a limited number of children using the [^{13}C]xylose breath test and report 100% sensitivity and 67% specificity.

Pediatric Experience The glucose test has been used to explain the origin of abdominal complaints[88] and has confirmed the suspected high incidence of SBO in CF patients.[87]

Colonic Bacterial Metabolism The [^{13}C]LUR breath test assesses OCTT based on hydrolysis of the marker by cecal bacteria. We found that children with Crohn's disease in remission did not yield a response and concluded that handling of [^{13}C]LUR by their colonic flora is different.[18] This led to the working hypothesis that the combined use of [^{13}C]LUR for breath analysis and (^{15}N(LUR for urinary analysis could shed some light on colonic flora metabolism.[89,90] Given the importance of colonic fermentation in the pathophysiology of chronic intestinal diseases, these tests will be developed further.

CONCLUSION

Breath tests offer a realm of possibilities for exploring digestion, metabolism, and physiology in children. Their advantages—noninvasiveness, specificity, and safety—are balanced by their burden for adherence to strict methodology by a skilled investigator, variability, and price. In some cases, breath analysis can be simplified and performed at lower cost. Test meals and analyzers can then be broadly commercialized, as is the case for HP diagnosis. Most stable isotope studies remain time consuming and require a specialized laboratory for analysis and calculations. Comparison to controls or reference values obtained in similar circumstances is crucial. Pediatric experience and literature are satisfactory on breath tests for carbohydrate and lipid digestion, gastrointestinal motility, and HP infection. Relevant pediatric clinical areas awaiting further expansion are hepatic dysfunction, such as in obesity and NASH, small bowel overgrowth, and colonic fermentation. Breath tests offer a good tool for collaborative projects and could be combined in certain disease entities: for example, H_2 lactulose test can be combined with [^{13}C]MTG test in cystic fibrosis to study transit and fat absorption simultaneously. Clinical endpoints in pediatric research can therefore be expanded as tools within the reach of creative scientists.

ACKNOWLEGMENT

The author is indebted to Professor Yvo Ghoos for a critical review and expert contributions to this chapter.

REFERENCES

1. Ghoos Y, Geypens B, Rutgeerts P. Stable isotopes and $^{13}CO_2$ breath tests for investigating gastrointestinal functions. *Food Nutr Bull.* 2002: 166–168.

2. Perman JA, Barr RG, Watkins JB. Sucrose malabsorption in children: noninvasive diagnosis by interval breath hydrogen determination. *J. Pediatr.* 1978;93:17–22.
3. Davidson G, Butler R. Breath analysis. In: Walker W, ed. *Pediatric Gastrointestinal Disease.* Hamilton, Ontario: BC Decker; 2000: 1529–1537.
4. Barr RG, Watkins JB, Perman JA, Schoeller DA. Mucosal function and breath hydrogen excretion: comparative studies in the clinical evaluation of children with nonspecific abdominal complaints. *Pediatrics.* 1981;68:526–533.
5. Perman JA, Modler S, Barr RG, et al. Fasting breath hydrogen concentration: normal values and clinical application. *Gastroenterology.* 1984;87:1358–1363.
6. Gilat T, Ben Hur H, Gelman-Malachi E, et al. Alterations of the colonic flora and their effect on the hydrogen breath test. *Gut.* 1978;19:602–605.
7. Strocchi A, Corazza G, Ellis CJ, Gasbarrini G, Levitt MD. Detection of malabsorption of low doses of carbohydrate: accuracy of various breath H_2 criteria. *Gastroenterology.* 1993;105:1404–1410.
8. Barth E, Tugtekin I, Weidenbach H, et al. Determination of $^{13}CO_2/^{12}CO_2$ ratio by IRMS and NDIRS. *Isotopes Environ Health Stud.* 1998;34:209–213.
9. Adamek RJ, Goetze O, Boedeker C, Pfaffenbach B, Luypaerts A, Geypens B. ^{13}C-methacetin breath test: isotope-selective nondispersive infrared spectrometry in comparison to isotope ratio mass spectrometry in volunteers and patients with liver cirrhosis. *Z Gastroenterologie.* 1999;37:1139–1143.
10. Boedeker C, Goetze O, Pfaffenbach B, Luypaerts A, Geypens B, Adamek RJ. ^{13}C mixed-triglyceride breath test: isotope selective non-dispersive infrared spectrometry in comparison with isotope ratio mass spectrometry in volunteers and patients with chronic pancreatitis. *Scand J Gastroenterol.* 1999;34:1153–1156.
11. Koletzko S, Haisch M, Seeboth I, et al. Isotope-selective non-dispersive infrared spectrometry for detection of *Helicobacter pylori* infection with ^{13}C-urea breath test. *Lancet.* 1995;345:961–962.
12. Germana B, Galliani E, Lecis P, Costan F. Diagnosis of *Helicobacter pylori* infections using isotope-selective non dispersive infrared spectrometry with ^{13}C-urea breath test. *Recenti Prog Med.* 2001;92:113–116.
13. Hauser B, De Schepper J, Caveliers V, Salvatore S, Salvatoni A, Vandenplas Y. Variability of the ^{13}C-acetate breath test for gastric emptying of liquids in healthy children. *J Pediatr Gastroenterol Nutr.* 2006;42:392–397.
14. Hauser B, De Schepper J, Caveliers V, Salvatore S, Salvatoni A, Vandenplas Y. Variability of the ^{13}C-octanoic acid breath test for gastric emptying of solids in healthy children. *Aliment Pharmacol Ther.* 2006;23:1315–1319.
15. Kasicka-Jonderko A, Kaminska M, Jonderko K, Setera O, Blonska-Fajfrowska B. Short- and medium-term reproducibility of gastric emptying of a solid meal determined by a low dose of ^{13}C-octanoic acid and nondispersive isotope-selective infrared spectrometry. *World J. of Gastroenterol.* 2006;12:1243–1248.
16. Schadewaldt P, Schommartz B, Wienrich G, Brosicke H, Piolot R, Ziegler D. Application of isotope-selective nondispersive infrared spectrometry (IRIS) for evaluation of [^{13}C]octanoic acid gastric-emptying breath tests: comparison with isotope ratio-mass spectrometry (IRMS). *Clin Chem.* 1997;43:518–522.
17. Morrison DJ, Dodson B, Slater C, Preston T. ^{13}C natural abundance in the British diet: implications for ^{13}C breath tests. *Rapid Commun Mass Spectrom.* 2000;14:1321–1324.
18. Van Den Driessche M. Study of gastro-intestinal motility in infants and children using ^{13}C breath tests. In: *Faculty of Medicine.* Leuven: Catholic University Leuven; 2001: 129.
19. Gudinchet F, Schutz Y, Micheli J, et al. Metabolic cost of growth in very low birth weight infants. *Pediatr Res.* 1982;16:1025–1030.

20. Altman P, Dittmer D. Metabolic rate of boys and girls according to weight in different age classes. In: *Biological Data Book*. 2nd ed. Bethesda, MD: FASEB. 1974;I:1527.
21. Schofield W. Predicting basal metabolic rate, new standards and review of previous work. *Hum Nutr Clin Nutr*. 1985;39:15–41.
22. Thomas JE, Dale A, Harding M, et al. Interpreting the ^{13}C-urea breath test among a large population of young children from a developing country. *Pediatr Rese*. 1999;46:147–151.
23. Koletzko B, Sauerwald T, Demmelmair H. Safety of stable isotope use. *Eur J Pediatr*. 1997;156 (Suppl 1):S12–S17.
24. Newcomer AD, McGill DB, Thomas PJ, Hofmann AF. Prospective comparison of indirect methods for detecting lactase deficiency. *Engl J Med*. 1975;293:1232–1236.
25. Davidson GP, Robb TA. Value of breath hydrogen analysis in management of diarrheal illness in childhood: comparison with duodenal biopsy. *J Pediatr Gastroenterol Nutr*. 1985;4:381–387.
26. Hiele M, Ghoos Y, Rutgeerts P, Vantrappen G, Carchon H, Eggermont E. $^{13}CO_2$ breath test using naturally ^{13}C-enriched lactose for detection of lactase deficiency in patients with gastrointestinal symptoms. *Lab Clin Med*. 1988;112:193–200.
27. Sibley E. Genetic variation and lactose intolerance: detection methods and clinical implications. *Am J Pharmacogenomics*. 2004;4:239–245.
28. Hiele M, Ghoos Y, Rutgeerts P, Vantrappen G. Measurement of the rate of assimilation of oligo- and polysaccharides by $^{13}CO_2$ breath tests and isotope ratio mass spectrometry. *Biomed Environ Mass Spectrom*. 1988;16:133–135.
29. Treem WR, McAdams L, Stanford L, Kastoff G, Justinich C, Hyams J. Sacrosidase therapy for congenital sucrase-isomaltase deficiency. *J Pediatr Gastroenterol Nutr*. 1999;28:137–142.
30. Lewindon PJ, Robb TA, Moore DJ, Davidson GP, Martin AJ. Bowel dysfunction in cystic fibrosis: importance of breath testing. *J Paediatr Child Health*. 1998;34:79–82.
31. Tooley KL, Saxon BR, Webster J, et al. A novel non-invasive biomarker for assessment of small intestinal mucositis in children with cancer undergoing chemotherapy. *Cancer Biol Ther*. 2006;5:1275–1281.
32. Hoekstra JH, van Kempen AA, Kneepkens CM. Apple juice malabsorption: fructose or sorbitol? *J Pediatr Gastroenterol Nutr*. 1993;16:39–42.
33. Hoekstra JH, van den Aker JH, Kneepkens CM, Stellaard F, Geypens B, Ghoos YF. Evaluation of $^{13}CO_2$ breath tests for the detection of fructose malabsorption. *J Lab Clin Med. Mar*. 1996;127 (3):303–309.
34. Lifschitz CH, Boutton TW, Carrazza F, et al. A carbon-13 breath test to characterize glucose absorption and utilization in children. *J Pediatr Gastroenterol Nutr*. 1988;7:842–847.
35. Douwes AC, van Caillie M, Fernandes J, Bijleveld CM, Desjeux JF. Interval breath hydrogen test in glucose-galactose malabsorption. *Eur J Pediatr*. 1981;137:273–276.
36. Hyams JS, Etienne NL, Leichtner AM, Theuer RC. Carbohydrate malabsorption following fruit juice ingestion in young children. *Pediatrics*. 1988;82:64–68.
37. Corazza GR, Strocchi A, Rossi R, Sirola D, Gasbarrini G. Sorbitol malabsorption in normal volunteers and in patients with coeliac disease. *Gut*. 1988;29:44–48.
38. Smith MM, Davis M, Chasalow FI, Lifshitz F. Carbohydrate absorption from fruit juice in young children. *Pediatrics*. 1995;95:340–344.
39. Christian MT, Amarri S, Franchini F, et al. Modeling ^{13}C breath curves to determine site and extent of starch digestion and fermentation in infants. *J Pediatr Gastroenterol Nutr*. 2002;34:158–164.
40. Dewit O, Prentice A, Coward WA, Weaver LT. Starch digestion in young children with cystic fibrosis measured using a ^{13}C breath test. *Pediatr Res*. 1992;32:45–49.

41. Amarri S, Harding M, Coward WA, Evans TJ, Weaver LT. ^{13}C and H_2 breath tests to study extent and site of starch digestion in children with cystic fibrosis. *J Pediatr Gastroenterol Nutr.* 1999;29:327–331.

42. Weaver LT, Dibba B, Sonko B, Bohane TD, Hoare S. Measurement of starch digestion of naturally ^{13}C-enriched weaning foods, before and after partial digestion with amylase-rich flour, using a ^{13}C breath test. *Br J Nutr.* 1995;74:531–537.

43. Hernell O. Assessing fat absorption. *J Pediatr.* 1999;135:407–409.

44. Borowitz D, Baker RD, Stallings V. Consensus report on nutrition for pediatric patients with cystic fibrosis. *J Pediatr Gastroenterol Nutr.* 2002;35:246–259.

45. Demmelmair H, Sauerwald T, Koletzko B, Richter T. New insights into lipid and fatty acid metabolism via stable isotopes. *Eur J Pediatr.* 1997;156 (Suppl 1):S70–S74.

46. Lembcke B, Braden B, Caspary WF. Exocrine pancreatic insufficiency: accuracy and clinical value of the uniformly labelled ^{13}C-hiolein breath test. *Gut.* 1996;39:668–674.

47. Watkins JB, Klein PD, Schoeller DA, Kirschner BS, Park R, Perman JA. Diagnosis and differentiation of fat malabsorption in children using ^{13}C-labeled lipids: trioctanoin, triolein, and palmitic acid breath tests. *Gastroenterology.* 1982;82:911–917.

48. McClean P, Harding M, Coward WA, Green MR, Weaver LT. Measurement of fat digestion in early life using a stable isotope breath test. *Arch Dis Child.* 1993;69:366–370.

49. McClean P, Harding M, Coward WA, Prentice A, Austin S, Weaver LT. Bile salt-stimulated lipase and digestion of non-breast milk fat. *J Pediatr Gastroenterol Nutr.* 1998;26:39–42.

50. Wutzke KD, Radke M, Breuel K, Gurk S, Lafrenz JD, Heine WE. Triglyceride oxidation in cystic fibrosis: a comparison between different ^{13}C-labeled tracer substances. *J Pediatr Gastroenterol Nutr.* 1999;29:148–154.

51. Braden B, Picard H, Caspary WF, Posselt HG, Lembcke B. Monitoring pancreatin supplementation in cystic fibrosis patients with the ^{13}C-hiolein breath test: evidence for normalized fat assimilation with high dose pancreatin therapy. *Z Gastroenterol. Feb.* 1997;35(2):123–129.

52. van Dijk-van Aalst K, Van Den Driessche M, van Der Schoor S, et al. ^{13}C mixed triglyceride breath test: a noninvasive method to assess lipase activity in children. *J Pediatr Gastroenterol Nutr.* 2001;32:579–585.

53. Ling SC, Amarri S, Slater C, Hollman AS, Preston T, Weaver LT. Liver disease does not affect lipolysis as measured with the ^{13}C-mixed triacylglycerol breath test in children with cystic fibrosis. *J Pediatr Gastroenterol Nutr.* 2000;30:368–372.

54. Vantrappen GR, Rutgeerts PJ, Ghoos YF, Hiele MI. Mixed triglyceride breath test: a noninvasive test of pancreatic lipase activity in the duodenum. *Gastroenterology.* 1989;96:1126–1134.

55. Weaver LT, Amarri S, Swart GR. ^{13}C mixed triglyceride breath test. *Gut.* 1998;43 (Suppl 3): S13–S19.

56. Slater C, Preston T, Weaver LT. Acetate correction for postabsorption metabolism does not improve the [^{13}C]mixed triacylglycerol breath test. *J Pediatr Gastroenterol Nutr.* 2006;43:666–672.

57. Slater C, Ling S, Preston T, Weaver L. Bulk and compound specific analysis of stool lipid confirm that the "missing" ^{13}C in the mixed triacylglycerol breath test is not in the stool. *Food Nutr Bull.* 2002;23:48–52.

58. Slater C, Preston T, Weaver LT. Improving the specificity of the [^{13}C]mixed triacylglycerol breath test by estimating carbon dioxide production from heart rate. *Eur J Clin Nutr.* 2006;60:1245–1252.

59. Amarri S, Harding M, Coward A, Evans TJ, Weaver LT. ^{13}Carbon mixed triglyceride breath test and pancreatic enzyme supplementation in cystic fibrosis. *Arch Dis Child. Apr.* 1997;76 (4):349–351.

60. De Boeck K, Delbeke I, Eggermont E, et al. Lipid digestion in cystic fibrosis: comparison of conventional and high-lipase enzyme therapy using the mixed-triglyceride breath test. *J Pediatr Gastroenterol Nutr*. 1998;408–411. 26(4):

61. Veereman-Wauters G, Behm M, Staelens S, Pelckmans P, Callens D, Mulberg A. ^{13}C mixed triglyceride breath test to assess pancreatic enzyme activity in infants and children with cystic fibrosis. *Pediatr Pulmonol*. 2006;29.

62. Matsumoto K, Suehiro M, Iio M, et al. [^{13}C]methacetin breath test for evaluation of liver damage. *Dig Dis Sci*. 1987;32:344–348.

63. Witschi A, Mossi S, Meyer B, Junker E, Lauterburg BH. Mitochondrial function reflected by the decarboxylation of [^{13}C]ketoisocaproate is impaired in alcoholics. *Alcohol Clin Exp Res*. 1994;18:951–955.

64. Festi D, Capodicasa S, Sandri L, et al. Measurement of hepatic functional mass by means of ^{13}C-methacetin and ^{13}C-phenylalanine breath tests in chronic liver disease: comparison with Child--Pugh score and serum bile acid levels. *World J Gastroenterol*. 2005;11:142–148.

65. Portincasa P, Grattagliano I, Lauterburg BH, Palmieri VO, Palasciano G, Stellaard F. Liver breath tests non-invasively predict higher stages of non-alcoholic steatohepatitis. *Clin Sci (Lond.)*. 2006;111:135–143.

66. Iikura Y, Iwasaki A, Tsubaki T, et al. Study of liver function in infants with atopic dermatitis using the ^{13}C-methacetin breath test. *Int Arch Allergy Immunol*. 1995;107:189–193.

67. Veereman-Wauters G. Neonatal gut development and postnatal adaptation. *Eur J Pediatr*. 1996;155: 627–632.

68. Gatti C, di Abriola FF, Dall'Oglio L, Villa M, Franchini F, Amarri S. Is the ^{13}C-acetate breath test a valid procedure to analyse gastric emptying in children? *J Pediatr Surg*. 2000;35:62–65.

69. Ghoos YF, Maes BD, Geypens BJ, et al. Measurement of gastric emptying rate of solids by means of a carbon-labeled octanoic acid breath test. *Gastroenterology*. 1993;104:1640–1647.

70. Barnett C, Snel A, Omari T, Davidson G, Haslam R, Butler R. Reproducibility of the ^{13}C-octanoic acid breath test for assessment of gastric emptying in healthy preterm infants. *J Pediatr Gastroenterol Nutr*. 1999;29:26–30.

71. Pozler O, Neumann D, Vorisek V, Bukac J, Bures J, Kokstein Z. Development of gastric emptying in premature infants. Use of the (13)C-octanoic acid breath test. *Nutrition*. 2003;19:593–596.

72. Veereman-Wauters G, Ghoos Y, van der Schoor S, et al. The ^{13}C-octanoic acid breath test: a noninvasive technique to assess gastric emptying in preterm infants. *J Pediatr Gastroenterol Nutr*. Aug. 1996;23(2):111–117.

73. Van Den Driessche M, Marien P, Ghoos Y, Devlieger H, Veereman-Wauters G. Gastric emptying in formula-fed and breast-fed infants measured with the ^{13}C-octanoic acid breath test. *J Pediatr Gastroenterol Nutr*. 1999;29:46–51.

74. Veereman-Wauters G, Staelens S, Van Den Driessche M, et al. Gastric emptying in newborns fed an intact protein formula, a partially and an extensively hydrolysed formula. *J Pediatr Gastroenterol Nutr*. 2004;39:84.

75. Wutzke KD, Schutt M. The duration of enzyme induction in orocaecal transit time measurements. *Eur J Clin Nutr*. 2007;61:1162–1166.

76. Morrison DJ, Dodson B, Preston T, Weaver LT. Gastrointestinal handling of glycosyl [^{13}C] ureides. *Eur J Clin Nutr*. 2003;57:1017–1024.

77. Heine WE, Berthold HK, Klein PD. A novel stable isotope breath test: ^{13}C-labeled glycosyl ureides used as noninvasive markers of intestinal transit time. *Am J Gastroenterol*. 1995;90:93–98.

78. Geypens B, Bennink R, Peeters M, et al. Validation of the lactose-[^{13}C]ureide breath test for determination of orocecal transit time by scintigraphy. *J Nucl Med*. 1999;40:1451–1455.

79. Van Den Driessche M, Van Malderen N, Geypens B, Ghoos Y, Veereman-Wauters G. Lactose-[^{13}C]ureide breath test: a new, noninvasive technique to determine orocecal transit time in children. *J Pediatr Gastroenterol Nutr.* 2000;31:433–438.

80. Peeters M, Ghoos Y, Geypens B, Rutgeerts P. Breath tests in *Helicobacter pylori* infection: methodological aspects. *J Physiol Pharmacol.* 1997;48 (Suppl 4):67–73.

81. Savarino V, Vigneri S, Celle G. The ^{13}C urea breath test in the diagnosis of *Helicobacter pylori* infection. *Gut.* 1999;45 (Suppl 1):I18–I22.

82. Kalach N, Briet F, Raymond J, et al. The ^{13}C urea test for the noninvasive detection of *Helicobacter pylori* in children: comparison with culture and determination of minimum analysis requirements. *J Pediatr Gastroenterol Nutr. Mar.* 1998;26(3):291–296.

83. Mohammad MA, Hussein L, Coward A, Jackson SJ. Prevalence of *Helicobacter pylori* infection among Egyptian children: impact of social background and effect on growth. *Public Health Nutr.* 2007;1–7.

84. Braga AB, Fialho AM, Rodrigues MN, Queiroz DM, Rocha AM, Braga LL. *Helicobacter pylori* colonization among children up to 6 years: results of a community-based study from northeastern Brazil. *J Trop Pediatr.* 2007;119:754–759.

85. Feydt-Schmidt A, Kindermann A, Konstantopoulos N, et al. Reinfection rate in children after successful *Helicobacter pylori* eradication. *Eur J Gastroenterol Hepatol.* 2002;14:1119–1123.

86. Dellert SF, Nowicki MJ, Farrell MK, Delente J, Heubi JE, The ^{13}C-xylose breath test for the diagnosis of small bowel bacterial overgrowth in children. *J Pediatr Gastroenterol Nutr.* 1997;25:153–158.

87. Fridge JL, Conrad C, Gerson L, Castillo RO, Cox K. Risk factors for small bowel bacterial overgrowth in cystic fibrosis. *J Pediatr Gastroenterol Nutr.* 2007;44:212–218.

88. de Boissieu D, Chaussain M, Badoual J, Raymond J, Dupont C. Small-bowel bacterial overgrowth in children with chronic diarrhea, abdominal pain, or both. *J Pediatr.* 1996;128:203–207.

89. Staelens S, Veereman G, De Preter V, Ghoos Y, Verbeke K. Could ^{13}C and ^{15}N lactose ureide be useful biomarkers to characterize the metabolism of intestinal bacteria in children? *Gastroenterology A.* 2005;169.

90. Staelens S, Veereman-Wauters G, De Preter V, Alliet P, Verbeke K. The effect of pre-, pro- and synbiotics on the intestinal flora of children with Crohn's disease. *J Pediatr Gastroenterol Nutr.* 2006;S53.

CHAPTER 39

Surrogate Endpoints: Application in Pediatric Clinical Trials

GEERT MOLENBERGHS, PhD

International Institute for Biostatistics and Statistical Bioinformatics, Universiteit Hasselt, Agoralaan 1, B-3590 Diepenbeek, Belgium

CAMILLE ORMAN, PhD

Johnson & Johnson Pharmaceutical Research and Development, 1125 Trenton-Harbourtown Road, Titusville, NJ 08560

INTRODUCTION

As part of the Critical Path Initiative,[1] the U.S. Food and Drug Administration (FDA) called for targeted research in six areas to stimulate the drug and device pipeline.[2] Highlighted was the development of biomarkers to more rapidly and/or more efficiently determine the benefit–risk profile of a new therapy. Potential uses of biomarkers that were listed included genomic tests to identify patients at high risk for serious toxicity, markers of drug metabolism to individualize drug dosage, and new imaging techniques to assess treatment efficacy. In addition, the qualification of new surrogate endpoints, a subset of biomarkers targeted at later-phase clinical trials, was identified as an important area to drive more rapid drug development. As recognized by the FDA, the use of surrogate endpoints holds great promise for improving the efficiency in clinical research. Given the pressing need for new pediatric therapies, incorporating surrogate endpoints could be a significant aid to accelerate development. However, the use of surrogate endpoints has been controversial; with the unique aspects of pediatric research, it is vital that surrogate endpoints be used appropriately, and potentially more frequently, in this population. In this chapter, we examine the use of surrogate endpoints in clinical research in general and the role that they can play in pediatric research in particular.

DEFINITION OF SURROGATE ENDPOINT

Although surrogate endpoints have been present in the scientific debate for over two decades, varying definitions have been used, the earliest one going back to Prentice.[3]

Pediatric Drug Development: Concepts and Applications
Edited by Andrew E. Mulberg, Steven A. Silber, and John N. van den Anker

In 2001, a National Institutes of Health (NIH) working group recommended the following terms and definitions.[4]

Biological Marker (Biomarker): A characteristic that is objectively measured and evaluated as an indicator of normal biologic process, pathogenic process, or pharmacologic responses to a therapeutic intervention.

Clinical Endpoint: A characteristic or variable that reflects how a patient feels or functions, or how long a patient survives.

Surrogate Endpoints: A biomarker intended to substitute for a clinical endpoint. A clinical investigator uses epidemiologic, therapeutic, pathophysiologic, or other scientific evidence to select a surrogate endpoint that is expected to predict clinical benefit, harm, or lack of benefit or harm.

A surrogate endpoint, therefore, does not directly measure clinical impact, but rather reflects the desired therapeutic treatment effect. For example, tumor shrinkage is an obvious candidate surrogate endpoint for long-term survival from carcinoma.

HISTORY OF SURROGATE ENDPOINTS

Assessment of the benefit and risk associated with a therapeutic intervention is the underpinning of a clinical development program. The traditional "gold standard" in assessing efficacy is the evaluation of treatment effect on a well-defined clinical endpoint. The choice of endpoint is a critical factor in determining the duration and complexity of the trial. However, often, the most sensitive and relevant clinical endpoint, the "true" endpoint, can pose severe challenges for evaluation. For example, the use of clinical endpoints such as survival in newly diagnosed patients with breast cancer, or short-term mortality in patients following acute myocardial infarction (a relatively infrequent event) would result in large, long, and expensive trials. An effective strategy in these situations is to identify alternative endpoints, or surrogates, that are less costly to measure, are more conveniently assessed, or occur earlier or more frequently than the true clinical endpoint.

In the 1980s, with the alarming rise in HIV infections and AIDS-related deaths, the approval process of new therapies relying on traditional clinical outcomes was questioned. The scientific and regulatory "gold standard" for a clinical trial endpoint was one where distinct clinical impact could be shown. In the case of demonstrating a treatment effect on HIV infections, this required assessing the clinical outcomes of either prevention of progression to AIDS or increased overall survival. However, the time and cost associated with trials using these outcome measures were unacceptable given the epidemic of HIV infection. In response to the demand for more rapid approval, the accelerated approval provisions were added to the U.S. new drug regulations[5] with a guidance following in 1998[6] (subsequently updated in 2004). This new regulation allowed for accelerated approval of drugs targeting serious or life-threatening diseases if the drug appeared to show a benefit over current therapy. In addition, the approval could be provisionally based on surrogate endpoints that were reasonably likely to predict clinical benefit (Subpart H). Under this regulation, sponsors must demonstrate long-term clinical benefit on the ultimate outcome measure if the association between the true clinical endpoint and the surrogate endpoint has not been demonstrated. Both Europe and Japan have similar regulatory provisions for accelerated approval based on surrogate endpoints. In 1998, the International Conference on Harmonisation (ICH) provided guidance on the use of surrogate endpoints that is also comparable.[7]

Regulatory authorization allowing approval based on surrogate endpoints coupled with the need to reduce the ever-rising costs of developing new therapies led to an increased use of surrogate endpoints as the regulatory basis for approval.[8] Examples include $CD4^+$ T-lymphocyte counts (CD4) in HIV-infected subjects rather than progression to AIDS or death, cholesterol levels in lieu of occurrence of myocardial infarction, and tumor response rates instead of reoccurrence.

CONTROVERSY WITH SURROGATE ENDPOINTS

In spite of the potential advantages, the use of surrogate endpoints has been controversial due to several dramatic failures of a surrogate endpoint to adequately substitute for a clinical endpoint.[9] During the early 1990s, the FDA approved the drugs encainide and flecainide after effective suppression of ventricular arrhythmias was demonstrated. It was believed that since arrhythmia is associated with a fourfold increase in sudden death following myocardial infarction, these drugs would reduce the death rate after myocardial infarction. Following approval, however, results from the Cardiac Arrhythmia Suppression Trial (CAST)[10] showed that the death rate among subjects treated with encainide and flecainide was more than twice that observed in the placebo subjects. In addition, relying on surrogate endpoints in smaller clinical trials has raised concerns about detecting safety issues that can only be detected in large randomized trials.[11]

In contrast, reliance on a surrogate endpoint may fail to show the true effect of a new therapy. For example, the evaluation of interferon gamma on recurrent infections in patients with chronic granulomatous disease, a surrogate endpoint for phagocytic function, did not show a therapeutic effect; but clinical benefit was demonstrated by substantial reduction in serious infections.[4,12]

These failures in the use of surrogate endpoints for evaluating therapeutic effect were fundamentally due to the misconception that an association between a true clinical endpoint and an observed biomarker is sufficient to declare a biomarker as a surrogate. The mere existence of an association between a candidate surrogate endpoint and the true endpoint is not sufficient for using the former as a surrogate: "a correlate does not a surrogate make."[9] Although an association between a potential and a true clinical endpoint is desirable, what is required is that the effect of the treatment on the surrogate endpoint reliably predict the effect on the true endpoint. Unfortunately, partly owing to the lack of appropriate methodology, this condition was not met in the early attempts to use surrogate endpoints and, consequently, negative opinions about the use of surrogates in the evaluation of treatment efficacy have been voiced.[9,13,14]

WHY CONTINUE WITH SURROGATE ENDPOINTS?

In spite of the failures, there are compelling reasons to continue the evaluation and use of surrogate endpoints. Technological advances have dramatically increased the number of new biomarkers fueling the pool of potential, new surrogate endpoints. Improved understanding of proposed therapies' mechanism of action at the molecular level facilitates the use of relevant biomarkers in the evaluation of benefit and risk.[15] Continued public pressure for fast approval of promising new drugs, particularly for serious illnesses where the effect on the true clinical endpoint is distant, encourages the use of surrogate endpoints that could reduce the time and cost of the required trials.[16]

Surrogate endpoints also can be used for early detection of safety issues that could point to toxicity problems of a new therapy. The duration and size of clinical trials designed to evaluate efficacy of a new drug are often insufficient to detect rare adverse events or events that occur after prolonged therapy.[17,18] The use of surrogate endpoints in this context of toxicity-related clinical endpoints might allow one to obtain information about such effects even during the clinical testing phase.

Furthermore, new discoveries in medicine and biology are creating an exciting range of possibilities for the development of many potentially effective treatments for a particular disease. This unquestionably is an achievement, but in turn it creates a challenge to rapidly evaluate a large number of new, promising treatments. Surrogate endpoints in the development program can offer an efficient route.

Finally, shortening the duration of a clinical trial using a surrogate endpoint not only can decrease the cost of the evaluation process, but also can limit potential problems with noncompliance and missing data, thereby increasing research effectiveness and reliability.[19,20] Benefits to the subject are also obvious in terms of reduced time and the number of potential studies related to the clinical trial burden.

The potential of surrogate endpoints to accelerate and improve the quality of clinical trials is clear. However, early experiences also demonstrate that only thoroughly evaluated surrogate endpoints should be used. The following section discusses this issue further.

STATISTICAL EVALUATION OF SURROGATE ENDPOINTS

Thus, while some of the past failures have led a number of researchers to the conclusion that surrogate endpoints should be avoided altogether, practice has clearly shown that sometimes surrogate endpoints are the only reasonable and plausible alternative to evaluate a new drug. Nevertheless, past attempts to use surrogate endpoints have made it clear that, before deciding on the use of a candidate surrogate endpoint, it is of the utmost importance to evaluate its validity. Developing a definition of a valid surrogate endpoint and operational criteria to assess a proposed surrogate were needed. Statistical methods to evaluate proposed surrogate endpoints have become the subject of intensive research since the 1980s. Note that, as in most clinical decisions, statistical arguments will play a major role but must be considered in conjunction with clinical and biological evidence.

The first formal statistical framework for the evaluation of potential surrogate endpoints dates back to 1989 when Prentice[3] proposed a formal definition of surrogate endpoints and outlined a set of evaluation criteria, all within a hypothesis testing paradigm. A perfect surrogate endpoint, as described by Prentice, can be represented as

$$X \Rightarrow S \Rightarrow T$$

where X is the treatment, S is the surrogate endpoint, and T is the true clinical outpoint. In this paradigm, the surrogate endpoint mediates all of the effect of the treatment on the true clinical endpoint. Prentice proposed four operational criteria to validate a proposed surrogate endpoint:

1. Treatment (X) has a significant effect on the surrogate endpoint (S).
2. Treatment (X) has a significant impact on the true endpoint (T).
3. Surrogate endpoint (S) has a significant impact on the true endpoint (T).
4. Full effect of treatment (X) on the true endpoint (T) is captured by the surrogate (S).

Although intuitively appealing, much debate ensued, for the criteria set out by Prentice are not straightforward to verify.[21,22] The fourth criterion is particularly challenging, as it requires that the surrogate must explain 100% of the treatment effect on the true clinical endpoint. In addition, Prentice's criteria could only be applied to binary endpoints (e.g., success vs. failure).[19,23]

Freedman et al.[22] supplemented Prentice's approach by introducing the term *proportion of treatment explained* (PE), aimed at measuring the proportion of the treatment effect mediated by the surrogate. This proposal was important, as it shifted the interest in the validation of surrogate endpoints from significance testing to estimation of the treatment effect explained by the surrogate. However, properties of the PE made it difficult to reliably estimate;[19,23] for example, the denominator of the proportion explained (the effect of treatment on the true clinical endpoint) usually cannot be estimated with precision.[24] Moreover, and fundamentally, the PE is flawed in the sense that it is not restricted to the unit interval. Attempting to further refine this approach, Buyse and Molenberghs[23] and Molenberghs et al.[25] showed that the PE can be decomposed into three different quantities: the ratio of the surrogate and true endpoint variances, the relative effect, and the adjusted association. This approach reflects the two dimensions of the problem of validating a surrogate endpoint. The first dimension is the capability of the surrogate to predict the treatment effect on the true clinical endpoint, while the second one is the capability to predict the outcome of the true clinical endpoint.[23,26]

The earlier proposals for a statistical framework to validate a surrogate endpoint have been based on utilizing data from a single trial. However, combining data from multiple studies, or meta-analyses, can lead to a more accurate assessment of a surrogate. Similar to the evaluation in single studies, meta-analyses examine the association between treatment effects on the surrogate endpoint and the true clinical endpoint. Based on the results of the association, the model assesses the reliability for predicting the treatment effect on the true clinical endpoint leading to an observed effect on the proposed surrogate. Daniels and Hughes,[28] Buyse and colleagues,[27] and Verbeke and Molenberghs[20] focusing on continuous response endpoints, employed linear mixed-effects models to predict the treatment effects on the true clinical endpoint based on data from the surrogate. In this approach, the quality of a surrogate is quantified using two coefficients of determination: R^2_{trial} and R^2_{indiv}. Both measures are unitless and can range from 0 to 1. Calculated from earlier trial results, R^2_{trial} measures how precisely the effect of treatment on the true clinical endpoint can be predicted based on the treatment effect on the surrogate endpoint in a new trial. If $R^2_{\text{trial}} = 1$, then the treatment effect on the true clinical endpoint can be predicted without error based on the treatment effect on the surrogate, whereas if $R^2_{\text{trial}} = 0$, then the treatment effect on the true clinical endpoint and the proposed surrogate are independent and therefore no meaningful prediction can be made. R^2_{indiv} has a very similar interpretation but it quantifies at the individual patient level how precisely the outcome on the true endpoint can be predicted using the outcome on the surrogate.

This meta-analytic method fully captures both dimensions of validation of a proposed surrogate. Nevertheless, a question that immediately arises in this setting is which of these two dimensions is the most important one in practice. There is no single answer to this question; it will depend on the context. For a trialist who wants to use the surrogate to predict the treatment effect on the true endpoint, the trial dimension will clearly be the most interesting one. However, for a treating physician who has observed a tumor response in a specific patient and wants to know how this can predict the survival of the patient, the individual dimension will be most useful.

Many extensions of this meta-analytic approach to surrogate validation have been developed to encompass a range of endpoints including binary, time-to-event, or repeated measures.[19] Each of these extensions has led to different ways of quantifying a proposed surrogate, which in turn could potentially lead to varying interpretations. Alonso and Molenberghs[29] have proposed a unifying framework for the validation of surrogate endpoints using information theory. This method applies to a wide variety of endpoints and reduces to the quantities previously introduced in the literature, providing a unified theoretical basis for the variety of statistical approaches to validate a surrogate endpoint. Recently, a new approach has been introduced based on causal inference concepts.[30,31] Although this approach is based on the single-trial setting and some strong assumptions are required, it appears to be a promising line of research, especially to evaluate potential surrogates in the initial stages of development when little information about the surrogate is available.

With the advent of new biomarkers and the increase in understanding of disease mechanism, there will be a continuing need for developing new statistical models for testing the validity of new proposed surrogates. To best assist in the evaluation of a biomarker as a potential surrogate endpoint, a statistical framework must be established during all stages of therapeutic development including the exploratory phases of a new compound.[32]

SURROGATE ENDPOINTS IN PEDIATRIC TRIALS

The recognition of the importance of conducting clinical trials in the pediatric population began in 1997 with the Food and Drug Administration Modernization Act (FDAMA) that allowed for market exclusivity based on pediatric clinical trials.[33] In 1999, the ICH issued a draft consensus guideline, *Clinical Investigation of Medicinal Products in the Pediatric Population* (ICH E11), that encouraged drug development in children while recognizing the unique challenges associated with the pediatric population.[34] Replacing the long-held belief that using children in clinical trials was unethical was the recognition that only through empirical evaluation in a clinical trial could the risk and benefit of a new therapy be accurately assessed.[35] The Best Pharmaceuticals for Children Act (BPCA), signed in 2002, reauthorized the FDAMA exclusivity provision to further encourage pediatric trials.[36] In 2003, the Pediatric Research Equity Act (PREA) required a review of all new active ingredients, dosage forms, routes of administration, indications, and dosing regimens for assessment of safety and efficacy in a pediatric population.[37,38] Through both requirements and incentives, regulatory authorities have attempted to encourage the study of new therapies in pediatrics, resulting in an understanding of the benefit–risk profiles that is equivalent to that required for adults. For further details regarding the current regulatory framework affecting trials in children and the incentives and requirements, please see Chapter 13 by Rose, Chapter 12 by Maldonado, and Chapter 14 by Nakamura and Ono for discussion of EU, U.S. and Japanese considerations.

Although ICH E11 and FDA regulations accept extrapolation of efficacy data from adults to children or from older to younger children, this is only suitable if the disease process and the outcome of therapy are comparable. However, as noted by the FDA, "children's bodies are not just small versions of adult bodies," so extrapolation is often inappropriate and clinical trials assessing both efficacy and safety are required.[39] Selecting clinical endpoints appropriate for a chronological or developmental age are critical in the design of a pediatric clinical trial. For example, specific assessment tools may be necessary to evaluate pain in infants and children.[39]

Similar arguments for using surrogate endpoints in adult studies can be applied to pediatric clinical trials. The FDA guidance on pediatric oncology studies states that "in the absence of available therapies to treat refractory stages of pediatric cancers, the FDA expects to use flexible regulatory approaches in approving drugs for pediatric research," including the use of surrogate endpoints such as the effect on tumor size in place of survival.[40] In addition to reducing the time to evaluation and the number of subjects, surrogate endpoints may be of value in pediatric research where patient-reported outcomes traditionally measured in adult studies are impossible to collect. Parents' resistance to invasive techniques to measure clinical outcome can drive the need for a surrogate endpoint that is more easily tolerated. Dosing may be better determined by sensitive surrogate measures.

Several statins (e.g., lovastatin, atorvastatin, simvastatin, and pravastatin) have been approved for familial hypercholesterolemia in pediatric patients. Similar to the approval process for these medications in adults, efficacy was based on the surrogate endpoint of lowering LDL cholesterol. Although pediatric patients with this diagnosis are at greater risk for coronary heart disease (CHD), no studies have examined the long-term safety of statin therapy or decrease in CHD morbidity and mortality with chronic exposure.

Vaccines are difficult to assess based on true clinical outcome due to the long duration of observation and, for some indications, the rarity of infection. Vaccines can be approved using a responder analysis demonstrating an immune response, for example, seroconversion in those subjects initially seronegative or the maintenance of an increase above prevaccination concentrations in subjects who were initially seropositive. The FDA will generally require Phase IV commitments to study adverse effects and long-term monitoring is required to ensure adequate protection. The debate over the association of autism and childhood immunizations highlights the difficulty in fully assessing the benefit–risk profile in pediatrics using a short-term surrogate endpoint. Even if true developmental safety issues exist, they can take years to assess.

As directed in the 2002 guidance for accelerated approval of antiretroviral drugs,[41] the five pediatric exclusivity approvals and one PREA approval to treat HIV infections have been based on the surrogate endpoints of HIV-1 RNA levels <400 copies/mL and increases in CD4 counts. Substantial scientific work has focused on HIV-1 RNA levels and CD4 counts as surrogate endpoints, including in-depth meta-analyses aimed at validating these measures as useful surrogates.[42] Such meta-analyses have focused on adult patients and it is not known whether these surrogate endpoints will prove reliable predictors for long-term outcome in the pediatric population treated early in life.

As new medications are developed with different proposed mechanisms of action, new surrogate endpoints must be identified and validated in the pediatric population. Gastroesophageal reflux disease (GERD) in infants can cause marked clinical problems including poor weight gain, failure to thrive, esophagitis, persistent irritability, pain, and feeding problems. Aspiration of refluxate in premature infants can lead to pneumonia, which can be life-threatening and lead to chronic respiratory problems. Although a few available medications are approved to treat GERD in the pediatric population, approvals have been based on bridging studies from adult clinical trials using the surrogate endpoint of control of gastric pH levels. However, for other classes of medications that are believed to alter gastrointestinal motility, there is little scientific basis to use gastric pH as an endpoint. Although two recent methods to detect reflux (intraluminal impedance and the ^{13}C-substrate breath testing) are promising as surrogate endpoints, neither has been validated.[43]

Numerous biomarkers have been associated with asthma, including measures of lung function (e.g., peak expiratory flow rate and FEV_1), exhaled nitrous oxide levels, and sputum eosinophils with FDA approvals based on FEV_1.[44] Although inhaled corticosteroids are the most effective current treatment for asthma, there are concerns about potential long-term effect on growth and final adult height.[45] However, conducting a study of sufficient duration to assess final adult height is logistically difficult. Various surrogate endpoints have been used to assess the impact corticosteroids may have on growth, including knemometry (to measure short-term changes in lower-leg length), growth velocity, and changes in height percentile.[45] Confounded with asthma-related height effects, measurement error and cyclical changes in height have made these short-term surrogate endpoints problematic as predictors of final height.

Although statistical methodology has been applied to qualifying surrogate endpoints in adult indications, little work has been done in the pediatric population. This may reflect the general lack of pediatric clinical trials and availability of data to perform the required analyses, but it is clearly an area in need of further investigation to adequately assess the value of proposed surrogate endpoints.

FUTURE AREAS FOR SURROGATE ENDPOINTS IN PEDIATRIC RESEARCH

There are several areas where surrogate endpoints might be particularly beneficial for developing new pediatric therapies. Surrogate measures of inflammation collected by noninvasive techniques such as biomarkers in induced sputum (e.g., total and differential cell counts, cytokines and neutrophil products) may prove useful to assess new cystic fibrosis therapies.[46] Biomarkers of the underlying inflammatory process of Crohn's disease (e.g., C-reactive protein, nitric oxide levels) may prove useful in identifying both subpopulations of patients as well as surrogate endpoints to evaluate treatment response.[47,48] Brain-imaging technology, allowing insight into the possible biological basis for psychiatric disease, may provide objective surrogate measures of drug response, as has recently been demonstrated in children with attention deficit hyperactivity disorder.[49] Developing surrogate measures of pain in babies in neonate intensive care units is needed to better evaluate the effectiveness and risks associated with analgesia, particularly in ventilated preterm infants.[50]

CONCLUSION

In this chapter, we have described the history of surrogate endpoints, the early misconceptions in assessing a proposed surrogate, and the development of statistical methodologies to accurately evaluate, in a quantitative fashion, surrogate endpoints. Surrogate endpoints in pediatric trials offer potential benefits in reducing the time to evaluation of a new therapy, as well as the number of exposed patients, thereby allowing for less invasive measurements, and collecting developmentally appropriate measures.

With the hoped for increase in pediatric clinical trials, a need for surrogate endpoints will likely increase as well. We must proceed cautiously to ensure that the selection of surrogates is based on sound scientific rationale. Noted failures of surrogate endpoints to adequately assess treatment effect in adults must not be repeated in pediatric clinical trials. The concern over inadequately assessing the benefit–risk ratio with a surrogate endpoint is of even greater concern in the pediatric population. Close collaboration between basic scientists, clinicians, and statisticians can facilitate the appropriate use of surrogate endpoints. Beginning early in the development cycle of a new drug, emphasis should be placed on identifying not only potential biomarkers associated with the proposed mechanism of action, but biomarkers that reflect developmental and chronological differences within the target patient population. Biomedical research will inevitably lead to new biomarkers that must be evaluated, and statistical models must be developed to assess new disease mechanisms and the association of surrogate endpoints. Trial designs should facilitate the evaluation of proposed surrogates, including standardizing collection methods and frequency of biomarker measurements to allow for across-study analyses to be performed. Careful and sound scientific methods will ensure that surrogate endpoints can effectively be used to assess new therapies in children.

REFERENCES

1. Food and Drug Administration. *Challenge and Opportunity on the Critical Path to New Medical Products.* Rockville, US Department of Health and Human Services, Food and Drug Administration; 2004.
2. Food and Drug Administration. *Critical Path Opportunities List.* Rockville, MD: US Department of Health and Human Services, Food and Drug Administration; 2006.
3. Prentice R. Surrogate endpoints in clinical trials: definition and operational criteria. *Stat Med.* 1989;8:431–440.
4. Biomarkers Definitions Working Group. Biomarkers and surrogate endpoints: preferred definitions and conceptual framework. *Clin Pharmacol Ther.* 2001;69:89–95.
5. Food and Drug Administration. Code of Federal Regulations 21 314.50 and 601.40. Rockville, MD: US Department of Health and Human Services, Food and Drug Administration.
6. Food and Drug Administration. *Guidance for Industry: Fast Track Drug Development Programs—Designation, Development, and Application Review.* Rockville, MD: US Department of Health and Human Services, Food and Drug Administration; 2004.
7. International Conference on Harmonisation. *General Considerations for Clinical Trials E8.* 1997.
8. Ellenberg SS, Hamilton JM. Surrogate endpoints in clinical trials: cancer. *Stat Med.* 1989; 8: 405–413.
9. Fleming TR, DeMets DL. Surrogate endpoints in clinical trials: Are we being misled? *Ann Intern Med.* 1996;125:605–613.
10. Echt BS, Liebson PR, Mitchell LB. Mortality and morbidity in patients receiving encainide, flecainide, or placebo. The cardiac arrhythmia suppression trial. *N Engl J Med.* 1991;324: 781–788.
11. Temple R. Are surrogate markers adequate to assess cardiovascular disease drugs? *JAMA.* 1999;282:790–795.

12. The International Chronic Granulomatous Diseases Cooperative Study Group. A controlled trial of interferon gamma to prevent infection in chronic granulomatous disease. *N Engl J Med.* 1991;324:509–516.
13. De Gruttola VG, Fleming TR, Lin DY, Coombs R. Perspectives: validating surrogate markers—Are we being naïve? *J Infect Dis.* 1997;175:237–246.
14. Fleming TR. Surrogate markers in AIDS and cancer trials. *Stat Med.* 1994;13:1423–1435.
15. Ferentz AE. Integrating pharmacogenomics into drug development. *Pharmacogenomics.* 2002; 3:453–467.
16. Lesko LJ, Atkinson AJ. Use of biomarkers and surrogate endpoints in drug development and regulatory decision making criteria, validation, strategies. *Ann Rev Pharmacol Toxicol.* 2000;41:347–366.
17. Dunn N, Mann RD. Prescription-event and other forms of epidemiological monitoring of side-effects in the UK. *Clin Exp Allergy.* 1999;29 (Suppl):217–239.
18. Jones TC. Call for a new approach to the process of clinical trials and drug registration. *BMJ.* 2001;322:920–923.
19. Burzykowski T, Molenberghs G, Buyse M, eds. *The Evaluation of Surrogate Endpoints.* New York: Springer; 2005.
20. Verbeke G, Molenberghs G. *Linear Mixed Models for Longitudinal Data.* New York: Springer-Verlag; 2000.
21. Fleming TR. Surrogate markers in AIDS and cancer trials. *Stat Med.* 1994;13:1423–1435.
22. Freedman LS, Graubard BI, Schatzkin A. Statistical validation of intermediate endpoints for chronic diseases. *Stat Med.* 1992;11:167–178.
23. Buyse M, Molenberghs G. Criteria for the validation of surrogate endpoints in randomized experiments. *Biometrics* 1998;54:1014–1029.
24. Molenberghs G, Buyse M, Burzykowski T. The history of surrogate endpoint validation. In: Burzykowski T, Molenberghs G, Buyse M, eds. *The Evaluation of Surrogate Endpoints.* New York: Springer; 2000: 67–82.
25. Molenberghs G, Buyse M, Geys H, et al. Statistical challenges in the evaluation of surrogate endpoints in randomized trials. *Controlled Clin Trials.* 2002;23:607–625.
26. Burzykowski T, Buyse M. Surrogate threshold effect: an alternative measure for meta-analytic surrogate endpoint validation. *Pharm Stat.* 2006;5:173–186.
27. Buyse M, Molenberghs G, Murzykowski T, et al. The validation of surrogate endpoints in meta-analysis of randomized experiments. *Biostatistics.* 2000;1:49–67.
28. Daniels MJ, Hughes MD. Meta-analysis for the evaluation of potential surrogate markers. *Stat Med.* 1997;16:1965–1982.
29. Alonso A, Molenberghs G. Surrogate marker evaluation from an information theoretic perspective. *Biometrics.* 2007;63:180–186.
30. Frangakis CD, Rubin DB. Principal stratification in causal inference. *Biometrics.* 2002;58:21–29.
31. Chen H, Geng Z, Jia J. Criteria for surrogate end points. *J R Stat Soc.* 2007;69:919–932.
32. De Gruttola VG, Clax P, DeMets DL, et al. Considerations in the evaluation of surrogate endpoints in clinical trials: summary of a National Institutes of Health workshop. *Controlled Clin Trials.* 2001;22:485–502.
33. Food and Drug Modernization Act of 1997, Public Law No 105–115.
34. International Conference on Harmonisation of Technical Requirements for Registration of Pharmaceuticals for Human Use. ICH Harmonized Tripartite Guideline: Clinical Investigation for Medicinal Products in the Pediatric Population. Brussels, Belgium: ICH; 2000.
35. Simar MR. Pediatric drug development: the International Conference on Harmonisation focus on clinical investigations in children. *Drug Information J.* 2000;34:809–819.

36. Best Pharmaceuticals for Children Act 2002, Public Law No 107–109.
37. Pediatric Research Equity Act 2003, Pubic Law No 108–155.
38. Mccune SK, Mathis LL, Cocchetta DM, et al. Safer, better more appropriate: clinical trial design for pediatric drug labels. *Drug Information J.* 2006;40:185–195.
39. Food and Drug Administration. *Critical Path Opportunities Reports.* Rockville, MD: US Department of Health and Human Services, Food and Drug Administration; 2006.
40. Food and Drug Administration. *Draft Guidance for Industry: Pediatric Oncology Studies in Response to a Written Request.* Rockville, MD: US Department of Health and Human Services, Food and Drug Administration; 2000.
41. Food and Drug Administration. *Guidance for Industry: Antiretroviral drugs Using Plasma HIV RNA Measurements—Clinical Considerations for Accelerated and Traditional Approval.* Rockville, MD: US Department of Health and Human Services, Food and Drug Administration; 2002.
42. Hughes MD. The evaluation of surrogate endpoints in practice: experience in HIV. In: Burzykowski T, Molenberghs G, Buyse M, eds. *The Evaluation of Surrogate Endpoints.* New York: Springer; 2000: 295–321.
43. Giacoia GP, Mattison DR. Newborns and drug studies: the NICHD/FDA newborn drug development initiative. *Clin Ther.* 2005;27:796–813.
44. Deykin A. Biomarker-driven care in asthma: Are we there? *J Allergy Clin Immunol.* 2006;118:565–568.
45. Price J, Hindmarsh P, Hughes S, Efthimiou J. Evaluating the effects of asthma therapy on childhood growth: principles of study design. *Eur Respir J.* 2002;19:1167–1178.
46. Sagel SD. Noninvasive biomarkers of airway inflammation in cystic fibrosis. *Curr Opin Pulm Med.* 2003: 516–521.
47. Ljung T, Axelsson L.-G, Heruf M, et al. Early changes in rectal nitric oxide and mucosal inflammatory mediators in Crohn's colitis in response to infliximab treatment. *Aliment Pharmacol Ther.* 2007;25:925–932.
48. Vermeire S, van Assche G, Rutgeerts P. C-reactive protein as a marker for inflammatory bowel disease. *Inflammatory Bowel Dis.* 2004;10:661–665.
49. Bush G, Spencer T, Holmes J, et al. Functional magnetic resonance imaging of methylphenidate and placebo in attention-deficit/hyperactivity disorder during the multi-source interference task. *Arch Gen Psychiatry.* 2008;65:102–114.
50. Boyle EM, Freer Y, Wong CM, et al. Assessment of persistent pain or distress and adequacy of analgesia in preterm ventilated infants. *Pain.* 2006;124:87–91.

CHAPTER 40

Patient-Reported Outcomes in Pediatric Drug Development

MARGARET ROTHMAN, PhD

World Wide Patient Reported Outcomes Center of Excellence, Johnson & Johnson Pharmaceutical Services, LLC, 301 South Alexander Avenue, Washington, Georgia 30673

LEAH KLEINMAN, DPH

Center for Health Outcomes Research, United BioSource Corporation, 2601 4th Ave, Seattle, Washington 98121

INTRODUCTION

General Considerations

Clinical trials with children are increasingly being conducted to obtain information about the safety and efficacy of new and existing products with this population. There are several sources of data for endpoints in clinical trials as shown in Figure 40.1. Traditionally, physiological and clinician-reported outcomes have been preferred over patient-reported outcomes (PROs). More recently, there has been an increased appreciation of the value of incorporating the perspective of the patient, including children, in evaluating new treatments. This has been reinforced by the publication of the FDA draft Guidance for Industry entitled *Patient-Reported Outcome Measures: Use in Medical Product Development to Support Labeling Claims.*[1] A PRO is defined in the guidance as "a measurement of any aspect of a patient's health status that comes directly from the patient (i.e., without the interpretation of the patient's response by a physician or anyone else) (p. 1). PROs are used to help determine the magnitude of treatment benefit defined in the guidance as "improvement in how a patient survives, feels or functions as a result of medical therapy." Data may be collected using many modalities including questionnaires, interviews, or diaries in paper and pencil or electronic format.

Collection of outcome data directly from patients is preferred in many cases because (1) some treatment effects are known only to the patient, (2) there is a desire to know the patient's perspective about the effectiveness of a treatment, or (3) systematic assessment of the patient's perspective may provide valuable information that can be lost when that

Pediatric Drug Development: Concepts and Applications
Edited by Andrew E. Mulberg, Steven A. Silber, and John N. van den Anker

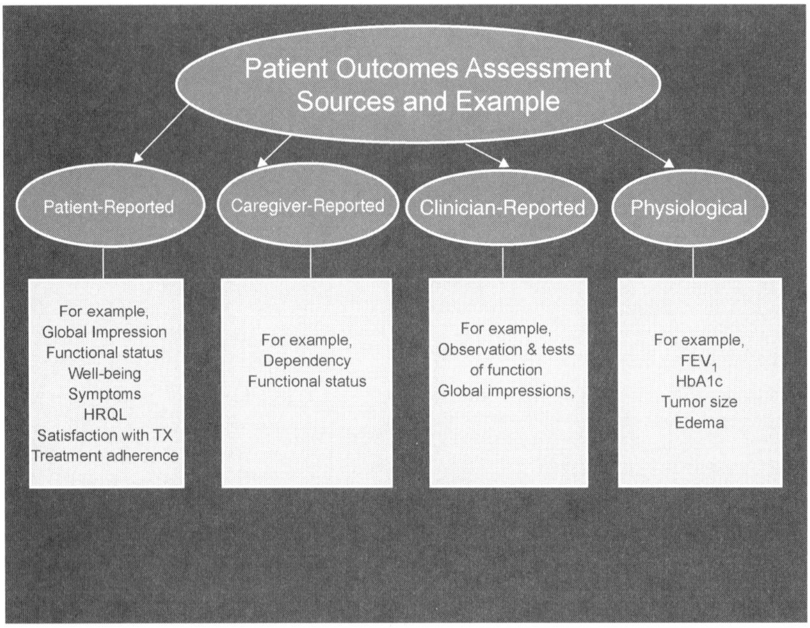

FIGURE 40.1 Sources of data in clinical trials.

perspective is filtered through a clinician's evaluation of the patient's response to clinical interview questions (see Ref. 1, p. 3). For some treatments, the patient is the only source of information. For example, it is not possible to evaluate the effectiveness of an analgesic product via observation or physical measures. For some measures, where there is an accepted physiological endpoint such as FEV_1 in studies of asthma, PRO data may be used to supplement the physiological endpoint because the physiologic measure does not necessarily reflect how the patient feels or functions. Good clinical practice has traditionally included asking patients about symptoms and functioning. In clinical trials, clinician perceptions are captured by structured questions. Obtaining the patient's perspective directly from the patient avoids interpretation by a third party and is potentially more reliable because it is not affected by interobserver variability.

The draft PRO guidance specifically addresses issues related to data collected directly from patients, but also has bearing on other types of subjective assessments that may be obtained from other sources such as caregivers or clinicians. This draft PRO guidance describes how the FDA evaluates PRO instruments used as effectiveness endpoints in clinical trials (Ref. 1, p. 1). Further thinking on this topic is described in a more recent paper by Patrick et al.[2] The following section references the PRO guidance as well as other documents; however, the statements below represent the opinions of the authors. We strongly advise that researchers review the available regulatory guidances including the FDA draft PRO guidance and discuss issues related to instrument selection or development with the appropriate health authority reviewing division before embarking on any studies that are aimed at obtaining labeling claims based on subjective assessments.

As noted earlier, the FDA's draft PRO guidance and subsequent publications provide direction to industry regarding use of PRO measures in medical product development specifically to support labeling claims. Claims in the U.S. regulatory context refer to "statements or implications of treatment benefit that appear in any section of a medical product's FDA-approved labelling." The draft PRO guidance is not intended to provide direction or suggestions in any other context.

There is some confusion in the outcomes literature regarding terminology that the draft PRO guidance attempts to clarify. For example, the terms *quality of life* (QOL), *health-related quality of life* (HRQL), and *patient-reported outcome* are sometimes used interchangeably; however, each has a specific meaning. As noted earlier, a PRO instrument or questionnaire encompasses any report coming directly from the patient and can include, for example, reports of functional status, well-being, and symptoms.

Evaluation of QOL encompasses all aspects of life that bear on general well-being including nonhealth aspects such as environment or financial status as well as health-related aspects of life. It is considered too broad to serve as the basis for a medical product claim. Health-related quality of life represents an individual's overall perception of the impact of an illness and its treatment. It is a type of PRO. By definition, HRQL is a multidimensional concept. A measure of HRQL would at a minimum include an assessment of physical, psychological, and social functioning. To support a claim for improvement in HRQL, the following criteria must be met: (1) the instrument measures all HRQL domains that are important to interpreting change in how the study population feels or functions as a result of treatment; and (2) improvement was demonstrated in all of the important domains.[1]

The Committee for Medical Products (CHMP) for Humn Use has also issued a "reflection paper" focusing on assessment of health-related quality of life that addresses similar measurement issues.[3] The purpose of the HRQL reflection paper is to "discuss the place that HRQL may have in drug evaluation and to give some broad recommendations on its use in the context of already existing documents." The CHMP reflection paper indicates that HRQL improvement as a claim implies that the most important and clinically relevant health-related domains of functioning that impact the patient's quality of life are known and measured. The document lays out some of the requirements for making such a claim, including demonstrating robust improvement in all prespecified domains, evidence of adequate development and validation of the instrument on which the claim is based, and a priori specification of a clinically meaningful improvement.

The PRO guidance and the CHMP HRQL reflection paper are similar in their recommendations regarding requirements for instrument development, validation, implementation, and statistical analysis. The primary way in which they differ is that the CHMP appears to be more open to the added value of HRQL data in some drug development programs and including such information in the summary of product characteristic (SmPC) as a secondary endpoint when requirements are met.

It should be noted that the PRO guidance does not specifically address the evaluation of pediatric outcomes, but the general principles described in the guidance are applicable across all subjective outcomes whether elicited from adults, children, caregivers, or clinicians. Data regarding health outcomes in pediatric trials has most frequently come from clinicians or parents. It is increasingly recognized, however, that children may have a unique perspective regarding their health status and the effectiveness of new medicines and products that may differ from their caregivers and clinicians. Thus, collecting data from children may have relevance as clinical trial endpoints. Collection of health status data directly from children, however, raises a number of unique issues related to child development and the limited reporting abilities of younger children.

The purpose of this chapter is to provide an overview of the measurement and child development issues that must be considered in the assessment of pediatric health outcomes in a clinical trial setting. A detailed summary of how to develop and evaluate a new or existing measure for use in pediatric clinical trials is outside the scope of this chapter and the reader is referred to one of the many textbooks that discuss such issues in detail.[4–6] While much of the focus of the next section is on patient-reported measures, it should be recognized that to a large extent, when speaking of rules of measurement, this terminology serves as a placeholder for subjective assessments provided by clinicians and caregivers as well as patients. For example, assessment instruments used to collect data concerning gastroesophageal reflux disease (GERD) symptoms in infants during a clinical trial are subject to the same rules and review processes as a patient-reported GERD symptom measure in older children or adults. The term PRO is used throughout the document for ease in communication.

Regulatory Guidance on Measurement of PRO Endpoints

The FDA has indicated its intent to evaluate carefully all PRO instruments intended to support labeling claims. Sponsors are encouraged to think about the structure of the desired product label claim in the beginning stages of development. A successful claim based on a PRO endpoint depends on alignment of PRO concepts, choice of PRO instrument, application, analysis, and interpretation in the context of adequate and well-controlled clinical trials.[2] Thinking through this process early in development allows sufficient time to identify an existing measure of the concept to be included in the claim or to determine the need to develop a new instrument or modify an existing one. Development of a new instrument is a resource intensive activity that may include iterative steps. These activities may only be performed adequately if sufficient time is available. Modification of an existing instrument may also require considerable time depending on the degree of modification required. The time required to develop a new instrument is variable, depending on a number of factors, including the complexity of the instrument, but may take anywhere from 12 months to several years.

The FDA guidance describes several aspects of the PRO instrument development process that will be considered as part of the evaluation of whether it is adequate for labeling purposes. These aspects focus on the validity of the instrument. Validity is an often misunderstood concept in constructing and evaluating health outcome instruments, although it has a long history in the social sciences. Within the social science tradition, characterization of validity has evolved from focusing on specific instruments and defining specific types of validity (e.g., face, content, construct) to thinking about validation as evaluation of the integrated evidence that supports the sound scientific basis for the proposed interpretation of scores generated by an instrument in a specific context.[7–10] To have confidence in the proposed interpretation of scores in a particular application, evidence based on instrument content, response processes, internal structure, and relations to other variables must be available for review. One aspect of PRO assessment that is highlighted in the regulatory context is the meaningfulness of a specific score difference (clinical meaningfulness), since this, in addition to statistical significance, is the standard against which the success or failure of a new compound will be evaluated.

Evaluation of Content

Instrument content refers to the themes, wording, and format of items or tasks; guidelines for procedures regarding administration and scoring and specification of the content domain. Item content for health outcome instruments may be identified using several sources including review of literature and other instruments evaluating similar concepts, focus groups, in-depth interviews, clinicians, family members, researchers, and other sources. Content of PRO instruments must be supported by evidence of input from the target population in addition to other sources since it is the experience of these patients that the instrument is designed to assess; that is, the content of the instrument must reflect those concepts that are important to patients with the condition and the items must adequately cover those concepts. Thus, it is important to describe the population for which the instrument was developed and tested as well as the population in which it is to be used. To the extent that these are not congruent, additional evidence of the appropriateness of instrument content to score interpretation may be required. The guidance notes the agency's intent to compare the patient population used in instrument development to the study population enrolled in the clinical trial with respect to patient age, sex, ethnic identity, and cognitive ability (Ref. 1, p. 9). Unfortunately, many instrument developers fail to clearly specify the concepts that the instrument is intended to measure and the source of the items, which makes evaluation of its usefulness as an endpoint measure challenging. Instruments that are not supported by such evidence may be unacceptable for labeling purposes.

Empirical analyses of the response processes of patients or other respondents can provide useful evidence concerning the fit between the concept being measured and scores on the instrument. For example, an instrument should measure the intended content and not reflect reading ability or the health status of the caregiver respondent.[11] The issue of the subject's interpretation of the items is most often evaluated through cognitive debriefing (a one-on-one interview with the expected respondents to determine how patients are interpreting the questions asked in a questionnaire). Evidence that respondents interpreted the instructions, items, and response scales as intended by the developer should be reported as should changes in the instrument as a result of the cognitive debriefing process.

Measurement involves using rules to assign numbers to characteristics of objects or entities.[12] Examination of the measurement properties of an instrument in a particular circumstance or setting is crucial to understanding the appropriateness of its use as an endpoint in clinical trials. As noted earlier, measurement characteristics are not properties of an instrument across any target population or setting. If the instrument is to be used in a different population or in a different setting than the one in which it was developed, measurement characteristics may need to be reevaluated. The extent of the reevaluation is dependent on the degree of the difference between the intended application and the available information. For example, the use of an instrument that was developed for children 9–10 years old may require minimal testing to extend it to children aged 11–12 years. However, the same instrument may require more extensive testing and adaptation for use in children aged 16–17 years because of the extensive developmental maturation occurring in the intervening years. The measurement properties of most interest in instruments used as endpoints in clinical trials are reliability and validity.

Evaluation of Reliability Reliability refers to that component of the observed score that does not reflect the concept of interest. This is referred to as measurement error. One of the goals of measurement is to reduce this component of the score because it reduces the usefulness of information obtained from measurement.[7] For example, it is difficult to show differences between treatments if the scores on the outcome measure include a large measurement error component. Three sources of variability that are often used to evaluate measurement error are test–retest, internal consistency, and interrater reliability. In the case of PRO assessments, the latter is referred to as interinterviewer reliability.[1] Reliability estimates are generally stated as coefficients that theoretically range from 0.00 to 1.00 with larger coefficients reflecting greater reliability or less measurement error. The reliability coefficients noted above reflect different sources of error and are not interchangeable. Evaluation of specific sources of error may be more appropriate for some types of assessments than others; for example, internal consistency reflects the relationship of individual items within an instrument, but is only appropriate for multiple item instruments. Interrater reliability reflects the consistency of measurement among raters. Interrater reliability would not be appropriate for a self-report instrument, but may be very important for clinician assessments where the same clinician does not always perform the assessment at each time point. Test–retest reflects the stability of scores over time in the absence of a change in the underlying condition and may be assessed for single or multiple item instruments as well as ratings made by patients or other raters such as clinicians or caregivers. All appropriate coefficients should be reported for any application of an instrument used in clinical trials. Standards for reliability coefficients have been reported,[4,13] although it is the tolerance for measurement error in a given situation that should drive determination of whether the reliability of scores in a specific context is adequate for the intended purpose. For a more thorough discussion of reliability and alternative approaches to evaluation, the reader is referred to one of the many textbooks that discuss this issue in detail.[4–6]

Evaluation of Validity Based on Relations with Other Variables As noted earlier, validity is not a property of a measuring instrument, but refers to the justification of a particular interpretation of scores resulting from a specific application of the instrument. It is incorrect to refer to an instrument as "valid." Establishing validity is an ongoing process. The validation process involves specifying hypotheses regarding how scores will perform based on knowledge of the concept and the target population, and collecting and analyzing data to evaluate those hypotheses. Validity is supported to the extent that the hypotheses are confirmed. Generally, two types of evidence are used to support measures used as endpoints in clinical trials: (1) evidence based on relationships between scores on the measure of interest and other variables and (2) evidence based on relationships between scores and a criterion. Evidence based on the relationship of scores to variables external to the measure is an important and common source of validity evidence. External variables may include measures of other concepts that are either similar to or different from the concept reflected in the measure of interest (referred to as convergent and discriminant validity). For example, scores on a measure of mobility would be expected to be more strongly associated with another measure of mobility or physical function than with a measure of anxiety. Categorical variables such as group membership are also used to support validity; for example, scores on a measure of anxiety would be expected to be different for respondents with and without a diagnosis of anxiety.

Evidence based on the relationship of scores to a criterion addresses the issue of how accurately the score on the new instrument predicts scores on the criterion either in the future or concurrently depending on the purpose of the measurement. Concurrent validation is generally of greater interest in evaluation of measures of health status used in clinical trials. The value of the relationship of the instrument scores for which validation evidence is being generated to the criterion depends on the reliability and validity of the criterion. Thus, it would not be appropriate to compare scores on two measures that both lack evidence of reliability and validity.

If a measure is expected to reflect change in a concept over time as in clinical trials, it is important to establish the responsiveness of scores. There are basically two methods of assessing responsiveness. The first is distributional, which evaluates the change in the instrument score relative to various measures of variability.[14] The second approach is to compare change scores to a clinically meaningful external anchor of change. The external anchor must be a valid measure of change in the target population.

Interpretation of Scores There is increasing focus among regulatory reviewers on the interpretation of scores in clinical trials; that is, observed score changes are expected to be "clinically meaningful" as well as "statistically" significant. It is expected that this difference be established prior to initiation of the pivotal trials. This may be difficult to accomplish unless there is considerable experience with an instrument and such information is available in the literature, for example, pain intensity numerical rating scale,[15] or studies to support a meaningful difference have been conducted during the Phase II trials. Two basic approaches are used to establish a clinically meaningful difference: (1) identification of a minimally important difference MID and (2) comparison of responders.[1] The MID is a theoretical treatment effect size that is presumed to represent a meaningful treatment benefit when the treatment and control groups each are considered as a group. One concern about this approach is that point estimates based on the group means may obscure important changes for individuals in either group. A second concern is that clinicians may find it difficult to interpret a group mean change in terms of an individual patient.[2]

The second basic approach to interpretation of scores is to focus on the within individual patient change in each treatment group and determine the proportion of patients who respond at this threshold. One of the issues with this approach is that some definition of a responder at the individual patient level must be identified. This responder criteria must be shown to be of perceptible importance and meaningful to patients. While various methods have been proposed for identifying and evaluating a responder definition,[16,17] appropriate standards have yet to be developed.[2]

ISSUES IN CHILD DEVELOPMENT RELATED TO INSTRUMENT DESIGN

The use of questionnaires to measure disease/therapy impact in a pediatric population poses an additional set of concerns beyond what is considered in an adult population, although the need for the instrument to be appropriately developed and to demonstrate reliability and validity as discussed earlier remains the same. Additional pediatric-specific concerns include differences in illness symptoms as children age, use of proxy respondents, age at which children can appropriately and adequately self-report, appropriate formatting, recall period and other instrument design issues, and ethics of children participating in research studies.[18,19]

Instrument Content

When designing or selecting an instrument appropriate for use in pediatric clinical trials, content is critical. An instrument with demonstrated validity and reliability in the adult population cannot simply be implemented "as is" into pediatric studies. Pediatric illnesses often present with different symptoms than the same illness in the adult population. For example, infants with gastroesophageal reflux disease (GERD) present with feeding and regurgitation issues while older children's symptoms may be more like adults and include abdominal pain, choking, refusal to eat, heartburn, and reflux whereas heartburn and reflux are considered the two main symptoms in adults. Thus, it is imperative that appropriate qualitative research such as focus groups with either parents/caregivers or the age appropriate pediatric population be conducted in order to identify instrument content. Furthermore, in order to meet the FDA draft guidance recommendations, available instruments must demonstrate evidence that such qualitative research has been conducted. Transcripts of focus groups are often requested by the FDA as well as evidence to support how items have been identified as important to the pediatric population. Evidence must also be provided regarding how the items have been worded and tested in the appropriate population to reflect the developmental capabilities of the pediatric respondent.

Use of Proxy Respondents and Agreement Between Proxy and Child

A PRO instrument is by definition designed to be completed by the patient. Many of the underlying concepts of a PRO (e.g., symptoms, psychological or physical function) are subjective and internal (known) to the respondent. However, many well-designed pediatric questionnaires require the use of a parent/caregiver proxy report rather than obtaining information directly from the child himself. Reasons for this include children's ability to read and follow instructions as well as the cognitive ability needed to respond to subjective questions using answers with shaded meanings rather than simple yes or no. While the use of proxies may be appropriate for infants and younger age groups, the decision to use proxy respondents with older children is the subject of much debate within the pediatric PRO literature.[18,20,21]

There are no straightforward guidelines as to the chronological age cutoff where a child can provide valid and reliable self-reports of the impact of disease and therapy. In general, in order to be able to complete a PRO instrument, there must be some comprehension on the part of the respondent about health and disease, and the respondent must be able to pay attention, understand the concepts inherent in the questions, have some recall of experience and understand gradients in responses (e.g., difference between "a little" and "some"), and be able to mark a response.[21] Furthermore, for a self-administered questionnaire, there must be a rudimentary reading ability. Some general age-range guidelines are available that incorporate stages of cognitive development as alluded to earlier.

It is generally accepted that children can report on concrete domains starting anywhere from 4 to 6 years of age. Ross and Ross[22] report that children this young can reliably report on pain. Because children that age are not necessarily able to read, such information must be gathered using an interview format. As long as the questionnaire has been developed specifically for the pediatric population, the reliability and validity of pediatric self-reports increases after age 7. As part of a larger study of health concepts, cognitive interviewing was conducted to examine children's understanding of health concepts and response options.[23] The 8–11-year-old group of children performed best, understanding about 74% of the terms while children at age 5 only understood about 27% of the content.

With the use of proxies, another question arises and that is whether or not the proxy can report accurately and adequately on the concept being measured. In the adult PRO literature, there is substantial evidence demonstrating that adult proxies such as spouses, adult children, and other caregivers report better (i.e., higher agreement statistically) on observable concepts such as physical function and less well on more subjective feelings such as emotional well-being. Examination of agreement between child and parental responses has yielded mixed results. There does seem to be increased agreement for the more observable domains as compared to emotional domains.[24]

However, agreement based on age demonstrated more equivocal results. In a review of health-related quality of life in children, Matza et al.[18] summarize studies by demonstrating that while some studies show higher agreement between older children and their parents, other studies demonstrate the converse. Similarly, the role of illness severity is unclear, with some studies demonstrating greater parent–child concordance for sicker children,[24] while others demonstrate that in fact agreement between parent and healthy child is higher.[25]

An additional concern once the decision to use a proxy has been made is the assumption that the proxy, most often the parent, is the person who spends enough time with the child to really be able to report accurately on the health concept in question. For infants, the appropriate proxy is almost always a parent: however, in some families a caretaker other than the parent may spend more daytime hours with the infant. As children age, daycare providers, teachers, or other family members may have insight into the frequency and severity of symptoms as well as the impact on behavior and functioning. If a parent is used as the proxy, there must be clear inclusion criteria that help specify which parent to use and why. Care must be taken to ensure that parents of research participants have equivalent direct experience with their children in order to make comparable ratings.

Some researchers have resolved the question of when to use proxies through the development of age-appropriate versions of different measures. For example, The Child Health Questionnaire contains a version where children aged 10–19 self-report and another version where parents self-report on their children; this version can be used for children aged 4–19. Another example is The Pediatric Quality of Life Questionnaire (PedsQL), which has multiple versions including one for children aged 8–12, adolescents aged 13–18, and parents reporting for children aged 8–18. Disease-specific measures in many therapeutic areas are also available and often have separate versions for proxies and children-respondents. However, although the FDA draft guidance does not address pediatrics specifically, there is no doubt that pediatric instruments will be held to similar standards when used in clinical trials for labeling and/or promotional purposes.

Instrument Formats and Designs

The format and comprehensibility of an instrument must be considered when designing or selecting an instrument for use in pediatric clinical trials. When selecting an existing instrument for use in a clinical trial, the researcher should ensure that there is good evidence that the developer considered the following areas during the development process: type of response scale, recall period, length of the instrument, ability to self-administer or amount of assistance required during administration, and formatting details.[18] Justification and documentation of decisions made in all these areas must be available for review. Evidence of qualitative research and cognitive debriefing with subsequent instrument modifications should be accessible to the researcher.

Examples of response scales used in instruments designed for use in the pediatric population include Likert-type scales (none, a little, some, most, and all), faces that range from a smile to a frown, and Likert-scales with visual aids such as increasingly larger circles as the child moves from never to always. The appropriate response options are critical when designing or selecting an age-appropriate questionnaire because of the possibility of response bias. This is the tendency to select a certain response (i.e., always selecting the extremes) regardless of what is being asked. This in turn can cause bias in the data and the possibility of inaccurate findings. Children from about age 8 and up appear able to use the full range of a 5-point or 7-point Likert-type scale to describe their health; however, younger children may tend to use the extreme ranges.[18]

Recall period can be troublesome in the pediatric population for several reasons. First, the FDA draft PRO guidance has expressed concern with the degree to which patients can remember and provide accurate recall and interest is growing in the use of daily diaries and ecological momentary assessment to obtain PRO data. However, there is little evidence available regarding use of children as respondents to diaries or ecological momentary assessment (i.e., the symptom in question is assessed at random moments throughout the day using electronic data gathering devices). Just as with comprehension, there may be age-related cognitive differences with children's ability to remember or to even understand the concept of one week or one month.[18]

Age also plays a role in the length of an instrument, with older children probably being able to complete a greater number of items than younger children simply because of attention span and ease or greater skill in language comprehension.

CONCLUSION

In summary, there is increasing attention to the measurement characteristics and interpretation of scores used in clinical trials intended for making claims about medical products. The issues raised in the draft FDA PRO guidance and the EMEA HRQL reflection paper are not new, but reflect best practices for measurement of subjective outcomes. These documents, however, do require that clinical trial researchers focus on issues that may not have been in the forefront of planning in the past. Attention to these issues may be one of several factors in greater acceptance of label claims based on subjective assessments. Three issues of special note are the increased emphasis on establishing content as well as other aspects of validity, a priori specification of a clinically meaningful difference in scores, and the need for documenting each step in developing the validation evidence to support the outcome measure used for making the claim.

As the number of pediatric clinical trials increases, the need for instruments specifically designed to measure subjective concepts such as functioning or symptoms in this population will continue to grow. Although it is recommended that these instruments follow the development path laid out for adult measures, there are several specific issues in this population that must be considered. These include identifying the appropriate age at which a child can self-report on a concept, identification of the appropriate proxy if the age is such that a child cannot self-report, and paying close attention to the format and content of the questionnaire including appropriate recall period, type of response option, and reading level.

The desire to include outcomes in clinical trials other than those based on physiologic measures is increasing due the recognition that subjective assessments often convey additional or more relevant information about the treatment. In order to ensure that such information is supportive of the intended claim, it is important to follow the guidance provided by regulatory authorities as well as best practices in the science of measurement. There are additional considerations that must be addressed when collecting data from children. Careful attention to these issues will improve the measurement of the concept for which a claim is desired. Each claim is very specific to the product and population being studied; therefore, it is essential that researchers discuss these issues with health authorities at the earliest possible stage of drug development.

Finally, it is important that researchers document all steps in the development and evaluation of any instrument used as an endpoint in clinical trials and that this documentation be available to health authorities upon request. Publication of results of instrument development and evaluation activities will not only enhance credibility of findings, but also help advance the field of measurement of pediatric outcomes in clinical trials.

REFERENCES

1. FDA Guidance for Industry. *Patient-Reported Outcome Measures: Use in Medical Product Development to Support Labeling Claims. Available at* http://www.fda.gov/cder/guidance/index.htm. (Accessed February 2006.)
2. Patrick DL, Burke LB, Powers JH, et al. Patient-reported outcomes to support medical product labeling claims. *Value in Health.* 2007;10(Suppl 2):S125–S137.
3. EMEA/CHMP/EWP/139391/2004. Came into effect January Available at www.emea.eu.int.
4. Nunnally JC, Bernstein IH. *Psychometric Theory*, 3rd ed. New York: McGraw-Hill; 1994.
5. McDowell I, Newell C. *Measuring Health*, 2nd ed. New York: Oxford University Press; 1996.
6. Steiner DL, Norman GR. *Health Measurement Scales: A Practical Guide to Their Development and Use*, 2nd ed. Oxford, UK: Oxford University Press; 1995.
7. Standards for Educational and Psychological Testing. Prepared by the American Educational Research Association, American Psychological Association, and the National Council on Measurement in Education. 2002.
8. Messick S. Validity of psychological assessment: validation of inferences from persons' responses and performances as scientific inquiry into score meaning. *Am Psychol.* 1995; 50(9):741–749.
9. Cronbach LJ. Test validation. In: Thorndike RL, ed. *Educational Measurement*, 2nd ed. Washington DC: American Council on Education; 1971.
10. Cronbach LJ, Meehl PE. Construct validity in psychological tests. *Psychol Bull.* 1955; 52:281–302.
11. Rothman ML, Hedrick SC, Bulcroft KA, Hickam D, Rubenstein LZ. The validity of proxy-generated scores as measures of patient health status. *Med Care.* 1991;29:115–124.
12. Thorndike RM, Cunningham GK, Thorndike RL, Hagen EP. *Measurement and Evaluation in Psychology and Education*, 5th ed. New York: Macmillan; 1991.
13. Scientific Advisory Committee of the Medical Outcomes Trust. Assessing health status and quality-of-life instruments: attributes and review criteria. *Qual Life Res.* 2002;11:193–205.
14. Kazis LE, Anderson JJ, Meenan RF. Effect sizes for interpreting changes in health status. *Med Care.* 1989;27:S178–S189.
15. Farrar JT, Young JP, La Moreaux L, et al. Clinical importance of changes in chronic pain intensity measured on an 11-point numerical pain rating scale. *Pain.* 2001;94:149–158.
16. Guyatt GH, Juniper EF, Walter SD, et al. Interpreting treatment effects in randomized trial. *BMJ.* 1998;316:690–693.
17. Guyatt GH, Norman GR, Juniper EF, et al. A critical look at transition ratings. *J Clin Epidemiol.* 2002;55(9):900–908.
18. Matza LS, Swensen AR, Flood E, et al. Assessment of health-related quality of life in children: a review of conceptual, methodological and regulatory issues. *Value in Health.* 2004; 7:79–92.

19. Weise KL, Smith ML, Maschke KJ, et al. National practices regarding payment of research subjects for participating in pediatric research. *Pediatrics*. 2002;110:577–582.
20. De Civita M, Regier D, Alamgir A, et al. Evaluating health-related quality of life studies in paediatric populations. Some conceptual, methodological and developmental considerations and recent applications. *Pharmacoeconomics*. 2005;23:659–685.
21. Riley AW. Evidence that school-age children can self-report on their health. *Ambulatory Pediatr*. 2004;4:371–376.
22. Ross DJ, Ross SA. Childhood pain: the school-aged child's viewpoint. *Pain*. 1984;20:179–191.
23. Rebok G, Riley A, Forrest C, et al. Elementary school-aged children's reports of their health; a cognitive interviewing study. *Qual Life Res*. 2001;10:59–70.
24. Eiser C, Morse R. Can parents rate their child's health-related quality of life? Results of systematic review. *Qual Life Res*. 2001;10:347–357.
25. Levi RB, Drotar D. Health-related quality of life in childhood cancer; discrepancy in parent–child reports. *Intl J Cancer Suppl*. 1999;12:58–64.

CHAPTER 41

Safety Monitoring in Pediatric Clinical Trials

SETH V. HETHERINGTON, MD

Icagen, Inc., Durham, North Carolina 27703 and Department of Pediatrics, University of North Carolina School of Medicine, Chapel Hill, North Carolina 27579

INTRODUCTION

In this section, we focus on safety monitoring during clinical trials of a pediatric age population. We start this discussion assuming that we are going to initiate a clinical trial of a new chemical entity (drug) that has not been granted marketing approval by regulatory agencies, at least for any pediatric age group. In fact, the concepts to be discussed are applicable for a study involving a new formulation such as a liquid formulation of a drug for which a solid form has previously been tested in adults, or even older children. These concepts are also applicable for studies of a drug previously tested in one pediatric age group (e.g., ages 6 and older), which will now be tested in a younger age group; and they are equally pertinent to the study of a device. New chemical entities, formulations, or devices are all medical interventions that, when tested in a pediatric population, require special considerations for safety monitoring. While the objective of any pediatric development plan is to show that the medical intervention under study is effective, the other main objective is to show that it is safe for use in a pediatric population. The testing of a new medical treatment applicable to a pediatric population, like any other clinical development program, requires an evaluation of the benefit-to-risk ratio of that treatment or intervention.

All clinical trials require a safety monitoring plan. The conventions used to assess safety for pediatric patients are the same as those for trials in any age group and are derived from a combination of legal regulations, regulatory guidance documents (national and international), and conventions drawn from the medical literature. The specific components of a safety monitoring plan will reflect the level of risk to the participants in the clinical trial.[1,2] Definitions of categories of risk have been created, as shown in Table 41.1.

The application of these conventions to the analysis of clinical trial data leads to the ultimate determination of whether an intervention is "safe" for use in a given population.

Pediatric Drug Development: Concepts and Applications
Edited by Andrew E. Mulberg, Steven A. Silber, and John N. van den Anker

TABLE 41.1 Categories of Risk in Pediatric Clinical Trials

21 CFR 50.51, 45 CFR 46.404	Not involving greater than minimal risk
21 CFR 50.52, 45 CFR 46.405	Involving greater than minimal risk but presenting the prospect of direct benefit to individual subjects
21 CFR 50.53, 45 CFR 46.406	Involving greater than minimal risk and no prospect of direct benefit to individual subjects, but likely to yield generalizable knowledge about the subject's disorder or condition
21 CFR 50.54, 45 CFR 46.407	Not otherwise approvable, which presents an opportunity to understand, prevent, or alleviate a serious problem affecting the health or welfare of children

The key here is that a clinical trial quantifies safety for the population under study. If that population is 6–12 year olds, the assessment of safety may or may not reflect the safety of the intervention in 2–5 year olds, just as the safety in adult populations may not predict safety in pediatric populations. The complexity of pediatrics is that it is composed of multiple age groups, with differences among them that may be greater than the differences to adult populations. Six to 12 year olds are more like adults that they are like neonates. This chapter focuses on selective topics for the development of a data monitoring safety plan for a clinical trial in pediatric patients.

OBJECTIVE OF THE SAFETY MONITORING PLAN

The goal of a safety monitoring plan is to minimize risk to health among the population under study. This starts with the development of the protocol itself. The investigator(s) or author of the protocol should plan to enroll only enough subjects to achieve the primary objective, be it a study of pharmacokinetics, safety, or efficacy. A study that is appropriately powered statistically will limit exposure of an undue number of children to a yet-to-be proven intervention. The safety monitoring plan for the protocol details what will be monitored, by whom, when and how the data is to be analyzed, and how results are communicated. It should also include actions to be taken with regard to modifying or stopping a study if questions of safety arise. The plan is usually summarized in the protocol, but could be a separate document and may include multiple parts, such as the protocol, a safety monitoring plan document with greater detail, and documents related to a Data Safety Monitoring Board, if one is established for the clinical trial. The plan may include guidance on how investigators are to manage adverse events experienced by subjects, including rules for interrupting, restarting, or permanently discontinuing study medication. Review of a protocol by an institutional review board (IRB) provides a check on these measures. The IRB should include members familiar with pediatric health issues (or alternatively have pediatric expertise accessible), in order to assess risks and potential benefits to subjects.

The safety monitoring plan will specify who performs the review of safety data, how often, and what data is to be reviewed. The reviewer may be the investigator or an independent data safety monitoring board (DSMB). Guidance from the FDA is available for further information in requirements and issues related to establishing a DSMB in Guidance for Clinical Trial Sponsors Establishment and Operation of Clinical Trial Data

Monitoring Committees (available at http://www.fda.gov/cber/guidelines.htm). This guidance discusses the roles, responsibilities, and operating procedures of Data Monitoring Committees (DMCs) (also known as Data and Safety Monitoring Boards (DSMBs) or Data and Safety Monitoring Committees (DSMCs)) that may carry out important aspects of clinical trial monitoring. This guidance is intended to assist clinical trial sponsors in determining when a DMC may be useful for study monitoring, and how such committees should operate.

It is important that those involved with the review of safety be familiar with pediatric issues in general and optimally the disease or condition under study. The individual investigator or healthcare provider or investigator who enrolls, treats, examines, collects data from the individual subject, and carries out the procedures of the protocol is also integral to minimize risk to patients enrolled in clinical trials. Nothing in the protocol or safety monitoring plan should interfere with best medical practice in delivering care to the patient.

Adverse Events

The definition of an adverse event (AE) has been standardized internationally for many years as "any untoward medical occurrence in a patient or clinical investigation subject" administered a pharmaceutical product and which does not necessarily have to have a causal relationship with this treatment. An AE can therefore be any unfavorable and unintended sign (including an abnormal laboratory finding), symptom, or disease temporally associated with the use of a medicinal product, whether or not considered related to the medicinal product.[3]

Two elements are key to this definition: the event is "bad" (e.g., not beneficial), and it may or may not have any relationship to the drug or intervention under study. Laboratory values out of the normal range may not necessarily be adverse or considered an AE. Those abnormal laboratory results related to clinical symptoms or that place the subject at risk are classified as AEs. Use of age-appropriate laboratory values is critical to assessment of clinical laboratory studies.

During the course of a clinical trial, information on each AE is collected, including the following: start date, stop date, severity, relatedness (to the drug under study), action taken, and outcomes. Three of these elements deserve additional discussion: severity (how bad is it?), seriousness (different from severity in that it denotes a level of importance that may include severity, but that reaches a predefined set of conditions), and relatedness (is it due to the intervention or not?).

Serious Adverse Event Of the three traits mentioned previously, regulatory authorities specify standards only for seriousness.[3] A serious adverse event (SAE) is defined as one that results in death, is life-threatening, prolongs or results in hospitalization, results in a significant or persistent disability, or is a congenital anomaly. In addition, events that are medically important may meet the definition of an SAE. While the regulations specify a definition of an SAE, there may be circumstances where modifications of this definition are justified. The definition may be expanded when there is a specific AE of interest, for example, one that has been identified in prior clinical trials, and where safety signal detection would be improved by adding specificity to a definition because of the complexity of the medical condition under study. For example, in the studies of drotrecogin in children with sepsis, the definition of serious bleeding was expanded to include episodes where transfusion was required.[4,5] In an extremely complex patient population, there may be

TABLE 41.2 Simple Severity Categories for Adverse Events

Mild	No limitation of usual activities
Moderate	Some limitation of usual activities
Severe	Inability to carry out usual activities

circumstances where an event that would be considered serious for an otherwise healthy population would occur commonly, and only those meeting a threshold of severity would be included in the definition of seriousness. Modifications of the definition of SAEs should be discussed with regulatory authorities.

Severity: Grading of Adverse Events For each AE reported in a clinical trial, the severity of that event will be quantified. This is true for all AEs, but it is especially considered a key data element in the description of an event subject to the rules of expedited reporting.[3] While there is no standard method for grading the severity of an AE, there are several common systems, each designating a level and providing a definition. In the simplest system, events are graded mild, moderate, or severe. Note that "severe" and "serious" are not the same, as the latter is a separate definition specified in the FDA regulations (CFR 312.32) and the ICH guidelines.[3] In the three-part grading scale, simple definitions are applied as shown in Table 41.2.

Other grading scales use up to five categories, often by adding a grade 4 (life-threatening) and grade 5 (death). Both fulfill the criteria for SAEs. Some, to designate no AE, add a grade "0" but for the purposes of Good Clinical Practice, this is unnecessary.[6]

Several collaborative clinical research groups have constructed standard severity grade tables for use across numerous multicenter clinical trials. One advantage is that it provides multiple investigators participating in a single trial a standard scale by which to compare AEs. This brings consistency across clinical trial sites. Second, when clinical trials are part of a larger effort to develop drugs for a specific therapeutic area, it allows comparison of trial results by encouraging uniformity in reporting and data analysis. The third advantage is that clinical laboratory results may be graded according to agreed upon ranges that represent severity and are relevant for the age group and disease under study. One example is the Common Terminology Criteria for Adverse Events (CTCAE), in its third version as of 2006, used by the National Cancer Institute.[7,8] This uses a five-level grading system, an example of which is shown in Table 41.3.

First, note that not all events have designations for all grades. Myelodysplasia grades start at grade 3 (severe), while bone marrow cellularity has no grade 4. Second, note the example of a laboratory finding (hemoglobin) assigned severity grades by range of values. Finally, SAEs may be identified by attributes in the table (as in the criteria of death), but that if any event fulfills any SAE criterion, regardless of grade, it would be recorded as an SAE. For example, an event of grade 2 that results in hospitalization would be considered an SAE.

Other examples of grading scales include those drafted by the Division of Microbiology and Infectious Diseases of the NIH and by the AIDS Clinical Trial Group.[9,10] All provide grading scales for pediatric patients using age-appropriate norms for clinical and laboratory events. Severity grades are also available for pediatric patients less than 3 months of age.[11] All or part of these severity grades may be incorporated into clinical trials. For events not listed, a general grading scheme can be applied, mapping a descriptive term to the numerical grade (grade 1 = mild, grade 2 = moderate, grade 3 = severe, etc.). Other modifications

TABLE 41.3 Severity of Blood/Bone Marrow Adverse Events

	Grade				
Adverse Event	1	2	3	4	5
Bone marrow cellularity	Mildly hypocellular or ≤25% reduction from normal for age	Moderately hypocellular or >25 to ≤50% reduction from normal for age	Severely hypocellular or >50 to ≤75% reduction in cellularity from normal for age	—	Death
Hemoglobin	<LLN to 10.0 g/dL	<10.0 to 8.0 g/dL	<8.0 to 6.5 g/dL	<6.5 g/dL	Death
Myelodysplasia	—	—	Abnormal marrow cytogenetics (marrow blasts ≤5%)	RAEB or RAEB-T (marrow blasts >5%)	Death

Source: http://www.fda.gov/cder/cancer/toxicityframe.htm.

to severity scales may be adopted as appropriate for the population under study. In a recently conducted clinical trial in premature infants with birth weights of 500–1250 g (<32 weeks gestational age), these tables were further adapted using published scales for adverse events specific to the premature infant population.[12]

Relatedness Relatedness as defined in 21 CFR 312.32 (revised April 1, 2006) is as follows:

> *Associated with the use of the drug*. There is a reasonable possibility that the experience may have been caused by the drug.

There is, however, no standard definition of "reasonable possibility." Some situations are clearly obvious; for example, a patient who experiences an anaphylactic reaction during the intravenous infusion of a biologic agent has undoubtedly experienced a drug-related adverse event. On the other hand, special populations with chronic diseases will have medical complications that are AEs and may be related to a drug or investigational agent. These events could be reasonably expected to occur in the population under study, but a drug could increase frequency or severity. Further complicating the assessment of relatedness is the factor of age in a pediatric population. Some diseases have different phenotypes by age. If a phenotype appears in a patient that is unusual for age during participation in a clinical trial, is that merely a variant of the disease, or has the investigational drug under study influenced the disease expression?

In clinical trials, collection of information on an AE included an assessment of relatedness by the investigator. In the simplest approach, the investigator chooses between "likely related" and "unlikely related." This is the recommendation of the CIOMS Working Group VI.[13] In practice, this is reasonable, because the regulatory reporting requirements make no other distinction by degree of likelihood. In other reporting strategies, events are reported by degree of likelihood for relatedness into one of up to five categories, such as not related, unlikely related, possibly related, probably related, and definitely related.

Although there are no standard definitions, several have been developed in an attempt to provide guidance to investigators and increase the consistency of reporting across multiple clinical sites. The Division of Microbiology and Infectious Diseases of the National Institute of Allergy and Infectious Diseases use one reproduced here in Table 41.4. For each category, criteria are listed and, in this case, specific guidance on how many of each may need to be present to qualify.

For blinded clinical trials, the assessment of relatedness must be made prior to any unblinding of treatment assignment if this is to be done at all. Follow-up information may become available that impacts the assessment of relatedness and thus the report of the event may be updated accordingly.

TABLE 41.4 Categorization of Relatedness for Adverse Events

Unrelated	Adverse event is clearly due to extraneous causes (e.g., underlying disease, environment).
Unlikely (must have 2)	1. Does not have temporal relationship to intervention. 2. Could have been due to environmental or other interventions. 3. Does not reappear or worsen with reintroduction of intervention. 4. Could readily have been produced by the subject's clinical state. 5. Does not follow known pattern of response to intervention.
Possible (must have 2)	1. Has a reasonable temporal relationship to intervention. 2. Could not readily have been due to environmental or other interventions. 3. Could not readily have been produced by the subject's clinical state. 4. Follows a known pattern of response to intervention.
Probable (must have 3)	1. Has a reasonable temporal relationship to intervention. 2. Could not readily have been produced by the subject's clinical state or have been due to environmental or other interventions. 3. Follows a known pattern of response to intervention. 4. Disappears or decreases with reduction in dose or cessation of intervention.
Definite (must have all 4)	1. Has a reasonable temporal relationship to intervention. 2. Could not readily have been produced by the subject's clinical state or have been due to environmental or other interventions. 3. Follows a known pattern of response to intervention. 4. Disappears or decreases with reduction in dose or cessation of intervention and recurs with reexposure.

Source: http://www.niaid.nih.gov/ncn/sop/adverseevents.htm.

Early in the conduct of clinical trials of a new drug, the ability to distinguish relatedness may be limited. Preclinical studies may identify AEs to be anticipated in human trials, but studies in animals are often poor predictors of toxicity in humans, especially with regard to cutaneous reactions. Also, drugs with significant safety signals in preclinical studies often do not proceed to human studies. Prior studies in adults may provide a safety profile, but age group differences in response to medications are well known. Finally, if similar drugs have been studied, those known safety profiles may give investigators an idea of what to expect for AEs with a new investigational agent. As clinical experience grows, especially with placebo-controlled trials, the ability to analyze data by treatment assignment allows better characterization of drug-related AEs, with less reliance on a point-of-contact assessment of likelihood of relatedness.

Expedited Reporting of Adverse Events

During the conduct of a clinical trial, certain adverse events have required reporting to regulatory authorities in countries where the compound is under investigation and to the IRBs of the participating clinical trial sites. These reports, referred to in the United States as IND safety reports (http://www.accessdata.fda.gov/scripts/cdrh/cfdocs/cfcfr/CFRSearch.cfm?fr=312.32) or expedited reports (http://www.emea.europa.eu/pdfs/human/ich/013595en.pdf), fulfill three characteristics: (1) they meet the definition of "serious," (2) they are associated with the use of the drug, and (3) they are "unexpected." The first two characteristics have been discussed. "Unexpected" means that the event is not listed on the product label (for an approved product) or in the investigator's brochure (for an investigational agent). Reporting requirements at the international level are well defined. The sponsor of the clinical trial is responsible for the submission of expedited reports to regulatory agencies and all investigators. There may be local requirements set by the IRB; compliance with these requirements is the responsibility of the investigator.

Adverse Event Detection and Sample Size

Clinical trials basically count events, whether safety or some measure of health benefit, and compare the frequencies of those events by assignment to some medical intervention. In most clinical trials, the sample size is driven by the objectives of the study, the expected frequency of the events to be counted, and some target for statistical power. Sizes of trials are also influenced by the stage of clinical development. In initial clinical trials (Phase I), the population size will be small as the goal will be to assess pharmacokinetics and initial safety. In later clinical trials (Phases II and III), the numbers of subjects studied are larger, but the determination of sample size usually rests with an efficacy endpoint. This often limits the ability to quantify risks, especially for rare and severe AEs. It is important to understand the limitations of AE detection and estimation of frequencies for each clinical trial phase.

First Time in Pediatric Patients: Phase I The timing of the first study in pediatric patients will depend on a number of factors, as identified in the ICH Guideline E11.[14] Assuming that the first Phase I study in pediatric patients will study pharmacokinetics, the number of dose groups and starting dose will depend on previous experience, if any, in adult age groups. Similarly, AEs likely to be observed will depend on prior experience in older age groups. What AEs may occur and with what frequency will be very much unknown, however, because (1) events unique to pediatric patients may occur, (2) events observed among adult populations in earlier studies of the compound may occur at a different frequency or severity among other age groups, and (3) numbers of patients studied previously may have been limited, and therefore would not have identified infrequent or rare AEs. Therefore, to minimize risk to pediatric patients, the safety monitoring plan for a Phase I study should include rules for termination of individual subjects, cohorts, and/or the clinical trial. For specific issues related to oncology trials in pediatric subjects, please see Chapter 24 by Norris and colleagues in this book devoted to considerations for trials in pediatric oncology with specific emphasis on Phase I trial development and safety considerations.

Phase I Studies: Stopping Rules Phase I dose escalation studies that seek to determine a maximum tolerated dose (MTD) should include a definition of MTD as well as rules for discontinuation of study drug and cessation of the dose escalation study. The North American Brain Tumor Consortium Phase I/II study of imatinib mesylate in children with recurrent malignant glioma is an example of these principles in practice.[15] Cohorts of three patients were enrolled beginning with a dose of 400 mg/d. Adverse events were graded according to the National Cancer Institute Common Toxicity Criteria (v 2.0), and a dose-limiting toxicity (DLT) was defined in part as one of the following: grade 4 anemia, grade 3 thrombocytopenia, or grade ≥ 3 nonhematologic toxicity. If one patient developed a dose-limiting toxicity, three additional patients were enrolled at that dose level. If at least one more patient developed a dose-limiting toxicity, the MTD was said to have been achieved (and defined as the next lower dose where only 0 or 1 of 6 patients had a DLT). On the surface, the authors are saying that, for this serious disease, a do32 se that results in up to one grade 3 (severe) event in six patients (17%) is acceptable. While this may be reasonable given the disease and alternative therapies available (or not), the precision of the estimated frequency of grade 3 or 4 events is low. For this example, the 95% confidence intervals for the estimates for frequency of grade 3 or 4 AEs are demonstrated in Table 41.5.

Thus, if 1 of 6 patients develops a grade 3 or 4 (severe or life-threatening) AE, the next cohort will be enrolled at a higher dose, but the 95% confidence interval for the true frequency of the observed event ranges from 1% to 64%. If the next dose cohort demonstrates a DLT for 2 patients, the 95% confidence interval for the true frequency of grade 3 or 4 AEs ranges from 4% to 78%, and the true MTD could be lower than estimated. This should be remembered as larger Phase I or II clinical trials are carried out: doses may need to be adjusted downward. Furthermore, given a true rate of occurrence for DLT events, the likelihood of stopping for MTD is lower than one might expect, as shown in Table 41.6.

TABLE 41.5 The 95% Confidence Intervals for Estimates of Frequencies of AEs

Number of Subjects with grade 3 or 4 AE	Lower 95% Confidence Interval	Upper 95% Confidence Interval
0	0.00	0.46
1	0.01	0.64
2	0.04	0.78
3	0.12	0.88
4	0.22	0.96
5	0.36	1.00
6	0.54	1.00

TABLE 41.6 Likelihood that a Dose Escalation Will Not Occur by Rate of Grade 3/4 AE[a]

True Rate of Grade 3 or 4 AE	Probability of Observing MTD
5%	0.04
10%	0.11
25%	0.47
50%	0.89

[a]Assumes that maximum tolerated dose is defined as observing 2 of 6 subjects with a dose-limiting toxicity.

From Table 41.6, if the rate of grade 3 or 4 events (DLTs) is 25%, the probability that at least 2 of 6 patients will demonstrate a DLT, thereby stopping any further dose escalation, is 0.47. If the true rate is 50%, there is still an 11% (1.00 – 0.89) chance that dose escalation will occur.

Phase II and III Studies Phase II and III clinical trials build a larger safety database from which to define the safety profile of the drug under study. In the case of a double-blind, placebo-controlled study, the frequencies of AEs among patients assigned to active study drug (the drug under study) will be compared to those among patients assigned to placebo or active control/comparator arms. Data collected will include a wide range of clinical and laboratory assessments. Although data management is not the subject of this chapter, data entry and classification of events ("coding") may have profound impact on the analysis. An understanding of age-related behaviors may lead investigators to merge certain AEs into one common category. For example, infants may express pain by crying, irritability, or lethargy; but if each of these terms is analyzed separately, true differences between treatment groups may be obscured. Adoption of standardized severity scales or common terminology may avoid confusion and ambiguity.

The limitations of analysis of safety data should not be underestimated. Data are often presented in a tabular form, listing the most common events by treatment group with statistical tests applied for each pairwise comparison, and often without accounting for multiple comparisons. Since clinical trials are not generally powered to detect differences in individual AE frequencies, the lack of a difference between two groups is not evidence that no difference exists. Many Phase II and III pediatric trials span a wide range of age groups. Analysis by commonly used age ranges (neonate, infant, toddler, child, adolescent) usually has limited power to detect significant differences by age group. Finally, the formulation used may have an impact on dose-related safety events. If solid dosage forms are used (tablet or capsule), the actual dose based on weight (i.e., per kilogram) may vary within the studypopulation. Where there is a narrow therapeutic index, such differences could significantly affect frequencies of adverse events.

Quantifying What Is Known and Not Known

At the end of a clinical trial, or pediatric development program, the safety for pediatric patients can be quantified by two components: severity and frequency of each observed AE. Generally, mild to moderate AEs are common (drugs with common SAEs will not progress in development except in unusual circumstances such as life-threatening illnesses with no therapeutic alternatives). Although the FDA Guidance to Industry discourages the use of categories of frequencies, the definition in 21 CFR 201.57 (g)(2) (revised as of April 1, 2006) categorizes AEs as "common" (occurring in 1–10% of patients or more), infrequent (0.1–1%), or rare (<0.1%). The lower the frequency of an event, the more difficult it is to detect. Thus, many rare and SAEs are not detected until the postmarketing period, when tens or hundreds of thousands of patients would be exposed. To put this in perspective, in a small, 30-patient pediatric study, if no SAEs were detected, one would have 90% confidence that the frequency of an undetected SAE would be less than 10%, which is still a considerable number. If the clinical trial were to enroll 300 patients and still no SAE were to be detected, there would be a 90% likelihood that the frequency of a SAE would be less than 1%. Depending on the target indication of the drug under study, this could still be considered a considerable risk. Few pediatric clinical trials exceed this number of exposed patients; thus, an adequately powered, well-controlled study of a new medication in a pediatric population could miss detection of an infrequent, but serious adverse event.

Minimizing Risk

In the design of all clinical trials, the objective is to minimize risk and maximize potential benefit to patients. The assessment of risk is made complex because of the different age groups encompassing the pediatric spectrum, from premature infant to adolescent. The depth of accumulated risk–benefit data from trials in adult populations and relevance for a pediatric population will vary. For example, data from adults of pharmacokinetics, safety, and potential for efficacy may allow a reasonable assessment of risk and benefit for 13–16 year olds, but may have no relevance for neonates. Thus, every safety monitoring plan should attempt to minimize risk to the individual patient and the population to be enrolled.

GENERAL PROTOCOL DESIGN

The first step is to limit the number of subjects who will be exposed to an intervention or series of procedures anticipated by the protocol design. To achieve this, only the number of participants required to achieve the stated objectives of the protocol should be enrolled. Concurrent with this, the authors of the protocol should make a realistic assessment of the stated protocol objectives and restrict these to objectives that add value to knowledge of use of the therapeutic agent. Additional blood sampling for tests that increase the knowledge of the disease may be desirable, but could increase the volume of blood required beyond a reasonably safe limit (see later discussion), or create additional sampling times with accompanying pain and discomfort. Even the addition of extra clinic visits for such procedures may create discomfort for the child and inconvenience for the caretaker. All clinical trial objectives should be critically evaluated within the context of the number of patients required for the primary study objective. If this number is insufficient to support a reasonable chance of answering an additional question, then that objective should be discarded.

There are study designs and design features that limit patient exposure and hence are particularly attractive for studies in pediatric patients. Once such study type is the group sequential design.[16] In this study, patients are initially randomized to multiple doses of a study drug or a placebo arm. At protocol-specified intervals, data is analyzed according to a prespecified data analysis plan that compares outcomes for each treatment arm against the outcome of the placebo arm. A futility criterion is set that assesses the likelihood that a given treatment arm would be found to be superior to the placebo arm by the end of the trial. If this criterion is not met, that treatment arm is closed to future enrollment and subsequent patients are randomized to the remaining treatment arms or placebo. As the trial progresses, the amount of information available for analysis increases and the criterion for futility becomes stricter. It is also possible to add a criterion for treatment success, such that if a treatment arm exceeds a predefined level, the study may be terminated early for efficacy. There are many caveats to this approach, but the key point is that it limits exposure of patients to treatment arms that are extremely unlikely to show benefit, and thus avoids unnecessary risk. As an example, this approach was used in a Phase II clinical trial to determine the dose of an antistaphylococcal immune globulin for a Phase III trial. [12] Results from this trial were used to determine a dose for a subsequent Phase III trial.

Monitoring Clinical Laboratory Studies During Clinical Trials

The types and frequency of laboratory studies will be determined by the time course expected for efficacy or endpoint determination (assuming that a laboratory study is part of the objective), the duration of therapy, and any expected safety signal anticipated based on preclinical studies or information obtained from studies in adults. In a pediatric clinical study, steps may be taken to limit risk and discomfort when associated with clinical laboratory testing.

Blood Sampling from Pediatric Patients

The number and frequency of laboratory studies and phlebotomies will determine the total volume of blood to be obtained during the course of the study. Pediatric blood volume is limited, especially in infants, raising a common question: "How much blood is too much for clinical testing?" Several guidelines have been promulgated and the investigator should consult local IRBs, regulatory authorities, or institutions for applicable rules. The Partners Human Research Protection Program provides the following guidelines for investigators:[17]

> Blood volume taken from children must be less than 3 cc/kg body weight per 8-week period. In studies where the direct benefit far outweighs this volume restriction, a full protocol must be submitted for review of the full committee, and the following guidelines will apply:
>
> - If more than 3 cc / kg body weight per 8-week period is required and justified by the potential benefits, up to 9 cc venous blood/kg body weight / 8-week period may be considered in older children (e.g., not neonates, toddlers), with the latter figure being the absolute upper limit.
> - Any child involved in a study involving removal of venous blood in the range of 3–9 cc / kg body weight per 8-week period should be placed on iron supplementation therapy. It is recommended that a dose of 30 mg ferrous sulfate / kg / day in three divided doses be given. Such therapy should continue for at least 8 weeks and should be monitored by hemoglobin measurements.

For studies involving life-threatening illnesses where treatment options are limited or absent, exceeding this limit could be justified. In order to remain within any phlebotomy limit, the investigator should critically examine the proposed schedule of clinical and/or research laboratory studies for relevance to the key objective(s) of the study and realistic expectation for the ability to obtain meaningful information within the context of the study and number of patients planned for enrollment. Additionally, use of a local laboratory, rather than a central laboratory, should be considered. If this is possible, timing of the laboratory studies should coincide as much as possible with clinical laboratory studies obtained in the course of general medical care, especially for children with chronic diseases or diseases that required prolonged hospital care. For example, in a study of an immune globulin in premature neonates, all data collected for periodic safety laboratory studies were derived from routine clinical laboratory data available during the care of the infants.[18] Thus, no additional blood sampling was required. In addition, all pharmacokinetic studies were performed on plasma samples scavenged from the hospital clinical laboratory, and no additional volume of blood was required for the entire study beyond that which was drawn during the routine course of the medical care. For logistical reasons, in cases where multiple clinical laboratory studies are obtained in the course of a day (or any time period), some rules may be necessary to decide which values would be collected for the purposes of the clinical trial, as collection and reporting of all values would not be reasonable. Assays utilizing the smallest amount of blood feasible should be used or developed for a clinical trial of pediatric patients. Development of microassays may be required prior to conducting a clinical trial of pharmacokinetics. Another method for reducing the requirement for blood sampling is the use of population pharmacokinetics. In this strategy, sparse sampling is conducted on a larger number of patients than would be utilized in a study of intensive pharmacokinetics with frequent sampling over a shorter period of time.[19–21] Thus, the number of samples per patient can be reduced to as few as two. A pharmacokineticist familiar with this technique and with the proper analysis and modeling of the data should be consulted. Samtani and Kimko further discuss use of population pharmacokinetics for pediatric studies in Chapter 22 in this book. For any study where blood sampling (or any potentially painful procedure) is performed, local anesthetics should be used to reduce pain.[22]

Minimizing Exposure for the Individual Patient

The methods described previously limit exposure of a population to risks associated with an investigational intervention. An additional way to limit the exposure for the individual patient is through an "escape clause" in the protocol design. This strategy has the additional benefit that it can provide treatment options available to patients in circumstances where currently available therapies are limited. In this strategy, each patient enrolled in the clinical trial is assessed for efficacy at a predefined interval (one that would allow sufficient time for any benefit to be realized). If the measured result does not meet a predefined criterion for benefit, that subject discontinues the assigned treatment. This strategy is particularly useful when surrogate markers for benefit are utilized. The patient then has the option to discontinue study drug (while still maintaining any blind) and initiate open label therapy with the active compound. Neither the patient nor the investigator will know the patient's treatment assignment until the end of the study. This design was utilized in a Phase III study of abacavir in children with HIV infection.[23] In this placebo-controlled, double-blind study, children with HIV infection were randomized to receive standard therapy or standard therapy plus abacavir. Plasma HIV-1 RNA was measured as a surrogate marker for treatment effect. Patients whose plasma HIV-1 RNA remained above a defined amount had the option of continuing in the study, receiving open label abacavir in addition to other antiretroviral therapy, or seeking other treatment alternatives. Similarly, for studies of new antiepileptic drugs, one common design is to randomize patients to add study drug (placebo or test drug) to their current treatment regimen. A criterion is established for "treatment failure" such as the time to *n*th seizure or onset of a new seizure type. Once a patient meets that criterion, he/she may elect to receive open label active drug or other appropriate treatment option, if available. This type of design has advantages. It minimizes the time during which a patient receives suboptimal therapy. It also allows open access to a new treatment option in instances where current treatment options are limited. It limits the amount of time that a patient receives placebo and allows the patient to seek other options outside the clinical trial. A variation of this design, for studies with more than one active treatment arm, is a rerandomization to another treatment option of the study.[24]

Adverse event evaluation and management guidelines built into clinical trials can limit the risk of toxicity for an individual patient, while minimizing premature discontinuation (due to an unrelated AE). Guidelines are provided for interruption or permanent discontinuation of the study drug, based on the severity of an AE. Typically, if a subject experiences a mild AE, study drug may be continued. If the severity is moderate, study drug administration is interrupted and the patient is monitored weekly. If the severity decreases, the study drug may be restarted. If not, or if the original AE was severe, the study drug is permanently discontinued. As knowledge of the drug safety profile increases, more specific guidance may be provided.

The Data Safety Monitoring Board

The safety monitoring plan for each clinical trial should specify not just what will be monitored, but by whom, how frequently, what types of actions could be taken, and how the results of that safety review will be communicated (and to whom). In certain circumstances the clinical trial will convene a Data Safety Monitoring Board (DSMB). These committees come under a number of names (e.g., Data Monitoring Committees), but their function is essentially the same.[25] Some groups, such as the NIH National Institute of Allergy and Infectious Diseases, make a distinction between DSMB, generally used for Phase II and III studies, and a Safety Monitoring Committee (SMC) for Phase I studies.[26]

Determining the Need for a DSMB There are circumstances when a DSMB is not required. Review of safety data on an ongoing basis may be adequately performed by the principal investigator, a qualified physician designated by the sponsor, or another individual with appropriate training relevant to pediatric patients and the disease under study. Types of studies where a DSMB is established or generally not required are shown in Table 41.7. Some institutions, notably the National Institutes of Health, and individual institutions may have policies that dictate when a DSMB is required.[27]

For Phase I studies, a small number of pediatric patients will be at risk. If the risk to patients is minimal or greater than minimal risk but presenting the prospect of direct benefit to individual subjects, a DSMB may not be necessary. The investigator(s) may choose an independent safety monitor or a Safety Monitoring Committee (SMC). Proposed guidelines for studies not requiring a traditional DSMB (Phase I and II studies) have been promulgated.[28] In contrast, if the population under study is complex, considered vulnerable, at high risk for complications due to a disease state, or if the study is placebo controlled with treatment assignment blinded, either a SMC or DSMB may be appropriate. In Phase II studies, where larger numbers of patients will be recruited, a DSMB is likely most

TABLE 41.7 Guidelines for Use of a Data Safety Monitoring Board in Clinical Trials

Use a DSMB	DSMB Generally Not Used
• Multicenter clinical trials, especially Phase II and III studies	• Early trials (Phase I) in limited numbers of patients
• Outcomes are significant morbidity or mortality	• Endpoints are lesser outcomes: symptoms, unless there is risk for more severe outcomes
• Greater than minimal risk to participants	• Minimal risk to participants
• For earlier phase trials (Phase I and Phase II), a DSMB may be appropriate if the studies involve multiple clinical sites, the studies are blinded or masked, or the studies involve high-risk interventions or vulnerable populations	

appropriate. In well-controlled adequately powered studies (Phase III), the DSMB has the primary function of monitoring the safety of patients enrolled in the clinical trial.

Establishing the DSMB Recent guidelines on the establishment and operations of DSMBs have been promulgated by the FDA and World Health Organization (http://www.who.int/tdr/cd_publications/pdf/operat_guidelines.pdf) and (http://www.fda.gov/cber/gdlns/clintrialdmc.pdf). In addition to providing guidance on determining the need for a DSMB, these documents identify the key components to the operations of a DSMB. These topics include membership, establishing the operating procedures, setting objectives and responsibilities, and the reporting requirements of the DSMB. Additional topics include interactions between the DSMB and the sponsor and/or regulatory agencies.

Membership of the DSMB will also vary depending on the trial. A DSMB may be as simple as three members, one of whom should be a biostatistician familiar with the design and analysis of clinical trials. For a pediatric clinical trial, the other two members should be clinicians familiar with the care of pediatric patients and the specific disease being studied. Additional membership may be considered and include a clinical trials design specialist, patient advocate and/or community member, and a bioethicist. For pediatric trials, experts in the care and perhaps in the behavioral aspects of children may be important, especially if multiple or prolonged procedures over an extended period of time are contemplated.

Each DSMB will require a charter that clearly states the objectives and procedures for the board. Examples and templates for DSMBs are available.[29,30] Some institutions have standing DSMBs for reviewing and monitoring safety of clinical trials conducted by their members (or faculty in the case of academic institutions). Studies funded by the NIH that are multisite and involve interventions that pose more than a minimal risk to the participants are required to have a DSMB to monitor the trial.

The DSMB charter will address several key issues: who is on the board, what their responsibilities will be, how the DSMB will discharge its responsibilities, and how it will report findings. Membership has been discussed earlier. The first responsibility of the DSMB is to review the protocol (in some instances the DSMB may have been part of the protocol development process), the statistical analysis plan (often a draft as the final analysis plan may not be finalized until just prior to final database lock), and the safety monitoring plan. The DSMB should establish how often meetings will occur and what data will be reviewed. The latter will be highly dependent on the data monitoring plan established by the sponsor and/or investigator(s). Analyses of interim data should be specified in the charter or a separate DSMB Interim Analysis Plan. The possible actions of the DSMB review should be specified and generally fall into three outcomes: continue the clinical trial as planned, continue with modifications, or stop the clinical trial. Issues related to reporting and acting upon review results are beyond the scope of this chapter, but many resources are available.[31,32] Many other roles of the DSMB have been suggested and are controversial, such as monitoring of enrollment, advising on modifications of the protocol, and recommendations for discontinuation of some participating centers.[33]

Conduct of DSMB Meetings Decisions need to be made as to what information will be available to the DSMB for review. Some understanding of the quantity of the data relative to the total amount that will be available at the end of the study may put the information into context. In addition, the quality of the data would be assessed: how much has been monitored by an independent source, how much if any has been queried for inconsistencies, what information is missing from reports of serious adverse events or other key safety data (particularly follow-up or information on resolution of events), and how much data if any has been quality assured. Based on expectations of availability, the DSMB may decide the appropriate intervals between meetings. There should be flexibility for ad hoc meetings to address issues that arise unexpectedly.

Reporting by the DSMB The methods by which the DSMB reports its findings should be specified in the board charter. In general, information is transmitted to a designated representative of the sponsor or the investigator (or protocol steering committee if there is one). The reporting must protect the blind of the study and as such should be brief, without providing rationale or potential information that would allow deciphering the deliberations and thinking of the DSMB. Interactions with regulatory agencies are also generally done in such a way as to prevent unblinding. It is up to the sponsor, investigator, or steering committee to transmit the conclusions of the DSMB to other investigators or regulatory agencies. Investigators are responsible for transmitting the outcomes of a DSMB review to their local institutional review boards.

CONCLUSION

Each clinical trial involving pediatric age patients should have a safety monitoring plan. Monitoring of safety in clinical trials of pediatric patients follows the same general principals of those performed in adults. Key differences are the potential for different responses to medical interventions under study and the heterogeneity of the pediatric population as a whole. In addition, attention should be paid to how differences in pediatric patients influence the operation of a data safety monitoring plan. Individuals with expertise in diseases and care of pediatric patients should be involved at all stages in the development, conduct, data collection, analysis, and communications of a pediatric clinical trial.

REFERENCES

1. Ethical Considerations for Clinical Trials Performed in Children. 2006. Available at http://ec.europa.eu/enterprise/pharmaceuticals/paediatrics/docs/paeds_ethics_consultation20060929.pdf. (Accessed November 15, 2007.)

2. Additional Safeguards for Children in Clinical Investigations. 2006. Available at http://www.accessdata.fda.gov/scripts/cdrh/cfdocs/cfCFR/CFRSearch.cfm?CFRPart=50&showFR =1&subpart Node = 21:1.0.1.1.19.4. (Accessed November 15, 2007.)
3. ICH Topic E2A Clinical Safety Data Management: Definitions and Standards for Expedited Reporting. 1995. Available at http://www.emea.europa.eu/pdfs/human/ich/037795en.pdf. (Accessed November 15, 2007.)
4. Goldstein B, Nadel S, Peters M, et al. ENHANCE: results of a global open-label trial of drotrecogin alfa (activated) in children with severe sepsis. *Pediatr Crit Care Med.* 2006;7(3): 200–211.
5. Nadel S, Goldstein B, Williams MD, et al. Drotrecogin alfa (activated) in children with severe sepsis: a multicentre phase III randomised controlled trial. *Lancet.* 2007;369(9564):836–843.
6. ICH Topic E6 (R1) Guideline for Good Clinical Practice. European Medicines Agency (EMEA). 2002. Available at http://www.fda.gov/cder/guidance/959fnl.pdf. (Accessed November 15, 2008).
7. Common Terminology Criteria for Adverse Events v3.0 (CTCAE). 2006. Available at http://ctep.cancer.gov/forms/CTCAEv3.pdf. (Accessed November 15, 2008).
8. Trotti A, Colevas AD, Setser A, et al. CTCAE v3.0: development of a comprehensive grading system for the adverse effects of cancer treatment. *Semin Radiat Oncol.* 2003;13(3):176–181.
9. Division of Microbiology and Infectious Diseases Pediatric Toxicity Tables. 2001. Available at http://www3.niaid.nih.gov/research/resources/DMIDClinRsrch/toxtables.htm. (Accessed November 15, 2008).
10. Division of AIDS Table for Grading the Severity of Adult and Pediatric Adverse Events. 2004. Available at http://rcc.tech-res.com/DAIDS%20RCC%20Forms/TB_ToxicityTables_DAIDS_AE_GradingTable_FinalDec2004.pdf. (Accessed November 15, 2008).
11. Division of AIDS (DAIDS) Table for Grading Severity of Pediatric (<3months of age) Adverse Experiences. 1994. Available at http://hsc.unm.edu/som/gcrc/ftp/ToxicityTables_Pediatric_Under3MonthsAge.pdf. (Accessed November 15, 2008).
12. De Jonge M., Burchfield D, Bloom B, et al. Clinical trial of safety and efficacy of INH-A21 for the prevention of nosocomial staphylococcal bloodstream infection in premature infants. *Pediatr.* 2007;151(3):260–265.
13. Woods D, Thomas E, Holl J, Altman S, Brennan T. Adverse events and preventable adverse events in children. Institute for Health Services Research and Policy Studies, Feinberg School of Medicine, Northwestern University, 339 E Chicago Ave, Chicago, IL 60611, USA woods@northwesternedu 2005;115 (1): 155–160.
14. Guidance for Industry E11. Clinical Investigation of Medicinal Products in the Pediatric Population. 2000. Available at http://www.fda.gov/cder/guidance/4099FNL.PDF. (Accessed November 15, 2008).
15. Wen PY, Yung WK, Lamborn KR, et al. Phase I/II study of imatinib mesylate for recurrent malignant gliomas: North American Brain Tumor Consortium Study 99-08 *Clin. Cancer Res.* 2006;12(16):4899–4907.
16. Whitehead J. *The Design and Analysis of Sequential Clinical Trials*, 2nd ed. New York: Wiley; 1997.
17. Blood Sampling Guidelines. Available at http://healthcare.partners.org/phsirb/bldsamp.htm. (Accessed November 15, 2008).
18. Capparelli EV, Bloom BT, Kueser TJ, et al. Multicenter study to determine antibody concentrations and assess the safety of administration of INH-A21, a donor-selected human staphylococcal immune globulin, in low-birth-weight infants. *Antimicrob. Agents Chemother.* 2005;49 (10):4121–4127.
19. Anderson BJ, Allegaert K, Holford NH. Population clinical pharmacology of children: general principles. *Eur. J. Pediatr.* 2006;165(11):741–746.

20. Meibohm B, Laer S, Panetta JC, Barrett JS. Population pharmacokinetic studies in pediatrics: issues in design and analysis. *AAPS J.* 2005;7(2):E475–E487.

21. Falcao AC, Fernandez de Gatta MM, Delgado Iribarnegaray MF, et al. Population pharmacokinetics of caffeine in premature neonates *Eur. J. Clin. Pharmacol.* 1997;52(3):211–217.

22. Joyce TH 3rd. Topical anesthesia and pain management before venipuncture. *Pediatrics.* 1993;122 (5 Pt 2):S24–S29.

23. Saez-Llorens X, Nelson RP Jr, Emmanuel P, et al. A randomized, double-blind study of triple nucleoside therapy of abacavir, lamivudine, and zidovudine versus lamivudine and zidovudine in previously treated human immunodeficiency virus type 1-infected children. The CNAA 3006 Study Team. Institute for Health Services Research and Policy Studies, Feinberg School of Medicine, Northwestern University, 339 E Chicago Ave, Chicago, IL 60611, USA woods@northwesternedu 2001;107(1):E4.

24. Englund JA, Baker CJ, Raskino C, et al. Zidovudine, didanosine, or both as the initial treatment for symptomatic HIV-infected children AIDS Clinical Trials Group (ACTG) Study 152 Team. *N, Engl. Med.* 1997;336(24):1704–1712.

25. Ellenberg SS, Independent data monitoring committees: rationale, operations and controversies *Stat Med.* 2001;20(17–18):2573–2583.

26. Policy and Guidelines for Data and Safety Monitoring. 2001. Available at http://www3.niaid.nih.gov/research/resources/DMIDClinRsrch/PDF/dsm_pol_guide.pdf. (Accessed November 15, 2008).

27. NIAID Principles for Use of a Data and Safety Monitoring Board (DSMB) for Oversight of Clinical Trials. 2006. Available at http://www3.niaid.nih.gov/research/resources/toolkit/attachments/DSMB_PolicyAug2006.pdf. (Accessed November 15, 2008).

28. Dixon DO, Freedman RS, Herson J, et al. Guidelines for data and safety monitoring for clinical trials not requiring traditional data monitoring committees *Clin Trials (London).* 2006;3(3):314–319.

29. A proposed charter for clinical trial data monitoring committees: helping them to do their job well. *Lancet.* 2005;365 (9460): 711–722.

30. University of New Mexico Health Sciences Center General Clinical Research Center (GCRC) Data and Safety Monitoring Board Charter. 2006. Available at http://hsc.unm.edu/som/gcrc/ftp/DSMB charter template v3.doc. (Accessed November 15, 2008).

31. Califf RM, Ellenberg SS. Statistical approaches and policies for the operations of Data and Safety Monitoring Committees. *Am. Heart J.* 2001;141(2):301–305.

32. DeMets D, Califf R, Dixon D, et al. Issues in regulatory guidelines for data monitoring committees. *Clin Trials (London).* 2004;1(2):162–169.

33. Tharmanathan P, Calvert MJ, Freemantle N. Striding forward or getting too big for their boots? The developing role of data monitoring committees in clinical trials. *J. Clin. Pharm. Ther.* 2006;31(2):111–118.

CHAPTER 42

Pharmacovigilance: Assessing Safety Postapproval

KAREN M. KAPLAN, MD

Johnson & Johnson Pharmaceutical Research & Development LLC, Titusville, New Jersey 08560

WHAT IS PHARMACOVIGILANCE?

Pharmacovigilance (PV) is a branch of the broader discipline of pharmacoepidemiology, and it has adverse drug reactions (ADRs)[1] as its primary focus. PV is fundamentally a public health function and comprises the full spectrum of safety evaluation after pharmaceutical products (drugs, vaccines, and devices) are introduced into the market. It serves to characterize the practical use of a product in an expanded population base and for its "life span" (i.e., as long as the drug, vaccine, or device is used as a therapeutic agent).

Safety monitoring during clinical trials is best described as "event" reporting: all adverse events that occur in the course of a trial are reported, only a subset of which are deemed to be true ADRs. Postapproval safety monitoring is based on suspicion reporting: healthcare providers, consumers, or others voluntarily report events that occur during treatment, based on the *belief* that they may be drug related (suspected ADRs). Adverse events may be related to drug exposure, progression of an underlying disease, comorbid conditions, concomitant drug exposures, or drug interactions, for example. A high index of suspicion is essential to recognize an ADR, particularly those that mimic comorbid conditions or those with a long latency. Nevertheless, the burden on the reporter is not to diagnose but to *suspect* based on what may be preliminary or relatively limited information. The centerpiece of PV methods is surveillance, (postmarketing or postapproval safety surveillance), a traditional public health tool; but PV includes all elements that contribute to assessing both the risk and benefit of a pharmaceutical product. Pharmacoepidemiology studies also are tools used in PV.

WHY IS PHARMACOVIGILANCE NECESSARY?

Safety information derived from preauthorization clinical trials is limited to events that occur frequently enough to be observed in the several thousand patients typically studied. In

Pediatric Drug Development: Concepts and Applications
Edited by Andrew E. Mulberg, Steven A. Silber, and John N. van den Anker

pediatric clinical trials used for registration, it is likely that the exposures will be even less than currently acceptable for adult clinical trials. Events with an incidence of at least 1/100 will be detected in clinical trials; events with an incidence of 1/1000 or less will not be reliably detected.[2] Until recently, very few drugs used to treat children had been studied in pediatric clinical trials; furthermore, those in whom they had been tested were generally a well-defined group with only the condition of interest. In clinical trials, surrogate measures often are used instead of the outcome of interest, particularly for longer-term outcomes (e.g., a trial of a diabetes drug measures blood glucose rather than the vascular complications of diabetes). Clinical trials also tend to be brief. The use of surrogate markers for pediatric clinical trials is discussed in Chapter 39 by Molenberghs and Orman in this book. Events that occur after prolonged exposure or events with a long latency (such as adult-onset events in children treated with cancer chemotherapy) would not be detected in clinical trials. Patients in clinical trials also have intensive safety monitoring with regularly scheduled visits and diagnostic testing. In clinical practice, patients are sicker, have other comorbid conditions, and take other medications. Office visits or diagnostic tests are scheduled as needed. Patients also may take the medicines for longer periods of time and for conditions other than those originally studied. Efficacy of a given drug is better evaluated in clinical trials than is safety. At the time a new product is approved, its safety profile has not yet been fully characterized.

BASIC PHARMACOVIGILANCE METHODS

The sources of risk from pharmaceutical products can be conceived as falling into four categories (Figure 42.1). Known events include the avoidable (often dose related) and those that are unavoidable or idiosyncratic. Medication errors and product quality defects also are sources of risk. Most important in the period after product approval are the remaining uncertainties (the unknown or unexpected). Although safety monitoring postapproval addresses all four categories, it is expected to be the primary source of information about remaining uncertainties.

Safety surveillance focuses on identifying signals that suggest a potential safety concern. These signals are analogous to signs or symptoms in clinical medicine (hypothesis generating), and they require additional tools for analysis (hypothesis testing). A signal is "an alert from any available source that a drug may be associated with a previously unrecognized hazard or that a known hazard may be quantitatively (e.g., more frequent) or qualitatively (e.g., more serious) different from existing expectations."[3] Signals can originate from multiple sources including from preclinical studies in animals, clinical trials (pre- or postapproval), the biomedical and scientific literature, epidemiological studies, data collected for other purposes such as vital statistics, disease registries, prescription or outcome databases, and postapproval spontaneous adverse event reporting.[4] In the case of very rare events, even a single (sentinel) case can represent a signal. In these situations, the SNIP criteria (strength of the association, newness of the event, clinical importance of the event, and potential for preventive measures) are useful in interpreting the clinical information.[5]

Spontaneous voluntary reporting of adverse events by healthcare providers, consumers, and others has been the foundation of traditional pharmacovigilance globally since its inception in the mid-20th century. Cases of suspected ADRs are reported to pharmaceutical manufacturers or health authorities or both. Case reporting to health authorities is mandatory

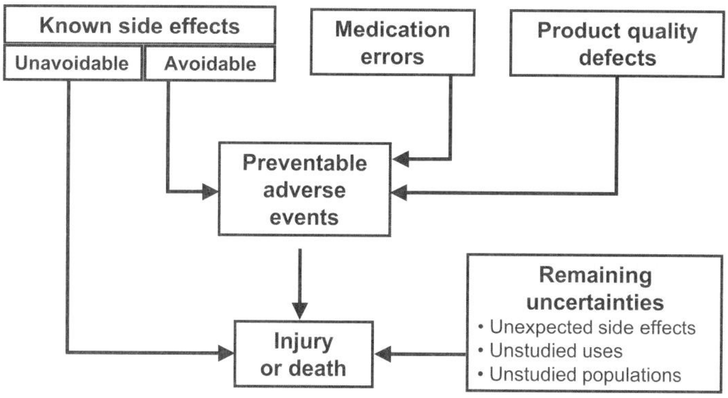

FIGURE 42.1 Sources of risk from drug products.[6]

for pharmaceutical companies. Of all cases in the Food and Drug Administration (FDA) Adverse Event Reporting System (AERS), 95% were first reported to pharmaceutical companies.[6] Approximately 460,000 cases believed related to drugs or devices are reported to the FDA via AERS annually:[6] as of 2006, the database totaled 2,885,276 cases.[7] Vaccine adverse events are reported to the FDA via the Vaccine Adverse Events Reporting System (VAERS) established in 1990. VAERS receives approximately 11,000 reports annually.[8]

The system of voluntary reporting has many weaknesses; including substantial underreporting, the magnitude of which is difficult to estimate and which varies depending on the product, the length of time a product has been on the market, and the nature of the event itself. Reports that are received represent only a fraction of occurrences. Databases based on spontaneous reporting have no denominator, making an estimate of true incidence impossible. Coding practices and even coding dictionaries vary among companies and between companies and health authorities. Numbers of reported cases often increase in the first years after a product enters the market and decline thereafter (the "Weber effect")[9] and can increase in response to publicity, direct to consumer advertising, and/or regulatory actions. How does one put the numbers of cases into a population context? Furthermore, the information received on individual cases may be so scanty as to preclude an in-depth understanding. Yet despite clear weaknesses, the system of passive reporting of adverse events *has* identified safety issues in marketed drugs and has lead to interventions by health authorities.

A supplement to traditional PV methods has been the use of data mining. Long used in many other venues, including retail sales and national security, data mining is an automated statistical technique in which large databases, such as adverse event databases, are searched or "mined" to detect strong consistent associations that occur at higher than expected frequencies. Data mining yields signal "scores" that reflect statistical rather than causal associations and is thus also a hypothesis-generating exercise. Unfortunately, the weaknesses and reporting biases that affect the interpretation of voluntary adverse event reports also affect the results of data mining the database. In the case of signal detection in children, cases in patients 18 years of age or younger represented 6.4% of cases in the FDA AERS database in 2006.[7] In 31% of cases, age was not reported.[7] Thus, whether data mining as it is currently used will be useful for the pediatric age group is yet to be determined.

Nevertheless, the potential exists for data mining to serve as an early warning system for the monitoring of pharmaceutical safety. The FDA issued a guidance document for the pharmaceutical industry in 2005 on pharmacovigilance practices, and included is a discussion of data mining for safety signal detection.[10]

Pharmacoepidemiologic studies represent the hypothesis-testing component of PV. The specific designs of the studies (case–control, cohort, or randomized clinical trial) depend on the potential safety signals to be addressed. Studies help answer the question: Does this signal represent a true safety risk with this product? As with any epidemiologic study, great care must be taken in the design, including case definition, and selection of population or existing data source.[11]

WHY PEDIATRIC PHARMACOVIGILANCE?

More children are living with chronic illnesses than at any time in history, and more children are using more therapeutic agents. More vaccines, one of the top 10 greatest public health achievements of the 20th century,[12] are available to prevent infectious diseases that once accounted for substantial childhood morbidity and mortality.

According to the 2005 National Health Interview Survey, more than 9 million U.S. children (13%) less than 18 years of age have "ever" had a diagnosis of asthma; 6.5 million (9%) still have it. Eleven percent suffered from hay fever in the previous 12 months. Nearly 4 million (7%) children 3–17 years of age have attention deficit hyperactivity disorder (ADHD),[13] of whom more than half are reported to take medication for the disorder.[14] The prevalence of epilepsy in the 0–15 age group is estimated to be 0.4%.[15] Children, less than the age of 18, with special healthcare needs ("those who have or are at increased risk for a chronic physical, developmental, behavioral, or emotional condition and who also require health and related services of a type or amount beyond that required by children generally"[16]) are estimated at 13.5 million in the United States.[17]

The 5-year relative survival rate for children with cancer has improved from less than 50% before the 1970s to nearly 80% in 2007. An important consequence is the potential for the development of late onset treatment-related adverse events, including cognitive impairment, progressive cardiotoxicity, endocrinopathies, and second malignancies.[18,19]

Children and adolescents also are taking more prescription medications: in 2005, 9.7 million children less than 18 years of age in the United States (13%) had a health problem for which a prescription medication had been taken regularly for at least 3 months. Sixteen percent of youths 12–17 were on regular medication, compared to 14% and 9% in the 5–11 and less than 5 year old age groups, respectively.[20] Whereas once children received brief courses of treatment for infectious and/or self-limited conditions, children are now being treated for chronic conditions for longer periods of time. Monitoring the safety of these products is essential: whether methods that address drug safety in an aging population also meet the needs of children is not clear.

WHAT IS UNIQUE ABOUT MONITORING DRUG SAFETY IN CHILDREN?

Because of the dramatic physical and developmental–behavioral changes that characterize the period from birth through adolescence, as well as the fact that some disease processes are unique to children, it is no surprise to find that ADRs in children differ from those observed in adults. Furthermore, treatment of a disease in childhood confers potential additional risks related to drug exposure during periods of accelerated growth and development (e.g., effects on linear growth or sexual maturation). Children also are at risk for events related either to long-term exposure to pharmaceutical products (for children with chronic illnesses) or events that emerge after treatment has ceased (e.g., long-term complications of treatment for respiratory disease in the premature infant or of cancer treatment in childhood).

Are traditional PV approaches sufficient for the monitoring of safety in children? Two important 20th century drug safety tragedies that disproportionately affected children, elixir of sulfanilamide in 1938 and thalidomide in the 1960s, catalyzed changes in the laws and regulations governing the testing and marketing of new medications. Paradoxically, adults were the principal beneficiaries of these changes: the evaluation of drugs in children was discouraged, and children became "therapeutic orphans."[21] Pediatric patients remain a vulnerable group with respect to what is known about the safety of pharmaceutical products.[22]

A number of ADRs are unique to *in utero* and childhood exposure. Thalidomide only causes phocomelia during limb formation *in utero*. Accumulated cloramphenicol, secondary to immature hepatic metabolic pathways of the newborn, causes gray baby syndrome.[23] Accumulated benzyl alcohol, used as a preservative in intravenous solutions, leads to "gasping syndrome" in small premature infants.[24] Tetracycline stains developing tooth enamel, and corticosteroids can compromise linear growth by affecting growing bones. Yet for the purposes of monitoring drug safety, despite the fact that children are not just small adults, some *perceived* differences also are a result of inadequate drug testing in children.[25]

Several considerations are unique to safety in children. First, drugs are likely to have been tested in few individuals, particularly drugs indicated for uncommon conditions. Thus, serious adverse events are unlikely to have been observed in clinical trials. Second, "off-label" use (i.e., use not consistent with the package insert/product labeling with respect to dosage, age, indication, or route) and unregistered use (extemporaneously created formulations for children, formulations that are modified, or drugs imported or used before a license is granted) have been common.

Products that have been developed specifically for children are subjected to rigorous evaluation and scrutiny by manufacturers and health authorities. The use of products off-label means that accurate information on dosing and safety is unavailable to either the prescriber or the patient/family.[26] The lack of age-appropriate formulations has led to the local development of liquids and suspensions not registered with any health authorities. Some pharmacies extemporaneously prepare liquid dosage forms for children, but these formulations have not been sufficiently tested to determine stability, efficacy, or expiration dating.[26]

A study in pediatric outpatients in France suggested an increased ADR risk associated with off-label use.[27] Studies done in different settings, different countries, and different age groups all have found high rates of off-label drug use.[28,29] Many, but not all, have found an increased ADR risk with off-label use.[30] Pandolfini and Bonati[30] regard off-label and unlicensed drug use as "uncontrolled experiments that do not capture information useful to understanding a drug's effects and in which patients are unknowingly being enrolled."

Medication errors are particularly important in pediatrics, not least because of the regular use of drugs off-label. They can occur during prescribing, dispensing, or administering drugs. Pediatric prescribing is particularly complex: For weight-based prescribing an accurate weight must be obtained and correctly transcribed (pounds or kilograms). In the course of a brief visit, the prescriber then must do the following: convert pounds to kilograms, make rapid weight-based calculations to determine total daily dose, divide daily dose into multiple doses based on the appropriate frequency for the medication, choose the correct preparation and concentration (liquid, chewables, tablets), and determine the amount of liquid/tablet to be taken for an individual dose. Some medications have both weight-based and age-based guidelines, and there are no established standards for switching from weight-based dosing (pediatric) to daily dosing (adult).[31] Because pediatric doses rely on weight-based calculations, the rate of tenfold errors owing to misplaced decimal points has been shown to be more common in children than in adults.[32]

A recent retrospective cohort study reviewed ADRs reported from 1995 to 2004 in a community-based tertiary care children's hospital. The authors found that although all healthcare providers are encouraged to report ADRS, 89% of the 1087 ADRs were reported by pharmacists, 10% by nurses, and <1% by physicians. In 93% of cases, the physician was notified of the ADR, but in only 29% was documentation made in the medical record.[33] Furthermore, healthcare providers often fail to inquire about the use of medications available over-the-counter (OTC) in their patients, and parents often do not perceive them as medications.[34]

Whether the reporting of adverse events is less frequent in children than in adults is not certain. Less is known overall about postapproval drug safety in children compared to adults, particularly in the United States, including reporting frequency, incidence, and preventability.[22] Many studies and commentary on the monitoring of postapproval drug safety in children have originated in Europe. In 2006, the European Medicines Agency (EMEA) issued a comprehensive guideline on pediatric pharmacovigilance that took effect in the European Union (EU) in January 2007.[35]

PEDIATRIC PHARMACOVIGILANCE FOR THE FUTURE

Ultimately, the goals of monitoring the safety of drugs in children are to ensure the safe use of pharmaceuticals and to target events that are preventable. Better quality information, systematically collected, is necessary to achieve both goals. Methods for surveillance of adverse events in children must be tailored more specifically to the pediatric population. All of the limitations of existing approaches to postapproval safety surveillance also apply to their use in pediatrics: the system relies on passive, voluntary reporting; information in case reports is scanty; reporting is not equivalent to incidence; and reports are difficult to interpret in a population context.

For children, better quality information related to age, age group, weight, and indication in reported cases is essential. The emphasis should be on issues important for children (off-label use, medication errors, poisoning) and serious events, as well as the effects of long-term exposure and late-onset effects of drug exposure in infancy or childhood. Active surveillance of pediatric healthcare providers, the use of registries (treatment registries and outcome registries as well as linked registries), and collaboration with existing networks of pediatric subspecialists are needed. Existing sources of data such as data from managed care organizations are potentially useful for capturing events with long latencies.

Studies have shown that training healthcare providers in the reporting of ADRs can be effective. A pilot Paediatric Regional Monitoring Centre (PRMC) established in the United Kingdom in 1998 used intensive education to promote pediatric ADR reporting. In its first year, reporting more than doubled: 25% of reports were for drugs used "off-label," while 15% of the total were regarded as medically significant.[36] A similar project funded by the FDA and carried out in Rhode Island in the late 1980s used physician education to improve the overall reporting of ADRs. Voluntary reporting of suspected ADRs increased 17-fold (including the reporting of serious events) in Rhode Island, and the state's physicians improved their knowledge of and attitudes toward ADR reporting.[37]

Better quality information begins with the reporters and their patients. Education of pediatric healthcare providers is needed that emphasizes adverse event "suspicion" reporting. Pediatricians, in particular, have considerable experience with the concept of suspicion reporting, since it forms the basis for the reporting of child abuse and neglect. Although the reporting of child abuse and neglect is intended to serve the needs of a child and family, and the reporting of ADRs is intended to serve the population: pediatrics has traditionally had a strong orientation toward community and public health. Prevention in general, the prevention of injury and poisoning in particular, and child advocacy are core elements of practice among both pediatric generalists and subspecialists.

Several recent studies on pediatric drug safety in the United States have been carried out by pharmacy professionals.[33,38] More routine collaboration between pediatric and pharmacy professionals would enrich the science of drug safety monitoring in children. A collaborative "culture of reporting" should become a core component of pediatric medical education and standard pediatric practice. Pediatricians are vital sources of information on ADRs and are well suited to play a central role in ensuring the safe use of pharmaceuticals in children.

REFERENCES

1. Mann RD, Andrews EB. Introduction. In: Mann R, Andrews EB, eds. *Pharmacovigilance.* Chichester, England: Wiley; 2002:3.
2. Strom BL. How the US drug safety system should be changed. *JAMA* 2006;295:2072–2075.
3. Waller PC, Lee EH. Responding to drug safety issues. *Pharmacoepidemiol Drug Saf.* 1999; 8:535–552.
4. Mann R, Andrews EB. Introduction. In: Mann R, Andrews EB, eds. *Pharmacovigilance.* Chichester, England: Wiley; 2002:106.
5. Waller PC, Lee EH. Responding to drug safety issues. *Pharmacoepidemiol Drug Saf.* 1999;8:540.
6. FDA. CDER 2005 Report to the Nation: Improving Public Health Through Human Drugs. Rockville, MD: FDA; 2006. Available at http://www.fda.gov/cder/reports/rtn/2005/rtn2005.pdf.
7. FDA. AERS Data Current Through Fourth Quarter, 2006. Released through The Freedom of Information Act. Accessed using WebVDME™ (Phase Forward, Inc).
8. Braun M. Vaccine event reporting system (VAERS): usefulness and limitations. Available at http://www.vaccinesafety.edu/VAERS.htm. (Accessed June 20, 2007.)
9. Weber JCP. Epidemiology of adverse reactions to nonsteroidal anti-inflammatory drugs. *Adv Inflam Res.* 1984;6:1–7.
10. US Department of HHS. *Guidance for Industry: Good Pharmacovigilance Practices and Pharmacoepidemiologic Assessment.* Rockville, MD: FDA; 2005.
11. Mann RD, Andrews EB. Introduction. In: Mann R, Andrews EB, eds. *Pharmacovigilance.* Chichester, England: Wiley; 2002: 8–9.
12. Centers for Disease Control and Prevention. Ten Great Public Health Achievements—United States, 1900–1999. *MMWR.* 1999;48:241–243.
13. Bloom B, Dey AN, Freeman G. Summary health statistics for US children: National Health Interview Survey, 2005. National Center for Health Statistics. *Vital Health Stat.* 2006;10(231).
14. Centers for Disease Control and Prevention. Prevalence of diagnosis and medication treatment for attention-deficit/hyperactivity disorder—United States, 2003. *MMWR.* 2005;54:842.
15. Centers for Disease Control and Prevention. Current trends: prevalence of self-reported epilepsy—United States, 1986–1990 *MMWR* 1994;43:810–811, 817–818.
16. McPherson M, Arango P, Fox H, et al. A new definition of children with special health care needs. *Pediatrics.* 1998;102:137–140.

17. Campaign for Children's Health Care. Available at http://www.childrenshealthcampaign.org/assets/pdf/Children-with-Special-Needs.PDF. (Accessed June 1, 2007.)
18. American Cancer Society. *Cancer Facts & Figures 2007*. Atlanta: American Cancer Society; 2007:11.
19. Alvarez JA, Scully RE, Miller TL, et al. Long-term effects of treatments for childhood cancers. *Curr Opin Pediatr*. 2007;19:23–31.
20. Bloom B, Dey AN, Freeman G. Summary health statistics for US children: National Health Interview Survey, 2005. National Center for Health Statistics. *Vital Health Stat*. 2006;10(23) 5,14.
21. Shirkey H. Editorial comment: therapeutic orphans. *J Pediatr*. 1968;72:119–120.
22. Impicciatore P, Choonara I, Clarkson A, et al. Incidence of adverse drug reactions in paediatric in/out-patients: a systematic review and meta-analysis of prospective studies. *Br J Clin Pharmacol*. 2001;52:77–83.
23. Stephenson T. How children's responses to drugs differ from adults. *Br J Clin Pharmacol*. 2005;59:670–673.
24. Gershanik J, Boecler B, Ensley H, McCloskey S, George W. The gasping syndrome and benzyl alcohol poisoning. *N Engl J Med*. 1982;307:1384–1388.
25. Stephenson T. How children's responses to drugs differ from adults. *Br J Clin Pharmacol*. 2005;59:670.
26. Christensen ML, Helms RA, Chesney RW. Is pediatric labeling really necessary? *Pediatrics*. 1999;104: 593–597.
27. Horen B, Montastruc JL, Lapeyre-Mestre M. Adverse drug reactions and off-label drug use in paediatric outpatients. *Br J Clin Pharmacol*. 2002;546:665–670.
28. Pandolfini C, Bonati M. A literature review on off-label drug use in children. *Eur J Pediatr*. 2005;164:552–558.
29. Shah SS, Hall M, Goodman DM, et al. Off-label drug use in hospitalized children. *Arch Pediatr Adolesc Med*. 2007;161:282–290.
30. Pandolfini C, Bonati M. A literature review on off-label drug use in children. *Eur. J Pediatr*. 2005;164:553.
31. McPhillips H. Medication safety in ambulatory pediatrics. In: Presentation to American Academy of Pediatrics, Safer health care for kids webinar. February 15, 2007.
32. Lesar TS. Tenfold medication dose prescribing errors *Ann Pharmacother*. 2002;36:1833–1839.
33. Le J, Nguyen T, Law AV, et al. Adverse drug reactions among children over a 10-year period. *Pediatrics*. 2006;118:555–562.
34. Gunn VL, Taha SH, Liebelt EL, Serwint JR. Toxicity of over-the-counter cough and cold medications. *Pediatrics*. 2000;108:e52.
35. European Medicines Agency Post-authorization Evaluation of Medicines for Human Use. Committee for Medicinal Products for Human Use (CHMP). Guideline on Conduct of Pharmacovigilance for Medicine Used by the Paediatric Population. London, UK. 2007. Available at http://www.emea.europa.eu/pdfs/human/phvwp/23591005en.pdf. (Accessed June 15, 2007.)
36. Clarkson A, Ingleby E, Choonara I, et al. A novel scheme for the reporting of adverse drug reactions. *Arch Dis Child*. 2001;84:337–339.
37. Scott HD, Thacher-Renshaw A, Rosenbaum SE, et al. Physician reporting of adverse drug reactions. Results of the Rhode Island adverse drug reaction reporting project. *JAMA*. 1990;263:1785–1788.
38. Temple ME, Robinson RF, Miller JC, et al. Frequency and preventability of adverse drug reactions in paediatric patients. *Drug Saf*. 2004;27:819–829.

PART IX

FORMULATION, CHEMISTRY, AND MANUFACTURING CONTROLS

CHAPTER 43

Formulation of Pediatric Dosage Forms

GERARD P. MCNALLY, PhD

McNeil Consumer Healthcare, 7050 Camp Hill Road, Fort Washington, Pennsylvania 19034

ANIRUDDHA M. RAILKAR, PhD

Johnson & Johnson Pharmaceutical Research and Development, LLC, Welsh and McKean Roads, Spring House, Pennsylvania 19477

INTRODUCTION: CHALLENGES TO A SUCCESSFUL PEDIATRIC FORMULATION

The development of pediatric dosage forms presents a significant challenge to the formulation scientist. Children are not "little adults" and, as such, a small percentage of an adult dose should not be administered to ensure that there are not safety or efficacy issues. Especially in the cases of children, having the right dosage form is as important as the drug being administered because the dose has to be successfully delivered. Since children go through various stages of growth and development, accurate administration of the dose based on age and weight is critical and the importance of dose changes needs to be understood. Development of age-appropriate dosage forms often involves consideration of the appropriate form, including liquids/suspensions, powders, chewable tablets, and tablets that can be swallowed, and understanding the impact on cost of development. However, since the market for pediatric dosage forms is small compared to the adult market, it is difficult for pharmaceutical companies to get a good return on their investment.

Most pharmaceutical actives are unpleasant tasting, and that taste can range from a lingering chemical taste to a harsh bitterness with intensities varying from moderate to high. This presents a challenge for the formulation scientist as the most common pediatric dosage forms, namely, liquids and chewable tablets, have the greatest potential for exposure of the drug in the oral cavity, resulting in bad taste and poor compliance. To deal with these taste issues, masking of bad tasting actives is seen as necessary and different

Pediatric Drug Development: Concepts and Applications
Edited by Andrew E. Mulberg, Steven A. Silber, and John N. van den Anker

technical approaches have been developed for different dosage forms and active drug compounds.[1,2]

The more complex nature of pediatric formulations presents other challenges in the development of a new drug product that must be highlighted. These include a more complicated analytical method development, primarily due to the presence of flavors and sweetening systems. Another related area that becomes more challenging is the stability program in which product changes, such as color, odor, and flavor, need to be monitored in addition to the normal chemical and physical stability. For pediatric liquid products, packaging and dosage administration present some unique requirements from component selection to the calibration of dose delivery devices.

The criticality of pediatric formulations has been strengthened by recent legislative changes in both the United States and European Union. Since 1997, three legislations, two in the United States and one in the EU (which was modeled after the U.S. legislation), have laid the groundwork to provide incentives to companies developing pediatric dosage forms. In the United States, the Best Pharmaceuticals for Children Act (BPCA) of 2002 provides a 6-month patent extension to a drug that is approved by the U.S. FDA for use in adults if the manufacturer has done its due diligence and conducted clinical trials to monitor the safety and efficacy of the dosage form in pediatric subjects and updated the drug label. The second legislation, in 2003, known as the Pediatric Research Equity Act (PREA), mandates that pharmaceutical companies testing new chemical entities (NCEs) carry out pediatric clinical trials if the drug is likely to be used in that population. The European legislation, referred to as Regulation (EC) No 1901/2006 of the European Parliament and of the Council on Medicinal Products for Paediatric Use and Amending Regulation (EEC) No 1768, also provides a 6-month patent extension to drugs that have been studied in children provided pharmaceutical companies adhere to strict standards for testing drugs in pediatric subjects. These topics are covered extensively in Chapter 12 by Maldonado and in Chapter 13 by Rose. Several notable differences between the U.S. and EU legislation applies to off-patent drugs. The EU legislation covers drugs that are not yet approved, patented medications currently on the market, and off-patent drugs. The EMEA has also recommended formulation options based on the age of pediatric patients.[3]

PHARMACOKINETIC ASPECTS OF PEDIATRIC DELIVERY

As children develop and mature, they go through many physical and maturational changes. Age-related changes also occur in drug disposition and metabolism. Differences between adults and children are seen in pH along the entire gastrointestinal tract (GIT), in gastric residence and emptying time, in intestinal transit time, in the nature of the bacterial population, and in P-glycoprotein. In addition, differences in the amount of total body water (resulting in changes in volume of distribution), clearance, plasma protein binding, fat content, blood flow, and enzymatic activity are also observed.[4] The FDA also has a draft guidance for industry on the conduct of clinical trials.[5] The American Academy of Pediatrics has also published a guideline on the ethical conduct of pediatric clinical studies.[6] The reader is referred to more extensive coverage of the pharmacokinetic principles underlying the distinctive differences in infants, children, and adolescents and their respective challenges in Chapters 18 and 19 on clinical pharmacology.

SPECIFICS OF FORMULATION DEVELOPMENT: INGREDIENT CONSIDERATIONS

Active Pharmaceutical Ingredients (APIs)

The issues with APIs can be discussed under two categories: new chemical entities (NCEs) or new molecular entities (NMEs) and existing APIs. A new chemical entity is the one that has never been approved before through a regulatory agency. The problems associated with formulation development for NCEs are that there is no existing clinical information so choosing a dose range is challenging. Similarly, there is very little information about safety and efficacy. At earlier stages of development, drug supply is limited, and preparing more complicated dosage forms is a challenge. It is easier to prepare simple capsule or tablet formulations. On the other hand, there is a lot of information for existing APIs and extrapolating that information to the development of pediatric formulations is much easier. There are also no issues with drug supply; thus, this allows for adequate experimentation in developing complex formulations.

Inactive Ingredient Selection

Typically, pediatric formulations contain more ingredients than the comparable adult formulation(s). Some of the additional ingredients are sweeteners, coloring agents, flavorants, wetting agents, thickening agents, pH modifiers, antioxidants, and preservatives. These ingredient categories are discussed in detail later in this chapter. As with any pharmaceutical formulation development program, preformulation studies need to be undertaken to determine the compatibility of the various excipients with the active pharmaceutical ingredient. These preformulation studies can become quite extensive for elegant pediatric liquid, suspension, or chewable tablet formulations. Other factors to be considered in ingredient selection for pediatric products, in particular, are side effects caused by the inactive ingredients, such as gastric irritation, diarrhea, allergy, and hypersensitivity. In general, the formulator,s challenge is to limit the number of inactive ingredients contained in a formulation while achieving a stable and aesthetically acceptable finished drug product.

PEDIATRIC DOSAGE FORMS

Since greater than 90% of pediatric dosage forms are for oral administration, formulation considerations for oral dosage forms only will be discussed. Age-appropriate dosage forms for children from 0 to 2 years usually constitute concentrated liquids and suspensions in order to reduce the volume of drug required per dose. For children 2–12 years old, the following dosage forms are appropriate: liquids and suspensions, granules and other multiparticulate dosage forms, and chewable tablets. Since there is little overlap in the formulation science of liquid and solid dosage forms, they will be discussed separately. The topic of taste masking is touched on under both liquid and solid dosage forms since different approaches may be taken based on the nature of the form. Developers of pediatric medicines need to be aware of children,s taste preferences, as the product taste will greatly affect the child,s willingness to take the medication repeatedly. This may be borne out in the parent or caregiver,s willingness to repurchase the product in the case of an over-the-counter drug product or comply with the dosing schedule if it is a prescription product such as an antibiotic. As such, compliance becomes a key driver of product formulation in the world of pediatric medicine. A number of studies investigating the impact of taste on compliance in the pediatric population have been published.[7–9]

Liquid Dosage Forms

Liquid preparations that are commonly prescribed for children under age 6 years include (1) various forms of solutions, such as simple aqueous solutions, elixirs, oils, and syrups; and (2) suspensions. In general, solutions are preferred over suspensions primarily because of the easier development and associated manufacturing process, formula stability, and assurance on dose uniformity and accuracy in dose delivery. Unfortunately, most liquid formulations, especially solutions, have an unpleasant taste that requires consideration of a variety of taste masking strategies.

Solutions The drug particles (the solute) are dissolved in a solvent to obtain a solution. Another description is that a solution is a homogeneous mixture of two or more ingredients forming a single phase. The primary advantages of solution formulations over suspensions are ease of formulation, homogeneity, and stability. The typical components of a solution are discussed next.

Solvents Water is the most commonly used solvent, but, in many instances, it may be necessary to use aqueous acidic or basic buffering or pH-modifying agents to maintain the drug particles in a dissolved state. Nonaqueous solvents such as alcohol, glycerin, polyethylene glycol and propylene glycol are occasionally used either on their own or in combination with water. In general, the use of alcohol is avoided where possible in pediatric products. Glycerin, propylene glycol, and polyethylene glycol are routinely used in combination with USP water in the preparation of solution formulations. The preferred grades of polyethylene glycol for use in pediatric liquids are PEG 1450, 4000, and 8000. There have been reports that propylene glycol can cause hyperosmolality in children.[10]

Bulk Sweeteners Sweeteners are used in the formulations to enhance palatability and, to some extent, increase viscosity. Commonly used bulk sweeteners are high fructose corn syrup, sucrose, maltitol, xylitol, and sorbitol. Since the 1970s, several artificial sweeteners have also become popular with formulators and are extremely useful in developing pleasant tasting pharmaceutical preparations; see Table 43.1 for a detailed list. Some sweeteners such as sorbitol can cause laxation (increased stool output) if taken in large enough amounts. The use of sugars needs to be considered in light of rising obesity among children and also in cases where long-term medication could have an impact on control of metabolic diseases, such as diabetes. There are also potential adverse reactions or events related to high fructose corn syrup such as contraindications for children with a certain metabolic disease, for example, hereditary fructose intolerance (HFI).

Thickening Agents In some instances, it may be necessary to increase the viscosity of an oral solution to improve the dosing accuracy, especially if the drug is dosed using a dropper.

TABLE 43.1 Sweeteners

Sweetening Agent Usage	Preferred Agent		No Concerns		Some Concerns		Not Preferred	
	Rx	OTC	Rx	OTC	Rx	OTC	Rx	OTC
Sucrose	X	X						
Sorbitol	X	X						
High fructose corn syrup					X	X		
Xylitol	X	X						
Maltitol	X	X						
Aspartame[a]							X	X
Sodium saccharine (saccharin)					X	X		
Sodium cyclamate[b]							X	X
Sucralose			X	X				
Potassium acesulfam			X	X				

[a]Unstable in liquid formulation especially at higher pH.
[b]Banned in United States.

Generally, adding low concentrations of gelling agents or gums, such as hypromellose, xanthan gum, or microcrystalline cellulose, increases viscosity.

Preservatives To prevent microbial growth in the solution, it is necessary to add preservatives. The most commonly used preservatives are methyl paraben, propyl paraben, and butyl paraben. Other systems include benzoic acid or benzoic acid/sodium benzoate combinations. Additionally, salts such as sodium edentate or potassium sorbate and acids such as sorbic acid are commonly used to preserve liquid formulations. The aforementioned preservatives have varying solubilities in water based on the pH of the system and therefore should be chosen appropriately.

Flavors, Dyes, and Colors Flavors are added to improve the palatability of the dosage form. Typical flavors used are fruit flavors and bubble gum flavor. One of the most common flavors used in adult medicine is mint; this is *not* a flavor that appeals to children. Colors improve the aesthetics and appearance of the dosage form, thus aiding compliance with the dosing regimen. However, it should be noted that this section references common excipients used in formulations manufactured in the United States. In general, choice of excipients having a sensory impact such as colors and flavors should be based on local cultural and aesthetic preferences.

These ingredients should be chosen on the basis of results from excipient compatibility and stability at normal and accelerated conditions. Care should be taken not to alter the safety, efficacy, and bioavailability of the dosage form. Another important component related to flavors and colors is the packaging materials. Certain packaging materials can have an impact on the loss of flavor and opacity of the components can cause fading of the formulation color over time and must be considered in developing the finished product.

Antioxidants In the case of certain drug compounds susceptible to degradation by oxidation, it may be necessary to add antioxidants such as butylated hydroxyanisole or butylated hydroxytoluene.

EXAMPLE OF EOLUTION FORMULATION

Ingredient	%Wt/Vol
Cosolvents	30–45
Bulk sweetening agent	30–45
pH modifier	0.1
Thickening agent	0.1–0.5
Preservative	0.05
Flavor	0.5–1.0
Colorant	0.05–0.10
Active ingredient	1–5.0
Purified water USP	q.s. to 100%

Suspensions Pharmaceutical suspensions are systems in which a solid drug compound is dispersed in a liquid. Thus, suspensions are a two-phase system and are more complex to formulate and manufacture than solutions. The rationale for formulating a suspension may stem from a drug being poorly soluble in a suitable solvent or from a drug having a tendency to degrade in solution. From the pediatric patient's perspective, another reason for developing the more complex suspension formulation of a particular drug is that a bad tasting solution can be made more palatable if the drug is presented in a suspension format.

The components of a suspension and solution are similar; however, suspensions always contain one or more suspending or thickening agents, which prevent drug particles from settling. Suspensions may also contain wetting agents that are used to aid in uniformly dispersing the drug particles. A well-formulated suspension should remain reasonably homogeneous while standing and at a minimum for periods between shaking the bottle and pouring or withdrawing a dose. If a suspension tends to settle over time then it should be easily and completely reconstituted with gently shaking of the container. In general, the active pharmaceutical ingredient should have a small particle size, preferably less than 100 microns, and ideally have a narrow size distribution. In general, the more concentrated the suspension the finer and more uniform the drug particles are required to be in order to produce a nongritty and palatable product. Suspensions almost always taste better than solutions and hence are particularly suitable for the pediatric patient. Even though the vast majority of the drug is not in solution, trace amounts may be soluble and thus result in a bad taste. Therefore, taste masking is also a consideration for suspensions as well as solutions albeit often an easier task to formulate a palatable suspension.

Thickening Agents Thickening agents or viscosity modifiers are an essential component of a suspension formulation. There are several categories of agents used to modify the rheology of a pharmaceutical suspension. The polysaccharides include natural gums, such as locust bean gum and acacia, and naturally derived materials, such as alginates, xanthan gum, and starches. The hydrated silicates include magnesium aluminum silicate (Veegum) and aluminum silicate (Bentonite) as the most commonly used ingredients. Water-soluble cellulosics are cellulose derivatives that are widely used in the preparation of suspensions; examples are the hypromeloses, the hydroxyethylcelluloses, microcrystalline cellulose, and sodium carboxymethylcellulose. All of these categories of suspending agent come in a variety of grades and may be selected based on the properties of the suspension and the compatibilities with the other selected ingredients.

Wetting Agents Because a suspension by definition contains insoluble solid drug particles, it is required that the liquid vehicle readily wets these particles, which in most cases is water. The wetted particles will then easily disperse into the suspending medium. However, many drug compounds are hydrophobic to some degree and are not easily wetted, which can be a significant problem in the manufacturing environment. Even on a small scale, such as a bottle of suspension, this phenomenon can be the cause of inadequate distribution of the drug throughout the liquid. This poor distribution may manifest itself as drug particles floating on the liquid surface and sticking to the walls of the container or forming clumps of drug particles within the suspension.

To ensure that drug particles are sufficiently wetted, the use of an appropriate wetting agent is recommended. These agents reduce the interfacial tension between the drug particles and the liquid so that the adsorbed air may be displaced from the solid surface, allowing the liquid to contact and wet the particles. The most commonly used wetting agents are the polysorbates, sorbitan esters, sodium lauryl sulfate, and sodium docusate.

EXAMPLE OF SUSPENSION FORMULATIONS

Ingredient	%Wt/Vol
Solvents/diluents	60–80
Thickening agent	0.5–1.0
Artificial sweetener	0.05
pH modifier	0.1
Preservatives	0.22
Wetting agent	0.01
Flavor	0.5–1.0
Colorant	0.05
Active ingredient	1–5.0
Purified water USP	q.s. to 100%

Taste Masking of Liquid Formulations For liquid formulations, which include solutions and suspensions, technical approaches to reduce bad taste may be categorized as taste masking, solubility modification, adsorption, complexation, and barrier systems.

Taste masking often involves the use of flavors and/or sweetening agents to literally cover up unpleasant tastes that might be present. Many pharmaceutical actives have characteristic medicinal flavor notes, which can be quite unpleasant to children. Good flavor work can reduce these medicinal notes by finding complementary flavors that mask or hide any medicinal tastes, thereby making the product more tolerable. Hence, the red berry flavors that are characterized by aldehydes, esters, and ionones are excellent complementary flavors to mask chemical tastes. Citrus flavors are also commonly used as their characteristic aldehydes, esters, ketones, and other flavor components provide excellent complementary masking of many chemical notes. When using flavors to mask unpleasant tastes, it is important not to overflavor the formulation, as these complex flavor systems can accentuate chemical notes if they are not balanced in the formulation.

An important aspect of flavor formulation, which is often given insufficient attention in pharmaceuticals, is flavor stability. Whereas flavors in food systems generally need only last a few weeks to 12 months at most, pharmaceutical products typically have shelf-life requirements of 2 years or more. Choosing flavor systems that are compatible with the pharmaceutical formulation is essential to avoid premature loss or unacceptable changes in the product flavor profile.

When considering a flavor system for a pharmaceutical product, an understanding of the chemical structure and reactivity of the drug component is critical to avoid chemical interaction between reactive flavor and drug functional groups. In addition, the dosage form needs to be taken into consideration as well since flavor stability and reactivity can vary significantly depending on whether the dosage form is a liquid or solid. In liquid formulations, degradation of flavors is generally the result of pH catalyzed, oxidation–reduction, and hydrolytic reactions.

Flavor companies can be of great assistance in helping to develop stable flavor systems for pharmaceutical dosage forms, as over the years they have developed an in-depth understanding of flavor chemistry and ways to mitigate changes. For example, most flavor companies offer fruit and citrus flavor systems with the reactive terpenes removed; also, flavor systems are often combined with carefully chosen antioxidants to reduce free radical autoxidation. In addition, many flavor companies offer various technologies, which greatly improve the shelf life of the flavored product. These technologies range from stabilized emulsions to spray dried encapsulates to extruded beadlets that provide extended flavor stability.[11]

Taste masking also uses sweetener systems to modify and bring out desired flavors as well as reduce the bitterness so often associated with drug compounds. Today, high fructose corn syrup is a widely used sweetener in liquid systems along with sucrose and high intensity artificial sweeteners as well. Inhibiting bitterness through the use of specific compounds that competitively inhibit bitter receptors on the tongue is an active area of research today and aside from a number of books and articles having been written on the subject, a number of companies offer products that have varying degrees of bitter blocking effectiveness.[12–14]

Another important technical approach for improving the taste of pediatric liquid formulations is to render the drug insoluble in the liquid base, thus creating a suspension. Rendering the drug insoluble in the liquid effectively removes the drug from ready access to taste receptors in the mouth. Typically, the drug is uncoated in the form of a fine powder with a particle size in the 30–50-micron range. Insolubilization of the drug substance can be achieved by adjusting the pH of the liquid for those drugs having a strong solubility dependence on pH. Insolubilization of the drug substance can also be achieved by lowering the water activity of the system or by limiting the amount of available water in the system for use in dissolving the drug. This approach works best for drugs that are not highly soluble, and often simple sugars, sugar alcohols, glycols, and polyols are used for this purpose.

Complexation is another technical approach for improving the taste of pediatric liquids and like solubility modification, complexation attempts to maintain the drug in a sequestered state while in the mouth, yet still allowing for dissolution and absorption in the stomach and gastrointestinal tract. Similar to solubility, modification of factors such as pH, water activity, and ionic strength of the solution affect the equilibrium between the bound drug and the free drug in solution. Simple adsorption of the drug onto the surface of a carrier such as clays or inorganic phosphates is often used and is relatively inexpensive. Other complexes include inclusions such as those formed by the interaction of a drug with cyclodextrin molecules wherein hydrophobic portions of a drug molecule reside within the center of the cyclodextrin ring. Ion exchange resins are yet another approach to complexation, where charged drug molecules can interact and bind with the oppositely charged surface of a resin particle. The most commonly employed ion exchange resins used for the purpose of taste masking are those based on polystyrene or poly(meth)acrylic polymers.[15]

Barrier systems are also used to improve the taste of liquid drug delivery dosage forms by attempting to physically remove the drug from exposure to the taste buds. In suspensions, viscosity modifiers can be added to create a thickened liquid, where drug particles are held more effectively in suspension and diffusion of solubilized drug is slowed so that the bulk of the drug is swallowed before the tongue discerns the full impact of an off-taste.

Coated or taste masked drug particles are also used in aqueous and nonaqueous liquid suspensions as well as reconstitutable formulations often in combination with viscosity modifiers. Due to the liquid nature of the suspension base, coated particles in these systems are typically finer in size for better suspendability and smoother mouthfeel. Because of the finer particle size, surface areas are higher than coated particles used in solid dosage forms and coating levels of 25% or more are common, creating the challenge of minimizing off-taste without compromising bioavailability and dissolution stability.

Dose Delivery Devices

An important consideration with regard to liquid dosage formulations is that the actual dose must be measured before administering to the patient. There are numerous options available that may be used for measuring liquid dosage forms. For the 0–2 year age group, calibrated droppers or more preferably calibrated syringes are typically the best delivery device. For the 2–12 year age group, graduated dose cups and spoons are more typical given the larger volume being administered. For these latter devices, it is important to use the one provided with the product as different methodologies can be used to calibrate such devices; that is, the dose cup may be calibrated to deliver a given volume accounting for residual to remain in the cup while others may be calibrated based on all of the contents being delivered. Another factor is that liquids or suspensions of different viscosities or surface properties can lead to inaccuracy in dosing. Thus, it is important not to use measuring cups or droppers interchangeably. The oral syringe is the most accurate method to measure children,s liquid formulations and assures that the correct dose of medicine is measured.

Some liquids come in unit dose packaging format; in addition to the convenience and portability, this also can ensure accurate measurement of the medication.

Solid Dosage Forms

The scope of solid dosage forms discussed in this section is limited to immediate release swallowable and chewable tablets. The use of tablets for the pediatric patient is generally limited to 6 years and up for chewable and over 12 years for swallowable tablets. The range of potential acceptance of swallowable tablets depends on multiple factors, including age, chronic disease, and tolerability of each individual child to handle this option. A recent literature review on safety of chewable tablets has concluded that they are a safe, well-tolerated alternative to traditional pediatric dosage forms and offer significant advantages in children 2 years of age and older.[15]

There are certain instances where children younger than 12 years may be required to take nonchewable tablets such as modified release forms that may not be crushed; these forms are discussed in more detail in Chapter 44 by Roche and Chapter 45 by Hoy and include products like Concerta™. Differences in product formulations have been shown to affect absorption rates, peak concentrations, and time to peak concentrations; however, relative bioavailability was not affected.[19] But this observation may not hold true for every API, so it is not advisable to use different formulations interchangeably.

Swallowable Tablets In instances where the solid dose is a swallowable tablet, the tablet may be coated externally, providing a barrier between the drug and the oral cavity. Typically, film coatings are used and these coatings may include high intensity sweeteners and flavors for improved palatability. Another dosage form useful in this respect is the sugar coated tablet; however, due to the relatively lengthy manufacturing process involved, the dosage form is not as common as it once was mostly due to the rise in popularity of the film coated tablet.

Chewable Tablets As children often have difficulty swallowing tablets, chewable tablets offer an attractive alternative that can improve compliance and ease of administration. The advantages of chewable tablets include palatability, stability over liquid forms, precise dosing, portability, and ease of delivery.[15] Over the past 10 years, manufacturers of chewable tablets have made them softer for ease of chewing and for the convenience of taking them without water. Flavors have tended to stay with the popular red berry and citrus types with sweetness levels ranging from moderate to high to help counter any bitterness present. As previously mentioned, bitter tasting drugs can lead to noncompliance in patients. Simple approaches to improving the taste of chewable tablets include in situ complexation between various adsorbates or complexing agents such as ion exchange resins.[14] Another example of a taste masking approach is the in situ complexation of polyvinyl acetate phthalate (PVAP) and polyvinylpyrrolidone (PVP), allowing entrapment of water-insoluble, bitter tasting drugs.[16] Use of hard fat matrices has resulted in chewable tablets of acetaminophen with suppressed bitterness and improved oral feel.[17] While these approaches work reasonably well for mildly bitter actives and may be tolerated by adults, invariably they are not well tolerated in the pediatric population. The basic fact that these tablets are designed to be chewed creates new challenges in taste masking the active drug compounds.

Chewable tablets that require taste masking of the active ingredient essentially involve two formulations. First, the taste-masked particle needs to be formulated; second, this particle is blended with the other ingredients that comprise a chewable tablet. The taste-masked particle usually requires the formation of drug granules that are subsequently coated with various polymers that prevent the drug from coming into contact with the oropharyngeal cavity.

Diluents For chewable tablets, typical diluents or fillers include sucrose, dextrates, mannitol, and occasionally calcium phosphates. If the drug compound does not have a particularly bad taste, it can be sprayed onto the diluent in a fluid bed processor or in a blender, ensuring better dose uniformity than if the drug is dry blended.[18,19]

Binders Binders are added to chewable tablet to ensure that there is adequate adherence between the granules to form a compact tablet. Examples of such dry binders include cellulose, polyvinylpyrrolidone, and polyethylene glycol.

Lubricants The primary function of the lubricant in a tablet formulation is to facilitate the tableting process by ensuring that there is little friction between the tablet and compression tools and die cavity in which that tablet is formed. Without a lubricant, the friction associated with the tableting process would result in broken and damaged tablets and ultimately limit production processes. The most widely used lubricants are magnesium stearate and stearic acid.

Flavors and Colorants As with pediatric liquid formulations, flavors are an essential component of chewable tablets, improving the overall palatability of the dosage form. Similar to the liquid formulas, the typical flavors used are fruit flavors and flavors that are popular in confectionery products. Colors are added to improve the aesthetics and appearance of the dosage form, thus aiding compliance with the dosing regimen. Flavor degradation in solid dosage forms is primarily caused by autoxidation and loss of volatile flavor components to evaporation.

Taste Masking of Drug Taste masking of bitter actives is equally important in pediatric solid dosage forms, as the off tasting or bitter pharmaceutical active can be directly exposed in the oral cavity.

To avoid excessive exposure of the drug in the mouth, the most widely used approach to improve the taste of chewable tablets is to use coated drug particles. In general, coatings should be relatively insoluble in the mouth, must release drug rapidly in the gastrointestinal tract, and must not interact chemically with the drug. Coatings for these particles range from water-soluble to water-insoluble polymers, latexes, as well as various lipids and waxes, many of which are sold commercially for this purpose.

Unlike particles used in liquid formulations, solid dose drug particles are usually granulated and are larger in size, typically in the range of 100–400 microns. These larger particles provide for good mixing and blend uniformity with other excipients, resistance to segregation, and good bulk flow from storage bins into and throughout the tablet press. Where high dose actives are of appropriate particle size, the particles can be coated directly without the need for granulation.

When developing taste-masked drug granulations, it is important to be aware of various physical parameters. Particle size distribution should be as narrow as possible to avoid issues with blend uniformity, blend stability, and even sample bias during thieving of the bulk blend. The general shape and morphology of the particle is important for good flow characteristics and spherical shaped particles are generally the ideal. To maintain good taste masking properties, the particles should have adequate physical integrity to avoid damage during blending (e.g., low friability or resistance to abrasion). Also, the particles should not be excessively brittle and should be strong enough to resist breakage during tablet compression and in the mouth during chewing.

EXAMPLE OF A CHEWABLE TABLET FORMULATION

Ingredient	%Wt/Wt
Primary diluent	50–80
Binder	5–10
Lubricant	0.25
Flavor	0.006
Artificial sweetener	0.003
Taste-masked API	5–20

EXTEMPORANEOUS FORMULATIONS

As mentioned before, age-appropriate dosage forms are not always available. Therefore off-label uses of adult dosage forms are often employed. However, for pediatric patients, the dose administered may be based on body weight so only a portion of an adult dosage form needs to be administered. Tablets and capsules don,t lend themselves to fractional dose administration. Methods such as breaking or splitting tablets and opening capsules are employed in hospital pharmacies or community pharmacies. Several references have addressed the issues related to dose inaccuracy and therapeutic failure of these methods.[20,21] Very few oral liquid formulations are commercially available, which requires a pharmacist to prepare extemporaneous formulations. But these formulations don,t take into account compatibility of the drug with the vehicle used and stability of the drug in the vehicle. Since there are no stability data available, these formulations tend to have a short shelf life. Extemporaneous formulations for ophthalmic use also need to be sterile. Several texts have been published which provide guidance in the field of extemporaneous formulation.[22,23]

CONCLUSION

Development of pediatric formulations encompasses all aspects of product development from preformulation through development and packaging. The pediatric population covers a broad age range in terms of physical capability, age considerations, and pharmacological considerations. Because of the broad age range within pediatrics, there is a challenge in developing a range of dosage forms that address the needs of the population. An additional challenge in developing dosage forms for the pediatric community is the fact that most pharmaceutical actives have an unpleasant taste. The taste can range from an unpleasant chemical taste to a harsh bitterness with intensities varying from moderate to high. Because the most commonly preferred pediatric dosage forms are liquids and chewable tablets, the challenges of developing aesthetically acceptable dosage forms for this population are more challenging than those faced with the adult population. A wide variety of technical solutions have been developed to minimize the impact of bad tasting actives in both liquid and chewable tablet formulations. The broadest variety of technical solutions have been developed for liquid formulations, which is not surprising given the flexibility of the dosage form in terms of weight-based dosing and the unique challenges it presents either in pure liquid format or as a suspension system.

REFERENCES

1. Lieberman HA, Lachman L, Schwartz JB, *Pharmaceutical Dosage Forms: Tablets*, 2nd ed. Vol.1. New York: Marcel Dekker; 1989;387–391.
2. Oas R. Taste masking—making bitter-tasting APIs palatable using the right combination of excipients. *Tablets Capsules*. July 2006;IV:12–18.
3. European Medicines Agency, Committee for Medical Products for Human Use; Reflection Paper: Formulations of Choice for the Paediatric Population. EMEA/CHMP/PEG/194810/2005, London, June 23, 2005.
4. Benedetti MS, Baltes ES. Drug metabolism and disposition in children. *Fund Clin Pharmacol*. 2003;17:281–299.

5. FDA. *Guidance for Industry: General Considerations for Pediatric Pharmacokinetic Studies for Drugs and Biological Products*. Rockville, MD: FDA; 1998.
6. American Academy of Pediatrics, Committee on Drugs. Guidelines for the ethical conduct of studies to evaluate drugs in pediatric populations. *Pediatrics*. 1995;95(2):286–294.
7. Bagger-Sjöbäck D, Bondesson G. Taste evaluation and compliance of two paediatric formulations of phenoxymethylpenicillin in children, VII. *Scand J Prim Health Care*. 1989;87–92.
8. Al-Shammari SA, Khoja T, Al-Yamani MJMS. Compliance with short-term antibiotic therapy among patients attending primary health centres in Riyadh, Saudi Arabia. *J R Soc Health*. August 1995;231–234.
9. El-Chaar GM, Mardy G, Wehlou K, Rubin LG. Randomized, double blind comparison of brand and generic antibiotic suspensions: II. A study of taste and compliance in children. *Pediatr Infect Dis J*. 1996;XV(1):18–22.
10. Glasgow AM, Boerckx RL, Miller MK, MacDonald MG, August CP, Goodman SI. Hyperosmolality in small infants due to propylene glycol. *Pediatrics*. 1983;72:353–355.
11. Risch SJ, Reineccius GA. Encapsulation and controlled release of food ingredients. *ACS Symposium Series*, 590, Washington DC, 1995.
12. Roy GM. *Modifying Bitterness: Mechanism, Ingredients, and Applications*. Basel: Lancaster; 1977.
13. Roy GM. Taste masking in oral pharmaceuticals. *Pharm Technol*. April 1994;84–99.
14. Elder DP. A complex solution. *Chem Ind*. April 2001;209–214.
15. Michele TM, Knorr B, Vadas EB, Reiss TF. Safety of chewable tablets for children. *J Asthma*. 2002;39(5):391–403.
16. Kumar V, Yang T, Yang Y. Interpolymer complexation II. Entrapment of ibuprofen by *in-situ* complexation between polyvinyl acetate phthalate (PVAP) and polyvinylpyrrolidone (PVP) and development of a chewable formulation. *Pharm Dev Tech*. 2001;6(1):71–81.
17. Suzuki H, Onishi H, Hisamatsu S, et al. Acetaminophen-containing chewable tablets with suppressed bitterness and improved oral feeling. *Int J Pharm*. 2004;278:51–61.
18. Maddi SS, Tandon S, Aithal KS. Clinical evaluation of sodium fluoride chewable tablets in dental caries. *Indian J Dental Res*. 1999;10:146–149.
19. Maas B, Garnett WR, Pellock JM, Comstoack TJ. A comparative bioavailability study of carbamazepine tablets and chewable tablet formulation. *Ther Drug Monit*. 1987;9:28–33.
20. Rosenberg JM, Nathan JP, Plakogiannis F. Weight variation of pharmacist-dispensed split tablets. *J Am Pharm Assoc*. 2002;42:200–205.
21. Teng J, Song CK, Williams RL, Polli JE. Lack of medication dose uniformity in commonly split tablets. *J Am Pharm Assoc*. 2002;42:195–199.
22. Jew RK, Mullen RJ, Winson S. *Extemporaneous Formulations*. American Society of Health-System Pharmacists, Inc, 2003.
23. Nahata MC, Hipple TF. *Pediatric Drug Formulations*, 3rd ed. Cincinnati, OH: Harvey Whitney Books Company; 1997.

CHAPTER 44

Drug Delivery Challenges for the Pediatric Patient: Novel Forms for Consideration

EDWARD J. ROCHE, PhD

McNeil, Johnson & Johnson, 420 Delaware Drive, Ft. Washington, Pennsylvania 19034

INHALATION DELIVERY

Key Issues for Pediatrics

Administration of pharmaceuticals to pediatric patients via the inhalation route presents a number of key challenges that are crucial to successful therapeutic outcomes. The area of asthma treatment via inhalation delivery clearly demonstrates these challenges. Three main types of devices are currently being used: nebulizers, metered dose inhalers (MDIs), and dry powder inhalers (DPIs). The dose of drug that reaches the airways depends on delivery device and the product formulation. Inhalation devices, many of which have been developed for adults, may be difficult for young children to use properly.[1] Difficulty in delivery may also lead to poor patient compliance and suboptimal outcomes.

Guidelines for choice of device have been provided by various groups but are not completely consistent. Guidance from the American Academy of Allergy, Asthma, and Immunology (AAAAI) recommends that for children under 2 years of age the use of either a nebulizer with a face mask or a pressurized MDI with a spacer/holding chamber and face mask.[2] The choice between a nebulizer and an MDI for infants remains controversial with some authors contending that use of MDIs should be discouraged due to the difficulty of coordination of activation and inhalation for these very young patients.[3] However, other authors cite evidence that nebulizers are less effective and more cumbersome to use.[4] It should be noted that nebulized cromolyn and nebulized budesonide are the only controller medications indicated for ages 1–4.[1] For school-age children, the AAAI recommends MDIs, dry powder inhalers, or nebulizers.[2]

Pediatric Drug Development: Concepts and Applications
Edited by Andrew E. Mulberg, Steven A. Silber, and John N. van den Anker

Nebulizers

Nebulizers are designed to deliver liquid medication in an extremely fine cloud or mist. Particle size is well known to play a key role in delivery with sizes in the 1–5-micron range generally considered to be inspirable. Particles in this range bypass the immunological defenses of the lung.[1] Nebulizers come in different types including conventional jet nebulizers, open-vent nebulizers, breath assisted devices, and adaptive aerosol delivery devices.[5] One of the key historic disadvantages of nebulizers has been their large size and lack of portability. Newer nebulizers are being developed that use a vibrating plate to generate a low velocity mist. One example is the eFlow device developed by Pari (Figure 44.1). This can produce advantages in terms of improved portability, battery operation, and minimal residual volume of medication in the device.[6]

Another disadvantage of nebulizer therapy has been the lengthy time of administration that can reduce compliance. Reducing the time of administration is a key focus for new product development in both asthma and other disease states such as cystic fibrosis, where nebulized products are common.

Patient factors such as face mask seal and breathing patterns are also key to successful nebulizer delivery. Young children and infants often try to escape the face mask, possibly leading to poor face mask seal. Optimal nebulizer delivery is achieved during quiet breathing; crying during administration can severely alter the amount of drug delivered.[7]

Metered Dose Inhalers

Pressurized MDIs are efficient, deliver medication in a short period of time, and are small and highly portable. The main disadvantage of these devices is the need to coordinate breathing

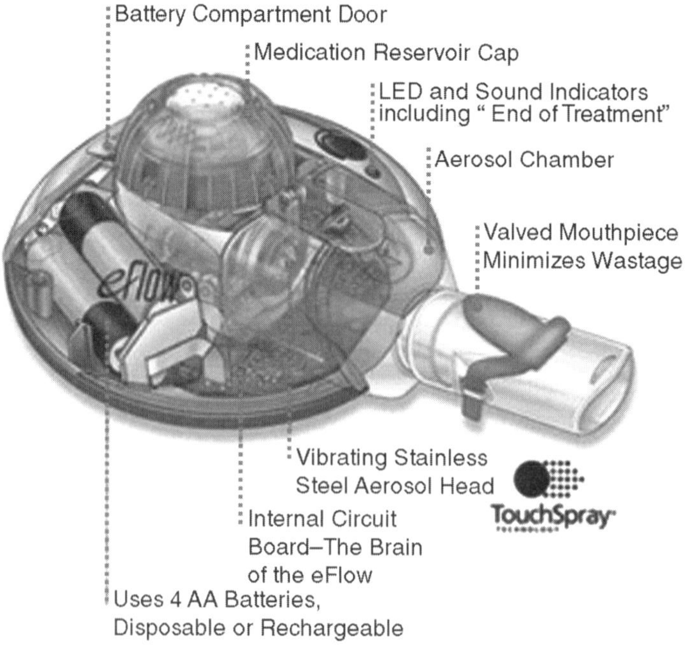

FIGURE 44.1 Pari eFlow portable nebulizer device.

with activation. In addition, when used without a spacer, the high aerosol velocity leads to oropharyngeal deposition, which is ineffective. The use of a spacer results in deceleration of the aerosol. Current recommendations support the use of a spacer regardless of patient age.[2] Some companies have started development of breath enhanced MDIs but there is still limited data at this time.[8]

Even with the use of a spacer or holding chamber, poor technique can compromise drug delivery with MDIs. Removing electrostatic charge by coating spacers with a detergent layer is also strongly recommended by some authors as a means of improving lung deposition.[9] In one study, common errors included not shaking the inhaler before use, not placing the mouthpiece between the teeth and lips, and not checking to see if the spacer valve was properly moving.[10] Instruction is generally an important factor in the successful use of inhalers by children.[11] Even with good instruction, patient follow-up is recommended as techniques may deteriorate over time.

Data on lung deposition for MDI delivery have been reported based on radiolabeled drug techniques.[12,13]

Dry Powder Inhalers

Dry powder inhalers (DPIs) now exist that can deliver multiple doses of drug over a prolonged period of time. The new DPIs are small, portable devices that can deliver medication quickly (Figure 44.2). However, all of the current DPIs require that a child breathe in forcefully with an inspiratory effort exceeding 30 L/min.[1] Generally, children older than 7 years of age can successfully use these devices.[10] DPI devices have been used increasingly for the concurrent administration of inhaled corticosteroids and long-acting bronchodilators. Some studies have been completed comparing the lung deposition with different DPIs (e.g., Diskus™ and Turbohaler™). Studies show that DPI use in children tends to deliver a lower percentage of drug than an MDI with a spacer.[14,15]

NASAL DELIVERY

The nasal route of administration has been far less used than other routes. However, with important medications such as some nasal corticosteroids (mometasone) now indicated to 2 years of age, this route is increasing in use. The nose is also a region of rich vasculature and can be used when more rapid onset of a medication is needed such as treatment of acute pain in a hospital setting.[16] In terms of delivery devices, the most common are the atomized spray

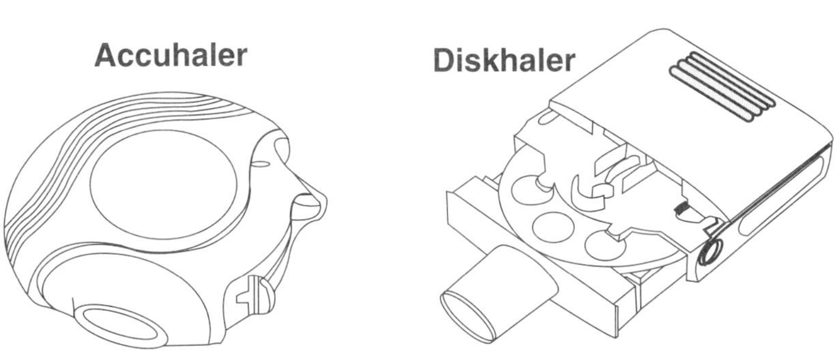

FIGURE 44.2 Examples of dry powder inhalers.

and nebulizer. The latter can be dosed using a mask that covers both the nose and mouth. A nasal adapter can be used in combination with a spray device to accommodate the smaller nasal openings in a child.

INTRAVENOUS/INJECTABLES/INTRAMUSCULAR DELIVERY

Special problems may also occur in these routes of delivery. Absorption of medication following an intramuscular (IM) injection is often erratic in neonates due to their small muscle mass and an inadequate perfusion rate at the intramuscular site.[17] The bioavailability of many drugs administered intramuscularly has not been evaluated in the pediatric population. Pediatric subjects may be less tolerant to the volume of fluid needed for injections. In some cases, the inclusion of a local anesthetic, such as lidocaine, may be advisable to reduce pain.[18] A recent analysis of ways to standardize infusion delivery in order to improve safety of pediatric patients was presented by Apkon et al.[19] The results showed that standardizing formulations for all infusions, developing database driven calculators, extending infusion hang times, and changes in other practices could improve patient safety and overall efficiency in an institutional setting.

TRANSDERMAL DELIVERY

There is some special consideration in transdermal drug delivery to pediatric patients and particularly in the case of neonates. Skin thickness and blood flow may vary with age in the children and have consequent effects on pharmacokinetics.[20] Central nervous toxicity occurred in neonates washed with hexachlorophene because their very thin skin and large body surface resulted in toxic blood levels.[21] Fatal toxicity was also encountered in early use of tetracaine/adrenaline/cocaine combinations applied for local anesthesia prior to suturing when excessive blood levels resulted in pediatric patients.[20] Examples of drugs currently or recently delivered by the transdermal route include scopolamine patches to prevent motion sickness, a eutectic mixture of local anesthetics (EMAs), corticosteroid creams for dermatology-related applications, fentanyl patches for treating cancer pain, and a methylphenidate patch for treatment of attention deficit hyperactivity disorder.

BUCCAL DELIVERY

The fentanyl Oralet™ is an example of successful buccal administration of a drug in children. Because the pK_a of fentanyl is 8.4, absorption through the oral mucosa is favored.[20] The product is approved for postoperative sedation and for painful procedures in a hospital setting. The buccal route may offer some protection from the adverse effects of intravenous fentanyl. The product offers the potential of sustained therapeutic blood levels resulting in prolonged analgesia.

CONCLUSION

There are multiple novel pathways for drug delivery that can be administered to the pediatric patient and the limitations are all specific for the type of delivery. What is clear is that the considerations for the formulation expert can be tailored depending on the uniqueness of the pediatric versus adult patient and their individual requirements for drug delivery.

REFERENCES

1. Berger WE. Paediatric pulmonary drug delivery: consideration in asthma treatment. *Expert Opin.* 2005;2(6):965–980.
2. American Academy of Allergy, Asthma, and Immunology: *Pediatric Asthma: Promoting Best Practices in Children;* 2004: 88–90.
3. Skoner D. Pharmacokinetics, pharmacodynamics, and the delivery of pediatric bronchodilator therapy, *J Allergy Clin Immunol.* 2000;106(3):S162.
4. Gillies J. Overview of delivery system issues in pediatric asthma. *Pediatr Pulmonol.* 1997;15 (Suppl):55–58.
5. Debendeictis F, Selvaggio D. Use of inhaler devices in pediatric asthma. *Pediatr Drugs.* 2003;5: 629–638.
6. Dhand R. New frontiers in aerosol delivery during mechanical ventilation. *Respir Care.* 2004;49:666–677.
7. Isles R, Lisfer P, Edmunds AT, et al. Crying significantly reduces absorption of aerosolized drug in infants. *Arch Dis Child.* 1999;81:163–165.
8. O' Callaghan C, Nerbrink P, Vidgrew MT, et al. *Drug Delivery to the Lung*, Vol. 162. New York: Marcel Decker; 2001: 337–370.
9. Pierart F, Wildhaber JH, Vrancken I, et al. Washing plastic spacers in household detergent reduces electrostatic charge and greatly improves delivery. *Eur Respir J.* 1999;13:673–678.
10. Kamps A. Poor inhalation technique, even after inhalation instructions, in children with asthma. *Pediatr Pulmonol.* 2000;29:39–42.
11. Brand P. Key issues in inhalation therapy in children. *Curr Med Res Opin.* 2006;21(Suppl 4): S27–S32.
12. Janssens H, Heijnen EM, DeJong VM, et al. Determining factors of aerosol deposition for four pMDI spacer combinations in an infant upper-airway model. *J Aerosol Med.* 2004;17:51–61.
13. Dubus J, Anhoj J. Inhaled steroid delivery from small-volume holding chambers, depends on age, holding chamber and interface in children. *J Aerosol Med.* 2004;17:225–230.
14. Wildhaber JH, Janssens HM, Pierart F, et al. High-percentage lung delivery in children from detergent based spacers. *Pediatr Pulmonol.* 2000;29:389–393.
15. Wildhaber JH, Deradason SG, Wilson JM, et al. Lung deposition of budesonide from turuhaler in asthmatic children. *Eur J Pediatr.* 1998;157:1017–1022.
16. Goldman R. Intranasal drug delivery for children with acute illness. *Curr Drug Ther.* 2006;1:127–130.
17. Loebstien R, Koren G. Clinical pharmacology and therapeutic monitoring in neonates and children. *Pediatr Rev.* 1998;19(12):423.
18. Danish M, Kottke M. Pediatric and geriatric aspects of pharmaceutics. In: Banker G, Rhodes C, eds. *Modern Pharmaceutics.* New York: Marcel Dekker; 2002: Chap 21.
19. Apkon M, Leonard J, Probst L, et al. Design of a safer approach to intravenous dug infusions: failure mode effects analysis. *Qual Safe Health Care.* 2004;13:265–271.
20. Alternative routes of drug administration—advantages and disadvantages. *Pediatrics.* 1997;100 (1): 143–152.
21. Tyrala EE, Hillman LS, Hillman RE, et al. Clinical pharmacology of hexachlorophene in newborn infants. *J Pediatr.* 1977;91:481–486.

CHAPTER 45

Oral Drug Delivery Challenges for the Pediatric Patient

MICHAEL R. HOY, PhD

Johnson & Johnson Pharmaceutical Research and Development, LLC, Raritan, New Jersey 08869

INTRODUCTION

Delivery into the oral cavity of the neonate, infant, child, or adolescent patient poses a range of challenges. Absorption, distribution, metabolism, and excretion (ADME) differ significantly across the continuum of pediatric ages. Assuming these components of physiological differences were well understood for new chemical entities and already marketed products, which they are not, a separate challenge confronts the drug development formulator. The challenge facing the expert in drug formulation is the timing and delivery of a drug efficiently for dose administration to minimize patient-to-patient variability and to maximize compliance. The outcome for the pediatric patient is effective treatment of the underlying disease under management.

Generally, pharmaceutical active ingredients have an unpleasant taste that can vary greatly from a harsh bitterness to a chemical aftertaste that can linger long after the dosage form is administered.[1] Additionally, taste preferences vary greatly from adults to children. The tastes that adults perceive as acceptable are likely much less acceptable to the infant or child. Furthermore, there are cultural differences that are critical for understanding their impact on tolerability. Therefore, taste testing in adults cannot be translated directly to children. Furthermore, the ethics associated with taste testing in children is questionable depending on the risk-to-benefit ratio of the drug being developed or if the drug is a new chemical entity (NCE) versus an already marketed prescription (Rx) or over-the-counter (OTC) medication. The safety of an existing OTC medication is generally well understood relative to that of an NCE. That is, an OTC medication in many cases has been marketed for years as an Rx drug before being switched from Rx to OTC status. Documentation of the Rx to OTC switch process and requirements can be found at the U.S. Food and Drug Administration (FDA) website for further information (http://www.fda.gov).[2] This switch process requires demonstration of the lowest efficacious dose and concomitant

Pediatric Drug Development: Concepts and Applications
Edited by Andrew E. Mulberg, Steven A. Silber, and John N. van den Anker

safety monitoring. Hence, taste testing of OTC medications can be done more safely. NCE rugs, on the other hand, have had significantly less testing across different populations; therefore, taste testing of NCE medications must be conducted carefully, especially in potentially higher risk patients such as children. Consideration for changing taste preferences must also be considered depending on the child's age as well as on the potential for the drug to impact the taste receptors in the mouth.[3] Clearly, the formulator is challenged to identify technologies that can address both ADME and taste-related issues.

IMMEDIATE VERSUS MODIFIED RELEASE: IMPACT ON RATE OF DRUG DELIVERY

Modified Release

Relative to medications labeled for adults, a limited number of immediate release (IR) dosage forms are currently labeled for the pediatric population beyond the OTC drug marketplace. Recently, these pediatric OTC drug products have been challenged to demonstrate clinical efficacy by FDA regulatory authorities. Specific challenge was brought forth most recently at an FDA Advisory Committee meeting regarding the use of cough and cold medications in children.[2] Modified release (MR) dosage forms are more limited in the marketplace. Literature sources highlight the need to develop MR forms for cardiovascular drug products specifically for children since the adult MR formulation dose is too high for the pediatric population.[4] Furthermore, adult MR forms cannot be altered through tablet splitting or crushing since the integrity of the dosage form would be destroyed, rendering the MR form useless where lower doses would be required for children. One does not need to think too long about which therapeutic classes of drugs would be useful in children. Many are the same as those already developed for adults including asthma, allergy, cardiovascular, antianxiety agents, analgesics, and antiepileptics.

MR technology has been developed and applied extensively in the adult prescription marketplace. Unfortunately, the same development has not been applied to pediatric prescription pharmaceuticals. The primary driver continues to be limited development of NCEs for use in pediatrics since bigger markets exist for adults. Since simple IR formulations are not routinely developed for children, the likelihood of MR forms being developed is further limited.

A number of technologies currently applied to the adult population may be considered in the development of MR formulations for children. They are described next.

Oros™ is an osmotic pump technology, which is currently used in numerous marketed products, including Concerta, Cardura XL, Covera HS, Ditropan XL, Glucotrol XL, Sudafed 24 hour, Procardia XL, and Volmax. Concerta™ is a product that has been developed initially for use in treating children for attention deficit hyperactivity disorder (ADHD) to control the release of methylphenidate for once-daily dosing. The advantage for the child is dosing only in the morning before leaving for school without the need to revisit the school nurse for redosing at midday. The Oros formulation cannot be modified once it leaves the manufacturer by splitting or crushing since it is a multilayer form that must remain intact. Therefore, the appropriate pediatric dose was predetermined by the manufacturer based on the age/weight of the child. The Oros drug delivery device is of limited use because of inherent drug load limitations; therefore, drugs used in the Oros form must be relatively potent. Furthermore, since Oros must be swallowed intact, it is best used for children who are capable of swallowing. Multiparticulates (beads) is another technology used for adult MR products. This technology is also used for children in another ADHD product called Adderall XL and has similar limitations to Oros. Multiparticulates are typically made up of beads or particles that are coated with polymers. The coat level can range up to approximately 25% depending on the level of drug release desired. Different coat levels may be combined in order to achieve the pharmacokinetic profiled desired. The type of polymer used can impart a pH-sensitive or pH-insensitive property depending on the region of the gastrointestinal tract where absorption is desired. Further control of drug release can be achieved by film coating the final tablet after compression. This film coating can be for both aesthetic purposes (mouth feel and taste) and/or MR purposes.

For treatment of children, modification of any bitter or chemical taste is critical to redosing and compliance.[3] Multiparticulates can be encapsulated or compressed into a tablet or caplet. In the case where multiparticulates are used for MR forms, the dosage form is usually encapsulated to ensure the integrity of the coated bead is maintained after it passes the mouth. If the MR beads are chewed, the coating is likely to be cracked, resulting in poorly controlled MR functionality. If the multiparticulate is used for taste-masking or mouth feel purposes, the dosage form can usually be compressed since some chewing can be withstood and still deliver a taste-masked drug through the mouth. Generally, a chewable is acceptable to many 3–4 year olds but this depends on each individual.[5] Encapsulated forms are also limited to children who can swallow.

Complexation is another delivery technique sometimes used. Both ion-exchange resins and cyclodextrins can be used. Ion-exchange resins used for the purpose of taste masking are those based on polystyrene or poly(meth)acrylic polymers. Ion exchange with these resins requires a positively charged molecule such as dextromethorphan in order to use it successfully. Delsym, an OTC cough suppressant, is available for both children and adults. Taste masking and MR are objectives for this product since it is dosed only twice daily relative to their IR counterparts, which are dosed four times daily. Considerations for the formulator are drug loading, pH, and ionic strength of the base.[6,7]

Cyclodextrin and drug complex formation involves fitting an active drug substance (drug load) into the cavity of a larger cyclodextrin receptor molecule (see Figure 45.1).

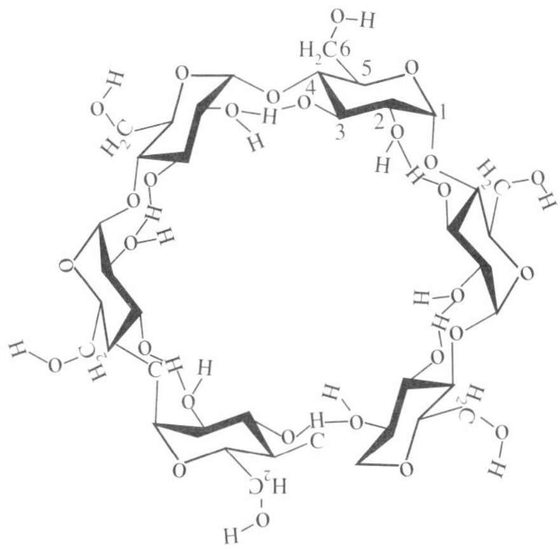

FIGURE 45.1 Cyclodextrin receptor molecule.

The availability of these receptor molecules is currently limited since their disposition in the body must be considered from a safety perspective. Extent and efficiency of drug load can be a limitation as well since the ratio of drug to cyclodextrin is usually 1:1. The large size of the cyclodextrin molecule limits the drug allowed for complex. If drug load is too high, the size of the final form will exceed the size that can be swallowed. Janssen uses hydroxypropyl-β-cyclodextrin in a formulation to solubilize itraconazole (Sporanox) for an intravenous infusion solution where each milliliter of solution contains 10 mg of itraconazole and 400 mg of hydroxypropyl-β-cyclodextrin. In this case, the rationale for use of cyclodextrin is enhanced solubilization of the itraconazole. Use of this delivery system would be unlikely since a clear advantage would need to be demonstrated relative to that of other delivery systems especially since use of cyclodextrins has been associated with concerns of renal toxicity in children.

Matrix formulations generally are used to modify the release rate of lower potency drugs since these delivery systems have the advantage there. Taste masking is not a property that is associated with matrix formulations unless an overcoat can be applied to the outer surface of the tablet or caplet to make the short residence time in the mouth more palatable or provide ease of swallowing. A disadvantage generally associated with a matrix formulation is food effect. Control of this effect can be adjusted to some extent by controlling the rate of diffusion versus erosion of the matrix. Generally, by increasing diffusion and minimizing erosion, the impact of food effect can be minimized. As erosion rate increases so does the impact of food effect. Osmotic devices, mentioned previously, overcome this disadvantage by maintaining zero order drug release.

Examples of modified release dosage forms currently on the market include Zyban-SR (bupropion), Sinemet-CR (carbidopa/levodopa), Madopar-CR (levodopa), Procardia XL (nifedipine),[8] Oxycontin (oxycodone), Tylenol® Arthritis Extended Release, and Contac®. Since the matrix formulations are limited to tablets or caplets, these forms would be useful in children who can swallow. One advantage of matrix formulation development is that it involves standard tablet/caplet manufacturing unit operations.

Multilayered formulations offer a developer the opportunity to modify drug release rates in each layer and/or protect multiple drugs from instability due to interaction with each other. The advantage is higher drug load and good control; however, complexity of development and equipment limit pragmatic use. Tylenol Arthritis Extended Release is based on both a bilayer and matrix approach.

Immediate Release

Immediate release technology for children has been developed and applied in the OTC marketplace over the past 20 years and has set the hurdle for prescription pharmaceuticals to break-in to pediatric drug delivery development. The ADME characteristics are well understood for OTC drugs in adults. Doses of OTC drugs for children have historically been based on an milligram/kilogram body weight (BW). The focus for the OTC formulator has been on developing the most age-appropriate dosage form acceptable to the child targeted to receive the medication. These formulation approaches include liquid solution/suspension as well as solid swallowable/chewable technologies mentioned by McNally and Railkar in Chapter 43. These OTC technologies can be applied to the pediatric prescription world as a means to make dosing easier for the parent and child by enhancing the aesthetic characteristics of the dosage form.

LESS TRADITIONAL DELIVERY SYSTEMS

More nontraditional drug delivery technologies are available for application in prescription drugs for the treatment of children. Each technology has its own attributes and drawbacks. These technologies are briefly described next. A prerequisite of the delivery system to be applied to the drug is that the form must have good mouth feel and taste good to the child. If these attributes are not acceptable, compliance will suffer.

As described earlier by McNally and Railkar in Chapter 43 suspensions, solutions, and chewable forms are very commonly used for children. Solutions and suspensions generally incorporate simple favor systems to overwhelm the taste receptors in the mouth versus that of the drug. Chewables can utilize flavor systems alone if the drug doesn't have a bad taste. If the drug is inherently bad tasting, the formulator must use additional approaches including application of polymer film coatings to drug-containing beads, which are then combined with traditional excipients to form a lightly compressed chewable tablet. Techniques to either block or competitively inhibit bitter receptors have also been investigated.

Fast dissolving dosage forms (FDDFs) or oral disintegrating tablets (ODTs) are forms that are sometimes mentioned interchangeably. These dosage forms generally dissolve quickly in the oral cavity above the tongue, whereas sublingual forms dissolve in the oral cavity under the tongue. Ideally, these forms should be small in size, simple to handle and insert in the mouth, taste good, remain stable in a bottle, and disintegrate/dissolve in less than 10 seconds. Interestingly, speed of disintegration seems to be more important to an adult than to a child. Children tend to prefer forms that taste good and remain in the mouth longer, similar to hard candy. An ongoing consideration is finding a balance between making a medication taste too good to the point where a child is no longer aware that it is a medication, which can cause safety concerns if ingested in candy-like quantities.

FDDFs generally are faster to dissolve than ODTs. FDDFs have historically been a lyophilized dosage form; therefore, when water from the mouth comes in contact with the molded lyophilized form, it instantaneously dissolves. Examples of prescription products on the market include Maxalt (rizatriptan by Merck) and Imitrex (sumatriptan by GlaxoSmithKline); both are for treatment of migraine. ODTs, on the other hand, utilize more standard pharmaceutical excipients, similar to those found in chewables, but are lightly compressed or otherwise molded. The result is a larger charge of powder that must be wetted by the mouth and will dissolve more slowly, perhaps in less than 30–60 seconds.

The FDDF and ODT dosage forms have drug load limitations, in the range of 50 mg or less. Furthermore, taste masking of an inherently fast dissolving form is challenging since quick dissolution of a bitter drug on the taste receptors of the tongue isn't a positive experience for children.

Oral thin films (OTFs), a dissolvable film technology, have evolved from a purely confectionary novelty form to a drug delivery platform capable of delivering up to 50 mg or so of active drug. As with the other technologies mentioned, OTFs have potential application beyond use in children. The OTFs, along with FDDFs and ODTs, provide very nice aesthetics for the child if formulated properly. Each of these forms can also have a mucoadhesive functionality added if desired. Inherently, the OTF is best suited since FDDFs and ODTs tend to be used most when quick disintegration is desired. Most consumers are familiar with the confectionary OTFs, including personal products like breath mints or OTC pharmaceuticals like cough/cold symptom and sore throat pain products.[9]

Sublingual forms are not useful in children since young children, under 2 years of age, cannot understand or apply directions for using a sublingual. Based on personal observation, the use of sublingual forms in children more than 3 years of age is dependent on the individual child but not ideal.[10] Generally, a sublingual form is used when very rapid delivery/absorption is required, such as in the case of nitroglycerin in the treatment of angina. In children, such an emergency is unlikely, except in a hospital setting, where alternative forms would be more desirable including injectables.

Effervescent forms are tablets that are put into a liquid and then swallowed. Effervescent forms are most useful when a liquid is inconvenient to carry for the parent or if the drug to be dosed can remain stable in liquid for relatively short periods of time. In such cases, an effervescent tablet containing bicarbonates, the drug, and other standard excipients is formulated. The effervescent should be flavored according to the age of the children receiving the doses. Generally, these flavors are berry or bubblegum-like flavors, which are most favored by the pediatric population.

GELS AND SPRAYS

These are mentioned here only for completeness. Pragmatic application in children generally has not been considered since a specific need has not been identified. Soft gel or gummy "candy-like" forms could be anticipated if active drug were soluble and did not impart poor taste. Sprays could be developed where swallowability is an issue in order to deliver drug bucally. However, oftentimes such cases are in-house, where intravenous or intramuscular dosage forms are applied.

MEDICATED LOZENGES

Lieberman, Lachman, and Schwartz[11] in Volume 1 of their *Pharmaceutical Dosage Forms* series discuss medicated lozenges, where they provide a comprehensive review of the formulation considerations for these delivery systems. Pragmatically, while these forms can be very appealing to children, they are challenging from a commercialization perspective since Good Manufacturing Process facilities to commercialize hard candy formulations do not exist. Actiq, an oral fentanyl lollipop, has been marketed.

RECONSTITUTABLE POWDERS

Reconstitutable powders are widely available dosage forms in the prescription pharmaceutical arena. The pharmacist adds water (diluent) to the powder prior to dispensing to the patient. The rationale for keeping the water separate from the powder is to maintain the stability of the powder in the dry form. Once water is added to the powder the stability of the reconstituted powder is limited to between 7 and 14 days. The most common reconstitutable powders are antibiotics, which are flavored in order to mitigate bad taste. Often, flavoring systems only cover up a bad taste and are minimally acceptable. The formulator should focus on using other techniques, including film coating the active drug substance in order to prevent contact between the drug and the taste receptors on the tongue. Addition of flavor will then significantly enhance patient acceptability. Compliance in children can be enhanced significantly when the dose is more palatable.[12,13]

It is important to note that many of the more unique delivery forms discussed earlier can require more specialized packaging since they are generally hygroscopic in nature. If standard packaging is used, these dosage forms will adsorb water and become unusable as a form and many times lead to drug instability. High moisture barrier packaging can have significant cost implications. Another consideration for specialized packaging is the inherent friability of many of the faster dissolving dosage forms. Since their inherent structure is porous, they are often physically weak and cannot withstand the normal rigors of transport from one unit operation to another or from manufacturer to the retail shelf unless specialized packaging is utilized. Therefore, a balance between the need for the unique form and cost must be considered.

REFERENCES

1. Ettner N, Grave A. Tastemasking: reducing the bitterness of drugs. *PFQ*. 2006;5:24–27.
2. Final Report from the Joint Meeting of the Nonprescription Drugs Advisory Committee and the Pediatric Advisory Committee, October 18–19, 2007. Available at http://www.fda.gov/ohrms/dockets/ac/07/minutes/2007-4323m1-Final.pdf.
3. Beauchamp GK, Moran M. Dietary experience and sweet taste preference in human infants. *Appetite*. 1982;3:139–152.
4. Standing JF, Tuleu C. Paediatric formulations—getting to the heart of the problem. *Int J Pharm*. 2005;300:56–66.
5. Michele TM, Knorr B, Vadas EB, Reiss TF. Safety of chewable tablets for children. *J Asthma*. 2002;39(5):391–403.
6. Pisal S, Zainnuddin R, Nalawade P, Mahadik K, Kadam S. Molecular properties of ciprofloxacin-indion 234 complexes. *AAPS PharmSciTech*. 2004;5(4):62.
7. *Drug Development and Industrial Pharmacy*. November 2007;11:1205–1215.
8. Abrahamsson B, Alpsten M, Bake B, Jonsson UE, Eriksson-Lepkowska M, Larsson A. Drug absorption from nifedipine hydrophilic matrix extended-release (ER) tablet-comparison with an osmotic pump tablet and effect of food. *J Controlled Release* 1998;52:301–310.
9. Frey P. Film strips and pharmaceuticals. *Pharm Manufacturing Packaging Sources*. Winter 2006: 92–93.
10. Passalacqua G, Baena-Cagnani CE, Berardi M, Canonica GW. Oral and sublingual immunotherapy in paediatric patients. *Curr Opin Allergy Clin Immunol*. 2003;3:139–145.
11. Lieberman HA, Lachman C, Schwartz JB, eds. *Pharmaceutical Dosage Forms: Tablets*, Volume 1, 2nd ed. New York: Information Health Care; 1990:419–565.
12. Tiwari SB, Murthy TK, Pai MR, Mehta PR, Chowdary PB. Controlled release formulation of tramadol hydrochloride using hydrophilic and hydrophobic matrix system. *AAPS PharmSciTech*. 2003;4(3):31.
13. Breitkreutz J, Boos J. Paediatric and geriatric drug delivery. *Expert Opin Drug Delivery*. January 2007;4(1):37–45.

CHAPTER 46

The Jelly Bean Test: A Novel Technique to Help Children Swallow Medications

ROBBYN E. SOCKOLOW, MD and ALIZA B. SOLOMON, DO

Division of Pediatric Gastroenterology and Nutrition, New York Presbyterian Hospital, Weill Cornell Medical College, New York, New York 10021

INTRODUCTION

As children grow, they learn to master each step of their development. Most of these milestones are achieved with patient encouragement from their caretakers. For example, let us consider walking. The scenario is usually toddlers awkwardly keeping themselves in balance on their legs so that they will move slowly one foot in front of the other to the encouraging voice of the caregiver just arm's length away. This setting is not pressured, and repeat attempts are made as the child acquires this skill. There is always a cheering person to encourage that first step. Positive reinforcement is given with hugs and kisses. Now, let us imagine a child who needs medication. The situation is no longer without pressure, no longer with the patience for several attempts, and the time frame to acquire this milestone is desperate. This is what happens to children who are sick or whose diagnosis requires certain medications that can only be ingested whole often due to the lack of appropriate pediatric formulations. This latter topic is addressed elsewhere in this book, specifically in Chapter 43 by McNally and Railkar.

Many parents are faced with this situation if they have children requiring medication. Their child has been diagnosed with an illness or condition that requires a medication that cannot be chewed, cannot have the capsule opened, and cannot be made into an appropriately palatable formulation (i.e., pleasant tasting liquid or suspension). Children are then asked to learn how to take a pill at a time of great stress and at the behest of their caregivers who are concerned about the inability to give their child the prescribed medicine. There is no longer patience in that encouraging voice and there is too much invested in the child achieving that new skill. Often a child will not take the medicine because it tastes bad or they are afraid of the size of the pill. It may also be at this time that a child is not feeling well and is too irritable to cooperate by trying something new. Parents may resort to bribing and threatening as a way to get the child to take his/her medication. Parents may try to hide the medications in food,

Pediatric Drug Development: Concepts and Applications
Edited by Andrew E. Mulberg, Steven A. Silber, and John N. van den Anker

which may lead to children avoiding certain foods and losing trust in the parents. In addition, whether the particular medication is indeed stable or bioavailable in other types of suspension vehicles often is never elucidated or described in the product label.

Pharmaceutical companies tend to develop many of their medications for the adult population and medications for populations with specialized needs are often developed as secondary product line extensions in the lifecycle of the individual drug. Although children may have obvious hurdles, even adults with certain issues such as neuromuscular diseases, swallowing disorders, need for enteral feeding tubes, and esophageal dysmotility must have their medications modified for their successful administration. The elderly population, who are without teeth and have muscular weakness, share similar needs of the pediatric population. Successful pill swallowing has been reported in healthy children as young as 18 months, and sustained long-term success with this skill has been achieved in most.[1,3] A majority of cognitively normal children ages 6–11 years old were followed in an observational cohort study conducted by Meltzer et al.[2] A total of 113 of the 124 children (91%) were able to swallow a tablet from a regular cup or a patented pill cup. Of these 124 patients, 57 who said they could swallow the pill were capable. Forty-seven children were taught how to swallow with a regular cup and nine learned how to swallow with the patented pill cup. The last 11 children did not learn how to swallow at that time.

Through behavioral training, young children can learn to swallow pills successfully. Children as young as 3 years old and even with developmental challenges and chronic conditions can be taught to swallow pills successfully.[4–7] Czyzewski et al.[5] followed 29 HIV-infected children ages 3–13 years old who were either naïve to swallowing pills or had difficulty. Seventeen of the children learned how to swallow large capsules that mimicked the size of their protease inhibitor medications and remained compliant for 6 months. Eighty-two percent of these children were trained how to swallow the capsules in one 30-minute teaching session.

Children with attention deficit hyperactivity disorder (ADHD) and children with autistic spectrum disorder may exhibit impulsive behavior and lack of attention, which may be an obstacle during training. Children with these special needs typically require medications in the form of a pill. Children with ADHD and autism have also been successfully taught to swallow pills.[6] Ghuman et al.,[4] in a pilot study, attempted to teach pill swallowing to four children with autistic spectrum disorder who carried a comorbid diagnosis of ADHD. By the end of the study, two learned how to swallow the capsules, one was able to successfully swallow with the behavioral therapist, and one was not able.

Beck et al.[6] taught eight children, four between 6 and 9 years old with ADHD and the other four, ages 4–6 years old, with autistic spectrum disorder (AD). Seven of the eight children were able to swallow the medication in the clinic with a therapist. Six of the eight were able to maintain this skill with their caregivers. In both of these studies, positive reinforcement was a successful tool, but these children may require several training sessions over a period of weeks to acquire this skill.

MECHANISM OF SWALLOWING

Understanding the typical swallowing mechanism and how to overcome the gag reflex are paramount for teaching children to swallow hard objects. If this skill is practiced in a supportive and calm environment, even young children can learn how to swallow medications. The key is getting children to start with tiny ingestible candies. The "trainer" then repetitively asks them to swallow the same size candy a few times and then slowly and methodically increases the size of these candies until the desired dimension is achieved.

The groups of children who may not be good candidates for this training may be those with oral motor problems, such as structural anomalies,[8,9] those who may not have the ability to understand the teacher, those with an anxious baseline, or those who have had a past negative experience with pill swallowing.

The ability to swallow is a complex, coordinated, and continuous act that involves both voluntary and involuntary muscles as well as striated and smooth muscles. In utero, infants are able to swallow at approximately 11 weeks gestation. They acquire the ability to have a coordinated suck–swallow mechanism at about 35 weeks gestation. The bolus of liquid nutrients through either the maternal nipple or by the bottle nipple is immediately situated after a successful suck on the back of the tongue. When the baby extracts the liquid bolus with sucking, the posterior tongue moves upward and backward. The liquid bolus is propelled back toward the pharynx. The pharyngeal muscles contract and the upper esophageal sphincter relaxes to allow the bolus to pass down to the esophagus.

In older infants and children, diets begin to change from liquid to purees to drier, lumpier foods. In addition, the foods are introduced by spoon to the anterior part of the mouth and tongue. The mechanism of swallowing alters in these older children by allowing this now thicker and drier consistency to pass from the front area of the mouth to the back part of the tongue. Coordination of the tongue, jaws, and cheeks along with closed lips create a pathway for more solid food to pass. Although the initiation of swallowing is under conscious control, most of the time it happens automatically. A sensation is created from contact of the bolus with the areas of the pharynx, base of the tongue, and soft palate. This is the pivotal point in permitting the contents to be swallowed. When there is acceptance of this bolus, the upper esophageal sphincter relaxes, breathing is inhibited, and the larynx moves forward and upward and the contents are propelled into the esophagus.

LEARNING TO SWALLOW PILLS

Simple Strategies

Children should be assessed for readiness for pill swallowing. A discussion with the child prior to learning how to swallow a pill is often helpful for understanding why they have to take the medication. There are child-friendly books that are available that can explain to children on their terms why it is important to take medication.[10,11] Parents often can teach their children how to swallow medication at home.

The best setting to start is a place free from distraction. Simple language should be used for each step. The caregiver begins by showing the child a glass of water, the straw if needed, and the medication. They should explain how to swallow the pill followed by actually demonstrating a pill-swallow. Next, the child is asked to take several practice swallows with large gulps of water. Straws can sometimes be a great distraction in the process, as they do not focus on the pill. Instruct the child to take the medication, place it on their tongue, and take a gulp of water. It may take several tries for the child to situate the pill so that it easily floats to the back of the mouth toward the throat. Some simple strategies are tilting the head up if the drug is a tablet and down if it is a capsule. Capsules tend to float to the top of the fluid. At the end of the attempt, the child should be praised whether the pill was swallowed or not. With a successful pill-swallow, positive reinforcement such as stickers may be given to the child.

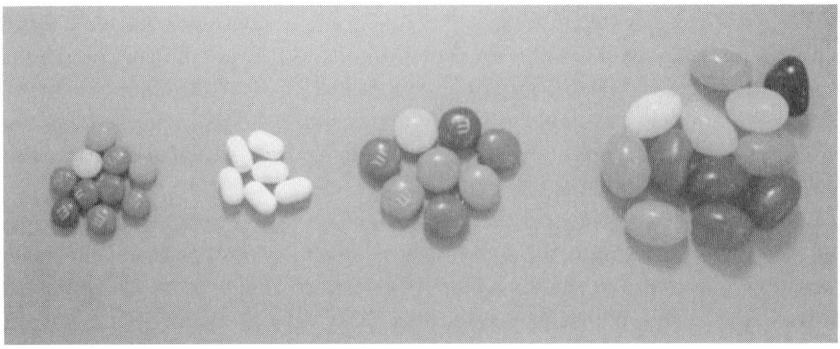

FIGURE 46.1 Progressive candy sizes in pill-swallowing training.

Some find that drinking a cold carbonated beverage is helpful in decreasing the sensation on the back of the throat. Another option for successful swallowing may be taking the pill with a semisoft food such as applesauce or ice cream. Compatibility with medications should be understood prior to using certain soft foods for reasons of bioavailability. It is important to review the product label for instructions.

Jelly Bean Test

At our institution, we have devised a simple strategy to prepare our children for wireless capsule endoscopy by having them practice with different sized candies. All patients are screened with the "Jelly Bean Test." We ask the families to start with a mini M&M®, advance to a TicTac® followed by a regular sized M&M®. If they could swallow all of these, the child is then asked to practice swallowing with a Brach's® jellybean (Figure 46.1) The 20 mm × 15 mm Brach's jellybean is comparable to the PillcamSB® capsule, which is 26 mm × 11 mm (Figure 46.2). We have found that "practicing" with the jellybeans reduces any apprehension prior to swallowing the actual PillcamSB.

Behavioral Training

For those children who cannot be successfully taught by their caregivers, another alternative is to teach them through behavioral training. This intervention is a series of well-defined and

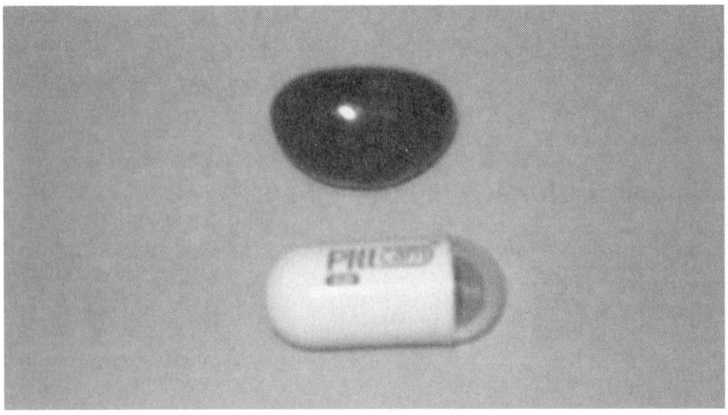

FIGURE 46.2 PillcamSB capsule compared to jellybean.

measurable steps that are repeated during practice sessions. The goal is to implement desensitization of the gag reflex. The process is time limited and a reward system is created with each successful trial. When there is not success, encouragement is given. As a part of the routine, children should be evaluated for their appropriateness for teaching pill swallowing. Those children with neuromuscular oral motor disorders, congenital anomalies of the oral motor cavity, and inability to understand language tend to be poor candidates. In addition, children with severe anxiety disorders or previous history of a severe negative experience of attempting to swallow pills may be quite difficult to teach. A practitioner for their ability to swallow should examine children, and their oral motor function must be assessed. Important as well is the caregiver's impression of the individual's readiness to learn.

Once the children are assessed for their ability to attempt pill swallowing, the trainer should meet with the families to discuss what the process will entail. As it may take several sessions, the consistency in the training is the key to the overall success.

Children should have the ability to understand simple verbal instructions and physical prompts. A trust should be established between the trainer and the child with a reward system clearly defined.

The child should sit down in a quiet place. Children should only see the smallest pills or candies with which they are practicing. After repeated successful attempts the next sized candy is introduced. Different edible cake decorations have been used by some investigators to teach children, effectively, in a sized sequential manner to swallow.[4] The trainer may begin by showing the child all the objects that will be used for training such as a glass of water, a straw, and the candies. Children may prefer to use another type of beverage or to use a straw. The trainer should give the child the control of opening and closing her mouth, sticking out her tongue, and independently placing the candy in her mouth. A glass of water or favorite beverage is offered with or without a straw. A few practice tries with swallowing the liquid may be helpful. Children should begin with the smallest hard candy that they are willing to swallow. With each successful attempt of the same sized candy, positive reinforcement is given to the child. After five consecutive attempts of swallowing the identical sized candy, children can progress to the next gradual size. Again, after five successful swallows with the increasing sized candy, the child may proceed. If there is hesitation, or unsuccessful trials with the larger candy, the last smaller sized candy should be swallowed again to give the final attempt a successful outcome. The next practice session should be on the next day. Caregivers should be informed of the last successful size of the candy. Children should be able to perform a successful swallow in front their caregiver of this last size. This demonstration shows the guardian that the child is willing and instructs the guardian ultimately how to do this at home. If comfortable, the family can practice with the last successfully swallowed candy at home. No attempts with increasing sizes should be carried out at home. Studies have shown that some children may be effectively trained after one session.

Patented Swallowing Cups

There is a patented cup called the *Pill Swallowing Cup* that was FDA approved in 2006 that is manufactured by Oraflo® Technologies. This cup facilitates swallowing of medications for children 4 years old and older. The cup is filled halfway with fluid, the lid is secured on top, and the pill is placed into the spout. The device allows water to wash into the back of the throat, carrying the pill with it. The angled mouthpiece is designed to allow the pill and liquid to be swallowed without tilting the head backward.

Discussion

Children can be trained to swallow pills at young ages. The opportunity to teach them younger could avoid future negative experiences when there are no choices of medication form or time to learn. Parents can teach simple steps of pill swallowing in a willing child. In more challenging situations, such as chronic illness or developmental delay, well-designed behavioral training sessions can achieve the goal of swallowing medications. A team of coordinated professionals who carry out the systematic desensitization of the gag reflex can teach young children with disabilities to take their medication. Medical professionals who take care of children should take a proactive role in encouraging the acquisition of this ability. This skill has lifelong benefits for the child in medication selection and ultimately adherence to therapy.

REFERENCES

1. Wright L, Woodcock JM, Scott R. Conditioning children when refusal oral medication is life threatening. *Pediatrics*. 1969;44:969–972.
2. Meltzer EO, Welch MJ, Ostrom NK. Pill swallowing ability and training in children 6 to 11 years of age. *Clin Pediatr*. 2006;45:725–733.
3. Funk MJ, Mullins LL, Olson RA. Teaching children to swallow pills. *Child Health Care*. 1984;13:20–23.
4. Ghuman JK, Catalso MD, Beck MH, Slifer KJ. Behavioral training for pill-swallowing difficulties in young children with autistic disorder. *J Child Adolesc Psychopharmacol*. 2004; 14(4):604–611.
5. Czyzewski DI, Runyan RD, Lopez MA, Calles NR. Teaching and maintaining pill swallowing in HIV-infected children. *AIDS Read*. 2000;10(2):88–94.
6. Beck MH, Cataldo M, Slifer KJ, Pulbrook V, Guhman JK. Teaching children with attention deficit hyperactivity disorder (ADHD) and autistic disorder (AD) how to swallow pills. *Clin Pediatr*. 2005;44:515–526.
7. Garvie PA, Lensing S, Rai SN. Efficacy of a pill swallowing training intervention to improve antiretroviral medication adherence in pediatric patient with HIV/AIDS. *Pediatrics*. 2007;119: e893–e899.
8. Milla PJ. Feeding, tasting, and sucking. In: Walker WA, Durie P, Hamilton JR, Walker-Smith J, Watkins J, eds. *Pediatric Gastrointestinal Disease*, Volume 1. Philadelphia: BC Decker; 2000: 217–223.
9. Vandenplas Y. Esophageal dysfunction. In: Delvin E, Lentze MJ, eds. *Gastrointestinal Functions*. Philadelphia: Lippincott/Williams & Wilkins; 2001: 235–255.
10. Galvin M, Ferraro S. *Otto Learns About His Medicine: A Story About Medication for Children with ADHD*. Washington, DC: American Psychological Association.
11. Klein NC, Holden M. *Healing Images for Children: Teaching Relaxation and Guided Imagery to Children Facing Cancer and Other Serious Illnesses*. Watertown, WI: Inner Coaching.

PART X

CASE STUDIES: SUCCESSES FOR CHILDREN

CHAPTER 47

A Case Study of Psychiatric Research in Children: The Risperdal® Example

MAGALI HAAS, MD, PhD

Johnson & Johnson Pharmaceutical Research and Development, Turnhoutseweg 30, B-2340 Beerse, Belgium

INTRODUCTION

Childhood and adolescence are the stages of life in which most mental disorders begin, even if they are detected or diagnosed significantly later in life. Most of these mental disorders are recognized as severe psychiatric conditions (e.g., depression, bipolar disorder, schizophrenia) with potentially serious acute consequences (e.g., suicide-related mortality)[1] and long-term impact on the development of the individual with particular impact felt in various areas including school, social, and sexual.[2–5] Many of these conditions are further exacerbated in adulthood by the development of other comorbid conditions such as substance abuse, other psychiatric disorders, or chronic diseases.

A fundamental tenet of clinical research is that with greater scientific understanding we would develop targeted interventions to obviate the inherent suffering, functional impairment, risk of suicide, and stigma that is associated with these mental disorders in children and adolescents. Despite the obvious needs, conducting clinical trials in these populations remains controversial and challenging. Hence, most medications used in children have still not been thoroughly studied in children, and many are being used in different indications from those approved in adults.

In the absence of timely pediatric clinical trials, the only solution for pediatricians and psychiatrists who wish to provide pharmacologic interventions for their patients is to prescribe medications in an off-label manner. The consequences of this practice are well described[6] and are tantamount to experimentation in themselves:

> Off-label prescribing is in fact a form of experimentation with no consistent eligibility criteria, no consistent dosing regimens, no pre-defined response criteria or stopping boundaries, no data safety monitoring, and inadequate sample sizes to power informative analyses, which means that nothing is learned.

Pediatric Drug Development: Concepts and Applications
Edited by Andrew E. Mulberg, Steven A. Silber, and John N. van den Anker

For these reasons, and as described throughout this book, legislation has been enacted in both the United States (2002 Best Pharmaceuticals for Children Act (BPCA) and the 2003 Pediatric Research Equity Act (PREA)[7,8] and Europe (2007 EU Paediatric Regulation)[9] to create incentives as well as a permissive environment for the pharmaceutical industry to improve pediatric drug labeling by conducting well-controlled studies in these populations. More recently, the BPCA has been reapproved and is discussed by Maldonado in Chapter 12.

The operational challenges of conducting pediatric research have been presented in hypothetical terms in previous chapters. The purpose of this chapter is to provide a case study using risperidone, a globally prescribed antipsychotic agent, to highlight some of the real-world challenges facing researchers and regulators in the conduct of pediatric psychiatry studies. These challenges are described in sections ranging from diagnosis, to validation of efficacy endpoints, to ascertainment of safety in these populations.

THE CONUNDRUM AND THE PARADOX OF EARLY DETECTION

One of the immediate operational challenges of pediatric psychiatry studies, in contrast to adult studies, is the conundrum of early diagnosis. While many of the severe psychiatric diseases manifest early in childhood development, the challenge is to detect and interpret premorbid symptoms to establish a diagnosis before a diagnosis is typically establishable. Diagnoses are currently made using guidelines such as the DSM-IV and ICD-10,[10,11] which were originally developed for adults. When evaluating risperidone for schizophrenia in our first study of an adolescent population, we wrestled with this issue in regard to inclusion criteria for the population to be studied.

Schizophrenia is a devastating, chronic neurodevelopmental disorder that afflicts 1 in every 10,000 persons ages 12–60 every year.[12] Onset is typically during adolescence or young adulthood; however, most epidemiologic studies have found that first contact with mental health services occurs when patients are aged 25–35 years, regardless of gender. In 75% of cases, a prodromal phase and psychotic prephase, totaling approximately 6 years, precedes the index admission.[13] As we currently understand this disorder, a complex interplay of genetic, epistatic, and environmental factors[14,15] produce a temporally dynamic pattern of illness and progressive structural brain abnormalities.[16] The considerable delays in recognizing this syndrome in common practice are, therefore, understandable.

According to the DSM-IV, the diagnostic criteria (a 6-month history with at least 1 month of active-phase symptoms) for schizophrenia are the same for pediatric and adult populations.[17] The ICD-10 diagnostic criteria largely mirror the DSM-IV criteria, with the exception that the required duration of symptoms is shorter. However, it is well recognized that young people often have subthreshold symptomatology well before they meet the formal diagnostic criteria for the disorder.[8]

Therefore, the use of DSM-IV criteria for conduct of pediatric clinical trials is most conservative. With this standard, we would potentially identify a more homogeneous population, but perhaps fewer numbers of subjects, who met these criteria. In contrast, the ICD-10 standard might produce a more inclusive, yet heterogeneous cohort. In our clinical trials, we resolved to use the DSM-IV criteria in a highly conservative way, that is, additionally excluding schizophreniform, schizoaffective, psychosis NOS, bipolar, depressive, and anxiety diagnoses. Careful characterization of a subject's symptoms was achieved by using a semistructured interview, the K-SADS-PL,[18] conducted by a trained pediatric psychiatrist.

When we developed our protocol for juvenile bipolar disorders, we were fortunate that many, but not all, controversies were addressed head-on by a consortium established between the U.S. Food and Drug Administration (FDA), the National Institutes of Health (NIH), academia, and industry.

Bipolar disorder is defined by the presence in the lifetime history of at least one episode of mania (or hypomania), with or without major depression (mixed episode), with a subsequent lifetime course of recurrent mania or depression.[19,20] Bipolar disorder is a serious condition with a high lifetime risk of suicide and is associated with substantial impairment in social and occupational functioning. It is commonly associated with equally disabling comorbid disorders and high relapse rates. Most disturbing is a trend for earlier onset of symptoms in successive generations of offspring of those with bipolar disorder.[21]

The diagnosis of episodes of mania in adults is based on symptoms that include inflated self-esteem, pressure of speech, racing thoughts, increased activity, impulsive behavior, hypersexuality, and decreased need for sleep. Mixed episodes have mania with concomitant major depression. Manic episodes are frequently followed by depressive episodes and often have some subsyndromal depressive symptoms.

There is evidence that adolescent mania may represent a developmental subtype of bipolar disorder with distinctive characteristics; however, debate continues regarding diagnostic phenotypes.[22,23] The fundamental basis of controversy is whether the presentation of bipolar disorder is, or should be, defined similarly to adults, per DSM-IV. A narrow definition of phenotype would define juvenile bipolar disorder based on the emergence of euphoria, grandiosity, and other classic manic symptoms, whereas a broader phenotype would encompass irritability and mood lability.[24] Part of the controversy stems from the often complex and variable presentation in this age group, where more patients experience mixed episodes (mania accompanied by depressive features) as well as psychotic symptoms, and/or rapid changes in mood. Under any definition, it is a disorder that significantly impacts the scholastic, social, and personality development, as well as the overall functioning and life experience of the children affected.

In June 2002, the American Academy of Child and Adolescent Psychiatry, in collaboration with Best Practice, convened a working conference on methodological issues and controversies in clinical trials with child and adolescent patients with bipolar disorder.[25] The explicit purpose of the meeting was to develop a template for clinical trials of acute mania/bipolar disorder in children and adolescents. Invited participants included clinical researchers with expertise in childhood and adult bipolar illness, pharmaceutical industry sponsors with an interest in mood stabilizer products, staff of the FDA (and their counterparts from regulatory agencies in Canada and the European Union) and the National Institute of Mental Health, and representatives of families with affected children.

Conference participants reached agreement on 18 broad methodological questions. Key points of consensus were to assign priority to placebo-controlled studies of acute manic episodes in children and adolescents ages 10–17 years, who may or may not be hospitalized, and who may or may not suffer from common comorbid psychiatric disorders; to require that specialist diagnostic "gatekeepers" screen youths' eligibility to participate in trials; to monitor interviewer and rater competency over the course of the trial using agreed upon standards; and to develop new tools for assessment, including scales to measure aggression/rage and cognitive function, while using the best available instruments (e.g., Young Mania Rating Scale—YMRS) in the interim.

In the conduct of our own juvenile bipolar study, we adopted many of these recommendations; however, we were limited in our ability to execute the program outside the

United States. Despite considerable evidence that bipolar disorder in the 10–17-year-old population is thought to be relatively common and phenomenologically similar to bipolar disorder in adults,[26] feedback from European health authorities (via scientific advice), IRBs, and investigators all suggested that inclusion of subjects below age 13 would not be acceptable, nor the inclusion of subjects with a mixed episode of bipolar mania. The appropriateness of the adolescent version of the YMRS was questioned along with the notion that subjects could be outpatients. Some of the arguments were focused on the perception that this disorder was not as prevalent in the European Union, which, given these stricter criteria, might be borne out. In the future, in order to conduct larger bipolar studies on a global scale, the field will need to arrive at a common definition of disease and relevant design elements.

THE PARADOX OF DIAGNOSIS

Developing a consensus opinion to diagnose universally bipolar disorder does not mitigate a fundamental concern and dilemma for pediatric researchers—the stigma of early diagnosis. The paradox occurs because investigators are obligated to establish a definitive diagnosis in order to enroll a subject in most psychiatry studies. Ironically, once a patient is diagnosed, in our social systems they are essentially labeled for the purposes of educational, medical, and social programs and this can result in significant stigma for the patient and his/her family. In current medical practice, there is no need to diagnose in order to treat. The practical implication in trial conduct can be reluctance of the investigator to diagnose or of the family to participate in the trial. Therefore, in order to push science forward, the researcher has to consider the impact on the subject of conclusively establishing a diagnosis and should educate the family of the potential benefits provided by a clearer etiology and earlier intervention.

MEASURING BENEFIT

Having successfully diagnosed a cohort population and identified the target of treatment, the next major concern of clinical studies is measuring benefit. Within psychiatric studies, few direct assessments or biomarkers exist that assess specific domains of disease symptomatology. Lacking direct prognostic indicators, these studies typically utilize validated scales or patient-reported outcomes to assess benefit. In this section we explore the experience we had in this arena when designing our adolescent schizophrenia program.

Characteristic symptoms of early-onset schizophrenia include disorganization and a high frequency of poorly differentiated and nonspecific symptoms, severe conduct disorders, and functional impairment.[8] In adult studies of schizophrenia, benefit from antipsychotic agents has been extrapolated based on improvements measured with scales such as the Positive and Negative Symptom Scale (PANSS).[27] The PANSS is a scale that is composed of 30 items relating to three core aspects of schizophrenia: positive symptoms (Items 1–7), negative symptoms (Items 8–14), and general signs of psychopathology (Items 15–30). This scale has been validated extensively in adults and a reduction of 20% in scores has been established as clinically relevant.[28] The 30-item PANSS is rated on a scale of 1 (absent) to 7 (extreme) with total scores ranging from 30 to 210, higher scores representing a worse condition. The primary outcome to assess benefit in schizophrenia trials generally is the change from baseline to endpoint on the PANSS.

The PANSS is only validated for use in adults. Therefore, the scale was modified through successive field trials on the basis of developmental characteristics of children and adolescents. The Kiddie-PANSS is a version of the PANSS specifically designed to assess positive and negative symptoms in severely disturbed children and adolescents.[29] In our initial design of the schizophrenia program, we chose to incorporate the Kiddie-PANSS for the assessment of the primary endpoint. However, in subsequent implementation at investigator meetings, we found that investigators were generally unfamiliar with the tool and had difficulty rating it, despite extensive training attempts. Since this psychometric tool had not been extensively validated, needed further development on its anchor points, was difficult to administer, and had not received wide adoption, we reverted to use of the adult PANSS.

To ensure the consistent administration and scoring of the PANSS in the adolescent study population, in consultation with thought leaders, we developed and implemented PANSS administration guidelines, a training program, and a certification process for investigators.

The main features of the revised administration guidelines were (1) interview of both patient and parent or guardian, (2) rearrangement of the order of the questions to make them less threatening to an adolescent and potentially more revealing, (3) rewording of the questions to make them age-appropriate and comprehensible to an adolescent, and (4) adjustments in scoring to account for developmental factors associated with age, such as uncooperativeness. The final product was a guidance document that accompanied the study protocol as well as training tapes that utilized the new techniques and adapted questions. The revised instrument was well accepted and produced high interrater reliability during the conduct of the subsequent schizophrenia trials. For global use, further modifications were made to address interpretation of several proverbs in foreign languages, a key component of the interview for cognitive understanding.

SAFETY ASSESSMENT IN PEDIATRIC PSYCHIATRY RESEARCH

A very important manifestation of the legislation on pediatric research is the establishment of a more expansive safety database for products used in children. In the conduct of all clinical trials, there are a variety of required and standard assessments that are collected on each subject (e.g., adverse event data, electrocardiograms, and laboratory tests). In pediatric research, investigators must also consider effects on growth and development and the impact of chronic treatment for an expanded lifetime period.

In our programs, a seemingly "simple" solution was to add these parameters (height, weight, body mass index (BMI)) to our studies and also extend the duration of long-term open-label observation following double-blind study. With the use of common instruments (e.g., same type of calibrated scales) and procedures (e.g., clothed; without shoes) objective measures of height and weight were obtainable. But, what about development? In current practice, physicians assess pubertal development via the sexual maturity rating developed by Marshall and Tanner.[30,31] In this scale, sexual maturity is defined according to five stages of genital and pubic hair development assessed by visual inspection and graded independently (please see Chapter 28 by Rovner and Zemel for further description).

For pediatric psychiatry trials, where psychiatrists typically have limited experience conducting these types of examinations, and where patients tend to be intolerant to the procedure, obtaining accurate data can be challenging. To address this, some trials have tried to allow subjects to provide self-assessments of developmental status by matching themselves to the Tanner stages. This has been demonstrated to be a highly unreliable practice.[32]

When we initiated our programs in the late 1990s to early 2000s, a paucity of data was available to assess the limits of normal development. The first national cohort studies, in a representative population, were emerging in this same period.[33,34] Determinations of the impact of drug treatment on normal growth and development were thereby restricted. Even the relatively objective measures of height and weight are hampered because "normal" growth varies by region and nutritional status and few data exist that provide assessments across all regions in which studies are conducted today.[35]

Similar challenges are faced in the assessment of hormonal and metabolic parameters. Metabolic disorders have not been adequately studied in children, with only a handful of studies evaluating the long-term impact of maintained antipsychotic treatment on metabolic parameters in this population. The advent of the atypical antipsychotics in the mid-1990s, with their broader spectrum efficacy and lower risk of movement disorders (including tardive dyskinesia), brought important improvements in the treatment of many psychiatric disorders. However, more recently, the atypicals have been linked to medical morbidity in adults, in particular, obesity and concomitant hyperglycemia, and abnormal lipid levels. This cluster of metabolic symptoms and obesity, which was first described by Reaven as Syndrome X[36] and is now known as the metabolic syndrome, comprises a serious propensity for diabetes and cardiovascular morbidity.[37] Insulin resistance appears to play an important role in the development of this syndrome and its complications.[36,37] Few systematic assessments of metabolic parameters in children and adolescents receiving antipsychotics have been reported.[38]

There are currently no universally accepted criteria for diagnosing hyperinsulinemia in children.[39–41] Assessment of normal insulin levels is further confounded by the observation that normal pubertal development is associated with changes in insulin sensitivity along with multiple other physical and hormonal changes.[42] The decrease in insulin sensitivity during puberty is associated with a compensatory increase in insulin secretion, which results in increased levels of plasma insulin. For instance, transition from Tanner stage I to Tanner stage III is associated with a 32% reduction in insulin sensitivity and increases in fasting glucose and insulin.[43] Thus, any assessments of metabolic alterations in children and adolescents must take into account the changes naturally associated with altering developmental status or puberty. A priori, however, there needs to be consensus about the criteria for defining the metabolic syndrome in children[44] (and even adults)[45] if its use as a diagnosis is to be helpful.

As most severe psychiatric disorders diagnosed in children and adolescents are lifelong in duration, chronic treatment is a mainstay of psychiatric therapeutic intervention. Most regulatory and ICH guidances require that long-term data, typically 1 year, be acquired for the studied population, to confirm long-term durability of benefit as well as impact on safety. For youths with psychiatric disorders, providing continuity of care is a challenge, since they are often noncompliant. In a clinical trial setting, this results in early termination of treatment and impacts the ability of researchers to analyze safety endpoints. Moreover, the previously standard 1-year durations of open-label safety studies are now being recognized by regulators as being insufficient to address topics of growth and pubertal development in the face of chronic treatment.

To further pediatric research within drug development, we must first develop universal, standard, age-based growth and development charts in the normal population as well as reference standards for important hormones and laboratory parameters. Furthermore, these standards have to account for longitudinal changes occurring in individuals as they mature. This type of very long-term safety and efficacy assessment in psychiatric populations will be one of the main operational challenges for the next generation of studies.

A VISION OF THE FUTURE

One vision of the future for treatment of psychiatric disorders holds that at-risk cohorts would be identified prior to any disease manifestation and treated prophylactically. In order for psychiatric research in children and adolescents to pioneer along this projection, technology platforms, design, and operational implementation will also need to advance. Most psychiatric disorders are complex multifactorial diseases with highly genetic and phenotypic heterogeneity. As the research community continues to identify susceptibility genes, the capability to develop high-prediction diagnostics will emerge. Each genetic variant in combination with nongenetic or environmental factors produces a unique phenotype of disease, resulting in the heterogeneity observed in many psychiatric disorders. Future studies will likely combine diagnostics and endophenotyping to create a homogeneous cohort and thereby enhance the power to detect and then assess novel molecular targets to treat disease and possibly prevent expression. The operational details of this schema will require a wholesale shift in the approach taken by regulators, physicians, and researchers in the design, conduct, and analysis of clinical trials conducted in pediatric populations.

A Final Word: On June 20, 2007, Johnson & Johnson Pharmaceutical Research & Development received an approvable letter from the FDA in support of its file for adolescent schizophrenia and juvenile bipolar disorders.

REFERENCES

1. Pelkonen M, Marttunen M. Child and adolescent suicide: epidemiology, risk factors, and approaches to prevention. *Paediatr Drugs*. 2003;5(4):243–265.
2. Froehlich TE, Lanphear BP, Epstein JN, Barbaresi WJ, Katusic SK, Kahn RS. Prevalence, recognition, and treatment of attention-deficit/hyperactivity disorder in a national sample of US children. *Arch Pediatr Adolesc Med*. September 2007;161(9):857–864.
3. Moreno C, Laje G, Blanco C, Jiang H, Schmidt AB, Olfson M. National trends in the outpatient diagnosis and treatment of bipolar disorder in youth. *Arch Gen Psychiatry*. 2007;64:1032–1039.
4. Olson M, Blanco C, Liu L, Moreno C, Laje G. National trends in the outpatient treatment of children and adolescents with antipsychotic drugs. *Arch Gen Psychiatry*. 2006;63(6):679–685.
5. Olfson M, Marcus SC, Druss B, Elinson L, Tanielian T, Pincus HA. National trends in the outpatient treatment of depression. *JAMA*. 2002;287(2):203–209.
6. Schachter AD, Ramoni MF. Paediatric drug development. *Nat Rev*. 2007;6:429–430.
7. US FDA Center for Drug Evaluation and Research. Guidance for Industry. Qualifying for Paediatric Exclusivity Under Section 505A of the Federal Food, Drug, and Cosmetic Act. 1999 Available at http://www.fda.gov/cder/guidance/2891fnl.htm.
8. US FDA. Pharmaceuticals Research Equity Act, 2003. Available at http://www.fda.gov/cder/pediatric/S-650-PREA.pdf.

9. European Medicines Agency. Medicines for Children—The EU Paediatric Regulation, 26 January 2007. Available at http://www.emea.europa.eu/htms/human/paediatrics/regulation.htm.
10. World Health Organization. *International Statistical Classification of Diseases and Related Health Problems*, Tenth Revision (IDC-10). 1992.
11. American Psychiatric Association. *Diagnostic and Statistical Manual of Mental Disorders*, 4th ed, Text Revision. Washington DC: American Psychiatric Association; 2002.
12. Hafner H, an der Heiden W. Epidemiology of schizophrenia.*Can J Psychiatry*. 1997;42: 139–151.
13. Hafner H, Maureer K, Loffler W, et al. Onset and early course of schizophrenia. In: Hafner H, Gattaz WF, eds. *Search for the Causes of Schizophrenia*, Vol III. Berlin:Springer-Verlag; 1995: 43–66.
14. Harrison PJ, Weinberger DR. Schizophrenia genes, gene expression, and neuropathology: on the matter of their convergence. *Mol Psychiatry*. 2005;10:40–68 (image 5).
15. Rapoport JL, Addington AM, Frangou S, Psych MR. The neurodevelopmental model of schizophrenia: update 2005. *Mol Psychiatry*. 2005;10:434–449.
16. Woods BT. Is schizophrenia a progressive neurodevelopmental disorder? Toward a unitary pathogenetic mechanism. *Am J Psychiatry*. 1998;155:1661–1670.
17. American Psychiatric Association. Practice guideline for the treatment of patients with schizophrenia. *Am J Psychiatry*. 1997;154(4 Suppl):1–63.
18. Kaufman J, Birmaher B, Brent D, Rao U, Ryan N. Kiddie-SADS—Present and Lifetime Version 1.0 (K-SADS-PL). In: American Psychiatric Association Task Force for the Handbook of Psychiatric Measures. *Handbook of Psychiatric Measures*. Washington DC: American Psychiatric Association; 2000.
19. Kessler RC, Berglund P, Demler O, Jin R, Merikangas KR, Walters EE. Lifetime prevalence and age-of-onset distributions of DSM-IV disorders in the National Comorbidities Survey Replication. *Arch Gen Psychiatry*. 2005;62(6):593–602.
20. Merikangas KR, Akiskal HS, Angst J, et al. Lifetime and 12-month prevalence of bipolar spectrum disorder in the National Comorbidity Survey replication. *Arch Gen Psychiatry*. 2007;64 (5):543–552.
21. Rice J, Reich T, Andreasen NC, et al. The familial transmission of bipolar illness. *Arch Gen Psychiatry*. 1987;44(5):441–447.
22. Masi G, Perugi G, Toni C. The clinical phenotypes of juvenile bipolar disorder: toward a validation of the episodic-chronic-distinction. *Biol Psychiatry*. 2006;59(7):603–610.
23. Ghaemi SN, Martin A. Defining the boundaries of childhood bipolar disorder. *Am J Psychiatry*. 2007;164(2):185–188.
24. McClellan J, Kowatch R, Findling RL, et al. Practice parameter for the assessment and treatment of children and adolescents with bipolar disorder. *J Am Acad Child Adolesc Psychiatry*. 2007;46 (1):107–125.
25. Carlson GA, Jensen PS, Findling RL, et al. Methodological issues and controversies in clinical trials with child and adolescent patients with bipolar disorder: report of a consensus conference. *JCAP*. 2003;13(1):13–27.
26. Geller B, Luby J. Child and adolescent bipolar disorder. A review of the past 10 years. *J Am Acad Child Adolesc Psychiatry*. 1997;36:1168–1176.
27. Kay SR, Fiszbein A, Opler LA. The positive and negative syndrome scale (PANSS) for schizophrenia. *Schizophr Bull*. 1987;13:261–276.
28. Peralta V, Cuesta MJ. Psychometric properties of the PANSS in schizophrenia. *Psychiatry Res*. July 1994;53(1):31–40.
29. Fields JH, Grochowski S, Lindenmayer JP, et al. Assessing positive and negative symptoms in children and adolescents. *Am J Psychiatry*. 1994;151(2):249–253.

30. Marshall WA, Tanner JM. Variations in the pattern of pubertal changes in boys. *Arch Dis Child.* 1970;45:13–23.
31. Tanner JM, Whitehouse RH. Clinical longitudinal standards for height, weight, height velocity, weight velocity, and stages of puberty. *Arch Dis Child.* 1976;51:170–179.
32. Desmangles JC, Lappe JM, Lipaczewski G, Haynatzki G. Accuracy of pubertal Tanner staging self-reporting. *J Pediatr Endocrinol Metab.* 2006;19(3):213–221.
33. Herman-Giddens ME, Wang L, Koch G. Secondary sexual characteristics in boys: estimates from the national health and nutrition examination survey III, 1988–1994. *Arch Pediatr Adolesc Med.* 2001;155(9):1022–1028.
34. Sun SS, Schubert CM, Chumlea WC, et al. National estimates of the timing of sexual maturation and racial differences among US children. *Pediatrics.* 2004;113(1 Pt 1):177–178.
35. National Center for Health Statistics: National Health and Nutrition Examination Survey. Available at www.cdc.gov/nchs/about/major/nhanes/growthcharts/datafiles.htm.
36. Reaven GM, Banting lecture 1988. Role of insulin resistance in human disease. *Diabetes.* 1988;37:1595–1607.
37. Eckel RH, Grundy SM, Zimmet PZ. The metabolic syndrome. *Lancet* 2005;365:1415–1428.
38. Jensen P, Buitelaar J, Pandina G, Binder C, Haas M. Management of psychiatric disorders in children and adolescents with atypical antipsychotics. *Eur Child Adolesc Psychiatry.* 2006;16 (2):104–120.
39. Sullivan CS, Beste J, Cummings DM, et al. Prevalence of hyperinsulinemia and clinical correlates in overweight children referred for lifestyle intervention. *J Am Diet Assoc.* 2004;104:433–436.
40. Travers SH, Jeffers BW, Bloch CA, Hill JO, Eckel RH. Gender Tanner stage differences in body composition and insulin sensitivity in early pubertal children. *J Clin Endocrinol Metab.* 1995;80:172–178.
41. Roemmich JN, Clark PA, Lusk M, et al. Pubertal alterations in growth and body composition. VI. Pubertal insulin resistance: relation to adiposity, body fat distribution and hormone release. *Int J Obes Relat Metab Disord.* 2002;26:701–709.
42. Caprio S, Plewe G, Diamond MP, et al. Increased insulin secretion in puberty: a compensatory response to reductions in insulin sensitivity. *J Pediatr.* 1989;114:963–967.
43. Goran MI, Gower BA. Longitudinal study on pubertal insulin resistance. *Diabetes.* 2001;50:2444–2450.
44. Li C, Ford ES. Is there a single underlying factor for the metabolic syndrome in adolescents? A confirmatory factor analysis. *Diabetes Care.* 2007;30(6):1556–1561.
45. Assmann G, Guerra R, Fox G, et al. Harmonizing the definition of the metabolic syndrome: comparison of the criteria of the Adult Treatment Panel III and the International Diabetes Federation in United States American and European populations. *Am J Cardiol.* 2007;99 (4):541–548.

CHAPTER 48

Topiramate Case Study

SETH L. NESS, MD, PhD

Johnson & Johnson Pharmaceutical Research & Development LLC, 1125 Trenton-Harbourton Road, Titusville, New Jersey 08560

URSULA MERRIMAN, RN

Ortho-McNeil Janssen Scientific Affairs, LLC, Titusville, New Jersey 08560

JEFFREY S. NYE, MD, PhD

Johnson & Johnson Pharmaceutical Research & Development, Welsh and McKean Roads, Springhouse, Pennsylvania 19477

INTRODUCTION

The following is a case study of a global pediatric drug development program focused on infants with epilepsy. The trial dynamics are specific, but many points may be generalized and useful for other pediatric programs that are in development, or being contemplated for execution in various therapeutic areas. The studies involved topiramate as adjunctive therapy in infants from 1 month to 2 years old with refractory epilepsy and who were being treated. The main components in the program consisted of PEP-1002, a pharmacokinetics and tolerability study; PEP-3001, an efficacy and long-term safety study; and two additional pharmacokinetic studies in healthy adult volunteers. There was also an oral liquid formulation developed especially for infants, who are not yet ready for solid foods.

To our knowledge, this has been one of the largest global clinical research programs in infants done by a pharmaceutical company to date. As the regulatory environment in the United States and European Union (EU) shifts to requiring pediatric investigational plans for neonates, infants, young children, older children, and adolescents, the lessons contained here should be applicable to development programs in other therapeutic areas.

LESSONS FROM THE REGULATORY HISTORY IN THE UNITED STATES

A proposed pediatric study request (PPSR) was submitted in 2000 to the U.S. Food and Drug Administration (FDA) but a final response was received only in July 2004. The importance

Pediatric Drug Development: Concepts and Applications
Edited by Andrew E. Mulberg, Steven A. Silber, and John N. van den Anker

of this delay is that the exclusivity period of topiramate was near to expiration when the written request (WR) was issued, limiting the amount of time for successful execution of the proposed studies. In addition, surprisingly, the written request included two indications, primary generalized tonic–clonic seizures and adjunctive therapy of partial complex epilepsy. These indications were clearly based on the issued and granted indications for topiramate down aged 2 and older, which in the mind of the FDA, were logical extensions for children below 2 years as well as infants below 2 months.

To comply with the regulations, the FDA issued its written request in conformity with the granted indications for topiramate. However, after consulting with a number of therapeutic area experts, the clinical team soon learned that primary generalized tonic–clonic seizures were an entity that is either not well known or entirely nonexistent in infants under age 2 months. This was not widely known in the literature and has not been clarified by the existing pediatric neurologist community. Experts in the field and practitioners have long tolerated the use of the terms generalized tonic–clonic seizures in infants, despite the fact that their electroencephalograms (EEGs) never showed generalization and their movements were rarely tonic and clonic.

This difference between infants and older children and adults is well documented and a general theme of this book. There are diseases that may resemble the disorders of older children and adults but, because of major developmental differences, are really not the same and may not require the same therapy. After providing adequate documentation, written expert opinion, and the testimony of a world recognized expert in pediatric epileptology, the team was able to convince the FDA of the nonexistence of primary generalized tonic–clonic seizures and was granted approval to proceed with only a single indication. Importantly, regulatory agencies must have a productive discourse with the marketing authorization holder and appreciate diagnostic nuances of pediatric disorders at the various ages to ensure scientific credibility of studies and eventual labeling.

The FDA division that issued the written request apparently did not have any pediatric neurology experience, and general pediatricians probably are not aware of the differences in epilepsy subtypes in infants and older children, a gap that could possibly be filled when regulatory agencies are making requests for additional studies for infants and, in the future, for neonates. An automatic assumption that diseases for adult indications have exact replicas in a pediatric age cohort needs to be substantiated and challenged if there is any significant scientific doubt of the relationship.

BUILDING THE TEAM

Pharmaceutical companies are highly motivated to complete pediatric written request studies, but that financial motivation does not necessarily extend to a motivation to develop large pediatric resources within the company. In addition, the existing employees with pediatric experience within the company may not be accessible to the specific team or have the same motivation to be available for these studies for various competing reasons. Because of these dynamics, the operational, though not the medical, aspects of this global program were outsourced to a partner with expertise in pediatric research. After amassing the available staff with experience in pediatric medicine, pediatric clinical research, or written request regulatory filings, the team began to develop a program to comply.

STUDY DESIGN

The design of the studies for the program was particularly challenging, both because of the ill patient population, the need to study infants, and the global nature of the trial. Previous trials of epilepsy in infants had not used a placebo arm nor had an enrichment design. The written request limited the design options, specifying many features of the trial. An advisory board with experts in pediatric epileptology was convened in order to discuss key elements of the protocol design. It was also important to retain an academic ethicist to assist us with the particular ethical issues inherent in a study in infants.

Once the protocols were complete, it was decided to submit them to the FDA for a special protocol assessment (SPA). Particularly in a written request program, the SPA is important and should be seriously considered, but it must be kept in mind that because of the underresourcing at the agency, the FDA may be unable to meet its mandated timelines. The SPA response, which was due after 90 days, took approximately 6 months. Because of the short timeline, global regulatory submission had already taken place, forcing the team to make amendments to accommodate changes requested in the delayed SPA response. Despite well-intentioned regulators, delays in providing timely responses can put programs at risk of failure to meet the required submission timing and, consequently, could abrogate the motivation of pharma teams who are planning important but expensive and difficult studies in pediatric populations.

Three points in the design of the protocol deserve special mention. The issues of particular relevance for this program include the use of a placebo arm, dosing regimen for infants, and the issue of the central laboratory and phlebotomy requirements.

Placebo

The written request issued by the FDA mandated the use of a placebo comparator in a well-designed trial. It was anticipated that use of placebo therapy in seriously ill infants would pose ethical challenges and may not receive approval by regulatory agencies, institutional review boards (IRBs), and ethics committees. In this regard, consultation with a medical ethicist was particularly helpful. Because there was no approved medications for the treatment of epilepsy in this age, and because the trial utilized topiramate as adjunctive therapy—and thus all subjects, even those on placebo, were also on another antiepileptic medication—the trial ultimately secured approval worldwide. The important ethical consideration of equipoise was achieved. For further considerations of ethics in pediatric clinical trials, please see Chapter 9 by Nelson and Chapter 10 by Baer in this book.

Related to the issue of the placebo, and an issue that was of strong concern to the advisory board, was the duration of time that the subjects could remain on placebo treatment. The need to titrate topiramate up to a target dose was an impetus toward a longer double-blind period. However, the requirement for the subject to be on placebo made a short period more desirable. The duration of the double-blind phase of the infant epilepsy efficacy study was ultimately set at only 20 days. It was thought by the advisors and by the company that this period of time was the maximum amount of time that it would be ethical to keep these infants on placebo without having the opportunity to alter their other antiepileptic drugs (AEDs) or to add another AED. Another factor in the decision to allow 20 days of therapy was the community standard for adjustments of anticonvulsant medications, typically requiring a trial period of approximately 20 days, even in infants. However, the short duration of the double-blinded period and the need to titrate topiramate mandated that the maximum dose reached in patients treated with topiramate was limited, the speed of titration was unusually rapid, and the duration of treatment at the plateau highest dose was extremely short.

Dosing Scheme and Availability of Formulation

At the time the program was started, a sprinkle formulation for use in children was already available. An oral liquid solution was therefore developed for infants in the program. Both of these formulations were found to be bioequivalent, and both were therefore used in the study. During the double-blind phase and the initial phase of the pharmacokinetics (PK) study, subjects were required to stay on one formulation, the choice of which was made based on weight, development, and physical ability to ingest the sprinkles formulation. During the open label extension of the studies, subjects were permitted to switch back and forth between formulations.

The dosing scheme for the studies was particularly challenging due to two formulations, four dosing arms, the need for blinded up-titration of the doses, the need for a blinded transition taper between the blinded drug in the double-blind phase and the open label drug in the open label extension phase to avoid unblinding of investigators and parents, and the dosing of infants and children on a milligram per kilogram (mg/kg) basis rather than a classical milligram basis.

In addition, to allow maximum flexibility to the investigators in changing doses in a naturalistic fashion, as they would in their standard practice, the protocol attempted to minimize the number of capsules that were required for dosing. Because of all these factors, the dosing scheme ended up extremely complex, requiring tremendous efforts by the clinical supplies unit and the factories producing the drug supply. For a complex scheme like this, it is critical to involve the clinical supplies unit from the very beginning.

Large amounts of effort and time were also devoted to designing the interactive voice response system (IVRS). Proper design and user testing of the IVRS is critical to success. Trials have been compromised due to faulty IVRS functioning and the problem becomes more acute with complex dosing, as may be expected in infant trials. Ultimately, despite the complexity, the dosing scheme was successful, and the IVRS system functioned well. However, the cost and benefits of such a complex approach, compared to a more simple, perhaps rigid, dosing scheme in infants should be carefully weighed.

Central Laboratory

A key concern from the beginning of the program was the issue of laboratory samples, particularly blood tests, in infants. The team knew from the start that it would be very difficult to use a central laboratory that could only handle adult or adolescent volumes of blood. Drawing several milliliters of blood from a 2-month-old baby at each visit would have presented an almost insurmountable problem. Fortunately, we were able to find an academic children's hospital with a reference laboratory that was willing to serve as a central laboratory for our trial. This central laboratory was able to use microsampling, small volumes of blood of several hundred microliters coupled with the infant type collection tubes that are typically used for infants by pediatricians. Because we were able to minimize the blood volumes drawn, we were able to obtain ethics committee approval for all the many required blood draws in this study.

Nevertheless, the use of this children's hospital as a central laboratory did create unexpected problems. The percentage of samples that were clotted or hemolyzed was certainly greater than would be typically found using larger samples with the major central laboratories. The problems that arose with shipping from many countries around the globe were also greater than a typical trial. Database issues arising from the unfamiliarity of an academic children's hospital with the database structures used in the pharmaceutical industry were also a problem.

Unfortunately, it appears that this academic hospital will no longer be working as a central laboratory for global trials run by the pharmaceutical industry. For a successful outcome in infants and neonates, it will be important for a major central laboratory vendor to allow the use of reduced blood volumes and to accept the special collection tubes. Alternatively, local laboratories could be used, perhaps as part of one of the "virtual laboratories" currently being marketed by the major central laboratories, although this approach is not ideal.

For urgent follow-ups that could only be done on fresh samples, local laboratories were required. These results and their distinct normal ranges needed to be integrated into the clinical database. In some cases, there were no pediatric normal ranges available at the local laboratory and generic ranges from our central laboratory were used.

OTHER SAFETY ISSUES

In any trial concerning infants, growth should be taken into account. Standardized procedures and equipment for measuring weight, height, and length need to be established. For length, a rigid measuring board should be used and for all parameters multiple measurements should be taken at each time point and averaged. Ideally, the same person should perform the measurements at each visit. For further details, please review Chapter 29 by Rovner and Zemel in this book.

Safety monitoring of growth is best done using programmatically generated growth charts. Analysis of growth parameters is best conducted using Z-scores. Growth charts and tables of percentiles and Z-scores can be obtained from the U.S. Centers for Disease Control website. Corrected age for premature infants should be used for growth analysis.

The impact of the growth of subjects over the course of the trial must be taken into consideration in the analyses of many safety parameters. For instance, in adults, the lack of any mean change in weight over a year would be unremarkable. In infants, it might indicate a major safety concern. In addition, the differences in vital signs and laboratory tests need to be considered. In addition, the special considerations for evaluation of trial-related safety and determination of ADRs is found in Chapter 41 by Hetherington and discussion of laboratory parameters in Chapter 37 by Mulberg and Maldonado in this book.

OPERATIONAL AND LOGISTICAL CHALLENGES

The resources needed to successfully implement and complete a clinical research program in this population cannot be underestimated. The rule of thumb was to estimate the time and cost for each aspect of the study, such as regulatory submissions and approvals, IVRs setup, and vendor training, and double it at the very least. Thus, the original estimate of the cost of the study was doubled. This was due to the slow recruitment rate and the extension of regulatory timelines (the regulatory approvals in the 30 countries proposed took nearly 3 times as long as an adult study). One country obtained approval 18 months after submission. When performing feasibility at both the country and site levels, it is wise to anticipate a 20% nonapproval rate. This of course would impact recruitment and timeline estimates and increase cost.

Proper training of clinical research organization (CRO) staff is essential as well. Monitoring experience in pediatric studies varies, especially when the study is conducted on a global scale. Increased resources for site training must be taken into consideration as well. Many of the sites participating in this program were academic centers; the clinicians were highly regarded pediatric epileptologists. However, the experience in Good Clinical Practice (GCP) global research studies and adherence to GCP and strict study procedures varied widely. Thus, it was necessary to create a monitoring plan that was more intensive and frequent than is typically utilized.

In the end, the take-home lesson of this experience was that clinical research in the very young is not business as usual, in any aspect of the study. While the processes employed were typical of most GCP clinical research studies, the intensity and degree to which these processes were implemented required extreme dedication and commitment on the part of all involved. The success of this study was due to the team keeping the ultimate goal in mind: the protection of these very young, vulnerable patients who rely on the compassion and dedication of those around them to act in their best interests.

CHAPTER 49

Ascent Pediatrics: A Model for Successful Pediatric Drug Development

EMMETT CLEMENTE, PhD

Manchester Consulting, Inc., Manchester, Massachusetts 01944

INTRODUCTION

Before discussing the specifics of my personal experience in starting and constructing a successful pediatric company, I believe that it will be useful to first describe some important and fundamental concepts in building a business.

Much of what was important in building Ascent Pediatrics was gained in my working experience. Although I was technically trained and was employed in the pharmaceutical field as a bench scientist in two large pharmaceutical companies, I had the good fortune to have joined a start-up pharmaceutical division of a multinational organization in 1972 that provided a different set of principles that proved indispensable.

My years at the bench in the large pharmaceutical companies (Warner–Lambert and Richardson–Merrill) were focused on developing medications in both the prescription (Rx) and the over-the-counter (OTC) market. This was important as it gave me some experience in the nuances of the market dynamics in developing an Rx from an OTC drug, including not only the technical issues but the regulatory aspects as well. The commercial necessities really came into focus during my experience in the start-up pharmaceutical division of the multinational company, Fisons Corporation. I was remarkably fortunate to have been exposed to various commercial parts of the business, including product acquisitions, reviewing and analyzing market audit data, among other important commercial aspects of the business. During this period I had various technical responsibilities too, which included regulatory, medical, and pharmaceutical development. Those various duties provided the appreciation for the importance of the different disciplines in drug development. Functions that I was not exposed to or proficient in, such as financial, accounting, and legal, were matters that I realized needed to be attended to. During my years at Fisons, I had mentors who provided the guidance in developing the skills to negotiate licensing agreements and to progress the development of products that were important for our local

Pediatric Drug Development: Concepts and Applications
Edited by Andrew E. Mulberg, Steven A. Silber, and John N. van den Anker

market with a minimum of financial and personnel resources; an attribute that cannot be overstated.

SOME THINGS LEARNED ALONG THE WAY

I am not the most gifted of individuals, but I was and am knowledgeable enough to believe that it is very important to know what you do not know rather than what you believe you know. Some of the principles that I learned prior to founding Ascent Pediatrics was that a company could be created based on some protected technology. Additionally, a company had several options in entering the market, depending on the resources that were available. Most of these commercial considerations, although learned, were only fully appreciated during my tenure at Ascent. In fact, the experiences at Fisons were similar to what was to come at Ascent. Although Fisons was a multinational organization at the time I joined the company, it had no pharmaceutical presence in the United States. The flagship product was an asthma medication, Intal®, which was going through the regulatory review process. The plan was to build a sales force, acquire some marketed products from third parties, and enter the market, at hopefully, the most opportune time to limit the costs of a waiting sales force with the introduction of a new asthma treatment. There were many things to accomplish while the regulatory aspects played out. This was true at Fisons and at Ascent.

In general, it was necessary to establish the company with the physician group(s) that were to be called on and the trade groups (wholesalers, retailers, managed care), as well as to build the infrastructure within the organization to support the sales and marketing efforts. All of these functions were to be coordinated and managed while the research and development effort was ongoing; the short- and long-term viability of the organization depends not only on the commercial aspects of the business, but on how well the company identifies and executes the product development effort. I learned that if an organization is focused on a particular therapeutic area, for example, respiratory disease treatment, it will be committed to that field in its research and development (R&D) strategy as well as in its sales and marketing effort, at least in the short term. This need to stay focused and working with "old" drugs was an important aspect in building a successful start-up such as Ascent. Some R&D scientists find it difficult to understand that working with off-patented drugs and applying proprietary technology to improve patient compliance or some other commercial benefit is technically challenging and important therapeutically. However, it is crucial to the success of the organization to hire scientists who share this passion. This principle cannot be emphasized enough as paramount in attracting the right commercial people.

Reviewing the business plan on a regular basis is a prudent exercise and keeps one current on any developments in the targeted market.

FACTORS ASSOCIATED WITH THE ADULT AND PEDIATRIC FOCUS

An organization that develops medications for the adult is usually committed to certain therapeutic categories, as it is very costly in detailing large numbers of physicians in different specialties. It may not be apparent, however, but resources are always being competed for as there are more programs identified than there is available money to develop. Why do companies with a focus in the adult market not develop pediatric products? There are really several reasons: first, the pediatric market is economically small as compared to the adult market ($20 vs. $300 billion). Twenty billion is a sizable amount of money but must be analyzed from the perspective of a company developing adult medications. There are only a few therapeutic categories that are large enough to support the effort required to develop a pediatric medication. There are other important hurdles, including the necessity of developing multiple product presentations, conducting clinical studies, and the reluctance of pediatricians to prescribe a new chemical entity. Pediatricians, in my experience, would rather prescribe a time-honored drug that has been modified to solve a problem such as its bitter taste, or a product which can be administered on a less frequent basis. All of these issues lead to the fact that we are dealing with a category that requires medications that are for the most part off-patented, and have a projected modest return with a high risk of competition. Additionally, the dilution of resources will continue with the need to divert the organization's sales and marketing effort, to promote the pediatric product, which may require the sales representatives to call on pediatricians. This will obviously take them away from their primary physician universe, reconstruct the sales territories, or force them to hire a contract sales organization. The list is impressive and not appealing to companies that need to focus on their main business.

It has been my observation that those organizations that have had some presence in the pediatric market have either had a presence historically or progressed a medication for the adult market and in the latter part of the development program, when the resources became available, formulated a pediatric dosage form for a particular indication which is lucrative enough to make the effort. Notably the major therapeutic area that fits this model is the anti-infectives and, in particular, antibiotics for the treatment of respiratory diseases, including middle ear infection in children. Some of the large pharmaceutical corporations focused in the adult market may from time to time develop a dosage form of an adult drug if the economics are acceptable. That is to say, they cannot focus in the pediatric market, but will be opportunistic providing their business criteria are met. This scenario will not be viable if the organization does not have a way to protect the product (usually by patents) from competition, which severely rules out the off-patented drugs. Until now, we have raised the issues that face the large pharmaceutical companies, which focus on the adult market. The pediatric prescription drug market worldwide,[1] which is growing at a rate of 6.2% per year, is projected to exceed $46 billion by 2009.

ASCENT PEDIATRICS, INC.

I knew from my experiences in developing products for both the adult and pediatric markets, and having established and kept good working relationships with quality manufacturers, that the company had a reasonable chance in successfully developing pediatric products. Part of the business strategy was to outsource our manufacturing and development activities with FDA approved manufacturers that had development laboratories. Ascent would identify the product to be developed, including the product specifications, and manage the process. This strategy was key to conserving cash. We entered into agreements with manufacturers which provided the manufacturer rights to produce the product that met their needs. Ascent was responsible for all other parts of the New Drug Application (NDA), and also owned the NDA. Our approach to clinical trial requirements was to outsource this activity with clinical research organizations (CROs) and to manage the process similar to our manufacturing strategy. With these applications we could focus our long-term commitment of building the sales and marketing effort when required.

In 1989, with seed round funding from a then recently established venture capital group at Harvard University, Medical Science Partners, Ascent Pediatrics was established.

A friend and former mentor allowed me to use the lobby of his company and their telephones, so that I could contact various people to establish and confirm that Ascent Pediatrics was an entity. He also provided an office when one became available, until I could locate space for the company. I think this story is important for several reasons, including the fact that we need more than ourselves to succeed and to conserve as much cash as possible. We found some office space that was convenient for potential employees. The initial task was to write the business plan and to conduct market focus studies with pediatricians. During this phase, which was to last about a year, my wife, Lorraine, and I spent long hours, she establishing the office management side and setting up the company, assuring that we projected a professional image. I focused on drafting and redrafting the business plan, which was to become our selling document for additional capital. During this time period, several important activities were pursued including the hiring of a number of services such as an accounting and law firm. We put together the board of directors, an advisory board, and outside consultants in both the technology and business sectors of the company. Additionally, we conducted some exploratory development with potential products that I thought would be useful in pediatrics and required modest resources, such as a specially formulated nasal saline product, Pediamist®, which did not require FDA approval. The concept was to develop products that had development timelines that were short, medium, and long term. The longer the timeline, the higher the risk for success, but the larger the return. After the first year Ascent had raised a few million dollars from venture organizations both in the United States and Europe.

The first several years focused on developing products identified as needed by the pediatric community, including an antibiotic, trimethoprim; a bronchodilator, albuterol, in a controlled release oral liquid presentation; and an antipyretic/analgesic, acetaminophen, in an oral liquid controlled release dosage form, among others. We had at that time about six employees, most in the regulatory and technology area, but our first hire was on the commercial side as I knew that we needed to stay glued to the products identified in the focus studies. Some potential products were terminated, an oral liquid antihistamine, terfenadine, and an oral laxative, polycarbophil. Others were added, such as an oral liquid steroid, prednisolone sodium phosphate, Orapred®, which contained our patented taste masking system. All of our effort was in developing improved products from off-patented drugs. The challenge was not only to technically succeed but to protect the product from competition. We did this by developing our own patented technology, and by in-licensing patented technology. Additionally, we were engaged in performing clinical trials for market approval, even for the acetaminophen product, due to the fact that the dosage form was a controlled release presentation and required FDA approval. These factors, proprietary technology, clinical studies, and required FDA approval, did result in barriers to entry from competitors but do not provide the protection that a patented drug affords.

By 1996, we had 12 employees, received one ANDA, and were anticipating an approval for the indication of middle ear infection for the antibiotic trimethoprim, our brand, Primsol®. Based on this, the company decided to put together a public offering. Prior to the public offering, we had raised about $19 million and were adding additional people mainly in the commercial area.

GOING PUBLIC

The preparation of the documents for the public offering and the "Road Show" were real learning experiences. The process took about 9 months and effectively removed me from the day-to-day running of the business. By this time we had hired a new CEO, with a commercial background, and split the duties of the scientific and commercial aspects of the company. We had identified a pediatric product for acquisition, Feverall®, which was a suppository dosage form of acetaminophen. This would serve as one of our products that the sales force, which was being recruited, would promote to pediatricians along with Primsol. Ascent raised about $20 million and became a public company in May 1997. This process was time consuming and difficult. The entire effort took 1 year and cost about a million dollars. The Road Show took us to many countries in Europe and throughout the United States. I believe we made approximately forty presentations.

The acquisition was concluded, and our sales force along with the supporting functions were in place and ready for the introduction of our antibiotic. The product approval was significantly delayed and by the time the approval was received important changes in the market had occurred. These market dynamics had a very negative effect on our ability to carry our expenses, especially the sales force. The board was reluctant to dissolve the sales force and chose to borrow additional capital.

After a year in the market with Primsol, it became obvious that this product would be marginal to the company. Our second prescription product, Orapred, was approved in December 2000 and was launched in January 2001. Ascent was transformed into a leaner organization with some strategic managerial changes. Additionally, we had the personnel to put into place and execute the marketing plan for the new product, Orapred.

I cannot overstate the effort and execution that our marketing and sales force performed with a limited budget and a truncated timeline to get distribution into the wholesaler and retail trade channels along with the acceptance of the managed care organizations. We had significant challenges with product supply shortly before we were to introduce the product as the outside product stability laboratory became financially insolvent, putting our launch at risk. However, with a Herculean effort by our regulatory director and myself, we were able to make the necessary technical transfers to another FDA approved site and meet the FDA requirements, thus preserving the launch timetable.

Orapred was an immediate success, with sales rising each week and the product capturing significant market share: by all measures, a real advance in treating children with asthma and a commercial success. The product was tracking at $16–19 million in sales in its first year and projected to climb to about $40–45 million by year three, which was achieved.

CONCLUSION

Although we were a public company, the board was constituted with a majority of our financial investors, who understandably wanted to close up their investment. In an effort to keep Ascent a dedicated pediatric company, I spoke with a number of pharmaceutical companies that expressed an interest in entering the pediatric market. The focus was to keep the sales force intact and utilize the relationship that was constituted with pediatricians, and to continue to develop the products in the pipeline, two of which had successful clinical trial results. This approach, although difficult from a personal standpoint, was in my view best for the shareholders, and included the employees. The company entered into an agreement with Medicis Pharmaceutical Corporation and merged Ascent into that organization as a wholly owned division in 2001.

Ascent Pediatrics consisted of some intelligent and dedicated people with the purpose of developing improved off-patented medications for the pediatric patient and serving the pediatrician. We made our mistakes and accomplished much as well. Important lessons are to have the product approval in hand before building a sales force and the attendant support functions and to be prudent in taking on debt.

A dedicated pediatric pharmaceutical company is feasible, as the Ascent model has shown. This is a market that has many unmet needs and if approached intelligently can provide improved medications from off-patented drugs and a reasonable return for its shareholders.

REFERENCE

1. *The Worldwide Market for Prescription Drugs*, 2nd ed. New York, NY: Kalorama Information; 2006.

CHAPTER 50

Case Study for the Development of an Enteric-Coated and High-Buffered Pancreatic Enzyme Product

TIBOR SIPOS, PhD

Digestive Care, Inc., Bethlehem, Pennsylvania 18017

INTRODUCTION

In 2003, the Pediatric Research Equity Act (PREA) was signed into law in the United States, as a further measure supporting the Pediatric Rule (ca.1994).[1] In passing the PREA, Congress countermanded the suspension of provisions of the Pediatric Rule by court order. The Pediatric Rule provided for the inclusion of more complete information in prescription drug labeling about the use of a drug in the pediatric population. Under PREA, human drug applications were required to include pediatric assessments for indications for which sponsors were receiving or seeking approval in adults, unless the requirement was waived or deferred.

The pediatric assessment shall contain data that are adequate to assess the safety and effectiveness of the drug or the biological product for the claimed indication(s) in all relevant pediatric subpopulations; and to support dosing and administration for each pediatric subpopulation for which the drug or the biological product is safe and effective. The pediatric subpopulations are defined by the following age groups in completed days, months, or years:

- Term newborn infants (0 to 27 days)
- Infants and toddlers (28 days to 23 months)
- Children (2 to 11 years)
- Adolescents (12 to 16–18 years, dependent on region)

The pediatric assessment data shall be gathered using appropriate formulations for each age group for which the assessment is required. Many adult dosage forms such as tablets and capsules are inappropriate for certain age groups. In the absence of a suitable commercial

Pediatric Drug Development: Concepts and Applications
Edited by Andrew E. Mulberg, Steven A. Silber, and John N. van den Anker

age-appropriate formulation, extemporaneous formulations (e.g., dissolving tablets in a beverage, or mixing the capsule contents in food) may be an acceptable drug delivery system.

In accordance with the above stated pediatric assessment requirements, this chapter provides background information on the disease manifestation of exocrine pancreatic insufficiency (EPI), the unmet medical need in the currently available pancreatic enzyme product (PEP) treatment options, the criteria to be met by an age-appropriate pediatric formulation, the development efforts undertaken by Digestive Care, Inc. resulting in its enteric coated (EC) and high-buffered PEP, and specific examples for the age-appropriate delivery and clinical evidence for the safety and efficacy of PEPs in the EPI pediatric cystic fibrosis (CF) population.

BACKGROUND

Under normal conditions, the pancreas secretes a sufficient amount of bicarbonate-rich digestive fluids into the intestine to aid in the breakdown of food into readily absorbable nutrients. When the pancreas is not functioning properly, or is partially removed surgically, lesser amounts of pancreatic digestive enzymes (i.e., lipase for fat digestion, protease for protein digestion, and amylase for starch digestion) are released into the intestine, resulting in the manifestation of EPI. EPI does not occur until the pancreatic enzyme output level is reduced by more than 90% because of the large reserve capacity of the pancreas.[2] This level of reduction in pancreatic enzyme output, however, causes insufficient digestion and absorption of fats and lipids, proteins, and carbohydrates and poor absorption of fat-soluble vitamins A, D, E, and K, iron, folic acid, and other micronutrients. The incomplete digestion of fats induces osmotic diarrhea (steatorrhea) that causes malabsorption of nutrients, which consequently results in growth retardation in children and adolescents. Adequate pancreatic enzyme replacement therapy is obviously most crucial during the formative years. Afflicted children should receive an intake of 120–150% of the recommended daily allowance of kilocalories and fat-soluble vitamin supplementation to compensate for these inherent metabolic discrepancies.[3] Treatment with pancreatic enzymes is necessary to relieve the symptoms of EPI and to reduce malnutrition.[3–5]

Cystic fibrosis (CF) is the second most common life-shortening, childhood onset inherited disorder in the United States. Approximately 30,000 people in the United States have CF.[6] CF is characterized by the production of abnormally thick and sticky mucus, which most frequently obstructs the lungs and exocrine glands, leading to progressive chronic and life-threatening lung infections and destruction of the secretory capacity of the pancreatic gland resulting in EPI. Approximately 90% of all individuals with CF have EPI and are treated with pancreatic enzyme replacement therapy (PERT).[7] The majority of individuals with CF are diagnosed before the age of 1 year and the diagnosis in this age group is often made due to signs and symptoms associated with EPI.

The effective use of PERT in infants can be extrapolated from older children since the course of disease and management of EPI are similar in all age groups. Guidelines for dosing and administration of pancreatic enzymes to infants with CF have been established by consensus groups with expertise in the treatment of CF, and are available to clinicians who treat CF.[8]

Prior to the introduction of PERT for CF, the major cause of death for infants with CF was malnutrition. PERT and advances in the management of respiratory complications have significantly improved the life expectancy of pediatric CF patients over the past three decades and death from malnutrition is uncommon. Complications from lung disease, however, still remain to be the primary cause of death in CF.[9]

TABLE 50.1 Total Enzyme Activity Released from the Pancreas in Response to Hormone Stimulation in Health and Disease[a]

Group	Lipase	Trypsin	Chymotrypsin	Amylase	Protease	Proteins	Bicarbonate (mEq/L)
Healthy adult	496,638	40,284	56,946	48,844	3,846	2,402	34.23
Chronic pancreatitis	62,644	12,808	8,544	11,664	1,723	1,283	7.21
CF Aadult	121	80	26	145	65	370	1.74

[a] During a 60-minute stimulation period with secretin. Enzyme activities are expressed in international units—IU (total IU/hour).[14]

CF patients are deficient in pancreatic fluid volume, enzymes, and bicarbonate secretion.[10] As a result of the insufficiency in bicarbonate secretion, most CF patients lack the ability to neutralize the acidic chyme and the pH at the duodenaljejunal junction will be in the acidic range (pH 3.4–6.6).[11,12] Under this low pH range, pancreatic lipases are minimally active to exert their enzymatic (digestive) activity.[2] The low intraduodenal pH is due to a defect by the bicarbonate/chloride exchanger protein to transport bicarbonate into the duodenal lumen.[13] (See Table 50.1)

Treatment for EPI in CF requires the use of exogenous pancreatic enzymes. The intraluminal activity of pancreatic enzymes (lipase, amylase, and protease) is dependent on the intestinal pH. Table 50.2 illustrates the pH dependency of lipase activity. The enzymatic activity of pancreatic enzyme-containing products, however, is variable, especially lipases, the most essential enzymes in EPI. There are several reasons for this variable activity: enzymes are partially inactivated by gastric conditions, even though some are coated with an acid-resistant polymeric film, that is, enteric coated (EC); they are physically too large (>2.4 mm in diameter) to uniformly mix with food and pass through the pylorus into the duodenum; they lack a favorable basic microenvironment (pH > 7.0) for optimized lipase activity; they have delayed dissolution in the lower intestine (ileum) due to too much enteric coating; or they have an overall acidic composition due to the acidity of the EC polymer and failure to dissolve at low intestinal pH values (Table 50.2). In addition, the intraluminal activity of pancreatic enzymes (lipase, amylase, and protease) is dependent on the intestinal pH, lipase having maximal activity at pH 9.0, amylase having maximal activity at pH 6.9, and protease having maximal activity at pH 7–9.[15]

TABLE 50.2 In Vitro Lipase Activity of Pancreatic Lipase at Different pH Values[a]

pH	USP Unit/mg	Activity (%)
9.0	20.6	100
8.5	18.1	88
8.0	14.4	70
7.5	10.5	51
7.0	4.9	24
6.5	0.7	3
6.0	0.1	<1

[a]Maximal lipase activity was obtained at pH 9.0. The same amount of lipase showed only 51% activity at pH 7.5, 24% activity at pH 7.0, 3% activity at pH 6.5, and <1.0% activity at pH 6.0.[14]

TABLE 50.3 In Vitro pH (Acidity) of EC Polymers Employed in EC Enzyme Products[a]

pH	EC Polymers
3.5	Cellulose acetate phthalate polymer (Eastman Kodak Company)
3.5	Hydroxypropyl methyl cellulose phthalate (Eastman Kodak Company)
3.4	Acrylic acid copolymer (Eudragit®) (Rohm Pharm, GmBH, Germany)

[a] Technical brochures are available from Eastman Kodak Company and Rohm Pharm, GmBH.

Table 50.3 illustrates the acidity (pH) of the conventionally employed, EC polymers that are used to coat the active drug composition in order to protect the pancreatic enzymes against inactivation by gastric acid and pepsin during gastric transit into the duodenum.

Due to the low pH (acidity) of the EC polymers that were employed to protect the pancrelipase composition during gastric passage into the upper intestine, the total pancrelipase composition is rendered acidic unless a buffer is included in the composition to neutralize the acidic enteric coating. Table 50.4 illustrates the differences in overall pH of the

TABLE 50.4 In Vitro pH of EC Pancreatic Enzyme-Containing Products

Product	Active Ingredients	Inactive Ingredients	pH
Cotazyme® - S (Organon)	Lipase—8,000 USP units/capsule Amylase—30,000 USP units/capsule Protease—30,000 USP units/capsule	Cellulose acetate phthalate, dimethyl phthalate, calcium carbonate, gelatin, sodium glycolate, cornstarch, talc, FD&C Green #3 and Yellow #10, titanium dioxide, magnesium stearate	5.7
Pancrecarb® (Digestive Care, Inc.)	Lipase—8,000 USP units/capsule Amylase—35,000 USP units/capsule Protease—35,000 USP units/capsule	Cellulose acetate phthalate, dimethyl phthalate, sodium carbonate, sodium bicarbonate, povidone, gelatin, sodium starch glycolate, ursodiol, and talc	9.0
Creon® 10 (Solvay)	Lipase—10,000 USP units/capsule Amylase—33,200 USP units/capsule Protease—37,500 USP units/capsule	Dibutyl phthalate, dimethicone, hydroxypropyl methylcellulose phthalate, light mineral oil, polyethylene glycol, gelatin, titanium dioxide, black iron oxide, red iron oxide, yellow iron oxide, FD&C Yellow #10, and FD&C Red #40	5.5
Pancrease® (McNeal)	Lipase—4,000 USP units/capsule Amylase—20,000 USP units/capsule Protease—25,000 USP units/capsule	Cellulose acetate phthalate, diethylphthalate, gelatin, povidone, sodium starch glycolate, cornstarch, sugar, talc, titanium dioxide	5.9
Ultrase® MT12 (Scandipharm)	Lipase—12,000 USP units/capsule Amylase—39,000 USP units/capsule Protease—39,000 USP units/capsule	Hydrogenated castor oil, silicon dioxide, sodium Carboxymethylcellulose, magnesium stearate, cellulose microcrystalline, methacrylic acid copolymer (type C), simethicone, triethyl citrate, iron oxides, titanium dioxide	5.7

composition with buffers (Pancrecarb® pH 9.0) as compared to nonbuffered compositions (pH 5.5–5.9).

THE UNMET MEDICAL NEED IN THE TREATMENT OF EPI

There is an unmet need for new treatments for EPI. No new PEPs have been approved by the U.S. Food and Drug Administration (FDA) for EPI since Cotazym in 1996 (NDA 20-580); Cotazym, a non-EC pancrelipase powder filled into gelatin capsules is no longer marketed in the United States.

The enzymes in older PEP formulations, such as Cotazym, were largely inactivated by gastric acidity, with less than 10% of the lipolytic and 20% of the tryptic activity reaching the ligament of Treitz in the duodenum.[2,16] The introduction of pH-controlled EC enzyme preparations has improved the effectiveness[17] of PEPs. Use of a pH-controlled polymer coating that resists dissolution of the preparation in the stomach but releases the enzyme in the more alkaline duodenum (normal subjects) substantially increases fat absorption with utilization of fewer capsules than required with uncoated pancreatic enzymes.[18]

At present, treatment for EPI consists of administration of exogenous pancreatic enzyme extracts in an attempt to normalize digestion. Unfortunately, this treatment is usually only partially successful and some patients continue to suffer from maldigestion.[19] Fat digestion rarely reverts to normal, and up to one-quarter of patients lose in excess of 25% of their fat intake despite seemingly adequate doses of enzymes.[5] Incorrect timing of enzyme delivery with food passing from the stomach into the duodenum, together with the problem of acid–pepsin inactivation of enzymes due to proteolytic hydrolysis of the enzymes within the environment of the stomach and duodenum, are clearly contributing factors to this problem. Attempts to overcome these barriers by neutralizing the gastric environment with antacids, by coadministration of antacids and histamine antagonists and proton pump inhibitors, or alternatively coating the enzyme with acid-resistant films have not been successful uniformly. Many patients with CF may have gastric acid hypersecretion, which, together with a deficiency of bicarbonate secretion from the pancreas, may result in a highly acidic intestinal environment. As a result, the intestinal milieu may be below the optimal conditions for enzyme activity.[5]

Effective treatment of fat malabsorption resulting from EPI requires that active pancreatic enzymes are present in a high enough concentration in the small intestine to ensure adequate digestion of complex foods and absorption of the liberated nutrients. Protective coating of the enzymes to prevent inactivation by gastric acid and pepsin is a critical advance in this treatment.

PHARMACEUTICAL DEVELOPMENT OF PEP FORMULAS

Recognizing the importance for the need of high-lipase potency, the acid–pepsin inactivation of the unprotected and exogenously administered pancrelipase, the limitation of particle size passage from the stomach into the duodenum (>2.4 mm particles are retained in the stomach while <1.5 mm particles are passed freely into the duodenum), and the need for uniform distribution of the enzyme with the acid chyme in the stomach led to the invention and development of the EC pancrelipase microspheres[20] that were marketed as Pancrease in 1978. The EC pancrelipase microspheres were a major advancement in the nutritional management of CF and have contributed to the increased life expectancy of CF patients by at least a decade.[21,22]

The next major advancement in the development of improved digestive enzyme formulation was the recognition that lipase activity is pH dependent with maximal activity at pH 9, and the fact that CF patients are deficient in both lipase and bicarbonate secretions, which results in low duodenal and intestinal pH values. By exploiting this scientific knowledge, one theoretically could achieve a more efficient fat digestion at lower lipase doses per capsule if the pH of the composition is maintained in the pH range of 8–9.0.

The appreciation of this scientific information led to the invention and commercial development of the high-buffered (pH 9) and EC pancrelipase microspheres product[14] in the mid-1990s, which was marketed as Pancrecarb in 1996. The microspheres contain both pancrelipase and bicarbonate buffer as a homogeneous composition and are enteric coated to protect both the enzyme and the bicarbonate buffer components from acid and pepsin hydrolysis during gastric passage. Lipases are inactive at low duodenal and intestinal pH < 5.5.[23,24] However, the bicarbonate in Pancrecarb establishes a suitable microenvironment surrounding the microspheres in the pH range of 8.5–9.0 that theoretically provides optimal lipase activity for digestion of fats and lipids compared to nonbuffered formulations.

The following experiments were carried out to demonstrate the technical feasibility of this concept:

- Investigate the compatibility and stability of pancrelipase with buffers and other formulation excipients.
- Develop prototype buffered pancrelipase compositions that are processed into microspheres and enteric coated.
- Characterize the EC and buffered pancrelipase microspheres for integrity, enzyme potency, and stability under controlled in vitro environmental conditions.
- Initiate accelerated stability studies under elevated temperatures and humidity conditions to establish the long-term stability of the EC composition.
- Establish product specifications and test methods for the buffered and EC pancrelipase microspheres.
- Prepare a large-scale batch of the EC and bicarbonate-buffered pancrelipase microspheres and drug product with 4000 (MS-4), 8000 (MS-8), and 16,000 (MS-16) USP units of lipase per capsule under controlled conditions. Characterize the clinical batch for total lipase, amylase, and protease activities and dissolution of lipase from the gastric fluid exposed EC microspheres in simulated intestinal fluid at pH 6.0. Determine loss on drying (LOD), pH of the composition after dissolution, and microbial content.
- Perform preclinical and clinical studies for safety and efficacy.

Results of the in vitro studies demonstrated compatibility and stability of pancrelipase with excipients as illustrated in Table 50.5. The buffered pancrelipase composition, in powder formulation or in EC microsphere form, was stable, had a pH of 9.0, and retaisned >90% of lipase activity, while the nonbuffered composition had a low pH of 5.2 and had only 15% lipase activity.

The in vitro enzymatic activities of the EC microspheres of Formula 2 (Table 50.5) on dissolution in simulated intestinal fluid demonstrated >90% recovery of lipase, amylase, and protease activities within 20 minutes.

TABLE 50.5 Summary of Enzymatic Activities of Buffered Pancrelipase Composition During Manufacture of the EC Microspheres

Formulation	Lipase (%)	Amylase (%)	Protease (%)	pH
Formula 1 (enzyme powder blend)[a]	100	100	100	6.5
Formula 2 (enzyme powder blend with buffer)	109	103	111	9.0
Formula 9 (enzyme powder blend[b] with excipients[c] without buffer)	15	27	42	5.2
EC microspheres of Formula 2	90	114	95	9.0

[a]The activity of Formula 1 was taken as 100%. All the other formulas' enzymatic activities were expressed relative to Formula 1.

[b]Formula 9 contained no buffer and the dissolved composition in water had a pH of 5.2. At this low pH, lipase, amylase, and protease were only partially active.[14]

[c]Excipients contained sodium starch glycolate, NF; ursodiol, USP; polyvinyl pyrrolidone, USP; cellulose acetate phthalate, NF; talc, USP; and buffers—sodium carbonate anhydrous, NF; and sodium bicarbonate, NF.[14]

The buffered and EC pancrelipase formulas at 25 °C/60% relative humidity (RH) were stable for 24–30 months. The pancrelipase compositions of Formulas 1 and 9 without buffers when exposed to 30 °C and 40 °C accelerated stability conditions for 3 months in high-density polyethylene containers showed a rapid loss of lipase activity at 40 °C while the buffered formula retained >80% lipase activity at 40 °C over a 3-month accelerated stability period.

Collectively, the results presented support preservation of the lipase, amylase, and protease activity and performance of the encapsulated EC and buffered pancrelipase microspheres at room temperature (25 °C) when exposed to accelerated stability conditions at 30 °C and 40 °C.

The outcome of this development work resulted in the marketing of Digestive Care, Inc.'s PEP. Pancrecarb (pancrelipase) capsules contain EC microspheres comprised of buffered pancreatic enzymes (lipase, amylase, and protease), isolated and concentrated from porcine pancreatic glands. Pancrecarb (pancrelipase) capsules are provided in three strengths designated as MS-4 (4000 USP units of lipase), MS-8 (8000 USP units of lipase), and MS-16 (16,000 USP units of lipase) per capsule.

The enzyme-containing microspheres are coated with an acid-resistant but base-soluble cellulose-based polymeric film to provide protection against inactivation of the buffered enzymes during gastric passage. The inclusion of sodium carbonate/sodium bicarbonate buffer excipients in the formulation affords an optimized microenvironment (pH 9) for the lipases to exert maximized enzymatic activity.

PEP AGE-APPROPRIATE DELIVERY SYSTEM

In order for an EC pancrelipase product to satisfy age-appropriate pediatric dosage requirements, it must meet certain criteria to be acceptable by children and be convenient for administration by caregivers. Following are key criteria to be considered when developing an age-appropriate pediatric formulation:

- Physical Considerations
    - Not to be retained in the mouth
    - Not to cause mouth and oral mucosal irritation and ulceration

  - o Need to mix easily with baby food
  - o Convenience to administer the medication to infants and toddlers by caregiver
  - o Child needs to be able to swallow medication
- Palatability
  - o Taste and odor of pancrelipase needs to be masked
  - o Medication must be compatible with food to be coadministered at feeding
- Texture of Food
  - o Food must be coarse enough to be compatible with the microspheres
- pH of Food Used as Vehicle for Drug Administration
  - o Must be on the acidic side to prevent premature activation of the microspheres and inactivation of lipase
- Particle Size
  - o Must be small enough to mix uniformly with food

Since some infants cannot swallow capsules, the capsules are opened and the microspheres are sprinkled and mixed into a soft food, most commonly applesauce. The mixture of enzymes/applesauce is then administered to the infant via a small spoon or with other alternative feeding devices (i.e., bottle nipple) several times throughout breast or bottle-feeding. Mixing Pancrecarb EC microspheres in applesauce and administering it to infants as described is a practical and age-appropriate method of administration.

Applesauce satisfies the criteria of palatability, texture, pH, and particle size of food that is compatible with the EC microspheres and can be coadministered with food during feeding. Infants and toddlers (1–23 months) are able to swallow the applesauce-dispersed medication due to the coarse texture of applesauce and its good taste. The offensive odor and unpleasant taste of pancrelipase are masked by the polymeric enteric coating. An additional advantage of Pancrecarb is that the small particle size of the MS-4 microspheres (0.4–0.7 mm) are compatible with the coarse texture of applesauce. Therefore, applesauce is considered to be an age-appropriate delivery vehicle for Pancrecarb.

CLINICAL EVIDENCE FOR THE SAFETY AND EFFICACY OF THE AGE-APPROPRIATE DOSAGE IN PEDIATRIC POPULATION

Individual Case Studies

Table 50.6 summarizes 11 individual case studies supporting pediatric dosage efficacy of Pancrecarb that were provided by physicians and caregivers of CF pediatric patients from CF centers throughout the United States. The individual case studies document enzyme treatment, reduction of enzyme dose administered, and clinical response observed prior and after switchover to Pancrecarb in children from 1 to 16 years of age.

Pediatric and Adult Clinical Study Results: Comparative Clinical Study at ~50% Reduced Lipase Dose

A prospective, open label, randomized active-controlled two-way crossover metabolic study with each subject as his/her own control was conducted with 20 CF patients recruited

TABLE 50.6 Individual Patient Case Studies Supporting Pediatric Dosage Efficacy of Pancrecarb

Subject Number	Gender	Age	Age at CF Diagnosis	CF Mutation	Pancreatic Enzyme Replacement Therapy (PERT) History	Pancrecarb Dose	Treatment Effects of Pancrecarb
1	F	12 mo	7 mo	Heterozygous (F508 and unidentified)	Ultrase regular: 1400 U lipase/kg/meal and snack	Pancrecarb MS-4: 700 U lipase/kg/meal and snack (50% reduction compared to previous PERT)	Improved stool consistency. Physician reported that the small microspheres in Pancrecarb MS-4 lead to better compliance and ease of swallowing for this patient.
2	M	13 mo	2 weeks	F508/1717-1G>A	Ultrase regular: 621 U lipase/kg/meal to 2625 U lipase/kg/meal	Pancrecarb MS-4: 1500–2265 U lipase/kg/meal (44% to 15% reduction compared to previous PERT)	Patient had an increased appetite, weight gain, and a decrease in the volume and number of stools. Stools were also normal in appearance. Smaller microspheres were easier for patient to take and for parent to administer.
3	M	14 mo	9 mo	F508/F508 genotype	Creon 5: 1 capsule before meals	Pancrecarb MS-4: 1 capsule before meals	Improved weight gain and stool consistency. Patient experienced fewer numbers of stools per day with no grease or mucus noted in the stool. He continued to thrive and tolerates his current regime well.
4	M	9 yr	2 mo	Negative/G542X	Creon 10: maximum dose of 2631 U lipase/kg/meal + nocturnal gastrostomy tube feedings	Pancrecarb MS-8: 2105 U lipase/kg/meal (20% reduction compared to previous PERT) Pancrecarb MS-4 administered by nocturnal gastrostomy tube feedings	Bowel movements were more formed and patient complained less of abdominal discomfort. Weight has increased. Patient tolerated nighttime gastronomy tube feedings with Pancrecarb MS-4 better as he would not have to be awakened to administer the enzymes. Physician noted that this route of administration is only possible with Pancrecarb MS-4. Patient has been able to maintain his weight and has had very few complaints of abdominal discomfort.

(Continued)

TABLE 50.6 *(Continued)*

Subject Number	Gender	Age	Age at CF Diagnosis	CF Mutation		Pancrecarb Dose	Treatment Effects of Pancrecarb
5	F	6 yr 5 mo	9.5 mo	F508, 2183 del AA-G	Ultrase MT20: 2521 U lipase/kg/meal and 1680 U lipase/kg/snack	Pancrecarb MS-8: 2689 U lipase/kg/meal and 2017 U lipase/kg/snack (lipase dose has since been titrated down to 85% starting)	Weight and height have increased. Patient has not experienced any more episodes of rectal prolapse, which were frequent before switching. At her last clinical visit, she had formed stools, no abdominal complaints, no oil in stool, and stooling 1–2 times per day.
6	M	9 yr	Not provided by physician	Not provided by physician	Pancrease MT4 → Pancrease MT10 → Creon 10 → Ultrase MT12 (2483–3310 U lipase/kg/meal)	Pancrecarb MS-8: 2 capsules with meals and nocturnal feeding (842 U lipase/kg/meal) and 1 capsule with snacks	Patient's weight has increased. He has continued without recurrence of abdominal distention or abdominal pain.
7	F	12 yr	Not provided by physician	Not provided by physician	Creon 20: maximum dose of ~5000 U lipase/kg/meal	Pancrecarb MS-8: ~2000 U lipase/kg/meal (~60% reduction compared to previous PERT)	Patient's height and weight increased. Physician stated that switching to Pancrecarb MS-8 allowed for a significant reduction in lipase dose, which placed patient at significantly less risk of developing complications from high dose pancreatic enzyme supplements.
8	F	13 yr	19 mo	Not provided by patient	Ultrase MT20 (6207 U lipase/kg/meal) → Creon 20 (5647 U lipase/kg/meal)	Combination of Pancrecarb MS-8 + Creon 20 (3145 U lipase/kg/meal) (45% reduction compared to previous PERT)	Patient stated that the combination therapy has made a difference in gaining and maintaining weight. She had gained 8 lb over a 3-month period and is currently taking 2445 U lipase/kg/meal.

9	F	14 yr	During first year	Not provided by physician	Pancrease MT7 → Ultrase MT12 (1384 U lipase/kg/meal to 1463 U lipase/kg/meal)	Pancrecarb MS-8: 585 U/lipase/kg/meal to 600 U/lipase/kg/meal (~57-59% reduction compared to previous PERT)	Symptoms of abdominal distention, poorly formed and greasy bowel movements, and abdominal pain had completely resolved since switching to Pancrecarb. Weight has increased.
10	F	15 yr	During infancy	Not provided by physician	Creon 25 (1800 U lipase/kg/meal) → Ultrase MT20 (2200 U lipase/kg/meal)	Pancrecarb MS-8: 800 U lipase/kg/meal (~64% reduction compared to previous PERT)	Patient experienced resolution of intermittent abdominal pain as well as less flatulence. Weight had increased.
11	M	16 yr	3 mo	Homozygous for F508	Ultrase MT20: 3205 U lipase/kg/meal	Pancrecarb MS-8: 5 capsules/meal and 2 capsules/snack to 7 capsules/meal and 5 capsules/snack (~40–55% reduction compared to previous PERT)	In 3 months, weight increased from 36.5 kg to 40.2 kg. Bowel movements decreased from 5–8 to 2–4 daily. Abdominal symptoms (large, loose stools, persistent flatus, and abdominal discomfort) were nearly resolved.

and 18 patients (12–27 years of age) completing the study.[25] This study determined the safety and efficacy of an EC high-buffered pancreatic enzyme replacement therapy (EC, high-buffered PERT) as compared to the patient's usual brand of EC, non buffered PERT, Pancrease (McNeil Pharmaceutical), Creon (Solvay Pharmaceutical), and Ultrase (Axcan Pharma) for the treatment of fat malabsorption in EPI. The study was conducted in the clinical research center under controlled conditions of dietary intake of fat (30–40%), protein (15–20%), and carbohydrates (45–55%). Both EC, high-buffered PERT and the usual pancrelipase products (Pancrease, Creon, and Ultrase) were administered at ~50% reduced lipase dose, but not lower than ~1800 units of lipase per gram of fat intake per day. After the screening visit, eligible patients were initially assigned randomly to one of two treatment groups. After 1 week of treatment, they were crossed over to the alternative treatment. The primary efficacy variable was the coefficient (percent) of fat absorption in the 72-hour stool collections using stool markers. The secondary efficacy variable was the percent of nitrogen malabsorption in the 72-hour stool collection using stool markers. Other efficacy variables included fat intake, protein intake, carbohydrate intake, energy, total fat excretion, stool weight, and calculated kilocalories excreted in stool from fat. The results of this study showed significantly increased fat absorption ($p = 0.01$) for EC, high-buffered PERT, 81.8% versus 75.1% for the usual EC, nonbuffered PERT. Thirteen of 18 subjects (72%) excreted less fat when receiving EC, high-buffered PERT whereas five subjects did not respond. The results of this study showed that EC, high-buffered PERT at ~50% reduced lipase dose is safe and effective in reducing steatorrhea and may represent a clinical advantage with respect to improvement in fat absorption and utilization of additional kilocalories from the increased absorption of fat compared to other enzyme treatments.

CONCLUSION

This chapter recounts the development and clinical evidence for the successful age-appropriate delivery of an EC, high-buffered PERT (Pancrecarb) for the treatment of EPI in pediatric CF patients.

Based on the totality of the in vitro data and the in vivo clinical study results, it is concluded that the above-described EC, high-buffered PERT (Pancrecarb) and the applesauce delivery vehicle represents a safe and effective age-appropriate pediatric dosage delivery system for the treatment of pancreatic enzyme and bicarbonate insufficiencies associated with pediatric CF patients. Such a delivery system leads to better patient compliance with recommended PEP dosages and thus improves the effectiveness of the treatment. Especially in the pediatric population, adherence to a PEP regimen is important to supply the nutrients necessary for a patient to grow and thrive.

ACKNOWLEDGMENTS

Special thanks and appreciation go to Eve Damiano, independent consultant; Jack Alhadeff of Lehigh University; and Timothy A. Halton and Robin LaPadula of Digestive Care, Inc., for their assistance in the preparation and editing of this chapter.

REFERENCES

1. US Food and Drug Administration, Pediatric Research Equity Act (PREA), December 3, 2003.
2. DiMagno EP, Malagelada JR, Go VLW, et al. Fate of orally ingested enzymes in pancreatic insufficiency. *N Engl J Med.* 1977;296:1318–1322.
3. Littlewood JM, Wolfe SP, Conway SP. Diagnosis and treatment of intestinal malabsorption in cystic fibrosis. *Pediatr Pulmonol.* 2006;41:35–49.
4. Keller J, Layer P. Human pancreatic exocrine response to nutrients in health and disease. *Gut.* 2005;54:1–28.
5. Durie P, Kalnins D, Ellis L. Uses and abuses of enzyme therapy in cystic fibrosis. *J R Soc Med.* 1998;91:2–13.
6. Cystic Fibrosis Foundation, Patient Registry 2005 Annual Report, Bethesda, MD.
7. Durie P, Forstner G. Pathophysiology of the exocrine pancreas in cystic fibrosis. *J R Soc Med.* 1989;82:2–10.
8. Borowitz DS, Grand RJ, Durie PR for the Consensus Committee (Cystic Fibrosis Foundation). Use of pancreatic enzyme supplements for patients with cystic fibrosis in the context of fibrosing colonopathy. *J Pediatr.* 1995;127:681–684.
9. Borowitz DS, Durie PR, Clarke LL, et al. Gastrointestinal outcomes and confounders in cystic fibrosis. *J Pediatr Gastroenterol Nutr.* 2005;41:273–285.
10. Lagerlof HO, Schutz HB, Holmer S. A secretin test with high doses of secretin and correction for incomplete recovery of duodenal juice. *Gastroenterology.* January 1967;52(1):67–77.
11. Youngberg CA, Berardi RR, Howatt WF, et al. Comparison of gastrointestinal pH in cystic fibrosis and healthy subjects. *Dig Dis Sci.* 1987;32:472–480.
12. Dutta SK, Sudhir K, Hubbard VS, et al. Critical examination of therapeutic efficacy of a pH-sensitive enteric-coated pancreatic enzyme preparation in treatment of exocrine pancreatic insufficiency secondary to cystic fibrosis. *Dig Dis Sci.* 1988;33:1237–1244.
13. Kopelman H, Corey M, Gaskin K, et al. Impaired chloride secretion, as well as bicarbonate secretion, underlines the fluid secretory defect in the CF pancrease. *Gastroenterology.* 1988;95:349–355.
14. Sipos T. US Patent #5, 750,104. High buffer containing enteric coated digestive enzyme bile acid compositions and method of treating digestive disorders herewith. May 12, 1998.
15. Bergmeyer HU. *Methods of Enzymatic Analysis*, Volume 2. New York: Academic Press; 1974: 814, 885, 1000.
16. Graham DY. Enzyme replacement therapy of exocrine pancreatic insufficiency in man: relation between *in vivo* enzyme activities and *in vivo* potency in commercial pancreatic extracts. *N Engl J Med.* 1977;296:1314–1317.
17. Khaw K, Adeniyi-Jones S, Gordon D, et al. Efficacy of pancreatin preparations on fat and nitrogen absorption in cystic fibrosis patients. Abstract presented at meeting sponsored by the Harvey R. Colton, MD Foundation, Boston, MA, 1981.
18. Nassif EG, Younoszai MK, Weinberger MM, et al. Comparative effects of antacids, enteric coating, and bile salts on the efficacy of oral pancreatic enzyme therapy in cystic fibrosis. *J Pediatr.* 1981;98:320–323.
19. Lapey A, Kattwinkel J, di Sant' Agnese PA, et al. Steatorrhea and azotorrhea and their relation to growth and nutrition in adolescents and young adults with cystic fibrosis. *J Pediatr.* 1974;84:328–334.
20. Sipos T. US Patent #4, 079,125. Preparation of enteric coated digestive enzyme composition. March 14, 1978.

21. Shepherd R, Cooksley WG, Cooke WD. Improved growth and clinical, nutritional, and respiratory changes in response to nutritional therapy in cystic fibrosis. *J Pediatr.* 1980;97:351–357.

22. Dodge JA, Turck D. Cystic fibrosis: nutritional consequences and management. *Best Practice Res Clin Gastroenterol.* 2006;20:531–546.

23. Go VL, Poley JR, Hofmann AF, Summerskill WH. Disturbances in fat digestion induced by acidic jejunal pH due to gastric hypersecretion in man. *Gastroenterology.* 1970;58:638–646.

24. Heizer WD, Cleaveland CR, Iber FL. Gastric inactivation of pancreatic supplements. *Bull Johns Hopkins Hospital.* 1965;116:261–270.

25. Brady MS, Garson JL, Krug SK, et al. An EC high-buffered pancrelipase reduces steatorrhea in patients with cystic fibrosis: a prospective, randomized study. *J Am Diet Assoc.* 2006;106:1181–1186.

APPENDIX

Reference Tables and Figures

Pediatric Drug Development: Concepts and Applications
Edited by Andrew E. Mulberg, Steven A. Silber, and John N. van den Anker

RELEVANT CODE OF FEDERAL REGULATIONS (CFR) AND FDA GUIDANCES FOR CONDUCT IN PEDIATRIC CLINICAL TRIALS

Nonclinical Safety Evaluation

- U.S. FDA, Center for Drug Evaluation and Research, Guidance for Industry, Nonclinical Safety Evaluation of Pediatic Drug Products, U.S. Department of Health and Human Services, Rockville, MD, February 2006:
  - http://www.fda.gov/cder/guidance/5671fnl.htm
  - http://www.fda.gov/cder/guidance/s5a.pdf
  - http://www.fda.gov/cder/guidance/6748fnl.htm
- Maintenance of the ICH Guideline on Non-clinical Safety Studies for the Conduct of Human Clinical Trials for Pharmaceuticals M3 (R1).
- EMEA/CHMP/SWP/169215/2005: Draft Guideline on the Need for Non-clinical Testing in Juvenile Animals on Human Pharmaceuticals for Paediatric Indications. September 2005.
- Committee for Human Medicinal Products (CHMP): Guideline on the Need for Non-Clinical Testing in Juvenile Animals on Human Pharmaceuticals for Paediatric Indications (Draft). Doc. Ref. EMEA/CHMP/SWP/169215/2005, September 29, 2005:
  - http://www.emea.org/
  - Frequently asked questions on regulatory aspects of Regulation (EC) No 1901/2006 (Paediatric Regulation) amended by Regulation (EC) No 1902/2006. Doc. Ref. EMEA/520085/2006; January 12, 2007.
- ICH Harmonized Tripartite Guideline: Maintenance of the ICH Guideline on Non-Clinical Safety Studies for the Conduct of Human Clinical Trials for Pharmaceuticals M3 (R1), November 9, 2000.
  - http://www.emea.europa.eu/htms/human/paediatrics/sci_gui.htm

Population Pharmacokinetics

- Guidance for Industry, Population Pharmacokinetics:
  - http://www.fda.gov/cder/guidance/1852fnl.pdf

General Guidances

- American Academy of Pediatrics. Guidelines for the Ethical Conduct of Studies to Evaluate Drugs in Pediatric Populations:
  - http://aappolicy.aappublications.org/cgi/content/abstract/pediatrics;95/2/286?maxtoshow=&HITS=&hits=&RESULTFORMAT=&fulltext=Guidelines+for
    +the+Ethical+Conduct+of+Studies+to+Evaluate+Drugs+in+Pediatric+Populations&searchid=1169123798554_184.
- 45 CFR 46, Subpart D:
  - http://www.hhs.gov/ohrp/humansubjects/guidance/45cfr46.htm.
- Guidance for Industry E11. Clinical Investigation of Medicinal Products in the Pediatric Population. 2000:
  - http://www.fda.gov/cder/guidance/4099FNL.PDF.

- Ethical Considerations for clinical trials performed in children. 2006:
  - http://ec.europa.eu/enterprise/pharmaceuticals/paediatrics/docs/paeds_ethics_consultation20060929.pdf.
- ICH Topic E 6 (R1) Guideline for Good Clinical Practice. European Medicines Agency (EMEA). 2002:
  - http://www.fda.gov/cder/guidance/959fnl.pdf.
- Blood Sampling Guidelines.
  - http://healthcare.partners.org/phsirb/bldsamp.htm.
- International Conference on Harmonization Guidance: Good Clinical Practice (ICH E6).
  - http://www.fda.gov/cder/guidance/959fnl.pdf.
- United States Code of Federal Regulations, Title 21, Part 50—Protection of Human Subjects:
  - http://www.accessdata.fda.gov/scripts/cdrh/cfdocs/cfcfr/CFRSearch.cfm?CFRPart=50.
- United States Code of Federal Regulations, Title 21, Part 312—Investigational New Drug Application:
  - http://www.accessdata.fda.gov/scripts/cdrh/cfdocs/cfcfr/CFRSearch.cfm?CFRPart=312.
- United States Code of Federal Regulations, Title 21, Part 314-Applications for FDA Approval to Market a New Drug:
  - http://www.accessdata.fda.gov/scripts/cdrh/cfdocs/cfcfr/CFRSearch.cfm?CFRPart=314.
- Promoting Safe Medicines in Children, World Health Organization:
  - https://www.who.int/medicines/publications/essentialmedicines/Promotion_-safe_med_childrens.pdf.
  - http://eurex.europa.eu/LexUriServ/site/en/oj/2006/l_378/l_37820061227en00200021.pdf: Editorial

Risk–Benefit and Ethical Considerations in the Neonate

- 21 CFR 50.51, 21 CFR 50.53: Subpart D deals specifically with protecting child research subjects. In contrast to Subpart A (which pertains to adults), allowable risk is restricted to either no more than minimal risk or a minor increase over minimal risk for research that does not offer the prospect of direct benefit to its participants.
- 21 CFR 50.52: risk of the therapeutic component should be justified by the prospect of direct benefit to the participant.
- 21 CFR 50.54: federal panel for possible approval.
- 21 CFR 50.55: parent permission and child assent.

Safety

- Promoting Safety of Medicines for Children. World Health Organization, ISBN 978-92-4-156343-7 (NLM classification: WS 366).

- Additional Safeguards for Children in Clinical Investigations. 2006:
  - http://www.accessdata.fda.gov/scripts/cdrh/cfdocs/cfCFR/CFRSearch.cfm?CFRPart=50&showFR=1&subpartNode=21:1.0.1.1.19.4.
- ICH Topic E2A Clinical Safety Data Management: Definitions and Standards for Expedited Reporting. 1995:
  - http://www.emea.europa.eu/pdfs/human/ich/037795en.pdf.
- Common Terminology Criteria for Adverse Events v3.0 (CTCAE). 2006.
  - http://ctep.cancer.gov/forms/CTCAEv3.pdf.
- European Medicines Agency Post-authorization Evaluation of Medicines for Human use. Committee for Medicinal Products for Human Use (CHMP). Guideline on conduct of pharmacovigilance for medicine used by the paediatric population. London, UK 2007:
  - http://www.emea.europa.eu/pdfs/human/phvwp/23591005en.pdf
- Data Safety Monitoring Board:
  - http://www.who.int/tdr/cd_publications/pdf/operat_guidelines.pdf.
  - http://www.fda.gov/cber/gdlns/clintrialdmc.pdf.
  - Policy and Guidelines for Data and Safety Monitoring. 2001. http://www3.niaid.nih.gov/research/resources/DMIDClinRsrch/PDF/dsm_pol_guide.pdf.
  - NIAID Principles for Use of a Data and Safety Monitoring Board (DSMB) for Oversight of Clinical Trials. 2006. http://www3.niaid.nih.gov/research/resources/toolkit/attachments/DSMB_PolicyAug2006.pdf.
- Categorization of Relatedness for Adverse Events:
  - http://www.niaid.nih.gov/ncn/sop/adverseevents.htm.
- Severity of Hematologic Adverse Events:
  - http://www.fda.gov/cder/cancer/toxicityframe.htm.
- IND safety reports:
  - http://www.accessdata.fda.gov/scripts/cdrh/cfdocs/cfcfr/CFRSearch.cfm?fr=312.32.
- Expedited reports:
  - http://www.emea.europa.eu/pdfs/human/ich/013595en.pdf.

Developing Nations and Drug Development

- World Medical Association Declaration of Helsinki. Ethical Principles for Medical Research Involving Human Subjects:
  - http://www.wma.net/e/policy/b3.htm
- World Health Organization. Ethical considerations arising from vaccine trials conducted in pediatric populations with high disease burden in developing countries:
  - http://www.who.int/vaccine_research/documents/en/manu774_.pdf.

GENERAL REGULATIONS GOVERNING HUMAN SUBJECT PROTECTIONS

Subpart C--IRB Functions and Operations

21 CFR 56.111 Criteria for IRB approval of research.

(a) In order to approve research covered by these regulations the IRB shall determine that all of the following requirements are satisfied:

(1) Principle of "Scientific Necessity"

(1) Risks to subjects are minimized:

(i) By using procedures which are consistent with sound research design *and*...

which do not unnecessarily expose subjects to risk, *and*

(ii) whenever appropriate, by using procedures already being performed on the subjects for diagnostic or treatment purposes.

(3) Selection of subjects is equitable. In making this assessment the IRB should take into account the purposes of the research and the setting in which the research will be conducted and should be particularly cognizant of the special problems of research involving vulnerable populations, such as children.

(2) Appropriate Balance of Risk and Potential Benefit

(2) Risks to subjects are reasonable in relation to anticipated benefits, *if any*, to subjects, and the importance of the knowledge that may be expected to result.

Apply Additional Safeguards of 21 CFR 50, Subpart D

(b) When some or all of the subjects, such as children, ... are likely to be vulnerable to coercion or undue influence additional safeguards have been included in the study to protect the rights and welfare of these subjects.

(c) In order to approve research in which some or all of the subjects are children, an IRB must determine that all research is in compliance with part 50, subpart D of this chapter.

FDA ALGORITHM FOR DETERMINING NEED FOR PEDIATRIC STUDIES USING THE PRINCIPLE OF SCIENTIFIC NECESSITY/EXTRAPOLATION (UNDER BPCA OR PREA)

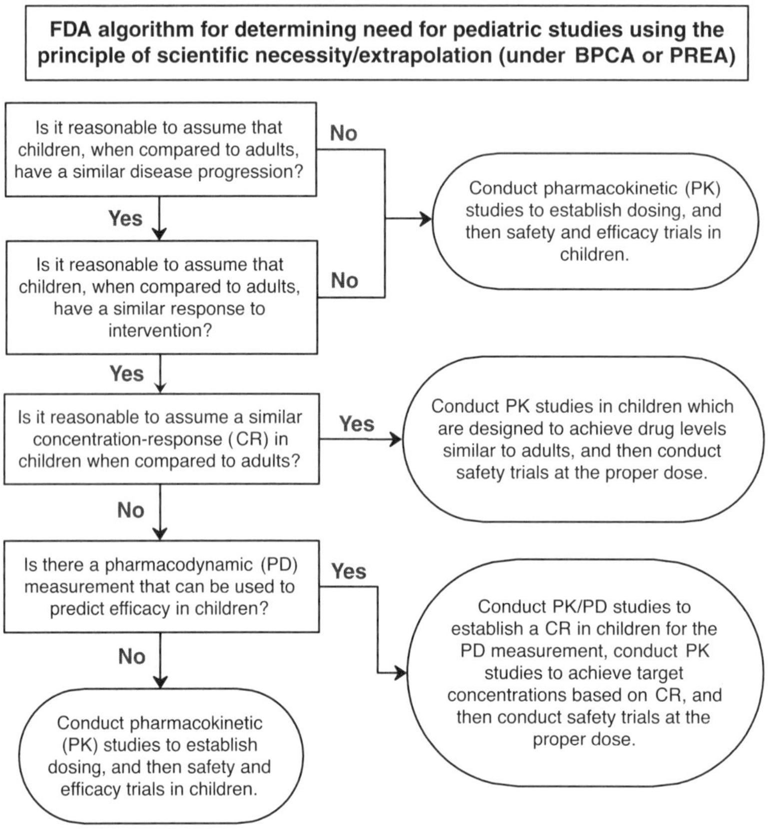

ANALYSIS OF A PROPOSED CLINICAL INVESTIGATION UNDER 21 CFR 50, SUBPART D

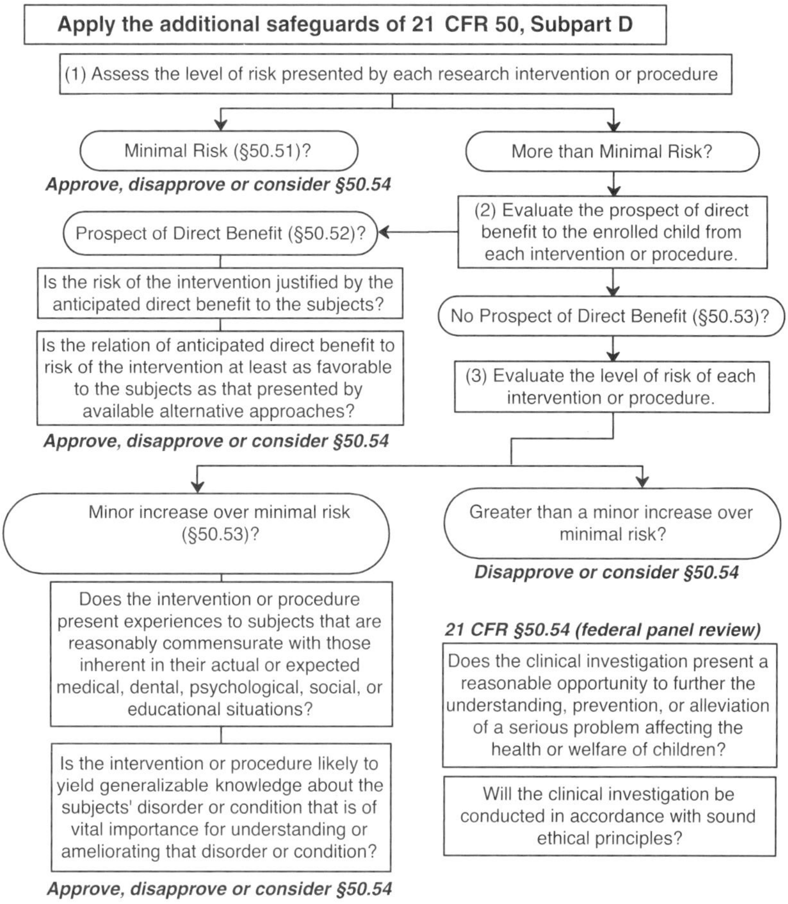

DEVELOPMENTAL EVENTS THAT DIFFER POSTNATALLY ACROSS SPECIES

Organ System	Morphological and Functional Markers
Central nervous	Locomotor and fine motor development, sensory and reflexive development, cognitive development, social play, myelination, and blood–brain barrier
Male reproductive	Preputial separation, testes descent, blood–testis barrier, sex hormone production, puberty
Female reproductive	Vaginal opening, sex hormone production, puberty
Renal	Urine volume control, concentration ability, acid–base equilibrium, glomerular filtration rate, and tubular secretion
Pulmonary	Lung volume, alveolar and saccular maturation
Cardiovascular	Electrophysiology (ECG), cardiac output and hemodynamics, coronary vasculature, cardiac innervation
Skeletal	Longitudinal bone growth (at epiphyseal growth plate long bones), diametric bone growth, fusion of secondary ossification centers, vascularity
Immune	Functional immunocompetence (B and T cell, NK cell), T-dependent antibody response, IgG levels
Digestive	Gastrointestinal motility, gastric pH, gastric emptying, HCl and pepsin secretion, bacterial colonization

DOSING ROUTES IN NEONATAL/JUVENILE ANIMAL SPECIES

Administration Route (Earliest Day Post partum)	Species				
	Rat	Mouse	Dog	Minipig	Rabbit
Oral gavage	From D1	From D4	From D1	From D7	From D14
Subcutaneous	From D1		From D1	From D7	
Intravenous bolus (repeated)	From D15		From D1	From D7	From D14
Intravenous infusion	From D21		From D56	From D7	
Inhalation	From D4	From D21	From D10	From D7	
Dermal[a]	From D21	From D21	From D42		From D35

[a] Not recommended in preweaned animals.

REFLEX ONTOGENY AND BEHAVIORAL TESTING BATTERY

Parameter	Example Tests
Reflex ontogeny (assessed on an age basis)	Surface righting Air righting Negative geotaxis Auditory startle (Preyer response)
Sensorimotor function, locomotor activity, reactivity	Extensor thrust response Hopping Grip strength Swimming ontogeny Rotorod Hindlimb landing foot splay Auditory startle, or auditory startle with pre-pulse inhibition and evoked potentials
Learning and memory	Passive avoidance[30] Biel maze[31] Morris water maze[32] Radial arm maze Cincinnati T-maze[33]

Species	Isoform	Protein Expression or Enzyme Activity	Ontogeny	References
Mouse	B(a)P hydroxylase	Activity	Gradual increase from birth until weaning.	33
	EH	Activity	High levels at birth; decreased first days after birth and then increased to adult value.	
	GST	Activity	Low at birth; starts increasing on PND 6; adult level not reached at weaning.	
	FMO	Dimethylaniline N-oxidase activity	Rapid increase at birth to reach 60–80% of adult levels at PND 3; slow gradual increase thereafter.	10, 30
Rat	GST α en μ	Protein	No expression in fetal hepatocytes.	66, 67
	GST π	Protein	Present in fetal hepatocytes.	
	Microsomal GSTs	Activity	Very low in fetal liver; increasing continuously after birth; adult level at PND 50–150.	68, 69
	FMO	Dimethylaniline N-oxidase activity	Activity in liver microsomes doubles between 3 and 12 weeks of age.	10, 70
Rat (Wistar)	UGT1A1	Activity	UGT1A1 activity increased gradually from birth to PND 6, followed by a sharper increase until adult level.	71
Rat (Sprague–Dawley)	Carboxylesterase	Activity	Approximately 20% at PND 3 and 50% at PND 21 of adult level.	43
	A-esterase	Activity	20–50% at PND 3 and 90% at PND 21 of adult level.	
Rat	CYP1A	Protein	(Mainly CYP1A1) Sharp increase just before weaning; then fourfold decrease to adult levels by PND 60.	36
Rat (Wistar)	CYP3A	Activity (6OHT) and protein (CYP3A2)	Gradual increase up to PND 25 then decrease in female and further increase in male.	37
Rat	CYP3A	EROD activity	Activity increased approximately sixfold in males from PND 14 to PND 60; no increase observed in females.	72
	CYP2D	Protein	Sharp increase from PND 3 to PND 14.	73

Rat (Wistar)	CYP1A	EROD activity	Slow and minor increase from PND 10 to PND 20; strong increase (ca. fivefold) from PND 20 to PND 40.	38
	CYP2B	PROD activity	Strong (fourfold) increase from PND 10 to PND 40.	
	CYP2E1	CYP2B1	Gradual increase from PND 10 to PND 40.	
Rabbit	CYP3A	Demethylase activity	Sharp increase from PND 1 to PND 16; maximum activity at PND 30, then decrease to adult level.	40
	Esterase	DTZ acetylase activity	Gradual decrease in activity after birth until adulthood.	39
	CYP3A	DTZ demethylase acitvity	Increased with age at PND 30.	39
	FMO	Dimethylaniline N-oxidase activity	Activity levels at PND 4 are ca. 20% of adult levels; gradual increase.	10, 31
Dog	P450	Protein	Low at birth; increases fivefold at PND 28–42.	74
	G6P	Activity	Fourfold increase from birth to adult level at PND 28–42.	
	UGT	Activity	Fourfold increase from birth to PND 28–42; then slight decrease to adult.	
	CYP1A	EROD activity	Marked increase during first 8 weeks, then decrease to lower level.	75
	CYP2B	Aminopyrine N-demethylase	Increase during first 8 weeks, then more gradual further increase.	
	CYP2E1	Aniline hydroxylase	Increase from birth until 3 weeks of life, then constant.	
	CYP2A	Coumarin hydroxylase	Increase during first 5 weeks, then constant.	
Pig	CYP450 content		Low at birth and increasing to maximal levels at 4–6 weeks.	76
	CYP	*p*-Nitroanisole O-demethylation	Marked increase to reach maximal activity at 4 weeks of age.	76
	UGT	*p*-Nitrobenzoic acid reduction and Phenolphthalein	Marked increase from birth to reach maximum activity at 4–6 weeks of age.	

Abbreviations: EH, epoxide hydrolase; GST, glutathione-*S*-transferase; UGT, glucuronosyltransferase; CYP, cytochrome P450 enzyme; 6OHT, 6-OH testosterone; EROD, ethoxy-resorufine O-demethylation; PROD, pentoxy-resorufine O-deethylation; G6P, glucose-6-phosphatase; PND, postnatal day.

ONTOGENY OF TRANSPORTERS IN SEVERAL ORGANS OF VARIOUS ANIMALS AT DIFFERENT LEVELS OF EFFECT

Species	Organ	Isoform (Gene Symbol)	mRNA, Protein, or Activity	Ontogeny	References
Mouse	Brain	Mdr1a (*Abcb1a*)	mRNA	PND 1 level is 30% (females) to 50% (males) of adult level (PND 42).	25
		Mdr1b (*Abcb1b*)		Constant between PND 1 and PND 42.	
	Kidney	Mdr1a (*Abcb1a*)		Twofold (males) to sixfold (females) increase between PND 1 and PND 42.	
		Mdr1b (*Abcb1b*)		Constant between PND 1 and PND 42, except fivefold increase in females at PND 42.	
	Liver	Mdr1a (*Abcb1a*)		For females, PND 1 level is 1/6 of adult level, reached at PND 12; for males, PND 1 level is equal to adult level, with eightfold peak at PND 12.	
		Mdr1b (*Abcb1b*)		At PND 1 level is approximately twofold that of adult (PND 42) and PND approximately five fold for 12–19.	
	Brain	Mdr1a (*Abcb1a*)	Protein	PND 1 levels are 20% of adult levels, reached at PND 21.	20
	Intestine	Mdr1a/b (*Abcb1a/b*)	Protein	PND 1 levels are 20% of adult levels, gradual increase.	20, 21
	Liver			Expressed at adult levels from birth onwards.	
	Kidney			Expressed at adult levels from birth onwards.	
	Liver	Mrp2/4 (*Abcc2/4*)	mRNA	PND 1 level approximately equal to adult levels, limited age-dependency in between.	77
		Mrp3 (*Abcc3*)		PND 1 level is 25% adult level, reached at PND 30.	
		Mrp6 (*Abcc6*)		Not detected before PND 10, 50% reduction between PND 10 and PND 15.	
	Kidney	Mrp1 (*Abcc1*)	mRNA	PND 1 levels equal to adult level in females or twice the adult level in males.	77
		Mrp2 (*Abcc2*)		PND 1 level is 30% of PND 45 level; rapid increase at PND 15.	

		Mrp3 (*Abcc3*)		Sharp increase at PND 15, back to perinatal levels in males at PND 45.	
Mouse	Kidney	Mrp4 (*Abcc4*)		Constant in males, gradual increase by threefold in females (PND 1–PND 4, 5).	
		Mrp5 (*Abcc5*)		Highest levels at birth, approximately twofold reduction by PND 45.	
		Mrp6 (*Abcc6*)		Constant levels for PND 1–45.	
	Liver	Oct1 (*Slc22a1*)	mRNA	Level at birth is 10% of adult (PND 45) level, reached by PND 22.	78
	Kidney	Oct1 (*Slc22a1*)		Gradual increase between PND 0 and PND 45.	
		Oct2 (*Slc22a2*)		Constant levels in female mice for PND 0–45, 50% of adult levels in male mice for PND 0–22.	
		Octn1/2 (*Slc22a4/5*)		10–30% of adult levels PND 0–10, adult levels by PND 15.	
	Liver	Oatp1a1 (*Slco1a1*)	mRNA	Absent before PND 30; female = 30% of male.	46
		Oatp1a4 (*Slco1a4*)		Absent before PND 10; male = 30% of female.	
		Oatp1b2 (*Slco1b2*)		Level at birth is 30% of adult level, reached at PND 23.	
	Kidney	Oatp2b1 (*Slco2b1*)		Very low until PND 15, adult levels reached at PND 23.	
		Oatp1a1 (*Slco1a1*)		Absent until PND 30, then only expressed in males (PND 45).	
		Oatp1a4 (*Slco1a4*)		Low expression from before birth until adulthood.	
		Oatp1a6 (*Slco1a6*)		Level at birth is 25% of adult level, reached at PND 15–22.	
		Oatp2b1 (*Slco2b1*)		Level at birth corresponds to adult level.	
		Oatp3a1 (*Slco3a1*)		Constant in female PND 0–45, doubles in males PND 30–45.	
Rat	Brain	Mdr1a/b (*Abcc1a/b*)	Protein	Increased with postnatal development starting on PND 7.	27
	Intestine	Octn2 (*Slc22a5*)	mRNA	Perinatal levels are twofold higher than adult.	79
			Activity	Na^+-dependent L-carnitine uptake significantly higher in newborn compared to suckling rats; activity disappears after weaning.	
Rat	Kidney	Oat1 (*Slc22a6*)	mRNA	Large increase in expression directly after birth.	80
			Activity	Probenecid-sensitive PAH accumulation in renal cortical slices of neonatal rats is 75% of adult level.	

(*continued*)

Species	Organ	Isoform (Gene Symbol)	mRNA, Protein, or Activity	Ontogeny	References
Rat	Kidney	Oat1 (*Slc22a6*)	mRNA	Level at birth is 20% of adult level, gradual increase.	81
		Oat2 (*Slc22a7*)		Very low until PND 35, then rapid increase in female rats only.	
		Oat3 (*Slc22a8*)		Level at birth is 30% of adult level, gradual increase.	
	Liver	Ntcp (*Slc10a1*)	mRNA	PND 1 expression is 35% of adult level, reached after 1 week.	55
			Protein	PND 1 expression is 50% of adult level, reached after 1 week.	
	Liver	Bsep (*Abcb11*)	Immunodetectable protein at membrane surface	First detected at GD 19, adult levels reached by PND 12.	50
		Mrp2 (*Abcc2*)		First detected before GD 15, adult levels by PND 10.	
		Mrp6 (*Abcc6*)		First detected before GD 15, adult levels by PND 10.	
		Ntcp (*Slc10a1*)		First detected at GD 19, adult levels reached by PND 4.	
		Oatp1a1 (*Slco1a1*)		First detected at PND 19, adult levels reached by PND 29.	
		Oatp1a4 (*Slco1a4*)		First detected at PND10, adult levels reached by PND 29.	
		Oatp1b2 (*Slco1b2*)		First detected at PND 4, adult levels reached by PND 29.	
	Liver	Oatp1a4 (*Slco1a4*)	mRNA	In males, PND 1 level ~30% and ~75% of PND 30 and adult (PND 45) levels; in females, PND 1 level ~30% of adult level, reached at PND 25.	49
			Protein	PND 0–20: very low expression; expression peaks in males at PND 35, then 50% decline by PND 45; adult level in females reached at PND 30.	
	Liver	Oatp1b2 (*Slco1b2*)	mRNA	PND 1 level is 20% of adult (PND 45) level, gradual increase with age.	47
	Liver	Ntcp and/or Oatp(s)	Activity	±Twofold lower V_{max} for taurocholate uptake in hepatic basolateral membrane vesicles from suckling rat compared to adult rats.	51
	Liver	Ntcp and/or Oatp(s)	Activity	Progressively increased V_{max} for taurocholate uptake in freshly isolated hepatocyte suspensions for rats from PND 7 to PND 54.	51

Abbreviation: GD, gestation day.

SUMMARY OF DEVELOPMENTAL CHANGES IN DRUG DISPOSITION

Physiological System	Age-Related Trends	Pharmacokinetic Implications	Clinical Implications
Gastrointestinal tract	Neonates and young infants: reduced and irregular peristalsis with prolonged gastric emptying time.	Slower rate of drug absorption (e.g., increased T_{max}) without compensatory compromise in the extent of bioavailability; impaired retention of suppository formulations.	Potential delay in the onset of drug action following oral ingestion; potential for reduced extent of bioavailability from rectally administered drugs.
	Neonates: greater intragastric pH (>4) relative to infants.		
	Infants: enhanced lower GI motility.		
Integument	Neonates and young infants: thinner stratum corneum (neonates only), greater cutaneous perfusion, enhanced hydration, and greater ratio of total BSA to body mass.	Enhanced rate and extent of percutaneous drug absorption; greater relative exposure of topically applied drugs as compared to adults.	Enhanced percutaneous bioavailability and potential for toxicity; need to reduce amount of drugs applied to skin.
Body compartments	Neonates and infants: decreased fat, decreased muscle mass, increased extracellular and total body water spaces.	Increased apparent volume of distribution for drugs distributed to body water spaces and reduced apparent volume of distribution for drugs that bind to muscle and/or fat.	Requirement of higher weight-normalized (i.e., mg/kg) drug doses to achieve therapeutic plasma drug concentrations.
Plasma protein binding	Neonates: decreased concentrations of albumin and α_1-acid glycoprotein with reduced binding affinity for albumin bound weak acids.	Increased unbound concentrations for highly protein-bound drugs with increased apparent volume of distribution and potential for toxicity if the amount of free drug increases in the body.	For highly bound (i.e., >70%) drugs, need to adjust dose to maintain plasma levels near the low end of the recommended "therapeutic range."

(continued)

Physiological System	Age-Related Trends	Pharmacokinetic Implications	Clinical Implications
Drug metabolizing enzyme (DME) activity	Neonates and young infants: immature isoforms of cytochrome P450 and phase II enzymes with discordant patterns of developmental expression.	Neonates and young infants: decreased plasma drug clearance early in life with an increase in apparent elimination half life.	Neonates and young infants: increased drug dosing intervals and/or reduced maintenance doses.
	Children 1–6 years: apparent increased activity for selected drug metabolizing enzymes over adult normal values.	Children 1–6 years: increased plasma drug clearance (i.e., reduced elimination half life) for specific pharmacologic substrates of drug metabolizing enzymes.	Children 1–6 years: for selected drugs, need to increase dose and/or shorten dose interval in comparison to usual adult dose.
	Adolescents: attainment of adult activity after puberty.		
Renal drug excretion	Neonates and young infants: decreased glomerular filtration rates (first 6 months) and active tubular secretion (first 12 months) with adult values attained by 24 months.	Neonates and young infants: accumulation of renally excreted drugs and/or active metabolites with reduced plasma clearance and increased elimination half-life, greatest during first 3 months of life.	Neonates and young infants: increased drug dosing intervals and/or reduced maintenance doses during first 3 months of life.

VALUES OF MEAN (RANGE) eGFR FOR INFANTS AND CHILDREN OF VARIOUS AGES USING CLASSIC CLEARANCE TECHNIQUES

Age	eGFR ($mL/min/1.73\ m^2$)
Preterm neonates at birth[a]	
28 weeks gestation	0.35
32 weeks gestation	0.50
36 weeks gestation	1.21
Full-term neonates at birth[a]	2.24
2–8 days	39 (17–60)
4–28 days	47 (26–78)
37–95 days	58 (30–86)
1–6 months	77 (39–114)
6–12 months	103 (49–157)
12–19 months	127 (62–191)
2–12 years	127 (89–165)

[a]mL per minute uncorrected for body surface area.

EXAMPLES OF SUBSTANCES SECRETED OR REABSORBED BY ACTIVE MECHANISMS IN THE PROXIMAL TUBULE

Acids	Bases
Secreted	
Penicillins	Morphine
Cephalosporins	Tetraethylammonium
Acetazolamide	Choline
Thiazides	*N*-Methylnicotinamide
Furosemide	Triamterene
Salicylate	Thiamine
Phenylbutazone	Tolazoline
Uric acid[a]	Aminoglycosides
p-Aminohippuric acid[a]	Histamine
Taurocholic acid[a]	Catecholamines
Glucuronide and sulfate conjugates of metabolized drugs	
Reabsorbed	
Glucose	
Amino acids	
Urate	
Bile acids	
p-Aminohippuric acid	
Taurocholic acid	
Lithium	
Bromide	

[a]Secretion of acids and bases may not be demonstrable when large amounts are subsequently reabsorbed.

CDC BODY MASS INDEX (BMI) CATEGORIES FOR ADULTS[a]

BMI	Weight Status
Below 18.5	Underweight
18.5–24.9	Normal
25.0–29.9	Overweight
30.0 and above	Obese

[a]For adults 20 years old and older, BMI is interpreted using standard weight status categories that are the same for all ages and for both men and women.
Source: Kuczmarski et al.[52]

CDC BODY MASS INDEX CATEGORIES FOR CHILDREN[a]

BMI Percentile	Weight Status
Less than the 5th percentile	Underweight
5th to less than the 85th percentile	Normal
85th to less than the 95th percentile	At risk overweight
Equal to or greater than the 95th percentile	Overweight

[a]For children, the interpretation of BMI is both age and gender specific.
Source: Kuczmarski et al.[52]

SELF-ASSESSMENT QUESTIONNAIRE FOR ASSESSMENT OF THE TANNER STAGES OF SEXUAL MATURATION: (A) GIRLS AND (B) BOYS

GIRLS SELECT ONE FROM EACH SET OF DRAWINGS BELOW.

Name ______________________

D.O.B. ______________ **Age** ______

Medical Record No. ______________

SET ONE: The drawings below show 5 different stages of how the breasts grow. A girl can go through each of the 5 stages as shown. Please look at each drawing and read the sentences that match the drawings. Then, mark an"X" in the box above the drawing that you think is *closest* to your stage of breast growth.

Stage 1	Stage 2	Stage 3	Stage 4	Stage 5

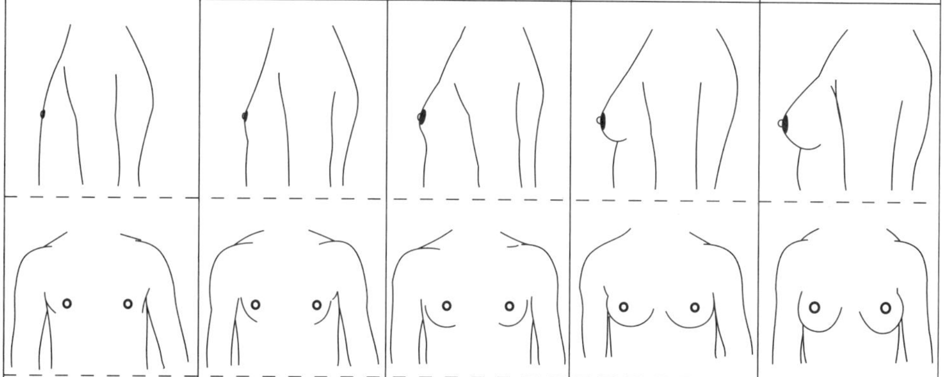

The nipple is raised a little. The rest of the breast is still flat.	This is the breast bud stage. In this stage, the nipple is raised more than in stage 1. The breast is a small mound. The areola is larger than stage 1.	The breast and areola are both larger than in stage 2. The areola does not stick out away from the breast.	The areola and the nipple make up a mound that stiks up above the shape sticks up above the shape of the breast. NOTE: This stage may not happen at all for some girls. Some girls develop from stage 3 to stage 5 with no stage 4.	This is the mature adult stage. The breast are fully developed. Only the nipple sticks out in this stage. The areola has moved back in the general shape of the breast.

SET TWO: The drawings below show 5 different stages of female public hair growth. A girl goes through each of the 5 stages as shown. Please look at each drawing and read the sentences below that match each drawing. Then, mark an "X" in the box above the drawing that you think is *closest* to the amount of your pubic hair growth.

Stage 1	Stage 2	Stage 3	Stage 4	Stage 5

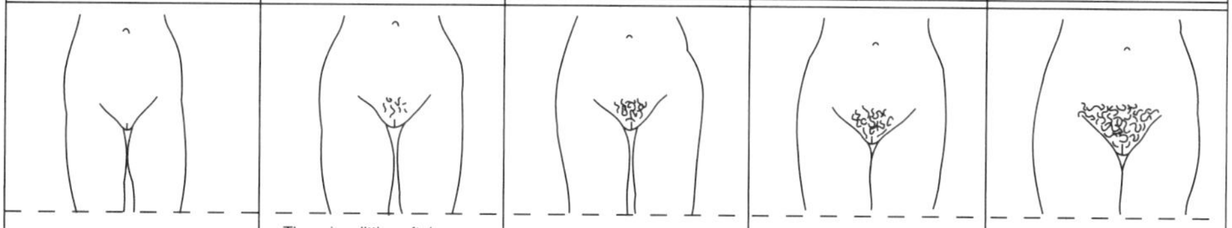

There is no pubic hair at all.	There is a little soft, long, lightly-colored hair. This hair may be straight or a little curly.	The hair is darker in this stage. It is coarser and more curled. It has spread out and thinly covers a bigger area.	The hair is now as dark, curly, and coarse as that of an adult female. The area that the hair covers is not as big as that of an adult female. The hair has NOT spread out to the legs.	The hair is now like that of an adult female. It covers the same area as that of an adult female. The hair usually forms a triangular (▽) pattern as it spreads out to the legs.

Adapted from: Morris, N.M., and Udry, J.R., (1980), **Validation of a Self-Administred Instrument to Assess Stage of Adolescent Development**. *Journal of Youth and Adoescence, Vol. 9. No. 3: 271-280.*

(a)

(continued)

BOYS **SELECT ONE FROM EACH SET OF DRAWINGS BELOW.**

Name	______________________
D.O.B	______________ **Age** ______
Medical Record No.	______________________

SET ONE: The drawings below show 5 different stages of testes, scrotum, and penis growth. A boy can go through each of the 5 stages as shown. Please look at each drawing and read the sentences that match the drawings. Then, mark an "X" in the box above the drawing that you think is *closest* to your stage of penis, scrotum, and testes growth.

Do not look at or select for pubic hair growth with this set of drawings.

Stage 1 ☐	Stage 2 ☐	Stage 3 ☐	Stage 4 ☐	Stage 5 ☐

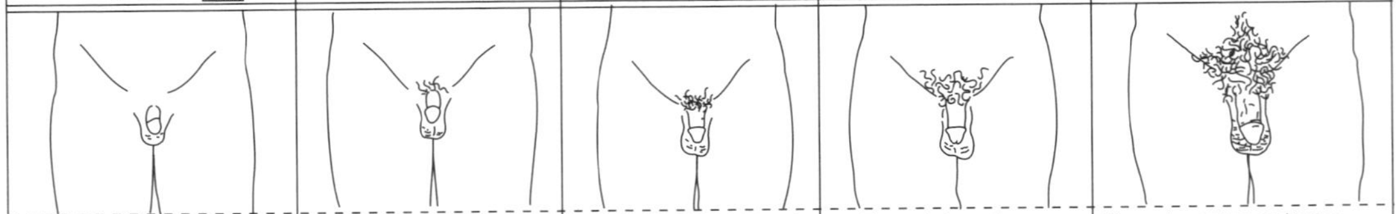

Stage 1	Stage 2	Stage 3	Stage 4	Stage 5
The testes, scrotum and penis are about the same size and shape as they were when you were a child.	The testes scrotum are bigger. The skin of the scrotum has changed. The scrotum, the sack holding the testes, has gotten lower. The penis has gotten only a little bigger.	The penis has grown in length. The testes and scrotum have grown and dropped lower than in drawing 2.	The penis has gotten even bigger. It is wider. The glans (the head of the penis) is bigger. The scrotum is darker than before. It is bigger because the testes are bigger	The penis, scrotum, and testes are the size and shape of an adult man.

SET TWO : The drawings below show 5 different stages of male pubic hair growth. A boy can pass through each of the 5 stages as shown. Please look at each drawing and read the sentences below that match each drawing. Then, mark an "X" in the box above the drawing that you think is *closest* to your stage of your pubic hair growth. *Do not look at or select for penis size with this set of drawings.*

Stage 1 ☐	Stage 2 ☐	Stage 3 ☐	Stage 4 ☐	Stage 5 ☐

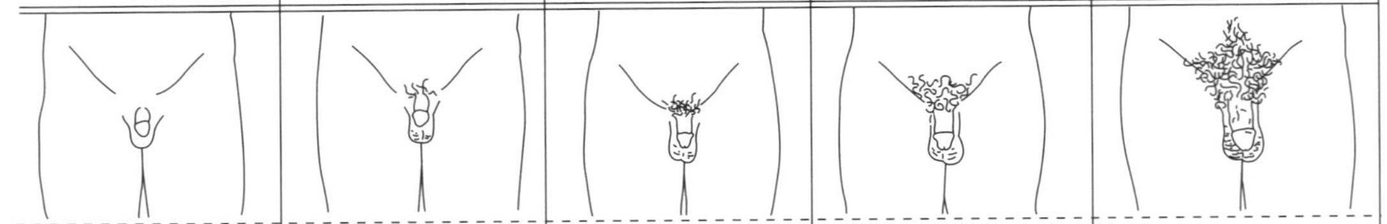

Stage 1	Stage 2	Stage 3	Stage 4	Stage 5
There is no pubic hair at all.	There is a little soft, long, lightly-colored hair. Most of the hair is at the base of the penis. This hair may be straight or a little curley.	The hair is darker in this stage. It is more curled. It has spread out and thinly covers a bigger area.	The hair is now as dark, curley, and coarse as that of an adult man. The area that the hair covers is not as big as that of an adult man. The hair has NOT spread out to the legs.	The hair has spread out to the legs. The hair is now like that of an adult man. It covers the same area as that of an adult man.

Adapted from: Morris, N.M., and Udry, J.R., (1980), **Validation of a Self-Administered Instrument to Assess Stage of Adolescent Development**. *Journal of Youth and Adolescence, Vol. 9. No. 3: 271-280.*

(b)

MEDIAN AGES IN YEARS AT ENTRY INTO TANNER STAGES

	Non-Hispanic White	Non-Hispanic Black	Mexican-American
		Girls	
Breast			
Stage 2	10.4	9.5	9.8
Stage 3	11.8	10.8	11.4
Stage 4	13.3	12.2	13.1
Stage 5	15.5	13.9	14.7
Pubic Hair			
Stage 2	10.6	9.4	10.4
Stage 3	11.8	10.6	11.7
Stage 4	13.0	11.9	13.2
Stage 5	16.3	14.7	16.3
		Boys	
Genitalia			
Stage 2	10.0	9.2	10.3
Stage 3	12.3	11.8	12.5
Stage 4	13.5	13.4	13.8
Stage 5	16.0	15.0	15.8
Pubic Hair			
Stage 2	12.0	11.2	12.3
Stage 3	12.7	12.5	13.1
Stage 4	13.6	13.7	14.1
Stage 5	15.7	15.3	15.8

Source: Sunn et al.[37]

PARADIGM FOR RECRUITMENT AND RETENTION OF PEDIATRIC SITES AND PATIENTS

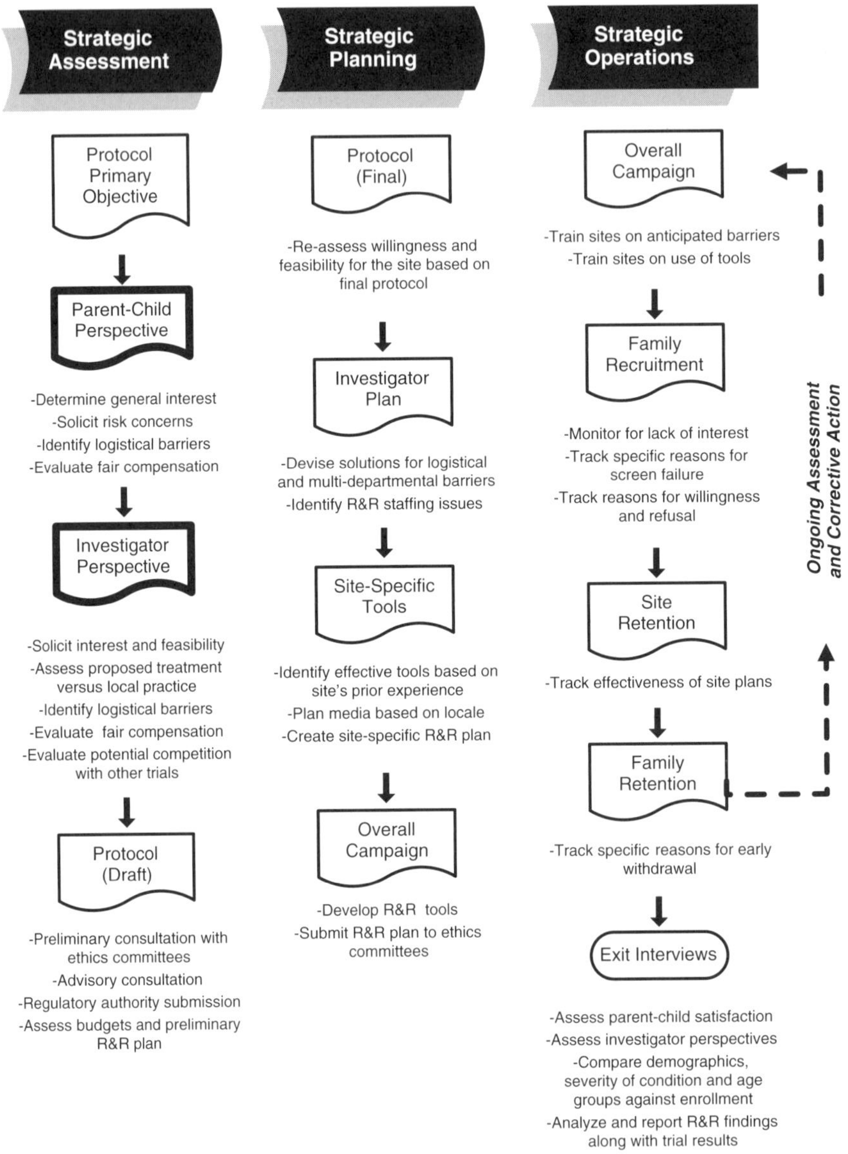

LOCAL AND NATIONAL NEONATAL AND PEDIATRIC CLINICAL RESEARCH NETWORKS LISTED ON IECRN THAT PARTICIPATE IN PHASE I–IV CLINICAL TRIALS

Network or Consortium	Website
NEONATAL	www.atnonline.org/
Biliary Atresia Research Consortium (BARC)	www.barcnetwork.org/
Neonatal Research Network (NRN)	http://neonatal.rti.org/
Vermont Oxford Network (VO)	www.vtoxford.org
PEDIATRIC	
Adolescent Trials Network for HIV/AIDS Interventions (ATN)	www.atnonline.org
American Spinal Muscular Atrophy Randomized Trials Consortium (AmSMART)	www.amsmart.org
Blood and Marrow Transplant Clinical Trials Network (BMTCTN)	www.bmtctn.net
CF[a] Foundation	
Child and Adolescent Psychiatry Trials Network (CAPTN)	www.captn.org
Childhood Arthritis & Rheumatology Research Alliance (CARRA)	www.carragroup.org
Childhood Asthma Research and Education Network (CARENET)	www.asthma-carenet.org
Children's Oncology Group (COG)	www.curesearch.org
Cincinnati Pediatric Research Group (CPRG)	www.cprg.org
Clinical Trials Network of Columbia University, Cornell University, and NY Presbyterian Hospital (CTN)	www.columbiaclinicaltrials.org/ctn
Cooperative Clinical Trials in Pediatric Transplantation (CCTPT)	Not available
Diabetes Research in Children Network (DirecNet)	http://public.direc.net
Emergency Medicine Network (EMNet)	www.emnet-usa.org
Glaser Pediatric Research Network	www.gprn.org
HIV Vaccine Trials Network (HVTN)	www.hvtn.org
International Maternal Pediatric Adolescent AIDS Clinical Trials (IMPAACT)	http://impaact.s-3.com
National Collaborative Pediatric Critical Care Research Network (CPCCRN)	http://www.cpccrn.org/
Pediatric Acute Lung Injury and Sepsis Investigators (PALISI)	http://pedsccm.wustl.edu/research/palisi/palisi.html
Pediatric AIDS Clinical Trials Group (PACTG)	http://impactg.s-3.com/
Pediatric Brain Tumor Consortium (PBTC)	www.pbtc.org
Pediatric Emergency Care Applied Research Network (PECARN)	www.pecarn.org
Pediatric European Network for Treatment of AIDS (PENTA)	www.Pentatrials.org
Pediatric Eye Disease Investigator Group (PEDIG)	http://public.pedig.jaeb.org
Pediatric Heart Network (PHN)	www.pediatricheartnetwork.com

(continued)

Network or Consortium	Website
Pediatric Neuromuscular Clinical Research Network	Not available
Pediatric Pharmacology Research Units Network (PPRU)	www.ppru.org
Pediatric Rheumatology International Trials Organization (PRINTO)	www.printo.it
Project Cure SMA	www.projectcuresma.org
Research Units on Pediatric Psychopharmacology Autism Network (RUPP Autism NW)	
Southern California Permanente Clinical Trials	http://xnet.kp.org/clinicaltrials/
The National Collaborative Pediatric Critical Care Research Network (CPCCRN)	www.cpccrn.org
Therapeutics Development Network (CF TDN)	Not available
Tuberculosis Trials Consortium (TBTC)	www.TBtrialsnetwork.org/TBTC
Type 1 Diabetes TrialNet (Trial Net)	www.diabetestrialnet.org
Vaccine and Treatment Evaluation Units (VTEUs)	Not available

[a] CF—cystic fibrosis.

Source: Available at https://www.clinicalresearchnetworks.org/srchnet.asp.

NORMAL BLOOD VALUES: NEONATAL, PEDIATRIC, AND ADULT POPULATIONS

	Total Leukocytes[a]		Neutrophils			Lymphocytes			Monocytes		Eosinophils	
Age	Mean	Range	Mean	Range	%	Mean	Range	%	Mean	%	Mean	%
Birth	18.1	9.0–30.0	11.0	6.0–26.0	61	5.5	2.0–11.0	31	1.1	6	0.4	2
12 hours	22.8	13.0–38.0	15.5	6.0–28.0	68	5.5	2.0–11.0	24	1.2	5	0.5	2
24 hours	18.9	9.4–34.0	11.5	5.0–21.0	61	5.8	2.0–11.5	31	1.1	6	0.5	2
1 week	12.2	5.0–21.0	5.5	1.5–10.0	45	5.0	2.0–17.0	41	1.1	9	0.5	4
2 weeks	11.4	5.0–20.0	4.5	1.0–9.5	40	5.5	2.0–17.0	48	1.0	9	0.4	3
1 month	10.8	5.0–19.5	3.8	1.0–9.0	35	6.0	2.5–16.5	56	0.7	7	0.3	3
6 months	11.9	6.0–17.5	3.8	1.0–8.5	32	7.3	4.0–13.5	61	0.6	5	0.3	3
1 year	11.4	6.0–17.5	3.5	1.5–8.5	31	7.0	4.0–10.5	61	0.6	5	0.3	3
2 years	10.6	6.0–17.0	3.5	1.5–8.5	33	6.3	3.0–9.5	59	0.5	5	0.3	3
4 years	9.1	5.5–15.5	3.8	1.5–8.5	42	4.5	2.0–8.0	50	0.5	5	0.3	3
6 years	8.5	5.0–14.5	4.3	1.5–8.0	51	3.5	1.5–7.0	42	0.4	5	0.2	3
8 years	8.3	4.5–13.5	4.4	1.5–8.0	53	3.3	1.5–6.8	39	0.4	4	0.2	2
10 years	8.1	4.5–13.5	4.4	1.8–8.0	54	3.1	1.5–6.5	38	0.4	4	0.2	2
16 years	7.8	4.5–13.0	4.4	1.8–8.0	57	2.8	1.2–5.2	35	0.4	5	0.2	3
21 years	7.4	4.5–11.0	4.4	1.8–7.7	59	2.5	1.0–4.8	34	0.3	4	0.2	3

[a] Numbers of leukocytes are in thousands per mm^3, ranges are estimates of 95% confidence limits, and percentages refer to differential counts.
[b] Neutrophils include band cells at all ages and a small number of metamyelocytes and myelocytes in the first few days of life.

HEMATOLOGIC VALUES IN CHILDREN AND ADULTS[a]

	Hemoglobin (g/dL)		Hematocrit (%)		Red Cell Count (10^{12}/L)		MCV (fL)		MCH (pg)		MCHC (g/dL)	
Age	Mean	−2 SD	Mean	−2 SD	Mean	−2 SD	Mean	−2 SD	Mean	−2 SD	Mean	−2 SD
Birth (cord blood)	16.5	13.5	51	42	4.7	3.9	108	98	34	31	33	30
1–3 days (capillary)	18.5	14.5	56	45	5.3	4.0	108	95	34	31	33	29
1 week	17.5	13.5	54	42	5.1	3.9	107	88	34	28	33	28
2 weeks	16.5	12.5	51	39	4.9	3.6	105	86	34	28	33	28
1 months	14.0	10.0	43	31	4.2	3.0	104	85	34	28	33	29
2 months	11.5	9.0	35	28	3.8	2.7	96	77	30	26	33	29
3–6 months	11.5	9.5	35	29	3.8	3.1	91	74	30	25	33	30
0.5–2 years	12.0	10.5	36	33	4.5	3.7	78	70	27	23	33	30
2–6 years	12.5	11.5	37	34	4.6	3.9	81	75	27	24	34	31
6–12 years	13.5	11.5	40	35	4.6	4.0	86	77	29	25	34	31
12-18 years												
Female	14.0	12.0	41	36	4.6	4.1	90	78	30	25	34	31
Male	14.5	13.0	43	37	4.9	4.5	88	78	30	25	34	31
18–49 years												
Female	14.0	12.0	41	36	4.6	4.0	90	80	30	26	34	31
Male	15.5	13.5	47	41	5.2	4.5	90	80	30	26	34	31

[a] The mean ± SD can be expected to include 95% of the observations in a normal population.

SERUM AND PLASMA CHEMISTRY VALUES IN CHILDREN AND ADULTS

Test	Age	Reference Value (USA)	Reference Value (SI)
Alanine aminotransferase (ALT, SGPT)	0–5 days	6–50 U/L	6–50 U/L
	1–19 years	5–45 U/L	5–45 U/L
Albumin	Premature 1 day	1.8–3.0 g/dL	18–30 g/L
	Full term <6 days	2.5–3.4 g/dL	25–34 g/L
	<5 years	3.9–5.0 g/dL	39–50 g/L
	5–19 years	4.0–5.3 g/dL	40–53 g/L
Ammonia	<30 days	21–95 μmol/L	21–95 μmol/L
	1–12 months	18–74 μmol/L	18–74 μmol/L
	1–14 years	17–68 μmol/L	17–68 μmol/L
	>14 years	19–71 μmol/L	19–71 μmol/L
Aspartate aminotransferase (AST, SGOT)	0–5 days	35–140 U/L	35–140 U/L
	1–9 years	15–55 U/L	15–55 U/L
	10–19 years	5–45 U/L	5–45 U/L
Bicarbonate	Premature	18–26 mEq/L	18–26 mmol/L
	Full term	20–25 mEq/L	20–25 mmol/L
	>2 years	22–26 mEq/L	22–26 mmol/L
Bilirubin (conjugated)	Any age	0–0.4 mg/dL	0–8 μmol/L
Bilirubin (total)	Cord		
	Premature	<2 mg/dL	<34 μmol/L
	Term	<2 mg/dL	<34 μmol/L
	0–1 day		
	Premature	<8 mg/dL	<137 μmol/L
	Term	<6 mg/dL	<103 μmol/L
	1–2 days		
	Premature	<12 mg/dL	<205 μmol/L
	Term	<8 mg/dL	<137 μmol/L

(continued)

Test	Age	Reference Value (USA)			Reference Value (SI)		
	3–5 days						
	Premature	<16 mg/dL			<274 µmol/L		
	Term	<12 mg/dL			<205 µmol/L		
	>5 days						
	Premature	<2 mg/dL			<34 µmol/L		
	Term	<1 mg/dL			<17 µmol/L		
	Adult	0.1–1.2 mg/dL			1.7–20.5 µmol/L		
Calcium (ionized)	Cord blood	5.0–6.0 mg/dL			1.25–1.50 mmol/L		
	<24 hours old	4.3–5.1 mg/dL			1.07–1.27 mmol/L		
	24–48 hours old	4.0–4.7 mg/dL			1.00–1.17 mmol/L		
	>2 days	4.8–4.92 mg/dL			1.12–1.23 mmol/L		
Calcium (total)	Cord blood	9.0–11.5 mg/dL			2.25–2.88 mmol/L		
	<24 hours old	9.0–10.6 mg/dL			2.30–2.65 mmol/L		
	24–48 hours	7.0–12.0 mg/dL			1.75–3.00 mmol/L		
	4–7 days	9.0–10.9 mg/dL			2.25–2.73 mmol/L		
	Child	8.8–10.8 mg/dL			2.20–2.70 mmol/L		
	Adult	8.4–10.2 mg/dL			2.10–2.55 mmol/L		
Chloride	Cord blood	96–104 mEq/L			96–104 mmol/L		
	Newborn	97–110 mEq/L			97–110 mmol/L		
	>Newborn	98–106 mEq/L			98–106 mmol/L		
Cholesterol (total)							
	1–3 years	45–182 mg/dL			1.15–4.70 mmol/L		
	4–6 years	109–189 mg/dL			2.80–4.80 mmol/L		
		Percentiles					
	Male	**5th**	**75th**	**95th**	**5th**	**75th**	**95th**
	6–9 years	126 mg/dL	172 mg/dL	191 mg/dL	3.26 mmol/L	4.45 mmol/L	4.94 mmol/L
	10–14 years	130 mg/dL	179 mg/dL	204 mg/dL	3.36 mmol/L	4.63 mmol/L	5.28 mmol/L
	15–19 years	114 mg/dL	167 mg/dL	198 mg/dL	2.95 mmol/L	4.32 mmol/L	5.12 mmol/L

	Female						
	6–9 years	122 mg/dL	173 mg/dL	209 mg/dL	3.16 mmol/L	4.47 mmol/L	5.41 mmol/L
	10–14 years	124 mg/dL	174 mg/dL	217 mg/dL	3.21 mmol/L	4.50 mmol/L	5.61 mmol/L
	15–19 years	125 mg/dL	175 mg/dL	212 mg/dL	3.23 mmol/L	4.53 mmol/L	5.48 mmol/L
	Adult	Desirable	Borderline	High	Desirable	Borderline	High
		<200 mg/dL	200–239 mg/dL	>240 mg/dL	<5.18 mmol/L	5.18–6.19 mmol/L	>6.2mmol/L
Cholesterol—HDL							
	1–13 years	35–84 mg/dL			0.9–2.15 mmol/L		
	14–19 years	35–65 mg/dL			0.9–1.65 mmol/L		
	Adult (desirable)	>45 mg/dL			>1.17 mmol/L		
Cholesterol—LDL							
		Male	Female		Male	Female	
	Cord blood	10–50 mg/dL	10–50 mg/dL		0.26–1.30 mmol/L	0.26–1.30 mmol/L	
	1–9 years	60–140 mg/dL	60–150 mg/dL		1.55–3.63 mmol/L	1.55–3.89 mmol/L	
	10–19 years	50–170 mg/dL	50–170 mg/dL		1.30–4.40 mmol/L	1.30–4.40 mmol/L	
	20–29 years	60–175 mg/dL	60–160 mg/dL		1.55–4.53 mmol/L	1.55–4.14 mmol/L	
	30–39 years	80–190 mg/dL	70–170 mg/dL		2.07–4.92 mmol/L	1.81–4.40 mmol/L	
	40–49 years	90–205 mg/dL	80–190 mg/dL		2.33–5.31 mmol/L	2.07–4.92 mmol/L	
	Adult	Desirable	Borderline	High	Desirable	Borderline	High
		<130 mg/dL	130–159 mg/dL	>160 mg/dL	<3.37 mmol/L	3.37–4.13 mmol/L	>4.14 mmol/L
Creatinine							
	Cord	0.6–1.2 mg/dL			53–106 μmol/L		
	Newborn	0.3–1.0 mg/dL			27–88 μmol/L		
	Infant	0.2–0.4 mg/dL			18–35 μmol/L		
	Child	0.3–0.7 mg/dL			27–62 μmol/L		
	Adolescent	0.5–1.0 mg/dL			44–88 μmol/L		
	Man	0.6–1.3 mg/dL			53–115 μmol/L		
	Woman	0.5–1.2 mg/dL			44–106 μmol/L		
Glucose							
	Cord blood	45–96 mg/dL			2.5–5.3 mmol/L		
	Premature	20–60 mg/dL			1.1–3.3 mmol/L		
	Full term	30–60 mg/dL			1.7–3.3 mmol/L		
	1 day old	40–60 mg/dL			2.2–3.3 mmol/L		

(continued)

Test	Age	Reference Value (USA)	Reference Value (SI)
	>1 day old	50–90 mg/dL	2.8–5.0 mmol/L
	Child	60–100 mg/dL	3.3–5.5 mmol/L
	Adult	70–105 mg/dL	3.9–5.8 mmol/L
Gammaglutamyltranspeptidase (GGT, GGTP)	Cord blood	19–270 U/L	19–270 U/L
	Premature	56–233 U/L	56–233 U/L
	0–3 weeks	0–130 U/L	0–130 U/L
	3 weeks–3 months	4–120 U/L	4–120 U/L
	>3 months, boys	5–65 U/L	5–65 U/L
	>3 months, girls	5–35 U/L	5–35 U/L
	1–15 years	0–23 U/L	0–23 U/L
	Adult male	11–50 U/L	11–50 U/L
	Adult female	7–32 U/L	7–32 U/L
Lactate dehydrogenase			
	Neonate	160–1500 U/L	160–1500 U/L
	Infant	150–360 U/L	150–360 U/L
	Child	150–300 U/L	150–300 U/L
	Adult	0–220 U/L	0–220 U/L
Lactate dehydrogenase isoenzymes			
LD_1 heart		24–34%	24–34%
LD_2 heart, erythrocytes		35–45%	35–45%
LD_3 muscle		15–25%	15–25%
LD_4 liver, trace muscle		4–10%	4–10%
LD_5 liver, muscle		1–9%	1–9%
Lipase	1–18 years	3–32 U/L	3–32 U/L
Magnesium	Any age	1.3–2.0 mEq/L	0.65–1.0 mmol/L
Manganese	Newborn	2.4–9.6 μg/dL	0.44–1.75 μmol/L
	2–18 years	0.8–2.1 μg/dL	0.15–0.38 μmol/L

Phosphatase, alkaline					
	1–9 years	145–420 U/L		145–420 U/L	
	10–11 years	130–560 U/L		130–560 U/L	
		Male	Female	Male	Female
	12–13 years	200–495 U/L	105–420 U/L	200–495 U/L	105–420 U/L
	14–15 years	130–525 U/L	70–130 U/L	130–525 U/L	70–130 U/L
	16–19 years	65–260 U/L	50–130 U/L	65–260 U/L	50–130 U/L
Phosphorus, inorganic					
	0–5 days	4.8–8.2 mg/dL		1.55–2.65 mmol/L	
	1–3 years	3.8–6.5 mg/dL		1.25–2.10 mmol/L	
	4–11 years	3.7–5.6 mg/dL		1.20–1.80 mmol/L	
	12–15 years	2.9–5.4 mg/dL		0.95–1.75 mmol/L	
	16–19 years	2.7–4.7 mg/dL		0.90–1.50 mmol/L	
Plasma volume					
	Male	25–43 mL/kg		0.025–0.043 L/kg	
	Female	28–45 mL/kg		0.028–0.045 L/kg	
Potassium					
	<2 months	3.0–7.0 mEq/L		3.0–7.0 mmol/L	
	2–12 months	3.5–6.0 mEq/L		3.5–6.0 mmol/L	
	>12 months	3.5–5.0 mEq/L		3.5–5.0 mmol/L	
Protein, (L) total					
	Cord	4.8–8.0 g/dL		48–80 g/L	
	Full term	4.4–7.6 g/dL		44–76 g/L	
	1–30 days	4.4–7.6 g/dL		44–76 g/L	
	1–3 months	3.6–7.4 g/dL		36–74 g/L	
	4–6 months	4.2–7.4 g/dL		42–74 g/L	
	7–12 months	5.1–7.5 g/dL		51–75 g/L	
	13–24 months	3.7–7.5 g/dL		37–75 g/L	
	25–36 months	5.3–8.1 g/dL		53–81 g/L	
	3–5 years	4.9–8.1 g/dL		49–81 g/L	
	6–8 years	6.0–7.9 g/dL		60–79 g/L	
	9–11 years	6.0–7.9 g/dL		60–79 g/L	
	12–16 years	6.0–7.9 g/dL		60–79 g/L	
	Adult	6.0–8.0 g/dL		60–80 g/L	

(continued)

Test	Age	Reference Value (USA)		Reference Value (SI)	
Sodium					
	Newborn	134–146 mEq/L		134–146 mmol/L	
	Infant	139–146 mEq/L		139–146 mmol/L	
	Child	138–145 mEq/L		138–145 mmol/L	
	Adolescent	136–146 mEq/L		136–146 mmol/L	
Triglycerides (after >12 hours of fast)					
		Male	Female	Male	Female
	Cord blood	10–98 mg/dL	10–98 mg/dL	0.10–0.98 g/L	0.10–0.98 g/L
	0–5 years	30–86 mg/dL	32–99 mg/dL	0.30–0.86 g/L	0.32–0.99 g/L
	6–11 years	31–108 mg/dL	35–114 mg/dL	0.31–1.08 g/L	0.35–1.14 g/L
	12–15 years	36–138 mg/dL	41–138 mg/dL	0.36–1.38 g/L	0.41–1.38 g/L
	16–19 years	40–163 mg/dL	40–128 mg/dL	0.40–163 g/L	0.40–1.28 g/L
	20–29 years	44–185 mg/dL	40–128 mg/dL	0.44–1.85 g/L	0.40–1.28 g/L
	Adults	40–160 mg/dL	35–135 mg/dL	0.40–1.60 g/L	0.35–1.35 g/L
Urea nitrogen (BUN)					
	Cord blood	21–40 mg/dL		7.5–14.3 mmol urea/L	
	Premature (1 week)	3–25 mg/dL		1.1–9.0 mmol urea/L	
	Full term	3–12 mg/dL		1.1–4.3 mmol urea/L	
	Infant/child	5–18 mg/dL		1.8–6.4 mmol urea/L	
	Adolescent	7–18 mg/dL		2.5–6.4 mmol urea/L	
Uric acid					
	0–2 years	2.4–6.4 mg/dL		0.14–0.38 mmol/L	
	2–12 years	2.4–5.9 mg/dL		0.14–0.35 mmol/L	
	12–14 years	2.4–6.4 mg/dL		0.14–0.38 mmol/L	
	Adult male	3.5–7.2 mg/dL		0.20–0.43 mmol/L	
	Adult female	2.4–6.4 mg/dL		0.14–0.38 mmol/L	

INDEX

Pediatric Drug Development: Concepts and Applications
Edited by Andrew E. Mulberg, Steven A. Silber, and John N. van den Anker